Edited by
Erick J. Vandamme and José Luis Revuelta

Industrial Biotechnology of Vitamins, Biopigments, and Antioxidants

Edited by
Erick J. Vandamme and José Luis Revuelta

Industrial Biotechnology of Vitamins, Biopigments, and Antioxidants

WILEY-VCH
Verlag GmbH & Co. KGaA

The Editors

Prof. Dr. Erick J. Vandamme
Ghent University
Department of Biochemical and Microbial Technology
9000 Gent
Belgium

Prof. Dr. José Luis Revuelta
Universidad de Salamanca
Departamento de Microbiologia y Genetica
Campus Miguel de Unamuno
37007 Salamanca
Spain

Cover courtesy
Henrik5000/iStockphoto; Brand X/Getty Images; David Marchal/iStockphoto; fusebulb/Shutterstock

Library of Congress Card No.: applied for

British Library Cataloguing-in-Publication Data
A catalogue record for this book is available from the British Library.

Bibliographic information published by the Deutsche Nationalbibliothek
The Deutsche Nationalbibliothek lists this publication in the Deutsche Nationalbibliografie; detailed bibliographic data are available on the Internet at <http://dnb.d-nb.de>.

Print ISBN: 978-3-527-33734-7
ePDF ISBN: 978-3-527-68176-1
ePub ISBN: 978-3-527-68177-8
Mobi ISBN: 978-3-527-68178-5
oBook ISBN: 978-3-527-68175-4

Typesetting SPi Global, Chennai, India
Printing and Binding Markono Print Media Pte Ltd, Singapore

Printed on acid-free paper

To Mireille (EJV)

To Ines (JLR)

Contents

List of Contributors

Spiros N. Agathos
Université Catholique de Louvain
Earth & Life Institute –
Bioengineering Laboratory
Place Croix du Sud 2
bte. L07.05.19
1348 Louvain-la-Neuve
Belgium

Christoph Albermann
Universität Stuttgart
Institut für Mikrobiologie
Allmandring 31
70569 Stuttgart
Germany

Luis Aliaga
University of Granada
Department of Medicine
Faculty of Medicine
Avenida de Madrid
s/n c.p.
18071 Granada
Spain

Alberto Amaretti
University of Modena and Reggio
Emilia
Department of Life Sciences
Fermentation Chemistry and
Biotechnology
via Campi 103
41125 Modena
Italy

Elmira Arab-Tehrany
Université de Lorraine
Laboratoire Ingénierie des
Biomolécules (LIBIO)
EA4367
54506 Vandoeuvre Les Nancy
France

Paula Arense
University of Murcia. Regional
Campus of International
Excellence 'Campus Mare
Nostrum'
Department of Biochemistry and
Molecular Biology B and
Immunology
Faculty of Chemistry
Campus of Espinardo
E-30100, Murcia
Spain

Vicente Bernal
University of Murcia. Regional
Campus of International
Excellence 'Campus Mare
Nostrum'
Department of Biochemistry and
Molecular Biology B and
Immunology
Faculty of Chemistry
Campus of Espinardo
E-30100, Murcia
Spain

and

Repsol Technology Center
Group of Biology
Repsol S.A. Autovía de
Extremadura
S/N. E-28935 Mostoles
Madrid
Spain

Holger Beuttler
Georgii-Gymnasium
Lohwasen 1
73728 Esslingen am Neckar
Germany

Tek Chand Bhalla
Himachal Pradesh University
Department of Biotechnology
Summer Hill
171 005 Shimla
India

Manuel Cànovas
University of Murcia. Regional
Campus of International
Excellence 'Campus Mare
Nostrum'
Department of Biochemistry and
Molecular Biology B and
Immunology
Faculty of Chemistry
Campus of Espinardo
E-30100, Murcia
Spain

Fusheng Chen
Huazhong Agricultural
University
College of Food Science and
Technology
Food Biotechnology and Food
Safety Laboratory
430070 Wuhan
Hubei Province
P.R. China

Wanping Chen
Huazhong Agricultural
University
College of Food Science and
Technology
Food Biotechnology and Food
Safety Laboratory
430070 Wuhan
Hubei Province
P.R. China

Isabelle Chevalot
Université de Lorraine
Laboratoire Réaction & Génie
des Procédés (LRGP)
UMR CNRS 7274
54506 Vandoeuvre Les Nancy
France

Luca Costantino
University of Modena and Reggio Emilia
Department of Life Sciences
Fermentation Chemistry and Biotechnology
via Campi 103
41125 Modena
Italy

Antoine Delhaye
Université Catholique de Louvain
Earth & Life Institute – Bioengineering Laboratory
Place Croix du Sud 2
bte. L07.05.19
1348 Louvain-la-Neuve
Belgium

Yanli Feng
Huazhong Agricultural University
College of Food Science and Technology
Food Biotechnology and Food Safety Laboratory
430070 Wuhan
Hubei Province
P.R. China

and

Hubei Key Laboratory of Edible Wild Plants Conservation & Utilization
435002 Huangshi
Hubei Province
P.R. China

Christine Gerardin
Université de Lorraine
Laboratoire Etude & Recherche Matériau Bois (LERMAB)
EA4370
54506 Vandoeuvre Les Nancy
France

Jesus Gonzalez-Lopez
University of Granada
Department of Microbiology
Faculty of Pharmacy
Campus universitario de la Cartuja, s/n c.p
18071 Granada
Spain

Alejandro Gonzalez-Martinez
University of Granada
Institute of Water Research
C/Ramón y Cajal, 4
18071 Granada
Spain

Ana M. González-Paramás
Universidad de Salamanca
Grupo de Investigación de Polifenoles (GIP-USAL)
Facultad de Farmacia
Campus Miguel de Unamuno
37007 Salamanca
Spain

Borhane S. Grama
Université Catholique de Louvain
Earth & Life Institute – Bioengineering Laboratory
Place Croix du Sud 2
bte. L07.05.19
1348 Louvain-la-Neuve
Belgium

Hans-Peter Hohmann
Biotechnology R&D
DSM Nutritional Products
Wurmisweg 576
CH-4303 Kaiseraugst
Switzerland

Edouard Husson
Université de Picardie Jules Verne
Unité de Génie Enzymatique et
Cellulaire
FRE CNRS 3580
80039 Amiens
France

Clayton Jeffryes
Université Catholique de Louvain
Earth & Life Institute –
Bioengineering Laboratory
Place Croix du Sud 2
bte. L07.05.19
1348 Louvain-la-Neuve
Belgium

Alberto Jiménez
Universidad de Salamanca
Metabolic Engineering Group
Department of Microbiology and
Genetics
Edificio Departamental Lab 323
37007 Salamanca
Spain

Hyun Y. Jung
Pukyong National University
Department of Biotechnology
and Bioengineering
45 Yongso-Ro
Nam-Gu
608-737 Busan
Korea

Eun J. Kim
Pukyong National University
Department of Biotechnology
and Bioengineering
45 Yongso-Ro
Nam-Gu
608-737 Busan
Korea

Joong K. Kim
Pukyong National University
Department of Biotechnology
and Bioengineering
45 Yongso-Ro
Nam-Gu
608-737 Busan
Korea

Christophe Lacroix
ETH Zurich
Laboratory of Food
Biotechnology
Institute of Food
Nutrition and Health
Schmelzbergstrasse 7
LFV C20 Zurich
Switzerland

Rodrigo Ledesma-Amaro
Universidad de Salamanca
Metabolic Engineering Group
Department of Microbiology and
Genetics
Edificio Departamental Lab 323
37007 Salamanca
Spain

Maria V. Martinez-Toledo
University of Granada
Department of Microbiology
Faculty of Science
Campus de Fuentenueva, s/n c.p
18071 Granada
Spain

Yalan Ni
Jiangnan University
State Key Laboratory of Food
Science and Technology
1800 Lihu Avenue
214122 Wuxi
China

and

Jiangnan University
Key Laboratory of Industrial
Biotechnology
Ministry of Education
School of Biotechnology
1800 Lihu Avenue
214122 Wuxi
China

and

Jiangnan University
Synergetic Innovation Center of
Food Safety and Nutrition
1800 Lihu Avenue
214122 Wuxi
China

Günter Pappenberger
Biotechnology R&D
DSM Nutritional Products
Wurmisweg 576
CH-4303 Kaiseraugst
Switzerland

Stefano Raimondi
University of Modena and Reggio
Emilia
Department of Life Sciences
Fermentation Chemistry and
Biotechnology
via Campi 103
41125 Modena
Italy

Colin Ratledge
University of Hull
Department of Biological
Sciences
Hull HU6 7RX
UK

José L. Revuelta
Universidad de Salamanca
Metabolic Engineering Group
Department of Microbiology and
Genetics
Edificio Departamental Lab 323
37007 Salamanca
Spain

Maddalena Rossi
University of Modena and Reggio
Emilia
Department of Life Sciences
Fermentation Chemistry and
Biotechnology
via Campi 103
41125 Modena
Italy

Celestino Santos-Buelga
Universidad de Salamanca
Grupo de Investigación de
Polifenoles (GIP-USAL)
Facultad de Farmacia
Campus Miguel de Unamuno
37007 Salamanca
Spain

Savitri
Himachal Pradesh University
Department of Biotechnology
Summer Hill
171 005 Shimla
India

Yanchun Shao
Huazhong Agricultural University
College of Food Science and Technology
Food Biotechnology and Food Safety Laboratory
430070 Wuhan
Hubei Province
P.R. China

Feng Shi
Jiangnan University
State Key Laboratory of Food Science and Technology
1800 Lihu Avenue
214122 Wuxi
China

and

Jiangnan University
Key Laboratory of Industrial Biotechnology
Ministry of Education
School of Biotechnology
1800 Lihu Avenue
214122 Wuxi
China

and

Jiangnan University
Synergetic Innovation Center of Food Safety and Nutrition
1800 Lihu Avenue
214122 Wuxi
China

Marc J.A. Stevens
ETH Zurich
Laboratory of Food Biotechnology
Institute of Food Nutrition and Health
Schmelzbergstrasse 7
LFV C18 Zurich
Switzerland

Janice Marie Sych
ZHAW Wädenswil
ZHAW Life Sciences und Facility Management
Institute for Food and Beverage Innovation
Einsiedlerstrasse 34
8820 Wädenswil
Switzerland

Hirohide Toyama
University of the Ryukyus
Department of Bioscience and Biotechnology
1 Senbaru
Nishihara-cho
Okinawa 903-0213
Japan

Erick J. Vandamme
Ghent University
Centre of Expertise-Industrial Biotechnology and Biocatalysis
Department of Biochemical and Microbial Technology
Faculty of Bioscience Engineering
Block B, 2nd floor
Coupure links 653
9000 Ghent
Belgium

Nannan Wang
Jiangnan University
State Key Laboratory of Food Science and Technology
1800 Lihu Avenue
214122 Wuxi
China

and

Jiangnan University
Key Laboratory of Industrial Biotechnology
Ministry of Education
School of Biotechnology
1800 Lihu Avenue
214122 Wuxi
China

and

Jiangnan University
Synergetic Innovation Center of Food Safety and Nutrition
1800 Lihu Avenue
214122 Wuxi
China

Hui Xu
Institute of Applied Ecology
Chinese Academy of Sciences
No. 72 Wenhua Road
110016 Shenyang
China

Weichao Yang
Institute of Applied Ecology
Chinese Academy of Sciences
No. 72 Wenhua Road
110016 Shenyang
China

Youxiang Zhou
Huazhong Agricultural University
College of Food Science and Technology
Food Biotechnology and Food Safety Laboratory
430070 Wuhan
Hubei Province
P.R. China

and

Hubei Academy of Agricultural Sciences
Institute of Agricultural Quality Standards and Testing Technology Research
430064 Wuhan
Hubei Province
P.R. China

Preface

Vitamins, provitamins and related compounds belong to the few chemicals that evoke a positive appeal to most people; even for a layman, the term vitamin sounds synonymous to vitality, health, physical and mental strength, fitness, well-being and so on. Indeed, each one of us needs his/her daily intake of vitamins, which should normally be provided by a balanced and varied diet. However, even today, this is not always the case. Current food habits or preferences, food availabilities, as well as food processing, cooking or preservation methodologies and technologies do not always assure a sufficient balanced natural daily vitamin supply to a healthy individual, let alone to a sick or stressed human being. Today, modern society is seldom confronted with the notorious avitaminoses of the past in the Western World, but they do still occur frequently in overpopulated, war-ridden, poverty- or famine-struck regions in many parts of the World. Apart from their *in vivo* nutritional–physiological roles as essential growth factors and coenzymes for human beings, animals, plants and microorganisms, vitamins and related compounds are increasingly being introduced as food and as feed additives, as medical–therapeutical agents, as health-promoting aids, and also as technical aids, for example, as antioxidants or biopigments. Today, an impressive number of processed foods, feeds, cosmetics, pharmaceutical and chemical formulations contain extra vitamins or vitamin-related compounds, and single and multivitamin preparations are commonly taken or prescribed. These considerations point towards an extra need for vitamin supply, other than those provided from microbial, plant and animal food sources. Most added vitamins and related compounds are indeed now industrially prepared via chemical synthesis, extraction technologies and/or biotechnological routes, such as fermentation and/or biocatalysis. This volume focusses on the use of industrial biotechnological principles and bioprocesses for the production of vitamins and related compounds such as biopigments and antioxidants.

Industrial biotechnology encompasses the exploitation of the genetic and biochemical machinery of useful microorganisms (bacteria, fungi, yeasts and microalgae) and of higher cells for the synthesis of bulk and fine chemicals (including vitamins and related factors), pharmaceuticals, enzymes, biomaterials and energy, using renewable resources rather than fossil ones. Two main types of microbiology-based enabling technologies are involved: fermentation-based technologies and enzyme-based technologies. Fermentation technology relates to the directed and controlled mass production of microbial or higher cells, their enzymes and/or their metabolites. Enzyme technology or biocatalysis deals with the use of microbial or higher cells for their enzyme systems (produced via fermentation processes) to catalyse desirable chemical chiral reactions. Both technologies were initially often rescued only when chemical processes failed to be successful or were uneconomical. Nowadays, they are often the first-choice technologies for several reasons: they are based on renewable resources, deliver simple as well as very complex molecules directly in a desirable chiral form and in an economically favourable way, and they are considered in the society as clean, sustainable and re-usable technologies. Industrial microbiology has its foundations based on knowledge of basic sciences and of technologies as well. It has always been a cornerstone of 'microbial biotechnology', even before this name was coined. Indeed, the discipline has attracted the interest of scientists and bioengineers for decades, but new developments in science, in technology, in industry and in society have made it an even more fascinating and indispensable field of research and application. Scientific breakthroughs in high-throughput screening methodologies, in molecular genetics of industrial microbial strains, in systems (micro)biology, in directed evolution, metabolic engineering and modelling, but equally in enzyme and cell engineering, in novel culture techniques, rapid sampling and sensor methodologies, in bioreactor design and in downstream processing, all have contributed to the growing interest and use and impact of industrial microbiology and biotechnology in the industry. The design-based engineering of industrial microbial strains is still hampered by incomplete knowledge of cell biochemistry, metabolic regulation and cell biology. Advances in systems biology technologies and in synthetic (micro)biology can now also contribute to fill this gap. Equally, microbial enzymes are increasingly being used in industry and are further optimised as to their characteristics for practical use in large-scale biocatalytic reactions; basic and applied studies of enzyme and protein engineering and of enzyme technology are essential here. Protein engineering of microbial enzymes is now an important tool to overcome the limitations of natural enzymes as useful biocatalysts; combination of directed evolution and rational protein design using computational tools has become significant to create even novel enzymes, expanding their application potential in industry. The asymmetric biocatalysis with microbial enzymes and cells has now achieved high efficiency, enantioselectivity and yield, such that – for a wide variety of chiral products, including vitamins, biopigments, antioxidants and related compounds – biocatalysis has become a preferred production alternative in organic synthesis and in the chemical industry for fine as well as bulk chemicals.

All the aforementioned developments have justified the timely publishing of a comprehensive book on industrial biotechnology of current vitamin production, biopigments, antioxidants and related compounds. Eighteen comprehensive chapters, all written by renown experts, focus on all aspects, from historical to the latest developments in both fields, fermentation science and enzyme technology, as applied to (pro)vitamins, biopigments, antioxidants and related compounds.

So far, such information is scattered widely in the scientific literature; for some compounds, only secrecy and sparse data are available. Some well-known vitamin compounds that are produced currently only chemically are deliberately not covered in this biotech-focussed volume, including B_1, B_6, B_7, D, E and K. For some of these molecules, biotechnological processes are being developed, although, indeed, not competitive as yet with chemical synthesis. Other published volumes cover only one or a few specific vitamin compounds or deal mainly with chemical synthesis, nutritional, biochemical, pharmaceutical or medical aspects.

This volume also aims at demonstrating the broad potential of industrial microbiology and biotechnology to produce these chemically quite complex molecules and its impact on society; it may awake the mind of the researchers – also in other fields of science and technology – to speed up the introduction of these clean biotechnologies in the industry and their products in society!

The help of several colleagues and friends in suggesting potential authors for difficult-to-get chapters has been invaluable to assemble a comprehensive volume. We want to mention especially Dr. Hans-Peter Hohmann, DSM Nutritional Products, Basel, Switzerland; Em. Prof. Yoshiki Tani, Faculty of Agriculture, Kyoto University, Japan; Dr. Hideo Kawabe and Dr. Hideharu Anazawa, Japan Bioindustry Association, Japan and Prof. K. Matsushita, Yamaguchi University, Japan; and Em. Prof. Colin Ratledge, Department of Biological Sciences, University of Hull, UK.

The positive interaction with all the contributing authors is highly appreciated as well. The editors are very much indebted to the staff of Wiley-VCH, Verlag GmbH & Co, Weinheim, Germany, especially to Dr. Reinhold Weber, Dr. Andreas Sendtko and Mrs. Lesley Fenske, who were extremely helpful at the different stages from the conception to the birth of this book! Most gratitude goes to our respective wives, Mireille and Ines, who could only have withstood our mental absence, strengthened with multivitamin preparations, although our sole vitamin shot was their encouraging and moral support during this biotechnological enterprise!

Belgium
Spain
2016

Erick J. Vandamme
José L. Revuelta

1 Vitamins, Biopigments, Antioxidants and Related Compounds: A Historical, Physiological and (Bio)technological Perspective

Erick J. Vandamme and José L. Revuelta

1.1 Historical Aspects of the Search for Vitamins

In hindsight, the history of organic compounds that are now called *vitamins* can be traced back to the ancient Egyptians; they experienced that feeding animal liver to a person would help cure night blindness, an illness now known to be caused by vitamin A deficiency. About 400 BC, the Greek physician – and father of Western medicine – Hippocrates of Kos (460 to 370 BC) reported via his 'Corpus Hippocraticum' that eating liver could cure the same vision problem. Indeed, the value of eating certain foods to maintain health was thus recognised long before vitamins were ever identified (Bender, 2003).

In the thirteenth century, the Crusaders frequently suffered from scurvy, now known to be caused by a lack of vitamin C in their food (Carpenter, 2012). Scurvy was a particular deadly disease in which the tissue collagen is not properly formed, causing poor wound healing, bleeding of the gums, severe pain and, finally, death. It had also long since been a well-known disease, appearing towards mid-winter in Northern European countries. Much later, in the sixteenth century, the therapeutic effects of lemon juice against scurvy (then named scorbut) became gradually known during long sea and ocean discovery voyages. The disease name, scorbut, seems to be derived from the Old Nordic 'skyr-bjugr', meaning 'sour milk-abscess', believed to be caused by continuous use of sour milk or 'skyr' as main food on long sea journeys; the Medieval Latin term was *scorbutus*, later known as *Sceurbuyck* in French, *Scheurbuyck* in Dutch and *scorbuicke* in English and then as *scorbut*, but it is now known as *scurvy*. The chemical name of vitamin C, L-ascorbic acid, is actually derived from these old names (Davies, Austin and Partridge, 1991). Scurvy had caused the loss of most ship crew members on Vasco da Gama's journey rounding the Cape of Good Hope in 1499 and those of Ferdinand Magellan during his first circumnavigation of our globe during 1519–1522. The Scottish physician James Lind, a pioneer in naval hygiene, studied this disease in 1747 and described, in 1753, in his book 'A treatise of the scurvy', the beneficial effect of eating fresh vegetables and citrus fruits in preventing it. He recommended that the British Royal Navy use lemons and limes to avoid scurvy; this led to the nickname 'limeys' for

Industrial Biotechnology of Vitamins, Biopigments, and Antioxidants, First Edition.
Edited by Erick J. Vandamme and José L. Revuelta.

British sailors at that time. However, these findings were not widely practiced even by the Royal Navy's Arctic expeditions in the nineteenth century, where it was believed that scurvy could be prevented by practising good hygiene and exercise, rather than by a diet of fresh food. (Ant)Arctic expeditions thus continued to be plagued by scurvy and other deficiency diseases further into the twentieth century. The prevailing medical theory was that scurvy was caused by tainted canned foods!

For another nutritional deficiency disease (vitamin B_3 or niacin deficiency) already described for its dermatological effects in 1735 by Gaspar Casal in Spain, the Italian medical doctor Francesco Frapoli used the name pellagra (pelle = skin; agra = rough), referring to a rough skin appearance. Pellagra was common in people who obtained most of their food energy from maize, notably in the Americas, but also in Africa and China. Its emergence also depended on neglecting the once common practice of the 'nixtamalisation' process – a special method of milling the whole dried corn kernel – making niacin, bound as niacytin, nutritionally available in the kernel.

In the nineteenth century, in Japan, the Hikan child diseases (keratomalacia or necrosis of the cornea and xerophthalmia or eye dryness) were successfully treated by including cod liver oil, eel fat or chicken liver, as a source of vitamin A, in the diet. It was also found that cod liver oil and also direct sunlight had a curing effect on rickets (vitamin D deficiency), a disease already well described by the English physician Daniel Whistler in 1645 and based on earlier observations of his colleague Francis Glisson. During the late eighteenth and early nineteenth centuries, the use of food deprivation studies, especially with mice and rats, but also with humans, allowed scientists gradually to isolate and identify a number of vitamins. Lipids from fish oil were successfully used to cure rickets in rats, and the fat-soluble nutrient was named 'antirachitic A or vitamin A'; this first vitamin 'bioactivity' ever isolated, which cured rickets, is now named vitamin D.

In 1881, the Russian surgeon Nikolai Lunin, while studying the effects of scurvy at the University of Tartu (now Estonia), compared the effects of feeding mice with milk versus an artificial mixture of then known milk constituents (proteins, fats, carbohydrates and salts); the mice that received only the individual milk constituents died, while those fed milk developed normally. He concluded that 'a natural food such as milk must therefore contain small quantities of unknown substances essential for life'.

In the Far East, when hulled rice was replaced by dehulled or polished white rice as the staple food of the middle class, a sharp increase in the occurrence of beriberi, a Sinhalese term meaning 'serious weakness' (due to lack of vitamin B_1), was observed, and it became an endemic disease. In 1884, Takaki Kanehiro, a British trained medical doctor of the Imperial Japanese Navy, observed that beriberi was endemic among the low-ranking crew, just eating rice, but not among officers who also consumed a Western-style diet. He experimented with using crews of two battleships: one was fed only white rice, and the other received a diet of meat, fish, barley, rice and beans. The group that ate only white rice reported 161 crew members with beriberi and 25 deaths, while the other group had only 14 cases of beriberi and no deaths. This convinced the Japanese Navy that the white rice diet was the cause of beriberi.

In 1897, the Dutch physician Christiaan Eijkman, working in what is now Indonesia, further observed that poultry fed with polished rice developed polyneuritis, a disease similar to human beriberi. This disease could also be prevented and cured by feeding rice and the silver fleece of the rice kernel; his co-worker, Gerrit Grijns hypothesised that beriberi was caused by a 'protecting factor' (later known as *vitamin* B_1) that was obviously lacking in dehulled rice. In 1898, the English biochemist Frederick G. Hopkins postulated that some foods contain 'accessory factors', in addition to proteins, carbohydrates, fats and nucleic acids that are necessary for the healthy functioning of the human body. Later, Hopkins and Eijkman were awarded the Nobel Prize for Physiology/Medicine in 1929 for their research on vitamins. Around 1910, F.G. Hopkins in the United Kingdom and T.B. Osborne and L.B. Mendel in the United States initiated research on modern vitamins with animal models and substantiated a theory, stating that diseases, such as night blindness, scurvy, pellagra, rickets, beriberi, hypocobalaminemia and paraesthesia, were the result of a lack of certain essential food components in the diet. We know now that all these aforementioned diseases are the result of nutritional vitamin deficiencies, that is, vitamin A, vitamin C, vitamin B_3 or niacin, vitamin D, vitamin B_1 or thiamine, vitamin B_{12} and vitamin B_5 deficiencies (Rosenfeld, 1997).

1.2
Vitamins: What's in a Name

The first vitamin complex was isolated in 1910 by the Japanese scientist Umetaro Suzuki, who succeeded in extracting a water-soluble complex of micronutrients from rice bran that prevented beriberi and named it 'aberic acid'. He published his discovery as an article in a Japanese scientific journal that, however, in a more accessible German translation failed to mention that it was a novel nutrient, thus gaining little attention! In 1912, the Polish biochemist Casimir Funk isolated the same beriberi-preventing complex of micronutrients from rice bran, displaying chemical properties of an amine; this led him in 1912 to coin the name 'vitamine' for this type of 'vital amine' compounds (Piro *et al.*, 2010). Funk also found in aqueous extracts of brewer's yeast a growth-promoting additive for the diet of young rats; it was called *vitamin B complex*. This vitamin B complex was, in the coming decades, to be resolved into its component vitamins: B_1, B_2, B_3, B_5, B_6, B_7, B_9 and B_{12}. The name 'vitamine' soon became synonymous with Hopkins' 'accessory factors', and by the time it was shown that not all vitamins are amines, this word was already in general use. In 1920, Jack Cecil Drummond proposed that the final 'e' be dropped to de-emphasise the 'amine' reference, as more researchers began to realise that not all vitamins have an amine moiety.

In 1913, American nutritional biochemists Elmer V. McCollum and M. Davis demonstrated a lipo-soluble factor A in butter fat and egg yolk, and in 1915, a

Table 1.1 Discovery years of vitamins and their original source.

Discovery year	Vitamin	Used food source
1910	B_1 (Thiamine)	Rice bran, yeast
1913	A (Retinol)	Cod liver oil
1920	D_3 (Calciferol)	Cod liver oil
1920	B_2 (Riboflavin)	Meat, dairy, eggs
1922	E (Tocoferol)	Wheat germ oil, unrefined vegetable oils
1926	B_{12} (Cobalamin)	Liver, animal products, eggs
1928	C (Ascorbic acid)	Citrus
1928	F (Essential fatty acids)	Plant oils
1929	K_1 (Phylloquinone)	Leaf vegetables
1930	F (Essential fatty acid)	Plant oils
1931	B_5 (Pantothenic acid)	Meat, whole grains
1931	B_7 (Biotin)	Meat, dairy products, eggs
1934	B_6 (Pyridoxine)	Meat, dairy products
1936	B_3 (Niacin)	Meat, grains
1941	B_9 (Folic acid)	Leafy vegetables
1957	Q_{10} (Ubiquinone)	Beef heart tissue

water-soluble factor B was found in wheat germ. It was Drummond who, in 1920, named the fat-soluble factor vitamin A; the water-soluble anti-beriberi factor was named vitamin B; the water-soluble anti-scorbut factor was first isolated in 1928 and named hexuronic acid, now vitamin C. In 1925, the fat-soluble anti-rickets factor was named vitamin D. After the 1920s, discovery and isolation of several other vitamins followed relatively quickly (see Table 1.1), and their structures, nutritional and chemical properties and chemical synthesis were studied in great detail in the following two decades.

In 1930, the Swiss chemist Paul Karrer elucidated the structure of beta-carotene, the main precursor of vitamin A and identified other carotenoids as pigments. Karrer and the British chemist Norman Haworth also made significant contributions to the chemistry of flavins, leading to the identification of riboflavin, for which they received the Nobel Prize in Chemistry in 1937. In 1931, the Hungarian physiologist Albert Szent-Györgyi and a fellow researcher, Joseph Svirbely, suspected that 'hexuronic acid' was actually vitamin C; they gave a sample to Charles Glen King, who proved its anti-scorbutic activity in his long-established guinea-pig scorbutic assay. In 1937, Szent-Györgyi received the Nobel Prize in Physiology/Medicine. In 1943, American biochemist Edward Albert Doisy and Danish biochemist Hendrik Dam were awarded the Nobel Prize in Physiology/Medicine for their discovery of vitamin K and for the elucidation of its chemical structure. In 1967, American George Wald became a Nobel laureate for his discovery that vitamin A participated directly in the physiological and chemical processes in the visual cycle.

Vitamin nomenclature was initially based on the use of letter symbols alphabetically arranged according to the time of discovery; soon it appeared that one-letter named vitamins were multiple complexes, and this led to the addition of an index to the original letters (B_1, B_2, …). Often, when the function of the vitamin became known, an appropriate letter symbol was chosen, that is, vitamin K, with K being the first letter of the German word 'Koagulation'; other names reflected deficiencies, that is, aneurin (B_1, now thiamine) for anti-polyneuritis vitamin; vitamin PP (B_3 or niacin) stood for 'pellagra-preventing' vitamin. Reasons that the list of vitamins skips certain letters of the alphabet are given as follows: certain compounds were discarded as false leads, were reclassified over time or were renamed because of being part of a complex. Letter names or trivial names are generally more in use than the IUPAC names. The division into fat-soluble and water-soluble vitamins as introduced about 100 years ago by McCollum and Davis is still universally in use today (Eggersdorfer *et al.*, 2012).

Another term that is often encountered in vitamin nomenclature is 'vitamer': by definition, a vitamer of a particular vitamin refers to any of a number of chemical compounds, generally having a similar molecular structure, each of which shows varying vitamin activity in a vitamin-deficient biological system (Table 1.2). As an example, vitamin A refers to at least six vitamer chemical structures, each displaying slightly differing properties: four of these are found naturally in plant foods and are carotenoids; the retinol and retinal forms occur in animal-based foods, and these are several times (up to six times) as effective in humans as the carotenoid forms; for example, the carotenoid forms of vitamin A cannot be absorbed by cats and ferrets and therefore display no vitamin A activity in them.

Table 1.2 List of vitamins by generic descriptor, with some of their vitamers including active forms.

Vitamin generic descriptor name	Vitamer chemical name(s) or chemical class of compounds
Vitamin A	Retinol, retinal and four carotenoids: the carotenes alpha-carotene, beta-carotene, gamma-carotene; and the xanthophyll, beta-cryptoxanthin
Vitamin B_1	Thiamine, thiamine pyrophosphate (TPP)
Vitamin B_{12}	Cyanocobalamin, hydroxycobalamin, methylcobalamin, adenosylcobalamin
Vitamin B_2	Riboflavin, flavin mononucleotide (FMN), flavin adenine dinucleotide (FAD)
Vitamin B_3	Niacin (nicotinic acid), niacinamide
Vitamin B_5	Pantothenic acid, panthenol, pantetheine
Vitamin B_6	Pyridoxine, pyridoxamine, pyridoxal, pyridoxal 5-phosphate
Vitamin B_9	Folic acid, folinic acid, 5-methyltetrahydrofolate
Vitamin C	Ascorbic acid, Dehydroascorbic Acid, calcium ascorbate, sodium ascorbate, other salts of ascorbic acid
Vitamin D	Calcitriol, ergocalciferol (D_2), cholecalciferol (D_3)
Vitamin E	Tocopherols (alpha, beta, gamma and delta-tocopherol), tocotrienols (alpha-, beta-, gamma-, delta-tocotrienols)
Vitamin F	Linoleic acid and alpha-linolenic acid
Vitamin K	Phylloquinone (K_1), menaquinones (K_2), menadiones (K_3)

1.3 Physiological Functions of Vitamins and Related Compounds

From a chemical point of view, vitamins are a very heterogeneous and diverse group of organic compounds, yet they can be considered as a particular single group of molecules. A vitamin is an organic chemical, an essential vital nutrient that an organism requires in limited amounts, one that the organism cannot synthesise on its own in sufficient quantities and that normally must be obtained through the diet. This implicates that the term 'vitamin' is conditional upon a particular organism and the food habits and situation. Vitamin C is a vitamin for humans, but not for most other animal organisms (except primates, guinea pigs, bats, some birds and fishes) nor for plants or microbes. Vitamin supplementation is important for the treatment of certain health conditions and for malnutrition as indicated earlier (Bender, 2003).

There are other essential nutrients such as dietary minerals and essential amino acids that are usually not considered as vitamins by convention. However, over the past decades, novel vitamin-like compounds have been discovered and studied in every detail and are now being considered as real vitamins; some are already commercialised. They include the essential fatty acids (EFAs), also called *vitamin F* or *polyunsaturated fatty acids* (PUFAs), the coenzyme ubiquinone (vitamin Q_{10}) and several quinoprotein factors such as pyrroloquinoline quinone (PQQ). Other molecules are still considered as growth factors and include inositol, glutathione, L-carnitine, carnosine, gamma-aminobutyric acid (GABA) and flavonoids. They perform diverse essential physiological functions or behave as antioxidants.

Most vitamins have to be provided via daily food/feed intake, but certain vitamins can be formed partially or indirectly within the body. Examples are:

- compounds – often called provitamins – with no apparent or low vitamin activity that can be converted into a vitamin within the body:
  - provitamin A or beta-carotene (in vegetables and fruits) converted into vitamin A
  - the amino acid tryptophan (in protein-rich food) converted into vitamin B_3 (niacin)
  - provitamin B_5 (panthenol) converted into vitamin B_5 (pantothenic acid)
  - provitamin D_2 or ergosterol (in yeasts, fungi, plants) converted into vitamin D_2 (ergocalciferol)
  - provitamin D_3 or 7-dehydrocholesterol (in our skin) converted into vitamin D_3 (cholecalciferol).
- other vitamins that are formed by the intestinal microbiota (Guarner and Malagelada, 2003; Le Blanc *et al.*, 2013), that is,
  - vitamin K_2 (menaquinone)
  - some B vitamins (B_1 or thiamin, B_2 or riboflavin, B_7 or biotin, B_{12} or cobalamin)

It is also well known that most fermented foods and drinks are enriched in their vitamin content derived from the beneficial microorganisms involved in their production by fermentation (Farnworth, 2003).

Vitamins have a catalytic role in the body, in enabling optimal biosynthesis, conversion and degradation of macromolecules, such as nucleic acids, proteins, lipids and carbohydrates or their building blocks. The physiological/biochemical function of most water-soluble vitamins is now well known: they are part of coenzymes, involved in enzymatic group transfer and thus responsible for specific biochemical reactions to occur (Padh, 2009). A survey is summarised in Table 1.3; see also Table 1.5.

The physiological functions of fat-soluble vitamins and water-soluble vitamin C are more varied and complex. Some examples are given in Table 1.4; see also Table 1.5.

Much debate exists about the positive effects of high doses of water-soluble vitamins on human and animal physiology; on the other hand, several hypervitaminoses of fat-soluble vitamins are well known. Compounds that specifically counteract the functioning of vitamins are known as *antivitamins* or *vitamin antagonists*; their negative action can be based on degradation of the vitamins or on the complexation of the vitamins into a non-resorbable complex, that is, avidin (in raw egg white) with biotin. Dicoumarin excludes vitamin K from the prothrombin synthesis system, and amethopterin is an antagonist of folic acid. Antivitamins present in our daily food are usually destroyed during food processing and cooking.

Table 1.3 Water-soluble vitamins and their corresponding coenzymes.

Vitamin	Coenzyme	Group transfer
B_1 (Thiamine)	Thiamine pyrophosphate (TPP)	C_2-aldehyde, decarboxylation
B_2 (Riboflavin)	Flavin adenine mononucleotide (FMN)	Hydrogen
	Flavin adenine dinucleotide (FAD)	Hydrogen
B_3 (Niacin)	Nicotinamide adenine dinucleotide (NAD+)	Hydrogen
	Nicotinamide adenine dinucleotide phosphate (NADP+)	Hydrogen
B_5 (Pantothenic acid)	Coenzyme A	Acyl
B_6 (Pyridoxine)	Pyridoxalphosphate	Amino, decarboxylation
B_7 (Biotin)	Biocytin	Carboxyl
B_9 (Folic acid)	Tetrahydrofolic acid	Formyl
B_{12} (Cyanocobalamin)	B_{12} coenzyme	Carboxyl, H-X rearrangements

Table 1.4 Physiological functions of vitamin C and fat-soluble vitamins.

Vitamin	Important physiological functions
Vitamin C	Cosubstrate of monooxygenases; role in redox reactions; hydroxylation of amino acids; hormone synthesis; iron absorption
Vitamin A	Active form (11-*cis*-retinol) is part of rhodopsin, the light-sensitive molecule in the eye; biosynthesis of proteoglycans; epithelial cell formation; immunostimulation
Vitamin D_3	Active form (1,25-dihydroxycholecalciferol) regulates Ca and P metabolism; bone and teeth formation; prevention of osteoporosis
Vitamin E	Antioxidant towards unsaturated compounds; protects membrane integrity
Vitamin F	Long-chain polyunsaturated fatty acids form prostaglandins, thromboxanes and related compounds having physiological effect in the body such as being anti-inflammatory, preventing platelet aggregation
Vitamin K_1	Formation of γ-carboxyglutamate residues in osteocalcin; bone formation

1.4 Technical Functions of Vitamins and Related Compounds

In addition to their nutritional, physiological and medical importance, vitamins and related compounds have also found large-scale technical applications, for example, as antioxidants (D-isoascorbic acid as the C_5-epimer of vitamin C, glutathione, vitamin E), as acidulants (vitamin C) and as biopigments (carotenoids, riboflavin) in the food, feed, cosmetic, chemical, nutraceutical and pharmaceutical sectors. There is a special need for natural pigments of (micro)biological origin to replace synthetic pigments and colourants; certain carotenoids (beta-carotene, lycopene, astaxanthin) and *Monascus* pigments have already been used in this respect (Vandamme, 2002, 2011; Patakova, 2013). Details about technical applications of vitamins, pigments, antioxidants and other related molecules are discussed in the corresponding chapters in this volume.

1.5 Production and Application of Vitamins and Related Factors

The staple food of humans, including cereals, rice, potato, vegetables, fruits, fish, meat, milk and eggs, forms the basic source of vitamins and related growth factors. Adequate nutrition should thus supply this daily need of vitamins. This need, however, increases with an unbalanced diet, physical exercise, pregnancy, lactation, active growth, re convalescence, drug abuse, stress, air pollution and so on. Pathological situations, such as intestinal malresorption, stressed intestinal microbiota, liver/gall diseases, treatment with drugs, antibiotics or hormones and

Table 1.5 Survey of the vitamins with main food sources, deficiency diseases, Recommended Dietary Allowance (RDA) and overdose diseases.

Vitamin A
Food sources: *retinol (in animal-derived food): liver, meat, butter, margarines, fatty fish, milk and derived products, cheese, egg yolk
*Provitamin A carotenoids (in plant-derived food): leafy vegetables, spinach, carrots, yellow and orange fruits
Deficiency diseases: night blindness, hyperkeratosis, keratomalacia, dry and scaly skin, brittle hair
Recommended dietary daily allowances (RDA): 350–750 μg (as retinol)
Overdose: >7500 μg retinol/day; fatigue, liver intoxication

Vitamin D
Food sources: *cholecalciferol (D_3): formed in our skin + UV in sunlight; also from animal-derived food: butter, margarines, fatty fish (herring, eel, salmon, mackerel), milk, cheese, egg yolk
*Ergocalciferol (D_2): yeast, wheat germ oil, cabbage, citrus fruits
Deficiency diseases: rickets, osteomalacia, osteoporosis
RDA: 10–15 μg
Overdose: >50 μg/day; hypercalcemia

Vitamin E
Food sources: plant oils rich in vitamin F, nuts, seeds, vegetables, fruits, bread, grains, cereals
Deficiency diseases: heamolytic anaemia, neurological disorders
RDA: 1–10 mg/day per gram
Overdose: >1 g/day

Vitamin F (EFAs)
Food sources : fish, especially oily fish (sardines, herring, salmon, etc.), egg yolks
Deficiency diseases: absence of long-chain PUFAs in neonatal children has adverse effects on brain and eye development. Therefore, they are now added to infant formula in over 70 countries
RDA: 1–2% of total daily calorie intake
Overdose: no known effects if consumed in humans at up to 7 g/day; also safe up to 30 g/kg body weight when fed to rats

Vitamin K
Food sources: *vitamin K_1 (phylloquinone): green leafy vegetables, fruits, milk, meat, egg yolk, cereals
*Vitamin K_2 (menaquinone): via gut microbiota
Deficiency diseases: impaired blood coagulation; haemorrhage
RDA: 10–35 μg/day
Overdose: not known

Vitamin B_1 (thiamine)
Food sources: bread, cereals, potatoes, vegetables, pork meat, milk products, eggs
Functions: synthesis of nucleic acids; essential in carbohydrate and energy metabolism; nerve impulse functioning
Deficiency diseases: beriberi; Wernicke–Korsakoff syndrome; depression; memory loss; neurological disorders; heart damage
RDA: 0.3–1 mg/day
Overdose: not known

(continued overleaf)

Table 1.5 (*Continued*)

Vitamin B_2 (riboflavin) Food sources: milk and other dairy products (to be stored in the dark), meat (liver), vegetables, fruits, bread, cereals Functions: essential role in metabolism of carbohydrates, proteins, lipids; promotes conversion of tryptophan into niacin; conversion of vitamin B_6 and vitamin B_9 into active forms; mobilisation of iron Deficiency diseases: glossitis; inflammation of skin, mucous membranes, seborrhoeic dermatitis, vision problems, secondary iron deficiency due to intestinal malabsorption, impairs B_6 and B_9 activation RDA: 0.4–1.6 mg/day Overdose: not known
Vitamin B_3 (niacin) Food sources: meat (liver), fish, whole meal bread, vegetables, potatoes, yeast, nuts Functions: essential role in energy metabolism; involved in numerous enzymatic reactions (synthesis of fatty acids and cholesterol); DNA repair and stress responses Deficiency diseases: pellagra, via diet mainly based on maize RDA: 8–18 mg/day; expressed as niacin equivalents, NE: 1 NE = 1 mg nicotinic acid or nicotinamide and = 60 mg food source tryptophan (B_2 is involved in this conversion) Overdose: >500 mg nicotinic acid per day; liver and eye damage; blood vessel dilatation
Vitamin B_5 (pantothenic acid) Food sources: meat, eggs, whole grain cereals, vegetables, pulses, fruits, milk products Functions: role in carbohydrate and fatty acid metabolism; synthesis of cholesterol and fatty acids; formation of red blood cells, formation of sex and stress-related hormones Deficiency diseases: burning feeling in extremities, depression, irritability, vomiting, stomach pains RDA: 2–12 mg/day Overdose: diarrhoea, increase the risk of bleeding
Vitamin B_6 (pyridoxine) Food sources: meat (chicken, beef liver, pork and veal), eggs, bread, grain products, potatoes, pulses, vegetables, milk and products, cheese Functions: important role in energy metabolism, in polyunsaturated fatty acids, phospholipids and amino acid metabolism; production of hormones, red blood cells and cells of the immune system; controls (along with vitamin B_{12} and vitamin B_9) homocysteine levels in the blood; improves conversion of tryptophan into niacin and into serotonin Deficiency diseases: anaemia, depression and nervous system disorders; impairment of the immune system; inflammation of skin and mucosa RDA: 0.4–1.7 mg/day Overdose: >50 mg/day; irreversible neuropathy of limbs
Vitamin B_7 (biotin) Food sources: yeast, kidney, eggs, liver, milk and milk products, nuts, pindas Functions: role in energy metabolism and formation of fatty acids; maintaining healthy skin and hair Deficiency diseases: seldom; anaemia; depression; cracking in the corners of the mouth, swollen and painful tongue; dry eyes; loss of appetite; fatigue; insomnia RDA : 10–100 μg/day; also formed by intestinal microbiota Overdose: not known

Table 1.5 (*Continued*)

Vitamin B_9 (folic acid) Food sources: whole grain products, bread, cereals, green vegetables, fruits, milk and dairy products Functions: red blood cell formation; involved in metabolism of histidine, glycine, methionine, DNA and RNA synthesis in the presence of B_6 or B_{12}; maintenance of cells; development of the brain and spinal marrow in foetus Deficiency diseases: macrocytic anaemia; birth defects (spina bifida, harelip, cleft palate); growth retardation; increased homocysteine levels in the blood RDA: 50–400 μg/day; prevents spina bifida (neural tube defects) in babies Overdose : overdose can mask B_{12} deficiency
Vitamin B_{12} (cyanocobalamin) Food sources: meat and other animal products (milk and dairy, cheese, eggs); not present in plant-derived food Functions: formation of red blood cells; nerve system functioning; controls, together with vitamin B_6 and vitamin B_9, homocysteine levels in the blood; production of nucleic acids Deficiencies: pernicious anaemia; neurological disorders; memory loss; deficiency risk also caused by stomach surgery (insufficient secretion of 'intrinsic factor', IF) or intestinal diseases (Crohn's disease); heart disease RDA: 0.5–2.0 μg/day; uptake depends on level of IF, secreted by parietal stomach gland cells Overdose: >200 μg/day
Vitamin C (ascorbic acid) Food sources: many fruits (citrus, kiwi, raspberry, strawberry, guava, mango), vegetables (Brussels sprouts, cabbage, paprika, potatoes) Functions: formation of collagen; metabolism of sugars, proteins and lipids; muscle and brain metabolism; bone formation; hormone synthesis; iron uptake from food; immune defence; antioxidant Deficiencies: scurvy; fatigue; retarded wound healing; dry and splitting hair; inflammation of the gums; decreased ability to ward off infection RDA: 35–110 mg/day Overdose: extremely high doses (>2–5 g/day) increase the risk of kidney stones; diarrhoea and gastrointestinal disturbances

enzyme deficiencies, can also lead towards vitamin shortages despite sufficient intake. Malnourishment in many underdeveloped countries but equally wrong food habits in developed countries also ask for direct nutritional and medical remediation, combined with daily diet adjustment. Vitamin-enriched and medicated feed are used worldwide to procure healthy livestock. Overdose of vitamins, especially fat-soluble ones, but also some water-soluble ones (high doses of C, B_3, B_5, B_6, B_9), can lead to hypervitaminoses and diseases. Table 1.5 presents a survey of vitamins with main food sources, deficiency diseases, recommended dietary daily allowance (RDA) and overdose diseases.

Concentrates or extracts derived from these vitamin-rich natural staple food products (of plant, animal or microbial origin), however, find relatively little use in the food, feed, pharmaceutical or cosmetic sector. Some of the reasons are:

- the level of vitamins in the natural plant/animal source is usually relatively low and fluctuates drastically (i.e. exceptions are PUFAs in plant oils and fish oils, vitamin D in fish oils).
- their organoleptic presentation and shelf life are often not optimal.
- vitamins are labile molecules during the process of harvest, preservation, storage or preparation of foodstuffs and are generally sensitive to pH, heat (B_2, B_5, B_6, B_9, C, E), light (B_2, B_6, B_9, B_{12}, C, D), oxygen (B_9, C, D, F); water-soluble vitamins are easily lost by aqueous extraction or other manipulations of these natural food-vitamin sources.

These drawbacks have led to the industrial manufacturing of most vitamins and related factors. Currently, several vitamins are produced chemically (A, D_3, E, K and B_1, B_5, B_6, B_7, B_9), although microbiological/biotechnological methods exist or are being developed, though not economically profitable as yet (Demain, 2000, 2007; Laudert and Hohmann, 2011). Others are produced (exclusively) by microbial fermentation with bacteria and/or fungi (C, D_2, B_2, B_{12}, EFAs). Some are produced by a combination of chemical steps and microbial/enzymatic steps (B_3, B_5, C) (Vandamme, 1989, 1992; Eggersdorfer *et al.*, 1996; De Baets, Vandedrinck and Vandamme, 2000; Shimizu, 2008; Laudert and Hohmann, 2011). Some are produced via microalgal culture in ponds or fermenter vessels (beta-carotene, PUFAs) (Cadoret, Garnier and Saint-Jean, 2012; Borowitzka, 2013).

The detailed biosynthetic pathways (and their metabolic regulation and controls) used by those microorganisms have been almost fully elucidated for most vitamins and similar compounds but only over the past two decades, mainly by studying model microbial strains and/or producer microorganisms, such as bacteria (*Escherichia coli, Serratia, Bacillus, Lactobacillus, Pseudomonas, Gluconobacter, Sinorhizobium, Agrobacterium, Propionibacterium, Rhodobacter, Arthrospira*), yeasts (*Saccharomyces, Candida, Xanthophyllomyces, Yarrowia*), fungi (*Blakeslea, Ashbya, Mortierella, Mucor, Monascus*), as well as green microalgae (*Dunaliella, Euglena, Haematococcus*), marine non-photosynthetic dinoflagellates (*Crypthecodinium*) and marine non-photosynthetic thraustochytrid–microalgae (*Schizochytrium*) (Laudert and Hohmann, 2011; Borowitzka, 2013; Ledesma-Amaro *et al.*, 2013; Bellou *et al.*, 2014). For some of the vitamins and related factors, microbial overproduction still remains a challenge. In the future, the advent of synthetic biology will allow for the complete construction of tailor-made microbial vitamin producer strains (Wang, Chen and Quinn, 2012).

This volume focuses especially on the biotechnological aspects of vitamins and related compounds – on biosynthesis and on their production processes. Apart from obtaining these vitamins and related compounds via a natural process – which is what microbial fermentation, biocatalysis and algal culture are all about, fermentation-based or enzymatic biocatalytic processes furthermore yield the desired enantiomeric compound, and they can be redirected via genetic and biotechnological modification of the involved bacteria, yeast and fungi or microalgae into high-yielding production systems. Especially, the advancement of genetic engineering techniques and the introduction of metabolic engineering have recently allowed high-yielding microbial strains that are suitable for industrial production of vitamins and related compounds to be constructed, and this has led to their wider application.

A broad range of applications now exists for these vitamin preparations (and similarly for related factors) in the food, feed, cosmetic, technical and pharmaceutical sectors:

- *Revitamination*: restoring the original vitamin level of a foodstuff.
- *Standardisation*: addition of vitamins to compensate for natural fluctuations.
- *Vitamin enrichment*: further addition of vitamins to a level higher than the original one.
- *Vitamination*: addition of vitamins to products lacking them.
- *Technical additive*: beta-carotene as pigment, vitamins C and E as antioxidants, riboflavin as yellow pigment.
- *Medical applications*: to alleviate hypo- or even avitaminoses.

1.6 Outlook

Vitamins and related compounds belong to those few chemicals with a strong positive appeal to most people worldwide. How they were discovered and how they are produced are not very well known nor understood by most people, layman as well as even academics, as long as they are widely available!

This volume hopes to contribute this understanding, not only to scientists, microbiologists, biochemists, nutritionists and medical people, but also to process biochemists and industrial biotechnologists already involved in – or attracted to – the production enigmas and application potential of vitamins, biopigments, antioxidants and related molecules.

Even today, vitamins remain to be seen as fascinating yet still elusive molecules!

References

Bellou, S., Baeshen, M.N., Elazzazy, A.M., Aggelis, D., Sayegh, F., and Aggelis, G. (2014) Microalgal lipids biochemistry and biotechnological perspectives. *Biotechnol. Adv.*, **32**, 1476–1493.

Bender, D.A. (2003) *Nutritional Biochemistry of the Vitamins*, Cambridge University Press, Cambridge.

Borowitzka, M.A. (2013) High value products from microalgae -their development and commercialization. *J. Appl. Phycol.*, **25**, 743–756.

Cadoret, J.P., Garnier, M., and Saint-Jean, B. (2012) Microalgae functional genomics and biotechnology. *Adv. Bot. Res.*, **64**, 285–341.

Carpenter, K.J. (2012) The discovery of vitamin C. *Ann. Nutr. Metab.*, **61**, 259–264.

Davies, M.B., Austin, J., and Partridge, D.A. (1991) *Vitamin C: its Chemistry and Biochemistry*, Royal Society of Chemistry Paperbacks, p. 134.

De Baets, S., Vandedrinck, S., and Vandamme, E.J. (2000) in *Encyclopedia of Microbiology*, Vol. 4, 2nd edn (ed J. Lederberg), Academic Press, New York, pp. 837–853.

Demain, A.L. (2000) Small bugs, big business: the economic power of the microbe. *Biotechnol. Adv.*, **18**, 499–514.

Demain, A.L. (2007) The business of biotechnology. *Ind. Biotechnol.*, **3** (3), 269–283.

Eggersdorfer, M., Adam, G., John, M.W.H., and Labler, L. (1996) in *Biotechnology*, vol. 4 (eds H. Pape and H.-J. Rehm), VCH, Weinheim, pp. 114–158.

Eggersdorfer, M., Laudert, D., Létinois, U., McClymont, T., Medlock, J., Netscher, T., and Bonrath, W. (2012) One hundred years of vitamins – a success story of the natural sciences. *Angew. Chem. Int. Ed.*, **51** (52), 12960–12990.

Farnworth, E.R. (2003) *Handbook of Fermented Functional Foods*, CRC Press LLC.

Guarner, T. and Malagelada, J.R. (2003) Gut flora in health and disease. *Lancet*, **36** (9356), 512–519.

Laudert, D. and Hohmann, H.-P. (2011) in *Comprehensive Biotechnology*, Vol. 3, 2nd edn (ed M. Moo-Young), Elsevier B.V, pp. 583–602.

Le Blanc, J.G., Milani, C., de Giori, G.S., Sesma, F., van Sinderen, D., and Ventura, M. (2013) Bacteria as vitamin suppliers to their hosts: a gut microbiota perspective. *Curr. Opin. Biotechnol.*, **24**, 160–168.

Ledesma-Amaro, R., Santos, M.A., Jimenez, A., and Revuelta, J.L. (2013) *Microbial Production of Food Ingredients, Enzymes and Nutraceuticals*, vol. 246, Woodhead Publishing Ltd, pp. 571–594.

Padh, H. (2009) Vitamin C: newer insights into its biochemical functions. *Nutr. Rev.*, **49** (3), 65–70.

Patakova, P. (2013) *Monascus* secondary metabolites: production and biological activity. *J. Ind. Microbiol. Biotechnol.*, **40**, 169–181.

Piro, A., Tagarelli, G., Lagonia, P., Tagarelli, A., and Quattrone, A. (2010) Casimir Funk: his discovery of the vitamins and their deficiency disorders. *Ann. Nutr. Metab.*, **57**, 85–88.

Rosenfeld, L. (1997) Vitamine-vitamin. The early years of discovery. *Clin. Chem.*, **43** (4), 680–685.

Shimizu, S. (2008) in *Biotechnology: Special Processes*, Vol. 10, 2nd edn (eds H.-J. Rehm and G. Reed), Wiley-VCH Verlag GmbH, Weinheim, pp. 320–340.

Vandamme, E.J. (ed) (1989) *Biotechnology of Vitamins, Pigments and Growth factors*, Elsevier Applied Science, London, New York.

Vandamme, E.J. (1992) Production of vitamins, coenzymes and related biochemicals by biotechnological processes. *J. Chem. Technol. Biotechnol.*, **53**, 313–327.

Vandamme, E.J. (2002) (Micro) biological colors. *Agron. Food Hi-Technol.*, **13** (3), 11–16.

Vandamme, E.J. (2011) Natural colors but of course!. *SIM-News*, **61** (5), 121–128; Society for Industrial Microbiology.

Wang, X., Chen, J., and Quinn, P. (eds) (2012) *Reprogramming Microbial Metabolic Pathways*, Springer.

Part I
Water-Soluble Vitamins

Industrial Biotechnology of Vitamins, Biopigments, and Antioxidants, First Edition.
Edited by Erick J. Vandamme and José L. Revuelta.

2
Industrial Production of Vitamin B_2 by Microbial Fermentation

José L. Revuelta, Rodrigo Ledesma-Amaro, and Alberto Jiménez

2.1
Introduction and Historical Outline

Vitamin B_2 (riboflavin) is an essential component of the human and animal diet, since it is the precursor of the flavocoenzymes flavin mononucleotide (FMN) and flavin adenine dinucleotide (FAD), which are mostly involved in redox reactions. Riboflavin FMN and FAD are also referred to as *flavins*. Flavin cofactors (FMN and FAD) participate in the metabolism of carbohydrates, fats, ketone bodies and proteins. Additionally, riboflavin is closely related to the metabolism of other vitamins such as B_3, B_6 and A, glutathione recycling and homocysteine metabolism. Since humans and animals lack the capacity to synthesise riboflavin, it must be obtained through the diet. Thus, riboflavin is industrially produced and commercialised both for animal feeding and as a food additive (Ledesma-Amaro *et al.*, 2013).

Riboflavin was initially discovered in 1879 as lactoflavin, a yellow pigment obtained from milk. In 1932, Warburg and Christian purified a yellow enzyme from aqueous yeast extracts that could be fractioned into two enzymatically inactive components: a protein and a yellow dye (Warburg and Christian, 1932). Soon after, Paul Karrer and Richard Kuhn independently elucidated the molecular structure of the yellow prosthetic group, which was called *riboflavin* (Karrer, Schöpp and Benz, 1935; Kuhn and Weygand, 1934). These findings were awarded with the Nobel Prizes in chemistry for Karrer and Kuhn in 1937 and 1938, respectively.

2.2
Occurrence in Natural/Food Sources

The main natural sources of riboflavin are milk, dairy products, eggs and lean meat. Green leafy vegetables, fish, legumes, cereals and nuts are also important sources of the vitamin. Since vitamin B_2 is light-sensitive, it is important that these food sources are stored in dark environments (Powers, 2003).

Industrial Biotechnology of Vitamins, Biopigments, and Antioxidants, First Edition.
Edited by Erick J. Vandamme and José L. Revuelta.

The Recommended Daily Allowance (RDA) of this vitamin varies with age and gender and is between 0.3 and 0.4 mg/day for infants, between 0.5 and 0.9 mg/day for children, 1.3 mg/day for adult males and 1.1 mg/day for adult females. The RDA for women during pregnancy and lactation is increased to 1.4 and 1.6 mg/day, respectively (Eldridge, 2004; Powers, 2003). Riboflavin deficiencies in humans (ariboflavinosis) often occur in association with multiple nutrient deficits, and they are associated with increased risk of cardiovascular diseases and impairment in iron metabolism (anaemia). Ariboflavinosis can also result in developmental abnormalities and growth retardation. In contrast, because riboflavin is a water-soluble vitamin, excess uptake is easily eliminated through the urine. Some diseases such as cancer, cardiac disease and diabetes mellitus can exacerbate riboflavin deficiency, and, therefore, higher intakes of riboflavin are recommended for these risk groups (Eldridge, 2004; Institute of Medicine (US) Standing Committee on the Scientific Evaluation of Dietary Reference Intakes and its Panel on Folate, 1998; Powers, 2003).

2.3
Chemical and Physical Properties; Technical Functions

Riboflavin crystallises as yellow–orange needles with poor solubility in water (5–15 mg dissolves in 100 ml of water). In general, riboflavin is thermostable and its melting point is 290 °C. It is highly sensitive to light, and it is incompatible with strong oxidising agents, reducing agents, bases, calcium and metallic salts (Eggersdorfer *et al.*, 2000).

Flavins (riboflavin, FMN and FAD) are all derivatives of the dimethyl isoalloxazine skeleton (7,8-dimethyl-10-(2,3,4,5-tetrahydroxypentyl)benzo[g]pteridine-2,4(3H,10H)-dione), with a substitution in the position 10, which in the case of riboflavin is a D-ribityl group (Figure 2.1). In nature, riboflavin occurs as a free vitamin, as 5′-phosphate (FMN) and as 5′-adenosine diphosphate (FAD) (Figure 2.1). In acidic solutions, flavin nucleotides are hydrolysed to free riboflavin (Merrill *et al.*, 1981; Rivlin, 1975, 2007).

Riboflavin and other flavins show a characteristic yellow–green fluorescence under UV light at 535 and 565 nm in a pH range of 3–8. Neutral aqueous solutions of riboflavin exhibit absorption peaks at 220, 265, 372 and 445 nm. The irradiation of flavins with visible light leads to their decomposition, producing either lumiflavin (7,8,10-trimethyl isoalloxazine) or lumichrome (7,8-dimethyl isoalloxazine) in alkaline or neutral acid solutions, respectively. Riboflavin can form metal chelates when it binds to heavy metals (e.g. Fe, Mo, Cu, Ag, Cd, Ni, Zn, Co) (Merrill *et al.*, 1981; Rivlin, 2007).

2.4
Assay Methods and Units

Light absorbance and fluorescence are usually employed for the detection and quantification of riboflavin and other flavins. When different flavins are to be

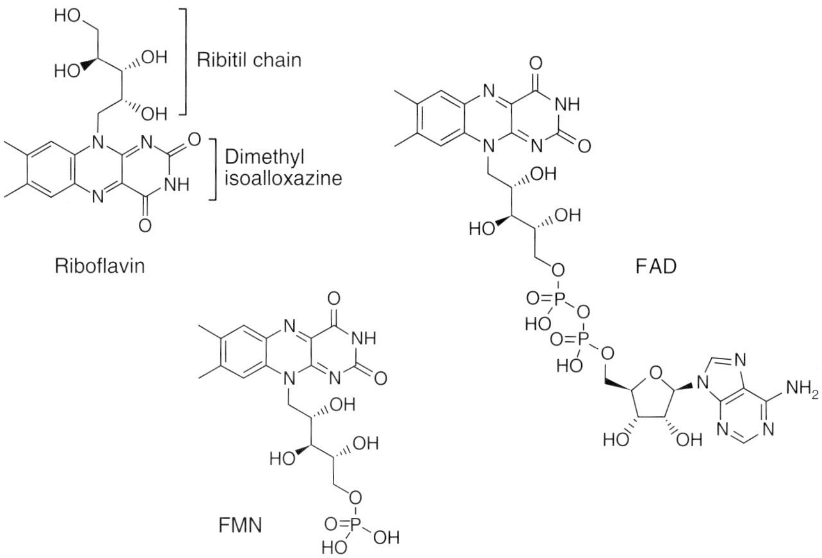

Figure 2.1 Chemical structure of riboflavin and the two coenzymes derived from riboflavin, FMN and FAD. FMN is formed from riboflavin by the addition of a phosphate group derived from adenosine triphosphate. FAD is formed from FMN after addition of an adenosine monophosphate group derived from a second molecule of adenosine triphosphate.

separated from a mixture, the method of choice is high-performance liquid chromatography (HPLC) with fluorescence detection (Vinas *et al.*, 2004; Zandomeneghi, Carbonaro and Zandomeneghi, 2007). Other techniques have also been used successfully to assay flavins such as thin-layer or ion exchange chromatography, electrophoresis and extraction by 2-phenylethanol (Abbas and Sibirny, 2011; Gliszczynska and Koziolowa, 1998).

2.5 Biological Role of Flavins and Flavoproteins

The biological role of riboflavin is mostly connected with the function of the FMN and FAD coenzymes and of the covalently bound flavins. Both FMN and FAD are cofactors of the so-called flavoproteins or flavoenzymes, which are estimated to be 3%, on average, of the gene products in eukaryotic genomes (De Colibus and Mattevi, 2006; Gudipati *et al.*, 2014; Lienhart, Gudipati and Macheroux, 2013).

The isoalloxazine tricyclic ring system of flavins is able to undergo redox reactions through the oxido-reduction of the nitrogen atoms. Hence, the isoalloxazine system can exist in three different redox states: the fully oxidised quinone state, the one-electron reduced state (semiquinone state) and the two-electron reduced state (hydroquinone state) (De Colibus and Mattevi, 2006; Massey, 2000). Accordingly, flavins are very versatile electron transfer molecules that are able to transfer both one and two electrons. They can, therefore, participate in redox reactions as one- or two-electron mediators, while most of the redox cofactors can transfer exclusively either one or two electrons. The high versatility of flavins for participating in many different biochemical functions has contributed to the ubiquity of flavoenzymes in most biological systems (Massey, 2000; Miura, 2001).

Flavoenzymes can be classified into five different classes according to the redox reaction that they catalyse (i) *transhydrogenase,* where two-electron equivalents are transferred, along with the corresponding hydrogen ions, from one substrate to another; (ii) *dehydrogenase–oxidase,* where two-electron equivalents are transferred from an organic substrate to the flavin, where molecular oxygen is the oxidising substrate, being reduced to H_2O_2; (iii) *dehydrogenase–monooxygenase,* where the flavin is reduced by a reduced pyridine nucleotide and where, on oxidation, with O_2 in the presence of a cosubstrate, one atom of oxygen is inserted into the cosubstrate, while the other is reduced to H_2O; (iv) *dehydrogenase–electron transferase,* where the flavin is reduced by two-electron transfer from a reduced substrate and then reoxidised in sequential single electron transfers to acceptors, such as cytochromes and iron–sulfur proteins and (v) electron transferase, where the flavin is reduced and reoxidised in one-electron steps (Abbas and Sibirny, 2011; Mattevi, 2006; Miura, 2001; Webb, 1989).

Additionally, a few flavoproteins that are not oxidoreductases catalyse reactions with no net redox change. Some of these flavoenzymes use the redox power of flavin directly in catalysis with either two-electron chemistry (*N*-methylglutamate synthase and 5-hydroxyvaleryl-CoA dehydratase) or free-radical chemistry (chorismate synthase, DNA photolyase, (6–4) photolyase and 4-hydroxybutyryl-CoA dehydratase) (Bornemann, 2002).

Flavins are also involved in light-sensing processes. There are three major classes of flavin photosensors: light–oxygen–voltage (LOV) domains, blue-light sensor using FAD (BLUF) proteins and cryptochromes (CRYs). FMN is the absorbing chromophore of the blue-light-sensing photoreceptors, which comprise LOV domains and are involved in phototropism and chloroplast movement, among other functions (Briggs and Christie, 2002; Conrad, Manahan and Crane, 2014). In contrast, FAD is the cofactor of CRYs and BLUF proteins that also participate in sensory transduction and other related processes such as circadian timekeeping (Conrad, Manahan and Crane, 2014; Gomelsky and Klug, 2002; Krishnan *et al.*, 2001; Lin *et al.*, 1995; Linden, 2002).

Flavins also play a role in bioluminescence, since some fluorescent proteins (mostly lumazine proteins) use riboflavin, FMN or the riboflavin biosynthetic precursor 6,7-dimethyl-8-ribityllumazine (DRL) as prosthetic groups (Chatwell *et al.*, 2008). In addition, the luminescence operons of some bacteria also contain genes involved in the synthesis of riboflavin (Dunlap, 2014).

Another biological function of FMN is to participate as a substrate in the biosynthesis of vitamin B_{12} (cobalamins). The 5,6-dimethylbenzimidazole moiety of vitamin B_{12} is formed from FMN in aerobic and some aerotolerant bacteria (Eggersdorfer *et al.*, 2000; Lingens *et al.*, 1992).

Although most of the biological functions of flavins can be attributed to both FMN and FAD as coenzymes, free flavins secreted from cells can also play important roles in some biological systems. For example, riboflavin plays a role in iron intake through Fe^{++} reduction in several species such as *Helicobacter pylori* and *Campylobacter jejuni* (Crossley *et al.*, 2007; Worst *et al.*, 1998). Flavins are also required for avian embryonic development, being transported through the bloodstream and stored within the avian egg by a specific riboflavin-binding protein (White, 1987).

2.6 Biotechnological Synthesis of Riboflavin

2.6.1 Riboflavin-Producing Microorganisms

Most microorganisms are able to synthesise riboflavin, but there are only a few that are able to produce high amounts of this vitamin, typically more than 10 mg/l (Demain, 1972, 2007). These microorganisms are referred to as *flavinogenic microorganisms*, and they can be classified into three groups according to their capacities of riboflavin production: weak overproducers (able to produce around 10 mg/l), medium or moderate overproducers (with riboflavin yields up to 600 mg/l) and highly flavinogenic overproducers (producing more than 10 g/l) (Demain, 1972).

There are flavinogenic bacteria as well as flavinogenic yeast and fungi. Among bacteria, *Clostridium acetobutylicum, Shewanella oneidensis* and some species of *Mycobacterium* and *Corynebacterium* are the highest natural overproducers (Coursolle *et al.*, 2010; Crossley *et al.*, 2007; Hickey, 1945; Worst *et al.*, 1998). *Bacillus subtilis* is not a flavinogenic microorganism; however, *B. subtilis* industrial strains that are overproducers of riboflavin have been isolated and are currently used for the biotechnological production of riboflavin (Perkins *et al.*, 1999; Stahmann, Revuelta and Seulberger, 2000).

The flavinogenic eukaryotes belong to different families of yeast and fungi, although several species of the genus *Candida* and *Pichia* have been described as being flavinogenic. *Candida famata*, currently known as *Candida flareri*, is one of the most flavinogenic yeasts capable of riboflavin overproduction during iron starvation (Dmytruk and Sibirny, 2012; Stahmann, Revuelta and Seulberger, 2000). Also, the yeasts *Candida (Pichia) guilliermondii, C. ghoshii, C. parapsilosis, Debaryomyces subglobosus, Sarcoscypha occidentalis, Torulopsis candida* and the filamentous fungi *Aspergillus niger* are flavinogenic (Abbas and Sibirny, 2011; Demain, 1972).

High-riboflavin overproducers also include two yeast-like molds, *Eremothecium ashbyii* and *Ashbya gossypii,* which synthesise riboflavin in concentrations greater than 20 g/l (Demain, 2007). A riboflavin overproducer such as *A. gossypii* produces 40 000 times more vitamin than it needs for its own growth. *E. ashbyii and A. gossypii* are filamentous hemiascomycetes that belong to the Saccharomycetaceae family. *E. ashbyii* was first identified as a natural producer of riboflavin in 1935 (Guilliermond, Fontaine and Raffy, 1935), but its genetic instability hinders its use in industrial applications. In contrast, *A. gossypii*, which was originally isolated from cotton plants as the pathogen agent causing stigmatomycosis, shows a more homogeneous production of riboflavin due to high genetic stability (Stahmann, Revuelta and Seulberger, 2000; Wickerham, Flickinger and Johnston, 1946). Throughout this book chapter, we mainly focus on the industrially exploited vitamin-B_2-producer microorganisms *A. gossypii* and *B. subtilis*, although some aspects of flavin nucleotides and riboflavin production by *C. flareri* are also reviewed.

2.6.2
Biosynthesis of Riboflavin

The discovery by MacLaren (1952) that the production of riboflavin can be increased by the addition of purine derivatives to the culture medium of *E. ashbyii* suggested a connection between purine and riboflavin. Further studies subsequently described that the pyrimidine moiety of riboflavin is biosynthetically related to guanine nucleotides (Bacher and Lingens, 1969; Bacher and Mailander, 1973). Soon after, guanosine triphosphate (GTP) was identified as the committed precursor of riboflavin, supplying the pyrimidine ring and the nitrogen atoms of the pyrazine ring, as well as the ribityl side chain of the vitamin (Foor and Brown, 1975; Mailander and Bacher, 1976). Early work on the riboflavin biosynthetic pathway has been reviewed comprehensibly elsewhere (Bacher *et al.*, 2000).

The biosynthesis of one riboflavin molecule requires one molecule of GTP and two molecules of ribulose-5-phosphate as substrates. Hence, riboflavin biosynthesis comprises a two-branched pathway with two main precursors: GTP and ribulose-5-phosphate (Figure 2.2). The two branches converge in the reaction catalysed by the product of the gene *RIB4* (DRL synthase) (Garcia-Ramirez, Santos and Revuelta, 1995). Seven enzymes are involved in riboflavin biosynthesis from the aforementioned precursors, constituting the so-called riboflavin pathway (Figure 2.2).

The first step in riboflavin biosynthesis is catalysed by the enzyme GTP cyclohydrolase II (EC 3.5.4.25), which is different from GTP cyclohydrolase I (EC 3.5.4.16), an enzyme that participates in the biosynthesis of folic acid. GTP cyclohydrolase II was first isolated from cell extracts of *Escherichia coli* (Foor and Brown, 1975). The enzyme catalyses both the release of C-8 of the imidazole ring of GTP as formate and the release of pyrophosphate from the phosphoribosyl side chain. The product of the GTP cyclohydrolase II is 2,5-diamino-6-ribosylamino-4-(3H)-pyrimidinone-5′-phosphate. GTP cyclohydrolase II from *E. coli* is encoded by the gene *ribA*, whereas the yeast genes coding for GTP cyclohydrolase II are designated *RIB1* (Buitrago *et al.*, 1993; Richter *et al.*, 1993). Plants and other bacteria, such as *B. subtilis,* comprise bifunctional proteins with the activities of GTP cyclohydrolase II and 3,4-dihydroxy-2-butanone 4-phosphate (DHBP) synthase

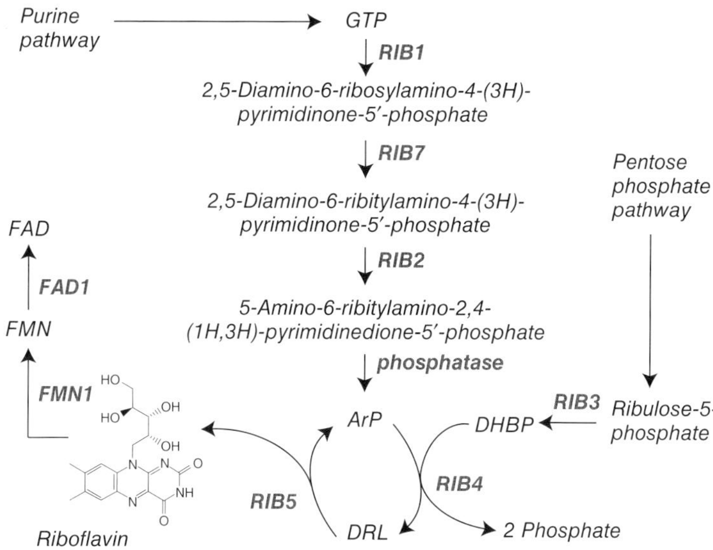

Figure 2.2 Biosynthesis of riboflavin and flavocoenzymes. *RIB1*: GTP cyclohydrolase II; *RIB7*: 2,5-Diamino-6-ribosylamino-4-(3H)-pyrimidinone-5′-phosphate reductase; *RIB2*: 2,5-Diamino-6-ribitylamino-4-(3H)-pyrimidinone-5′-phosphate deaminase; *RIB3*: 3,4-dihydroxy-2-butanone-4-phosphate synthase; *RIB4*: 6,7-dimethyl-8-ribityllumazine synthase; *RIB5*: riboflavine synthetase; *FMN1*: riboflavin kinase; *FAD1*: Flavin adenine dinucleotide synthetase.

that catalyse the initial steps of both branches of the riboflavin pathway (Hümbelin *et al.*, 1999; Singh *et al.*, 2013).

The product of GTP cyclohydrolase II must be reduced and deaminated in a series of two reactions consisting of the hydrolytic cleavage of the amino group at position 2 of the heterocyclic ring and the reduction of the ribosyl side chain to ribityl (Figure 2.2). In bacteria and plants, the deamination reaction occurs prior to the reduction of the ribosyl group (Richter *et al.*, 1997; Roje, 2007). In contrast, in yeast and fungi, the activity of an NADPH-dependent reductase encoded by the *RIB7* gene in *Saccharomyces cerevisiae* precedes the subsequent deamination of the pyrimidine ring, which is controlled by the *RIB2* gene in *S. cerevisiae* (Buitrago *et al.*, 1993; Chen *et al.*, 2013; Lv, Sun and Liu, 2013; Revuelta, Buitrago and Santos, 1995).

The product of the first three steps of the riboflavin pathway (5-amino-6-ribitylamino-2,4-(1H,3H)-pyrimidinedione-5′-phosphate) cannot serve as substrate for DRL synthase (Harzer *et al.*, 1978). Accordingly, the compound must be dephosphorylated prior to further conversion (Figure 2.2); however, both the mechanisms of dephosphorylation and phosphatase that catalyse the reaction remain to be elucidated (Haase *et al.*, 2014).

As mentioned earlier, the riboflavin pathway comprises two branches that converge to form the pteridine compound DRL in a reaction catalysed by the DRL synthase (also known as *lumazine synthase*) (Figure 2.2). The conversion of the pyrimidine ring into two condensed-ring pteridines requires the joining of a 4-carbon compound (Volk and Bacher, 1990). Hence, one of the substrates of the DRL synthase reaction is the dephosphorylated 5-amino-6-ribitylamino-2,4(1H,3H)-pyrimidinedione (ArP) from the GTP branch. The other substrate for the DRL synthase comes from the ribulose-5-phosphate branch of the pathway, where the 4-carbon precursor is converted to DHBP by the product of the gene *RIB3* in *S. cerevisiae* (Garcia-Ramirez, Santos and Revuelta, 1995; Revuelta, Buitrago and Santos, 1995; Volk and Bacher, 1990). Then, both ArP and DHBP are conjugated by DRL synthase (encoded by the *RIB4* gene in *S. cerevisiae*) to produce DRL (Garcia-Ramirez, Santos and Revuelta, 1995). The DHBP synthase of *E. coli* is a homodimer of 46 kDa encoded by the *ribB* gene (Richter *et al.*, 1999). However, in *B. subtilis* and plants, the DHBP synthase is part of a bifunctional protein, also containing GTP cyclohydrolase II activity. The DHBP synthase domain is located at the N-terminal part of the protein, whereas the GTP cyclohydrolase II domain occupies the C-terminal end of the protein (Hümbelin *et al.*, 1999).

The last step in the riboflavin pathway consists in the dismutation of DRL, which is catalysed by the enzyme riboflavin synthase. The second product of the dismutation is ArP, which is in turn the substrate of DRL synthase and is recycled in the biosynthetic pathway. In *S. cerevisiae,* the riboflavin synthase is a 25-kDa homotrimer that is encoded by the *RIB5* gene (Santos, Garcia-Ramirez and Revuelta, 1995). The riboflavin synthases from eubacteria and plants are also homotrimers (Fischer *et al.*, 2005; Liao *et al.*, 2001; Ritsert *et al.*, 1995). In contrast, riboflavin synthases from archaea are homopentamers (Ramsperger *et al.*, 2006).

Stoichiometrically, the biosynthesis of riboflavin requires 1 equiv. of GTP and 2 equiv. of ribulose-5-phosphate. Hence, of the 17 carbon atoms in the riboflavin molecule, 13 are derived from the pentose phosphate pathway (Bacher *et al.*, 2000).

Although riboflavin biosynthesis occurs only in plants, fungi and prokaryotes, the transformation of riboflavin into FMN or FAD can be achieved in all organisms. Most eubacteria comprise bifunctional riboflavin kinase/FAD synthetases that catalyse the formation of both FMN and FAD (Mack, van Loon and Hohmann, 1998; Manstein and Pai, 1986). In contrast, monofunctional riboflavin kinases are found in archaea, fungi, plants and animals (Clarebout, Villers and Leclercq, 2001; Karthikeyan *et al.*, 2003; Mashhadi *et al.*, 2008; Santos, Jiménez and Revuelta, 2000). In these organisms, monofunctional FAD synthetases are also found (McCormick *et al.*, 1997; Wu *et al.*, 1995).

The six genes encoding the riboflavin biosynthetic enzymes in *A. gossypii* (*AgRIB1, AgRIB2, AgRIB3, AgRIB4, AgRIB5* and *AgRIB7*) are structurally similar to those of the yeast *S. cerevisiae* and are scattered throughout chromosomes IV, V and VII (Revuelta, Buitrago and Santos, 1995).

In *B. subtilis,* the four genes encoding the catalytic enzymes for riboflavin synthesis (*ribGBAH*) are clustered in an operon (the *rib* operon), which contains an additional open reading frame (*ribT)* of unknown function. The untranslated leader region (designated *ribO*) was originally proposed to be the operator region. Although some mutations leading to riboflavin overproduction mapped into the *ribO* locus (Kil *et al.*, 1992), the identification of the gene that should code for the repressor protein was unsuccessful. Mutations in two *trans*-acting genes, *ribC* and *ribR*, which provoke riboflavin oversynthesis, were also identified (Mack, van Loon and Hohmann, 1998; Solovieva *et al.*, 1999). However, *ribC* and *ribR* encode bifunctional riboflavin kinase/FAD synthetase and monofunctional riboflavin kinase, respectively, which are not *ribO*-binding repressors but control *rib* operon expression by producing FMN and/or FAD. Analysis of the leader region of the *rib* operon reveals the presence of a riboswitch sequence that could fold into a characteristic and evolutionarily well-conserved RNA structure known as *RFN element* (Mironov *et al.*, 2002).

2.6.3
Regulation of the Biosynthesis of Riboflavin

Riboflavin biosynthesis is mainly regulated at enzymatic level; however, transcriptional regulation has also been described. Among the effectors of the regulation of riboflavin biosynthesis that have been identified are flavins, nucleotides, iron ions and other metals such as cobalt, chromium, magnesium and zinc (Abbas and Sibirny, 2011; Fischer and Bacher, 2005). Some of the enzyme activities are modulated by their end products; in the flavinogenic yeast *P. guilliermondii*, GTP cyclohydrolase II is inhibited allosterically by FAD (Shavlovsky *et al.*, 1980), while DRL synthase is inhibited by riboflavin but not FAD (Abbas and Sibirny, 2011; Logvinenko *et al.*, 1973).

Flavins do not exert a repression on the transcription of the *RIB* genes in yeast and fungi. In contrast, iron starvation has been documented to be important for riboflavin overproduction in most flavinogenic yeasts and also in some bacteria and plants (Coursolle *et al.*, 2010; Crossley *et al.*, 2007; Tanner, Vojnovich and Vanlanen, 1945; Vorwieger *et al.*, 2007). A role of riboflavin as an electron donor for iron reduction has been suggested (Abbas and Sibirny, 2011), but the exact mechanisms of riboflavin overproduction in response to iron-restrictive conditions are still unknown. Iron deficiency triggers the transcriptional activation of most genes of the riboflavin pathway in some flavinogenic yeasts (Abbas and Sibirny, 2011; Boretsky *et al.*, 2005; Wang *et al.*, 2008). In this regard, mutations in the *SEF1* gene, which codes for a transcription factor in *C. flareri*, have been linked to iron metabolism and riboflavin overproduction (Abbas and Sibirny, 2011).

One of the most important mechanisms of transcriptional regulation of the biosynthesis of riboflavin in bacteria is the riboswitch mechanism, such as the *RFN* element in *B. subtilis.* The riboflavin operon (*rib* operon) of *B. subtilis* contains a leader region that can fold into a conserved structure called the *RFN element*, which is involved in the mechanism of transcriptional termination (Gelfand *et al.*, 1999; Winkler, Cohen-Chalamish and Breaker, 2002). FMN can bind directly to the *RFN* element, thus leading to the formation of a terminator hairpin, which attenuates transcription. *RFN* elements have been found in the leader regions of genes involved in either the biosynthesis or the transport of riboflavin in many bacteria (Vitreschak *et al.*, 2002). Riboswitches are able to sense cellular metabolites directly and are involved in the transcriptional regulation of several vitamins and coenzymes in bacteria (Abbas and Sibirny, 2011; Bobrovskyy and Vanderpool, 2013). Riboswitches are also involved in the regulation of translation initiation. For example, the *rib* genes of *E. coli* are scattered throughout the chromosome, and they contain 5′-UTR regions that probably participate at the translation level, although the mechanisms involved remain unknown (Barrick and Breaker, 2007; Vitreschak *et al.*, 2002).

In *E. ashbyii* and *A. gossypii,* riboflavin overproduction occurs during the late stationary phase when vegetative growth has mainly ceased and both sporulation and lysis occur (Mateos *et al.*, 2006; Schlosser *et al.*, 2007; Stahmann, Revuelta and Seulberger, 2000). Riboflavin production resembles the production kinetics of secondary metabolites in these flavinogenic fungi. Accordingly, riboflavin overproduction in the filamentous fungi *E. ashbyii* and *A. gossypii* has been called *pseudosecondary biosynthesis* (Cerletti *et al.*, 1965). The physiological role of riboflavin overproduction in the late stationary phase in these fungi has been discussed in terms of its protective role against UV light and also as a mechanism involved in spore spreading by insects (Abbas and Sibirny, 2011; Stahmann *et al.*, 2001). The detoxification of an excess of purines cannot be excluded (Jiménez *et al.*, 2005; Jiménez, Santos and Revuelta, 2008). Nonetheless, the precise role of the natural overproduction of riboflavin in these fungi is unknown.

In *A. gossypii,* riboflavin biosynthesis is tightly regulated at both enzymatic and transcriptional levels, as previously described for the purine pathway, which provides the immediate precursor GTP (Jiménez *et al.*, 2005; Jiménez, Santos and Revuelta, 2008; Mateos *et al.*, 2006; Revuelta, Buitrago and Santos, 1995). In this regard, several reports have shown a relationship between the transcriptional regulation of *RIB* genes and an increase in riboflavin biosynthesis in several organisms (Karos *et al.*, 2004; Marx, Mattanovich and Sauer, 2008). Also, the transcriptional regulation of metabolic flux changes occurring during the riboflavin production phase has been confirmed with a genome-scale metabolic reconstruction. Interestingly, most *RIB* genes were predicted to be up-regulated during the riboflavin production phase in *A. gossypii* (Ledesma-Amaro *et al.*, 2014a).

2.7
Strain Development: Genetic Modifications, Molecular Genetics and Metabolic Engineering

As mentioned earlier, *A. gossypii* is currently used for the industrial production of riboflavin. The capability of *A. gossypii* to overproduce riboflavin was first reported in 1946 (Wickerham, Flickinger and Johnston, 1946). The wild-type strain of this filamentous fungus, originally discovered as a severe but today negligible cotton pathogen, produces up to 2 mg/g of riboflavin of its cell dry weight, possibly as part of a natural mechanism involved in the protection of its spores against UV radiation (Stahmann *et al.*, 2001). In contrast to *C. famata*, *A. gossypii* and *E. ashbyii* do not show iron-responsive repression of riboflavin oversynthesis. Moreover, unlike *E. ashbyii*, *A. gossypii* is genetically stable and the production of riboflavin is fairly homogeneous. Thus, *A. gossypii* shows better characteristics than other flavinogenic microorganisms as a cell factory for the biotechnological production of riboflavin.

In *A. gossypii*, improvement of strains for riboflavin productivity has been achieved by applying methods of classic mutagenesis and selection and, more recently, by modern strategies of metabolic engineering. Random, chemical and radiation mutagenesis is a rapid and efficient approach in the initial steps of strain development for metabolite production. However, this strategy soon reaches a limit and no further improvements can be achieved after several rounds of mutagenesis and selection, which is probably due to the accumulation of detrimental mutations in addition to the beneficial ones. Thus, gene-target metabolic engineering is the strategy to follow for further strain development.

The pioneer isolation and molecular characterisation of the genes involved in riboflavin biosynthesis (*RIB* genes) in the yeast *S. cerevisiae* allowed the isolation by functional complementation of the six corresponding homologous *RIB* genes of *A. gossypii* (Garcia-Ramirez, Santos and Revuelta, 1995; Revuelta, Buitrago and Santos, 1995; Revuelta *et al.*, 1994; Santos, Garcia-Ramirez and Revuelta, 1995). This achievement, together with the development of an efficient electrotransformation method by our group, the recycling of selectable markers and the isolation of appropriate promoters for *A. gossypii*, paved the way for the application of metabolic engineering techniques in riboflavin production in this fungus (Forster *et al.*, 1999; Ledesma-Amaro *et al.*, 2014b; Santos *et al.*, 2004).

Rational metabolic design has focused on different processes, which *a priori* seem relevant for riboflavin production. Since GTP is one of the committed precursors for riboflavin biosynthesis, the *de novo* purine biosynthetic pathway has attracted considerable attention. Purine biosynthesis is a tightly regulated pathway at the transcriptional and metabolic levels (Figure 2.3). Two enzymes, PRPP amidotransferase (encoded in *A. gossypii* by *AgADE4*), which transforms phosphoribosyl pyrophosphate (PRPP) to phosphoribosylamine (PRA), and PRPP synthetase (encoded by *AgPRS1*, *AgPRS2,4*, *AgPRS3* and *AgPRS5*), which catalyses the formation of PRPP, are subjected to feedback inhibition by their end products and are major control steps of the pathway. Accordingly, strains overexpressing inhibition-resistant forms of PRPP amidotransferase (*AgADE4*^{V,K,W}) or PRPP synthetase (*AgPRS2,4*^{I,Q} and *AgPRS3*^{I,Q}) were constructed and showed 10-fold and 2-fold increases, respectively, in riboflavin production (Jiménez *et al.*, 2005; Jiménez, Santos and Revuelta, 2008). Insertional mutagenesis has been also developed for *A. gossypii*, and its use has permitted the isolation of several mutants with improved production yields (Santos *et al.*, 2004). One of these mutants proved to be affected in a transcription factor (*BAS1*) reported to control the purine biosynthetic pathway transcriptionally in *S. cerevisiae*. By mimicking the insertional mutant, the construction of a strain expressing a *BAS1*

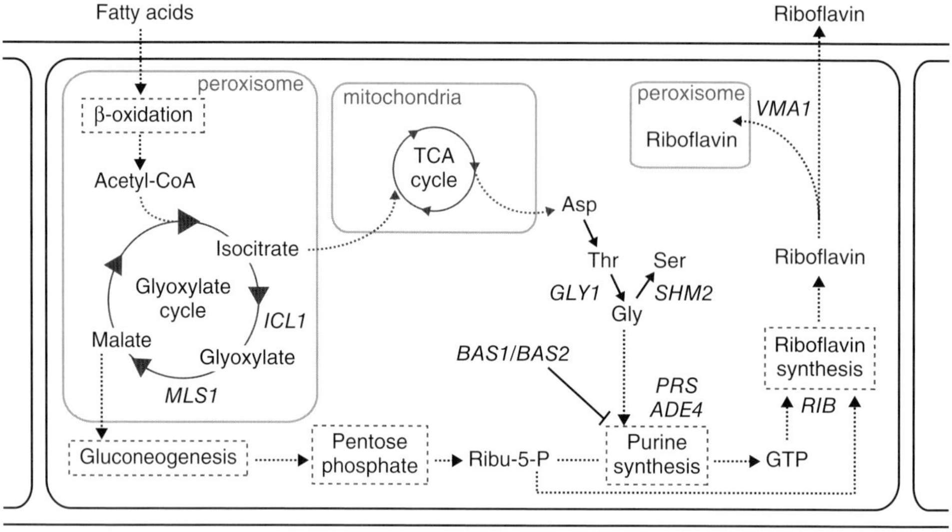

Figure 2.3 Modified genes to increase riboflavin production in *A. gossypii*. *ICL1*, isocitrate lyase; *MLS1*, malate synthase; *GLY1*, threonine aldolase; *SHM2*, serine hydroxymethyltransferase; *BAS1nBAS2*, Myb-related transcription factors; *PRS*, phosphoribosyl pyrophosphate synthetases; *ADE4*, phosphoribosyl pyrophosphate amidotransferase; *RIB*, genes of the Rib pathway; and *VMA1*, vacuolar ATPase.

truncated factor lacking the regulatory domain led to a deregulated, constitutive transcription of the genes involved in the purine biosynthetic pathway and a 10-fold enhanced riboflavin overproduction (Mateos *et al.*, 2006).

Glycine, which also participates in the biosynthesis of purines, stimulates riboflavin production in *A. gossypii*, and several examples illustrate how increasing the levels of intracellular glycine can enhance riboflavin production. In an effort to improve the supply of the purine precursor glycine, the *AgGLY1* gene, which encodes the glycine biosynthetic enzyme threonine aldolase, was overexpressed under the control of the strong P_{GPD1} promoter, resulting in a strong enhancement of riboflavin production in the engineered strain (Monschau, Sahm and Stahmann, 1998). Similarly, the disruption of the *AgSHM2* gene, encoding one of two isoenzymes of serine hydroxymethyltransferase, also increased the glycine supply by decreasing its conversion into serine and, accordingly, increased the production of riboflavin (Schlupen *et al.*, 2003). Heterologous expression of the alanine:glyoxylate aminotransferase encoding gene (*AGX1*) from *S. cerevisiae* was also used to enlarge the pool of glycine precursor (Kato and Park, 2006).

Since oils are the preferred carbon source for riboflavin fermentation in *A. gossypii*, an efficient glyoxylate cycle is required for acetyl-CoA to be converted into the carbohydrate precursors needed for riboflavin biosynthesis (Stahmann, Revuelta and Seulberger, 2000). Improvement of riboflavin production was achieved by the isolation of mutants resistant to itaconate, an inhibitor of the key isocitrate lyase enzyme that exerts the main control of the glyoxylate shunt

(Schmidt, Stahmann and Sahm, 1996). Introduction of an additional copy of the *ICL1* gene, encoding isocitrate lyase, enhanced riboflavin production in a medium containing soybean oil (Boeddecker *et al.*, 1997; Maeting *et al.*, 1999). Overexpression of the second enzyme of the glyoxylate pathway, malate synthase, was also performed in an attempt to improve the efficiency of oil consumption and riboflavin production (Sugimoto *et al.*, 2009).

Attention has also been paid to the riboflavin transport processes. In *A. gossypii*, a hitherto unidentified high-activity efflux carrier capable of maintaining a concentration gradient of at least 2 orders of magnitude over several hours exports riboflavin out of the cell. Riboflavin is also stored in the vacuolar compartment, leading to product retention and thereby reducing the excretion yields and requiring the disruption of cells to obtain the full amount of the product (Forster, Revuelta and Kramer, 2001). Knock-out of the *AgVMA1* gene, encoding vacuolar ATPase, which energises active riboflavin transport from the cytoplasm to the vacuole, resulted in complete excretion of riboflavin synthesised into medium and high total riboflavin production (Forster *et al.*, 1999).

Recent approaches guided by computational metabolic modelling have led to the overexpression of *RIB* genes (Ledesma-Amaro *et al.*, 2014a). Although *RIB1* and *RIB3* were the major limiting steps in riboflavin production, the strain overexpressing all the *RIB* genes showed the highest production yield (Althofer and Revuelta, 2003; Ledesma-Amaro *et al.* 2015).

It has been reported that *A. gossypii* industrial riboflavin producers accumulate more than 15 g/l of riboflavin (Bigelis, 1989). However, these data do not consider the improvements achieved by recent metabolic engineering approaches, and current industrial producer strains could surely accumulate much higher titres.

B. subtilis is also a good candidate to develop a bacterial process for riboflavin production by fermentation. Several useful features of *B. subtilis* include its classification as a GRAS (generally regarded as safe) microorganism, the deep knowledge on its physiology and recombinant DNA technology and its capability to produce large amounts of the riboflavin precursors inosine and guanosine (20–40 g/l) (Shiio, 1989), which could subsequently be converted metabolically into riboflavin, making *B. subtilis* a good candidate for developing a bacterial process of riboflavin production by fermentation.

The first riboflavin production strain of *B. subtilis* (VNIIGenetika 304/pMX45, Russian Institute for Genetics and Selection of Industrial Microorganisms, Moscow) obtained by genetic engineering methods consists of a host strain (VNIIGenetika 304) deregulated in the purine and the riboflavin biosynthetic pathway that harbours a plasmid (pMX45) containing the entire *rib* operon from a *ribO* mutant strain. This strain produced 4.5 g/l riboflavin after 25 h fermentation but was genetically unstable due to the presence of repeated chromosomal and episomal copies of the *rib* operon (Zhdanov and Stepanov, 1984). Genetically stable riboflavin production strains were obtained using integrative vector constructs. In its chromosome, the riboflavin production strain GM41/pMX4557 harbours a deregulated *Bacillus amyloliquefaciens rib* operon and the plasmid pMX45. Under small-scale fed-batch fermentation conditions, the strain accumulated 21 g/l riboflavin. The optimal copy number dose of deregulated *rib* operon sequences integrated into the chromosome of *B. subtilis* host strains for highest riboflavin production has recently been estimated at seven to eight copies (Hohmann and Stahmann, 2010).

Riboflavin production strains constructed by precise genetic engineering approaches have only been possible after the sequencing of the *B. subtilis rib* operon. Gene amplification and replacement of wild-type promoters and regulatory regions by a strong constitutive promoter from the *Bacillus* bacteriophage *SPO1* have resulted in a strain with remarkably improved riboflavin productivity. The engineered strain contains multiple copies of a modified *B. subtilis rib* operon integrated at two different sites in the host chromosome. The modified *rib* operons are expressed constitutively from strong *SPO1* phage promoters located at the 5′ end and in an internal region of the operon. The host strain also contains purine analogue-resistant mutations (azaguanine-, decoyinine- and psicofuranine-resistant mutations) known to deregulate the purine pathway and a riboflavin analogue-resistant mutation (roseoflavin-resistant) in *ribC* that deregulates the riboflavin biosynthetic pathway (Perkins *et al.*, 1999).

Further attempts to increase riboflavin production suggested that the bifunctional cyclohydrolase II-DHBP synthase protein encoded by *ribA* is the rate-limiting enzyme in the industrial riboflavin-producing strain. The introduction of an additional single copy of *ribA* under the control of the constitutive, medium-strength *vegI* promoter into the *sacB* locus of the riboflavin production strain led to 25% improvements in riboflavin titres and yields (Hümbelin *et al.*, 1999).

A detailed analysis of the *rib* leader sequence as a regulatory and mRNA stabilising element has recently allowed the construction of a new class of riboflavin production strains with a single copy of a precisely deregulated *rib* operon and (Lehmann *et al.*, 2011).

GTP and ribulose-5-phosphate are the two committed precursors for riboflavin biosynthesis, which are required at a 1 : 2 stoichiometric ratio. Hence, a sufficiently high intracellular concentration of ribulose-5-phosphate is needed to ensure a sufficient precursor supply for riboflavin oversynthesis. *B. subtilis* mutants defective in transketolase, a key enzyme in the pentose phosphate pathway, show high intracellular C_5 carbon sugar pools at levels that lead to the excretion of excess ribose into the fermentation broth, and an improved riboflavin was described in one *tkt* (transketolase) mutant (Gershanovich *et al.*, 2000; Wulf and Vandamme, 1997). Although transketolase knock-out mutations negatively affect riboflavin formation, *tkt* mutations, which impair but do not completely inactivate transketolases, showed increased riboflavin yield in riboflavin producer strains. For example, a *B. subtilis* strain expressing the R357A transketolase mutant showed a 43% riboflavin yield increase in glucose-limited fed-batch fermentation runs (Lehmann *et al.*, 2008).

Several efforts to redirect the central metabolic flux towards the riboflavin building blocks have been made, including expression modification of glycolysis,

TCA cycle or pentose phosphate pathway enzymes. With the exception of the transketolase modification described earlier, these attempts resulted in very limited success (Hohmann and Stahmann, 2010). This lack of success might be explained by the presence of an unusually rigid metabolic network in *B. subtilis* as revealed by comprehensive 'fluxome' analyses of a large number of *B. subtilis* knock-out mutants (Fischer and Sauer, 2005).

2.8
Fermentation Process

Initial studies to develop a fermentation process to produce riboflavin were begun around 1940 and are now well established (Perlman, 1979; Wickerham, Flickinger and Johnston, 1946). The fermentative production of riboflavin is carried out in submerged culture, and yield is strongly influenced by different factors such as the microbial strain, carbon source, minerals and pH. Inoculum preparation, the fermentation medium and the optimisation of fermentation conditions are key aspects of the *A. gossypii* fermentation process. The generally accepted inoculation procedure is based on the use of low-concentration (2–10% v/v) inoculum broths containing young, undifferentiated mycelium devoid of spores and sporiferous sacs.

Although the initial studies found that *A. gossypii* shows a healthy growth pattern in a medium containing glucose, corn steep liquor and animal stick liquor or meat scraps, the stimulatory effect on riboflavin production of peptones, lipids and accessory factors present in corn steep liquor was soon recognised (Perlman, 1979). Several amino acids and vitamins such as methionine, glycine, inositol, biotin and thiamine have been found to enhance growth and riboflavin production. Aerobic submerged fermentation of *A. gossypii* with a nutrient medium containing plant oil as the major carbon source, which yields more than 15 g/l, is currently the preferred method (Bigelis, 1989). Industrial waste materials such as activated bleaching earth (containing oil discharged from oil refinery factories) and agroindustrial by-products such as grape must, beet molasses, peanut seed cake and whey have also been assayed for the production of riboflavin, but with limited success (Kalıngan and Liao, 2002; Ming, Lara Pizarro and Park, 2003; Park and Ming, 2004). Fermentation conditions have been extensively studied to determine the optimum pH, aeration, temperature and substrate concentration (Kutsal and Özbas, 1989). The optimum temperature range for the *A. gossypii* fermentation is 27–30 °C, and it requires an airflow rate of at least 0.25 vvm, an initial pH of 6.5 and an initial carbon concentration of 50 g/l. Fermentations are run for about 120–140 h when yield usually peaks.

The fermentation process developed for the industrial production of riboflavin by *B. subtilis* strains is generally based on the common method of carbon-source-limited fed-batch. After an initial growth phase at its maximal growth rate, with unlimited carbon source to reach a high cell density, the bioreactor is fed with controlled limiting amounts of carbon source to restrict the microbial growth rate to the rate of substrate supply. Riboflavin is synthesised and excreted into the culture broth at low growth rates under the strictly glucose-limited conditions of the feeding phase. The *Bacillus*-based fed-batch process for commercial riboflavin production could be optimised using model-predictive control based on artificial neural networks (Hohmann and Stahmann, 2010).

A single-step fermentative riboflavin production process was also developed for a *B. subtilis* recombinant strain, which effectively produces riboflavin directly from glucose in fed-batch operation. The process is based on fuzzy control system featuring the identification of culture phases by fuzzy logic and was applied to large-scale vitamin B_2 production (Horiuchi and Hiraga, 1999).

2.9 Downstream Processing

Riboflavin can be recovered at different purification grades (Kutsal and Özbas, 1989). Since riboflavin is rather insoluble at neutral pH in aqueous solutions, the vitamin accumulates in the fermentation broth as needle-shaped crystals, facilitating the purification steps. Riboflavin crystals are recovered from the broth by centrifugation after inactivation of the microorganism by pasteurisation to ensure that no viable cells or spores of the fungus are present in the final product. This heating step also induces cell autolysis and helps to recover a significant part of the product, which is stored inside the vacuoles of the cells. After heating, the cell mass is separated from the fermentation broth by differential centrifugation, allowing the separation of cells and riboflavin crystals because of differences in their size and sedimentation behaviour. Riboflavin is then recovered from cell-free broth by evaporation and vacuum drying (Faust *et al.*, 1991; Kurth, 1992).

For some applications, additional purification steps can be accomplished in order to remove lipids with ether, or salt and glycogen, or by fractionate precipitation with alcohol or acetone. Other methods described for riboflavin recovery are based on bacteriological reduction of the fermentation broth, which produces a reddish-brown product (Hickey, 1946). Alternative chemical precipitation methods have been proposed, such as the use of sodium dithionite, stannous and chromium chlorides (Kutsal and Özbas, 1989). Thereafter, the crude precipitate can be readily converted to crystalline riboflavin using a hot polar solvent such as isopropyl alcohol. These needle-like crystals can be separated by filtration or centrifugation. Additionally, riboflavin can be purified by absorption methods such as the use of Fuller's earth in acid solution (Kutsal and Özbas, 1989).

Downstream process in *B. subtilis* consists of pasteurisation of the complete fermentation broth containing riboflavin crystals and biomass, followed by differential centrifugation to harvest and partially purify the crystals. After a washing step with hot mineral acid to remove impurities, a feed quality product (>96% purity) that does not contain any recombinant DNA can be obtained (Bretzel *et al.*, 1999).

2.10 Chemical Synthesis

Even though most of the riboflavin produced currently comes from microbial fermentation, the chemical process for the synthesis of this vitamin was established many years ago. Both Kuhn and Weygand (1934) and Karrer, Schöpp and Benz (1935) independently developed a method consisting of the reductive condensation of D-ribose with 3,4-xylidine (Goldberg and Williams, 1991). The D-ribose moiety can be produced chemically from either D-arabinose or glucose. D-ribose can also be obtained by microbial fermentation using *Bacillus* species, and this led to the semi-synthetic method to produce riboflavin, which was exploited industrially for many years (Ernst, Leininger and Paust, 1989; Wolf *et al.*, 1982).

2.11 Application and Economics

Riboflavin production at industrial scale by microbial fermentation has proved to be both cost-effective and environmentally friendly in comparison with conventional chemical synthesis: carbon dioxide emissions and use of non-renewable resources are reduced by 80% each and water emissions by 66%. Currently, therefore, riboflavin is mainly produced biotechnologically by engineered strains of *A. gossypii* and *B. subtilis* (Stahmann, Revuelta and Seulberger, 2000). Previously, *E. asbhyii* and *C. famata* were also used in riboflavin production, but both processes were discontinued due to the emergence of instabilities. Major worldwide producers are Aventis, BASF, Daicel, DSM, Hubei Guangji Pharmaceuticals, Kyowa, Mitsui, Roche, Shanghai Desano Vitamins Co. and Takeda (Abbas and Sibirny, 2011). More than 4/5 of total riboflavin is used as an additive for animal feeding, while the remaining 1/5 is used both for medical applications (vitamin) and as yellow colourant E-101 for beverages. The total riboflavin market in 2012 was around 9000 t, and the final price was about $15/kg for the feed-grade product and $35–50/kg for the food-grade product (Kato and Park, 2012). FMN is synthesised chemically and has 30% of impurities (Nielsen, Rauschenbach and Bacher, 1983), and this form is preferred for pharmaceutical applications since it is more soluble than riboflavin; FAD is used in medicine, and it is also produced biotechnologically (Shimizu, 2008).

References

Abbas, C.A. and Sibirny, A.A. (2011) Genetic control of biosynthesis and transport of riboflavin and flavin nucleotides and construction of robust biotechnological producers. *Microbiol. Mol. Biol. Rev.*, **75**, 321–360.

Althofer, H. and Revuelta, J.L. (2003) Genetic strain optimization for improving the production of riboflavin. Patent WO03048367 A1, BASF AG, DE.

Bacher, A., Eberhardt, S., Fischer, M., Kis, K., and Richter, G. (2000) Biosynthesis of vitamin B_2 (riboflavin). *Annu. Rev. Nutr.*, **20**, 153–167.

Bacher, A. and Lingens, F. (1969) The structure of the purine precursor in riboflavin biosynthesis. *Angew. Chem., Int. Ed. Engl.*, **8**, 371–372.

Bacher, A. and Mailander, B. (1973) Biosynthesis of riboflavin. The structure of the purine precursor. *J. Biol. Chem.*, **248**, 6227–6231.

Barrick, J.E. and Breaker, R.R. (2007) The distributions, mechanisms, and structures of metabolite-binding riboswitches. *Genome Biol.*, **8**, R239.

Bigelis, R. (1989) *Industrial Products of Biotechnology: Application of Gene Technology*, VCH Verlag, Weinheim.

Bobrovskyy, M. and Vanderpool, C.K. (2013) Regulation of bacterial metabolism by small RNAs using diverse mechanisms. *Annu. Rev. Genet.*, **47**, 209–232.

Boeddecker, T., Kaesler, B., Sahm, H., Schmidt, G., Seulberger, H. and Stahmann, K.P. (1997) Riboflavin production process by means of microorganisms with modified isocitrate lyase activity. Patent WO1997003208A1.

Boretsky, Y.R., Kapustyak, K.Y., Fayura, L.R., Stasyk, O.V., Stenchuk, M.M., Bobak, Y.P., Drobot, L.B., and Sibirny, A.A. (2005) Positive selection of mutants defective in transcriptional repression of riboflavin synthesis by iron in the flavinogenic yeast *Pichia guilliermondii*. *FEMS Yeast Res.*, **5**, 829–837.

Bornemann, S. (2002) Flavoenzymes that catalyse reactions with no net redox change. *Nat. Prod. Rep.*, **19**, 761–772.

Bretzel, W., Schurter, W., Ludwig, B., Kupfer, E., Doswald, S., Pfister, M., and van Loon, A.P.G.M. (1999) Commercial riboflavin production by recombinant *Bacillus subtilis*: down-stream processing and comparison of the composition of riboflavin produced by fermentation or chemical synthesis. *J. Ind. Microbiol. Biotech.*, **22**, 19–26.

Briggs, W.R. and Christie, J.M. (2002) Phototropins 1 and 2: versatile plant blue-light receptors. *Trends Plant Sci.*, **7**, 204–210.

Buitrago, M.J., Gonzalez, G.A., Saiz, J.E., and Revuelta, J.L. (1993) Mapping of the *RIB1* and *RIB7* genes involved in the biosynthesis of riboflavin in *Saccharomyces cerevisiae*. *Yeast*, **9**, 1099–1102.

Cerletti, P., Strom, R., Giordano, M.G., Barra, D., and Giovenco, S. (1965) Flavin coenzymes, flavinogenesis and reproduction in *Ashbya gossypii*. *J. Biochem.*, **57**, 773–786.

Chatwell, L., Illarionova, V., Illarionov, B., Eisenreich, W., Huber, R., Skerra, A., Bacher, A., and Fischer, M. (2008) Structure of lumazine protein, an optical transponder of luminescent bacteria. *J. Mol. Biol.*, **382**, 44–55.

Chen, S.C., Shen, C.Y., Yen, T.M., Yu, H.C., Chang, T.H., Lai, W.L., and Liaw, S.H. (2013) Evolution of vitamin B_2 biosynthesis: eubacterial *RibG* and fungal *Rib2* deaminases. *Acta. Crystallogr. D: Biol. Crystallogr.*, **69**, 227–236.

Clarebout, G., Villers, C., and Leclercq, R. (2001) Macrolide resistance gene mreA of *Streptococcus agalactiae* encodes a flavokinase. *Antimicrob. Agents Chemother.*, **45**, 2280–2286.

Conrad, K.S., Manahan, C.C., and Crane, B.R. (2014) Photochemistry of flavoprotein light sensors. *Nat. Chem. Biol.*, **10**, 801–809.

Coursolle, D., Baron, D.B., Bond, D.R., and Gralnick, J.A. (2010) The Mtr respiratory pathway is essential for reducing flavins and electrodes in *Shewanella oneidensis*. *J. Bacteriol.*, **192**, 467–474.

Crossley, R.A., Gaskin, D.J., Holmes, K., Mulholland, F., Wells, J.M., Kelly, D.J., van Vliet, A.H., and Walton, N.J. (2007) Riboflavin biosynthesis is associated with assimilatory ferric reduction and iron acquisition by *Campylobacter jejuni*. *Appl. Environ. Microbiol.*, **73**, 7819–7825.

De Colibus, L. and Mattevi, A. (2006) New frontiers in structural flavoenzymology. *Curr. Opin. Struct. Biol.*, **16**, 722–728.

Demain, A.L. (1972) Riboflavin oversynthesis. *Annu. Rev. Microbiol.*, **26**, 369–388.

Demain, A.L. (2007) Reviews: the business of biotechnology. *Ind. Biotechnol.*, **3**, 269–283.

Dmytruk, K.V. and Sibirny, A.A. (2012) *Candida famata* (*Candida flareri*). *Yeast*, **29**, 453–458.

Dunlap, P. (2014) Biochemistry and genetics of bacterial bioluminescence. *Adv. Biochem. Eng. Biotechnol.*, **144**, 37–64.

Eggersdorfer, M., Adam, G., John, M., Hähnlein, W., Labler, L., Baldenius, K.-U., von dem Bussche-Hünnefeld, L., Hilgemann, E., Hoppe, P., Stürmer, R., Weber, F., Rüttimann, A., Moine, G., Hohmann, H.-P., Kurth, R., Paust, J., Pauling, H., Weimann, B.J., Kaesler, B., Oster, B., Fechtel, U., Kaiser, K., de Potzolli, B., Casutt, M., Koppe, T., Schwarz, M., Weimann, B.-J., Hengartner, U., de Saizieu, A., Wehrli, C., and Blum, R. (2000) Vitamins, in *Ullmann's Encyclopedia of Industrial Chemistry*, Wiley-VCH Verlag GmbH & Co. KGaA.

Eldridge, A.L. (2004) Comparison of 1989 RDAs and DRIs for water-soluble vitamins. *Nutr. Today*, **39**, 88–93.

Ernst, H., Leininger, H. and Paust, J. (1989) Preparation of ribitylxylidine. Patent US 4806686, BASF AG, USA.

Faust, T., Meyer, J., Wellinghoff, G., Goesele, W., Martin, C. and Grimmer, J. (1991) Process for the separation of riboflavin from a fermentation suspension. Patent EP0438767, BASF AG, DE.

Fischer, M. and Bacher, A. (2005) Biosynthesis of flavocoenzymes. *Nat. Prod. Rep.*, **22**, 324–350.

Fischer, M., Haase, I., Feicht, R., Schramek, N., Kohler, P., Schieberle, P., and Bacher, A. (2005) Evolution of vitamin B_2 biosynthesis: riboflavin synthase of *Arabidopsis thaliana* and its inhibition by riboflavin. *Biol. Chem.*, **386**, 417–428.

Fischer, E. and Sauer, U. (2005) Large-scale in vivo flux analysis shows rigidity and suboptimal performance of *Bacillus subtilis* metabolism. *Nat. Genet.*, **37**, 636–640.

Foor, F. and Brown, G.M. (1975) Purification and properties of guanosine triphosphate cyclohydrolase II from *Escherichia coli*. *J. Biol. Chem.*, **250**, 3545–3551.

Forster, C., Revuelta, J.L., and Kramer, R. (2001) Carrier-mediated transport of riboflavin in *Ashbya gossypii*. *Appl. Microbiol. Biotechnol.*, **55**, 85–89.

Forster, C., Santos, M.A., Ruffert, S., Kramer, R., and Revuelta, J.L. (1999) Physiological consequence of disruption of the *VMA1* gene in the riboflavin overproducer *Ashbya gossypii*. *J. Biol. Chem.*, **274**, 9442–9448.

Garcia-Ramirez, J.J., Santos, M.A., and Revuelta, J.L. (1995) The *Saccharomyces cerevisiae RIB4* gene codes for 6,7-dimethyl-8-ribityllumazine synthase involved in riboflavin biosynthesis. Molecular characterization of the gene and purification of the encoded protein. *J. Biol. Chem.*, **270**, 23801–23807.

Gelfand, M.S., Mironov, A.A., Jomantas, J., Kozlov, Y.I., and Perumov, D.A. (1999) A conserved RNA structure element involved in the regulation of bacterial riboflavin synthesis genes. *Trends Genet.*, **15**, 439–442.

Gershanovich, V.N., Kukanova, A.I.A., Galushkina, Z.M., Stepanov, A.I., and Stepanov, A.I. (2000) Transketolase mutation in riboflavin-synthesizing strains of *Bacillus subtilis*. *Mol. Gen. Mikrobiol. Virusol.*, **3**, 3–7.

Gliszczynska, A. and Koziolowa, A. (1998) Chromatographic determination of flavin derivatives in baker's yeast. *J. Chromatogr. A*, **822**, 59–66.

Goldberg, I. and Williams, R. (1991) *Biotechnology and Food Ingredients*, Van Nostrand Reinhold, New York.

Gomelsky, M. and Klug, G. (2002) BLUF: a novel FAD-binding domain involved in sensory transduction in microorganisms. *Trends Biochem. Sci*, **27**, 497–500.

Gudipati, V., Koch, K., Lienhart, W.D., and Macheroux, P. (2014) The flavoproteome of the yeast *Saccharomyces cerevisiae*. *Biochim. Biophys. Acta*, **1844**, 535–544.

Guilliermond, A., Fontaine, M., and Raffy, A. (1935) Sur l'existence dans l'*Eremothecium ashbyii* d'un pigment jaune se rapportant au groupe des flavines. *C. R. Hebd. Seances Acad. Sci.*, **201**, 1077–1080.

Haase, I., Grawert, T., Illarionov, B., Bacher, A., and Fischer, M. (2014) Recent advances in riboflavin biosynthesis. *Methods Mol. Biol.*, **1146**, 15–40.

Harzer, G., Rokos, H., Otto, M.K., Bacher, A., and Ghisla, S. (1978) Biosynthesis of riboflavin. 6,7-Dimethyl-8-ribityllumazine 5′-phosphate is not a substrate for riboflavin synthase. *Biochim. Biophys. Acta*, **540**, 48–54.

Hickey, R.J. (1945) The inactivation of iron by 2,2′-bipyridine and its effect on riboflavin synthesis by *Clostridium acetobutylicum*. *Arch. Biochem.*, **8**, 439–447.

Hickey, R.J. (1946) Precipitation of riboflavin from aqueous solution by bacteriological reduction. *Arch. Biochem.*, **11**, 259–267.

Hohmann, H.-P. and Stahmann, K.-P. (2010) *Comprehensive Natural Products II Chemistry and Biology*, Elsevier, Oxford.

Horiuchi, J. and Hiraga, K. (1999) Industrial application of fuzzy control to large-scale recombinant vitamin B_2 production. *J. Biosci. Bioeng.*, **87**, 365–371.

Hümbelin, M., Griesser, V., Keller, T., Schurter, W., Haiker, M., Hohmann, H.P., Ritz, H., Richter, G., Bacher, A., and van Loon, A.P.G.M. (1999) GTP cyclohydrolase II and 3,4-dihydroxy-2-butanone 4-phosphate synthase are rate-limiting enzymes in riboflavin synthesis of an industrial *Bacillus subtilis* strain used for riboflavin production. *J. Ind. Microbiol. Biotech.*, **22**, 1–7.

Institute of Medicine (US) Standing Committee on the Scientific Evaluation of Dietary Reference Intakes and its Panel on Folate, Other B Vitamins, and Choline (1998) *Dietary Reference Intakes for Thiamin, Riboflavin, Niacin, Vitamin B_6, Folate, Vitamin B_{12}, Pantothenic Acid, Biotin, and Choline*, National Academies Press (US) CTI- The National Academies Collection: Reports Funded by National Institutes of Health.

Jiménez, A., Santos, M.A., Pompejus, M., and Revuelta, J.L. (2005) Metabolic engineering of the purine pathway for riboflavin production in *Ashbya gossypii*. *Appl. Environ. Microbiol.*, **71**, 5743–5751.

Jiménez, A., Santos, M.A., and Revuelta, J.L. (2008) Phosphoribosyl pyrophosphate synthetase activity affects growth and riboflavin production in *Ashbya gossypii*. *BMC Biotech.*, **8**, 67.

Kalingan, A.E. and Liao, C.M. (2002) Influence of type and concentration of flavinogenic factors on production of riboflavin by *Eremothecium ashbyii* NRRL 1363. *Bioresour. Technol.*, **82**, 219–224.

Karos, M., Vilarino, C., Bollschweiler, C., and Revuelta, J.L. (2004) A genome-wide transcription analysis of a fungal riboflavin overproducer. *J. Biotechnol.*, **113**, 69–76.

Karrer, P., Schöpp, K., and Benz, F. (1935) Synthesen von Flavinen IV. *Helv. Chim. Acta*, **18**, 426–429.

Karthikeyan, S., Zhou, Q., Mseeh, F., Grishin, N.V., Osterman, A.L., and Zhang, H. (2003) Crystal structure of human riboflavin kinase reveals a beta barrel fold and a novel active site arch. *Structure*, **11**, 265–273.

Kato, T. and Park, E.Y. (2006) Expression of alanine: glyoxylate aminotransferase gene from *Saccharomyces cerevisiae* in *Ashbya gossypii*. *Appl. Microbiol. Biotechnol.*, **71**, 46–52.

Kato, T. and Park, E.Y. (2012) Riboflavin production by *Ashbya gossypii*. *Biotechnol. Lett.*, **34**, 611–618.

Kil, Y.V., Mironov, V.N., Gorishin, I., Kreneva, R.A., and Perumov, D.A. (1992) Riboflavin operon of *Bacillus subtilis*: unusual symmetric arrangement of the regulatory region. *Mol. Gen. Genet.*, **233**, 483–486.

Krishnan, B., Levine, J.D., Lynch, M.K., Dowse, H.B., Funes, P., Hall, J.C., Hardin, P.E., and Dryer, S.E. (2001) A new role for cryptochrome in a *Drosophila* circadian oscillator. *Nature*, **411**, 313–317.

Kuhn, R. and Weygand, F. (1934) Synthetisches vitamin B_2. *Ber. Dtsch. Chem. Ges. (A/B Ser.)*, **67**, 2084–2085.

Kurth, R. (1992) Process for the enhancement of riboflavin levels in spray-dried riboflavin fermentation extracts. Patent EP0487985, BASF AG, DE.

Kutsal, T. and Özbas, M.T. (1989) in *Biotechnology of Vitamins, Pigments and Growth Factors* (ed E. Vandamme), Springer, pp. 149–166.

Ledesma-Amaro, R., Jimenez, A., Santos, M., and Revuelta, J. (2013) Microbial production of vitamins, in *Microbial Production of Food Ingredients, Enzymes and Nutraceuticals*, Technology and Nutrition, vol. **246**, Woodhead Publishing Series in Food Science, Sawston, Cambridge.

Ledesma-Amaro, R., Kerkhoven, E.J., Revuelta, J.L., and Nielsen, J. (2014a) Genome scale metabolic modeling of the riboflavin overproducer *Ashbya gossypii*. *Biotechnol. Bioeng.*, **111**, 1191–1199

Ledesma-Amaro, R., Santos, M.A., Jimenez, A., and Revuelta, J.L. (2014b) Strain design of *Ashbya gossypii* for single-cell oil production. *Appl. Environ. Microbiol.*, **80**, 1237–1244.

Ledesma-Amaro, R., Serrano-Amatriaín, C., Jimenez, A., and Revuelta, J.L. (2015)

Metabolic engineering of riboflavin production in *Ashbya gossypii* through pathway optimization. *Microb. Cell Fact.*, **14**, 163 doi:10.1186/s12934-015-0354-x.

Lehmann, M., Hohmann, H.-P., Laudert, D. and Hans, M. (2008) Modified transketolase and use thereof. Patent EP 1957640 A1, DSM.

Lehmann, M., Hohmann, H.-P., Laudert, D. and Hans, M. (2011) Improved production of riboflavin. Patent US20110312025, DSM.

Liao, D.I., Wawrzak, Z., Calabrese, J.C., Viitanen, P.V., and Jordan, D.B. (2001) Crystal structure of riboflavin synthase. *Structure*, **9**, 399–408.

Lienhart, W.D., Gudipati, V., and Macheroux, P. (2013) The human flavoproteome. *Arch. Biochem. Biophys.*, **535**, 150–162.

Lin, C., Robertson, D.E., Ahmad, M., Raibekas, A.A., Jorns, M.S., Dutton, P.L., and Cashmore, A.R. (1995) Association of flavin adenine dinucleotide with the Arabidopsis blue light receptor CRY1. *Science*, **269**, 968–970.

Linden, H. (2002) Circadian rhythms. A white collar protein senses blue light. *Science*, **297**, 777–778.

Lingens, B., Schild, T.A., Vogler, B., and Renz, P. (1992) Biosynthesis of vitamin B_{12}. Transformation of riboflavin 2H-labeled in the 1'R position of 1'S position into 5,6-dimethylbenzimidazole. *Eur. J. Biochem.*, **207**, 981–985.

Logvinenko, E.M., Shavlovskii, G.M., Trach, V.M., and Sibirnyi, V.A. (1973) Role of flavins in regulating riboflavin synthetase synthesis in *Pichia guilliermondii* and *Candida utilis*. *Mikrobiologiia*, **42**, 1008–1014.

Lv, Z., Sun, J., and Liu, Y. (2013) Structural and functional insights into *Saccharomyces cerevisiae* riboflavin biosynthesis reductase RIB7. *PLoS One*, **8**, e61249.

Mack, M., van Loon, A.P., and Hohmann, H.P. (1998) Regulation of riboflavin biosynthesis in *Bacillus subtilis* is affected by the activity of the flavokinase/flavin adenine dinucleotide synthetase encoded by *ribC*. *J. Bacteriol.*, **180**, 950–955.

MacLaren, J.A. (1952) The effects of certain purines and pyrimidines upon the production of riboflavin by *Eremothecium ashbyii*. *J. Bacteriol.*, **63**, 233–241.

Maeting, I., Schmidt, G., Sahm, H., Revuelta, J.L., Stierhof, Y.D., and Stahmann, K.P. (1999) Isocitrate lyase of *Ashbya gossypii*--transcriptional regulation and peroxisomal localization. *FEBS Lett.*, **444**, 15–21.

Mailander, B. and Bacher, A. (1976) Biosynthesis of riboflavin. Structure of the purine precursor and origin of the ribityl side chain. *J. Biol. Chem.*, **251**, 3623–3628.

Manstein, D.J. and Pai, E.F. (1986) Purification and characterization of FAD synthetase from *Brevibacterium ammoniagenes*. *J. Biol. Chem.*, **261**, 16169–16173.

Marx, H., Mattanovich, D., and Sauer, M. (2008) Overexpression of the riboflavin biosynthetic pathway in *Pichia pastoris*. *Microb. Cell Fact.*, **7**, 23.

Mashhadi, Z., Zhang, H., Xu, H., and White, R.H. (2008) Identification and characterization of an archaeon-specific riboflavin kinase. *J. Bacteriol.*, **190**, 2615–2618.

Massey, V. (2000) The chemical and biological versatility of riboflavin. *Biochem. Soc. Trans.*, **28**, 283–296.

Mateos, L., Jiménez, A., Revuelta, J.L., and Santos, M.A. (2006) Purine biosynthesis, riboflavin production, and trophic-phase span are controlled by a Myb-related transcription factor in the fungus *Ashbya gossypii*. *Appl. Environ. Microbiol.*, **72**, 5052–5060.

Mattevi, A. (2006) To be or not to be an oxidase: challenging the oxygen reactivity of flavoenzymes. *Trends Biochem. Sci.*, **31**, 276–283.

McCormick, D.B., Oka, M., Bowers-Komro, D.M., Yamada, Y., and Hartman, H.A. (1997) Purification and properties of FAD synthetase from liver. *Methods Enzymol.*, **280**, 407–413.

Merrill, A.H. Jr., Lambeth, J.D., Edmondson, D.E., and McCormick, D.B. (1981) Formation and mode of action of flavoproteins. *Annu. Rev. Nutr.*, **1**, 281–317.

Ming, H., Lara Pizarro, A.V., and Park, E.Y. (2003) Application of waste activated bleaching earth containing rapeseed oil on riboflavin production in the culture of *Ashbya gossypii*. *Biotechnol. Prog.*, **19**, 410–417.

Mironov, A.S., Gusarov, I., Rafikov, R., Lopez, L.E., Shatalin, K., Kreneva, R.A., Perumov, D.A., and Nudler, E. (2002) Sensing small molecules by nascent RNA: a mechanism to control transcription in bacteria. *Cell*, **111**, 747–756.

Miura, R. (2001) Versatility and specificity in flavoenzymes: control mechanisms of flavin reactivity. *Chem. Rec.*, **1**, 183–194.

Monschau, N., Sahm, H., and Stahmann, K. (1998) Threonine aldolase overexpression plus threonine supplementation enhanced riboflavin production in *Ashbya gossypii*. *Appl. Environ. Microbiol.*, **64**, 4283–4290.

Nielsen, P., Rauschenbach, P., and Bacher, A. (1983) Phosphates of riboflavin and riboflavin analogs: a reinvestigation by high-performance liquid chromatography. *Anal. Biochem.*, **130**, 359–368.

Park, E.Y. and Ming, H. (2004) Oxidation of rapeseed oil in waste activated bleaching earth and its effect on riboflavin production in culture of *Ashbya gossypii*. *J. Biosci. Bioeng.*, **97**, 59–64.

Perkins, J.B., Sloma, A., Hermann, T., Theriault, K., Zachgo, E., Erdenberger, T., Hannett, N., Chatterjee, N.P., Williams, V. II, Rufo, G.A. Jr., Hatch, R., and Pero, J. (1999) Genetic engineering of *Bacillus subtilis* for the commercial production of riboflavin. *J. Ind. Microbiol. Biotech.*, **22**, 8–18.

Perlman, D., 1979. In: H.J Peppler, D. Perlman, *Microbial Processes*, Vol. **1**. Academic Press, New York, pp. 521–527.

Powers, H.J. (2003) Riboflavin (vitamin B-2) and health. *Am. J. Clin. Nutr.*, **77**, 1352–1360.

Ramsperger, A., Augustin, M., Schott, A.K., Gerhardt, S., Krojer, T., Eisenreich, W., Illarionov, B., Cushman, M., Bacher, A., Huber, R., and Fischer, M. (2006) Crystal structure of an archaeal pentameric riboflavin synthase in complex with a substrate analog inhibitor: stereochemical implications. *J. Biol. Chem.*, **281**, 1224–1232.

Revuelta, J.L., Buitrago, M.J. and Santos, M.A. (1995) Riboflavin synthesis in fugi. Patent WO9526406 A3, BASF AG.

Revuelta, J.L., Santos, M.A., Garcia-Ramirez, J.J., Gonzalez-Hernandez, M.A. and Buitrago, M.J. (1994) Riboflavin synthesis in yeast. Patent WO1994011515, BASF AG.

Richter, G., Fischer, M., Krieger, C., Eberhardt, S., Luttgen, H., Gerstenschlager, I., and Bacher, A. (1997) Biosynthesis of riboflavin: characterization of the bifunctional deaminase-reductase of *Escherichia coli* and *Bacillus subtilis*. *J. Bacteriol.*, **179**, 2022–2028.

Richter, G., Kelly, M., Krieger, C., Yu, Y., Bermel, W., Karlsson, G., Bacher, A., and Oschkinat, H. (1999) NMR studies on the 46-kDa dimeric protein, 3,4-dihydroxy-2-butanone 4-phosphate synthase, using 2H, 13C, and 15N-labelling. *Eur. J. Biochem.*, **261**, 57–65.

Richter, G., Ritz, H., Katzenmeier, G., Volk, R., Kohnle, A., Lottspeich, F., Allendorf, D., and Bacher, A. (1993) Biosynthesis of riboflavin: cloning, sequencing, mapping, and expression of the gene coding for GTP cyclohydrolase II in *Escherichia coli*. *J. Bacteriol.*, **175**, 4045–4051.

Ritsert, K., Huber, R., Turk, D., Ladenstein, R., Schmidt-Base, K., and Bacher, A. (1995) Studies on the lumazine synthase/riboflavin synthase complex of *Bacillus subtilis*: crystal structure analysis of reconstituted, icosahedral beta-subunit capsids with bound substrate analogue inhibitor at 2.4 A resolution. *J. Mol. Biol.*, **253**, 151–167.

Rivlin, R.S. (1975) *Riboflavin*, Plenum Press, New York.

Rivlin, R.S. (2007) in *Handbook of Vitamins*, 4th edn (eds R.B. Rucker, J.W. Suttie, and D.B. McCormick), CRC Press, Boca Raton, FL, pp. 233–251.

Roje, S. (2007) Vitamin B biosynthesis in plants. *Phytochemistry*, **68**, 1904–1921.

Santos, M.A., Garcia-Ramirez, J.J., and Revuelta, J.L. (1995) Riboflavin biosynthesis in *Saccharomyces cerevisiae*. Cloning, characterization, and expression of the RIB5 gene encoding riboflavin synthase. *J. Biol. Chem.*, **270**, 437–444.

Santos, M.A., Jiménez, A., and Revuelta, J.L. (2000) Molecular characterization of FMN1, the structural gene for the monofunctional flavokinase of *Saccharomyces cerevisiae*. *J. Biol. Chem.*, **275**, 28618–28624.

Santos, M.A., Mateos, L., Stahmann, K.P., and Revuelta, J.L. (2004) *Methods in Biotechnology*, vol. **18**, Humana Press Inc., Totowa, NJ.

Schlosser, T., Wiesenburg, A., Gatgens, C., Funke, A., Viets, U., Vijayalakshmi, S., Nieland, S., and Stahmann, K.P. (2007)

Growth stress triggers riboflavin overproduction in *Ashbya gossypii*. *Appl. Microbiol. Biotechnol.*, **76**, 569–578.

Schlupen, C., Santos, M.A., Weber, U., de Graaf, A., Revuelta, J.L., and Stahmann, K.P. (2003) Disruption of the SHM2 gene, encoding one of two serine hydroxymethyltransferase isoenzymes, reduces the flux from glycine to serine in *Ashbya gossypii*. *Biochem. J.*, **369**, 263–273.

Schmidt, G., Stahmann, K.P., and Sahm, H. (1996) Inhibition of purified isocitrate lyase identified itaconate and oxalate as potential antimetabolites for the riboflavin overproducer *Ashbya gossypii*. *Microbiology*, **142**, 411–417.

Shavlovsky, G., Logvinenko, E., Benndorf, R., Koltun, L., Kashchenko, V., Zakalsky, A., Schlee, D., and Reinbothe, H. (1980) First reaction of riboflavin biosynthesis - Catalysis by a guanosine triphosphate cyclohydrolase from yeast. *Arch. Microbiol.*, **124**, 255–259.

Shiio, I. (1989) Production of primary metabolites, in *Bacillus Subtilis: Molecular Biology and Industrial Application*, Kodansha Ltd. and Elsevier Science Publishers, Tokyo.

Shimizu, S. (2008) Vitamins and related compounds: microbial production, in *Biotechnology: Special Processes*, vol. **10**, 2nd edn (eds H.-J. Rehm and G. Reed), Wiley-VCH Verlag GmbH, Weinheim. doi: 10.1002/9783527620937.ch11.

Singh, M., Kumar, P., Yadav, S., Gautam, R., Sharma, N., and Karthikeyan, S. (2013) The crystal structure reveals the molecular mechanism of bifunctional 3,4-dihydroxy-2-butanone 4-phosphate synthase/GTP cyclohydrolase II (Rv1415) from *Mycobacterium tuberculosis*. *Acta. Crystallogr. D: Biol. Crystallogr.*, **69**, 1633–1644.

Solovieva, I.M., Kreneva, R.A., Leak, D.J., and Perumov, D.A. (1999) The ribR gene encodes a monofunctional riboflavin kinase which is involved in regulation of the *Bacillus subtilis* riboflavin operon. *Microbiology*, **145** (Pt. 1), 67–73.

Stahmann, K.P., Arst, H.N. Jr., Althofer, H., Revuelta, J.L., Monschau, N., Schlupen, C., Gatgens, C., Wiesenburg, A., and Schlosser, T. (2001) Riboflavin, overproduced during sporulation of *Ashbya gossypii*, protects its hyaline spores against ultraviolet light. *Environ. Microbiol.*, **3**, 545–550.

Stahmann, K.P., Revuelta, J.L., and Seulberger, H. (2000) Three biotechnical processes using *Ashbya gossypii*, *Candida famata*, or *Bacillus subtilis* compete with chemical riboflavin production. *Appl. Microbiol. Biotechnol.*, **53**, 509–516.

Sugimoto, T., Kanamasa, S., Kato, T., and Park, E.Y. (2009) Importance of malate synthase in the glyoxylate cycle of *Ashbya gossypii* for the efficient production of riboflavin. *Appl. Microbiol. Biotechnol.*, **83**, 529–539.

Tanner, F.W. Jr., Vojnovich, C., and Vanlanen, J.M. (1945) Riboflavin production by Candida species. *Science*, **101**, 180–181.

Vinas, P., Balsalobre, N., Lopez-Erroz, C., and Hernandez-Cordoba, M. (2004) Liquid chromatographic analysis of riboflavin vitamers in foods using fluorescence detection. *J. Agric. Food. Chem.*, **52**, 1789–1794.

Vitreschak, A.G., Rodionov, D.A., Mironov, A.A., and Gelfand, M.S. (2002) Regulation of riboflavin biosynthesis and transport genes in bacteria by transcriptional and translational attenuation. *Nucleic Acids Res.*, **30**, 3141–3151.

Volk, R. and Bacher, A. (1990) Studies on the 4-carbon precursor in the biosynthesis of riboflavin. Purification and properties of L-3,4-dihydroxy-2-butanone-4-phosphate synthase. *J. Biol. Chem.*, **265**, 19479–19485.

Vorwieger, A., Gryczka, C., Czihal, A., Douchkov, D., Tiedemann, J., Mock, H.P., Jakoby, M., Weisshaar, B., Saalbach, I., and Baumlein, H. (2007) Iron assimilation and transcription factor controlled synthesis of riboflavin in plants. *Planta*, **226**, 147–158.

Wang, L., Chi, Z., Wang, X., Ju, L., and Guo, N. (2008) Isolation and characterization of Candida membranifaciens subsp. Flavinogenie W14-3, a novel riboflavin-producing marine yeast. *Microbiol. Res.*, **163**, 255–266.

Warburg, O. and Christian, W. (1932) Ein zweites sauerstoffübertragendes Ferment und sein Absorptionspektrum. *Naturwissenschaften*, **20**, 688.

Webb, E.C. (1989) Enzyme nomenclature. Recommendations 1984. Supplement 2: corrections and additions. *Eur. J. Biochem.*, **179**, 489–533.

White, H.B. III, (1987) Vitamin-binding proteins in the nutrition of the avian embryo. *J. Exp. Zool. Suppl.*, **1**, 53–63.

Wickerham, L.J., Flickinger, M.H., and Johnston, R.M. (1946) The production of riboflavin by *Ashbya gossypii*. *Arch. Biochem.*, **9**, 95–98.

Winkler, W.C., Cohen-Chalamish, S., and Breaker, R.R. (2002) An mRNA structure that controls gene expression by binding FMN. *Proc. Natl. Acad. Sci. U.S.A.*, **99**, 15908–15913.

Wolf, R., Reiff, F., Wittmann, R. and Butzke, J. (1982) Process for the preparation of riboflavin. Patent US 4355158, BASF AG, USA.

Worst, D.J., Gerrits, M.M., Vandenbroucke-Grauls, C.M., and Kusters, J.G. (1998) *Helicobacter pylori* ribBA-mediated riboflavin production is involved in iron acquisition. *J. Bacteriol.*, **180**, 1473–1479.

Wu, M., Repetto, B., Glerum, D.M., and Tzagoloff, A. (1995) Cloning and characterization of FAD1, the structural gene for flavin adenine dinucleotide synthetase of *Saccharomyces cerevisiae*. *Mol. Cell. Biol.*, **15**, 264–271.

Wulf, P.D. and Vandamme, E.J. (1997) Production of D-ribose by fermentation. *Appl. Microbiol. Biotechnol.*, **48**, 141–148.

Zandomeneghi, M., Carbonaro, L., and Zandomeneghi, G. (2007) Biochemical fluorometric method for the determination of riboflavin in milk. *J. Agric. Food. Chem.*, **55**, 5990–5994.

Zhdanov, V.G., Stepanov and A.I. (1984) Riboflavin preparation. Patent 2546907 A1, Inst Genetiki I Selektsii, FR.

3
Vitamin B_3, Niacin

Tek Chand Bhalla and Savitri

3.1
Introduction

Vitamins are organic nutrients which perform specific and vital functions in various systems of the human body and are essential for maintaining optimal health (Bellows and Moore, 2012). These are classified as micronutrients and are present in food in minute quantities as compared to macronutrients (protein, carbohydrates and fat). The human body requires these nutrients to maintain normal metabolism, growth and good health. Most vitamins are not synthesised in the body or produced in insufficient amounts to meet our needs. They, therefore, have to be obtained mainly through the food we eat.

The two different types of vitamins are fat-soluble vitamins and water-soluble vitamins. Fat-soluble vitamins are vitamin A, D, E and K, which dissolve in fat before they are absorbed in the bloodstream to carry out their functions. Excess of these vitamins are stored in the liver and are not required every day in the diet. In contrast, water-soluble vitamins dissolve in water and are not stored in the body. Since they are eliminated in urine, we need a continuous daily supply of these in our food. The water-soluble vitamins include the vitamin B-complex group and vitamin C (Bellows and Moore, 2012).

Niacin (also known as *nicotinic acid*) and *niacinamide* (nicotinamide) are forms of vitamin B_3 (Figure 3.1). Niacin is an organic compound with the formula $C_6H_5NO_2$ and is one of the 20–80 essential human nutrients (Sharma, 2009). The term niacin is used for pyridine-3-carboxylic acid (nicotinic acid) and derivatives that exhibit the biological activity of nicotinic acid. It may refer either specifically to nicotinic acid or to the total amount of nicotinic acid and nicotinamide in the diet (Food and Nutrition Board, 1998).

Nicotinic acid and nicotinamide are vitamins of the vitamin-B group exhibiting equivalent vitamin activity. Both are precursors in the synthesis of the pyridine coenzymes NAD (nicotinamide adenine dinucleotide) and NADP (nicotinamide adenine dinucleotide phosphate) involved in numerous cell metabolic reactions. The major function is the removal of hydrogen from certain substrates and the transfer of hydrogen to another coenzyme. Reactions in which NAD

Industrial Biotechnology of Vitamins, Biopigments, and Antioxidants, First Edition.
Edited by Erick J. Vandamme and José L. Revuelta.

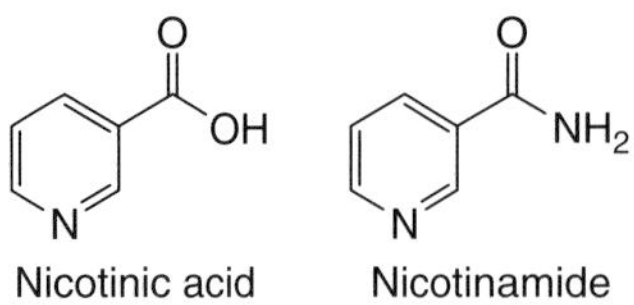

Figure 3.1 Chemical structure of nicotinic acid and nicotinamide.

and NADP are involved include the metabolism of carbohydrates, lipids and proteins.

Vitamin B_3 is required for cell respiration and helps in the release of energy and metabolism of carbohydrates, fats and proteins, proper circulation, maintenance of healthy skin, functioning of the nervous system and normal secretion of bile and stomach fluids. It is used in treatment of pellagra (it is a vitamin B_3 deficiency disease characterised by dermatitis, diarrhoea and mental disturbance), schizophrenia and other mental disorders and as a memory enhancer. Nicotinic acid administered in drug dosage improves the blood cholesterol profile and thus used in treatment of cardiovascular diseases.

3.2 History

Niacin has been known to organic chemists since 1867, long before its importance as an essential nutrient was recognised. The history of niacin and repercussions of its deficiency have been reported by many authors (Harris, 1919; McCollum, 1957; Darby, McNutt and Todhunter, 1975; Hankes, 1984; Loosli, 1991). As early as 1911–1913, Funk had isolated it from yeast and rice polishing in the course of an attempt to identify the water-soluble anti-beriberi vitamin. The interest in niacin for some time was subsequently lost as it was ineffective in curing pigeons of beriberi. Although Funk found that niacin did not cure beriberi, cures were more rapid when it was administered in conjunction with the concentrates containing the anti-beriberi vitamin (thiamin). Warburg and co-workers, for the first time, demonstrated a biochemical function of nicotinic acid when they isolated it from an enzyme in 1935 and showed that it is part of the hydrogen transport system. Before the discovery of its pellagra-preventing property, it was one of the products obtained during oxidation of nicotine. In 1867, Huber prepared nicotinic acid through potassium-dichromate-based oxidation of nicotine, and subsequently, many workers isolated this compound from various natural products. In 1894 and 1912, nicotinic acid from rice bran and yeast was isolated by Suzuki and Funk, respectively (van Eys, 1991). Following discovery that a crude extract of liver was effective in curing pellagra, and therefore was a source of the preventive factor, Elvehjem *et al.* (1937) isolated nicotinamide from the liver as the factor that would cure black tongue in dogs. Reports on the dramatic therapeutic effects of niacin in human pellagra quickly followed from several clinics. In 1945, Krehl and co-workers found that tryptophan was as active as niacin in the treatment of pellagra. Heidelberger reported that tryptophan is a precursor in the synthesis of nicotinic acid in animals and, through meticulous experimentation, demonstrated that the L-[^{14}C] tryptophan was converted to ^{14}C-labelled nicotinic acid in the rat. The conversion of tryptophan to niacin explained why foods rich in animal protein (e.g. milk) prevent and cure pellagra.

3.3 Occurrence in Nature/Food Sources

Free nicotinic acid and nicotinamide are present in nature in only small amounts. Nicotinic acid is mainly bound to the macromolecules in plants, while nicotinamide is usually a component of NADP in the animal world. Nicotinic acid can be formed in humans from the metabolism of dietary tryptophan, and so, it is not necessary to be present in food, provided that adequate tryptophan is available in the foods being consumed.

Niacin is widely distributed in foods of both plant and animal origin (Table 3.1). Large quantities of niacin are found in brewer's yeast and meat. Animal and fish by-products, distillers grains and yeast, various distillation and fermentation broth and certain oil meals are good sources of vitamin B_3. Niacin is present in uncooked foods mainly as the pyridine nucleotides NAD and NADP, but enzymatic hydrolysis of the coenzymes can occur during the course of food preparation. Most organisms use the essential amino acid tryptophan and synthesise niacin from it. However, since there is a preferential use of tryptophan for protein synthesis before, it becomes available for conversion to niacin (Kodicek

Table 3.1 Niacin in food and feedstuffs (milligram per kilograme on dry mass basis).

Alfalfa hay, sun cured	42	Molasses, sugarcane	49
Alfalfa leaves, sun cured	53	Oat, grain	16
Barley, grain	94	Pea seeds	36
Bean, navy (seed)	28	Peanut meal, solvent extracted	188
Blood meal	34	Potato	37
Brewers' grains	47	Rice, bran	330
Buttermilk (cattle)	9	Rice, grain	39
Chicken broilers (whole)	230	Rice, polished	17
Citrus pulp	23	Rye, grain	21
Clover hay, ladino (sun cured)	11	Sorghum, grain	43
Copra meal (coconut)	28	Soybean meal, solvent extracted	31
Corn, gluten meal	55	Soybean seed	24
Corn, yellow grain	28	Spleen, cattle	25
Cottonseed meal, solvent extracted	48	Timothy hay, sun-cured	29
Fish meal, anchovy	89	Wheat, bran	268
Fish meal, menhaden	60	Wheat, grain	64
Fish, sardine	81	Whey	11
Linseed meal, solvent extracted	37	Yeast, brewer's	482
Liver, cattle	269	Yeast, torula	525

Adapted from McDowell (2000).

et al., 1974). It seems unlikely that tryptophan conversion greatly contributes to the niacin supply as majority of the feedstuffs have low tryptophan content.

Important sources of preformed niacin include beef, pork, wheat flour, maize (corn) flour, eggs and cow's milk. Human milk contains a higher concentration of niacin than cow's milk. In the United Kingdom, there is mandatory fortification of flour with nicotinic acid at a level of not less than 1.6 mg/100 g flour for supplementation purposes. Vitamin B_3 is found in various animal and plant tissues including yeast, meat, fish, milk, eggs, green vegetables, beans and cereal grains.

3.4 Chemical and Physical Properties

3.4.1 Chemical Properties

Chemically, niacin ($C_6H_5O_2N$) is one of the simplest vitamins. The two forms of niacin – nicotinic acid and nicotinamide – correspond to 3-pyridinecarboxylic acid and its amide. Niacin (vitamin B_3) is the generic term for nicotinic acid (pyridine-3-carboxylic acid) and nicotinamide (nicotinic acid amide) and the coenzyme forms of the vitamin. Nicotinamide is the active form, which functions as a constituent of two coenzymes, namely NAD and NADP. These coenzymes, in their reduced states (NADH/NADPH), are the principal forms of niacin that exist in animal tissues. Synonyms include nicacid, nicangin, niconacid, nicotinipca, nicyl, PP (pellagra-preventive) factor, pellagra-preventive vitamin, anti-pellagra vitamin, pyridine-B-carboxylic acid, 3-pyridinecarboxylic acid, akotin, nicobid, nico-400 and nicotinex (Informatics, 1974). Nicotinic acid readily forms salts with metals such as aluminium, calcium, copper and sodium. When in acid solution, niacin readily forms quaternary ammonium compounds, such as nicotinic acid hydrochloride, which is soluble in water. When in a basic solution, nicotinic acid readily forms carboxylic acid salts.

3.4.2 Physical Properties

Both nicotinic acid and nicotinamide are white, odourless, crystalline solids soluble in water and alcohol. They are highly resistant to heat, air, light and alkaline conditions and thus are stable in foods. However, they will undergo decarboxylation at a high temperature when in an alkaline medium. The molecular weight of nicotinamide is 122.14 g/mol and that of nicotinic acid is 123.12 g/mol. Niacin is also stable in the presence of the usual oxidising agents. The melting point of nicotinamide and nicotinic acid is 129 and 236 °C, respectively, and the boiling point of nicotinamide is 150–160 °C. Nicotinic acid sublimes at high temperature. Nicotinamide is soluble in water, ether and glycerin; however, nicotinic acid is soluble in water with a solubility of 16.7 g/l. Both the compounds are insoluble in

Table 3.2 Chemical and physical properties of nicotinamide.

Chemical description	Colourless needles or white crystalline powder; odourless with a bitter taste	Lewis (1993)
Molecular weight	122.14	Lewis (1993)
Octanol/water partition coefficient	log-0.37	Unilever (1998)
Empirical formula	$C_6H_6N_2O$	Informatics (1974)
Density	1.40	Lewis (1993)
Solubility	Soluble in water, ether and glycerin	Lewis (1993)
Melting point	129 °C	Lewis (1993)
Boiling range	150° – 160°	National Toxicology Program (NTP) (2000)

Elmore and Cosmetic Ingredient Review Expert Panel (2005).

Table 3.3 Chemical and physical properties of nicotinic acid.

Chemical description	Colourless needles or white crystalline powder with a slight odour	Lewis (1993)
Molecular weight	123.12	Lewis (1993)
Octanol/water partition coefficient	0.63	Bronaugh and Stewart (1985)
Water solubility	16.7 g/l	Bronaugh and Stewart (1985)
Octanol solubility	10.5 g/l	Bronaugh and Stewart (1985)
Empirical formula	$C_6H_5NO_2$	Informatics (1974)
Density	1.473	Lewis (1993)
Solubility	Soluble in water and alcohol, insoluble in most lipid solvents	Lewis (1993)
Melting point	236 °C	Lewis (1993)
Boiling range	Sublimes	National Toxicology Program (NTP) (2000)

Elmore and Cosmetic Ingredient Review Expert Panel (2005).

lipid solvents. The chemical and physical properties of nicotinamide and nicotinic acid are given in Tables 3.2 and 3.3.

3.5 Vitamin B_3 Deficiency Disease (Pellagra)

Vitamin B_3 deficiency causes a condition called *pellagra* (Italian *'pelle agra'*; 'rough skin'), which is characterised by photosensitive dermatitis, diarrhoea, dementia and death (Karthikeyan and Thappa, 2002). The most common symptoms of niacin deficiency are changes in the skin and mucosa of the mouth, stomach and intestinal tract and the nervous system. The changes in the skin are among the most characteristic in human beings. Other signs and symptoms include dizziness, vomiting, constipation or diarrhoea and inflammation of the tongue and gastric mucosa. The neurological symptoms can include fatigue, sleeplessness, depression, memory loss and visual impairment.

This disease was demonstrated as a dietary deficiency disease by Goldberger in 1913. In the beginning of the twentieth century, Elvehjem and co-workers identified nicotinamide isolated from liver extract as PP factor (Sharma, 2009). People with poor diet, alcoholism and some types of slow-growing tumours called *carcinoid tumours* might be at risk for niacin deficiency.

It was thought in the beginning that the clinical manifestations of pellagra arise from the deficient NAD^+ and $NADP^+$ levels in maintaining energy for cellular functions (Hendricks, 1991). However, understanding of these multiple symptoms has progressed with the finding of NAD^+ as a substrate for poly (ADP-ribose) polymerases (PARPs) (Chambon, Weill and Mandel, 1963). PARP has been recognised to play multitude roles in DNA damage responses, including DNA repair, maintenance of genomic stability, transcriptional regulation, signalling pathways involving apoptosis, telomere functions and other multiple cellular functions (Oliver, Menissier-de Murcia and De Murcia, 1999).

3.6 Methods Used for Determination of Vitamin B_3

3.6.1 Microbiological Methods

The most sensitive method for the determination of niacin and related compounds is microbiological. *Lactobacillus plantarum* responds to both forms of the vitamin, whereas *Leuconostoc mesenteroides* measures only nicotinic acid. Niacin must be freed from bound forms before assay. Since niacin is very stable to strong acids, it can be released by acid hydrolysis.

3.6.2 Chemical Methods

Chemical methods of analysis are less sensitive than microbiological procedures and generally require more extensive extraction methods. The cyanogen bromide method of analysis is based on the reaction of pyridine derivatives with cyanogen bromide to form a coloured compound which can be measured quantitatively. The active coenzyme forms of niacin (NAD and NADP) can be determined by an enzyme-cycling colorimetric procedure (Nisselbaum and Green, 1969) or high-performance liquid chromatography (Stocchi *et al.*, 1987). Niacinamide and niacin have also been analysed by infrared and ultraviolet spectroscopy (Committee of Revision of the United States Pharmacopeial Convention (USP), Inc, 1995). Gas–liquid chromatography was utilised after converting niacin and niacinamide into ethylnicotinate and *N*-ethylnicotinamide (Prosser and Sheppard, 1968).

Bioassay procedures for niacin present two major difficulties, that is, (i) tryptophan in the diet is converted into niacin in the tissues and (ii) niacin is synthesised by intestinal bacteria to varying degrees. Chicks, puppies and weanling rats have been used for biological niacin assay.

3.7 Synthesis

Vitamin B_3 differs from the other vitamins of B-complex group in that an essential amino acid, tryptophan, serves as its precursor. Humans can synthesise nicotinamide cofactors from tryptophan. However, the process is somewhat inefficient; synthesis of 1 mg of niacin requires 60 mg of tryptophan. Niacin deficiency, therefore, is usually the result of a diet deficient in both niacin and tryptophan. However, some diets contain tryptophan or niacin in a biologically unavailable form. In corn, the niacin is poorly absorbed unless the corn is treated with alkali prior to ingestion.

3.7.1 Chemical Process Used for Nicotinic Acid Production

3-Picoline (3-methyl pyridine) is used as an ideal starting material for the production of nicotinic acid or nicotinamide. In 3-picoline, the methyl group can be selectively and readily oxidised to the carboxyl derivative with few side products. The present chemical industries utilise 3-picoline, 2-methyl-5-ethylpyridine and 3-cyanopyridine for the synthesis of nicotinic acid through chemical routes. 2-Methyl-5-ethyl-pyridine (MEP) is used as a starting material for the high-temperature and high-pressure liquid-phase oxidation with nitric acid, but it is not a good option (Chuck, 2005). 2-Methyl-5-ethylpyridine is synthesised from ethane/acetaldehyde and ammonia under high temperature and pressure. 3-Methylpyridine (3-picoline) is obtained as a by-product (20–40%) during the synthesis of pyridine from acetaldehyde, formaldehyde and ammonia. The chemical route used in the synthesis of these materials (3-picoline, 2-mentyl-5-ehtylpyridine and 3-cyanopyridine) is depicted in Figure 3.2.

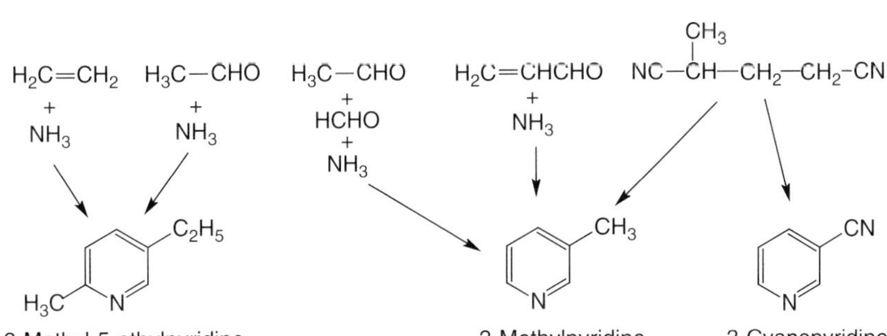

Figure 3.2 Chemical synthesis of starting materials required for nicotinic acid manufacture.

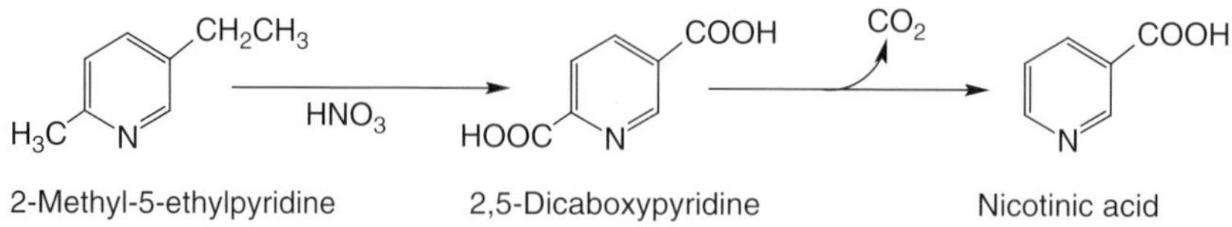

Figure 3.3 Synthesis of nicotinic acid from 2-methyl-5-ethylpyridine.

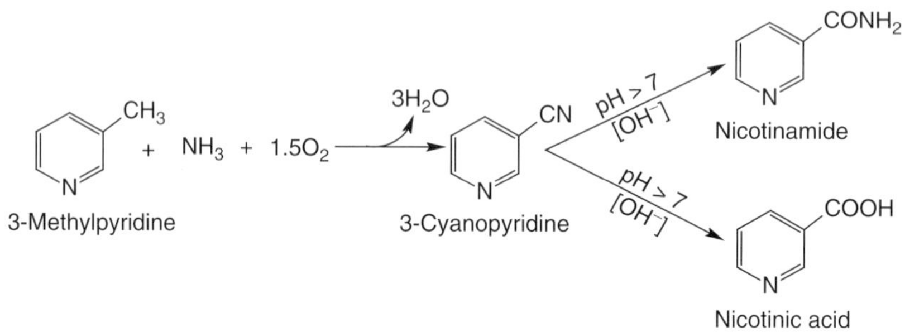

Figure 3.4 Ammoxidation of 3-picoline and hydrolysis of cyanopyridine to niacinamide and nicotinic acid.

Most of the nicotinic acid is manufactured by liquid-phase oxidation of 2-methyl-5-ethylpyridine. The dialkyl pyridine is subjected to oxidation with nitric acid followed by selective decarboxylation of the carboxyl group at the 2-position (Figure 3.3) (Weissermel and Arpe, 1997).

In an alternative process, 3-picoline is first converted into 3-cyanopyridine by gas-phase ammoxidation followed by hydrolysis either to nicotinamide or to nicotinic acid as shown in Figure 3.4. Vanadium oxide is the key catalyst used in the ammoxidation process. The catalyst is supported over oxides of silicon, aluminium, titanium and zirconium (Offermanns *et al.*, 1984).

The production of nicotinamide or nicotinic acid through ammoxidation reaction has received greater attention in the past two decades, both in the industry (Lukas, Neher and Arntz, 1996; Saito *et al.*, 1989; Dicosimo, Burrington and Grasselli, 1991) and in academic institutions (Luecke *et al.*, 1987; Suvorov *et al.*, 1991; Manohar and Reddy, 1998; Narayana *et al.*, 2002).

The oxidation of 3-picoline with stoichiometric or excess quantities of oxidising agents, such as permanganate, nitric acid or chromic acid, has severe environmental implications. Thus, a direct method of nicotinic acid production through gas-phase oxidation of 3-picoline in air has been developed (Chuck and Zacher, 1999). The reaction scheme for this direct gas-phase oxidation of 3-picoline is shown in Figure 3.5.

This process faces considerable difficulties in obtaining a selective and efficient reaction in gas phase. Also, nicotinic acid is less stable than 3-cyanopyridine and decarboxylates at the temperatures normally encountered in the gas-phase reaction. In addition, nicotinic acid desublimes at temperatures below 200 °C and thus can create plugging difficulties in the equipment. Picoline can also be

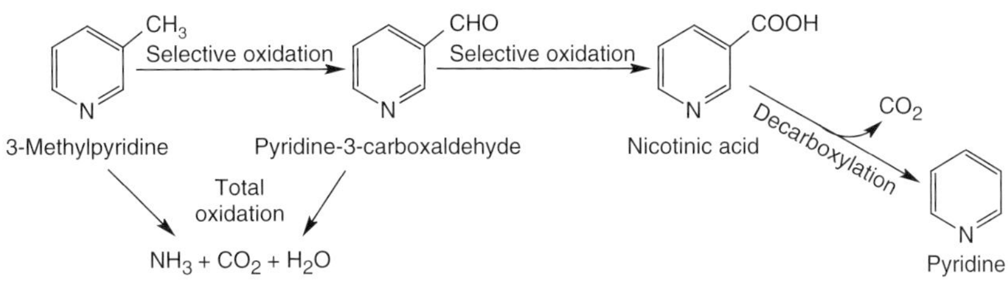

Figure 3.5 Reactions in the gas-phase oxidation of picoline to nicotinic acid.

selectively oxidised with air in the liquid phase to produce niacin (Asamidori, Hashiba and Takigawa, 1994). A combination of catalysts such as cobalt and manganese acetate and/or bromide is usually used in an acetic acid medium, and the air oxidation takes place under elevated temperatures and pressures. Through this method, 32% conversion of picoline and 19% of nicotinic acid are obtained (Hatanaka and Tanaka, 1993). 3-Cyanopyridine is also hydrolysed to nicotinic acid by refluxing with $Ba(OH)_2$ for 8–10 h. 3-Methylpyridine can be directly converted to nicotinic acid, with an yield of 57%, by heating to 200 °C and at 21 atm pressure in the presence of acetic acid as catalyst (Mathew *et al.*, 1988).

Electrochemical oxidation of alkylpyridine to nicotinic acid has also been developed at laboratory scale. Chemical selectivity (80%) and electrical efficiency (up to 90%) have been achieved during oxidation of 3-picoline using lead cells (Toomey, 1984, 1991).

3.7.2 Biosynthesis

The liver can synthesise niacin from the essential amino acid tryptophan, requiring 60 mg of tryptophan to produce 1 mg of niacin. The five-membered aromatic heterocycle of tryptophan is cleaved and rearranged with the alpha amino group of tryptophan into the six-membered aromatic heterocycle of niacin. Riboflavin, vitamin B_6 and iron are required in some of the reactions involved in the conversion of tryptophan to NAD. The pathway involved in the synthesis of nicotinic acid and NAD is shown in Figure 3.6.

3.7.2.1 Biological Processes Used for Nicotinic Acid Production

An alternative to chemical processes is to use biological systems (fermentative route) or their enzymes (enzymatic route) for organic synthesis.

Fermentative Routes for Production of Niacin A fermentative method for the preparation of nicotinic acid has been developed in which *Escherichia coli* was grown in culture medium to produce quinolinic acid, which was further decarboxylated to produce nicotinic acid (Kim *et al.*, 2014). The formation of B vitamins (nicotinic acid and nicotinamide, thiamine, vitamin B_6 and vitamin B_{12}) during the

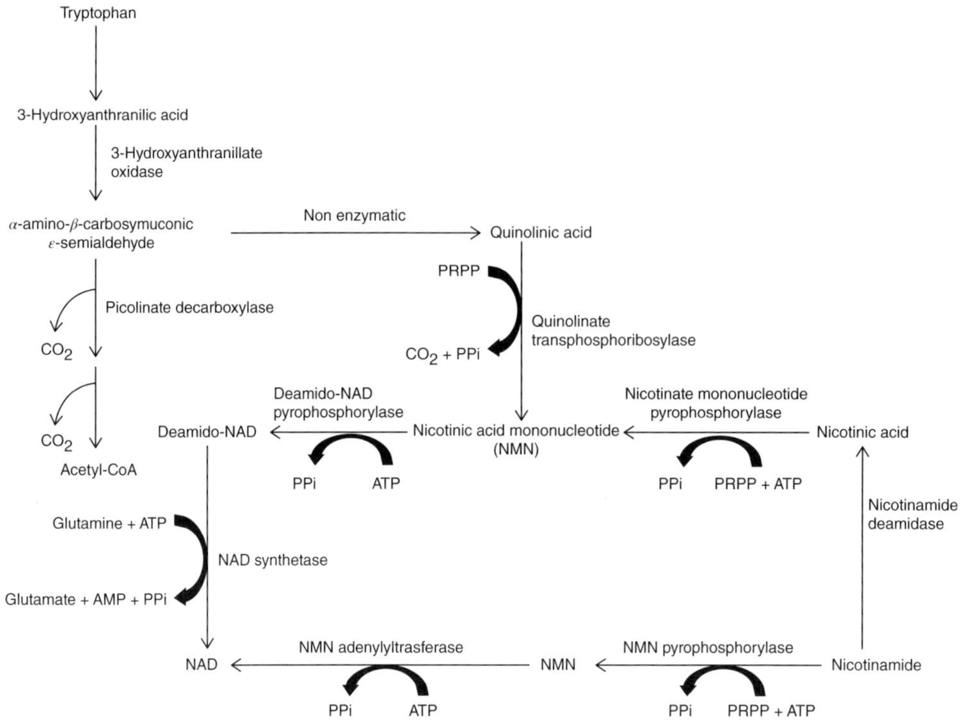

Figure 3.6 Pathways of nicotinic acid and NAD synthesis. (Modified from Greenbaum and Pinder, 1968.) (NMN): Nicotinic acid mononucleotide, NMN: Nicotinamide mononucleotide, PRPP: 5-phosphoribosyl 1-pyrophosphate, PPi: Pyrophosphates.

soaking of soybeans by bacteria isolated from *tempeh* was investigated (Denter and Bisping, 1994). Nicotinic acid and nicotinamide were produced by the species of *Lactobacillus* and *Citrobacter freundii* (Survase, Bajaj and Singhal, 2006).

A fermentative method of NAD synthesis was developed in which a large amount of NAD accumulated with AMP, ADP and ATP in the culture broth when *Brevibacterium ammoniagenes* ATCC 6872 was incubated in the medium containing adenine and nicotinic acid or nicotinamide. A large amount of nicotinic acid mononucleotide with a small amount of NAD accumulated when nicotinic acid or nicotinamide was singly added. NAD was isolated from the culture broth by ion-exchange chromatography and identified by paper chromatography, ultraviolet and infrared spectra, analyses of ribose and phosphate and reduction by alcohol dehydrogenase from yeast (Nakayama *et al.*, 1968).

Biocatalytic Routes for Production of Niacin The application of enzymes to organic chemical processing has attracted increasing attention of academia and industry for the past several decades. Microbial nitrilases or nitrile hydratases and amidases have been reported to hydrolyse 3-cyanopyridine to nicotinic acid or 3-cyanopyridine is hydrated to nicotinamide (Mathew *et al.*, 1988; Vaughan,

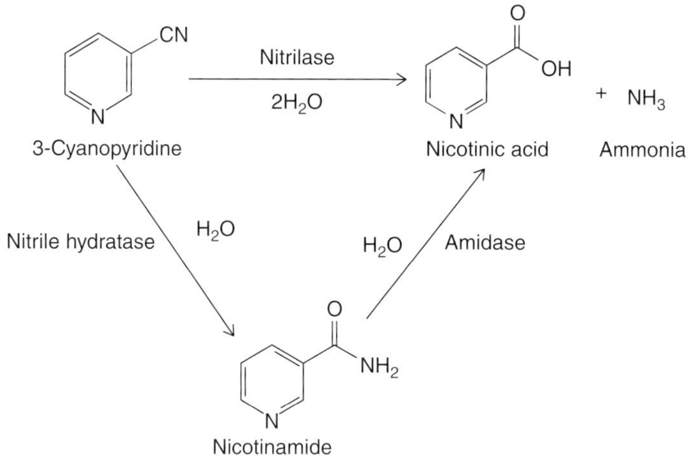

Figure 3.7 Biocatalytic route for synthesis of nicotinamide and nicotinic acid.

Knowles and Cheetham, 1989; Sharma *et al.*, 2006; Sharma, Sharma and Bhalla, 2010; Prasad *et al.*, 2007; Cantarella *et al.*, 2008). Enzymes are operated under mild conditions suitable for the synthesis of labile organic molecules and are efficient in terms of specificity. Compared to chemical methods, enzymatic conversion of 3-cyanopyridine is an advantageous alternative for the production of nicotinic acid and nicotinamide. Enzyme-catalysed conversion of 3-cyanopyridine to nicotinic acid is shown in Figure 3.7.

Asano *et al.* (1982) proposed an enzymatic production process for acrylamide involving nitrile hydratase as a catalyst. Mathew *et al.* (1988) attempted the microbial conversion of 3-cyanopyridine to nicotinic acid by using resting *Rhodococcus rhodochrous* J1 cells containing high benzonitrilase activity. Vaughan, Knowles and Cheetham (1989) developed column reactor using calcium-alginate-immobilised cells of *Nocardia rhodochrous* LL100-21 for the synthesis of nicotinic acid. A total of 96 g of nicotinic acid was formed through this reactor in 150 h. A thermostable nitrilase produced by *Bacillus pallidus* Dac521 catalysed the direct hydrolysis of 3-cyanopyridine to nicotinic acid without detectable formation of nicotinamide. Under optimised conditions, 100% of the 3-cyanopyridine substrate could be converted to nicotinic acid at a conversion rate of 76 nmol/min/mg dry cell mass (Almatawah and Cowan, 1999). One extremely valuable advantage of conducting biotechnological processes at higher temperatures is minimising the risk of contamination by common mesophiles.

Kaplan *et al.* (2006) also performed biotransformation of 3-cyanopyridine into nicotinic acid by fungal nitrilases. Prasad *et al.* (2007) used free cells of *Rhodococcus* sp. NDB 1165 in a fed-batch reaction, and a total of 1.6 M nicotinic acid was formed in 11 h at a rate of 72 mM nicotinic acid g/dcw/h. Nitto (Yamada and Nagasawa, 1989; To and Fujita, 1991) developed a biocatalytic process of selective hydrolysis of 3-cyanopyridine to niacinamide. BASF (Ress-Loeschke, Hauer and

Table 3.4 Microorganisms used as source of enzymes or whole cell biocatalysts for synthesis of nicotinic acid.

Microorganism	Biocatalyst	References
Arthrobacter sp.	Free cells	Asano *et al.* (1982)
Aspergillus niger K10	Purified biocatalyst	Kaplan *et al.* (2006)
Bacillus pallidus Dac521	Calcium alginate immobilised biocatalyst	Almatawah and Cowan (1999)
Nocardia globerulla NHB-2	Free cells	Sharma *et al.* (2006)
Nocardia rhodochrous LL100-21	Calcium alginate immobilised biocatalyst	Vaughan, Knowles and Cheetham (1989)
Rhodococcus sp. NDB 1165	Free cells	Prasad *et al.* (2007)
Rhodococcus rhodochrous J1	Free cells	Mathew *et al.* (1988)
Saccharomyces cerevisiae	Yeast cells	Ahmad and Moat (1966)

Mattes, 2001) and Lonza (Robins and Nagasawa, 1999) produced nicotinamide at industrial scale using biocatalytic route. This technology has also been patented for niacinamide production in China (Heveling *et al.*, 1997). The enzymatic production of nicotinic acid becomes the most efficient alternative. A list of microorganisms reported for the synthesis of nicotinic acid is given in Table 3.4.

3.8 Downstream Processing of Nicotinic Acid

The fermentation technology for the production of organic acids in particular has been known for more than a century, and these have been produced in the form of aqueous solutions. These bioconversions and recovery from fermentation broth are severely inhibited by the products (Kumar and Babu, 2009). Several separation methods such as liquid extraction, ultrafiltration, reverse osmosis, electrodialysis, direct distillation, liquid surfactant membrane extraction, anion exchange, precipitation and adsorption have been used for the recovery of carboxylic acids from fermentation broth. Spray drying, crystallisation and thermal decomposition of ammonium nicotinate have been used for nicotinic acid separation. Degussa (Moeller, Friedrich and Winkler, 1987) developed a crystallisation process to obtain large nicotinic acid crystals. This involves the total hydrolysis of 3-cyanopyridine with a strong base. Lonza (Chuck and Zacher, 1999) and Nippon Soda (Hayakawa and Hatayama, 2002) utilised the conversion of ammonium nicotinate at elevated temperatures. Transformation of concentrated solution of ammonium nicotinate to pure nicotinic acid has been carried out by spray drying, which also ensures a free flow of materials (Chuck and Zacher, 2002). The nicotinic acid can be freed from residual ammonium nicotinate by a thermal post-treatment in a fluidised bed or under reduced pressure. Boreskova (Andrushkevich *et al.*, 1998) developed a process which incorporates the desublimation of nicotinic acid out of the gas stream. All these separation techniques have several limitations requiring high energy and material consumption (Kumar and Babu, 2009).

3.9 Reactive Extraction

Reactive extraction is also used for the separation of nicotinic acid, which increases separation by solvent extraction and represents a link between chemical and physical phenomena. This method allows the production and recovery of fermentation products in one continuous step and reduces the recovery costs. The mechanism of the biosynthetic product separation by reactive extraction depends upon the extraction system used. The extraction can be achieved by means of a chemical reaction between the solute and the extractant (Kumar and Babu, 2009).

3.10 Physiological Role of Vitamin B_3 (Niacin)

Niacin exerts its major physiological effects through its role in the enzyme system for cell respiration. Niacin and niacinamide are required for the proper function of fats and sugars in the body and to maintain cellular health. At high doses, niacin and niacinamide can have different effects such as lowering of cholesterol, improvement in the levels of desired triglycerides in the blood and fibrinolytic effects. Still higher doses (50 mg or more) can cause side effects, most commonly 'niacin flush', which is a burning, tingling sensation in the face and chest, and red or flushed skin.

3.10.1 Coenzyme in Metabolic Reactions

Niacin is required by the human body for the formation of coenzymes NAD and NADP which has pellagra preventive/curative, vasodilating and antilipidemic properties (National Toxicology Program (NTP), 2000) and function in dehydrogenase–reductase systems requiring transfer of a hydride ion (McCormick, 1996, 1997). These coenzymes act as intermediate in most of the H^+ transfers in metabolism, including metabolism of carbohydrates, fatty acids and amino acids. Niacin has important roles as part of oxidation/reduction reaction involving energy metabolism, amino acid metabolism and detoxification reactions for drugs and other substances. NAD is also required for non-redox adenosine diphosphate–ribose transfer reactions involved in DNA repair (Berger, 1985) and calcium mobilisation. NAD functions in intracellular respiration and with enzymes involved in the oxidation of fuel substrates such as glyceraldehyde-3-phosphate, lactate, alcohol, 3-hydroxybutyrate and pyruvate.

NADP functions in reductive biosynthesis such as fatty acid and steroid synthesis and in the oxidation of glucose-6-phosphate to ribose-5-phosphate in the pentose phosphate pathway. NAD is primarily involved in catabolic reactions where it accepts electrons during the breakdown of molecules for energy. In contrast, NADPH (the reduced form of NADP) is primarily involved in biosynthetic reactions where it donates electrons required for synthesising new molecules. In most cells, NAD levels are much higher than NADH levels, while NADPH levels are much higher than those of NADP.

Important metabolic reactions catalysed by NAD and NADP are summarised as follows:

1) **Carbohydrate metabolism:**
   a. *Glycolysis:* NAD serves as an electron acceptor in glycolysis. When NAD accepts electrons, it also acquires a proton (H^+) and is converted into NADH. NADH is a reduced electron carrier. The overall process of glycolysis can be summarised in the following reaction:

$$\text{Glucose} + 2\ \text{NAD}^+ + 2\ \text{ADP} + 2\ \text{P}_i \rightarrow 2\ \text{Pyruvate} + 2\ \text{NADH} + 2\ \text{H}^+ + 2\ \text{ATP} + 2\ \text{H}_2\text{O}$$

   If glycolysis were to continue indefinitely, all of the NAD^+ would be used up and glycolysis would stop. To allow glycolysis to continue, organisms must be able to oxidise NADH back to NAD^+. NADH then becomes oxidised in the first step of electron transport by mitochondrial complex I or NADH dehydrogenase. NADH contains flavin mononucleotide (FMN) as a bound prosthetic group, which is responsible for catalysing the following reaction.

$$\text{NADH} + \text{H}^+ + \text{FMN} \longrightarrow \text{NAD}^+\ \text{FMNH}_2$$

   ATP is produced in this oxidation reaction by the enzymes of the respiratory chain.

   b. *Krebs cycle:* The Krebs cycle is the central metabolic pathway in all aerobic organisms. The cycle is a series of eight reactions that occur in the mitochondrion. These reactions take two carbon molecules (acetate) and completely oxidise them to carbon dioxide. The cycle is summarised in the following chemical equation:

$$\text{acetyl CoA} + 3\ \text{NAD} + \text{FAD} + \text{ADP} + \text{HPO}_4^{-2} \rightarrow 2\ \text{CO}_2 + \text{CoA} + 3\ \text{NADH}^+ + \text{FADH}^+ + \text{ATP}$$

   One example of the role of NAD as coenzyme in the Krebs cycle is oxidative decarboxylation of pyruvate to acetyl-CoA catalysed by a three-enzyme complex known as *pyruvate dehydrogenase.*

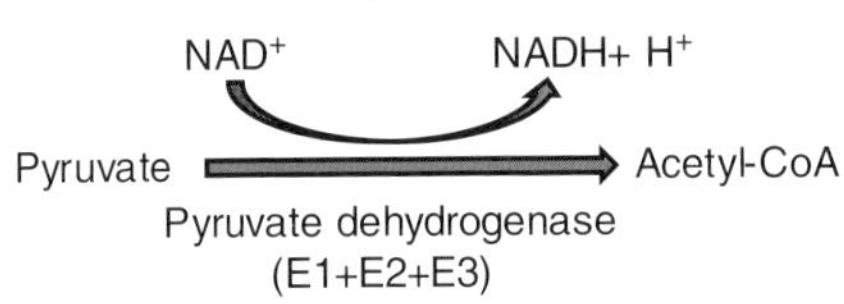

2) **Lipid metabolism:** $NADP^+$ carries reducing power for fatty acid biosynthesis and oxidation. Some examples of oxidation–reduction reactions occurring during fatty acid oxidation and synthesis are given as follows:
   a. Oxidation of L-3-hydroxyacyl CoA to 3-ketoacyl-CoA by NAD^+ and enzyme hydroxyacyl-CoA dehydrogenase. This converts the hydroxyl group into a keto group.

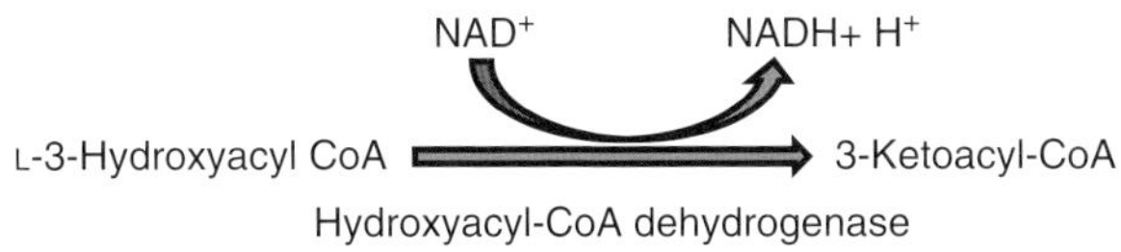

   b. Reduction of crotonyl-ACP to butyryl-ACP catalysed by enoyl-ACP reductase and NADPH during fatty acid synthesis.

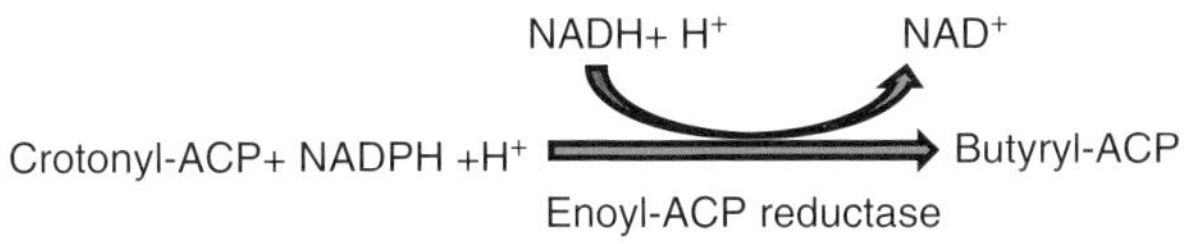

3) **Protein metabolism:** High level of ammonium ion (NH_4^+) produced as a result of biological nitrogen fixation is toxic to the cells, which must be eliminated by incorporating it into various organic forms. Reactions leading to three compounds, namely, glutamic acid, glutamine and carbamoyl phosphate, play key role in assimilating ammonium ion. The two amino acids (glutamic acid, glutamine) further participate in the synthesis of other amino acids and nitrogen-containing compounds. In plants and bacteria, ammonia is used in the synthesis of glutamic acid through a coupled reaction catalysed by glutamine synthetase and glutamic acid synthase in the presence of NADPH as shown in the following reaction.

$$\alpha\text{-ketoglutarate} + NH_4^+ + NADPH + H^+ + ATP \longrightarrow \text{Glutamic acid} + NADP^+ + ADP + Pi$$

4) **Photosynthesis:** In photosynthetic organisms, NADPH is produced by ferredoxin-$NADP^+$ reductase in the last step of the electron chain of the light reactions of photosynthesis. It is used as reducing power for the biosynthetic reactions in the Calvin cycle to assimilate carbon dioxide. In this pathway, the free energy of cleavage of ~P bonds of ATP and the reducing power of NADPH are used to fix and reduce CO_2 to form carbohydrate.
5) **Rhodopsin synthesis:** Rhodopsin, also known as *visual purple* is a light-sensitive receptor protein present in the photoreceptor cells of the retina. They are extremely sensitive to light, enabling vision in low-light conditions. Rhodopsin synthesis is an irreversible reaction in which NAD^+ and NADH take part. A list of enzymes that require $NAD^+/NADP^+$ and their functions is given in Table 3.5.

Table 3.5 Enzymes that require NAD and NADP as coenzyme.

	Enzyme	Function
Enzymes that use $NAD^+/NADH$	Alcohol dehydrogenase	Metabolizes alcohol
	Glyceraldehyde phosphate dehydrogenase	Catalyses important step in glycolysis
	Lactate dehydrogenase	Catalyses reactions in muscle and liver cells
	Pyruvate dehydrogenase	Catalyses reactions connecting glycolysis to the Krebs cycle
	α-Keto-glutarate dehydrogenase, isocitrate dehydrogenase, malate dehydrogenase	Catalyses reactions in the Krebs cycle, aerobic metabolism
	NADH dehydrogenase	Catalyses oxidative phosphorylation reactions
	Hydroxy-acyl-SCoA dehydrogenase	Important in fat catabolism
Enzymes that use $NADP^+/NADPH$	Glucose 6-phosphate dehydrogenase	Catalyses reactions in the pentose phosphate pathway
	β-Ketoacyl-ACP reductase β-enoyl-ACP reductase	Catalyses reactions in fatty acid synthesis
	Chloroplast glyceraldehyde phosphate dehydrogenase	Catalyses reactions in the Calvin cycle, glucose synthesis

Adapted from Metzler (1977).

3.10.2 Therapeutic Molecule

Nicotinic acid and nicotinamide find their applications in formulation of drugs for control/treatment of pellagra in pharmacologic doses (1–5 g/day), hypercholesterolemia and hypertriglyceridemia (McCormack and Keating, 2005), cardiovascular diseases, for formation of coenzyme (such as NAD and NADP), for treatment of cancer, diabetes, arthritis and for detoxification (Kirschmann and Kirschmann, 1996). The most commonly cited use of vitamin B_3 (niacin/nicotinic acid and niacinamide/nicotinamide) is for the treatment of pellagra.

The recommended dietary intake of niacin is 14–16 mg/day (Food and Nutrition Board, 1998). Currently, niacin therapy as a replacement or in combination with the existing cholesterol-reducing prescription drugs is under evaluation by various drug regulatory authorities around the world.

3.10.2.1 Treatment of Pellagra

Pellagra is a disease caused by a cellular deficiency of the nicotinamide coenzymes due to inadequate dietary supply of tryptophan and vitamin B_3 characterised by

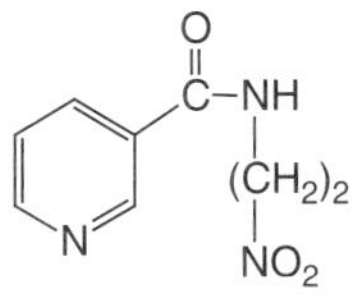

Figure 3.8 Nicorandil.

three Ds, that is, diarrhoea, dermatitis and dementia. Although it is not usually fatal, yet when the three Ds are present, death can occur. The adult intake of vitamin B_3 necessary to prevent pellagra is around 20 mg/day. The body can manufacture approximately 1 mg of niacin equivalents from 60 mg of tryptophan obtained mostly from dietary protein.

3.10.2.2
Treatment of Cardiovascular Diseases

Numerous clinical trials have demonstrated that niacin reduces the risk of coronary artery disease and is the most potent lipid-regulating agent for increasing levels of HDLC (high-density lipoprotein cholesterol) (Ganji *et al.*, 2006). Niacin is considered as a very effective and inexpensive agent for improving health outcomes in persons with elevated lipid levels at risk for heart disease (Keenan *et al.*, 1991; Morgan, Capuzzi and Guyton, 1998; Vogt *et al.*, 2006). Nicotinic acid is a precursor for the synthesis of nicorandil, a cardiovascular drug (Figure 3.8).

Antihyperlipidemic effect and inhibition of cholesterol synthesis discussed next are mainly responsible for reduction in cardiovascular diseases in persons taking niacin.

3.10.2.3
Antihyperlipidemic Effect

Niacin/nicotinic acid at an intake of 1000 mg or higher is an effective antihyperlipidemic agent. It is particularly effective in lowering the blood concentrations of low-density lipoprotein (LDL) and very low density lipoprotein (VLDL) cholesterol and in increasing the concentration of HDLC. The beneficial effects of nicotinic acid in the treatment of hyperlipidemia are attributed to four interrelated effects on lipid and lipoprotein metabolism such as (i) inhibition of lipolysis in adipose tissue; (ii) inhibition of the synthesis and secretion of VLDL by the liver; (iii) lowering of serum levels of lipoprotein(a), a variant form of LDL; and (iv) an increase in serum levels of HDL (DiPalma and Thayer, 1991). Intakes at quantities of 1 g or more, however, not only provide pharmacological benefits but also carry significant risk of adverse effect, thus requiring medical supervision and monitoring (Hathcock, 2004).

3.10.2.4
Treatment of Hypercholesterolemia

A number of studies in animals as well as in hyperlipidemic patients have indicated that nicotinic acid inhibits endogenous synthesis of cholesterol (Hotz,

1983). Plasma levels of squalene, an intermediate in the cholesterol biosynthesis, have been reported to be decreased in response to niacin intake (Kudchodkar *et al.*, 1978). Thus, at least a part of the antihyperlipidemic action of nicotinic acid appears to arise from inhibition of endogenous cholesterol biosynthesis. Niacinamide may be used in clinical treatment of hypercholesteremia, and niacin may be used in the prevention of pellegra and treatment of certain psychological disorders. Both ingredients are readily absorbed from the skin, blood and the intestines and widely distribute throughout the body (Elmore and Cosmetic Ingredient Review Expert Panel, 2005).

3.10.2.5
Diabetes

It has long been known that there are several compounds, which may produce acute insulin deficiency causing diabetes, by destructing ß-cells of the pancreas in experimental animals. A consistent finding has been that this experimental damage is closely related to a sharp decrease in intracellular NAD levels. The active coenzyme forms of niacinamide (NAD, NADP) are essential for normal carbohydrate, lipid and protein metabolism. Pharmacological doses of niacinamide are therefore studied for their potential benefit in the prevention and treatment of diabetes (Vague *et al.*, 1989; Kolb and Burkart, 1999; Shah *et al.*, 2013).

3.10.2.6
Fibrinolysis

Parenteral doses of nicotinic acid cause a significant fibrinolytic effect; however, oral dose of nicotinic acid does not cause fibrinolysis. The fibrinolytic effect only occurs with the first dose, and subsequent/continuous intravenous infusions are inactive. However, clinically effective fibrinolytic agents such as tissue plasminogen activator, streptokinase and urokinase limit its use in fibrinolysis (DiPalma, 1988).

3.10.2.7
Treatment of Neurodegenerative Disorders

People who consume higher amounts of niacin from food and multivitamin sources seem to have a lower risk of getting Alzheimer's disease than people who consume less niacin.

Niacin is also used for treating schizophrenia, anxiety, depression and chronic alcoholism (Ban, 1971), hallucinations due to drugs, age-related loss of thinking skills, chronic brain syndrome, motion sickness and oedema. Some people use niacin or niacinamide for treating acne, leprosy, attention deficit hyperactivity disorder (ADHD), memory loss, arthritis, preventing premenstrual headache, improving digestion, protecting against toxins and pollutants, reducing the effects of aging, lowering blood pressure, improving circulation, promoting relaxation and preventing cataracts.

3.11
Safety of Niacin

Niacin and niacinamide are likely safe for most people when taken orally and are authorised for use in food. A common minor side effect of niacin is a flushing reaction causing burning, itching and redness of the face, arms and chest, as well as headaches. Usually, this reaction goes away as the body gets used to the medication. Other minor side effects of niacin and niacinamide are intestinal gas, upset stomach, dizziness, pain in the mouth and so on. Niacin when taken over 3 g/day results in serious side effects including liver problems, gout, ulcers of the digestive tract, loss of vision, high blood sugar and irregular heartbeat (European Food Safety Authority (EFSA), 2009, 2012).

3.12
Toxicity of Niacin

Although therapeutically useful in lowering serum cholesterol, administration of chronic high oral doses of nicotinic acid has been associated with hepatotoxicity as well as dermatologic manifestations most typically skin flushing and itching. An upper limit (UL) of 35 mg/day is proposed by the US Food and Nutrition Board (1998).

3.12.1
Hepatotoxicity

Niacin has been associated with abnormal liver tests and causes significant liver toxicity. This has only been seen with slow release formulation of niacin and virtually never with immediate or extended release of niacin. Severe and potentially life-threatening hepatotoxicity has been observed in patients taking 3–9 g niacin per day for periods of months or years for the treatment of hypercholesterolaemia. Several cases show liver dysfunction and fulminant hepatitis and may even proceed to encephalopathy requiring liver transplantation (Scientific Committee on Food, 2002).

3.12.2
Vasodilation/Niacin Flush

High intakes of niacin produce a vasodilative effect known as the *niacin flush*. The vasodilatation is associated with an unpleasant sensation of intense warmth and itching that commonly starts in the face and neck and can proceed down through the body. The visible skin flush lasts only about 1–2 min, but vasodilation can be measured even in the lower limbs for about 30 min. Some individuals may experience a rash, hypotension and/or dizziness (Capuzzi *et al.*, 2000). Flushing is initiated via prostaglandin D_2-mediated vasodilatation of small subcutaneous blood vessels. The flush reaction disappears after weeks of continued daily nicotinic acid therapy.

3.12.3 Glucose Intolerance

Nicotinic acid (3 g/day) has been reported to impair glucose tolerance in otherwise healthy individuals treated for hypercholesterolaemia (Scientific Committee on Food, 2002).

3.13 Derivatives of Niacin

Apart from nicotinic acid and nicotinamide, niacin is also available in other derivative forms, for example, inositol hexanicotinate, which exhibit the biological activity of nicotinamide (Food and Nutrition Board, 1998). Inositol hexaniacinate (IHN) is the hexanicotinic acid ester of meso-inositol. This compound consists of six molecules of nicotinic acid (niacin) with an inositol molecule in the centre (Figure 3.9). It is described as 'no flush niacin'. These derivatives may be converted into nicotinic acid or may contain nicotinic acid, nicotinamide or their releasable moieties. Whether these compounds should be referred to as *niacin* depends on their biological effects, the rates of uptake and metabolism and the release of the chemical components that produce biological effects similar to the other forms of niacin.

In 2009, the European Food Safety Authority (EFSA) Scientific Panel on Food Additives and Nutrient Sources Added to Food concluded that nicotinate from IHN is a bioavailable source of niacin (European Food Safety Authority (EFSA), 2009). IHN, similarly to extended-release nicotinic acid, has been investigated for potential beneficial effects on serum lipids while minimising the flushing effect

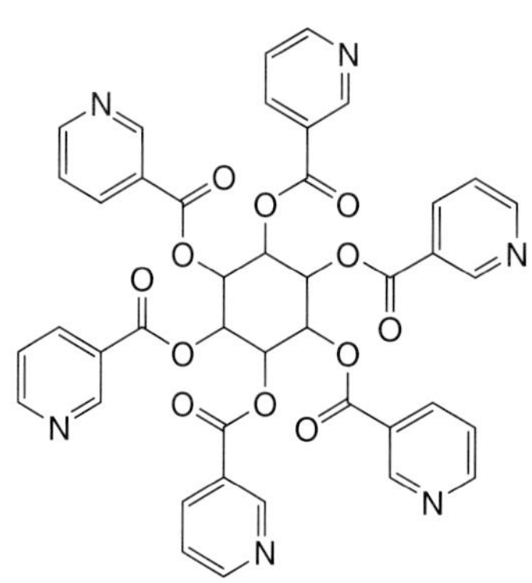

Figure 3.9 Chemical structure of inositol hexaniacinate (European Food Safety Authority (EFSA), 2009).

(Norris, 2006). The mechanisms of action of IHN are believed to be the same as those for niacin.

3.14 Application in Cosmetics, Food and Feed

Niacin and niacinamide have their application in cosmetics primarily as hair and skin conditioning agents. Niacinamide is used in around 30 cosmetic formulations including shampoos, hair tonics, skin moisturizers and cleansing formulations in the concentration range staring from as low as 0.0001% in night creams to a high of 3% in body and hand creams, lotions, powders and sprays. Nicotinic acid is also used as additives in food and animal feed and considered GRAS (Generally Recognized as Safe). Nicotinamide derived from nicotinic acid is also used as a brightener in electroplating baths and stabiliser for pigmentation in cured meat (Arum, 1998).

3.15 Future Prospects

Vitamin B_3 (nicotinic acid/nicotinamide) is considered to be the major B vitamin required for a healthy life. This vitamin has great potential to be used in the treatment of pellagra and cardiovascular diseases, as lipid-modifying drug, in the treatment of diabetes, which has resulted in an increased interest in the pharmacological properties of this drug. The clinical use of nicotinic acid, however, has been limited by unpleasant side effects, primarily the flushing problem. So, research is needed to better understand the mechanisms of physiological benefits of this vitamin and to minimise the adverse effects when used for pharmacological purposes. The growing demand of nicotinic acid further draws attention to intensify the production of nicotinic acid via fermentative or biocatalytic route in order to reduce the risks posed by chemical processes to the environment. Further, the rapid growth in the field of nanotechnology can be utilised for development of new and efficient delivery systems for vitamin B_3 which will improve the pharmacological profile of this molecule without any side effect.

References

Ahmad, F. and Moat, A.G. (1966) Nicotinic acid biosynthesis in prototrophs and tryptophan auxotrophs of Saccharomyces cerevisiae. *J. Biol. Chem.*, **241** (4), 775–780.

Almatawah, Q.A. and Cowan, D.A. (1999) Thermostable nitrilase catalyzed production of nicotinic acid from 3-cyanopyridine. *Enzyme Microb. Technol.*, **25**, 718–724.

Andrushkevich, T.V., Makarenko, M.G., Prokhorov, V.P., Vasilev, E.V., Zenkovets, G.A. and Al'kaeva, E.M. (1998) Industrial production of nicotinic acid. RU21097341-998, Boreskova Inst Kataliza Sibir.

Arum, S.D.V. (1998) Niacin, nicotinamide, and nicotinic acid, in *Encyclopaedia of Chemical Technology*, vol. **25**, 4th edn (eds J.I. Kroschwitz and M. Howe-Grant), John Wiley & Sons, Inc., New York.

Asamidori, Y., Hashiba, I. and Takigawa, S. (1994) Production of nicotinic acid. Patent JP 9,426,603, Nissan Chemical Industries Ltd., Japan.

Asano, Y., Yasuda, Y., Tani, Y., and Yamada, H. (1982) A new enzymatic method of acrylamide production. *Agric. Biol. Chem.*, **46**, 1183–1189.

Ban, T.A. (1971) Nicotinic acid and psychiatry. *Can. Psychiatr. Assoc. J.*, **16**, 413–431.

Bellows, L. and Moore, R. (2012) *Water-Soluble Vitamins: B-Complex and Vitamin C*, Food and Nutrition Series. Fact Sheet No. 9.312, Health Colorado State University.

Berger, N.A. (1985) Poly(ADP-ribose) in the cellular response to DNA damage. *Radiat. Res.*, **101**, 4–15.

Bronaugh, R.L. and Stewart, R.F. (1985) Methods for *in vitro* percutaneous absorption studies: permeation through damaged skin. *J. Pharm. Sci.*, **74**, 1062–1066.

Cantarella, M., Cantarella, L., Gallifuoco, A., Intellini, R., Kaplan, O., Spera, A., and Martínková, L. (2008) Amidase-catalyzed production of nicotinic acid in batch and continuous stirred membrane reactors. *Enzyme Microb. Technol.*, **42** (3), 222–229.

Capuzzi, D.M., Morgan, J.M., Brusco, O.A. Jr., and Intenzo, C.M. (2000) Niacin dosing: relationship to benefits and adverse effects. *Curr. Atheroscler. Rep.*, **2**, 64–71.

Chambon, P., Weill, J.D., and Mandel, P. (1963) Nicotinamide mononucleotide activation of a new DNA-dependent polyadenylic acid synthesizing nuclear enzyme. *Biochem. Biophys. Res. Commun.*, **11** (1), 39–43.

Chuck, R. (2005) Technology development in nicotinate production. *Appl. Catal., A*, **280**, 75–82.

Chuck, R.J. and Zacher, U. (1999) Verfahren zur Herstellung von Nicotinsäure. Patent EP 919,548 A1, May 2,1999.

Chuck, R.J. and Zacher, U. (2002) Process for the preparation of nicotinic acid. US Patent 6,376,677.

Committee of Revision of the United States Pharmacopeial Convention (USP), Inc (1995) *The United States Pharmacopeia*, 23rd edn, USP Convention, Inc., Rockville, MD.

Darby, W.J., McNutt, K.W., and Todhunter, E.N. (1975) Niacin. *Nutr. Rev.*, **33**, 289–297.

Denter, J. and Bisping, B. (1994) Formation of B vitamins by bacteria during the soaking process of soybeans for *tempe* fermentation. *Int. J. Food Microbiol.*, **22**, 23–31.

Dicosimo, R., Burrington, J.D. and Grasselli, R.K. (1991) Ammoxidation of methyl substituted heteroaromatics to make heteroaromatic nitriles. US Patent 5,028,713, Standard Oil Co.

DiPalma, J.R. (1988) Tissue plasminogen activator for coronary thrombosis. *Am. Fam. Physician*, **37**, 291–295.

DiPalma, J.R. and Thayer, W.S. (1991) Use of niacin as a drug. *Annu. Rev. Nutr.*, **11**, 169–187.

Elmore, A.R. and Cosmetic Ingredient Review Expert Panel (2005) Final report of the safety assessment of niacinamide and niacin. *Int. J. Toxicol.*, **24** (5), 1–31.

Elvehjem, C.A., Madden, R.J., Strong, F.M., and Wooley, D.W. (1937) Relation of nicotinic acid and nicotinic acid amide to canine black tongue. *J. Am. Chem. Soc.*, **59**, 1767–1768.

European Food Safety Authority (EFSA) (2009) Scientific opinion of the Panel on Food Additives and Nutrient Sources added to food on inositol hexanicotinate (inositol hexaniacinate) as a source of niacin (vitamin B3) added for nutritional purposes in food supplements. *EFSA J.*, **949**, 1–20.

European Food Safety Authority (EFSA) (2012) Scientific opinion on the safety and efficacy of niacin (nicotinamide) as feed additive for all animal species based on a dossier submitted by agrinutrition BV1, 2 EFSA panel on additives and products or substances used in animal feed (FEEDAP). *EFSA J.*, **10** (6), 2731.

van Eys, J. (1991) in *Handbook of Vitamins* (ed L. Machlin), Marcel Dekker, New York.

Food and Nutrition Board (1998) *Dietary Reference Intakes: Thiamin, Riboflavin, Niacin, Vitamin B-6, Vitamin B-12, Pantothenic Acid, Biotin, and Choline*, National Academy Press, Washington, DC, 123–149.

Ganji, S.H., Zhang, L.H., Kamanna, V.S., and Kashyap, M.L. (2006) Effect of niacin on lipoproteins and atherosclerosis. *Future Lipidol.*, **1**, 549–557.

Greenbaum, A.L. and Pinder, S. (1968) The pathway of biosynthesis of nicotinamide-adenine dinucleotide in rat mammary gland. *Biochem. J.*, **107**, 55.

Hankes, L.V. (1984) Nicotinic acid and Nicotinamide. in *Handbook of Vitamins* (ed L.J. Machlin), Marcel Dekker, New York pp. 329–377.

Harris, H.F. (1919) *Pellagra*, The Macmillan Co., New York.

Hatanaka, M. and Tanaka, N. (1993) Process for producing pyridinecarboxylic acid. World Patent 93.5022 A1.

Hathcock, J.N. (2004) *Vitamin and Mineral Safety*, 2nd edn, Council for Responsible Nutrition (CRN), Washington, DC.

Hayakawa, K., Hatayama, M. (2002) Process for producing free acids from ammonium carboxylates. Patent WO 9,900,350, (Nippon Soda Co., Ltd., Japan), Jan.7, 1999.

Hendricks, W.M. (1991) Pellagra and pellagra like dermatoses: etiology, differential diagnosis, dermatopathology, and treatment. *Semin. Dermatol.*, **10** (4), 282–292.

Heveling, J., Armbruster, E., Utiger, L., Rohner, M., Dettwiler, H.R. and Chuck, R.J. (1997) Process for the preparation of nicotinic acid amide. Patent EP 770,687, Lonza AG, Switzerland.

Hotz, W. (1983) Nicotinic acid and its derivatives: a short survey. *Adv. Lipid Res.*, **20**, 195–217.

Informatics (1974) Monograph on Nniacin. (Submitted by FDA in response to an FOI request–2000, 416 pages).2.

Kaplan, O., Vejvoda, V., Plíhal, O., Pompach, P., and Kavan, D. (2006) Purification and characterization of a nitrilase from *Aspergillus niger* K10. *J. Appl. Microbiol. Biotechnol.*, **73**, 567–575.

Karthikeyan, K. and Thappa, D.M. (2002) Pellagra and skin. *Int. J. Dermatol.*, **41** (8), 476–481.

Keenan, J.M., Fontaine, P.L., Wenz, J.B., Myers, S., Huang, Z.Q., and Ripsin, C.M. (1991) Niacin revisited. A randomized, controlled trial of wax-matrix sustained-release niacin in hypercholesterolemia. *Arch. Intern. Med.*, **151**, 1424–1432.

Kim, S.Y., Shin, Y.U., Kyung, H.E.O. *et al.* (2014) Method for the preparation of nicotinic acid. US Patent 20140315263 A1.

Kirschmann, G.J. and Kirschmann, J.D. (1996) Nicotinic acid, in *Nutrition Almanac*, McGraw Hill publication, New York. ISBN: 0-443-043051..

Kodicek, E., Ashby, D.R., Muller, M., and Carpenter, K.J. (1974) The conversion of bound nicotinic acid to free nicotinamide on roasting sweet corn. *Proc. Nutr. Soc.*, **33**, 105A–106A.

Kolb, H. and Burkart, V. (1999) Nicotinamide in type 1 diabetes, mechanisms of action revisited. *Diabetes Care*, **22** (2), B16–B20.

Kudchodkar, B.J., Sodhi, H.S., Horlick, L., and Mason, D.T. (1978) Mechanisms of hypolipidemic action of nicotinic acid. *Clin. Pharmacol. Ther.*, **24**, 354–373.

Kumar, S. and Babu, B.V. (2009) Process intensification of nicotinic acid production via enzymatic conversion using reactive extraction. *Chem. Biochem. Eng. Q.*, **23** (3), 367–376.

Lewis, R.J. Sr. (1993) *Hazardous Chemicals Desk Reference*, 3rd edn, Van Nostrand Reinhold, New York.

Loosli, J.K. (1991) in *Handbook of Animal Science* (ed P.A. Putnam), Academic Press, San Diego, CA.

Luecke, B., Martin, A., Seeboth, H., Ladwig, G., Parlitz, B., and French, J. (1987) German Patent *DD*, **241**, 903.

Lukas, V.H., Neher, A. and Arntz, D. (1996) Cyanopyridine preparation process and catalysts therefor. Patent EP 726,092, Degussa AG, Germany.

Manohar, B. and Reddy, B.M. (1998) Ammoxidation of 3-picoline to nicotinonitrile over vanadium phosphorus oxide-based catalysts. *J. Chem. Technol. Biotechnol.*, **71** (2), 141–146.

Mathew, C.D., Nagasawa, T., Kobayashi, M., and Yamada, H. (1988) Nitrilase catalyzed production of nicotinic acid from 3-cyanopyridine in *Rhodococcus rhodochrous* J1. *Appl. Environ. Microbiol.*, **54**, 1030–1032.

McCollum, E.V. (1957) *A History of Nutrition*, Houghton Mifflin, Boston, MA.

McCormack, P.L. and Keating, G.M. (2005) Prolonged-release nicotinic acid: a review of its use in the treatment of dyslipidaemia. *Drugs*, **65**, 2719–2740.

McCormick, D.B. (1996) in *Encyclopedia of Molecular Biology and Molecular Medicine*, vol. **1** (ed R.A. Meyers), VCH, Weinheim, DC, pp. 396–406.

McCormick, D.B. (1997) in *Encyclopedia of Human Biology*, 2nd edn (ed R. Dulbecco), Academic Press, San Diego, MA, pp. 847–864.

McDowell, L.R. (2000) *Vitamins in Animal and Human Nutrition*, 2nd edn, Iowa State University Press, Ames, IA, ISBN: 0-8138-2630-6.

Metzler, D.E. (1977) *Biochemistry: The Chemical Reactions of Living Cells*, Academic Press, New York.

Moeller, A., Friedrich, H.K. and Winkler, H.K. (1987) Verfahren zur herstellung grobkristalliner nicotinsaeure hoher reinheit. DE 3,614, 019, Degussa-Huels AG, Fed. Rep. Germany.

Morgan, J.M., Capuzzi, D.M., and Guyton, J.R. (1998) A new extended-release niacin (Niaspan): efficacy, tolerability, and safety in hypercholesterolemic patients. *Am. J. Cardiol.*, **82**, 29U–34U.

Nakayama, K., Sato, Z., Haruo, H., and Kinoshita, S. (1968) Production of nucleic acid-related substances by fermentative processes. *Agric. Biol. Chem.*, **32** (11), 1331–1336.

Narayana, K.V., Masthan, S.K., Rao, V.V., Raju, B.D., and Rao, P.K. (2002) Influence of V_2O_5 content on ammoxidation of 3-picoline over V_2O_5/AlF_3 catalysts. *Catal. Commun.*, **3** (4), 173.

National Toxicology Program (NTP) (2000) Entry on Niacinamide and Niacin. NTP Database, National Library of Medicine, Bethesda.

Nisselbaum, J.S. and Green, S. (1969) A simple ultramicro method for determination of pyridine nucleotides in tissues. *Anal. Biochem.*, **27**, 212–217.

Norris, R.B. (2006) Flush-free niacin": dietary supplement may be "benefit free. *Prev. Cardiol.*, **9**, 64–65.

Offermanns H, Kleeman A, Tanner H, Beshke H and Friedrich H (1984) *Krik-Othmer Encyclopaedia of Chemical Technology* (Mark HF, Othmer DF, Overberger CG and Seaborg GT), vol. **24**, John Wiley & Sons, Inc., New York p. 1.

Oliver, F.J., Menissier-de Murcia, J., and De Murcia, G. (1999) Poly(ADP-Ribose) polymerase in the cellular response to DNA damage, apoptosis, and disease. *Am. J. Hum. Genet.*, **64** (5), 1282–1288.

Prasad, S., Misra, A., Jangir, V.P., Awasthi, A., Raj, J., and Bhalla, T.C. (2007) A propionitrile-induced nitrilase of *Rhodococcus* sp. NDB 1165 and its application in nicotinic acid synthesis. *World J. Microb. Biot.*, **23**, 345–353.

Prosser, A.R. and Sheppard, A.J. (1968) Gas-liquid chromatography of niacin and niacinamide. *J. Pharm. Sci.*, **57**, 1004–1006.

Ress-Loeschke, M., Hauer, B.and Mattes, R. (2001) Nitrilase aus *Rhodococcus rhodochrous.* NCIMB 11216. DE 10,010,149, BASF AG, Germany.

Robins, K.T. and Nagasawa, T. (1999) Procede de fabrication d'amides. Patent WO 9,905,306, Lonza AG, Switzerland.

Saito, M., Tsukahara, K., Yamada, K. and Imai, H. (1989) Process for producing cyanopyridines. Patent EP 339,680, Mitsubishi Gas Chemical Co., Inc., Japan.

Scientific Committee on Food (2002) Opinion of the Scientific Committee on Food on the Tolerable Upper Intake Levels of Nicotinic Acid and Nicotinamide (Niacin). European Commission Health & Consumer Protection Directorate-General SCF/CS/NUT/UPPLEV/39: 6.

Shah, T.Z., Ali, A.B., Jafri, S.A., and Qazi, M.H. (2013) Effect of nicotinic acid (vitamin B3 or niacin) on the lipid profile of diabetic and non-diabetic rats. *Pak. J. Med. Sci.*, **29** (5), 1259–1264.

Sharma, N.N. (2009) Hyper induction of nitrilase in *Nocardia globerula* NHB-2 and its application in biotransformation of N-substituted aromatic nitriles. PhD thesis. Himachal Pradesh University, Shimla.

Sharma, N.N., Sharma, M., and Bhalla, T.C. (2010) An improved nitrilase-mediated bioprocess for synthesis of nicotinic acid from 3-cyanopyridine with hyper-induced*Nocardia globerula* NHB-2. *J. Ind. Microbiol. Biot.*, **38** (9), 1235–1243.

Sharma, N.N., Sharma, M., Kumar, H., and Bhalla, T.C. (2006) Nocardia globerula NHB-2: Bench scale production of nicotinic acid. *Process Biochem.*, **41** (9), 2078–2081.

Stocchi, V., Cucchiarini, L., Canestrari, F., Piacentini, M., and Fornaini, G. (1987) A very fast ion-pair reversed-phase HPLC method for the separation of the most significant nucleotides and their degradation products in human red blood cells. *Anal. Biochem.*, **167**, 181–190.

Survase, S.A., Bajaj, I.B., and Singhal, R.S. (2006) Biotechnological production of vitamins. *Food Technol. Biotechnol.*, **44** (3), 381–396.

Suvorov, B.V., Sembaev, D.K., Afanas'eva, T.A., Tol-macheva, T.P., Saurambaeva, L.I., Gostev, V.I., Kan, I.I. and Glubokovskikh, L.K. (1991) A method for producing cyanopyridine. Patent SU 1,014,228, Institute of Chemical Sciences, Academy of Sciences, Kazakh S.S.R., USSR.

To, F. and Fujita, C. (1991) JP 03,280,891, Nitto Chemical Industry Co., Ltd., Japan.

Toomey, J.E. Jr, (1984) Electrochemical oxidation of pyridine bases. Patent US 4,482,439 A, Nov. 13, 1984.

Toomey, J.E., Jr, (1991) Electrochemical synthesis of niacin and other N-heterocyclic compounds. Patent US 5,002,641 A, May 26, 1991.

Unilever (1998) Niacinamide: Safety Assessment (Document Number D97/059), Section 5. Safety Assessment of Topically Applied Niacinamide, CTFA, Washington, DC.

Vague, P., Picq, R., Bernal, M., Lassmann-Vague, V., and Vialettes, B. (1989) Effect of nicotinamide treatment on the residual secretion in type I (insulin dependent) diabetic patients. *Diabetologia*, **32**, 316–321.

van Eys, J (1991) Nicotinic acid. In: Handbook of Vitamins (Machlin L, ed.), *Marcel Dekker*, New York, pp. 311–340.

Vaughan, P.A., Knowles, C.J., and Cheetham, P.S.J. (1989) Conversion of 3-cyanopyridine to nicotinic acid by *Nocardia rhodochrous* LL 100-21. *Enzyme Microb. Technol.*, **11**, 815–823.

Vogt, A., Kassner, U., Hostalek, U., Steinhagen-Thiessen, E., and NAUTILUS Study Group (2006) Evaluation of the safety and tolerability of prolonged-release nicotinic acid in a usual care setting: the NAUTILUS study. *Curr. Med. Res. Opin.*, **22**, 417–425.

Weissermel, K. and Arpe, H.J. (1997) *Industrial Organic Chemistry*, Wiley-VCH Verlag GmbH, Weinheim.

Yamada, H. and Nagasawa, T. (1989) Process for biological production of amides. Patent EP 307,926, Nitto Chemical Industry Co., Ltd., Japan.

4
Pantothenic Acid

Jesus Gonzalez-Lopez, Luis Aliaga, Alejandro Gonzalez-Martinez, and Maria V. Martinez-Toledo

4.1
Introduction and Historical Outline

Pantothenic acid (also known as *pantothenate* or vitamin B_5, see Figure 4.1) is a water-soluble B-complex vitamin (Rucker and Bauerly, 2007) that was discovered in 1931 by chemist Roger J. Williams during his studies of the vitamin B complex (Williams, 1939). Williams observed that an acidic substance was capable of stimulating the growth of strains of the yeast *Saccharomyces cerevisiae.* In 1933, he named the substance pantothenic acid from the Greek word *panthos*, meaning 'from all sides', because of its widespread presence in food. Pantothenic acid was isolated and extracted from a sheep liver by Williams and colleagues in 1939 as an impure substance (about 40% pure). The initial isolation produced 3 g of pantothenic acid from 250 kg of sheep liver (Lanska, 2012). In 1939, a partial synthesis of pantothenic acid was carried out by Williams in Oregon and Conrad A. Elvehjem in Wisconsin, independently (Williams, 1939). Finally, the synthesis of pantothenic acid was performed in 1940 by American biochemist Karl Folkers and colleagues at Merck and Company in Rahway, NJ. The structure of pantothenic acid was determined by stepwise degradation and synthesis (Lanska, 2012).

Pantothenic acid is pantoic acid linked to β-alanine through an amide bond (Leonardi *et al.*, 2005). Pantothenic acid is of biologic importance because of its incorporation into coenzyme A (CoA) (Figure 4.2) and acyl carrier protein (ACP), on which acetylation and acylation, respectively, and other interactions depend. CoA is an indispensable cofactor in all living organisms, where it functions in over 70 enzymatic pathways, including fatty acid oxidation, carbohydrate metabolism, pyruvate degradation, amino acid catabolism, haem synthesis, acetylcholine synthesis and phase II detoxification acetylation. On the other hand, ACP is an essential component of the fatty acid synthase (FAS) complex required for fatty acid elongation.

CoA (also known as *CoASH*) itself is a complex and highly polar molecule, consisting of adenosine 3′,5′-diphosphate linked to 4″phosphopantethenic acid (vitamin B_5) and thence to β″mercaptoethylamine, which is directly involved in

Industrial Biotechnology of Vitamins, Biopigments, and Antioxidants, First Edition.
Edited by Erick J. Vandamme and José L. Revuelta.

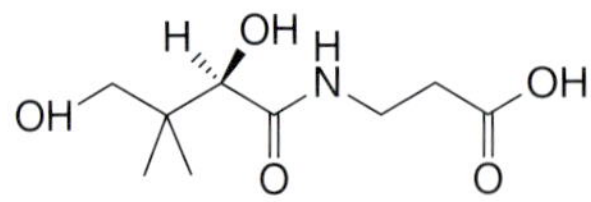

Figure 4.1 Chemical structure of pantothenic acid.

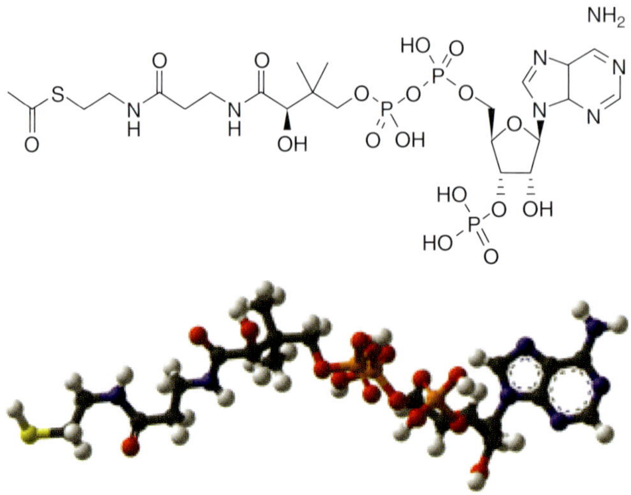

Figure 4.2 Chemical structure of coenzyme A (CoA).

acyl transfer reactions. The adenosine 3′,5′-diphosphate moiety functions as a recognition site, increasing the affinity of CoA binding to enzymes. While acyl-dephospho-CoAs lacking the 3′-phosphate group on the ribose moiety have been detected in tissues, their function is unknown.

Not only is CoA intimately associated with most reactions of fatty acids, but it is also a key molecule in the catabolism of carbohydrates via the citric acid cycle in which acetyl-CoA is a major end product. The genes encoding the enzymes for CoA biosynthesis have been identified, and the structures of many proteins in the pathway have been determined. Although there are substantial sequence differences between prokaryotes and eukaryotes, CoA is assembled in five steps from pantothenic acid in essentially the same way in both groups. However, pantothenic acid *per se* can only be synthesised by microorganisms and plants and must be acquired largely from the diet by animals. In animals, the process is believed to occur entirely in the cytosol of cells and the first and rate-limiting step involves the enzyme pantothenate kinase, several isoforms of which are known.

It is interesting that the 4′-phosphopantetheine moiety, linked via its phosphate group to the hydroxyl group of serine, is the active component in another important molecule in lipid metabolism, ACP. This is a small (8.8 kDa) but ubiquitous and highly conserved carrier of acyl groups during the synthesis of fatty acids. In yeast and mammals, it forms a separate region within a multifunctional FAS complex, but in bacteria and plastids, it remains as a small monomeric protein, though closely associated with the other elements of the FAS. The phosphopantetheine moiety in effect provides a long flexible chain, which permits the intermediates to remain covalently linked to the synthases in an energy-rich linkage with access to spatially distinct enzyme active sites in a manner that resembles an assembly line. However, the final step in fatty acid synthesis in many types of organism is the transfer of the fatty acyl group from ACP to CoA.

Intracellular free fatty acids arising from *de novo* synthesis or from the diet must be activated by a fatty acyl-CoA synthetase before they can be utilised for the synthesis of triacylglycerols, wax esters, long-chain aldehydes and alcohols or complex lipids or for the covalent modification of proteins by myristoylation or palmitoylation. In addition, the fatty acyl-CoA synthetases are essential to many aspects of intermediary metabolism.

Acyl-CoA synthetases activate fatty acids through a process that is energy-dependent and requires ATP (adenosine triphosphate) and CoA. It is a two-stage process, requiring magnesium ions in the first step, which involves the formation of an acyl-AMP intermediate. ATP is consumed and AMP (adenosine monophosphate) and pyrophosphate are produced.

$$\text{ATP} + \text{fatty acid} \rightarrow \text{Acyl}''\text{AMP} + \text{PPi}$$

$$\text{Acyl}''\text{AMP} + \text{CoA} \rightarrow \text{Acyl}''\text{CoA} + \text{AM}$$

At least five families of acyl-CoA synthetases are known in humans (Lopez Martinez, Tsuchiya and Gout, 2014) with specificities for fatty acids in groups with different chain lengths. The enzymes are distinguished by two highly conserved sequence elements, that is, an ATP/AMP-binding motif, which is common to enzymes that form an adenylated intermediate, and a fatty-acid-binding motif. Multiple isoforms of these enzymes are known to be present in animals and other life forms, and 6 have been identified in the yeast genome while there are at least 26 in the human genome, for example. They are generally believed to be membrane-bound, and each isoform appears to be at a unique subcellular location, where it may contribute acyl-CoA to different metabolic pools or where it can participate in the transport of fatty acyl moieties across membranes. For example, there is appreciable sequence homology between the very long chain acyl-CoA synthetases and certain fatty acid transport proteins in animals, and the significance of this is under active investigation. Acetyl-CoA derived via the citric acid cycle or from acetate via a CoA synthetase is of course the primary precursor for FASs. In addition, short-chain acyl-CoAs, including free CoA, acetyl-CoA and malonyl-CoA, are well-known regulators of metabolic flux, with the ratio of acetyl-CoA to free CoA tightly regulating glycolysis and fatty acid oxidation. As well as its role in fatty acid synthesis, malonyl-CoA decreases fatty acid oxidation by inhibiting the transport of acyl-CoA into mitochondria. In addition to their role in lipid biosynthesis and catabolism, CoA esters have been shown to regulate the activities of a variety of enzymes, including that of acetyl-CoA carboxylase, an essential enzyme in fatty acid biosynthesis. Many genes and enzymes are regulated by deacylation and acylation via various short-chain acyl-CoAs, such as acetyl and succinyl-CoA.

Long-chain acyl-CoA esters also bind to certain hormone receptors and have a signalling function. Many of the effects observed for free fatty acids in nuclear signalling may also be attributable to acyl-CoA esters.

Many bacterial species, both Gram-negative and Gram-positive, synthesise long-chain acyl-CoA esters for lipid synthesis, and this enables them to make efficient use of exogenous fatty acids. However, other bacterial species do not make use of CoA in this way but instead utilise newly synthesised acyl groups linked via the thiol bond to the ACP. Some species, including *Escherichia coli*, use both acyl-CoA esters and acyl ACPs for *de novo* synthesis of phosphatidic acid. Many other bacterial species activate fatty acids in a very different way, that is, as the fatty acyl phosphates.

CoA esters are required for a number of processes in addition to esterification. During fasting or starvation, intracellular long-chain fatty acids mobilised from adipose tissue reserves are catabolised as fuel by the mitochondrial β-oxidation pathway, and they must first be converted into CoA esters prior to synthesis of carnitine derivatives for translocation into the mitochondrion. Medium-chain fatty acids can enter mitochondria without carnitine transport, but they still must be activated before β-oxidation can occur.

Similarly, peroxisomes in animal cells have a distinct fatty acid β-oxidation system with a separate set of enzymes, including as many as three acyl-CoA oxidases. Acyl-CoA oxidase 1 catalyses the β-oxidation of straight-chain acyl-CoAs, while acyl-CoA oxidase 2 is involved in the oxidation of the side chain of bile acid precursors, and acyl-CoA oxidase 3 catalyses the oxidation of methyl-branched-chain CoA esters. Activation is also needed for α-oxidation in tissues. In addition, most other biological reactions of fatty acids, including chain elongation and desaturation (plants are an exception), require their activation. As they have both polar and hydrophobic molecular components, CoA esters of long-chain fatty acids have strong detergent-like physical properties and have the potential to be disruptive towards cells.

The intracellular concentration of free acyl-CoA esters is tightly controlled by feedback inhibition of the acyl-CoA synthetase and is buffered by specific acyl-CoA-binding proteins in the cytoplasm, which in effect reduce the concentration of free acyl-CoA by up to 104-fold. Mitochondrial acyl-CoA concentrations are 10-fold higher than in the cytoplasm. At high concentrations, acyl-CoA is a non-specific inhibitor of innumerable enzyme systems, and it must be removed from cells in part of their acyl-carnitine derivatives.

Only the dextrorotatory (D) isomer of D-pantothenic acid possesses biologic activity (Kelly, 2011). The reactive component of both CoA and ACP is not the pantothenic acid molecule but rather the sulfhydryl (SH) group donated from cysteine (Kelly, 2011). Pantethine is the stable disulfate form of pantetheine, the metabolic substrate that constitutes the active part of CoA and ACP (Anonymous, 2010). Thus, the disulfide form of pantothenic acid – pantethine – is considered the most active form of vitamin B_5 because it contains the SH group needed for biological activity in CoA and ACP (Anonymous, 2010; Kelly, 2011). Because D-pantothenic acid is relatively unstable, the more stable calcium pantothenate is the form of vitamin B_5 usually found in dietary supplements and used for study purposes (Kelly, 2011).

4.2 Occurrence in Natural Food Sources and Requirements

Animals and some microbes lack the capacity to synthesise pantothenate and are totally dependent on the uptake of exogenous pantothenic acid. However, most bacteria, such as *E. coli* and *Azotobacter chroococcum*, plants and fungi synthesise pantothenic acid and so, pantothenate is found virtually everywhere in biology (Leonardi *et al.*, 2005; Lopez Martinez, Tsuchiya and Gout, 2014). Pantothenic acid is found both free and conjugate in virtually all plant and animal cells (Institute of Medicine, 1998). However, data on the pantothenic acid content of food is very limited. Chicken, beef, potatoes, oats, tomato products, liver, kidney, peanuts, almonds, yeast, egg yolk, broccoli, cheese, lobster and whole grains are reported to be major sources of pantothenic acid (Institute of Medicine, 1998; Kelly, 2011). Royal bee jelly and the ovaries of tuna and cod have very high levels of pantothenic acid (Institute of Medicine, 1998). Other meats, vegetables, milk and fruits also contain moderate amounts of pantothenic acid. Processing and refining of grains produce a loss of pantothenic acid content (Lopez Martinez, Tsuchiya and Gout, 2014). Ordinary cooking does not cause excessive losses of pantothenic acid (Lopez Martinez, Tsuchiya and Gout, 2014). However, freezing and canning of vegetables, fish, meat and dairy products have been shown to decrease the pantothenic acid content of foods (Lopez Martinez, Tsuchiya and Gout, 2014).

The Food and Nutrition Board of the U.S. Institute of Medicine regularly updates dietary guidelines that define the quantity of each micronutrient that is 'adequate to meet the known nutrient needs of practically all healthy persons'. This Recommended Dietary Allowance (RDA) was revised between 1998 and 2001 (Lopez Martinez, Tsuchiya and Gout, 2014). As was stated in this revision, due to lack of suitable data, an Estimated Average Requirement and, thus, the RDA for pantothenic acid in humans of any age, cannot be established. The available information on pantothenic acid can only be used to support Adequate Intake (AI), the amount needed to prevent a state of deficiency in the vitamin. The usual pantothenic acid intake is 4–7 mg/day, as reported in small groups of adolescents and adults of various ages (Lopez Martinez, Tsuchiya and Gout, 2014). There is no evidence suggesting that this range of intake is inadequate. Thus, the approximate midpoint – 5 mg/day – is set as the AI for adults. The AIs in other age groups have usually been calculated by extrapolating from adult values (Table 4.1).

Except during pregnancy and lactation, there is no basis for determining a separate recommendation based on gender, so the AIs for men and women are the same. Curiously, a study reported that pantothenic acid levels in blood and urine were significantly lower in females using oral contraceptives (nine

Table 4.1 Adequate Intake of pantothenic acid in humans according to life stage groups.

Stage group	Adequate intake (mg/day)
Infants	
0–6 months	1.7
7–12 months	1.8
Children	
1–3 years	2.0
4–8 years	3.0
9–13 years	4.0
Adolescents	
14–18 years	5.0
Adults	
19–70+ years	5.0
Pregnancy	
(Any age)	6.0
Lactation	
(Any age)	7.0

Modified from *Dietary Reference Intakes: Vitamins* (U.S. Food and Nutrition Board, released 12 June 2000)

women) as compared with four females who were not (Lewis and King, 1980). Some studies have shown that certain subsets of the population might consume insufficient pantothenate in their diets (Kelly, 2011; Kolahdooz, Spearing and Sharma, 2013). However, it is possible that intestinal microbiota contribute to the overall vitamin B_5 status in humans (Kelly, 2011). Intestinal bacteria would produce enough pantothenate to ward off signs of a deficiency state in humans. However, the contribution of bacterial synthesis to body pantothenic acid levels or faecal losses in humans has not been quantified.

Whole blood and urine concentrations of pantothenate are indicators of status (Institute of Medicine, 1998). Although it is theoretically possible that erythrocyte concentrations are a more accurate representation of status than whole-blood concentrations because of the contribution of serum pantothenic acid to the latter, no clear advantage of using erythrocyte values has been shown (Institute of Medicine, 1998). Plasma or serum levels are not thought to be accurate for measuring pantothenate status.

As a consequence of the ubiquitous nature of pantothenic acid, a naturally occurring vitamin deficiency in humans either has not occurred or has not been recognised (Leonardi *et al.*, 2005). Presumably even in very poor diets, other vitamin deficiencies are limiting factors before pantothenic acid deficiency causes definite trouble (Kelly, 2011). Actually, our knowledge about pantothenic acid deficiency in humans comes from some studies on the burning-feet syndrome, a disorder considered a natural state of deficiency (Bibile *et al.*, 1957;

Gopalan, 1946), and investigations on inducing pantothenate deficiency in healthy volunteers fed with a diet devoid of pantothenic acid along with the administration of vitamin antagonists (Bean *et al.*, 1955; Hodges *et al.*, 1959).

In the Spanish Civil War and among malnourished prisoners held by the Japanese in the South Pacific during World War II, common complaints were numbness and burning pain in the feet (Lanska, 2012). Patients were reported to improve after adding rice polishings and yeast to their diet, although they were not cured completely. This finding suggests that a deficiency of some vitamin B-complex factor was responsible for the disease (Gopalan, 1946; Lanska, 2012). The underlying nutritional disorder was variously attributed to a deficiency in pantothenic acid, riboflavin, nicotinic acid, thiamine or some combinations of these. Pantothenic acid deficiency is now often considered responsible for these symptoms on the basis of the report by Gopalan (1946), in which the symptoms were remedied with calcium pantothenate supplementation, but not when other B-complex vitamins were given. However, a later controlled trial carried out on 56 patients from a rural area of Sri Lanka did not support these findings (Bibile *et al.*, 1957).

In the mid-to-late 1950s, internists William Bean and Robert Hodges and their colleagues at the University of Iowa induced an experimental pantothenic acid deficiency in men through the administration of a vitamin antagonist in combination with a pantothenic-acid-deficient diet (Hodges *et al.*, 1959). These studies were undertaken on a few healthy volunteers. After taking the drug omega-methyl pantothenic acid (a pantothenate kinase inhibitor), along with a partly synthetic diet deficient in pantothenate, serious clinical symptoms appeared within a few weeks. The triad of fatigue (including apathy and malaise), headaches and weakness was the most consistent finding. Other symptoms included emotional lability, impaired motor coordination, paraesthesia, burning sensations in the hands and feet, muscle cramps and gastrointestinal disturbance such as nausea, vomiting and abdominal cramps. Some subjects had tachycardia, orthostatic hypotension and fluctuations in arterial blood pressure. In some individuals, upper respiratory infections were common, in others, they were not. One subject who had many infections had a decrease in gamma globulins, but in other subjects, they were normal. Other lab abnormalities included a reduction of urinary 17-ketosteroids, a loss of the eosinopenic response to ACTH (adrenocorticotropic hormone), abnormal glucose tolerance and increased sensitivity to insulin (Hodges, Ohlson and Bean, 1958). Secretion of gastric hydrochloric and pepsin was reduced in these subjects (Thornton, Bean and Hodges, 1955). Unfortunately, these studies were performed on a small number of individuals and also with a considerable variation in clinical manifestations among them. Moreover, the clinical symptoms were non-specific, and some artefacts were introduced in the experiment due to the nature of the experimental plan. For example, subjects were isolated in a ward during the experiment and fed by gastric tube, and these conditions may explain some of the emotional alterations that the individuals suffered. In addition, one cannot rule out that some of these symptoms were not adverse side effects of the administered drug. Finally, prompt and complete recovery did not always follow

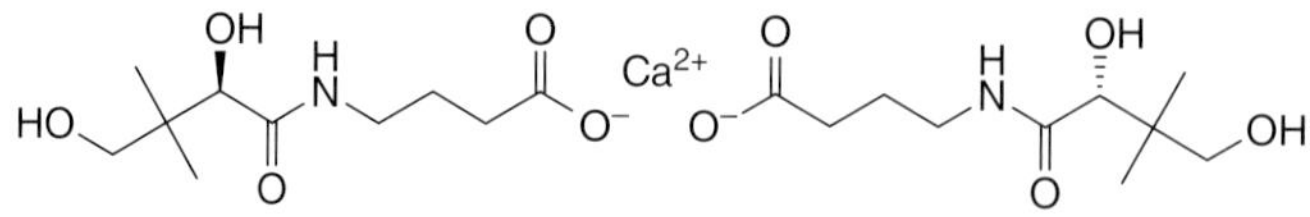

Figure 4.3 Chemical structure of calcium hopantenate.

pantothenic acid administration. Improvement of the paraesthesia and muscle weakness usually followed the administration of the vitamin, but fatigue and some degree of irritability persisted.

Calcium hopantenate has a structural formula (Figure 4.3) similar to that of pantothenic acid and is obtained by substituting the β-alanine moiety of pantothenic acid for γ-aminobutyric acid (GABA). So, it has a GABA-ergic effect on the central nervous system. Since 1978, this drug has been available only in Japan for the treatment of diminished reactivity in organic brain diseases in children and adults. This compound is also a pantothenic acid antagonist, with potency three times higher than that of ω-methyl pantothenic acid (Noda *et al.*, 1988).

Between 1983 and 1985, 11 Japanese children, aged between 9 months and 10 years, suffered from Reye-like syndrome during calcium hopantenate therapy and 7 of them died (Noda *et al.*, 1988). The duration of the administration of hopantenate was varied, ranging from 15 days to 15 months, and the dosage from 0.5 to 3 g/day. Noda *et al.* (1988) have reported three additional senile patients who developed fatal Reye-like syndrome coincident with the treatment of hopantenate for 120–124 days, at a dose of 33–58 mg/kg/day. Serum levels of pantothenic acid were measured in one patient and low levels were found. On the basis of these data, the authors speculated with the possibility that the pathogenesis of the Reye-like syndrome could be due to pantothenic acid deficiency produced by calcium hopantenate.

Finally, Leonardi *et al.* (2005), in their excellent and exhaustive review article on CoA, pointed out that extremely low CoA resulting from pantothenate deficiency, in either animals models or humans, is associated with hypoglycaemia, increased sensitivity to insulin, elevated serum triglycerides and hepatic steatosis (consistent with an inability to degrade fatty acids).

4.3 Physiological Role as Vitamin or as Coenzyme

Pantothenic acid is used in CoA and ACP, which carry and transfer acetyl and acyl groups, respectively (Shimizu *et al.*, 2001). *In vivo* effects of pantothenic acid are generally thought to be a result of its incorporation into these molecules. CoA is an essential cofactor in fatty acid oxidation, lipid elongation and fatty acid synthesis. It is involved in the production of many secondary metabolites such as polyisoprenoid-containing compounds (e.g. dolichol, ubiquinone (CoQ10), squalene and cholesterol), steroid molecules (e.g. steroid hormones, vitamin D and bile acids), acetylated compounds (e.g. acetylated derivatives of amino sugars (e.g. *N*-acetylglucosamine), acetylated neurotransmitters (e.g. *N*-acetylserotonin, acetylcholine) and prostaglandins and prostaglandin-like compounds.

Biosynthesis of phospholipids as well as plasmalogen, sphingenin and ceramide requires CoA. Directly or indirectly, CoA is related to the breakdown of the carbon skeleton of most of the amino acids. The breakdown of the pyrimidine bases, cytosine, uracil and thymine is also dependent on CoA. ACP is involved in fatty acid, polyketide and lysine synthesis and also in nonribosomal peptide synthetases.

To evaluate, the main roles of B_5 as a vitamin are basically linked to many different biochemical processes that utilise CoA as a substrate and cosubstrate, specifically given that the bulk of 4-phosphopantotheine incorporated into ACP also derives from transfer reactions that need CoA as a substrate. The following descriptions (Table 4.2) underscore how B_5 vitamin as a component of CoA and ACP can be considered as essential to virtually all aspects of cell metabolism.

According to Srivastava and Bernhard (1987), different intermediates obtained from the transfer reactions catalysed by CoA and 4′-phosphopantetheine in ACP can be considered as 'high-energy' substances. Consequently, for most reactions involving CoA or ACP, no additional energy is required for transfer of the acetyl or acyl group. CoA is also important for the balance between carbohydrate and fat metabolisms. Carbohydrate metabolism needs some CoA for the citric acid cycle to continue, and fat metabolism needs a larger amount of CoA for breaking down fatty acid chains during β-oxidation (Leonardi *et al.*, 2005). Finally, acetyl-CoA is also involved in Claisen condensations, which is the basis for the biosynthesis of several molecules, such as fatty acids, polyketides, phenols, terpenes and steroids.

CoA is also mainly involved in a broad spectrum of acyl and acetyl transfer reactions and processes implicated in primarily oxidative metabolism and catabolism

Table 4.2 Main functions of CoA (coenzyme A) and ACP (acyl carrier protein) in cell metabolism.

Function	Metabolic significance
Carbohydrate-related citric acid cycle transfer reactions	Oxidative metabolism
Acetylation of sugars (e.g. *N*-acetylglucosamine)	Production of carbohydrates
Phospholipid biosynthesis	Cell membrane formation
Isoprenoid biosynthesis	Cholesterol and bile salt synthesis
Steroid biosynthesis	Steroid hormone synthesis
Fatty acid elongation	Ability to switch membrane fluidity
Acyl fatty acid and triacylglyceride biosynthesis	Energy source
Protein acetylation	Altered protein conformation; activation of hormones, enzymes and transcriptional regulation
Protein acylation (e.g. myristic and palmitic acid and phenyl moiety additions)	Compartmentalisation and phenyl moiety additions activation of hormones and transcription cofactor

reactions. However, ACP is mainly related with synthetic processes. Thus, it has been suggested that adenosyl moiety of CoA represents an excellent site for tight binding to CoA-requiring enzymes, while allowing the 4′-phosphopantetheine portion to serve as a flexible arm to transfer substances from one catalytic centre to another (Tahiliani and Beinlich, 1991; Leonardi *et al.*, 2005). In the same way, when B_5 (as 4′-phosphopantetheine) in ACP is used in transfer reactions, it also functions as a flexible arm that allows for an orderly and systematic presentation of thiol ester derivatives to each of the active centres of the FAS (also denominated as FAS complex).

Lysine residues are also a target for acetylations (Yang, 2004). Lysine acetylations also occur post-translationally, although in this case, the lysine acetylation that occurs on internal lysine residues is balanced by the action of a large number of deacetylases which are nicotinamide adenine dinucleotide (NAD)-dependent. In the same way, non-histone proteins and transcription fractions that are reversibly acetylated have been associated in protein–protein interactions and have been shown to facilitate specific binding of regulatory proteins. Moreover, such chemical modifications can affect and produce some changes in protein structures.

Acetylation of protein facilitated by CoA can be produced as a post-translational modification. In this case, protein acylation can occur by covalent attachment of lipid groups to modify the polarity and strengthen the association of an acylated protein with cell membranes, both intracellularly and extracellularly. Actually, one of the best identified acylation pathways is that involving S-acyl linkages to proteins. In this context, many signalling proteins, such as receptors, G-proteins, protein tyrosine kinases and other cell-membrane 'scaffolding' molecules, are normally acylated.

In general terms, it has been reported that the addition of an acetyl group into an amino acid can drastically alter its chemical properties. The same is true for other biomolecules such as biogenic amines, carbohydrates, complex lipids and hormones, xenobiotics and drugs (Rose and Hodgson, 2004). It can be established that acetylation is critical to cell–cell surface and cell surface protein–protein interactions, for instance, antigenic sites and determinants, and it could be suggested that acetylation is the most common mechanism of protein modification. More specifically, acetylations are produced by a wide range of acyltransferases that transfer acetyl groups from acetyl-CoA to amino groups. As a consequence of this covalent modification into proteins, the enzymatic activities or other biological properties can be altered.

Amino-terminal acetylations occur co-translationally and post-translationally, as indicated earlier. Proteins containing serine and alanine termini are the most usually acetylated, although other amino acids such as methionine and glycine can be also acetylated. This type of acetylation is frequently irreversible and occurs right away after the initiation of translation. The biological significance of this chemical modification varies from one protein to another. Thus, while in some proteins, the acetylation is essential for its biological functions, in others, the acetylation is not required at all.

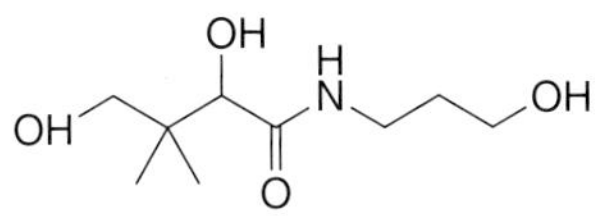

Figure 4.4 Chemical structure of pantothenol.

4.4 Chemical and Physical Properties

The D-isomer of pantothenic acid (Figure 4.1) is the only naturally occurring vitamin that has biological activity. Pantothenol (Figure 4.4) also has biological activity, which is a chemical synthetic compound, because it can be oxidised to pantothenic acid. Pantothenic acid, which has a molecular weight of 219.2, is composed by pantoic acid (butyric acid) linked to an amino acid (β-alanine) through a peptide bound. The free form of pantothenic acid and its sodium salts is too unstable and hygroscopic for commercial application. Therefore, the form for human supplements and therapeutic use is normally calcium D-pantothenate, which has a molecular weight of 474.5. The United States Pharmacopeial Convention (USP) standard is D-(+)-pantothenic acid. Other commercial forms include sodium and calcium salts and the alcohol pantothenol (Chonan *et al.*, 2014).

Pantothenic acid is yellow, viscous, oily and readily soluble in water, alcohol and dioxane but is rarely soluble in diethyl ether and acetone. It is insoluble in benzene and chloroform (De Leenheer, Lambert and van Bocxlaer, 2000). Calcium pantothenic acid is a colour- and odour-free microcrystalline powder and has a bitter taste. Pantothenic acid is highly hygroscopic, while calcium pantothenate is moderately hygroscopic and melts at 195–196 °C. The solubility of calcium salts is 40 g/100 ml in water and is slightly soluble in ethyl acetate and insoluble in diethyl ether. The pKa value of the calcium salt is 4.4 (dissociation of the carboxyl group). A 5% solution of the calcium salt has a pH of 7.2–8.0 (Eitenmiller, Lin and Landen, 2008). Pantothenic acid and other isomer compounds do not have a chromophore. It is an acid and has a marked tendency to absorb water from the air. Under alkaline hydrolysis, it breaks down into β-alanine and pantoic acid. The latter readily forms a lactone, D-(−)-pantolactone, in acid solution or on heating. Acid hydrolysis of pantothenic acid gives β-alanine and pantolactone.

The structure of pantothenic acid contains a single asymmetric centre, so that it is optically active; only the natural D-(−)-isomer has vitamin activity (Salunke and Vijayan, 1984). The calcium salt of pantothenic acid, which can be obtained as needle crystals from methanol, is moderately hygroscopic and is rather more stable to heat, air and light than the free acid is. It is soluble in water and glycerol and slightly soluble in alcohol and acetone. A review by Wagner and Folkers (1964) summarised early studies on the chemistry of pantothenic acid.

The naturally occurring derivatives of pantothenic acid (Table 4.3) can be grouped into three types on the basis of their chemical structures: simple pantothenate derivatives, pantetheine derivatives in which cysteamine (or its analogues) attaches by an amide linkage, and CoA derivatives in which the

Table 4.3 Pantothenic acid and its naturally occurring derivatives (Shimizu and Kataoka, 1999).

D-Pantothenic acid C9H17NO5 MW: 219.23	Unstable, viscous oil. Extremely hydroscopic, easily decomposed by acids, bases and heat. Soluble in water, ethyl acetate, dioxane, glacial acetic acid; moderately soluble in ether, amyl alcohol; insoluble in benzene, chloroform. Solutions are stable between pH 5 and 7
Calcium D-pantothenate C16H32CaN2O10 MW: 476.53	White needles. Moderately hygroscopic. Soluble in water, glycerol; slightly soluble in alcohol, acetone; insoluble in ether, benzene, chloroform. Decomposed by bases. Solutions are stable between pH 5 and 7
Sodium D-pantothenate C9H16NaNO5 MW: 241.21	White, hygroscopic crystals. Decomposed by acids and bases. Solutions are stable between pH 5 and 7. For solubility, see D-calcium pantothenate
4′-Phosphopantothenic acid (Ba salt) C9H16NO8P MW: 313.27	Soluble in water; insoluble in ethanol. Unstable to bases. Free acid is unstable
Pantothenoyl-L-cysteine (Ba salt) C12H22N2O6S MW: 322.38	Soluble in water, methanol; moderately soluble in ethanol; insoluble in ether. Unstable to acids and bases
4′-Phosphopantothenoyl-L-cysteine (Ba salt) C12H23N2O4PS MW: 416.42	Soluble in water; slightly soluble in alcohol. Unstable to acids and bases. Easily oxidised in air
Pantetheine C11H22N2O4S MW: 278.37	Syrup or glass. Soluble in water; slightly soluble in alcohol; insoluble in ether, benzene, chloroform, ethyl acetate. Unstable to acids and bases. Easily oxidised in air
Pantetheine C22H42N4O5S2 MW: 554.72	Disulfide form of pantetheine. Glassy, colourless to light yellow substance. Unstable to acids
4′-Phosphopantetheine (Ba salt) C11H23N2O7PS MW: 358.35	Soluble in water; slightly soluble in ethanol; insoluble in ether. Unstable to acids and bases. Easily oxidised in air
Dephospho-coenzyme A (Li salt) C21H35N7O12P2S MW: 687.56	Soluble in water, methanol; insoluble in acetone. Unstable to acids and bases
Coenzyme A C21H36N7O16P3S MW: 767.55	Soluble in water; insoluble in ethanol, ether, acetone. Decomposed to pantetheine-2′,4′-cyclic phosphate and 3′,5′-ADP in 1 N NaOH (100°, 2 min). Decomposed to pantetheine-4′-phosphate and adenine in 1 N HCl (100°, 5 min). Easily oxidised in air

pantetheine is adenosylylated. Pantothenyl alcohol, an alcohol analogue of pantothenic acid, is also a pharmaceutically important unnatural derivative.

The stability of pantothenic acid and calcium pantothenate is highly pH-dependent and moderately stable under light and atmospheric oxygen, if protected from moisture. Pantothenic acid is most stable at pH 4.0–5.0, while the calcium pantothenate is stable at pH 5.0–7.0. Therefore, because of greater stability under near-neutral conditions, calcium pantothenate, as opposed to the free pantothenic acid, is more often utilised in fortified foods and for pharmaceutical use. Although calcium pantothenate is more stable, it is easily degraded during autoclaving or under similar conditions. For example, when vegetables are cooked in water, the pantothenate salt is lost to a large extent (Ball, 2006; Eitenmiller, Lin and Landen, 2008). In aqueous solutions with pH values under 5.0 or above 7.0, calcium pantothenate becomes thermally labile and will undergo hydrolytic cleavage to produce pantoic acid, its salts and β-alanine. When compared with other B vitamins, pantothenic acid is more stable at higher pH values (Ball, 2006).

In plant or animal foods, pantothenic acid is present as both free and bound forms, but most of them are present in the bound form in food, because of CoA and ACP. Pantothenic acid is the most stable form in food storage; however, it is potentially leached 15–50% from cooked meats and 37–78% from blanched vegetables (Combs, 2008). In the processing industry, pantothenic acid may be destroyed through freezing, canning and refining processes (Whitney and Rolfes, 2011).

The bioavailability of pantothenic acid in foods and feedstuffs has not been fully investigated. It was reported that the bioavailability of pantothenic acid ranged from 40% to 61%, with a mean of 50%, based on urinary excretion from male subjects and tested by microbiological assay. In previous research for the study of bioavailability of pantothenic acid in five different types of foods, that is, wheat, coarse wholemeal bread, steamed potatoes, boiled pork and boiled beef, which were fortified with minerals, fat-soluble vitamins, enriched soybean oil and amino acids, the results showed that 65–81% of pantothenic acid was digestible in studied animals (e.g. pig), and indicated that the feed did not yield significantly different levels in the pigs. The bioavailability of pantothenic acid decreased in the following order: wheat diet > pork diet > potato diet > beef diet, and the coarse wholemeal bread diet only reached 28% (Combs, 2008).

4.5 Assay Methods

The traditional analytical methods and the current official methods for the determination of water-soluble vitamins are based on spectroscopic, chemical, enzymatic and microbiological assays. However, some of these methods are usually tedious and time-consuming, because various indispensable steps for sample preparation are required to remove the interfering chemicals. Regardless of the shortcomings of the aforementioned methods, both non-bioavailability and bioavailability measurements often give overestimated amounts of vitamins and lead to inaccurate assessment results (Bird and Thompson, 1967; Rychlik and Roth-Maier, 2005).

In the past decade, investigators in food laboratories have shown great interest in development of simultaneous determination methods for water-soluble vitamins. As a result, many techniques, including capillary electrophoresis (CE), UV-Vis spectrophotometry, fluorimetry, chemiluminescence, atomic absorption, micellar electrokinetic chromatography, micellar liquid chromatography, thin-layer chromatography (TLC), high-performance liquid chromatography (HPLC), gas chromatography (GC) and high-performance liquid chromatography/mass spectrometry (LC/MS), have been investigated and reported (Rychlik and Roth-Maier, 2005; Wang *et al.*, 2004). Among them, one of the most common methods for vitamin determination is HPLC (Havlíková *et al.*, 2006; Wang *et al.*, 2004), because improved quality and properties of stationary phases and chromatographic equipment have enabled significant improvement in chemical separation (resolution) and signal detection (sensitivity). Although many HPLC methods are still not officially approved as standard methods by the Association of Official Analytical Chemists (AOAC) because of different column conditions, the HPLC technique can obviously provide many benefits for determining vitamins in various products. For example, it is fast, sensitive, accurate, precise and can minimise required solvent and sample quantities. As a result, it is better than traditional methods (Tsuda, Matsumoto and Ishimi, 2011). In this sense, simultaneous determinations of water-soluble vitamins, including B_5, have been developed with an HPLC system installed with a ZORBAX Eclipse XDB-C18 (250 mm × 4.6 mm, 5 μm particle size, Aglient Technologies, Inc., Loveland, CO, USA) with a guard column (12.5 mm × 4.6 mm, 5 μm particle size). According to the results, LC-MS is the best methodology for simultaneous determination of soluble vitamins in light of its analytical accuracy, precision, sensitivity and versatility.

Many methods actually used in the determination of pantothenic acids in all kind of samples including foods are microbiological bioassay. In this sense, many different bioassay methods can be found in the scientific literature both at a laboratory scale and industrial scale. From this point of view, test microorganisms normally used for the microbiological assay of pantothenic acid are auxotrophic bacteria or yeast such as *Lactobacillus plantarum* ATCC 8014, *Lactobacillus casei* (ATCC 7469) and *Saccharomyces uvarum* (ATCC 9080; *Saccharomyces carlsbergensis*).

L. plantarum is suitable for determining unconjugated pantothenate in many different samples. It should be noted that pantetheine, when simultaneously present in a molar ratio to pantothenate of more than 0.5, yields positive errors in the determination. *S. uvarum* also shows almost specific growth response to free pantothenate, but β-alanine stimulates its growth. Hence, an assay procedure employing this organism is also the one chosen for determining the pantothenic acid that occurs in natural products together with other pantothenate forms. *L. casei* responds not only to pantothenate but also to several conjugated forms of pantothenate. *Lactobacillus helveticus* (ATCC 12046) and *Lactobacillus bulgaricus* B_1 have been recommended for the determination of pantetheine (or pantethine) because both these organisms require more than 100 times as much pantothenic acid as pantethine to produce the same response.

The enzymatic assay method using pantothenase has been reported (Airas, 1986), but the enzyme is not commercially available, and therefore, the methodology is not fully standardised for all sample types. In the same way, chemical and physical methods have also been assayed. These are often used in determining pantothenic acid in pharmaceutical products but are not suitable for the determination of natural samples because of their low sensitivity. Consequently, most of the standardised methods for the quantification of pantothenic acid consider the utilisation of bioassays as the most standardised method for the detection and quantification of this soluble vitamin in natural sources. However, more experimental research in this analytical field must be developed in the future.

4.6 Chemical and Biotechnological Synthesis

At present, commercial production of pantothenate depends basically on chemical synthesis. The conventional chemical process involves reactions yielding racemic pantolactone from isobutyraldehyde, formaldehyde and cyanide; optical resolution of the racemic pantolactone to D-(−)-pantolactone with quinine, quinidine, cinchonidine, brucine and so on; and condensation of D-(−)-pantolactone with β-alanine. This is followed by isolation of the calcium salt and drying to obtain the final product. A problem associated with this chemical process, apart from the use of poisonous cyanide, is the troublesome resolution of the racemic pantolactone and the racemisation of the remaining L-(−)-isomer. Therefore, most of the recent studies in this area have concentrated on the development of an efficient method to produce D-(−)-pantolactone.

Enzymatic resolution (Figure 4.5) of racemic pantolactone can be carried out by specific fungal lactonohydrolases. Shimizu *et al.* (2001) reported that many fungal strains belonging to the genera *Fusarium*, *Gibberella* and *Cylindrocarpon* stereospecifically hydrolyse D-(−)-pantolactone to D-(−)pantoic acid (Kataota *et al.*, 1999). If racemic pantolactone is used as a substrate for the hydrolysis reaction by the microbial lactonohydrolase, only the D-(−)-pantolactone might be converted to D-(−)-pantoate and the L-(−)-enantiomer might remain intact. Consequently, the racemic mixture could be resolved into D-(−)-pantoate and L-(−)-pantolactone. After the removal of L-(−)-pantolactone from the reaction mixture by solvent extraction and so on, the remaining D-(−)-pantoate could be easily converted to D-(−)-pantolactone by heating in an acidic environment. The reverse reaction, that is, lactonisation of D-(−)-pantoate, might also be possible for the resolution. In this case, D-(−)-pantoate in a racemic mixture of pantoate is specifically lactonised into D-(−)-pantolactone. When *Fusarium oxysporum* mycelia are incubated in 700 g/l aqueous solution of racemic pantolactone for 24 h

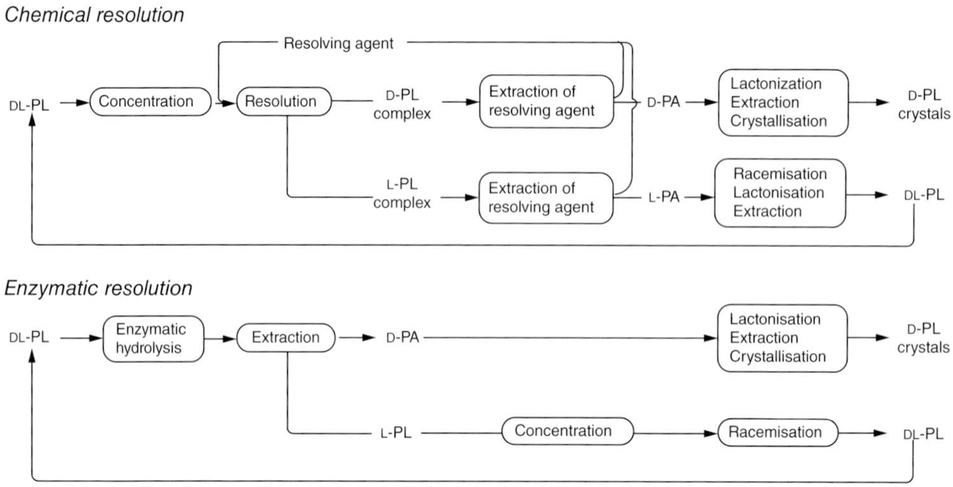

Figure 4.5 Comparison of enzymatic and chemical resolution processes for DL-PL (Shimizu *et al.*, 2001).

at 30 °C with automatic pH control (pH 6.8–7.2), about 90% of the D-(−)-isomer can be hydrolysed. The resultant D-(−)-pantoic acid in the reaction mixture shows a high optical purity (96%), and the coexisting L-(−)-isomer remained without any modification.

Practical hydrolysis of the D-(−)-isomer in a racemic mixture can be performed using immobilised mycelia of *F. oxysporum* as the catalyst. A stable catalyst with high hydrolytic activity can be prepared by entrapping the fungal mycelia in calcium alginate gels. When the immobilised mycelia are incubated in a reaction mixture containing 350 g/l racemic pantolactone for 21 h at 30 °C under automatic pH control (pH 6.8–7.2), 90–95% of the D-(−)-isomer is hydrolysed (optical purity, 90–97%). After the reaction is repeated 180 times (i.e. for 180 days), the immobilised mycelia retain more than 90% of their initial biological activity. The enzymatic process allows several tedious steps that are necessary in chemical resolution to be skipped and is highly advantageous for practical and industrial purposes. The production methods for CoA roughly fall into chemical and microbial categories. However, the chemical methods, which have been reviewed by Shimizu *et al.* (2001), can be considered too complex to be practical. Therefore, commercial production is carried out by different microbiological methods. Extraction of CoA from yeast cells has been performed since the early 1950s. Cells of baker's or brewer's yeasts, which are relatively rich in CoA, have usually been used as the CoA source. Later, an efficient enzymatic method using *Brevibacterium ammoniagenes* cells as the catalyst was developed, showing excellent results at real scale.

A successful enzymatic method using the biosynthetic route of CoA from pantothenic acid, L-cysteine and ATP has been reported in *B. ammoniagenes* (Jackowski, 1996). This microorganism has all five enzymes necessary for the biosynthesis of CoA in high activities. These three substrates, when added to a reaction mixture containing the bacterial cells, are converted to CoA with a satisfactory yield (2–3 g/l). Additionally, *B. ammoniagenes* can accumulate CoA directly in the culture medium on addition of: pantothenic acid, L-cysteine and AMP; adenosine or adenine in the presence of a surfactant, cetylpyridinium chloride and high levels of glucose (usually 10%); K_2HPO4; and $MgSO_4$. Thus, under optimal conditions, the amount produced can be approximately 5.5 g/l. Most CoA in the medium is produced in the disulfide form because of the vigorous shaking during the reaction. After treatment of the culture filtrate with Duolite S-30, charcoal and Dowex 1, a significant reduction of the disulfide is detected and a very pure thiol form can be obtained. However, the biosynthesis of CoA in *B. ammoniagenes* is controlled mainly by the feedback inhibition of pantothenate kinase by CoA, and consequently, it can be concluded that this is a major problem for industrial production, because the overproduced CoA itself stops the biosynthesis.

To improve the biosynthesis of CoA, the mechanism for regulation of biosynthesis has been investigated (Jackowski, 1996). Thus, it has been concluded that the biosynthesis is controlled mainly by the feedback inhibition of pantothenate kinase by CoA. Obviously, this is the main problem in practical production, because the overproduced CoA itself stops the biosynthesis. However, two methods to abolish this feedback inhibition have been developed. A synthetic scheme has been investigated in which the reaction is initiated by the condensation of 4′-phosphopantothenic acid and L-cysteine or the transadenosylylation of 4′-phosphopantetheine, because these routes do not involve phosphorylation of pantothenic acid or pantetheine by pantothenate kinase. Replacement of the enzymatic phosphorylation of pantothenate or pantetheine with chemical phosphorylation followed by the enzymatic reaction increased the yield of CoA 10- to 20-fold. Yields from 4′-phosphopantothenic acid and 4′-phosphopantetheine are 33 g/l and 115 g/l, respectively (Shimizu and Kataoka, 1999). This method is applicable to CoA production under ATP-generating conditions. 4′-phosphopantothenic acid (25 g/l), L-cysteine (15 g/l) and AMP (33 g/l), when added to the culture broth of *B. ammoniagenes*, are converted to CoA with a yield of 23 g/l.

Another way to improve the yield is to use microbial mutants derepressed for the feedback inhibition or those showing elevated pantothenate kinase activity. For example, a mutant of *B. ammoniagenes* that is resistant to oxypantetheine (the corresponding oxygen analogue of pantetheine) has been found to have a high activity of pantothenate kinase. Under ATP-generating conditions, the yields of CoA from pantothenic acid (3.6 g/l), L-cysteine (1.8 g/l) and AMP (6 g/l) or from pantetheine (5 g/l) and AMP (6 g/l) are 9.3 or 11.5 g/l, respectively. These values are about threefold higher than those obtained with the wild-type strains, and 70–100% of the added AMP was converted to CoA. 4′-phosphopantetheine together with other intermediates in CoA biosynthesis can be effectively synthesised by using microorganisms such as *B. ammoniagenes* cells as the catalyst and by modifying the reaction conditions. The amounts of these intermediates obtained by this method are summarised in Table 4.4.

Table 4.4 Production of the intermediates in CoA (coenzyme A) biosynthesis by *Brevibacterium ammoniagenes* (Shimizu and Kataoka, 1999).

Product	Substrate		Dried cells	Culture broth	Immobilised cells
Productivity enzyme source (mg/ml)					
4′-Phosphopantothenic acid	Pantothenic acid	ATP	3–4	—	1.5–2.5
4′-Phosphopantothenic acid	Pantothenic acid	AMP	—	4–5	—
4′-Phosphopantotheine	4′-Phosphopantothenic acid and L-cysteine	CTP	3–4	—	1.8
4′-Phosphopantotheine	4′-Pantothenic acid and L-cysteine	ITP and CTP	2–3	—	0.3
4′-Phosphopantotheine	Pantothenic acid and L-cysteine	GMP and CMP	—	3–4	—
4′-Phosphopantotheine	Pantetheine	ITP	2–3	—	0.9
4′-Phosphopantotheine	Pantetheine	GMP	—	4–5	—
3′-Dephospho-coenzyme A	Pantothenic acid and L-cysteine	ATP	1–2	—	—

Pantothenic acid is produced all over the world in amounts of several thousand tons per year. It is used, *inter alia*, in human medicine, in the pharmaceutical industry and in the foodstuff industry. A high proportion of the pantothenic acid produced is used for feeding economically useful animals such as poultry and pigs. The demand for this material is increasing every year. In view of the increasing demand for D-pantothenic acid, there remains a need for new methods for producing this material.

Pantothenic acid can be prepared by chemical synthesis or biotechnically by the fermentation of specific microorganisms in selected nutrient media. In the case of chemical synthesis, DL-pantolactone is an important precursor. This compound is prepared in a multi-step process from formaldehyde, isobutyl aldehyde and cyanide, the racemic mixture is resolved in a subsequent process step, D-pantolactone is condensed with β-alanine and D-pantothenic acid is obtained in this way. The typical commercial form is the calcium salt of D-pantothenic acid. The calcium salt of the racemic mixture D,L-pantothenic acid is also commonly available.

The advantage of fermentative preparation by microorganisms (Shimizu *et al.*, 2001) is the direct formation of the desired stereoisomeric form that is the D-pantothenic acid form, which contains no L-pantothenic acid. In this context, various species of bacteria, such as *E. coli, Arthrobacter ureafaciens, Corynebacterium erythrogenes, B. ammoniagenes, Azospirillum* sp. *Azotobacter vinelandii* and also yeasts such as *Debaryomyces castellii*, can produce D-pantothenic acid in a nutrient growth medium which contains glucose, D,L-pantoic acid and β-alanine. Furthermore, in the case of *E. coli*, the formation of D-pantothenic acid is increased by the amplification of pantothenic acid biosynthesis genes from *E. coli* which are contained in the plasmids pFV3 and pFV5, in a nutrient medium which contains glucose, DL-pantoic acid and β-alanine. However, the production of vitamin B_5 is directly affected not only by the selected strain but also by the culture conditions such as temperature, growth medium, oxygen concentration and incubation time.

Different examples of microorganisms used for the production of vitamin B_5 can be found in the literature. Thus, U.S. Patent no. 5518906 describes mutants derived from *E. coli* strain IF03547, such as FV5714, FV525, FV814, FV521, FV221, FV6051 and FV5069, which carry resistance to various antimetabolites such as salicylic acid, α-ketobutyric acid, β-hydroxyaspartic acid, O-methylthreonine and α-ketoisovaleric acid. They produce pantoic acid in a nutrient medium which contains glucose, and D-pantothenic acid in a glucose- and β-alanine-containing nutrient medium. Furthermore, in EP-A 0 590 857 and U.S. Patent no. 5518906, it is stated that the production of D-pantoic acid is improved in a glucose-containing nutrient media and the production of D-pantothenic acid is improved in a nutrient medium which contains glucose and β-alanine after amplification, in the aforementioned strains, of the pantothenic acid biosynthesis genes panB, panC and panD, which should be present in the plasmid pFV3 1. Furthermore, WO 97/10340 reports on the beneficial effect of enhancing the ilvGM operon on the production of D-pantothenic acid. Finally, EP-A-1001027 reports on the effect of enhancing the panE gene on the formation of D-pantothenic acid.

According to known procedures, D-pantothenic acid or the corresponding salt can be isolated from the fermentation broth and purified and then used in purified form or the entire D-pantothenic-acid-containing broth and used in particular as a foodstuff additive (Hüser *et al.*, 2005). However, a method for the fermentative preparation of D-pantothenic acid and/or its salts or feedstuff additives containing these by the fermentation of microorganisms from microorganisms of the Enterobacteriaceae family, in particular those which already produce D-pantothenic acid, wherein at least one of the nucleotide sequence(s) coding for the genes gcvT, gcvH and gcvP is enhanced in the microorganisms, in particular is overexpressed. In general terms, it could be proposed that fermentation technologies provide an alternative to chemical processes in the production of pantothenic acid. Different methods such as media optimisation, mutation and screening, genetic engineering and biocatalyst conservation must be used for improvement of the production of vitamin B_5.

Most plants and microorganisms accomplish biosynthesis of pantothenic acid by enzymatically combining pantoic acid with β-alanine. Mammals and some microbes lack the enzyme for this synthetic step, so are unable to synthesise pantothenic acid and need to obtain it from the external environment. Three routes to β-alanine have been described. Several microorganisms have been reported to form β-alanine by α-decarboxylation of L-aspartic acid. Confirmatory evidence for this conversion was provided by Willamson and Brown (1979), who purified (to apparent homogeneity), from extracts of *E. coli*, an enzyme that catalyses the α-decarboxylation of L-aspartic acid to yield β-alanine and CO_2. Willamson and Brown (1979) also reported that the enzyme is missing in a mutant of *E. coli* that requires either β-alanine or pantothenate as a nutritional factor, but is present in the wild-type strain and in a revertant strain of

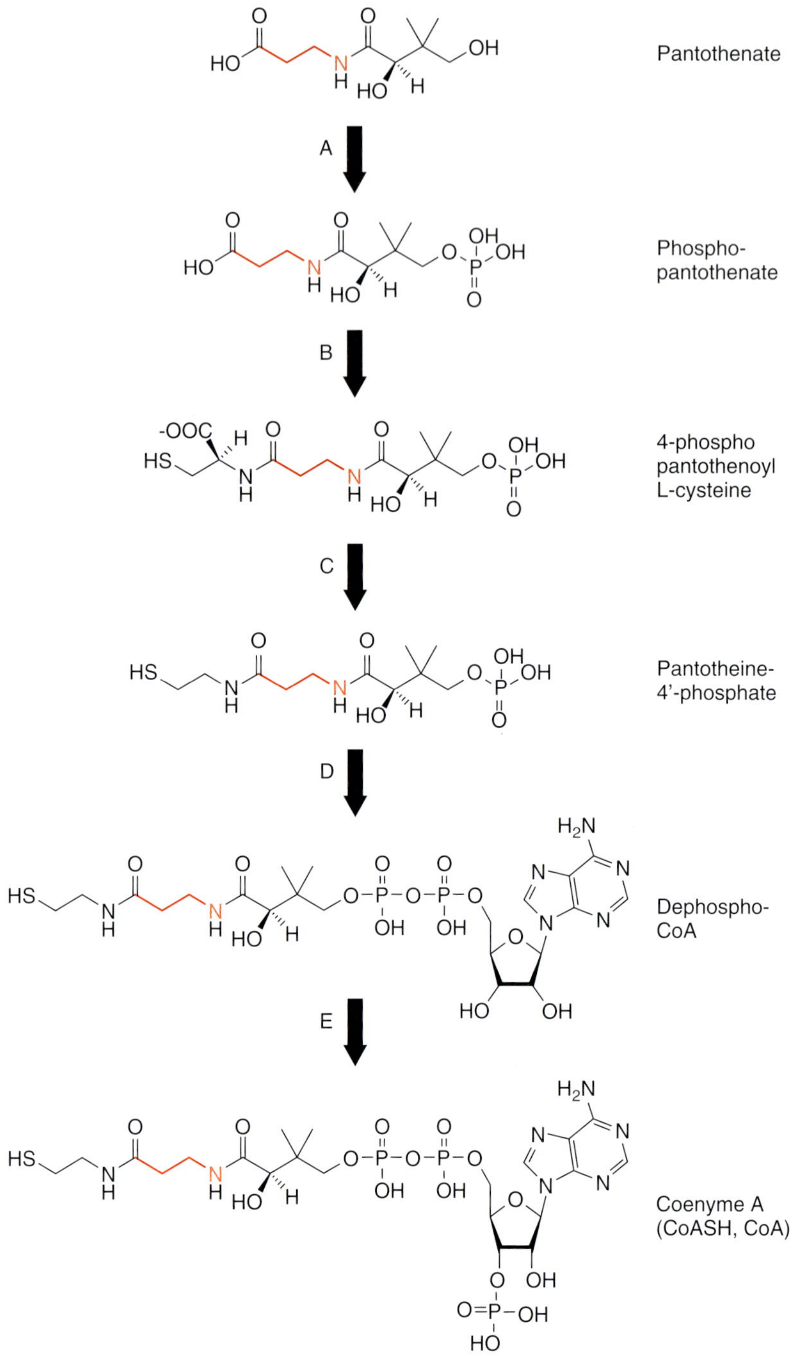

Figure 4.6 Pathway of biosynthesis of CoA from pantothenate (Combs, 2008).

the mutant. It has also been suggested, on the basis of the observation that mutants of *Salmonella typhimurium* lacking the ability to degrade uracil require *N*-carbamoyl-β-alanine, β-alanine or pantothenate, as a nutritional factor, that β-alanine is produced by decarboxylation of *N*-carbamoyl-β-alanine formed from uracil (West *et al.*, 1985). β-alanine may also be produced by transamination of malonyl semialdehyde produced from propionic acid, because enzyme activity catalysing this conversion has been detected in several microorganisms. However, there have been no further studies concerning this reaction.

The route to pantoic acid from pyruvate has been elucidated mainly in *E. coli* and *Neurospora crassa* (Shimizu *et al.*, 2001). Two enzymes catalysing the conversion of pyruvate to α-ketoisovalerate in this route are shared by the route for the biosynthesis of the branched chain amino acids. In *E. coli*, two enzyme activities have been detected for the conversion of α-ketoisovalerate to ketopantoic acid: one is dependent on tetrahydrofolate and the other is not. The physiological significance of tetrahydrofolate-independent activity seems to be questionable because of its high Km values for formaldehyde and α-ketoisovalerate. As a mutant lacking tetrahydrofolate-dependent activity requires pantothenate for growth, although the same amount of tetrahydrofolate-independent activity is found in the same mutant, concrete evidence is provided to support the theory that the tetrahydrofolate-dependent enzyme is responsible for the ketopantoate needed for the biosynthesis of pantothenate. The tetrahydrofolate-dependent enzyme (i.e. ketopantoate hydroxymethyltransferase) has been purified and characterised in some detail. The observation that pantoate, pantothenate and CoA are all allosteric inhibitors of this enzyme also supports this conclusion.

The reduction of ketopantoic acid to D-pantoic acid is catalysed by an NADPH-dependent enzyme, ketopantoic acid reductase. This enzyme has been studied in *S. cerevisiae* and *E. coli*. The same reduction is also catalysed by α-acetohydroxy acid isomeroreductase, which is the enzyme responsible for the transformation of α-acetolactate to α-ketoisovalerate (Primerano and Burns, 1983). Later, Shimizu *et al.* (2001) isolated ketopantoic acid reductase in a crystalline form from *Pseudomonas maltophilia* and characterised it in some detail. They also demonstrated that this reductase is the enzyme for D-pantoic acid formation, necessary for the biosynthesis of pantothenic acid, because mutants lacking this enzyme require either D-pantoic acid or pantothenate for growth and the revertants regain this activity.

The biosynthesis of CoA from pantothenic acid is an essential and universal pathway in prokaryotes and also in eukaryotes, which requires cysteine and ATP (Figures 4.6 and 4.7). CoA is generated from pantothenate through a series of five synthetic reactions (Jackowski, 1996). In the synthetic pathway of CoA, pantothenate is first phosphorylated to 4′-phosphopantothenate by the enzyme pantothenate kinase (CoaA). This step is considered the most important control step in the biosynthesis of pantothenate-dependent enzymes, and it is subjected to feedback regulation by CoA itself or its thioester derivatives. The next step is a condensation reaction with cysteine at the expense of ATP (or CTP in bacteria) yielding 4′-phosphopantothenoylcysteine, which

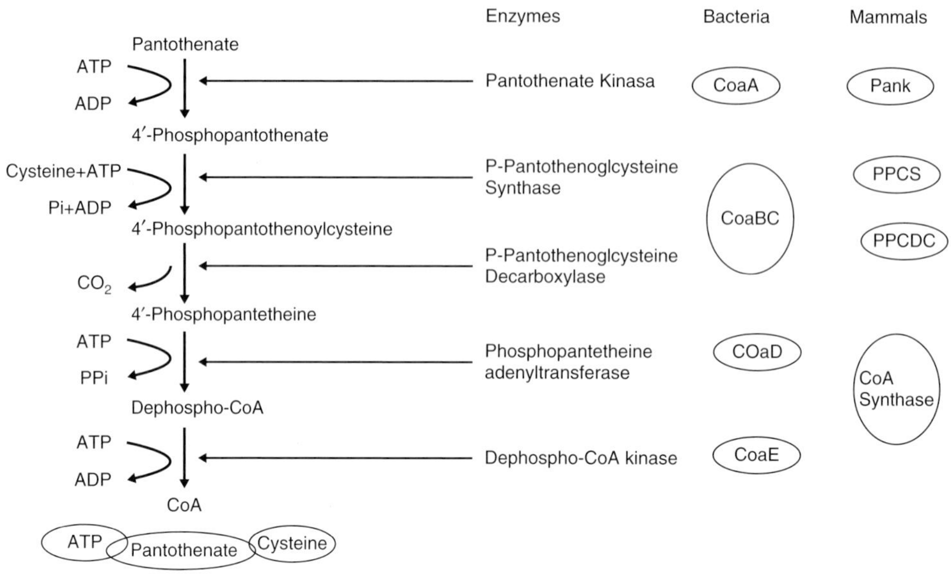

Figure 4.7 The CoA biosynthetic pathway and its key players in bacteria and mammals. ATP, adenosine triphosphate; ADP, adenosine diphosphate; CO_2, carbon dioxide; PPi, pyrophosphate (Lopez Martinez, Tsuchiya and Gout, 2014).

is decarboxylated to form 4′-phosphopantetheine. These two reactions are catalysed by the 4′-phosphopantothenoylcysteine synthase (CoaB) and 4′-phosphopantothenoylcysteine decarboxylase (CoaC) domains of a bifunctional enzyme in prokaryotes (CoaBC) and by two distinct proteins in eukaryotes (PPCS and PPCDC). 4′-phosphopantetheine is subsequently converted to dephospho-CoA by phosphopantetheine adenyltransferase (CoaD), a second rate-limiting reaction in the pathway. Afterwards, dephospho-CoA is phosphorylated by dephospho-CoA kinase (CoaE) at the 3′-OH of the ribose to form CoA. The CoaD and CoaE activities are associated with two separate enzymes in prokaryotes and plants, but fused in a bifunctional enzyme, also termed the CoA synthase (COASY), in mammals.

In bacteria, the nomenclature for the biosynthetic enzymes is CoaA, CoaBC (bifunctional enzyme), CoaD and CoaE for each of the steps presented in Figure 4.7. In mammals, the corresponding biosynthetic enzymes are PanK, PPCS, PPCDC and COASY, encompassing phosphopantetheine adenyltransferase and dephospho-CoA kinase as a unique enzyme.

CoA accounts for a large proportion of cellular pantothenic acid, although ACP also contains the pantothenic acid molecule. The synthesis of ACP is not yet completely elucidated. CoA catabolism occurs as the reverse of the biosynthetic pathway except that 4′-phosphopantetheine is converted to pantetheine followed by conversion to pantothenic acid by the pantetheinase enzyme (Lopez Martinez, Tsuchiya and Gout, 2014). In the metabolic pathway, CoA is dephosphorylated at the 3′ position of ribose to form dephospho-CoA. Dephospho-CoA is then degraded to 4′-phosphopantetheine and 5′-AMP. Dephosphorylation of 4′-phosphopantetheine forms pantetheine. In the final step in the metabolic pathway, pantetheine is hydrolysed into pantothenic acid and cysteamine by the enzyme pantetheinase (also called *vanin*). Pantothenic acid generated during the CoA degradation is recycled for another biosynthesis of CoA or can be excreted intact in urine.

The allosteric inhibition of ketopantoic acid hydroxymethyltransferase of *E. coli* by D-pantoic acid, pantothenic acid or CoA may be involved as a control mechanism in pantothenate biosynthesis. On the other hand, such inhibition is not observed in the case of ketopantoic acid reductase of *P. maltophilia* (Shimizu *et al.*, 2001).

In the pathway to CoA from pantothenate, the involvement of the feedback inhibition of pantothenate kinase by CoA and 4′-phosphopantetheine as a control mechanism in the biosynthesis has been demonstrated (Vallari and Jackowski, 1988). Because this inhibition has generally been observed regardless of species and the other four steps following this reaction are not significantly inhibited by CoA or 4′-phosphopantetheine, this may be one of the most important mechanisms in the control of cellular levels of CoA. No other mechanism, such as repression, has been observed in either pantothenate or CoA biosynthesis.

Pantetheinase, which specifically degrades pantetheine to pantothenic acid and cysteamine, may also be an important enzyme because CoA can be degraded to pantetheine enzymatically and pantetheine can be reused as a precursor of CoA after phosphorylation by pantothenate kinase. Cellular CoA levels may be affected by competition between pantetheinase and pantothenate kinase towards their substrate, pantetheine.

The spectrum of products that can be synthesised by biotechnological processes seems endless, provided that the corresponding metabolic genes are available. However, in order to produce commodity biochemicals with competitive costs, and thus expand the industrial scope and societal impact of biotechnology, the economics of biotechnological processes need to be further improved to compete with the apparent cost achieved by conventional chemical synthesis. As a result, there is still a need to optimise the efficiency of microbial technology, even if these improvements offer only marginal cost-effectiveness, as the impacts of small incremental improvements are greatly amplified by large production scales. In this context, pantothenate can be synthesised by a wide range of bacteria, fungi and microalgae. Thus, a 'first generation' of microorganisms developed directly from wild-type strains has been applied to the industrial production of this vitamin. However, most amino-acid-producing bacterial strains currently used have been constructed by random mutagenesis. A significant disadvantage of this approach is the possibility that the random distribution of mutations in regions not directly related to amino acid biosynthesis can cause unwanted changes in physiology and growth retardation. Moreover, although various large-scale analytical techniques such as transcriptome and proteome analysis are now available, it is difficult to apply these techniques to the randomly mutated industrial strains for further strain improvement because of unknown mutations in their genome. Rational metabolic engineering by specifically targeted modifications can overcome this disadvantage. The recent development of omic technology, combined with computational analysis, now provides a new avenue for strain improvement by providing new information extracted from a large number of data, which is termed 'systems biotechnology'. Thus, a 'second generation' derived from a pantothenate producer by rational design to assess its potential to synthesise and accumulate the vitamin pantothenate by batch cultivation has recently been developed, as is the case of biotechnological approaches using recombinant *E. coli* strains (Song and Jackowski, 1992).

Pantothenate biosynthesis and production have also been investigated in Gram-positive soil bacteria such as *Corynebacterium glutamicum*, which is widely used for large-scale fermentative production of amino acids, such as L-glutamate and L-lysine (Hüser *et al.*, 2005). In *C. glutamicum* ATCC 13032 (Figure 4.8), four enzymes are involved in the biosynthesis of pantothenate from the precursors ketoisovalerate and aspartate. The first reaction in the biosynthesis of pantothenate is catalysed by the panB-encoded ketopantoate hydroxymethyltransferase that converts ketoisovalerate (a precursor of valine and leucine biosynthesis) into ketopantoate using 5,10-methylenetetrahydrofolate as a cofactor. Subsequently, ketopantoate is reduced to pantoate by ketopantoate reductase activity. In *C. glutamicum*, the ketopantoate reductase activity is encoded solely by the ilvC gene, which is also involved in the common pathway for the synthesis of the branched-chain amino acids isoleucine and valine and encodes acetohydroxy acid isomeroreductase. Aspartate is converted into β-alanine by the product of the panD gene encoding aspartate-α-decarboxylase. The biosynthetic pathway is completed by the ATP-consuming condensation of β-alanine with pantoate.

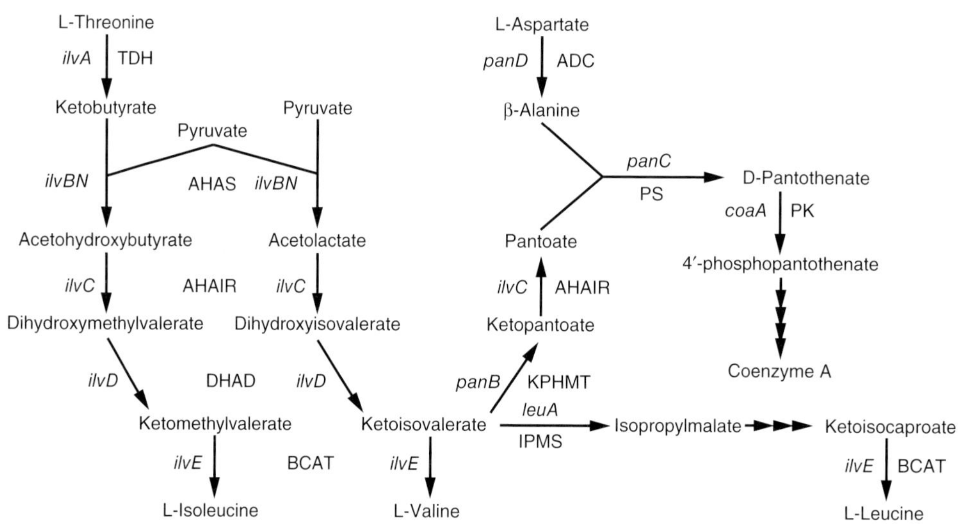

Figure 4.8 Biosynthesis pathway of pantothenic acid in *Corynebacterium glutamicum* ATCC 13032 (Hüser *et al.*, 2005).

C. glutamicum is a promising microorganism for the examination of pantothenate overproduction since not only the molecular physiology of amino acid biosynthesis in general but also the accumulation of valine has been analysed in detail. The biosynthesis of valine involves many enzymes required for the production of pantothenate (Figure 4.8), and the enzymatic activities and their regulation have been studied extensively in a valine-producing *C. glutamicum* strain. Furthermore, a 'first-generation' pantothenate producer was developed directly from the wild-type strain *C. glutamicum* ATCC 13032 (Hüser *et al.*, 2005). Two important genetic features for obtaining substantial pantothenate accumulation by *C. glutamicum* were a chromosomal deletion of the ilvA gene, encoding threonine dehydratase and combined overexpression of the ilvBNCD and panBC genes on two compatible plasmids. Using this type of production strain, up to 1 g/l of pantothenate accumulated in the culture supernatant.

Analysis of different production strains suggested that increased availability of ketoisovalerate is mandatory for enhanced pantothenate synthesis by *C. glutamicum*. The successful use of ilvBNCD overexpression to obtain pantothenate production is due to increased ketoisovalerate availability, since only then does panBC overexpression result in substantial accumulation of pantothenate. Carbon flux analysis of a 'first-generation' production strain of *C. glutamicum* during batch cultivation with β-alanine supplementation revealed that the flux towards valine was 10-fold higher than that directed to pantothenate, indicating that significant improvements of strain design could be obtained only if the carbon flux at the ketoisovalerate branch point of the pathway was modulated efficiently (Chassagnole *et al.*, 2003).

C. glutamicum is an industrial organism with a long history of use for the production of various fine chemicals. The advent of molecular biology enabled a new wave of development in which this industrial know-how was leveraged, not only to improve the performance of the existing lysine and threonine production processes, but also to enable the production of other amino acids and vitamins such as pantothenic acid. The utilisation of recombinant DNA techniques, combined with metabolic and carbon flux analyses (Chassagnole *et al.*, 2003), facilitated the identification of metabolic bottlenecks and their bypassing by expressing or repressing the corresponding genes to develop further improved industrial amino-acid production processes. The intrinsic characteristics of this food-grade microbial workhorse include its lack of pathogenicity and its lack of spore-forming ability, both of which are desirable traits, as well as its high growth rate, its relatively limited growth requirements, the ability of several strains not to undergo autolysis under conditions of repressed cell division, the absence of native extracellular protease secretion that makes corynebacteria suitable hosts for protein expression and the relative stability of the corynebacterial genome itself. These intrinsic attributes, combined with an up-to-date set of genetic engineering tools, make this organism ideal for the development of robust industrial processes that are increasingly competitive in comparison with *E. coli*, *Bacillus subtilis* or yeast-based processes. As a result, corynebacterial fermentations have become increasingly relevant to a wide range of industrial sectors, including food, feed, cosmetic, pharmaceutical and chemical companies.

Although systems biology approaches have only recently been established for various microorganisms, impressive progress has been made, especially with respect to fundamental insight into principles of metabolic regulation. Current and future efforts focus on the integration of quantitative data from genomics, transcriptomics, proteomics, metabolomics and flux analysis to build and evaluate metabolic and regulatory models of vitamin producers. Eventually, the potential of these approaches for the rational improvement of microbial strains for amino acid production and, more broadly, for white biotechnology will fully be harnessed. Moreover, this strategy not only successfully improved pantothenate production via genetically modified microbial strains but also revealed new constraints in attaining high productivity. However, after successful fermentation or enzyme reactions, desired products must be separated and purified. This final step is commonly known as *downstream processing* or *bioseparation*, which can account for up to 60% of the total production costs, excluding the cost of the purchased raw materials.

4.7 Application and Economics

The current world capacity of calcium pantothenate production and its demand are presumed to be about 4000 and 3600–4000 t/year, respectively. It is mainly used as an additive to animal feed (about 3000 t/year) and as a pharmaceutical product (about 600 t/year). Pantothenyl alcohol is used as a source of pantothenate activity for pharmaceutical vitamin products. Pantothenyl alcohol itself has no pantothenate activity; in fact, it is a competitive growth inhibitor of several pantothenate-requiring lactic acid bacteria. However, it has been demonstrated to be quantitatively converted to pantothenic acid in the animal body and to be equivalent to pantothenic acid in humans.

Pantethine, the disulfide of pantetheine and CoA are also used as pharmaceutical products in several countries. They have been suggested to be effective in reducing cholesterol levels, curing fatty liver and treating related diseases.

Some sulfonate derivatives of pantetheine or CoA (*Bifidus* factors), such as 4′-phosphopantetheine-*S* sulfonate, which were originally isolated from carrot roots have been shown to be growth factors of *Bifidobacterium* (Kolahdooz, Spearing and Sharma, 2013). Addition of the *Bifidus* factors to dried milk for infants has been suggested to be useful in improving the quality of the milk. A carbapenem antibiotic, OA-6129A produced by *Streptomyces* sp. OA-6129, may be an interesting example suggesting a new use of the vitamin as a building block for its synthesis (Bibile *et al.*, 1957).

Several clinical trials have been undertaken on humans using pantothenic acid supplementation and its derivatives in various medical fields, such as hyperlipidaemia, obesity, acne vulgaris, alopecia, hepatitis A, lupus erythematosus, osteoarthritis, rheumatoid arthritis and wound healing (Horváth and Vécsei, 2009). Unfortunately, firm conclusions regarding therapeutic effectiveness cannot be drawn from many of these studies, given the nonrandomised design, statistical biases, confounding variables and small sample size of participating patients. Moreover, specific cysteamine treatment has dramatically changed the course of cystinosis (Horváth and Vécsei, 2009). Furthermore, recent research on CoA metabolic enzymes has led to the discovery of uniquely non-metabolic roles for both enzymes and their metabolites, opening a broad field of investigation (Nitto and Onodera, 2013).

The effects of pantethine on the treatment of hyperlipoproteinaemia have been investigated in numerous studies. Pantethine is a dimer of pantothenic acid linked by a disulfide cystamine. McRae (2005) has reviewed 28 clinical trials from the literature on this topic, which provided a pooled population of 646 hyperlipidaemic patients. All but six of these investigations were conducted in Italy. Only 4 of the 28 published studies used a randomised double-blind study design, and only one of these was controlled with a placebo. Oral supplementation of pantethine resulted in a tendency towards normalisation of lipid values during a study period of 4 months. Only one study of these 28 clinical trials showed results in 9 and 12 months. Administration of pantethine resulted in a progressive decrease in total cholesterol, triglycerides and low-density lipoprotein cholesterol, along with an increase in high-density lipoprotein cholesterol, as is shown in Table 4.5. The doses of pantethine used ranged from 300 and 600 mg twice daily. The most common dosage administration was 300 mg three times a day. The mechanism of action of pantethine in normalising parameters associated with dyslipidaemia is unknown, although one can assume to be secondary to increased levels of intracellular pantothenate coenzymes. However, two recent papers have not confirmed these data in North American people (Evans *et al.*, 2014; Rumberger *et al.*, 2011). Thus, future randomised, double-blind and placebo-controlled trials with longer intervention are needed to clarify the possible therapeutic effect of pantethine on lipids, because some methodological shortcomings and narrow regional population involved in these investigations.

Table 4.5 Percentage mean change from baseline for serum lipids at months 1–4 (Yang *et al.*, 2014).

First month	Second month	Third month	Fourth month
Total serum cholesterol (%)			
↓8.7	↓11.6	↓12.6	↓15.1
Low-density lipoprotein cholesterol (%)			
↑6.1	↑7.8	↑10.7	↑8.4
Triglycerides (%)			
↓14.2	↓15.8	↓23.7	↓32.9

↑: Increase and ↓: Decrease.

A reduction in very low-density lipoprotein cholesterol and apolipoprotein A has also been reported in patients treated with pantethine. However, no studies have investigated whether pantothenic acid has lipid-lowering effects. Extrinsic factors, such as the plasma lipids, play a major role in the regulation of the platelet lipid pattern (Horváth and Vécsei, 2009). The platelet membrane lipids modulate certain important platelet functions, such as platelet aggregation and thromboxane A_2 synthesis. Therefore, a number of investigations have been performed on the effects of pantethine treatment on the platelet function. It has been shown that oral treatment with pantethine leads to significant decreases in total cholesterol and total phospholipids not only in plasma, but also in platelets, without any change in their ratio. The effects of pantethine on the membrane platelet composition may influence the fluidity of the cell membranes (Anonymous, 2010). Therefore, it has been suggested that pantethine supplementation might prevent atherogenesis in humans through its effect on serum lipid profile and platelet aggregability.

The most common serious complication of *Plasmodium falciparum* infection in man is cerebral malaria, with a case fatality rate of 20–50%. The pathogenesis of cerebral malaria is currently viewed in relation to the process of sequestration of parasitised erythrocytes in the cerebral microvasculature (Fairhurst and Wellems, 2010). Among other pathogenic mechanisms, the process is accompanied by platelet and endothelial cell activation. An investigation has shown that mice infected with *Plasmodium berghei* did not develop the cerebral syndrome when pantethine was administered (Penet *et al.*, 2008). The protection was associated with down-regulation of platelet responsiveness and impairment of endothelial cell activation. In this experiment, parasite development was unaffected by pantethine, as those infected mice that escaped from cerebral malaria died of high parasitaemia. Unfortunately, neither this nor any other experimental malaria model provides a reliable representation of human cerebral malaria.

For now, therapy of cystinosis relies on the aminothiol cysteamine (Bertholet-Thomas *et al.*, 2014). Cystinosis is a rare autosomal recessive disorder with an estimated incidence of 1 case per 100 000–200 000 live births. It is caused by mutations in the CTNS gene, mapped to chromosome 17p13, which encodes cystinosin, a lysosomal cystine transporter (Wilmer, Emma and Levtchenko, 2010). Defects in this transporter lead to the accumulation of intralysosomal cystine crystals and widespread cellular destruction. The predominant pathological finding in cystinosis is the presence of cystine crystals in almost all cells and tissues, including the conjunctivae, corneas, liver, spleen, lymph nodes, kidneys, thyroid, intestines, rectal mucosa, muscle, brain, macrophages and bone marrow. The disease is manifested as a multisystem disorder that affects the kidneys (Fanconi syndrome, renal failure), eyes, muscles, central nervous system, lungs and various endocrine organs; however, kidney involvement remains the earliest and foremost clinical characteristic of the disorder.

The mainstay of cystinosis therapy is oral cysteamine bitartrate (trade name Cystagon®, Mylan Pharma, USA), an aminothiol that can lower intracellular cystine content by 95% (Ghal, Thoene and Schneider, 2002). Cysteamine is administered in four divided doses; however, newly long-acting enteric-coated formulations are available (called RP103) from Raptor Pharmaceutical Corp. (USA), which can be administered twice a day. The mechanism of intralysosomal cystine depletion involves entry of cysteamine into the lysosomal compartment through a specific transporter, a disulfide reaction with cystine, resulting in the equimolar generation of a cysteine–cysteamine molecule and a molecule of cysteine. Both compounds can exit lysosomes via 'system c' transporters, bypassing the defective cystinosin pathway. In well-treated children, cysteamine has proven efficacy in delaying renal glomerular deterioration, enhancing growth and preventing hypothyroidism. Corneal cystine crystals do not dissolve with oral cysteamine therapy but do respond to the administration of cysteamine eye drops (Besouw *et al.*, 2013). The U.S. Food and Drug Administration (FDA) has approved a formulation for this purpose.

Approximately 14% of patients are unable to tolerate cysteamine therapy because of nausea and vomiting (Besouw *et al.*, 2013). Intracellular cysteamine is produced by the action of the enzyme pantetheinase on pantetheine. Pantethine, the disulfide dimer of pantetheine, was proven in the treatment of four cystinotic children. Pantethine is non-toxic and more palatable than cysteamine. The authors concluded that pantethine had less efficacy than cysteamine in depleting leukocytes from cystine, and therefore, it should only be considered in cases of cysteamine intolerance (Wittwer *et al.*, 1985).

An open-label study including 100 patients with acne to be treated with high doses of pantothenic acid (10 g/day in four divided doses) for 8 weeks or longer showed that the disease was usually controlled by 8 weeks in cases of moderate severity. A recent paper has indicated that the administration of a pantothenic-acid-based dietary supplement in healthy adults with facial acne lesions reduced total facial lesions after 12 weeks of supplementation with 2.2 g/day of pantothenic acid (Yang *et al.*, 2014). Also, 100 obese patients of Chinese descent treated with pantothenic acid (10 g/day in four divided doses) along with a calorie restricted diet noted an average weight loss of 1.2 kg/week and no side effects were observed (Kelly, 2011).

By the early 1960s, the effects of intramuscular daily injection of calcium-D-pantothenate into patients with rheumatoid arthritis had been tested (Baron-Wright and Elliott, 1963). The authors reported temporary alleviation of symptoms and no uniform success. Also, the efficacy of panthenol for paediatric post-tonsillectomy pain and wound healing has been evaluated (Wittwer *et al.*, 1985). Panthenol (synonyms: D-panthenol or dexpanthenol) is a stable alcoholic analogue of pantothenic acid. Post-operative administration of panthenol significantly accelerated the wound healing process and reduced tonsillectomy-related complaints, independent of the surgical technique used. Unfortunately, the mechanism(s) of action of panthenol for these effects remains elusive.

The effect of oral supplementation of 1.5 g/day each of D-pantothenic acid and L-cysteine on exercise performance was examined in eight healthy male volunteers aged 22.9 ± 1.4 years (Wall *et al.*, 2012). The conclusion of this study was that 'acute feeding with pantothenate and cysteine does not alter muscle CoA content and consequently does not affect muscle fuel metabolism or performance during exercise in humans'. Recently, cysteamine has been used in the treatment of non-alcoholic fatty liver disease in children with promising results (Besouw *et al.*, 2013). Moreover, some experimental and clinical data support the usefulness of cysteamine in the treatment of patients with Huntington's disease. Thus, a randomised, controlled, double-blind multicentre Phases II–III trial using RP103 cysteamine formulation is currently underway. In addition, some animal studies have suggested that cysteamine might also be beneficial in the treatment of Parkinson's disease, as well as in certain neuropsychiatric disturbances, such as schizophrenia and major depressive disorders.

Finally, during the past few decades, the inhibition of the metabolism of CoA is being extensively investigated in relationship to its potential antimicrobial effect. A breakthrough discovery was the molecular cloning of genes encoding CoA biosynthetic pathway enzymes, in particular pantothenate kinases. These enzymes have been divided into four groups based on their amino acid sequences: prokaryotic type I, II and III CoaAs and eukaryotic PanK (Leonardi *et al.*, 2005). Type I and III CoaAs are widely distributed among bacteria while the type II CoaA is limited to *Staphylococci*. The requirement for CoA in numerous metabolic bacterial processes and the diversity among the structure of pantothenate kinase in bacteria and mammals have made these enzymes an attractive drug target for the development of novel antimicrobial agents. Pantothenamides, amides derived from pantothenic acid, are substrates of the key rate-limiting enzyme pantothenate kinase (CoaA). *N*-pentylpantothenamide (*N*5-Pan) and *N*-heptylpantothenamide (*N*7-Pan) are the prototypes of pantothenamides and are active *in vitro* against Gram-negative and Gram-positive bacteria (Choudhry *et al.*, 2003), although only Gram-positive bacteria are sensitive to pantothenamide in pharmaceutically realistic concentrations. In addition, they have also been shown to possess activity against fungi and malaria parasites. It was thought that the mechanism of action of pantothenamides was the formation of CoA analogues that lead to the transfer of an inactive 4′-phoshopantothenamide moiety to ACP, which is the first step in the bacterial type II route of fatty acid synthesis (Jansen *et al.*, 2013). However, the precise mechanism by which pantothenamides act to inhibit bacterial viability is not yet completely resolved, because pantothenate supplementation does not revert this effect.

Another compound, the antibiotic CJ-15801, was discovered by Sugie *et al.* at Pfizer Laboratories in 2001, from the fermenting cultures of a *Seimatosporium* sp. fungus (Van der Westhuyzen *et al.*, 2012). Structural analysis showed the compound to resemble pantothenic acid, with the notable exception of a *trans*-substituted double bound in the β-alanine moiety. CJ-15801, similar to pantothenamides, selectively inhibits the growth of *Staphylococcus aureus* and the intraerythrocytic growth stage of the malaria parasite *P. falciparum*. The basis for CJ-15801's unique antimicrobial specificity may also be based on the type specificity of its pantothenate kinases, both of which have been characterised as atypical type II enzymes. Van der Westhuyzen *et al.* (2012) have demonstrated that CJ-15801 acts as an antimetabolite by using the first enzyme of the CoA biosynthesis to enter the pathway, after which it inhibits the second CoA biosynthesis enzyme, phosphopantothenoylcysteine synthase, by forming a tight-binding inhibitor *in situ*. This mode of action is reminiscent of the sulfonamide antibiotics, which block folic acid biosynthesis using a similar strategy.

Pantothenol is an alcoholic analogue of the pantothenic acid, which is widely used in health-care and cosmetic industries, as discussed next. Pantothenol has been reported to reduce the activities of type I and II CoaAs, with a more potent effect being observed on the type II CoaA than on the type I CoaA. In contrast, the type III CoaA is not inhibited by pantothenol (Chonan *et al.*, 2014). Thus, pantothenol has been shown to markedly inhibit the phosphorylation activity of pantothenate kinases *in vitro* of *E. coli* and *Staphylococci*. The growth of *Mycobacterium tuberculosis* is also inhibited by pantothenol (Kumar, Chhibber and Surolia, 2007). At the same time, Sabila, Ferru and Kirk (2005) showed that pantothenol inhibited the *in vitro* growth of *P. falciparum*. The authors speculated that the mechanism might involve competition with pantothenate, which could be attributed to inhibition of the parasite's pantothenate kinase. Therefore, the antimicrobial properties of the inhibitors of the metabolism of the CoA are an exciting field of research.

Tolerable Upper Intake Level (UL) is the maximum level of daily nutrient intake that is likely to pose no adverse health risks. The UL represents total intake from food, water and supplements. There is not sufficient scientific evidence on which to base a UL for pantothenic acid. Evidence available from clinical studies using high doses of pantothenic acid indicates that intake considerably in excess of 5 mg does not represent a health risk for the general population (Kelly, 2011). The existing clinical studies on pantothenic acid were not designed to monitor and assess side effects, so information of adverse effects in humans is limited. The most commonly reported side effect is mild transient gastrointestinal disturbance such as nausea, heartburn and diarrhoea (McRae, 2005). Adverse effects typically do not occur until doses exceed 1 g daily. In doses of 10 g/day, diarrhoea is reported to occur. There has been one case report of eosinophilic pleuropericardial effusion in a patient taking 300 mg/day of pantothenic acid in combination with 10 g/day of biotin for 2 months. The condition resolved after the vitamins were stopped (Debourdeau *et al.*, 2001).

Panthenol, the alcoholic analogue of pantothenic acid, is widely used in a variety of cosmetics and topical medical, over-the-counter and photoprotective products. Because of its moisturising and conditioning properties, it is mainly used in hair preparations, but is also added to other products. Allergic contact dermatitis caused by panthenol is considered to be rare, but has been occasionally reported following the use of medications, moisturisers and sunscreens (Fernandes *et al.*, 2012).

Pantothenic acid has an FDA Use-in-Pregnancy category A rating for doses at or below AI. What this means is that 'well-controlled studies have failed to demonstrate a risk to the foetus in the first trimester of pregnancy (and there is no evidence of risk in later trimester)' for doses at or below this level. Higher doses of pantothenic acid have a pregnancy category C rating (Besouw *et al.*, 2013). Cysteamine is also classified as a pregnancy category C drug. That is to say that 'animal reproduction studies have shown an adverse effect on the foetus and there are no adequate and well controlled studies in humans, but potential benefits may warrant use of the drug in pregnant women despite potential risks'.

Pantothenic acid is a naturally occurring physiological compound which offers potential effective therapeutic actions on diverse clinical conditions. The proper role of pantothenic acid in the therapeutic armamentarium is still challenging to answer properly. In light of the potential therapeutic benefits of pantothenic acid and its derivatives and the lack of adverse reactions to it, there is a need to address the issue of treatment with this compound. In our opinion, pantothenic acid (and its derivatives) deserves much more attention than it has received until now.

References

Airas, R.K. (1986) Pantothenase-based assay of pantothenic acid. *Methods Enzymol.*, **122**, 33–35.

Anonymous2010) Pantethine monograph. *Altern. Med. Rev.*, **15**, 279–282.

Ball, G.F.M. (2006) *Vitamins in Food: Analysis, Bioavailability, and Stability*, CRC Press, Boca Raton, FL.

Baron-Wright, E.C. and Elliott, W.A. (1963) The pantothenic acid metabolism of rheumatoid arthritis. *Lancet*, **282**, 862–863.

Bean, W.B., Hodges, R.E., Daum, K., Bradbury, J.T., Gunning, R., Manresa, J., Murray, W., Oliver, P., Routh, J.I., Schedl, H.P., Townsend, M., and Tung, I.C. (1955) Pantothenic acid deficiency induced in human subjects. *J. Clin. Invest.*, **34**, 1073–1084.

Bertholet-Thomas, A., Bacchetta, J., Tasic, V., and Cochat, P. (2014) Nephropatic cystinosis—a gap between developing and developed nations. *N. Engl. J. Med.*, **370**, 1366–1367.

Besouw, M., Masereeuw, R., van den Heuvel, L., and Levtchenko, E. (2013) Cysteamine: an old drug with new potential. *Drug Discov. Today*, **18**, 785–792.

Bibile, S.W., Lionel, N.D., Dunuwille, R., and Perera, G. (1957) Pantothenol and the burning feet syndrome. *Br. J. Nutr.*, **11**, 434–439.

Bird, O.D. and Thompson, R.Q. (1967) in *The Vitamins: Chemistry, Physiology, Pathology, Methods*, vol. 7, 2nd edn (eds P. György and W.N. Pearson), Academic Press, New York, pp. 209–241.

Chassagnole, C., Diano, A., Létisse, F., and Lindley, N.D. (2003) Metabolic network analysis during fed-batch cultivation of *Corynebacterium glutamicum* for pantothenic acid production: first quantitative data and analysis of by-product formation. *J. Biotechnol.*, **104**, 261–272.

Chonan, S., Murase, M., Kurikawa, K., Higashi, K., and Ogata, Y. (2014) Antimicrobial activity of pantothenol against staphylococci possessing a prokaryotic type II pantothenate kinase. *Microbes Environ.*, **29**, 224–226.

Choudhry, A.E., Mandichak, T.L., Broskey, J.P., Egolf, R.W., Kinsland, C., Begley, T.P., Seefeld, M.A., Ku, T.W., Brown, J.R., Zalacain, M., and Ratnam, K. (2003) Inhibitors of pantothenate kinase: novel antibiotics for staphylococcal infections. *Antimicrob. Agents Chemother.*, **47**, 2051–2055.

Combs, G.F. Jr., (2008) *The Vitamins: Fundamental Aspects in Nutrition and Health*, 3rd edn, Academic Press, Burlington, MA.

Debourdeau, P.M., Djezzar, S., Estival, J.L., Zammit, C.M., Richard, R.C., and Castot, A.C. (2001) Life-threatening eosinophilic pleuropericardial effusion related to vitamins B5 and H. *Ann. Pharmacother.*, **35**, 424–426.

De Leenheer, A.P., Lambert, W.E., and van Bocxlaer, J.F. (2000) *Modern Chromatographic Analysis of Vitamins*, 3rd edn, Marcel Dekker, New York.

Eitenmiller, R.R., Lin, Y., and Landen, W.O. Jr., (2008) *Vitamin Analysis for the Health and Food Sciences*, CRC Press, Boca Raton, FL.

Evans, M., Rumberger, J.A., Azumano, I., Napolitano, J.J., Citrolo, D., and Kamiya, T. (2014) Pantethine, a derivative of vitamin B5, favorably alters total, LDL and non-HDL cholesterol in low to moderate cardiovascular risk subjects eligible for statin therapy: a triple-blinded placebo and diet-controlled investigation. *Vasc. Health Risk Manag.*, **10**, 89–100.

Fairhurst, R.M. and Wellems, T.E. (2010) in *Mandell, Douglas, and Bennett's Principles and Practice of Infectious Diseases*, 7th edn (eds G.L. Mandell, J.E. Bennett, and R. Dolin), Churchill Livingstone Elsevier, Philadelphia, PA, pp. 3437–3462.

Fernandes, S., Macias, V., Cravo, M., Amaro, C., Santos, R., and Cardoso, J. (2012) Allergic contact dermatitis caused by dexpanthenol: report of two cases. *Contact Dermatitis*, **66**, 160–161.

Food and Nutrition Board (2000) *Dietary References Intakes*. Institute of Medicine of the National Academies. The National Academies Press. Washington DC. USA.

Ghal, W.A., Thoene, J.G., and Schneider, J.A. (2002) Cystinosis. *N. Engl. J. Med.*, **347**, 111–121.

Gopalan, C. (1946) The burning-feet syndrome. *Indian Med. Gaz.*, **81**, 22–25.

Havlíková, L., Matysová, L., Nováková, L., and Solich, P. (2006) HPLC determination of calcium pantothenate and two preservatives in topical cream. *J. Pharm. Biomed. Anal.*, **41**, 671–675.

Hodges, R.E., Bean, W.B., Ohlson, M.A., and Bleiler, R. (1959) Human pantothenic acid deficiency produced by omega-methyl pantothenic acid. *J. Clin. Invest.*, **38**, 1421–1425.

Hodges, R.E., Ohlson, M.A., and Bean, W.B. (1958) Pantothenic acid deficiency in man. *J. Clin. Invest.*, **37**, 1642–1657.

Horváth, Z. and Vécsei, L. (2009) Current medical aspects of pantethine. *Ideggyogy. Sz.*, **62**, 220–229.

Hüser, A.T., Chassgnole, C., Lindley, N.D., Merkamm, M., Guyonvarch, A., Elišáková, V., Pátek, M., Kalinowski, J., Brune, I., Pühler, A., and Tauch, A. (2005) Rational design of a *Corynebacterium glutamicum* pantothenate production strain and its characterization by metabolic flux analysis and genome-wide transcriptional profiling. *Appl. Environ. Microbiol.*, **71**, 3255–3268.

Institute of Medicine (1998) *Dietary Reference Intakes for Thiamin, Riboflavin, Niacin, Vitamin B6, Folate, Vitamin B12, Pantothenic Acid, Biotin, and Choline*, National Academy Press, Washington, DC, pp. 357–373.

Jackowski, S. (1996) in *Escherichia coli and Salmonella: Cellular and Molecular Biology*, 2nd edn (eds F.C. Neidhardt, R. Curtiss III,, J.L. Ingraham, E.C.C. Lin, K.B. Low, B. Magasanik, W.S. Reznikoff, M. Riley, M. Schaechter, and H.E. Umbarger), ASM Press, Washington, DC, pp. 687–694.

Jansen, P.A., Hermkens, P.H., Zeeuwen, P.L., Botman, P.N., Blaauw, R.H., Burghout, P., van Galen, P.M., Mouton, J.W., Rutjes, F.P., and Schalkwijk, J. (2013) Combination of pantothenamides with vanin inhibitos as a novel antibiotic strategy against Gram-positive bacteria. *Antimicrob. Agents Chemoter.*, **57**, 4794–4800.

Kataoka, M., Yamamoto, K., Kawabata, H., Wada, M., Kita, K., Yanase, H., and Shimizu, S. (1999) Steroseletive reduction of ethyl-4-chloro-3-oxo butanoate by *Escherichia coli* transformant cells coexpressing the aldehyde reductase and glucose dehydrogenase genes. *App. Micro. Bio.*, **51**, 486–490.

Kelly, G.S. (2011) Pantothenic acid monograph. *Altern. Med. Rev.*, **16**, 263–274.

Kolahdooz, F., Spearing, K., and Sharma, S. (2013) Dietary adequacies among South African adults in rural KwaZulu-Natal. *PLoS One*, **8**, e67184. doi: 10.1371/journal.pone.0067184

Kumar, P., Chhibber, M., and Surolia, A. (2007) How pantothenol intervenes in coenzyme-a biosynthesis of *Mycobacterium tuberculosis*. *Biochem. Biophys. Res. Commun.*, **361**, 903–909.

Lanska, D.J. (2012) The discovery of niacin, biotin, and pantothenic acid. *Ann. Nutr. Metab.*, **61**, 246–253.

Leonardi, R., Zhang, Y.-M., Rock, C.O., and Jackowski, S. (2005) Coenzyme a: back in action. *Prog. Lipid Res.*, **44**, 125–153.

Lewis, C.M. and King, J.C. (1980) Effect of oral contraceptives agents on thiamine, riboflavin, and pantothenic acid status in young women. *Am. J. Clin. Nutr.*, **33**, 832–838.

Lopez Martinez, D.L., Tsuchiya, Y., and Gout, I. (2014) Coenzyme a biosynthetic machinery in mammalian cells. *Biochem. Soc. Trans.*, **42**, 1112–1117.

McRae, M.P. (2005) Treatment of hyperlipoproteinemia with pantethine: a review and analysis of efficacy and tolerability. *Nutr. Res.*, **25**, 319–333.

Nitto, T. and Onodera, K. (2013) The linkage between coenzyme a metabolism and inflammation: roles of pantetheinase. *J. Pharmacol. Sci.*, **123**, 1–8.

Noda, S., Umezaki, H., Yamamoto, K., Araki, T., Murakami, T., and Ishii, N. (1988) Reye-like syndrome following treatment with the pantothenic acid antagonist, calcium hopantenate. *J. Neurol. Neurosurg. Psychiatry*, **51**, 582–585.

Penet, M.F., Abou-Hamdan, M., Coltel, N., Cornille, E., Grau, G.E., de Reggi, M., and Gharib, B. (2008) Protection against cerebral malaria by the low-molecular-weight thiol pantethine. *Proc. Natl. Acad. Sci. U.S.A.*, **105**, 1321–1326.

Primerano, D.A. and Burns, R.O. (1983) Role of acetohydroxy acid isomeroreductase in biosynthesis of pantothenic acid in *Salmonella typhimurium*. *J. Bacteriol.*, **153**, 259–269.

Rose, R.L. and Hodgson, E. (2004) in *A Textbook of Modern Toxicology*, 3rd edn (ed E. Hodgson), John Wiley & Sons, Inc., Hoboken, NJ, pp. 111–148.

Rucker, R.B. and Bauerly, K. (2007) in *Handbook of Vitamins*, 4th edn (eds J. Zempleni, R.B. Rucker, D.B. McCormick, and J.W. Suttie), CRC Press, Boca Raton, FL, pp. 289–305.

Rumberger, J.A., Napolitano, J., Azumano, I., Kamiya, T., and Evans, M. (2011) Pantethine, a derivative of vitamin B(5) used as a nutritional supplement, favorably alters low-density lipoprotein cholesterol metabolism in low- to moderate-cardiovascular risk North American subjects: a triple-blinded placebo and diet controlled investigation. *Nutr. Res.*, **31**, 608–615.

Rychlik, K.M. and Roth-Maier, D. (2005) Pantothenic acid quantification: method comparison of a stable isotope dilution assay and a microbiological assay. *Int. J. Vitam. Nutr. Res.*, **75**, 218–223.

Sabila, K.J., Ferru, I., and Kirk, K. (2005) Provitamin B_5 (pantothenol) inhibits growth of the intraerythrocytic malaria parasite. *Antimicrob. Agentes Chemother.*, **49**, 632–637.

Salunke, D.M. and Vijayan, M. (1984) Conformational analysis of D-pantothenic acid. *Biochim. Biophys. Acta. Gen. Subj.*, **798**, 180–186.

Shimizu, S. and Kataoka, M. (1999) in *Encyclopedia of Bioprocess Technology: Fermentation, Biocatalysis, and Bioseparation* (eds M.C. Flickinger and S.W. Drew), John Wiley & Sons, Inc., New York, pp. 1923–1934.

Shimizu, S., Kataoka, M., Honda, K., and Sakamoto, K. (2001) Lactone-ring-cleaving enzymes of microorganisms: their diversity and applications. *J. Biotechnol.*, **92**, 187–194.

Song, W.J. and Jackowski, S. (1992) Cloning, sequencing, and expression of the pantothenate kinase (coaA) gene of *Escherichia coli*. *J. Bacteriol.*, **174**, 6411–6417.

Srivastava, D.K. and Bernhard, S.A. (1987) Biophysical chemistry of metabolic reaction sequences in concentrated enzyme solution and in the cell. *Annu. Rev. Biophys. Biophys. Chem.*, **16**, 175–204.

Tahiliani, A.G. and Beinlich, C.J. (1991) Pantothenic acid in health and disease. *Vitam. Horm.*, **46**, 165–228.

Thornton, G.H.M., Bean, W.B., and Hodges, R.E. (1955) The effect of pantothenic acid deficiency on gastric secretion and motility. *J. Clin. Invest.*, **34**, 1085–1091.

Tsuda, H., Matsumoto, T., and Ishimi, Y. (2011) Biotin, niacin and pantothenic acid assay using lyophilized *Lactobacillus plantarum* ATCC 8014. *J. Nutr. Sci. Vitaminol. (Tokyo)*, **57**, 437–440.

Vallari, D.S. and Jackowski, S. (1988) Biosynthesis and degradation both contribute to the regulation of coenzyme a content in *Escherichia coli*. *J. Bacteriol.*, **170**, 3961–3966.

Van der Westhuyzen, R., Hammons, J.C., Meier, J.L., Dahesh, S., Moolman, W.J., Pelly, S.C., Nizet, V., Burkart, M.D., and Strauss, E. (2012) The antibiotic CJ-15,801 is an antimetabolite which hijacks and then inhibits CoA biosynthesis. *Chem. Biol.*, **19**, 559–571.

Wagner, A.F. and Folkers, K. (1964) *Vitamins and Coenzymes*, Intersciences Publishers, New York, pp. 93–137.

Wall, B.T., Stephens, F.B., Marimuthu, K., Constantin-Teodosiu, D., Macdonald, I.A., and Greenhaff, P.L. (2012) Acute pantothenic acid and cysteine supplementation does not affect muscle coenzyme a content, fuel selection, or exercise performance in healthy humans. *J. Appl. Physiol.*, **112**, 272–278.

Wang, T.-M., Chiu, Y.-M., Lin, L.-C., and Hsu, M.C. (2004) Liquid chromatographic method for determination of calcium pantothenathe preparations and related stability studies. *J. Food Drugs Anal.*, **12**, 1–6.

West, T.P., Traut, T.W., Shanley, M.S., and O'Donovan, G.A. (1985) A *Salmonella typhimurium* strain defective in uracil catabolism and beta-alanine synthesis. *J. Gen. Microbiol.*, **131**, 1083–1090.

Whitney, E. and Rolfes, S.R. (2011) *Understanding Nutrition*, Wadsworth Cengage Learning, Belmont, CA.

Willamson, J.M. and Brown, G.M. (1979) Purification and properties of L-Aspartate-alpha-decarboxylase, an enzyme that catalyzes the formation of beta-alanine in *Escherichia coli*. *J. Biol. Chem.*, **254**, 8074–8082.

Williams, R.J. (1939) Pantothenic acid – a vitamin. *Science*, **89**, 486.

Wilmer, M.J., Emma, F., and Levtchenko, E.N. (2010) The pathogenesis of cystinosis: mechanisms beyond cystine accumulation. *Am. J. Physiol. Renal Physiol.*, **299**, F905–F916.

Wittwer, C.T., Gahl, W.A., Butler, J.D., Zatz, M., and Thoene, J.G. (1985) Metabolism of pantethine in cystinosis. *J. Clin. Invest.*, **76**, 1665–1672.

Yang, X.J. (2004) The diverse superfamily of lysine acetyltransferases and their roles in leukemia and other diseases. *Nucleic Acids Res.*, **32**, 959–976.

Yang, M., Moclair, B., Hatcher, V., Kaminetsky, J., Mekas, M., Chapas, A., and Capodice, J. (2014) A randomized, double-blind, placebo-controlled study of a novel pantothenic acid-based dietary supplement in subjects with mild to moderate facial acne. *Dermatol. Ther. (Heidelb)*, **4**, 93–101.

5
Folate: Relevance of Chemical and Microbial Production

Maddalena Rossi, Stefano Raimondi, Luca Costantino, and Alberto Amaretti

5.1
Introduction

Folate is the term encompassing the different natural forms of the water-soluble B_9 vitamin and folic acid (FA), the synthetic oxidised form used in supplements. Humans depend on an adequate and constant intake of this essential nutrition component, which is a central cofactor of many metabolic reactions, required for biosynthetic and cellular processes, such as DNA, RNA and protein synthesis (Bailey and Caudill, 2012). At present, FA is industrially produced through chemical synthesis while, unlike other vitamins, microbial production of folate on industrial scale is not exploited. The sole application of microorganisms for folate production seems to be limited to the fortification of fermented dairy foods and to folate-producing probiotics.

This chapter provides an overview on biochemical and physiological role of this vitamin, natural occurrence and bioavailability, chemical synthesis for industrial production and applications of folate-producing lactic acid bacteria (LAB) and bifidobacteria.

5.2
Folates: Chemical Properties and Occurrence in Food

FA, or pteroyl-L-glutamate, is the synthetic form of the B_9 water-soluble vitamin (Figure 5.1). It consists of a pteridine ring bound to *p*-aminobenzoic acid and a glutamic acid moiety. It is not biologically active itself, but it is activated to tetrahydrofolate (THF) through two reduction reactions catalysed by dihydrofolate reductase (DHFR, EC 1.5.1.3) (Figure 5.1).

Dietary folates are reduced and usually linked to a polyglutamic chain, resulting in a tetrahydropteroyl heterocycle bound to a poly-L-glutamic acid side chain. The major ones are 5-methyltetrahydrofolate (5-MeTHF) and 10-formyltetrahydrofolate (10-formylTHF), in their polyglutamated forms (Figure 5.1). Reduction to THF generates a second chiral centre at C_6 of the

Industrial Biotechnology of Vitamins, Biopigments, and Antioxidants, First Edition.
Edited by Erick J. Vandamme and José L. Revuelta.

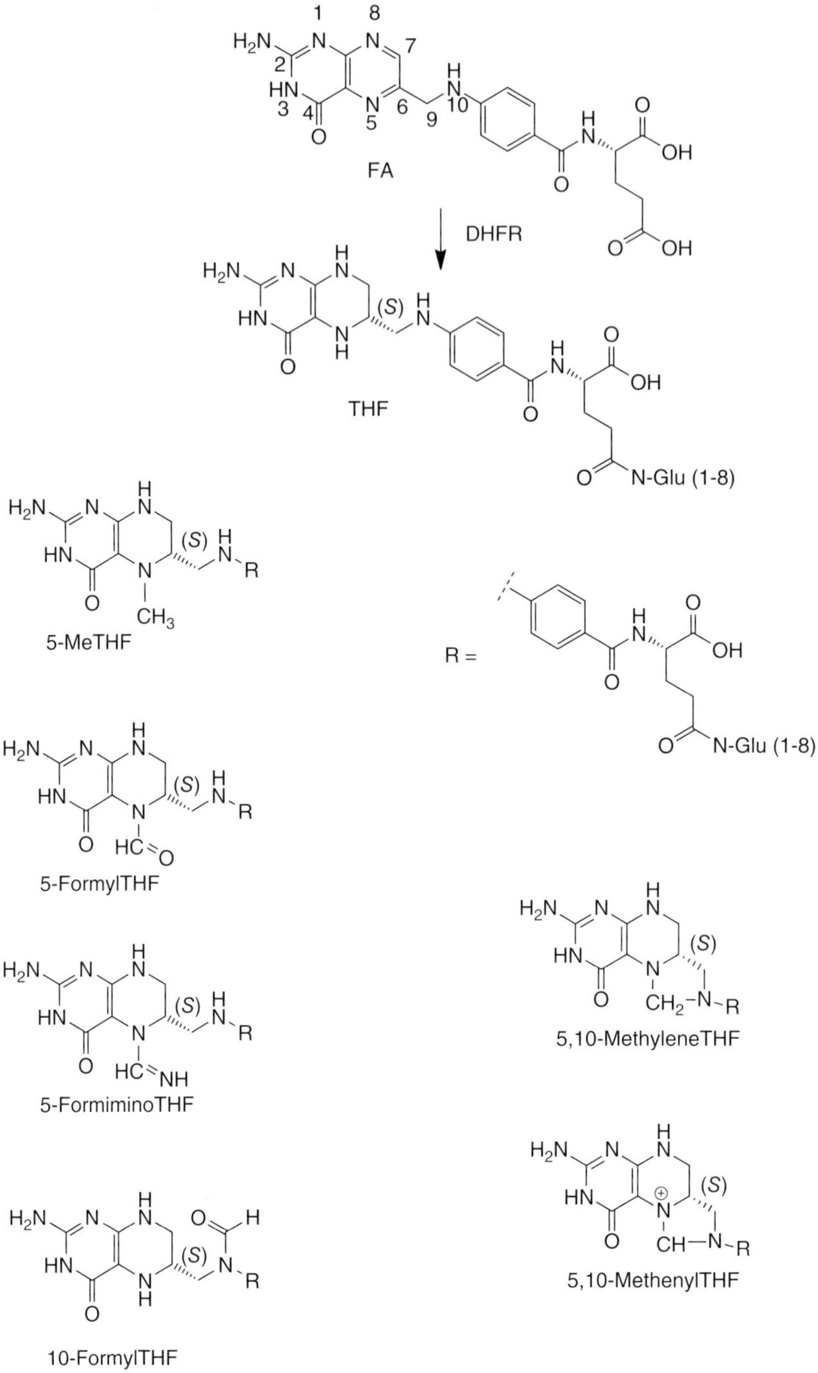

Figure 5.1 Structures of folic acid (FA), tetrahydrofolic acid (THF) and of its native, biologically active derivatives.

pteridine nucleus. Enzymatic reduction is stereoselective, yielding only the (6*S*) isomer, and all the biologically active naturally occurring folates are derivatives of the reduced form THF.

THF derivatives are cofactors of many enzymes that catalyse the transfer of one-carbon unit, with a highly compartmentalised metabolism between cytoplasm, mitochondria and nucleus (Tibbetts and Appling, 2010). The reactions involve the N_5 and N_{10} atoms (Figure 5.1) that bind an additional carbon (methyl, formyl, methylene or formimino group), derived from the catabolism of serine, glycine and histidine. Other intermediates occurring during folate interconversions that do not participate directly as C_1 group donors are 5-forminimoTHF, which is then converted to 5,10-methenylTHF, which in turn yields 10-formylTHF (Figure 5.1).

Folate occurs in a wide variety of foods, including vegetables (especially dark green leafy vegetables), fruits and fruit juices, nuts, beans, peas and other legumes, milk and dairy products, whole grains, poultry and meat, eggs, seafood. Spinach, liver, yeast, asparagus and Brussels sprouts are among the foods with the highest levels of folate (Jägerstad and Jastrebova 2012; USDA 2015).

Humans are unable to synthesise folate and depend on an adequate and constant intake. The U.S. Food and Drug Administration recommends an intake of 400 µg as daily folate equivalents (DFE) from natural foods, FA fortified foods and FA supplements, while 240 µg of DFE is advised by the European Union (EFSA Panel, 2014). A higher intake of folates and FA, up to 1 mg/day, is optimal for the pre-pregnant, pregnant and lactating women.

In order to meet micronutrient needs of the population, several countries introduced FA fortification (US FDA, 1996a). FA is more oxidised and more stable than natural folates, then it has been used for decades in fortification and nutritional supplements (Dwyer *et al.*, 2014). The U.S. FDA in 1996 authorised FA fortification of cereal grain products, which became mandatory in 1998 (US FDA, 1996b). This strategy is considered a major success of nutritional discoveries (Jägerstad, 2012), albeit European countries do not follow this way, since the scientific community is still debating on folate exerting a protective effect on cancer or increasing risk of malignancies (EFSA, 2009, EFSA, 2013).

5.3 Biosynthesis

Both prokaryotic and eukaryotic cells require reduced folate cofactors in a variety of biosynthetic processes. Albeit cellular requirement for folates is universal, animals cannot synthesise folates and assimilate them with the diet, while plants, fungi, certain protozoa and several archaea and bacteria can synthesise folates *de novo*, through the same general biosynthetic pathway with some modifications (White, 1988; Bermingham and Derrick, 2002; Hanson and Gregory, 2002; Levin *et al.*, 2004; De Crécy-Lagard *et al.*, 2007; Rossi, Amaretti and Raimondi, 2011).

Folate consists of a pterin moiety, originating from 6-hydroxymethyl-7,8-dihydropterin pyrophosphate (DHPP), bound to para-aminobenzoic acid (pABA, vitamin B_{10}). pABA is produced *de novo* by plants and bacteria (Figure 5.2).

In the shikimate pathway, erythrose 4-phosphate and phosphoenolpyruvate are transformed into chorismate, a branching point towards the biosynthesis of aromatic aminoacids and pABA. Chorismate is converted into 4-amino-4-deoxychorismate via aminodeoxychorismate synthase (EC 2.6.1.85), then pyruvate is cleaved by 4-amino-4-deoxychorismate lyase (EC 4.1.3.38) to give pABA.

The biosynthesis of DHPP proceeds via the conversion of guanosine triphosphate (GTP) in four consecutive reactions. GTP cyclohydrolase I (EC 3.5.4.16) carries out an extensive transformation of GTP, through Amadori rearrangement, resulting in the pterin ring structure. Following dephosphorylation, the pterin is transformed by aldolase and pyrophosphokinase into activated pyrophosphorylated DHPP.

Dihydropteroate synthase (EC 2.5.1.15) catalyses the formation of a C–N bond joining DHPP to pABA, yielding dihydropteroate (DHP). DHP is glutamylated by dihydrofolate synthase (EC 6.3.2.12) into dihydrofolate (DHF), which in turn is reduced by DHFR (EC 1.5.1.3) to the biologically active cofactor THF. THF-polyglutamate was generated by folylpolyglutamate synthase (EC 6.3.2.17), adding multiple glutamate moieties. Polyglutamylation may occur also before the reduction step, catalysed by DHF synthase or, in many bacteria, by a bifunctional enzyme which is responsible for both EC 6.3.2.12 and EC 6.3.2.17 activities (Levin *et al.*, 2004).

5.4
Physiological Role

Folate is essential for cell division and cell maintenance, acting as a coenzyme in the transfer and processing of one-carbon unit. In the mitochondria, it is involved in the catabolism of serine and glycine, generating formate, which in turn is utilised in the cytoplasm for the re-methylation of homocysteine to methionine and for the synthesis of nucleotides (Bailey and Caudill, 2012). In particular, folate plays an important role in thymidine synthesis, *de novo* synthesis of purines and repair of DNA (Liu and Ward, 2010; Nazki, Sameer and Ganaie, 2014). 5,10-MethyleneTHF is a cofactor in the reaction generating the thymidine monophosphate (Figure 5.3a), while 10-formylTHF provides one-carbon unit to synthetise the purine nucleotides adenine and guanine (Figure 5.3b).

5-MeTHF transfers methyl groups to homocysteine, producing methionine, which in turn participates in the synthesis of *S*-adenosylmethionine (SAM, Figure 5.3c), the universal donor of methyl groups in several methylation reactions of DNA, RNA and proteins (Blom and Smulders, 2011). Conversion of homocysteine into the powerful metabolite SAM requires as cofactors not only folates, but also B_{12} (cobalamin) and B_6 (pyridoxine). Folate deficiency can determine an insufficient availability of methyl groups, affecting methylation

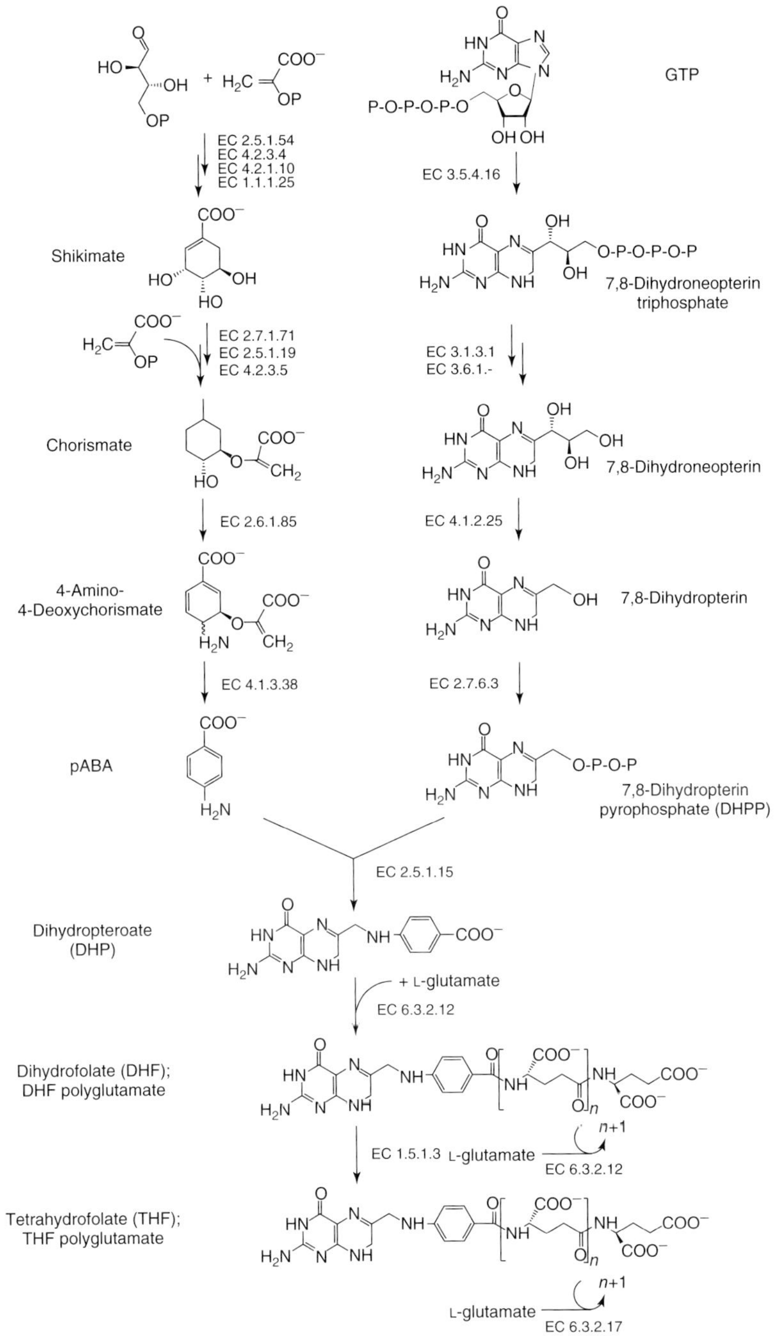

Figure 5.2 Pathway of *de novo* bacterial biosynthesis of folate. GTP, guanosine triphosphate; DHPP, 6-hydroxymethyl-7,8-dihydropterin pyrophosphate; pABA, para-aminobenzoic acid; DHP, dihydropteroate; DHF, dihydrofolate and THF, tetrahydrofolate. The figure is adapted from Rossi, Amaretti and Raimondi (2011).

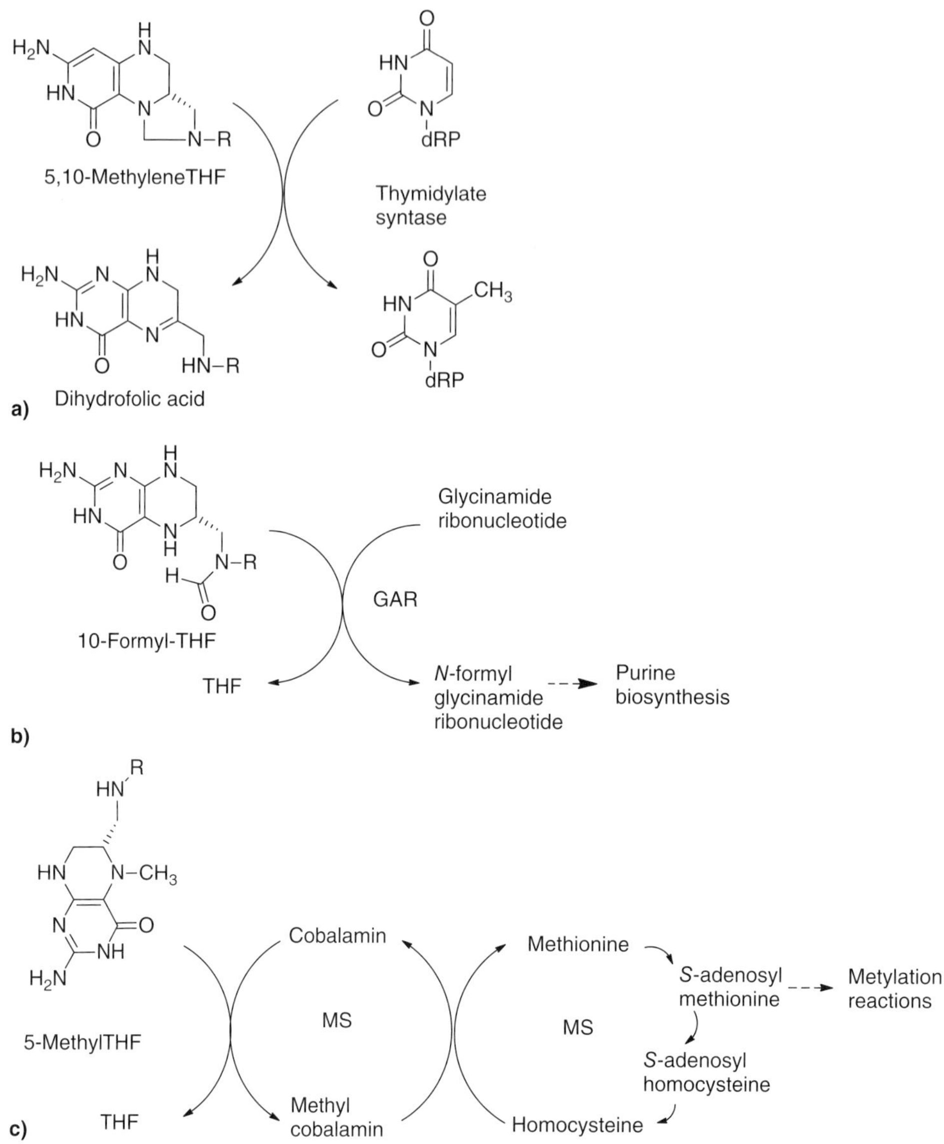

Figure 5.3 Structures of one-carbon derivatives of THF and example of reactions in which they are involved: (a) transfer of a C_1 moiety and reduction to DHF; (b) transfer of a formyl group (GAR, glycinamide ribonucleotide transformylase) and (c) transfer of a methyl (MS, homocysteine *S*-methyl transferase or methionine synthase).

efficiency and increasing the level of homocysteine. Hyperhomocysteinaemia exerts direct toxic effects on both the vascular and nervous systems, being responsible of several cardiovascular and neurologic disorders (Liu *et al.*, 2014; Ansari *et al.*, 2014; Litynski *et al.*, 2002).

Altered folate metabolism affects DNA synthesis, methylation, stability and integrity and can have a pivotal role in malignancies development (Liu and Ward, 2010; Duthie *et al.*, 2002). Folate deficiency reduces the division rate of all cells in the body, resulting, firstly, in megaloblastic anaemia. Low folate status, often affected by polymorphisms in the genes involved in its metabolism, is associated with increased risk of cancer of the colorectum, oropharynx, oesophagus, stomach, pancreas, lungs, cervix, ovary and breast and of neuroblastoma and leukaemia (Nazki, Sameer and Ganaie, 2014). In particular, low folate status may induce hypomethylation of DNA, promoting cancer on the proliferating cells of the colon–rectal mucosa that supports rapid and continuous renewal of the epithelium. However, while epidemiological studies suggest that dietary folate confers modest protection against colon–rectum cancer, high doses of FA used in intervention trials exerted adverse consequences, particularly in the case of an existing adenoma (Williams, 2012).

Summing up, folate deficiency is associated with leuco- and thrombocytopenia, cardiovascular disease, neurologic disorders, depression and cognitive impairment, embryonic defects (Blom and Smulders, 2011). Among the latter, neural tube defects can affect the brain (anencephaly) and the spine (spina bifida) during the first month of pregnancy. Folate prevents the majority of neural tube defects if consumed before and during early pregnancy (Imbard, Benoist and Blom, 2013). In order to reduce the risk of neural tube defect in pregnant women, the U.S. FDA introduced cereal grain products fortified with FA (US FDA, 1996b), a practice currently followed by over 60 countries (Jägerstad, 2012).

5.5 Bioavailability and Dietary Supplements

Dietary folates, most in the form of polyglutamates, need to be hydrolysed to monoglutamates to be transported. The glutamate carboxypeptidase II, anchored to the intestinal apical brush border, catalyses this reaction (Chandler *et al.*, 1991). Being hydrophilic anionic molecules that do not cross biological membranes by diffusion, they require specialised membrane transport systems for accumulation into mammalian cells and tissues. Monoglutamylated folates and FA are absorbed in the duodenum and upper part of the jejunum by the high-affinity proton-coupled folate receptor PCFT. Absorption also exploits other genetically and functionally distinct transporters, such as the folate receptors, the family of organic anion transporters and the reduced folate carrier, which is ubiquitously expressed (Qiu *et al.*, 2006). During transit through the intestinal mucosa, both natural folates and FA are converted to 5-MeTHF, which is the main circulating form of folate and can be transported into the cell with carrier- or receptor-mediated transport (Blom *et al.*, 2006).

FA and dietary folates differ in terms of both stability and bioavailability. FA is more oxidised and more stable than natural folates, which are vulnerable to a variable degree of degradation by cooking processes (Hawkes and Villota, 1989; Jägerstad and Jastrebova, 2013). 5-MeTHF, the major dietary folate and the most common form in body fluids, is moderately stable, while THF occurs in foods in minor amount and is the least stable folate.

The bioavailability of dietary folate depends on several factors, such as the food matrix, the efficiency of intestinal deconjugation of polyglutamic folates, the loss of certain unstable folates during digestion and the presence of dietary constituents that enhance its stability during digestion or prevent the attack of glutamate carboxypeptidase II. Food folate is less bioavailable than FA (~80%), taking into account that the bioavailability of FA is nearly 100% when it is consumed without food, whereas it decreases to approximately 85% when taken during a meal (Gregory, 2001; Winkels *et al.*, 2007). Therefore, FA is used for dietary supplements and food fortification due to the higher stability and bioavailability with respect to food folates (Dwyer *et al.*, 2014).

All the biological functions of the provitamin FA are dependent on the reduction to THF and other reduced derivatives. The activation of FA depends on the activity of DHFR. The physiological function of this enzyme is the reduction of DHF, generated during thymidylate synthesis, to THF (Figure 5.3a), but it also performs the reduction of synthetic FA to DHF and is necessary for FA to enter into one-carbon metabolism. However, FA is a poor substrate for DHFR, and relevant variability of enzyme activity occurs among individuals, underlying the importance of the administration of reduced cofactors with respect to the unnatural precursor FA (Bailey and Ayling, 2009).

Folate supplements, other than FA, are also available in the form of cofactors. Particularly, among the derivatives of the reduced form of FA, the diastereoisomer (6*S*)-5-MeTHF is commercially available as a medical food for the treatment of folate deficiency (Deplin) and depression symptoms (Shelton *et al.*, 2013). Moreover, the (6*S*)-5-MeTHF calcium (Metafolin) and glucosamine salt (Quatrefolic) are available as a commercial preparation, with a bioavailability similar to FA at equimolar doses (EFSA Panel, 2014).

5.6
Chemical and Chemoenzymatic Synthesis of Folic Acid and Derivatives

FA is produced industrially by chemical synthesis due to the low yields of fermentative processes (Miyata and Yonehara, 1999). Synthesis can be performed from 2,4,5-triamino-6-hydroxypyrimidine (**1**, Figure 5.4), a three-carbon compound of variable structure (**2.1**–**2.3**) and *N*-(*p*-aminobenzoyl)-L-glutamic acid (**3**). The reaction takes place in water-based solvent mixture, in the presence of sodium sulfite as reducing agent. Among the three-carbon compounds, 2-hydroxy malondialdehyde (**2.1**) was firstly used (Angier *et al.*, 1948), followed by a

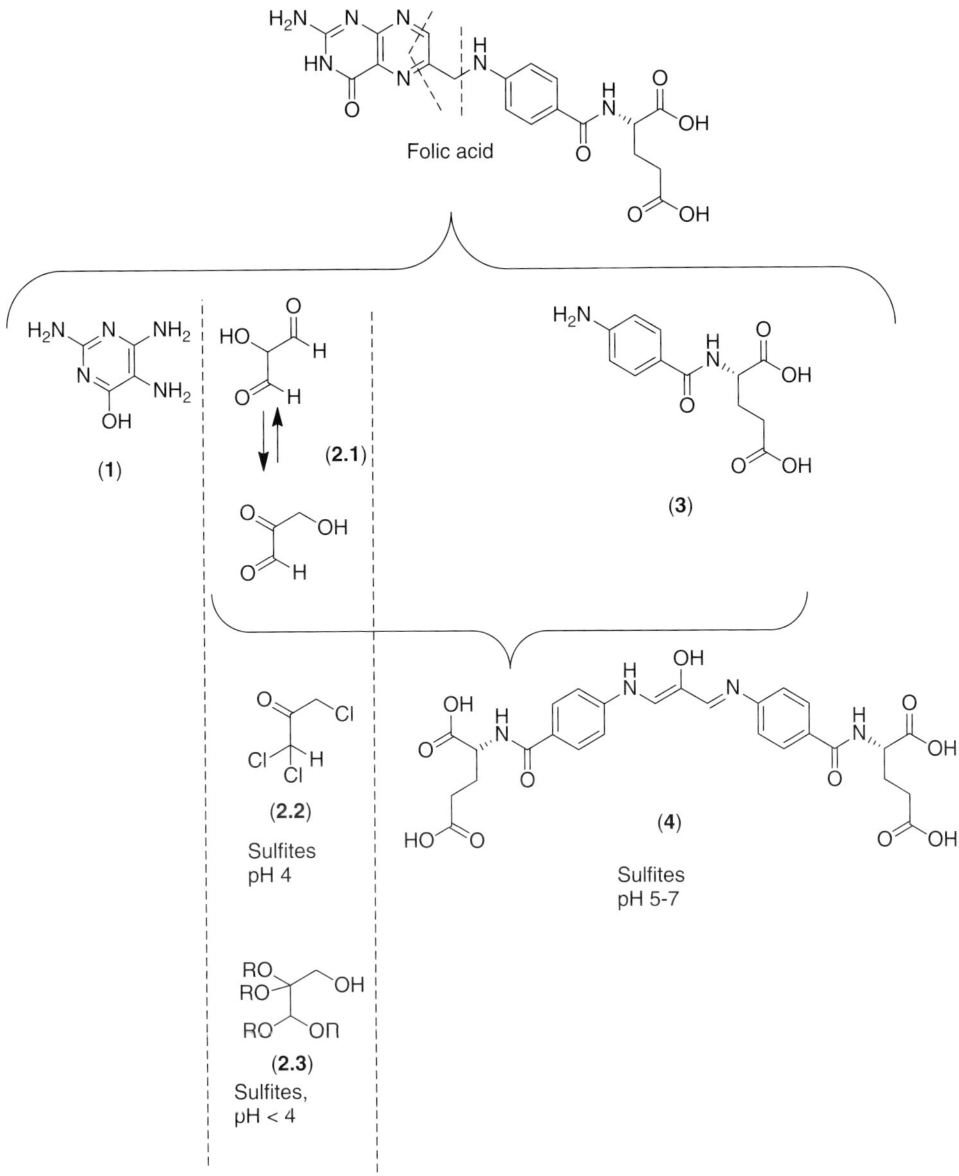

Figure 5.4 Chemical synthesis of FA.

haloacetone derivative (**2.2**) (Kawanishi, 1960) and a tetraalkoxypropanol derivative (**2.3**) (Botzem *et al.*, 2002). In another procedure, 2-hydroxymalondialdehyde (**2.1**) reacts firstly with *p*-aminobenzoylglutamic acid (**3**) forming a diimine (**4**), which subsequently reacts with 2,4,5-triamino-6-hydroxypyrimidine (**1**) in the presence of sodium sulfite. A one-pot reaction followed by crystallisation was also reported (Wehrli, 1996).

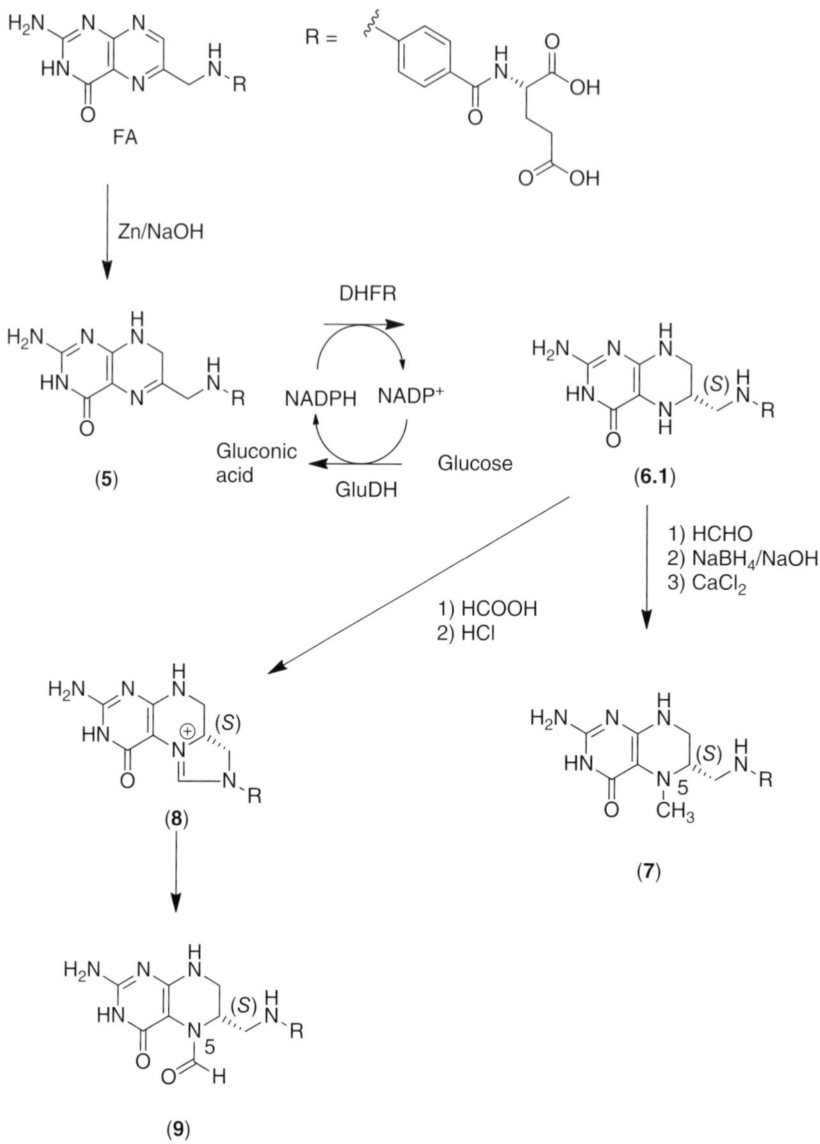

Figure 5.5 Chemoenzymatic synthesis of (6*S*)-methylTHF (**7**) and (6S)-5-formylTHF (**9**).

Eguchi *et al.* (1990) and Jones *et al.* (2012) disclosed the chemoenzymatic process for the stereoselective reduction of dihydrofolic acid (**5**, Figure 5.5) to (6*S*)-THF (**6.1**) with DHFR in the presence of $NADP^+$/NADPH, glucose and glucose dehydrogenase (GluDH). The availability of **6.1** allowed the production of the diastereoisomerically pure compounds, avoiding the pitfalls of previously reported processes, namely the purification by fractional crystallisation of the

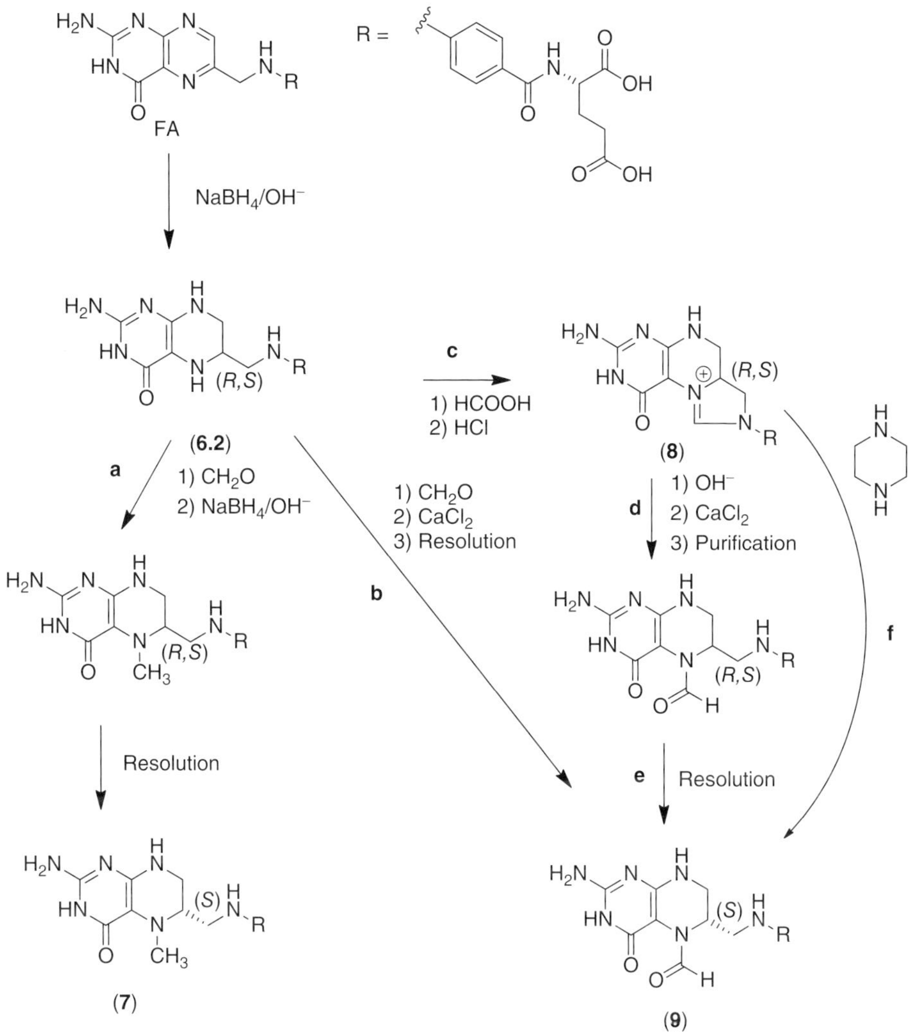

Figure 5.6 Chemical methods for the synthesis of (6*S*)-methylTHF (**7**) and (6*S*)-5-formylTHF (**9**).

diastereoisomeric mixture that originates from the chemical reduction of FA, by chromatographic separation of diastereomers or by stereoselective reduction (Rees *et al.*, 1986). (6*S*)-5-MeTHF (**7**) was then produced according to Jones *et al.* (2012). The intermediate **6.1** can also be used to produce 5-formylTHF (**9**) (Eguchi *et al.*, 1990). Due to intrinsic instability of (6*S*)THF, it was converted into 5,10-methylidene derivative (**8**), then into calcium (6*S*)-5-formylTHF (**9**).

Chemical methods for the synthesis of (6*S*)-5-MeTHF (**7**, Figure 5.6) and (6*S*)-5-formylTHF (**9**) have also been reported, based on the separation of the (6*R*,*S*) diastereoisomeric mixtures. During the classical synthesis, the intermediate

(6*R*,*S*)-THF (**6.2**) was obtained (Temple *et al.*, 1979). This compound is formylated and processed to obtain **7** and **9**, according to the procedures **a** and **b** (Vecchi, 1994). Moreover, Felder, Ripa and Distaso (1998) reported the synthesis of **9** starting from **6.2** through several steps (**c**, **d** and **e**) or directly (**c** and **f**) in the presence of a diamine such as piperazine.

5.7 Intestinal Microbiota, Probiotics and Vitamins

The human colon is colonised by up to 10^{11} microorganisms per gram of intestinal content, mostly represented by anaerobic bacteria. However, colonic microbiota also includes viruses (bacteriophages), archaea, yeasts and other eukaryotes. The dominating bacteria belong to *Firmicutes* and *Bacteroidetes*, followed by *Actinobacteria* and *Proteobacteria* (Eckburg *et al.*, 2005). Analysis of rRNA gene sequences indicates that over 1000 bacterial species can be found within the colonic microbial community. The gut microbiota benefits the host playing a pivotal role in nutrient digestion and energy recovery. Colonic bacteria produce enzymes that the host lacks, including those involved in breakdown of complex molecules, such as plant polysaccharides. The fermentation of the dietary components that escape digestion and absorption in the upper intestinal tract and of endogenous products (e.g. mucin) results in the production of organic acids (e.g. acetic, lactic, propionic and butyric acids), branched-chain fatty acids (e.g. isobutyric, isovaleric and 2-methylbutyric acids), H_2, CO_2, ammonia, amines and several other end-products (Louis, Hold and Flint, 2014). These fermentation products affect the gut environment and the host health, acting as energy sources, regulators of gene expression and cell differentiation and anti-inflammatory agents. They influence host–microbe interactions involved in the resistance to pathogens, gut development and epithelial homeostasis (O'Keefe, 2008; Russell *et al.*, 2013).

The microbiota also represents an important source of vitamins, which cannot be synthetised by the host. In fact, the colon microbiome contains a number of COGs (Clustered Orthologous Groups) involved in the synthesis of several essential vitamins (Gill *et al.*, 2006). Intestinal bacteria can produce vitamin K and several B group vitamins, such as biotin, cobalamin, folates, nicotinic acid, pantothenic acid, pyridoxine, riboflavin and thiamine. Unlike dietary vitamins, which are mainly absorbed in the proximal part of the small intestine, the uptake of microbial vitamins predominantly occurs in the colon through colonocytes, which are able to absorb biotin, thiamin, folates, riboflavin, pantothenic acid and menaquinones (Said and Mohammed, 2006; LeBlanc *et al.*, 2013). Vitamins from intestinal bacteria can contribute to the systemic vitamin levels and especially to the homeostasis of the vitamins in the localised epithelial cells (Ichihashi *et al.*, 1992; Said and Mohammed, 2006).

The colonic microbiota produces folate in the monoglutamylated form, which is absorbed at the highest rate (Zhao, Matherly and Goldman, 2009; Kim *et al.*,

2004; Aufreiter *et al.*, 2009). Several studies assessed the contribution of intestinal bacteria to the folate intake of animal host and demonstrated that intestinally synthesised folate is absorbed by the host (Kim *et al.*, 2004; Asrar and O'Connor, 2005; Sepehr *et al.*, 2003; Zimmerman, 1990; Pompei *et al.*, 2007b). The rate of absorption in the colon is lower than that in the small intestine, but the transit time in the distal portion of the gastrointestinal tract is longer than in the small intestine. Furthermore, the supply of folates by the colonic microbiota is expected to be constant and continuous, whereas their availability in the upper tract is discontinuous and mostly affected by food intake. Microbial production of folates in the colon provides the host with the reduced form of the vitamin, encompassing the pitfall synthetic provitamin FA, which requires activation by DHFR.

Bifidobacteria and LABs are natural colonisers of the gut, strictly gaining energy through the fermentation of carbohydrates. Commensal bifidobacteria and LABs exert a number of beneficial health effects through a variety of different mechanisms and are increasingly being used in functional foods and pharmaceutical products. As a consequence, they are generally regarded as probiotics, namely 'live microorganisms which when administered in adequate amounts confer a health benefit on the host' (FAO and WHO, 2006). The use of probiotics originated decades ago and is continuing to increase, contextually with the scientific understanding of the role of microbiota and of these specific microbial groups in health promotion.

The function of the probiotic bacteria comprises the reduction of potential pathogenic bacteria and/or harmful metabolites in the intestine, the normalisation of gastrointestinal functions and the production of bioactive or otherwise regulatory metabolites, including vitamins (Rossi and Amaretti, 2010; Williams, 2010). In this perspective, the use of folate-producing probiotic strains may represent a natural and efficient alternative to food fortification with chemically synthetised folate or provision of natural folates at colonic level.

5.8 Folate Production by Lactic acid Bacteria

LABs are a group of low G+C Gram-positive bacteria, classified within the phylum of the *Firmicutes* and the *Clostridium–Bacillus* subdivision. They are included in the order of *Lactobacillales,* comprising the families of *Enterococcaceae, Lactobacillaceae* and *Streptococcaceae.* They include a few hundred recognised species that present a wide phylogenetic, ecological and functional diversity. Common traits of LABs include being invariably anaerobic/microaerophilic, aciduric/acidophilic non-sporulating bacteria. They are saccharolytic and gain energy through the lactic fermentation of carbohydrates. Particularly, they are classified as obligate homofermentative (giving mainly lactic acid), obligate heterofermentative (giving mainly lactic acid, acetic acid and CO_2) or facultative heterofermentative (Makarova and Koonin, 2007).

LABs colonise different habitats: plants and plant-derived matrices, fermented foods (such as dairy products, meats and fermented dough, milk and vegetables) and diverse niches within the body of humans and animals. LABs of food origin, belonging to *Lactobacillus*, *Lactococcus* and *Streptococcus thermophilus*, are exploited in the production of fermented products and are increasingly utilised for the production of healthy functional foods. Furthermore, several species of *Lactobacillus* occur as human gut commensals and exert positive effects on the host's health status. Among them, the strains exhibiting specific beneficial properties relevant to human health are attracting considerable attention as probiotics. The species of *Lactobacillus* with the greatest relevance for the manufacturing of probiotics and functional foods are *L. acidophilus*, *L. casei*, *L. paracasei*, *L. plantarum*, *L. reuteri* and *L. salivarius* (de Vrese and Schrezenmeir, 2008).

Folate-fortified dairy products with improved nutritional value have potentially relevant applications. Similarly, strains from the human gastrointestinal tract could find application as folate-producing probiotics. With this perspective, lactobacilli and other LABs have been investigated as possible folate producers (LeBlanc *et al.*, 2007; Rao *et al.*, 1984; Sybesma *et al.*, 2003b; Santos *et al.*, 2008).

Lactococcus lactis and *S. thermophilus*, utilised in the manufacture of fermented dairy products, possess all the genes for *de novo* folate (Table 5.1) synthesis; thus, they accumulate the vitamin within the cells and excrete it into the medium. For these microorganisms, the extent of vitamin production, the partitioning between accumulation and excretion and the form in which the vitamin occurs (e.g. the number of glutamate residues and the association to formyl or methenyl groups) depend on the strain and, in some cases, are influenced by culture conditions, such as pH, growth rate and pABA concentration.

On the other hand, *Lactobacillus* species are generally unable to produce folate *de novo* and need vitamin supplement to grow. In particular, lactobacilli are generally unable to synthetise pABA due to the lack of the enzymes for chorismate conversion into pABA (Table 5.1). *L. acidophilus*, *L. brevis*, *L. casei*, *L. gasseri*, *L. johnsonii*, *L. rhamnosus* and *L. salivarius* also lack the genes of DHPP biosynthetic pathway and the gene encoding dihydropteroate synthase (EC 2.5.1.15), which condensates DHPP and pABA. However, these species possess the genes for DHP transformation into DHF, THF and THF-polyglutamate. Thus, they are expected to be auxotrophic for folates or DHP and incapable of folate production even in presence of pABA supplementation.

L. plantarum, *L. sakei*, *L. delbrueckii*, *L. reuteri*, *L. helveticus* and *L. fermentum* harbour a folate biosynthetic cluster that includes the gene encoding dihydropteroate synthase and all the genes for the biosynthesis of DHPP, with the exception of alkaline phosphatase (EC 3.1.3.1). Most lactobacilli harbour a number of genes encoding putative Nudix phosphohydrolases, but only *L. sakei*, *L. helveticus* and *L. delbrueckii* have, within the *fol* cluster, a homologue of the *L. lactis* gene encoding the Nudix enzyme performing dephosphorylation of dihydroneopterin triphosphate into the monophosphate (EC 3.6.1.-) (Klaus *et al.*, 2005). On the other hand, the *fol* cluster of *L. fermentum L. plantarum* and *L. reuteri* contains the gene of a putative non-Nudix purine nucleoside triphosphate pyrophosphatase, likely responsible for the hydrolysis of dihydroneopterin triphosphate. Thus, *L. plantarum*, *L. sakei*, and *L. delbrueckii*, *L. reuteri*, *L. helveticus* and *L. fermentum* can synthesise DHPP but are expected to necessitate pABA to produce folates. *L. reuteri* deserves a specific discussion. In fact, the genome sequences of 10 strains isolated from the intestine of three different mammals, including human, were subjected to multilocus sequence analysis and classified into six clades, which reflected adaptation to different hosts (Oh *et al.*, 2010). Interestingly, the two clades of human origin behaved differently with respect to folate production, even if the organisation of the *fol* genes is similar in both clades and comparable with the described *Lactobacillus* species. One of the human clades produced folate in the presence of pABA while the other did not (Spinler *et al.*, 2014).

Based on these evidences, the utilisation of *Lactobacillus* species as folate-producing probiotics seems to be excluded, since they are generally folate auxotrophs or produce the vitamin only in the presence of pABA. Similarly, lactobacilli are expected to reduce the folate levels of the fermented product (Crittenden, Martinez and Playne, 2003; Reddy, Shahani and Kulkarni, 1976; Kneifel *et al.*, 1992; Hoppner and Lampi, 1990), and they also seem unsuitable as bacterial starters for folate fortification. However, folate production and utilisation are complementary in mixed cultures of *S. thermophilus* and lactobacilli. Thus, proper selection of starters can result in increased folate levels in yoghurt and fermented milk, even though the folate levels remain relatively low in terms of recommended daily intake (Crittenden, Martinez and Playne, 2003).

Metabolic engineering was successfully utilised to increase folate production by *L. lactis* and to transform *L. gasseri* into a folate-producer cloning and overexpressing *fol* genes from *L. lactis* (Sybesma *et al.*, 2003a,c; Wegkamp *et al.*, 2004). Despite this approach being efficacious in improving the level of folates and in modulating the length of the polyglutamyl tail, at present, engineered microbes are not generally accepted by legislation for the use as probiotics or as starters for folate enrichment of fermented foods.

5.9 Folate Production by Bifidobacteria

Bifidobacterium is a genus of high G+C Gram-positive bacteria within the phylum of *Actinobacteria*. They are saccharolytic obligate anaerobes whose primary habitat is the gastrointestinal tract of animals, being among the first gut colonisers, and represent up to 8% of the total anaerobes in the gastrointestinal tract of adults. Among 42 species recognised so far, the most represented in the gastrointestinal tract of human adults or infants are *Bifidobacterium pseudocatenulatum*, *B. catenulatum*, *B. adolescentis*, *B. longum*, *B. infantis*, *B. breve*, *B. angulatum* and *B. dentium* (Mattarelli *et al.*, 2014; Biavati and Mattarelli, 2006). Bifidobacteria are one of the most important health-promoting groups of the colonic microbiota and are largely used as probiotics. Many commensal strains exert a number of beneficial effects through different mechanisms, such as immunostimulation, anticarcinogenic activity, pathogen growth inhibition, vitamin and amino acid production, reduction of the conversion of primary bile salts to secondary bile salts, bioconversion of a number of dietary compounds into bioactive healthy molecules.

Examination of *Bifidobacterium* spp. genomes shows that complete shikimate pathway is present in strains sequenced so far, which are all expected to produce chorismate (LeBlanc *et al.*, 2013). Even though all the available bifidobacterial genomes harbour a gene encoding the aminodeoxychorismate synthase (EC 2.6.1.85), only *B. adolescentis* and *B. dentium* possess the 4-amino-4-deoxychorismate lyase (EC 4.1.3.38) and should accomplish *de novo* biosynthesis of pABA. Conversely, *B. animalis* subsp. *lactis* lacks the gene encoding dihydropteroate synthase (EC 2.5.1.15) and all the genes for the biosynthesis of DHPP; thus, it should behave as auxotrophic for folates or DHP and remain incapable of folate production even in the presence of pABA supplementation (Table 5.1).

With the exception of *B. animalis* subsp. *lactis* and *B. animalis* subsp. *animalis*, bifidobacteria harbour a cluster of *fol* genes encoding dihydropteroate synthase (EC 2.5.1.15) and other enzymes for the biosynthesis of DHPP and for condensation to pABA. Unlike lactobacilli, the *fol* cluster of bifidobacteria encodes for neither pyrophosphohydrolase (EC 3.6.1.-) nor alkaline phosphatase (EC 3.1.3.1). Several putative Nudix pyrophosphatases and alkaline phosphatase are widespread in all the sequenced genomes of bifidobacteria, but it is not possible to argue whether they are involved in folate biosynthesis, because their function has not been investigated so far. As a whole, *B. adolescentis*, *B. dentium* and *B. longum* could accomplish the dephosphorylation of dihydroneopterin through an enzyme that is so far unidentified or an enzyme-independent chemical process and synthesise DHPP. Furthermore, it is conceivable that *B. adolescentis* and *B. dentium* are capable of *de novo* folate production, while *B. longum* needs to be provided with pABA, and *B. animalis* requires folates.

With the aim to screen them for folate production, bifidobacteria were cultured in folate-free, low-folate or complex media in several studies (Lin and Young, 2000; Crittenden, Martinez and Playne, 2003; Pompei *et al.*, 2007a; Strozzi and Mogna, 2008). Many strains were described to produce folate, and information about intracellular and secreted levels vitamins was reported. For instance, Pompei *et al.* (2007a) screened 76 bifidobacteria in a folate-free semisynthetic medium and observed that most of the strains were incapable of growth, while 17 of them belonging to 9 different species (*B. adolescentis*, *B. breve*, *B. pseudocatenulatum*, *B. animalis*, *B. bifidum*, *B. catenulatum*, *B. dentium*, *B. infantis* and *B. longum*) synthetised the vitamin in the folate-free medium. The level of folate production was not related to the extent of the growth and was not a characteristic of the species but seemed to be a trait of the single strains. Folate was mostly extracellular, with intracellular accumulation never exceeding the 38% of total vitamin. Strains belonging to *B. adolescentis* and *B. pseudocatenulatum* yielded the highest folate production, with the extracellular concentration reaching up to 82 ng/ml. Deguchi, Morishita and Mutai (1985) screened 24 strains of *B. bifidum*, *B. infantis*, *B. breve*, *B. longum* and *B. adolescentis* in a low-folate semisynthetic medium, obtaining significant differences in vitamin production among these species, since all the strains of *B. bifidum* and *B. infantis* were classified as high folate producers, while the strains of *B. breve, B. longum* and *B. adolescentis* gave significantly lower yield. In a complex medium such as reconstituted skim milk, the highest folate production was achieved by strains of *B. breve* and *B. infantis* or *B. longum* (Lin and Young, 2000).

The discrepancies among these studies, which failed to identify the same species as high producers, may be due to both strain-to-strain differences and diverse experimental designs (Pompei *et al.*, 2007a; Lin and Young, 2000; Deguchi, Morishita and Mutai, 1985). For instance, when folate-free medium was used, the cultures were passaged seven times in this medium to exhaust the vitamin before evaluating growth and net folate production (Pompei *et al.*, 2007a). It is also conceivable that several strains were taxonomically classified when the number of species was lower and need reclassification based on molecular phylogenetic analyses. Furthermore, it was shown that folate content in bifidobacteria is very dynamic and is particularly dependent on medium composition. In fact, intracellular folate reached up to 4000 μg/100 g of biomass if bifidobacteria were cultured in folate-containing complex medium but could increase above 9000 μg/100 g of biomass if autotrophs were cultured in folate-free medium (D'Aimmo *et al.*, 2012).

In the perspective to develop a probiotic based on folate-producing strains, it is important that vitamin biosynthesis is not affected by the environmental conditions occurring in the colon, and particularly by the level of exogenous vitamin, whose concentration range can be rather large depending by the dietary intake, absorption and excretion from the urine, skin and bile (Birn, 2006). Among the aforementioned strains, two *B. adolescentis* and one *B. pseudocatenulatum* were selected since they did not exhibit any feedback regulation of folate production, due to the presence of exogenous vitamin in the range between 0 and 50 ng/ml^{-1} (Pompei *et al.*, 2007a). Furthermore, neither pH nor the carbon source affected folate biosynthesis. These selected strains were administered to Wistar rats with induced folate deficiency, in order to investigate their effectiveness to improve folate status (Pompei *et al.*, 2007b). Lyophilised bifidobacteria were used alone or were added to bifidogenic fructans in a synbiotic formulation. At the end of the treatment, mean serum folate concentration in rats consuming both the probiotic and the synbiotic diet was significantly higher than in control. However, the simultaneous consumption of probiotics and prebiotic carbohydrates further increased the level of the probiotic strains in the intestine and resulted in the highest level of serum folate, confirming that the availability of a preferred indigestible carbon source is advantageous to the growth and the metabolic activity of probiotic bacteria.

These same strains of *B. adolescentis* and *B. pseudocatenulatum*, when given to 23 healthy volunteers in a pilot human study, significantly increased folate concentration in the faeces of the subjects (Strozzi and Mogna, 2008). These results corroborate the assumption that the increase of folate levels was markedly due to the effective growth of the folate-producing bifidobacteria. In this case, the levels of commensal bifidobacteria in the large intestine correlated with the vitamin availability, suggesting that bifidobacteria are capable of producing folate in the gut and that the folate synthesised in the large intestine can be absorbed and utilised by the host. In agreement with these results, in folate-depleted rats, the administration of diets containing bifidogenic ingredients (e.g. human milk solids or prebiotics) increased the folate concentration in the ceacum, colon, plasma and colonic tissue (Krause, Forsberg and O'Connor, 1996; Thoma, Green and Ferguson, 2003). These results support evidence that folate-producing probiotic strains may represent an endogenous source of vitamin preventing its deficiency in the colon. Localised folate production in the large intestine may provide the proliferating enterocytes with this essential vitamin with potential effects in reducing colonic carcinogenesis (Choi and Mason, 2002). Therefore, the trophic effects on colonocytes of folate-producing strains deserve to be evaluated. Moreover, the supply of folate by bifidobacteria may also contribute to lower hyperhomocysteinaemia, since the administration of folate-producing *B. longum* exerted beneficial effects on the homocysteine levels of hemodialysis patients (Taki, Takayama and Niwa, 2005).

Besides their exploitation as an endogenous source of vitamin, folate-producing bifidobacteria may also be used to fortify fermented dairy products, as milk is a poor source of folate. This concept was tested in a particular study, where seven strains of *Bifidobacterium* were evaluated for their capacity to enhance the folate concentration of reconstituted skim milk, resulting in a strain of *B. breve* being selected as the most promising (Crittenden, Martinez and Playne, 2003). Moreover, mixed-culture fermentations of reconstituted skim milk were successfully carried out using folate-producing strains of *Bifidobacterium* in conjunction with strains of *S. thermophilus* and/or *Lactobacillus delbrueckii* subsp. *bulgaricus* from conventional yoghurt, demonstrating that it is possible to increase folate levels in fermented milk products through appropriate selection of bacterial strains.

5.9.1
Conclusions

Various green plants and some microorganisms produce folate. However, vitamin supplements and food fortification utilise chemically synthetised FA or reduced derivatives. Numerous studies have shown that LAB such as *L. lactis* and *S. thermophilus* can produce folate *de novo*, but biotechnological production of folate has never been competitive with chemical synthesis, even though metabolic engineering of LAB may be promising.

Appropriate combination of selected LAB strains can be successfully used to enrich fermented foods in this vitamin, as demonstrated for dairy products. The use of folate-producing *Bifidobacterium* strains can be regarded as a specific use of probiotics. Human and animal trials proved that the administration of folate-producing bifidobacteria positively affected the plasmatic folate level, indicating that the vitamin is produced *in vivo* by the probiotic strains and absorbed.

Table 5.1 Genes and enzymes for the biosynthesis of DHPP, THF-polyglutamate, chorismate and pABA predicted from the sequenced genomes of genus *Bifidobacterium* (BLAST, www.gencme.jp/kegg/pathway.html).

Strain and origin[a)]				Chorismate			pABA			DHPP					THF-polyglu		
		aroF 2.5.1.54	*arcB* 4.2.3.4	*aroD* 4.2.1.10	*aroE* 1.1.1.25	*aroK* 2.7.1.71	*aroA* 2.5.1.19	*aroC* 4.2.3.5	*pabA* 2.6.1.85	*pabB* 4.1.3.38	*folE* 3.5.4.16	3.1.3.1	*folQ* 3.6.1.-	*folB* 4.1.2.25	*folK* 2.7.6.3	*folP* 2.5.1.15	*folC* 6.3.2.12/17	*dfrA* 1.5.1.3
Lactococcus lactis sbp. *cremoris* MG1363	f	○	▲	○	▲	▲	▲	▲	◆	◆	■	○	■	■	■	■	■	■
Streptococcus thermophilus CNRZ1066	f	○	▲	▲	▲	▲	▲	▲	◆	◆	■	○	○	■	■	■	■	○
Lactobacillus acidophilus NCFM	f																○	○
Lactobacillus brevis ATCC 367	v																○	○
Lactobacillus casei ATCC 334	f				○												○	○
Lactobacillus delbrueckii ATCC 11842	f										■		■	■	■	■	■	○
Lactobacillus fermentum IFO 3956	v					▼	▼	▼			■		■	■	■	■	■	○
Lactobacillus gasseri ATCC 33323	h																○	○

(*continued overleaf*)

Table 5.1 (*Continued*)

Strain and origin[a)]		Chorismate					pABA				DHPP					THF-polyglu		
		aroF 2.5.1.54	*aroB* 4.2.3.4	*aroD* 4.2.1.10	*aroE* 1.1.1.25	*aroK* 2.7.1.71	*aroA* 2.5.1.19	*aroC* 4.2.3.5	*pabA* 2.6.1.85	*pabB* 4.1.3.38	*folE* 3.5.4.16	3.1.3.1	*folQ* 3.6.1.-	*folB* 4.1.2.25	*folK* 2.7.6.3	*folP* 2.5.1.15	*folC* 6.3.2.12/17	*dfrA* 1.5.1.3
Lactobacillus helveticus DPC 4571	f										■		■	■	■	■	■	○
Lactobacillus johnsonii NCC 533	h																○	○
Lactobacillus plantarum WCFS 1	h	▲	▲	○	▲	▼	▼	▼			■		■	■	■	■	■	○
Lactobacillus reuteri DSMZ 20016	h										■		■	■	■	■	■	○
Lactobacillus rhamnosus GG	h				○												○	○
Lactobacillus sakei 23 K	f	▲	▲	▲	▲	▲					■		■	■	■	■	■	○
Lactobacillus salivarius UCC 118	h			▲	▲							○					○	○
B. adolescentis ATCC 15703	h	○	▲	▲	○	▲	○	▲	○	○	■	○	○	■	■	■	○	○

B. animalis subsp. *animalis* ATCC 25527	a	○	▲	▲	○	▲	○	▲	○			○	○				○	○
B. animalis subsp. *lactis* AD011	h	○	▲	▲	○	▲	○	▲	○			○	○				○	○
B. bifidum PRL2010	h	○	▲	▲	○	▲	○	▲	○		■	○	○	■	■	■	○	○
B. breve UCC2003	h	○	▲	▲	○	▲	○	▲	○		■	○	○	■	■	■	○	○
B. dentium Bd1	h	○	▲	▲	○	▲	○	▲	○	○	■	○	○	■	■	■	○	○
B. longum subsp. *infantis* ATCC 15697	h	○	▲	▲	○	▲	○	▲	○		■	○	○	■	■	■	○	○
B. longum subsp. *longum* JDM301	p	○	▲	▲	○	▲	○	▲	○		■	○	○	■	■	■	○	○

a) a, animal gastrointestinal tract; h, human gastrointestinal tract; p, probiotic; f, fermented food (meat or dairy products) and v, plant.
pABA, para-aminobenzoic acid; DHPP, 6-hydroxymethyl-7,8-dihydropterin pyrophosphate and THF, tetrahydrofolate. Within each strain, ▲, ▼, ◆ and ■ indicate genes organised within the same gene cluster; ○ indicates genes located elsewhere putatively encoding the enzyme.
The table is adapted from Rossi, Amaretti and Raimondi (2011) and Leblanc *et al.* (2013).

References

Angier, R.B., Stokstad, E.L.R., Mowat, J.H., Hutchings, B.L., Boothe, J.H., Waller, C.W., Semb, J., Subbarow, Y., Cosulich, D.B., Fahrenbach, M.J., Hultquist, M.E., Kuh, E., Northey, E.H., Seeger, D.R., Sickels, J.P., and Smith, J.M. (1948) Synthesis of pteroylglutamic acid. III. *J. Am. Chem. Soc.*, **70**, 25–26.

Ansari, R., Mahta, A., Mallack, E., and Luo, J.J. (2014) Hyperhomocysteinemia and neurologic disorders: a review. *J. Clin. Neurol.*, **10**, 281–288.

Asrar, F.M. and O'Connor, D.L. (2005) Bacterially synthesized folate and supplemental folic acid are absorbed across the large intestine of piglets. *J. Nutr. Biochem.*, **16**, 587–593.

Aufreiter, S., Gregory, J.F. III, Pfeiffer, C.M., Fazili, Z., Kim, Y.I., Marcon, N., Kamalaporn, P., Pencharz, P.B., and O'Connor, D.L. (2009) Folate is absorbed across the colon of adults: evidence from cecal infusion of (13)C-labeled [6S]-5-formyltetrahydrofolic acid. *Am. J. Clin. Nutr.*, **90**, 116–123.

Bailey, S.W. and Ayling, J.E. (2009) The extremely slow and variable activity of dihydrofolate reductase in human liver and its implications for high folic acid intake. *Proc. Natl. Acad. Sci. U.S.A.*, **106**, 15424–15429.

Bailey LB, Caudill MA. (2012) Folate. In (eds Erdman JW, Macdonald IA, Zeisel SH) *Present Knowledge in Nutrition*, 10th edn, Wiley-Blackwell, Oxford.

Bermingham, A. and Derrick, J.P. (2002) The folic acid biosynthesis pathway in bacteria: evaluation of potential for antibacterial drug discovery. *Bioessays*, **24**, 637–648.

Biavati B, Mattarelli P. (2006) In (eds Dworkin M, Falkow S, Rosenberg E, Schleifer KH, Stackebrandt E) *The Prokaryotes*, Chapter 3, vol. **3**, 3rd edn, pp. 322–382, Springer, New York.

Birn, H. (2006) The kidney in vitamin B12 and folate homeostasis: characterization of receptors for tubular uptake of vitamins and carrier proteins. *Am. J. Physiol. Renal Physiol.*, **291**, F22–F36.

Blom, H.J., Shaw, G.M., den Heijer, M., and Finnell, R.H. (2006) Neural tube defects and folate: case far from closed. *Nat. Rev. Neurosci.*, **7**, 724–731.

Blom, H.J. and Smulders, Y. (2011) Overview of homocysteine and folate metabolism. With special references to cardiovascular disease and neural tube defects. *J. Inherit. Metab. Dis.*, **34**, 75–81.

Botzem, J., Jaedicke, H., John, M. and Paust, J. (2002) Process for preparing folic acid. US Patent 6348593 B1.

Chandler, C.J., Harrison, D.A., Buffington, C.A., Santiago, N.A., and Halsted, C.H. (1991) Functional specificity of jejunal brush-border pteroylpolyglutamate hydrolase in pig. *Am. J. Physiol.*, **260**, G865–G872.

Choi, S.W. and Mason, J.B. (2002) Folate status: effects on pathways of colorectal carcinogenesis. *J. Nutr.*, **132**, 2413S–2418S.

Crittenden, R.G., Martinez, N.R., and Playne, M.J. (2003) Synthesis and utilisation of folate by yoghurt starter cultures and probiotic bacteria. *Int. J. Food Microbiol.*, **80**, 217–222.

D'Aimmo, M.R., Mattarelli, P., Biavati, B., Carlsson, N.G., and Andlid, T. (2012) The potential of bifidobacteria as a source of natural folate. *J. Appl. Microbiol.*, **112**, 975–984.

De Crécy-Lagard, V., El Yacoubi, B., de la Garza, R.D., Noiriel, A., and Hanson, A.D. (2007) Comparative genomics of bacterial and plant folate synthesis and salvage: predictions and validations. *BMC Genomics*, **8**, 245.

Deguchi, Y., Morishita, T., and Mutai, M. (1985) Comparative studies on synthesis of water-soluble vitamins among human species of bifidobacteria. *Agric. Biol. Chem.*, **49**, 13–19.

Duthie, S.J., Narayanan, S., Brand, G.M., Pirie, L., and Grant, G. (2002) Impact of folate deficiency on DNA stability. *J. Nutr.*, **132**, 2444S–2449S.

Dwyer, J.T., Woteki, C., Bailey, R., Britten, P., Carriquiry, A., Gaine, P.C., Miller, D., Moshfegh, A., Murphy, M.M., and Smith Edge, M. (2014) Fortification: new findings and implications. *Nutr. Rev.*, **72**, 127–141.

Eckburg, P.B., Bik, E.M., Bernstein, C.N., Purdom, E., Dethlefsen, L., Sargent, M.,

Gill, S.R., Nelson, K.E., and Relman, D.A. (2005) Diversity of the human intestinal microbial flora. *Science*, **308**, 1635–1638.

EFSA (2009) ESCO Report Prepared by the EFSA Scientific Cooperation Working Group on Analysis of Risks and Benefits of Fortification of Food with Folic Acid, http://www.efsa.europa.eu/it/supporting/pub/3e.

EFSA Panel on Dietetic Products, Nutrition and Allergies (2014) Scientific opinion on dietary reference values for folate. *EFSA J.*, **12**, 3893.

EFSA Panel on Food Additives and Nutrient Sources Added to Food (2013) Scientific Opinion on (6S)-5-methyltetrahydrofolic acid, glucosamine salt as a source of folate added for nutritional purposes to food supplements. *EFSA J.*, **11**, 3358.

Eguchi, T., Oshiro, T., Kuge, Y., Mochida, K. and Uwajima, T. (1990) Process for producing L(-)-tetrahydrofolic acid. US Patent 4929551.

FAO and WHO (2006) Probiotics in Food. Health and Nutritional Properties and Guidelines for Evaluation. FAO Food and Nutrition Paper 85.

Felder, E., Ripa, G. and Distaso, C. (1998) Process for the preparation and separation of diastereoisomeric salts of folinic acid. US Patent 5710271.

Gill, S.R., Pop, M., Deboy, R.T., Eckburg, P.B., Turnbaugh, P.J., Samuel, B.S., Gordon, J.I., Relman, D.A., Fraser-Liggett, C.M., and Nelson, K.E. (2006) Metagenomic analysis of the human distal gut microbiome. *Science*, **312**, 1355–1359.

Gregory, J.F. III, (2001) Case study: folate bioavailability. *J. Nutr.*, **131**, 1376S–1382S.

Hanson, A.D. and Gregory, J.F. III, (2002) Synthesis and turnover of folates in plants. *Curr. Opin. Plant Biol.*, **5**, 244–249.

Hawkes, J.G. and Villota, R. (1989) Folates in foods: reactivity, stability during processing, and nutritional implications. *Crit. Rev. Food Sci. Nutr.*, **28**, 439–538.

Hoppner, K. and Lampi, B. (1990) Total folate, pantothenic acid and biotin content of yogurt products. *Can. Inst. Sci. Technol. J.*, **23**, 223–225.

Ichihashi, T., Takagishi, Y., Uchida, K., and Yamada, H. (1992) Colonic absorption of menaquinone-4 and menaquinone-9 in rats. *J. Nutr.*, **122**, 506–512.

Imbard, A., Benoist, J.F., and Blom, H.J. (2013) Neural tube defects, folic acid and methylation. *Int. J. Environ. Res. Public Health*, **10**, 4352–4389.

Jägerstad, M. (2012) Folic acid fortification prevents neural tube defects and may also reduce cancer risks. *Acta Paediatr.*, **101**, 1007–1012.

Jägerstad, M. and Jastrebova, J. (2013) Occurrence, stability, and determination of formyl folates in foods. *J. Agric. Food. Chem.*, **61**, 9758–9768.

Jones, G.S., Saint Laurent, J.P., Goodrich, S.A. and Maguire, G. (2012) Synthesis of (6S)-5,6,7,8-tetrahydrofolic acid. US Patent 20120315679 A1.

Kawanishi, S. (1960) Process for the preparation of pteroylglutamic acid. US Patent 2956057.

Kim, T.H., Yang, J., Darling, P.B., and O'Connor, D.L. (2004) A large pool of available folate exists in the large intestine of human infants and piglets. *J. Nutr.*, **134**, 1389–1394.

Klaus, S.M., Wegkamp, A., Sybesma, W., Hugenholtz, J., Gregory, J.F. III,, and Hanson, A.D. (2005) A Nudix enzyme removes pyrophosphate from dihydroneopterin triphosphate in the folate synthesis pathway of bacteria and plants. *J. Biol. Chem.*, **280**, 5274–5280.

Kneifel, W., Kaufmann, M., Fleischer, A., and Ulberth, F. (1992) Screening of commercially available mesophilic dairy starter cultures: biochemical, sensory, and microbiological properties. *J. Dairy Sci.*, **75**, 3158–3166.

Krause, L.J., Forsberg, C.W., and O'Connor, D.L. (1996) Feeding human milk to rats increases *Bifidobacterium* in the cecum and colon which correlates with enhanced folate status. *J. Nutr.*, **126**, 1505–1511.

LeBlanc JG, de Giori GS, Smid EJ, Hugeholtz J, Sesma F. (2007) In (ed Mendez-Vilas A) *Communicating Current Research and Educational Topics and Trends in Applied Microbiology*, vol. **1**, pp. 329-339, Formatex, Badajoz.

LeBlanc, J.G., Milani, C., de Giori, G.S., Sesma, F., van Sinderen, D., and Ventura, M. (2013) Bacteria as vitamin suppliers to their host: a gut microbiota perspective. *Curr. Opin. Biotechnol.*, **24**, 160–168.

Levin, I., Giladi, M., Altman-Price, N., Ortenberg, R., and Mevarech, M. (2004) An alternative pathway for reduced folate biosynthesis in bacteria and halophilic archaea. *Mol. Microbiol.*, **54**, 1307–1318.

Lin, M.Y. and Young, C.M. (2000) Folate levels in cultures of lactic acid bacteria. *Int. Dairy J.*, **10**, 409–413.

Litynski, P., Loehrer, F., Linder, L., Todesco, L., and Fowler, B. (2002) Effect of low doses of 5-methyltetrahydrofolate and folic acid on plasma homocysteine in healthy subjects with or without the 677C → T polymorphism of methylenetetrahydrofolate reductase. *Eur. J. Clin. Invest.*, **32**, 662–668.

Liu, Y., Tian, T., Zhang, H., Gao, L., and Zhou, X. (2014) The effect of homocysteine-lowering therapy with folic acid on flow-mediated vasodilation in patients with coronary artery disease: a meta-analysis of randomized controlled trials. *Atherosclerosis*, **235**, 31–35.

Liu, J. and Ward, R.L. (2010) Folate and one-carbon metabolism and its impact on aberrant DNA methylation in cancer. *Adv. Genet.*, **71**, 80–121.

Louis, P., Hold, G.L., and Flint, H.J. (2014) The gut microbiota, bacterial metabolites and colorectal cancer. *Nat. Rev. Microbiol.*, **12**, 661–672.

Makarova, K.S. and Koonin, E.V. (2007) Evolutionary genomics of lactic acid bacteria. *J. Bacteriol.*, **189**, 1199–1208.

Mattarelli, P., Holzapfel, W., Franz, C.M., Endo, A., Felis, G.E., Hammes, W., Pot, B., Dicks, L., and Dellaglio, F. (2014) Recommended minimal standards for description of new taxa of the genera *Bifidobacterium*, *Lactobacillus* and related genera. *Int. J. Syst. Evol. Microbiol.*, **64**, 1434–1451.

Miyata, R. and Yonehara, T. (1999) Culturing *Candida famata*, *Candida guilliermondii*, *Yarrowia lipolytica*, *Pichia glucozyma*, or *Saccharomyces cerevisiae* in medium with glucose, fructose, sucrose, or maltose; recovering at least 0.3 milligrams folic acid per liter media. US Patent 5968788 A.

Nazki, F.H., Sameer, A.S., and Ganaie, B.A. (2014) Folate: metabolism, genes, polymorphisms and the associated diseases. *Gene*, **533**, 11–20.

Oh, P.L., Benson, A.K., Peterson, D.A., Patil, P.B., Moriyama, E.N., Roos, S., and Walter, J. (2010) Diversification of the gut symbiont *Lactobacillus reuteri* as a result of host-driven evolution. *ISME J.*, **4**, 377–387.

O'Keefe, S.J. (2008) Nutrition and colonic health: the critical role of the microbiota. *Curr. Opin. Gastroenterol.*, **24**, 51–58.

Pompei, A., Cordisco, L., Amaretti, A., Zanoni, S., Matteuzzi, D., and Rossi, M. (2007a) Folate production by bifidobacteria as a potential probiotic property. *Appl. Environ. Microbiol.*, **73**, 179–185.

Pompei, A., Cordisco, L., Amaretti, A., Zanoni, S., Raimondi, S., Matteuzzi, D., and Rossi, M. (2007b) Administration of folate-producing bifidobacteria enhances folate status in Wistar rats. *J. Nutr.*, **137**, 2742–2746.

Qiu, A., Jansen, M., Sakaris, A., Min, S.H., Chattopadhyay, S., Tsai, E., Sandoval, C., Zhao, R., Akabas, M.H., and Goldman, I.D. (2006) Identification of an intestinal folate transporter and the molecular basis for hereditary folate malabsorption. *Cell*, **127**, 917–928.

Rao, D.R., Reddy, A.V., Pulusani, S.R., and Cornwell, P.E. (1984) Biosynthesis and utilization of folic acid and vitamin B12 by lactic cultures in skim milk. *J. Dairy Sci.*, **67**, 1169–1174.

Reddy, K.P., Shahani, K.M., and Kulkarni, S.M. (1976) B-complex vitamins in cultured and acidified yogurt. *J. Dairy Sci.*, **59**, 191–195.

Rees, L., Valente, E., Suckling, C.J., and Wood, H.C.S. (1986) Asymmetric reduction of dihydrofolate using dihydrofolate reductase and chiral boron-containing compounds. *Tetrahedron*, **42**, 117–136.

Rossi M, Amaretti A. (2010) In (eds van Synderen D, Mayo B) *Bifidobacteria: Genomics and Molecular Aspects*, Chapter 6, pp. 97–123, Caister Academic Press, Norfolk.

Rossi, M., Amaretti, A., and Raimondi, S. (2011) Folate production by probiotic bacteria. *Nutrients*, **3**, 118–134.

Russell, W.R., Hoyles, L., Flint, H.J., and Dumas, M.E. (2013) Colonic bacterial metabolites and human health. *Curr. Opin. Microbiol.*, **16**, 246–254.

Said, H.M. and Mohammed, Z.M. (2006) Intestinal absorption of water-soluble vitamins: an update. *Curr. Opin. Gastroenterol.*, **22**, 140–146.

Santos, F., Wegkamp, A., de Vos, W.M., Smid, E.J., and Hugenholtzm, J. (2008) High-Level folate production in fermented foods by the B12 producer *Lactobacillus reuteri* JCM1112. *Appl. Environ. Microbiol.*, **74**, 3291–3294.

Sepehr, E., Peace, W., Storey, K.B., Jee, P., Lampi, B.J., and Brooks, S.P. (2003) Folate derived from cecal bacterial fermentation does not increase liver folate stores in 28-d folate-depleted male Sprague-Dawley rats. *J. Nutr.*, **133**, 1347–1354.

Shelton, R.C., Sloan Manning, J., Barrentine, L.W., and Tipa, E.V. (2013) Assessing effects of l-methylfolate in depression management: results of a real-world patient experience trial. *Prim. Care Companion CNS Disord.*, **15**, pii: PCC.13m01520.

Spinler, J.K., Sontakke, A., Hollister, E.B., Venable, S.F., Oh, P.L., Balderas, M.A., Saulnier, D.M., Mistretta, T.A., Devaraj, S., Walter, J., Versalovic, J., and Highlander, S.K. (2014) From prediction to function using evolutionary genomics: human-specific ecotypes of *Lactobacillus reuteri* have diverse probiotic functions. *Genome Biol. Evol.*, **6**, 1772–1789.

Strozzi, G.P. and Mogna, L. (2008) Quantification of folic acid in human feces after administration of *Bifidobacterium* probiotic strains. *J. Clin. Gastroenterol.*, **42**, S179–S184.

Sybesma, W., Starrenburg, M., Kleerebezem, M., Mierau, I., de Vos, W.M., and Hugenholtz, J. (2003a) Increased production of folate by metabolic engineering of *Lactococcus lactis*. *Appl. Environ. Microbiol.*, **69**, 3069–3076.

Sybesma, W., Starrenburg, M., Tijsseling, L., Hoefnagel, M.H., and Hugenholtz, J. (2003b) Effects of cultivation conditions on folate production by lactic acid bacteria. *Appl. Environ. Microbiol.*, **69**, 4542–4548.

Sybesma, W., Van Den Born, E., Starrenburg, M., Mierau, I., Kleerebezem, M., De Vos, W.M., and Hugenholtz, J. (2003c) Controlled modulation of folate polyglutamyl tail length by metabolic engineering of *Lactococcus lactis*. *Appl. Environ. Microbiol.*, **69**, 7101–7107.

Taki, K., Takayama, F., and Niwa, T. (2005) Beneficial effects of Bifidobacteria in a gastroresistant seamless capsule on hyperhomocysteinemia in hemodialysis patients. *J. Ren. Nutr.*, **15**, 77–80.

Temple, C., Elliott, R.D., Rose, J.D., and Montgomery, J.A. (1979) Preparation and purification of L-(+/-)-5-formyl-5,6,7,8-tetrahydrofolic acid. *J. Med. Chem.*, **22**, 731–734.

Thoma, C., Green, T.J., and Ferguson, L. (2003) Citrus pectin and oligofructose improve folate status and lower serum total homocysteine in rats. *Int. J. Vitam. Nutr. Res.*, **73**, 403–409.

Tibbetts, A.S. and Appling, D.R. (2010) Compartmentalization of Mammalian folate-mediated one-carbon metabolism. *Annu. Rev. Nutr.*, **21**, 57–81.

USDA (2015) USDA National Nutrient Database for Standard Reference, Release 28, ndb.nal.usda.gov/ndb/ (accessed 24 October 2015).

US Food and Drug Administration (1996a) Food additives permitted for direct addition to food for human consumption; folic acid (folacin). *Fed. Regist.*, **61**, 8798–8807.

US Food and Drug Administration (1996b) Food standards: amendment of standards of identity for enriched grain products to require addition of folic acid. *Fed. Regist.*, **61**, 8781–8797.

Vecchi, G. (1994) Process for the preparation of substituted tetrahydrofolic derivatives in the [6(R,S) (-)] forms and of their active [6(S)(-)] N5 diastereoisomers in form of alkali and alkaline earth metal salts. US Patent 5350850.

de Vrese, M. and Schrezenmeir, J. (2008) Probiotics, prebiotics, and synbiotics. *Adv. Biochem. Eng. Biotechnol.*, **111**, 1–66.

Wegkamp, A., Starrenburg, M., de Vos, W.M., Hugenholtz, J., and Sybesma, W. (2004) Transformation of folate-consuming *Lactobacillus gasseri* into a folate producer. *Appl. Environ. Microbiol.*, **70**, 3146–3148.

Wehrli, C. (1996) Process for the preparation of folic acid. EP Patent 0608693 B1.

White, R.H. (1988) Analysis and characterization of the folates in the non-methanogenic archaebacteria. *J. Bacteriol.*, **170**, 4608–4612.

Williams, N.T. (2010) Probiotics. *Am. J. Health Syst. Pharm.*, **67**, 449–458.

Williams, E.A. (2012) Folate, colorectal cancer and the involvement of DNA methylation. *Proc. Nutr. Soc.*, **71**, 592–597.

Winkels, R.M., Brouwer, I.A., Siebelink, E., Katan, M.B., and Verhoef, P. (2007) Bioavailability of food folates is 80% of that of folic acid. *Am. J. Clin. Nutr.*, **85**, 465–473.

Zhao, R., Matherly, L.H., and Goldman, I.D. (2009) Membrane transporters and folate homeostasis: intestinal absorption and transport into systemic compartments and tissues. *Expert Rev. Mol. Med.*, **11**, e4.

Zimmerman, J. (1990) Folic acid transport in organ-cultured mucosa of human intestine. Evidence for distinct carriers. *Gastroenterology*, **99**, 964–972.

6
Vitamin B_{12} – Physiology, Production and Application

Janice Marie Sych, Christophe Lacroix, and Marc J.A. Stevens

6.1
Introduction and Historical Outline

Cobalamin or B_{12} is an important group of water-soluble compounds belonging to the cobalt corrinoid family. The main active naturally occurring derivatives are adenosylcobalamin ($AdoB_{12}$), methylcobalamin (MeB_{12}) and hydroxycobalamin (OHB_{12}); cyanocobalamin is an important chemically transformed form with high stability. In this review, the term vitamin B_{12} refers to cyanocobalamin (CNB_{12}), as in standard terminology, whereas B_{12} (and cobalamin) alone refers generically to all forms of the vitamin; when required, specific names are given for other B_{12} forms.

Vitamin B_{12} research has a rich history of over 150 years (Castle, 1980; Okuda, 1999; Chanarin, 2000). Its discovery was driven by the need to treat a mysterious and life-threatening disease known today as pernicious *anaemia*, described by Addisson (1855). Over the years, a variety of symptoms were reported, such as sore mouth, numbness in the fingers and hands, large red blood cells, abnormalities of the stomach lining, reduced or absence of acid production in gastric juice and demyelinating spinal cord. In the early 1900s, an effective treatment was found, that is, the ingestion of an extrinsic factor present in the liver (and meat) whose action is dependent on the presence of an intrinsic factor (IF) secreted by the stomach. The search to isolate and purify the extrinsic factor, identified as B_{12}, led to isolation of the reddish, needle-like crystals of CNB_{12} in 1948 (Lester-Smith, 1948; Rickes *et al.*, 1948). Eight years later, Dorothy Hodgkin determined the structure of vitamin B_{12} (CNB_{12}); the structures $AdoB_{12}$ and MeB_{12} were confirmed several years later (Hodgkin *et al.*, 1956; Barker, Weissbach and Smyth, 1958; Lenhert and Hodgkin, 1961). The important role of microorganisms as the primary source of B_{12} also began to emerge in 1948. Using a microbiological assay with *Lactobacillus lactis* B_{12} (Shorb, 1947, 1948), significant quantities of B_{12} were found in different fermented media from cultures of *Streptomyces* sp., *Mycobacterium smegmatis*, *Lactobacillus arabinosus* and *Bacillus subtilis* (Darken, 1953). Another B_{12} source identified was manure and faeces (Spalla *et al.*, 1989), and increasingly, it became clear that bacteria were the original source of B_{12} in nature. The biochemical

Industrial Biotechnology of Vitamins, Biopigments, and Antioxidants, First Edition.
Edited by Erick J. Vandamme and José L. Revuelta.

pathway for B_{12} was first determined in *Pseudomonas denitrificans* in 1993 (reviewed in (Martens *et al.*, 2002)). B_{12} is industrially produced by a biotechnological process which has seen many developments over the years, such as medium optimisation, selection and improvement of strains, as reviewed in this chapter.

Today, the market of B_{12} is consolidated within a limited number of companies, most located in China. Currently, industrial vitamin B_{12} production uses strains of *P. denitrificans* and *Propionibacterium freudenreichii* to a lesser extent. Industrial strains have been selected for their rapid growth and high productivity and are eventually genetically engineered to enhance vitamin production levels. This chapter provides an overview of B_{12} chemistry and physical properties, its occurrence in natural sources and quantification of B_{12}, mainly in foods. A summary of key nutritional aspects of B_{12} in humans is also given, that is, absorption and uptake, metabolic cofactor functions and the current context of B_{12} deficiencies and their detection, with references to detailed reviews. The final part of the chapter presents the current state of B_{12} production by *Propionibacterium* and *Pseudomonas*, including genetic engineering of producing strains, downstream processing (DSP) of fermentation products, B_{12} purification and future directions for industrial B_{12} production.

6.2 Occurrence in Food and Other Natural Sources

Vitamin B_{12} is synthesised only by certain bacteria; therefore, it is found primarily in animal foods, originating from B_{12}-producing microbiota of ruminant animals or ingestion of B_{12}-fortified animal feed (Burgess, Smid and van Sinderen, 2009; Watanabe *et al.*, 2014). In foods of animal origin, the main naturally occurring B_{12} derivatives are $AdoB_{12}$, OHB_{12} and MeB_{12}, whereas CNB_{12} is the chemically derived form used in vitamin supplements and fortified foods. Highest levels of B_{12} are found in the liver and kidney, estimated at >23 μg/100 g and >13 μg/100 g, respectively (Souci, Fachmann and Kraut, 2008). Other good sources are shellfish, fish, muscle meats, fish and eggs (ranging from 9 to 2 μg/100 g, respectively). Despite its lower levels (0.3–0.4 μg/100 g) milk is an important B_{12} source, as it is highly consumed by many populations and also a major component of infant nutrition. Foods fermented by B_{12}-producing microorganisms contain higher B_{12} levels as in Swiss-type cheeses (3 μg/100 g, (Souci, Fachmann and Kraut, 2008)) or fermented fish (Watanabe *et al.*, 2014), prevalent in Japan and other areas of Asia. In nutrient data bases, B_{12} levels may be overestimated (5–30%) as they are based on the microbiological assay which erroneously includes B_{12}-analogues (Ball, 2006). Indeed, lower levels of active B_{12} were recently reported for certain edible shellfish and also meat products analysed by improved methods able to characterise different B_{12} forms (Guggisberg, Risse and Hadorn, 2012; Watanabe *et al.*, 2013).

Increasingly present on the market are vitamin B_{12}-fortified foods such as cereals, yoghurt, juices and other beverages with levels of CNB_{12} typically ranging from 0.2 to 2 μg B_{12}/100 g of product (Zeuschner *et al.*, 2013). These represent an important source of B_{12} for the general population, especially for vegans and the elderly.

Algal supplements are increasingly marketed as alternative B_{12} sources for vegetarians. The majority of these supplements are derived from *Spirulina* sp., which are, in fact, cyanobacteria. These organisms produce B_{12} analogue which lacks biological activity and may even block vitamin B_{12} metabolism (discussed in Herbert, 1988; Stupperich and Nexo, 1991). Recent studies based on physiological tests in rats suggested that edible algae (*Porphyra* sp., Korean purple laver) may be a potential source of active B_{12} (Miyamoto *et al.*, 2009; Watanabe *et al.*, 2014).

One fundamental nutritional effect of the gut microbiota is the production of vitamins for the host. Early studies emphasised the requirement of increased dietary B_{12} for the health of germ-free animals (Wostmann, 1981). However, the vitamin contribution of the human gut microbiota is considered insufficient for humans because it is largely taken up by other members of the microbiome (Degnan *et al.*, 2014).

Consequently, B_{12} is an essential nutrient with recommended levels of B_{12} for adults ranging from 2.0 to 3.0 μg recommended daily allowance (RDA) in most countries. To date, no adverse effects have been reported for high intake levels; therefore, no upper limit exists for B_{12} (Flynn *et al.*, 2003).

6.3 Physiological Role as a Vitamin or Coenzyme

6.3.1 Absorption and Transport

In humans, the uptake of B_{12} from food and its absorption and transport to cells constitute a complex but very efficient pathway, most of which has now been well described. The pathway involves three key proteins, i.e. haptocorrin (HC), intrinsic factor (IF) and transcobalamin (TC), mediated by several complex receptors (Green and Miller, 2014; Quadros, 2010; Said and Nexo, 2012; Nielsen *et al.*, 2012) (Figure 6.2). In the stomach, after the release of B_{12} from food proteins, HC binds not only to vitamin B_{12} but also to B_{12} analogues. In the upper part of the intestine, pancreatic proteases degrade HC, allowing B_{12} to combine with the IF. The IF binds specifically to active forms of B_{12}, providing an important first screening to limit the entry of B_{12} analogues, derived from food sources or degraded B_{12} in the body. At the apical brush border of the distal ileum, the complex IF-B_{12} is recognised and taken up by the receptor cubam, involving cubilin and amnionless protein (He *et al.*, 2005; Said and Nexo, 2012). After degradation of IF within the intestinal cell, released B_{12} is transported by TC to cells throughout the body. The larger portion of B_{12} (approximately 80%) remains bound to HC in circulation, as reserves. Transport protein TC also binds only to active B_{12}. Consequently, both IF and TC have a filtering role, to prevent B_{12} analogues from entry and intracellular access, respectively. The occurrence and role of B_{12} analogues detected in the body, as well as their possible interference with the reactions catalysed by B_{12}-dependent enzymes, remain unclear (Toporok, 1960; Stupperich and Nexo, 1991). However, they lack vitamin activity in humans since they cannot be converted into B_{12} coenzymes.

After release from TC, all active forms of B_{12} are processed in the cytosol into the coenzyme forms, 5′-deoxyadenosylcobalamin (AdoB_{12}) or methylcobalamin (MeB_{12}) and directed to their respective B_{12}-dependent enzymes, *L*-methylmalonyl-CoA mutase (MCM) and methionine synthase (MS), according to physiological needs (Banerjee, Gherasim and Padovani, 2009; Froese and Gravel, 2010; Quadros, 2010). The processing tasks include the removal of the ß-axial ligand; a series of conversion reactions; and stepwise reduction of the cobalt in B_{12} (from trivalent to monovalent), following complex mechanisms which are still under study (Figure 6.2) (Banerjee, Gherasim and Padovani, 2009; Froese and Gravel, 2010; Quadros, 2010).

The active absorption of B_{12} is restricted to about 1–1.5 μg/day, limited to the capacity of the IF-B_{12} receptor (Seetharam and Yammani, 2003). This low absorption is compensated by appreciable quantities of B_{12} stored in the body, mainly in the liver and kidney (about 5 mg), by enterohepatic recirculation (about 1 μg/day), and also limited losses from the body (1 μg/day, (Allen, 2010; Nielsen *et al.*, 2012; Said and Nexo, 2012)). Depending on B_{12} body reserves, a 5–10-year delay may occur between the onset of insufficient dietary intake and clinical symptoms of B_{12} deficiencies, which complicates their diagnostics (Zeuschner *et al.*, 2013).

Uptake of a small portion of B_{12} may also proceed by passive absorption through the gastrointestinal tract and possibly through oral and nasal membranes (Green and Miller, 2014).

6.3.2
Metabolic Functions

In humans, the main metabolic function of B_{12} is its cofactor role for two enzymes, that is, methionine synthase (MS) and MCM, in the cytoplasm and mitochondria, respectively. The reactions catalysed by these enzymes proceed by complex mechanisms (Kräutler, 2005; Kräutler, 2012) and affect the functioning of other key metabolic pathways in the body, mainly the folate and methylation cycles. In the cytoplasm, MeB_{12} acts as cofactor for MS in the folate-dependent methylation of homocysteine which regenerates methionine, allowing the synthesis of *S*-adenosylmethionine (Figure 6.2). This links B_{12} to many essential methylations in the body required for the synthesis of myelin, phospholipids, proteins and neurotransmitters. In the absence of MeB_{12}, these reactions are impaired and a wide range of neurological consequences may develop. In folate metabolism, MeB_{12} is essential for the conversion of 5-methyltetrahydrofolate (5-methyl THF) into tetrahydrofolate (THF), ensuring the availability of folate derivatives required for synthesis of purines and pyrimidines. Therefore, untreated B_{12} deficiency leads to impaired DNA synthesis and the development of megaloblastic anaemia, the major haematological symptom shared by both B_{12} and folate deficiencies.

In the mitochondria, $AdoB_{12}$ acts as cofactor for MCM, which catalyses the reversible isomerisation of *L*-methylmalonyl-CoA to succinyl-CoA (Takahashi-Iniguez *et al.*, 2012). This reaction is an intermediate step in the catabolism of ketogenic amino acids, odd-chain fatty acids and cholesterol. Absence of $AdoB_{12}$ results in an accumulation of methylmalonic acid or may lead to methylmalonic aciduria; both are associated with impaired myelination and cognitive functioning (Moore and Warren, 2012).

Depending on severity and duration, the main symptoms of B_{12} deficiencies are impaired neurological, haematological (Green and Miller, 2014; Allen, 2010) and cognitive functions (Moore and Warren, 2012). If left untreated, these disorders may cause irreversible damage and be life-threatening.

6.3.3
Main Causes and Prevalence of Deficiencies

Impaired absorption (malabsorption) and low dietary intake of B_{12} are the two main causes of B_{12} deficiencies but have different origins (Herrmann and Obeid, 2012; Zeuschner *et al.*, 2013; Green and Miller, 2014). Classical pernicious anaemia (PA) is a severe and irreversible form of malabsorption resulting from impaired synthesis of the IF, following an autoimmune reaction. As the IF-B_{12} absorption pathway is no longer available, treatment of these deficiencies requires regular and life-long intramuscular injections of CNB_{12} or OHB_{12}. Some advantages, such as improved retention, have been reported for OHB_{12} (Green and Miller, 2014).

Especially prevalent among the elderly is malabsorption caused by decreased stomach acidity (chronic or atrophic gastritis) and reduced levels of proteases, both preventing the release of B_{12} from food, as discussed in the literature (Andres *et al.*, 2004; Allen, 2010; Green and Miller, 2014). This type of B_{12} deficiency is typically treated with oral administration of CNB_{12} or OHB_{12} since unbound B_{12} can be absorbed to some degree. Many other factors may also interfere with B_{12} absorption, such as gastrectomy (partial or complete) or gastric bypass surgery; or certain drugs and other medical treatments which cause damage to the stomach or intestine (Green and Miller, 2014). Low intake of B_{12} is another major cause of B_{12} deficiencies, associated with inadequate diets or restriction of animal products. This is especially prevalent in developing countries; for example, in Latin America, an estimated 40% of the population may be affected (Allen, 2009). In developed countries, growing popularity for various forms of vegetarianism may contribute to inadequate B_{12} status. Indeed, the long-term effects of marginal B_{12} deficiencies within the general population are a concern, but especially for pregnant and lactating women, whose offspring may be at risk (Rush, Katre and Yajnik, 2014).

A variety of in-born errors associated with the B_{12} pathway may occur, affecting the absorption and assimilation, plasma transport or intracellular metabolism of B_{12} (Froese and Gravel, 2010; Nielsen *et al.*, 2012; Green and Miller, 2014) and leading to B_{12} deficiencies during infancy and childhood, often with very serious complications. To date, eight different defects of intracellular B_{12} metabolism have been identified and their corresponding genes were also characterised (Froese and Gravel, 2010). Study of these defects has led to major advances in understanding. Somewhat more prevalent are B_{12} deficiencies caused by single-nucleotide polymorphisms, occurring at any age and leading to a wide range of detrimental effects (Green and Miller, 2014).

6.3.4 Diagnosis of Deficiencies

The diagnosis of B_{12} deficiencies is complex and typically involves the assessment of cobalamin levels in serum (or plasma) and of one or more functional biomarkers in serum or urine (Carmel, 2011). Direct measurement of total serum B_{12} is still commonly used to detect B_{12} deficiency (cut-off <200 pg/ml), despite limitations of sensitivity and specificity (Herrmann and Obeid, 2012). Alternatively, serum holotranscobalamin (holoTC) may offer several advantages as a direct indicator, as it reflects quantities of active B_{12} transported by TC from intestinal cells into body tissues (Selhub *et al.*, 2008; Herrmann and Obeid, 2012; Green and Miller, 2014). The use of functional biomarkers such as methylmalonic acid (in urine or serum) or plasma total homocysteine leads to improved sensitivity, allowing the detection of marginal (or sub-clinical) B_{12} deficiencies. These deficiencies are challenging as they may occur without haematological or neurological clinical symptoms and may be highly prevalent in the general population, which raises concern about their possible long-term effects. Due to inherent weaknesses of each existing indicator, no golden standard exists in B_{12} diagnostics (Carmel, 2011). To detect B_{12} malabsorption, the Schilling test was used for several decades and required ingestion and subsequent determinations of radioactively labelled B_{12}, but is no longer used (Allen, 2010). A promising qualitative absorption test was recently introduced, that is, the C-CobaSorb test, based on quantification of holoTC after administration of vitamin B_{12} (Hardlei *et al.*, 2010).

6.4 Chemical and Physical Properties

The B_{12} molecule is large and complex, consisting of a corrin ring with cobalt (I) ion at its centre, which is coordinated with four nitrogen atoms in the corrin ring. A nucleotide moiety consisting of 5,6-dimethylbenzimidazole (DMBI) forms the lower axial α-ligand with cobalt but is also attached covalently to the ring by a side chain (Figure 6.1). The conformation of B_{12} is referred to as *base-on* when DMBI is attached or *base-off/His-on* when DMBI is replaced by a histidine residue, as occurs during binding of B_{12} cofactors with B_{12}-dependent enzymes. Cobalt is covalently bound to the upper axial β-ligand, which can be an adenine, methyl, hydroxyl or a cyano group (Figure 6.1b). The α-ligand is essential for biological

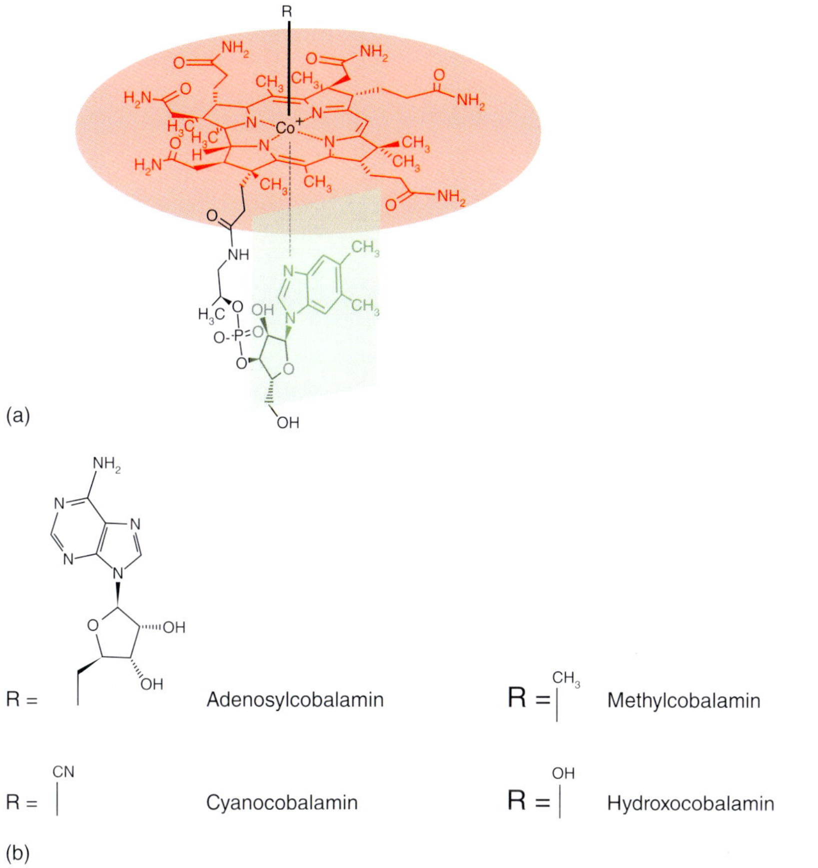

Figure 6.1 Structure of vitamin B_{12}. The DMBI group (green) is coordinated with the cobalt at the lower ligand (α) and linked to the corrin ring (red) by an amino-propanol-ribosyl side chain (base-on conformation). At the upper ligand (β), an R group is linked to the cobalt. (a) 3D structure: the planar corrin ring and the orientation of the ligands, (b) the different upper-or β-ligand occurring in food.

activity, whereas the β-ligand affects the form and function of B_{12} (Roth, Lawrence and Bobik, 1996; Schneider and Stroinski, 1987; Ball, 2006). The planar corrin ring is set at right angles to the nucleotide moiety (Figure 6.1b) and exhibits a degree of flexibility with a characteristic upward folding towards the β-ligand (Antonopoulos and Charalambos, 2013; Green and Miller, 2014).

Several unique characteristics of the cobalt–carbon (Co–C) bond of B_{12} affect the structure, reactivity and function of B_{12} forms and coenzymes. Under physiological conditions, the cobalt ion may occur as Co(III), Co(II) or Co(I), which determines the number and type of axial ligands present (Kräutler, 2005, 2012). The β-ligand is adenosyl or a methyl moiety in B_{12} coenzymes $AdoB_{12}$ and MeB_{12}, respectively, which are required for enzymes MCM and MS, respectively

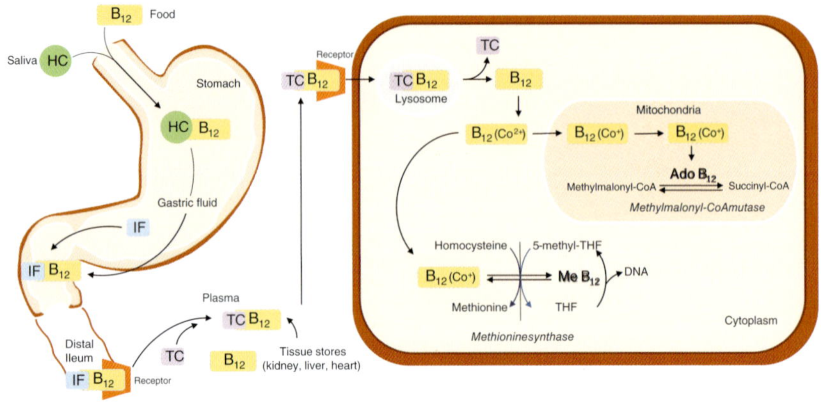

Figure 6.2 Absorption, cellular uptake and cofactor roles in human metabolism of vitamin B_{12}. B_{12}: active B_{12} form; Co: cobalt and oxidation state; HC: haptocorrin; IF: intrinsic factor; $AdoB_{12}$: 5′-deoxyadenosylcobalamin; MeB_{12}: methyl-cobalamin; THF: tetrahydrofolate; TC: transcobalamin (modified from Green and Miller, 2014; Randaccio *et al.*, 2010).

(Figure 6.2). In the reactions catalysed by these enzymes, both coenzymes $AdoB_{12}$ and MeB_{12} undergo cleavage of the Co–C bond but following two different mechanisms, haemolytic (radical) or heterolytic (non-radical) (Kräutler, 2005, 2012). Replacement of the β-ligand with a cyano group by KCN treatment leads to CNB_{12} with greatly improved chemical stability (Randaccio *et al.*, 2010; Antonopoulos and Charalambos, 2013). This B_{12} form has vitamin activity in the body after removal of the cyano group and coenzyme-conversion reactions (Herbert, 1988). When compared with alkyl forms of B_{12} ($AdoB_{12}$, MeB_{12}), the higher stability of CNB_{12} has been associated with higher dissociation energies and shorter distances of the axial β-ligand involving cobalt and carbon (Lodowski *et al.*, 2011) as well as with increased folding angle of its corrin ring (reviewed in Antonopoulos and Charalambos (2013) and Randaccio *et al.* (2010)).

An important structural criterion for vitamin B_{12} activity is maintenance of the intact DMBI at the α-ligand (Grasbeck, 1967). Distance between the nucleotide moiety and corrin ring may influence the selective binding of IF to B_{12} during absorption (Stupperich and Nexo, 1991), a characteristic which affects the uptake of active vitamin B_{12} (Section 6.3.1). Exchange of the DMBI group with adenine results in pseudovitamin B_{12}, for example, produced by the lactic acid bacterium *Lactobacillus reuteri* (Santos *et al.*, 2007). Although the biological relevance of this pseudovitamin in humans is still disputed, its activity in microorganisms has been demonstrated (Molina *et al.*, 2009; Santos *et al.*, 2009).

The chemical properties of CNB_{12} have been the most studied. In aqueous solution, it is stable at room temperature with highest stability at 4.5–5. Loss of vitamin activity may occur due to heat treatments under mild acid or alkaline conditions or due to contact with reducing agents, such as ascorbic acid, nitrous oxide, sulfite and iron (II) salts (Schneider and Stroinski, 1987; Ball, 2006). In pharmaceutical preparations, contact with ascorbic acid can lead to important losses of vitamin B_{12} activity. In a recent study, degradation rates of cyanocobalamin and hydroxocobalamin (OHB_{12}) in the presence of ascorbic acid were highest at approximately pH 5 (Ahmad *et al.*, 2014). Results suggested that CNB_{12} was first degraded to OHB_{12} followed by cleavage of the corrin ring and formation of oxidation products (Ahmad *et al.*, 2014). The alkyl forms MeB_{12} and $AdoB_{12}$ are the most sensitive to photodegradation and are rapidly converted to OHB_{12} which is considered to have good stability (Juzeniene and Nizauskaite, 2013). However, the presence of reactive oxygen species led to increased degradation rates of OHB_{12} exposed to UVA radiation (Juzeniene and Nizauskaite, 2013), suggesting the need for further studies of this important B_{12} form. The photostability of B_{12} in food or biological samples may be increased due to matrix effects, such as binding to proteins TC or HC (Allen, 2012; Juzeniene and Nizauskaite, 2013; Farquharson and Adams, 1976). Furthermore, light penetrates only slightly below the surface of foods, which would suggest that B_{12} photosensitivity is not problematic in most foods. However, treatment of foods by penetrating waves such as microwaves was shown to promote degradation of B_{12} into inactive forms, with losses estimated at 30–40% (Watanabe *et al.*, 1998).

In food matrices, B_{12} is generally considered to be rather heat stable compared with other water-soluble vitamins (Harris, 1988). Reported losses depend on the type of food, processing and storage conditions (Ball, 2006). Several studies have focused on B_{12} in milk, with reported losses following long durations of heat processing, microwave heating and exposure to fluorescent light (Ball, 1998; Watanabe *et al.*, 1998). However, pasteurised milk stored and refrigerated for 9 days and under retail or domestic conditions showed minimal losses (Andersson and Öste, 1994).

The stability of B_{12} in food products during processing, preparation and storage is an important parameter, directly affecting the supply of B_{12} to consumers. Furthermore, B_{12} losses lead to increased discrepancies between actual vitamin contents and food label declarations.

6.5
Assay Methods

The determination of active B_{12} levels is challenging for many reasons. It is indeed a large complex molecule with several possible structures, leading to differences in chemical reactivity which affects how it interacts with other substances in the matrix. In fermented broths or cell suspensions, B_{12} is intracellular and must be removed, whereas in biological samples, it may be tightly bound to matrix components; in both cases, many interfering substances may be present. Further, B_{12} in natural sources is typically present at only low concentrations, and therefore, a method should not only be specific but also highly sensitive. Reliable methods of extraction and analysis are therefore needed to analyse B_{12} in food and feed, in microbial vitamin productions, and also to identify B_{12} deficiencies in biological samples (Snow, 1999; Watanabe *et al.*, 2013; Chamlagain *et al.*, 2015).

A number of methods to determine B_{12} presence and concentrations have been developed in the past few decades (Kumar, Chouhan and Thakur, 2010). Unfortunately, many of these methods are expensive, slow or not suitable for foods (Pakin *et al.*, 2005). Furthermore, the presence of other molecules in samples can interfere with B_{12} analyses, for example, B_{12} analogues or B_{12}-binding proteins, and hence, a partial purification of the B_{12} is often needed. To prevent degradation of B_{12} during purification, a KCN treatment is performed which converts all B_{12} forms to stable CNB_{12} (Piao *et al.*, 2004). Consequently, most methods detect only B_{12}. A number of methods are addressed in this chapter, with focus on those applicable in food industry and microbiology.

The microbiological assay for B_{12} determination is based on the growth requirement of B_{12} by some bacteria. Samples containing B_{12} are added to a growth medium lacking B_{12}, and the maximum dilution of samples still enabling growth of the test organism is recorded. The parallel use of a CNB_{12} standard solution then allows the determination of B_{12} in the sample. This method is relatively simple but has some drawbacks. One important limitation is that the indicator organisms may also use B_{12} forms that have minimal or no activity in humans. For example, the test organism used for determining vitamin B_{12} concentration according to the U.S. Pharmacopeial Convention (USP) and the Cobalamin Assay in the Official Methods of Analysis of AOAC International (Horowitz, 2006; UPC, 2008) *Lactobacillus bulgaricus* subs. *lactis* ATTC 7830 (*Lactobacillus leichmannii*) is able to grow on pseudovitamin B_{12} (Santos *et al.*, 2009b). Although the method is labour-intensive and slow, it is used frequently because it is inexpensive, no additional equipment or purification is needed and the detection limit is relatively low (30 ng/ml). Therefore, it is still recommended by AOAC as a method for food product analysis (Horowitz, 2006). Furthermore, ready-to-use kits for microbiological assays are on the market with greatly improved accuracy and reliability (Vitafast).

Various human proteins that bind B_{12} have been identified, and they form the basis of currently used B_{12}-binding assays. These assays are highly sensitive but require highly specialised equipment. Two types of binding assays are used: competitive and inhibitory binding. Competitive binding is based on addition of labelled CNB_{12} molecules which compete for binding sites with non-labelled B_{12} present in samples; the ratio of labelled/non-labelled B_{12} bound to binding sites allows quantifying of the B_{12} concentration. The method of competitive binding is sensitive enough for detection in samples from vitamin-B_{12}-deficient patients, with a detection limit as low as 4 pg/ml when using radiolabelled B_{12} (Rothenberg, 1963; Breuel *et al.*, 1973). However, radiolabelled B_{12} is expensive, and a radioisotope counter is needed for the analysis. Therefore, chemical labels are frequently applied in competitive binding assays (Karmi *et al.*, 2011). The detection limit using chemical labels is slightly higher, about 30 pg/ml, but still sufficient for B_{12} analyses of blood serum (Roche, 2008). The binding factors used in the competitive assays can interact with proteins from food, thereby interfering with the test results (Muhammad, Briggs and Jones, 1993).

An alternative to the competitive binding assay is the inhibitory binding assay. Binding of a tester molecule to B_{12} inhibits the binding between the tester and a sensor; this reduction can then be quantified and expressed as B_{12} concentration. The sensing of the tester molecule has been engineered thoroughly in the past few years and is currently performed using highly sensitive surface plasmon resonance technology. The detection limit of inhibitory binding assay is 80 pg/ml, and commercial kits are available for food samples with quantification limits of 200 pg/ml (GE Healthcare Life Sciences, 2014).

Another principle for B_{12} detection is based on chemiluminescence. Chemiluminescence occurs after release of cobalt on acidification and the catalytic effect of the cobalt on chemiluminescence of luminol (Qin, Zhang and Liu, 1997). The method was optimised and can now quantify B_{12} levels as low as 10 pg/ml (Kumar, Chouhan and Thakur, 2009). Chemiluminescence-based methods were combined with immunity-based B_{12} enrichment to allow analyses of complex matrixes. This type of combined technology was used to analyse levels of B_{12} in human milk (Hampel *et al.*, 2014), where tight binding occurs between HC and B_{12} (Allen, 2012). Further, a dipstick-based method for liquids was developed with a detection limit of 1 ng/ml (Selvakumar and Thakur, 2012). The method is simple without the need for a photon-counter, fast (5 min) and robust. A standard assay based on chemiluminescence for B_{12} for serum samples is currently available on the market (Siemens Healthcare, 2014), and the method seems to be a good alternative to the microbiological method.

Fluorescence-based measurements are generally highly sensitive. Vitamin B_{12} is a non-fluorescent molecule, but one of its fragments, α-ribazole, is fluorescent. This feature has been used to measure B_{12} concentration in different food products (Pakin *et al.*, 2005). The B_{12} was enriched on an immunoaffinity column, the α-ribazole fragment was chemically released from the corrin ring and its fluorescence was measured. After extensive sample preparation, quantification limit was low (3 ng/g), but repeatability only reached 'satisfactory levels' (Pakin *et al.*, 2005; Chamlagain *et al.*, 2015).

Chromatography has played a pivotal role in identification and quantification of B_{12}. Today, high-performance liquid chromatography (HPLC) is probably the most frequently used method to determine B_{12} concentrations in food samples. Reversed-phase HPLC using C_{18}-column allows a clean separation of the main B_{12} forms CNB_{12}, OHB_{12}, $AdoB_{12}$ and MeB_{12} (Frenkel, Kitchens and Prough, 1979). The detection limit of HPLC methods without prior enrichment of B_{12} is estimated at 40 ng/ml (Moreno and Salvado, 2000). HPLC-based methods are precise and allow identification of different B_{12} forms. However, the detection limit of HPLC is above the vitamin B_{12} concentrations of 3–20 ng/ml found in most foods and the 70–900 pg/ml found in human milk (Heudi *et al.*, 2006; Hampel *et al.*, 2014). Enrichment of B_{12} over an immunoaffinity column followed by HPLC analyses lowers the detection limit in a sample to 3 ng/ml and the quantification limit to 10 ng/ml, allowing detection of B_{12} in food (Heudi *et al.*, 2006). This method has been used to determine vitamin B_{12} contents in meat products and infant formula (Guggisberg, Risse and Hadorn, 2012; Kirchner *et al.*, 2012).

Further development of HPLC techniques with higher pressure and smaller particles (ultra-HPLC) has led to increased resolution, sensitivity and speed of analyses (Swartz, 2005). Vitamin B_{12} can be quantified at levels as low as 2 ng/ml using UHPLC (Zironi *et al.*, 2013). Combination of UHPLC and immunoaffinity enrichment allowed detection of B_{12} in cereal-based products fermented by Propionibacteria (PAB) (Chamlagain *et al.*, 2015) and distinguishes between B_{12} and pseudovitamin B_{12}.

The microbiological method is still widely applied to determine B_{12} concentrations. However, its limitations and recent developments in HPLC techniques have resulted in the proposal of the latter as standard method (Campos-Gimenez *et al.*, 2012). HPLC-based techniques allow rapid detection and quantification of B_{12} at levels encountered in food products, as well as quantification of bioactive B_{12} forms. During sample preparation, the different B_{12} forms are converted to CNB_{12} which is subsequently determined by HPLC. One important advantage of HPLC is that it can distinguish between CNB_{12} and KCN-treated pseudovitamin B_{12} (Chamlagain *et al.*, 2015). Many research laboratories have an HPLC system, and, therefore, the method is widely used in science (Kumar, Chouhan and Thakur, 2010). Binding-based methods are favoured for B_{12} determinations in clinical samples (Roche, 2008). The inhibitory binding assay is currently being assessed by AOAC for acceptance as one of the standardised methods for food analyses (Vyas and O'Kane, 2011; Vyas, O'Kane and Dowell, 2012).

6.6 Biotechnological Synthesis

6.6.1 Producing Microorganisms

B_{12} is the most complex cofactor in nature and is exclusively synthesised by certain prokaryotes (Moore and Warren, 2012). The biosynthesis of B_{12} involves approximately 30 enzymatic reactions and intermediates, of which several are highly oxygen sensitive. The latter makes synthesis in eukaryotic cells problematic and might explain that B_{12} synthesis occurs exclusively in prokaryotes (Martens *et al.*, 2002). The chemical pathways involved in B_{12} synthesis have been extensively reviewed in the past, and these reviews provide detailed information on the synthesis (Martens *et al.*, 2002; Moore and Warren, 2012). Microbial production of B_{12} leads to a mixture of different analogues, some of which are inactive for humans (Krfilovfi and Rauch, 1985), which emphasises the need for both accurate quantitative and qualitative analyses of the different forms of B_{12} produced.

The main function of B_{12} in microbes is its cofactor role; $AdoB_{12}$ is required in dehydratase, methylase and ammonia lyase reactions (Martens *et al.*, 2002). In PAB, the reaction from succinyl-CoA to *R*-methylmalonyl-CoA is

$AdoB_{12}$-dependent, a reaction which occurs in the opposite direction in humans. Another important role of B_{12} in microbes is regulation of gene expression. B_{12} binds to some mRNAs, thereby impacting on translation efficiency and acts as cofactor for certain transcriptional regulator proteins (Klug, 2014). Generally, it is believed that B_{12} arose to support anaerobic fermentation of small molecules in microbes, later followed by a secondary acquisition of B_{12}-dependent reactions (Martens *et al.*, 2002). A recent study showed that human gut microbes use elaborate mechanisms to capture and differentiate corrinoids *in vivo* that are determinant for host colonisation (Degnan *et al.*, 2014).

Until now, only a few bacteria and archaea were believed to synthesise B_{12}, yet recent genomic studies suggest that B_{12} biosynthesis is more widespread. A systematic genome assessment of B_{12} biosynthesis of 256 common human gut bacteria indicated that about 45% of the analysed strains belonging to the dominant phyla *Bacteroidetes* and *Firmicutes* were predicted to produce cobalamin (B_{12}) by the anaerobic biosynthesis route (Magnusdottir *et al.*, 2015). Nevertheless, it was recently suggested that most of the corrinoids produced in the gut are taken up by non-producer bacteria. These findings suggest a *corrinoid economy* within the gut, where corrinoid-based compounds are required by at least 83% of over 300 sequenced microbiota-derived bacteria representing a form of currency which is exchanged between microbes (Degnan *et al.*, 2014). Further, all available genomes of Thaumarchaeota, which are ubiquitous and abundant in aquatic environments, possess cobalamin synthesis genes, predominantly from the anaerobic pathway, suggesting widespread genetic capacity for cobalamin synthesis (Doxey *et al.*, 2015). Moreover, B_{12} production has been observed in members of the Clostridia group, *Propionibacterium* sp., *Streptomyces* sp., *Bacillus* sp. and *Nocardia* sp. (Darken, 1953). Production of pseudovitamin B_{12} was observed in *Lactobacillus reuteri* (Santos *et al.*, 2007).

Production at industrial scale has been performed using *P. denitrificans* (aerobic pathway) and *P. freudenreichii* (anaerobic pathway), with maximum reported yields of approximately 200 mg/l. Both species, but especially *P. denitrificans*, are attractive for biotechnological production because of their rapid growth with simple nutrient demand, their genetic accessibility, highest production and volumetric productivity (Kang *et al.*, 2012). Moreover, *P. freudenreichii* has GRAS (generally regarded as safe) approval by the FDA and QPS (Qualified Presumption of Safety) by EFSA. This species has been successfully applied to the commercial production of B_{12}, with high reported production of about 200 and 300 mg/l for natural and engineered strains, respectively, reviewed by Kang *et al.* (2012). *Bacillus megaterium* (anaerobic pathway) has been used in the past for B_{12} industrial production and cobalamin production was recently revisited in *B. megaterium* DSM319. Its ability to grow on inexpensive carbon sources, such as raw glycerol from biodiesel production, makes this organism an ideal production host. Cobalt bioavailability was first optimised allowing to increase B_{12} yields from approximately 2–3 to 13 mg/l for the parent strain, and bypassing of the B_{12}-riboswitch by cloning led to even higher yields, ≥220 mg/l (Moore *et al.*, 2014).

6.6.1.1
Propionibacteria (PAB)

The genus *Propionibacterium* belongs to the family Propionibacteriaceae and to the phylum actinobacteria, a phylum that roughly consists of all Gram-positive bacteria with a high G + C content in their genome. Currently, the genus contains 16 species of which the human skin commensal and pathogen *Propionibacterium acne* is doubtless the most studied. PAB are found to be ubiquitous, ferment a wide range of carbon sources to gain energy and are mostly aerotolerant. Most PAB strains produce B_{12}, and especially, *P. freudenreichii* strains produce high quantities (Vorobjeva, 1999; Hugenschmidt *et al.*, 2010). *Propionibacterium* sp. are classically divided into two groups, 'dairy-' and 'cutaneous-' derived members (Patrick and McDowell, 2012). Dairy-related species include *P. freudenreichii* (ssp. *shermanii* and ssp. *freudenreichii*) and *Propionibacterium jensenii*, all originally isolated from Swiss-type cheese. *P. freudenreichii* subsp. *freudenreichii* cannot ferment lactose and has a nitrate reductase activity, whereas *P. freudenreichii* subsp. *shermanii* is capable of lactose fermentation but cannot reduce nitrate. However, phylogenetic analyses and the identification of *P. freudenreichii* strains possessing or lacking both features suggest that the division into subspecies is not justified anymore (Dalmasso *et al.*, 2011; Thierry *et al.*, 2011).

The name 'Propionibacterium' originates from their ability to produce large amounts of propionic acid, as well as acetate and carbon dioxide out of fermentable carbon sources. Remarkably, lactate can also be used as carbon source which also results in the production of propionic acid as fermentation end product. The latter conversion occurs in Swiss-type cheese, where the lactate produced by lactic acid bacteria is used by PAB, giving the cheese a nutty taste, whereas carbon dioxide production leads to the production of the typical large holes. The conversion of lactate produced by other fermenting bacteria, also known as *secondary fermentation*, is a distinguishing feature of the genus, and lactate as sole carbon source is a powerful method allowing the isolation of PAB from fermented dairy products (Drinan and Cogan, 1992).

PAB have a unique fermentative pathway to produce propionic acid. Under anaerobic or micro-aerobic conditions, PAB metabolise glucose by glycolysis to pyruvate which is then oxidised to acetate and carbon dioxide. The reducing power produced in this conversion is used in the methylmalonyl-CoA, randomising or Wood–Werkman cycle to produce propionic acid. The methylmalonyl-CoA pathway is dependent on two vitamins: biotin for transfer of CO_2 and $AdoB_{12}$ for conversion of succinyl-CoA to *R*-methylmalonyl. The pathway converts pyruvate by a reverse citrate cycle to succinyl-CoA, which is then converted to *R*-methylmalonyl and eventually to the end-product propionic acid. ATP is produced in the glycolysis and an additional ATP in the methylmalonyl-CoA pathway, with the latter ATP unique for PAB (Dalmasso *et al.*, 2011). The B_{12} dependency of the pathway explains the production of this vitamin by PAB.

PAB are capable of growth under micro-aerobic conditions and can use molecular oxygen as electron acceptor (Vorobjeva, 1999). All the genes needed to produce a respiratory chain are present in the genome of *P. freudenreichii* (Falentin *et al.*, 2010). Remarkably, if a culture is shifted from anaerobic to aerobic conditions, *P. freudenreichii* starts to use propionate and produce pyruvate by a reverse methylmalonyl-CoA pathway and using oxygen as final electron acceptor (Ye *et al.*, 1999). The production of B_{12} was shown to stop under aerobic conditions, but restarted rapidly when switched back to anaerobic conditions (Ye *et al.*, 1999). Other inorganic electron acceptors that can be used by PAB are sulfite, nitrate and iron (Vorobjeva, 1999). The latter case is an indirect use of inorganic electron acceptors; humic acid is used as electron acceptor which is oxidised by ferric iron (Benz, Schink and Brune, 1998).

6.6.1.2
Pseudomonades

Pseudomonas is a diverse genus of Gram-negative rods occurring in various environmental niches and belonging to the class of γ-proteobacteria and the family Pseudomonadaceae containing 191 validly described species (LPSN, 1998). Pseudomonades are straight or slightly curved rods which are flagellated, and hence motile, and accumulate poly-β-hydroxybutyrate as storage molecule (Kersters *et al.*, 1996). They grow aerobically and use oxygen as electron acceptor. Some species, such as *Pseudomonas stutzeri*, are able to use nitrate as electron acceptor. The genus has gone through some taxonomic reclassifications, and the genus *Pseudomonas* sensu stricto can only be differentiated based on 16S RNA sequence comparisons. Some *Pseudomonas* species are opportunistic pathogens such as *Pseudomonas aeruginosa* and the plant pathogen *Pseudomonas syringae*. *P. aeruginosa* and the biochemical versatile species *Pseudomonas putida* and *Pseudomonas fluorescens* are by far the best-studied members of the genus, whereas reports on other species are rare (Palleroni, 2005). The capability to produce B_{12} is present in many *Pseudomonas* species, for example, *P. aeruginosa* and *P. putida* (Cameron *et al.*, 1989; Lee *et al.*, 2012).

Pseudomonades have excellent properties for a broad range of biotechnology uses due to their metabolic versatility. They can grow rapidly on inexpensive and simple formula, are robust and genetically accessible and their metabolic features have been extensively studied. Nowadays, *Pseudomonas* species, especially *P. putida*, are widely used for the production of complex organic compounds, including B_{12}, which are used for bio-based polymers, pharmaceuticals and herbicides (Poblete-Castro *et al.*, 2012). However, the close relation of *P. putida* to the pathogen *P. aeruginosa* complicates its use as producer organism for medical and food applications.

Currently, *P. denitrificans* is almost exclusively used for industrial production of B_{12} because of its technical properties such as rapid aerobic growth and high productivity (Li *et al.*, 2008; Laudert and Hohmann, 2011). *P. denitrificans* was first described in 1903, but the species was placed on the list of nomina rejicienda by the Society for General Microbiology in 1982 (Bacteriology, 1982). The decision was based on observations that *P. denitrificans* isolates did not match the species description, whereas other *Pseudomonas* isolates clearly belonging to other species did (Doudoroff *et al.*, 1974). Nevertheless, three isolates were further considered in the literature as *P. denitrificans*. The neotype IAM 12023 designating from NCIB 10465 (=ATCC 19244) was proposed to be reclassified as *P. denitrificans* and is closely related to *Pseudomonas pertucinogena* (Anzai *et al.*, 2000). To our knowledge, derivatives from this isolate were never used for B_{12} research. The genome of NCIB 9496 (=ATCC 13867), clearly a different species than NCIB 10465 (Doudoroff *et al.*, 1974), was completely sequenced and published as *P. denitrificans* (Ainala, Somasundar and Park, 2013). Research on this strain focused mainly on nitrite reduction and not on B_{12} production. In addition to these two isolates, a B_{12} production strain owned and well-studied by the Merck company is commonly named *P. denitrificans* MB580 (Long and Parlin, 1962; Demain *et al.*, 1968). The species ambiguity is known, as explained in the following statement: 'although the taxonomic validity of the species *P. denitrificans* is questionable, we retain this taxonomic definition for strains derived from MB580' (Cameron *et al.*, 1989). Recent papers on B_{12} production by *P. denitrificans* use 'industrial production' strain without any further references. The exact species of the production strains is not clear, nor their relation to each other and to other strains assigned to the former species *P. denitrificans*.

General characteristics of *P. denitrificans* cannot be listed because the species does not exist anymore. The capability of *P. denitrificans* strains to produce B_{12} seems solely related to methionine biosynthesis, because growth of some methionine auxotrophic mutants can be restored by B_{12} addition (Lago and Demain, 1969). Moreover, genes encoding for the other B_{12}-dependent reactions, that is, glycerol dehydration and succinyl-CoA to *R*-methylmalonyl conversion, are absent in the *P. denitrificans* ATCC 13867 genomes (KEGG, 2015).

6.6.2
Biosynthesis and Metabolic Regulation

Synthesis of B_{12} starts with a set of reactions shared by the synthesis pathways of all tetrapyrrole containing molecules in living cells, including haem and chlorophyll. The biosynthesis of B_{12} is commonly divided into three parts: the synthesis of uroporphyrinogen III, the corrin ring formation and the corrin ring adenylation and nucleotide group assembly (Santos, 2008). The synthesis starts with 5-aminolevulinate, which can be produced from glutamate in the C_5-pathway or from glycine in the C_4-pathway or Shemin-pathway. The C_4-pathway occurs in humans and is the start of haem synthesis. Most bacteria use the C_5-pathway but α-proteobacteria are known to use C_4-pathway. Alternatively, 5-aminolevulinate can be transported into the cell (Martens *et al.*, 2002). In the next step, two molecules 5-aminolevulinate combine to form a heterocyclic aromatic ring known as *pyrrole*, and four pyrroles are then cyclised into uroporphyrinogen III, a large ring molecule. Uroporphyrinogen III can be decarboxylated as occurs in the synthesis of haems and chlorophylls, or it can be methylated to form precorrin-2, the last common metabolite for the synthesis of coenzyme F430, sirohaem and cobalamin. At precorrin-2, the cobalamin pathway diverges into an aerobic and anaerobic pathway (Martens *et al.*, 2002; Moore and Warren, 2012).

Two distinct pathways for corrin ring formation occur in nature, aerobic and anaerobic, with one main difference concerning the chelation of cobalt. Synthesis by the aerobic pathway was extensively researched in *P. denitrificans*, and the anaerobic pathway was elucidated in *Salmonella enterica* subsp. *enterica* serovar Typhimurium. In the anaerobic pathway, cobalt is inserted at an early stage and the reaction is not ATP-dependent. In the aerobic pathway, the insertion of cobalt occurs nine enzymatic steps later and is ATP-dependent. The corresponding cobalt chelatases that catalyse the entrapment of cobalt into the ring are not related. As a consequence, the intermediates of the anaerobic pathway are cobalt complexes, whereas those of the aerobic pathway are not, resulting in different substrate specificities (Martens *et al.*, 2002). However, for both pathways, a ring contraction occurs by the removal of the C-20 atom of the ring and the end product is Cob(II)yrinic acid a,c-diamide, the corrin ring of B_{12}. Genes of the anaerobic and aerobic synthetic pathways are designated *cobi*namide-biosynthesis (*cbi*) s and *cob*alaminbiosynthesis (*cob*) genes, respectively. The synthesis pathways are the aerobic pathway for *P. denitrificans* and *Rhodobacter capsulatus* (Heldt *et al.*, 2005) and the anaerobic pathway for *S.* Typhimurium, *B. megaterium* and *P. freudenreichii* ssp. *shermanii* (Moore and Warren, 2012).

In the final steps of synthesis, reduction of the corrin ring serves to stabilise the cobalt chelation. Then, adenylation of the cobalt occurs, forming the β-ligand. Finally, the α-ligand is produced by a set of reactions. First an amino-propanol arm is attached to the corrin ring and phosphorylated or, alternatively, an already phosphorylated amino-propanol arm is attached. The phosphor group is than activated by addition of an adenosine-GDP at the expense of one GTP. Concomitantly, an α-ribazole ((*α*)-D-ribosyl-5,6-dimethylbenzimidazole) is activated and a transferase reaction takes place which replaces adenosine-GDP by α-ribazole and results in the formation of coenzyme B_{12}.

6.6.3 Engineering of B_{12} Production

6.6.3.1 Propionibacteria

Knowledge of the biosynthetic pathway and the corresponding genes is highly beneficial in controlling and improving the biotechnological production of any compound. Synthesis of B_{12} by PAB has been extensively studied in *P. freudenreichii* subsp. *shermanii*. This microbe uses the C_4 and C_5 pathways for the production of the precursor 5-aminolevulinate concomitantly with a flux estimated to be 50–65% by the C_4 pathway (Iida and Kajiwara, 2000). The anaerobic pathway is used for ring contraction, independent of aerobic or anaerobic conditions by all PAB (Iida, Ohtaka and Kajiwara, 2007). Genes involved in the biosynthesis are organised into two clusters in *P. acne* and *Propionibacterium acidipropionici* (Parizzi *et al.*, 2012). A small cluster encodes for 5-aminolevulinate synthesis and uroporphyrinogen III formation and a large cluster for cobalt transport and B_{12} synthesis. In *P. freudenreichii*, the genes are organised in four clusters. One cluster encodes a cobalt transporter, a second for the anaerobic pathway, a third for corrin ring modification and the fourth cluster contains genes for uroporphyrinogen III formation (Falentin *et al.*, 2010). In general, the clusters for B_{12} differ on the presence or absence of some transporters and on the occurrence of fused genes. The operon *cbi*EGH is found in all PAB, but the organisation of *cob* genes is different (Parizzi *et al.*, 2012).

Genetic engineering has been applied to improve the production of B_{12} by *P. freudenreichii*. Overexpression of several genes involved in B_{12} biosynthesis resulted in 1.5- to 1.9-fold increased yield (Piao *et al.*, 2004). However, only seven genes were higher expressed, all encoding enzymes involved in the corrin formation. Possibly these enzymes do not catalyse a bottle-neck reaction in B_{12} synthesis, and therefore, higher production by genetic engineering is still feasible, as described for *P. denitrificans* (Blanche *et al.*, 1998). Genome shuffling is another genetic method and does not lead to genetically modified organisms. This method was used to produce a variant of *P. freudenreichii* subsp. *shermanii* which could produce 2.8 mg/l as compared with 1.8 mg/l by the parent strain (Zhang *et al.*, 2010).

Vectors for metabolic engineering in PAB have been developed and successfully applied to enhance propionic acid and 5-aminolevulinate production (Kiatpapan and Murooka, 2001; Zhuge *et al.*, 2013). High B_{12} production yields are, however, difficult to reach due to the complexity of the B_{12} pathway (Piao *et al.*, 2004) and also product-feedback inhibition (Biedendieck *et al.*, 2010). Consequently, medium and process optimisation seems to be the most promising strategy now.

6.6.3.2
Pseudomonades

Genetic methods to improve B_{12} production by *P. denitrificans* were also successful. Several random mutagenesis steps with the production strains MB580 resulted in the high-producing strain SC510 (Cameron *et al.*, 1989). Elucidation of the B_{12} biosynthesis pathway and cloning of the 22 cobalamin biosynthetic genes of the organism was used by researchers at Rhone Poulenc (now Sanofi-Aventis) to construct genetically engineered production strains (Blanche *et al.*, 1998). Although there are no published data, the resulting strains presumably have technological benefits in terms of productivity and yield on raw materials over strains solely obtained by classical random mutagenesis and selection. The general consensus is that these engineered production strains might produce up to 300 mg/l under optimised fermentation conditions (Martens *et al.*, 2002).

6.6.4
Fermentation Process

6.6.4.1
Propionibacteria

PAB exhibit a complex metabolism but have few nutritional requirements. *P. freudenreichii* is able to synthesise all amino acids, and all but a few vitamins, and can grow on chemically defined media containing a carbon and energy source; NH_4 as nitrogen source; minerals and vitamins (pantothenate, biotin, thiamine) (Thierry *et al.*, 2011). They primarily synthesise coenzyme forms of B_{12} (AdoB_{12} and MeB_{12}), however, they may synthesise small amounts of cobamides inactive for humans, for example, pseudovitamin B_{12}. B_{12} production by PAB is intracellular with an accumulation in the stationary phase (Yongsmith *et al.*, 1982); extracellular B_{12} has also been detected but is likely released after cell lysis. Production of B_{12} is highly strain-dependent and screening of large diversity of strains has been done to identify high producers (Hugenschmidt *et al.*, 2010). Additionally, many studies have been carried out to optimise the medium composition and to improve the cultivation process (suspended and planktonic cells, batch, fed-batch and continuous) in order to enhance the B_{12} yield and productivity of *P. freudenreichii* and also of *P. denitrificans* and decrease production costs (Spalla *et al.*, 1989; Survase, Bajaj and Singhal, 2006; Kang *et al.*, 2012).

PAB ferment a variety of substrates, including carbohydrates, polyols such as glycerol and organic acids such as lactic acid (Thierry *et al.*, 2011). In B_{12} synthesis, the culture-medium carbon source may consist of glucose, sucrose or a glucose and fructose mixture (inverted molasses) at concentrations of 50–100 g/l. The medium is supplemented with nitrogenous compounds which also supply growth enhancers and other nutrients including yeast extract, casein hydrolysate and corn steep liquor (50–70 g/l, (Spalla *et al.*, 1989)). Suitability of dairy waste such as whey or whey permeate (containing lactose and lactic acid) has also been studied for B_{12} production (Marwaha and Sethi, 1984; Hugenschmidt, Miescher Schwenninger and Lacroix, 2011). Other components are typically added to stimulate the fermentation: small amounts of ferrous, manganous and magnesium salts, in addition to cobalt salts (60–100 mg/l), and the B_{12} precursor DMBI (10–25 mg/l) (Marwaha, Sethi and Kennedy, 1983a,b; Spalla *et al.*, 1989). Betaine, and to some extent choline, and glutamic acid have shown stimulatory effects on B_{12} production (Marwaha *et al.*, 1983c).

Generally, the temperature of the culture is set at 30 °C, close to the optimum growth temperature for PAB. The pH of the culture has to be controlled, usually in the range 6.5–7, to neutralise the accumulated organic acids and prevent inhibition of the culture by low pH and undissociated acetic and propionic acids. PAB used in industrial production are microaerophilic and produce B_{12} in high yields only under very low oxygen concentrations. However, oxygen might be needed to stimulate B_{12} production or facilitate the attachment of DMBI. Although PAB are able to synthesise DMBI from riboflavin, at least one step seems to require oxygen (Gray and Escalante-Semerena, 2007). The B_{12} bioprocess is usually divided into two stages (Martens *et al.*, 2002; Hugenschmidt, Miescher Schwenninger and Lacroix, 2011). In the first 3 days of fermentation, the bacteria are incubated anaerobically to enhance growth while they produce a B_{12} intermediate missing the DMBI moiety. Subsequently, the fermentation is completed by gentle aeration of the whole culture for 1–3 days, allowing the bacteria to undertake the oxygen-dependent synthesis of DMBI and link it to cobamide (Martens *et al.*, 2002).

To reduce the concentration of inhibitory propionic acid, the culture can be switched to aerobic and back to anaerobic conditions. Propionic acid is utilised during the aerobic phase resulting in higher B_{12} production in a new anaerobic cycle. This switching sequence resulted in an increased yield from 6 to 12 mg/l (Miyano, Ye and Shimizu, 2000). In a mixed culture with the propionic acid utilising *Ralstonia eutropha*, the propionic acid concentration was reduced under aerobic conditions and the yield could be further increased to 19 mg/l (Miyano, Ye and Shimizu, 2000). Recently, an expanded bed absorption *in situ* product recovery (ISPR) process was developed to bind propionic acid with a resin, prevent the accumulation of inhibitory product for cell growth and B_{12} biosynthesis, and facilitated the DSP, but only resulted in an 18% increase in B_{12} yield (Wang *et al.*, 2012).

A process combining lactic acid bacteria and PAB was recently reported and patented for concurrent production of B_{12} and natural folate (B_9) in a food-grade fermentation process in whey permeate medium (Smid and Lacroix, 2013; Hugenschmidt, Miescher Schwenninger and Lacroix, 2011). This co-cultivation is a classic example of commensalism, with a trophic chain for the carbon source and synergistic growth effects. The two vitamins are co-metabolised in humans, and their ratio could be adjusted for optimal nutritional effects by manipulating fermentation conditions. The fermentate produced in a food-grade process, containing high vitamin concentrations in physiological ratio, could be directly used as a vitamin bioingredient without extraction and purification steps (Hugenschmidt, Miescher Schwenninger and Lacroix, 2011).

6.6.4.2
Pseudomonades

Studies on B_{12} production were mainly performed with the production strain *P. denitrificans* MB580, and its overproducing derivative SC510. *P. denitrificans* uses the aerobic pathway for B_{12} synthesis and genes encoding for B_{12} production are located on four loci on the genome (Cameron *et al.*, 1989). In contrast with PAB, cobalamin biosynthesis of *Pseudomonas* occurs parallel to growth under aerobic conditions (Spalla *et al.*, 1989). Production of B_{12} was optimised by medium optimisation, fermentation settings and genetic engineering. Fermentation is conducted with aeration and agitation, in the presence of nutrients, such as yeast extract, sucrose and several mineral salts in the growth medium, at pH of about 7 and 30 °C for a period of 6–7 days, giving a yield >150 mg/l (Spalla *et al.*, 1989). The fermentation is well aerated during the exponential growth phase, but B_{12} production is enhanced by controlling the dissolved oxygen concentration (DOC) and also the level of CO_2 in the inlet gas. Biomass and B_{12} production both are oxygen-dependent, but the latter is favoured by lower DOC as compared with the first. Therefore, a multi-stage DOC control strategy was recently validated at large fermentation scale of 120 m^3. The fermentation started with a high DOC period (8–10%) to allow biomass production followed by a progressive reduction to 2–5% (49–106 h) and below 2% (107–168 h) DOC, for an increase of B_{12} production of about 20% compared to DOC-stat strategy (Li *et al.*, 2012; Peng *et al.*, 2014). Furthermore, an optimal exhausted CO_2 fraction control strategy was also recently reported which included aeration of the culture with a defined mixture of air and CO_2 that enhanced the B_{12} yield by about 10% compared with the control (Wang *et al.*, 2014).

Components of the medium have been extensively investigated to enhance vitamin B_{12} production. The medium has to be supplemented at the beginning of the culture with 10–25 mg/l of DMBI and 40–200 mg/l of cobaltous nitrate (Daniels, 1970). Addition of betaine or choline as methyl donor stimulates production of B_{12} precursor δ-aminolevulinic acid and results in increased B_{12} production (Demain *et al.*, 1968; Fa, Kusel and Demain, 1984). In addition, B_{12} biosynthesis includes seven other methylation reactions which are probably stimulated by betaine addition (Laudert and Hohmann, 2011). The production of the precursors glutamate and glycine as well as methionine was also enhanced by betaine addition (Xia *et al.*, 2015). Sugar beet molasses is often used as carbon source because of its high betaine and glutamate content (Moine *et al.*, 2012; Li *et al.*, 2013). A set up with betaine-control feeding was tested to prevent inhibition of cell growth by elevated betaine concentrations. An effective and economical strategy was reported for B_{12} fermentation in a 120-m^3 fermenter, with continuous feeding of betaine to maintain its concentration of the broth in the range of 5–7 g/l during 50–140 h of fermentation, resulting in an increase of about 10% of B_{12} yield compared with the control (Li *et al.*, 2008). Under optimal fermentation conditions, around 200 mg/l B_{12} predominantly in the form of $AdoB_{12}$ accumulates in the fermentation medium during 7-day runs (Laudert and Hohmann, 2011).

6.7 Downstream Processing; Purification and Formulation

DSP is most important for commercial success of fermentation products. DSP includes the various stages of processing that occur after the completion of the fermentation or bioconversion stage, including separation, purification and packaging of the product. Fermentations produce a mixture of MeB_{12}, OHB_{12} and $AdoB_{12}$ which are not separately isolated, thus avoiding tedious manipulations. OHB_{12} is obtained when cobalamins are converted to the chloro, sulfate or nitrate form during extraction and then subjected to alkaline ion exchange (Moine *et al.*, 2012). $AdoB_{12}$ and MeB_{12} are isolated directly from the fermentation broth.

The major aim of DSP is to produce B_{12} with a high degree of purity, particularly when the producing organism (e.g. *Pseudomonas*) is non-food grade or genetically modified. General DSP conditions have been reported, and these apply for intracellular B_{12} (Spalla *et al.*, 1989; Survase, Bajaj and Singhal, 2006). Briefly described, DSP of intracellular B_{12} involves biomass separation by centrifugation to obtain a cell concentrate. Cell lysis is typically induced by heating the cell suspension, for example, 80–120 °C for 10–30 min at pH 6.5–8.5. The produced B_{12} forms are converted to CNB_{12}, by addition of potassium cyanide, usually in the presence of sodium nitrite and heat. The vitamin solution is clarified by filtration, treatment with zinc chloride and then precipitated by the addition of tannic acid or cresol to give a product with 80% purity, sufficient for use as an animal food additive. For pharmaceuticals or food supplements, the clarified solution requires several more extractions with organic solvents, such as carbon tetrachloride, and then with water and butanol. Adsorption processes such as using ion exchangers, aluminium oxide or activated carbon can also be used. Pure B_{12} is finally obtained by crystallisation after the addition of organic solvents, such as phenol, and water.

Production in a food-grade process using natural *P. freudenreichii* (GRAS status) may allow simplifying the DSP steps, therefore facilitating the production of natural supplements. Typical DSP steps include concentration and drying (e.g. spray drying) (Hugenschmidt, Miescher Schwenninger and Lacroix, 2011).

6.8 Application and Economics

The most important sectors for B_{12} are food/feed industry and supplement and pharmaceutical industry (Moine *et al.*, 2012; DSM, 2015). B_{12} is available in numerous forms (from chewable tablets to nasal spray and injection) and grades. The United States is by far the single largest market on a per capita basis, for B_{12} used in dietary supplement formulations. Pharmaceutical B_{12} is used for treating and preventing B_{12} deficiency and for treating pernicious anaemia. Furthermore, it can be used for treating cyanide poisoning (Hamel, 2011) and for lowering homocysteine levels (WebMD, 2015). Additional claims of B_{12} benefits are reductions in memory loss in Alzheimer's disease and stimulations of mood, energy, concentration and the immune system, but scientific evidence is lacking (WebMD, 2015). Animal feed is fortified with B_{12} worldwide, also for ruminants, accounting for fairly wide usage (DSM, 2015). Generally, it is dosed into all animal feeds in Europe and the United States with the exception of ruminants. The dosage levels are of 10–30 mg/t of feed for poultry, pigs and for calves as milk replacer (Spalla *et al.*, 1989).

CNB_{12} is the most produced B_{12} form and is mainly produced for the food sector as food additive (Spalla *et al.*, 1989; Moine *et al.*, 2012). OHB_{12} is produced mainly for the pharmaceutical industry due to its higher uptake and a more sustained serum level compared with CNB_{12} (Heinrich, 1970). For the same reasons, $AdoB_{12}$ and MeB_{12} are used in the pharmaceutical industry, whereas their application in food industry is planned for the future (Aguilar *et al.*, 2008).

More than 1000 patents on B_{12} production have been published, most of them no longer existing (Moine *et al.*, 2012). It is very difficult to get reliable published estimates for the overall world market, because information is either not sourced or carried over from older papers with no update. A high increase of B_{12} production occurred over the last 25 years, from approximately 3 metric tons per year in 1989 (Spalla *et al.*, 1989) to more than 10 t/year in 2005, for worldwide market worth €77 million (Kaesler, 2005) and 30 t/year estimated in 2011 (Laudert and Hohmann, 2011), with a tendency to increase. A selling price of several thousand euros per kilogram attracted a number of Chinese producers into the market that was dominated by French Sanofi-Aventis. China is dominant on this market with large producers such as CSPC Huarong Pharmaceutical Company, NCPC Victor, Yufeng Bioengineering and Duowei Pharmaceutical Company. As a result, the market is currently characterised by severe production overcapacities and concomitant price pressure (Laudert and Hohmann, 2011).

6.9 Conclusions and Outlook

B_{12} is a very complex water-soluble vitamin, produced by a limited number of microorganisms and found only in animal products. From food or supplements, it follows an intricate pathway of absorption and transfer into cells, where it assumes its main metabolic function as an essential coenzyme for two important enzymes in humans, methionine synthase and methylmalonyl-CoA. Nowadays, the supply of B_{12} relies almost exclusively on biotechnological production using *P. denitrificans* which has been developed to reach high-yield and large-scale production, using optimised processes and, in some cases, engineered strains. In addition to its classical role as an essential nutrient, B_{12} might also have additional unidentified roles. As recently proposed, B_{12} might be molecule of exchange among producers and users in natural ecosystems such as the gut or oceans. Moreover, B_{12} could be an important modulating factor impacting on gut microbiota and gut functions. Future demand for B_{12} will likely continue to increase in both developed and developing countries, mainly due to increases in human and animal populations, growing popularity of vegan and vegetarian diets, limited supply of animal foods combined with environmental issues for their production and the ageing population with higher needs for B_{12}-fortified food. In addition to supplementing food with purified forms of B_{12}, *in situ* production during food fermentation and addition of food-grade bioingredients produced only by *P. freudenreichii* may become increasingly important to the supply of B_{12} for humans and greatly simplifying DSP.

References

Addison, T. (1855) *On the Constitutional and Local Effects of Disease of the Suprarenal Capsules*, S. Highley, London.

Aguilar, F., Charrondiere, U., Dusemund, B., Galtier, P., Gilbert, J., Gott, D.M., Grilli, S., Guertler, R., Kass, G.E.N., Koenig, J., Lambré, C., Larsen, J.-C., Leblanc, J.-C., Mortensen, A., Parent-Massin, D., Pratt, I., Rietjens, I., Stankovic, I., Tobback, P., Verguieva, T., and Woutersen, R. (2008) On 5'-deoxyadenosylcobalamin and methylcobalamin as sources for vitamin B_{12} added as a nutritional substance in food supplements. *EFSA J.*, **815**, 1–21.

Ahmad, I., Qadeer, K., Zahid, S., Sheraz, M.A., Ismail, T., Hussain, W., and Ansari, I.A. (2014) Effect of ascorbic acid on the degradation of cyanocobalamin and hydroxocobalamin in aqueous solution: a kinetic study. *AAPS PharmSciTech*, **15** (5), 1324–1333.

Ainala, S.K., Somasundar, A., and Park, S. (2013) Complete genome sequence of *Pseudomonas denitrificans* ATCC 13867. *Genome Announc.*, **1** (3), e00257-13.

Allen, L.H. (2009) How common is vitamin B-12 deficiency? *Am. J. Clin. Nutr.*, **89** (2), 693S–696S.

Allen, L.H. (2010) Bioavailability of vitamin B_{12}. *Int. J. Vitam. Nutr. Res.*, **80** (4-5), 330–335.

Allen, L.H. (2012) B vitamins in breast milk: relative importance of maternal status and intake, and effects on infant status and function. *Adv. Nutr.*, **3** (3), 362–369.

Andersson, I. and Öste, R. (1994) Nutritional quality of pasteurized milk: vitamin B_{12}, folate and ascorbic acid content during storage. *Int. Dairy J.*, **4**, 161–172.

Andres, E., Loukili, N.H., Noel, E., Kaltenbach, G., Abdelgheni, M.B., Perrin, A.E., Noblet-Dick, M., Maloisel, F., Schlienger, J.L., and Blickle, J.F. (2004) Vitamin B_{12} (cobalamin) deficiency in elderly patients. *Can. Med. Assoc. J.*, **171** (3), 251–259.

Antonopoulos, A. and Charalambos, A. (2013) *The Chemistry of Cobalamins*, Royal Society of Chemistry.

Anzai, Y., Kim, H., Park, J.Y., Wakabayashi, H., and Oyaizu, H. (2000) Phylogenetic affiliation of the pseudomonads based on 16S rRNA sequence. *Int. J. Syst. Evol. Microbiol.*, **50** (Pt. 4), 1563–1589.

Judicial Commission of the International Committee on Systematic Bacteriology (1982) Opinion 54: rejection of the species name *Pseudomonas denitrificans* (Christensen) Bergey *et al* 1923. *Int. J. Syst. Evol. Microbiol.*, **32**, 466.

Ball, G.F.M. (1998) Vitamin B12, in *Bioavailability and Analysis of Vitamins in Foods*, Chapman & Hall, London.

Ball, G.F.M. (2006) *Vitamins in Foods – Analysis, Bioavailability and Stability*, Taylor & Francis Group, Boca Raton, FL.

Banerjee, R., Gherasim, C., and Padovani, D. (2009) The tinker, tailor, soldier in intracellular B_{12} trafficking. *Curr. Opin. Chem. Biol.*, **13** (4), 484–491.

Barker, H.A., Weissbach, H., and Smyth, R.D. (1958) A coenzyme containing pseudovitamin B(12). *Proc. Natl. Acad. Sci. U.S.A.*, **44** (11), 1093–1097.

Benz, M., Schink, B., and Brune, A. (1998) Humic acid reduction by *Propionibacterium freudenreichii* and other fermenting bacteria. *Appl. Environ. Microbiol.*, **64** (11), 4507–4512.

Biedendieck, R., Malten, M., Barg, H., Bunk, B., Martens, J.H., Deery, E., Leech, H., Warren, M.J., and Jahn, D. (2010) Metabolic engineering of cobalamin (vitamin B_{12}) production in *Bacillus megaterium*. *Microb. Biotechnol.*, **3** (1), 24–37.

Blanche, F., Cameron, B., Crouzet, J., Debussche, L., Levy-Schil, S. and Thibaut, D. (1998) Polypeptides involved in the biosynthesis of cobalamines and/or cobamides, dna sequences coding for these polypeptides, and their preparation and use. European Patent 0516647 B1.

Breuel, H.P., Nowrousian, M.R., Hesch, R.D., and Emrich, D. (1973) Clinical significance of serum vitamin B_{12} determination. Comparison with Schilling's test. *Blut*, **27** (2), 99–104.

Burgess, C.M., Smid, E.J., and van Sinderen, D. (2009) Bacterial vitamin B_2, B_{11} and B_{12} overproduction: an overview. *Int. J. Food Microbiol.*, **133** (1–2), 1–7.

Cameron, B., Briggs, K., Pridmore, S., Brefort, G., and Crouzet, J. (1989) Cloning and analysis of genes involved in coenzyme B_{12} biosynthesis in *Pseudomonas denitrificans*. *J. Bacteriol.*, **171** (1), 547–557.

Campos-Gimenez, E., Fontannaz, P., Trisconi, M.J., Kilinc, T., Gimenez, C., Andrieux, P., and Nelson, M. (2012) Determination of vitamin B_{12} in infant formula and adult nutritionals by liquid chromatography/UV detection with immunoaffinity extraction: First Action 2011.08. *J. AOAC Int.*, **95** (2), 307–312.

Carmel, R. (2011) Biomarkers of cobalamin (vitamin B-12) status in the epidemiologic setting: a critical overview of context, applications, and performance characteristics of cobalamin, methylmalonic acid, and holotranscobalamin II. *Am. J. Clin. Nutr.*, **94** (1), 348S–358S.

Castle, W.B. (1980) *The Conquest of Pernicious Anemia*, McGraw-Hill, New York.

Chamlagain, B., Edelmann, M., Kariluoto, S., Ollilainen, V., and Piironen, V. (2015)

Ultra-high performance liquid chromatographic and mass spectrometric analysis of active vitamin B_{12} in cells of *Propionibacterium* and fermented cereal matrices. *Food Chem.*, **166**, 630–638.

Chanarin, I. (2000) Historical review: a history of pernicious anaemia. *Br. J. Haematol.*, **111** (2), 407–415.

Dalmasso, M., Nicolas, P., Falentin, H., Valence, F., Tanskanen, J., Jatila, H., Salusjarvi, T., and Thierry, A. (2011) Multilocus sequence typing of *Propionibacterium freudenreichii*. *Int. J. Food Microbiol.*, **145** (1), 113–120.

Daniels, J.H. (1970) Some factors influencing vitamin B_{12} production by *Pseudomonas denitrificans*. *Can. J. Microbiol.*, **16**, 809–815.

Darken, M.A. (1953) Production of vitamin B_{12} by microorganisms and its occurrence in plant tissues. *Bot. Rev.*, **19**, 99–130.

Degnan, P.H., Barry, N.A., Mok, K.C., Taga, M.E., and Goodman, A.L. (2014) Human gut microbes use multiple transporters to distinguish vitamin B_{12} analogs and compete in the gut. *Cell Host Microbe*, **15** (1), 47–57.

Demain, A.L., Daniels, H.J., Schnable, L., and White, R.F. (1968) Specificity of the stimulatory effect of betaine on the vitamin B_{12} fermentation. *Nature*, **220** (5174), 1324–1325.

Doudoroff, M., Contopou, R., Kunisawa, R., and Palleron, N.J. (1974) Taxonomic validity of *Pseudomonas denitrificans* (Christensen) Bergey *et al*–request for an opinion. *Int. J. Syst. Bacteriol.*, **24** (2), 294–300.

Doxey, A.C., Kurtz, D.A., Lynch, M.D., Sauder, L.A., and Neufeld, J.D. (2015) Aquatic metagenomes implicate Thaumarchaeota in global cobalamin production. *ISME J.*, **9** (2), 461–471.

Drinan, F.D. and Cogan, T.M. (1992) Detection of propionic acid bacteria in cheese. *J. Dairy Res.*, **59** (1), 65–69.

DSM (2015) Vitamine B12, http://www.dsm.com/markets/anh/en_US/Compendium/poultry/vitamin_B12.html (accessed 26 May 2015).

Fa, Y.H., Kusel, J.P., and Demain, A.L. (1984) Dependence of betaine stimulation of vitamin B_{12} overproduction on protein synthesis. *Appl. Environ. Microbiol.*, **47** (5), 1067–1069.

Falentin, H., Deutsch, S.M., Jan, G., Loux, V., Thierry, A., Parayre, S., Maillard, M.B., Dherbecourt, J., Cousin, F.J., Jardin, J., Siguier, P., Couloux, A., Barbe, V., Vacherie, B., Wincker, P., Gibrat, J.F., Gaillardin, C., and Lortal, S. (2010) The complete genome of *Propionibacterium freudenreichii* CIRM-BIA1, a hardy actinobacterium with food and probiotic applications. *PLoS One*, **5** (7), e11748.

Farquharson, J. and Adams, J.F. (1976) The forms of vitamin B12 in foods. *Br. J. Nutr.*, **36** (1), 127–136.

Flynn, A., Moreiras, O., Stehle, P., Fletcher, R.J., Muller, D.J., and Rolland, V. (2003) Vitamins and minerals: a model for safe addition to foods. *Eur. J. Nutr.*, **42** (2), 118–130.

Frenkel, E.P., Kitchens, R.L., and Prough, R. (1979) High-performance liquid chromatographic separation of cobalamins. *J. Chromatogr.*, **174** (2), 393–400.

Froese, D.S. and Gravel, R.A. (2010) Genetic disorders of vitamin B_{12} metabolism: eight complementation groups–eight genes. *Expert Rev. Mol. Med.*, **12**, e37.

GE Healthcare Life Seciences (2014) The Qflex Kit Vitamin B_{12} PI http://www.gelifesciences.com/webapp/wcs/stores/servlet/productById/en/GELifeSciences/14100677 (accessed 26 May 2015).

Grasbeck, R. (1967) Intrinsic factor and the transcobalamins with reflections on the general function and evolution of soluble transport proteins. *Scand. J. Clin. Lab. Invest. Suppl.*, **95**, 7–18.

Gray, M.J. and Escalante-Semerena, J.C. (2007) Single-enzyme conversion of FMNH2 to 5,6-dimethylbenzimidazole, the lower ligand of B12. *Proc. Natl. Acad. Sci. U.S.A.*, **104** (8), 2921–2926.

Green, R. and Miller, J.W. (2014) Vitamin B_{12}, in *Handbook of Vitamins* (eds J. Zempleni and R.B. Rucker), Taylor & Francis Group, Boca Raton, FL.

Guggisberg, D., Risse, M.C., and Hadorn, R. (2012) Determination of vitamin B_{12} in meat products by RP-HPLC after enrichment and purification on an immunoaffinity column. *Meat Sci.*, **90** (2), 279–283.

Hamel, J. (2011) A review of acute cyanide poisoning with a treatment update. *Crit. Care Nurse*, **31** (1), 72–81; quiz 82.

Hampel, D., Shahab-Ferdows, S., Domek, J.M., Siddiqua, T., Raqib, R., and Allen, L.H. (2014) Competitive chemiluminescent enzyme immunoassay for vitamin B_{12} analysis in human milk. *Food Chem.*, **153**, 60–65.

Hardlei, T.F., Morkbak, A.L., Bor, M.V., Bailey, L.B., Hvas, A.M., and Nexo, E. (2010) Assessment of vitamin B(12) absorption based on the accumulation of orally administered cyanocobalamin on transcobalamin. *Clin. Chem.*, **56** (3), 432–436.

Harris, R.S. (1988) in *Nutritional Evaluation of Food Processing*, 3rd edn (eds E. Karmas and R.S. Harris), Van Nostrand Reinhold Company, New York, pp. 3–5.

He, Q., Madsen, M., Kilkenney, A., Gregory, B., Christensen, E.I., Vorum, H., Hojrup, P., Schaffer, A.A., Kirkness, E.F., Tanner, S.M., de la Chapelle, A., Giger, U., Moestrup, S.K., and Fyfe, J.C. (2005) Amnionless function is required for cubilin brush-border expression and intrinsic factor-cobalamin (vitamin B_{12}) absorption in vivo. *Blood*, **106** (4), 1447–1453.

Heinrich, H.C. (1970) Renale vitamin B_{12}-Ausscheidung. *Dtsch. Med. Wochenschr.*, **95**, 1134.

Heldt, D., Lawrence, A.D., Lindenmeyer, M., Deery, E., Heathcote, P., Rigby, S.E., and Warren, M.J. (2005) Aerobic synthesis of vitamin B_{12}: ring contraction and cobalt chelation. *Biochem. Soc. Trans.*, **33** (Pt. 4), 815–819.

Herbert, V. (1988) Vitamin B-12: plant sources, requirements, and assays. *Clin. Nutr.*, **48**, 852–858.

Herrmann, W. and Obeid, R. (2012) Cobalamin deficiency. *Subcell. Biochem.*, **56**, 301–322.

Heudi, O., Kilinc, T., Fontannaz, P., and Marley, E. (2006) Determination of Vitamin B_{12} in food products and in premixes by reversed-phase high performance liquid chromatography and immunoaffinity extraction. *J. Chromatogr. A*, **1101** (1-2), 63–68.

Hodgkin, D.C., Kamper, J., Mackay, M., Pickworth, J., Trueblood, K.N., and White, J.G. (1956) Structure of vitamin B_{12}. *Nature*, **178** (4524), 64–66.

Horowitz, W. (ed) (2006) *Official Methods of Analysis of AOAC International*, 18th edn, AOAC International, Gaithersburg, MD.

Hugenschmidt, S., Miescher Schwenninger, S., Gnehm, N., and Lacroix, C. (2010) Screening of a natural biodiversity of lactic and propionic acid bacteria for folate and vitamin B_{12} production in supplemented whey permeate. *Int. Dairy J.*, **20** (12), 852–857.

Hugenschmidt, S., Miescher Schwenninger, S., and Lacroix, C. (2011) Concurrent high production of natural folate and vitamin B_{12} using a co-culture process with *Lactobacillus plantarum* SM39 and *Propionibacterium freudenreichii* DF13. *Process Biochem.*, **46**, 1063–1070.

Iida, K. and Kajiwara, M. (2000) Evaluation of biosynthetic pathways to delta-aminolevulinic acid in *Propionibacterium shermanii* based on biosynthesis of vitamin B_{12} from D-[1-13C]glucose. *Biochemistry*, **39** (13), 3666–3670.

Iida, K., Ohtaka, K., and Kajiwara, M. (2007) Mechanism of the ring contraction process in vitamin B_{12} biosynthesis by the anaerobe *Propionibacterium shermanii* under aerobic conditions. *FEBS J.*, **274** (13), 3475–3481.

Juzeniene, A. and Nizauskaite, Z. (2013) Photodegradation of cobalamins in aqueous solutions and in human blood. *J. Photochem. Photobiol., B*, **122**, 7–14.

Kaesler, B. (2005) Vitamins: 9. Vitamin B_{12} (Cyanocobalamins), in *Ullmann's Encyclopedia of Industrial Chemistry*, Wiley-VCH Verlag GmbH, Weinheim.

Kang, Z., Zhang, J., Zhou, J., Qi, Q., Du, G., and Chen, J. (2012) Recent advances in microbial production of delta-aminolevulinic acid and vitamin B_{12}. *Biotechnol. Adv.*, **30** (6), 1533–1542.

Karmi, O., Zayed, A., Baraghethi, S., Qadi, M., and Ghanem, R. (2011) Measurement of vitamin B_{12} concentration: a review on available methods. *IIOAB J.*, **2** (2), 23–32.

KEGG (2015) KEGG: Kyoto Encyclopedia of Genes and Genomes, Release 75.0, http://www.genome.jp/kegg/ (accessed 10 July 2015).

Kersters, K., Ludwig, W., Vancanneyt, M., De Vos, P., Gillis, M., and Schleifer, K.H. (1996) Recent changes in the classification of the pseudomonads: an overview. *Syst. Appl. Microbiol.*, **19** (4), 465–477.

Kiatpapan, P. and Murooka, Y. (2001) Construction of an expression vector for propionibacteria and its use in production of 5-aminolevulinic acid by *Propionibacterium freudenreichii*. *Appl. Microbiol. Biotechnol.*, **56** (1-2), 144–149.

Kirchner, U., Degenhardt, K., Raffler, G., and Nelson, M. (2012) Determination of vitamin B_{12} in infant formula and adult nutritionals using HPLC after purification on an immunoaffinity column: first action 2011.09. *J. AOAC Int.*, **95** (4), 933–936.

Klug, G. (2014) Beyond catalysis: vitamin B_{12} as a cofactor in gene regulation. *Mol. Microbiol.*, **91** (4), 635–640.

Kräutler, B. (2005) Vitamin B_{12}: chemistry and biochemistry. *Biochem. Soc. Trans.*, **33** (4), 806–810.

Kräutler, B. (2012) Biochemistry of B_{12}-cofactors in human metabolism. *Subcell. Biochem.*, **56**, 323–346.

Krfilovfi, B. and Rauch, P. (1985) The effect of antimetabolites on the microbial biosynthesis of vitamin B_{12} and related compounds. *Appl. Microbiol. Biotechnol.*, **21**, 331–335.

Kumar, S.S., Chouhan, R.S., and Thakur, M.S. (2009) Enhancement of chemiluminescence for vitamin B_{12} analysis. *Anal. Biochem.*, **388** (2), 312–316.

Kumar, S.S., Chouhan, R.S., and Thakur, M.S. (2010) Trends in analysis of vitamin B_{12}. *Anal. Biochem.*, **398** (2), 139–149.

Lago, B.D. and Demain, A.L. (1969) Alternate requirement for vitamin B_{12} or methionine in mutants of *Pseudomonas denitrificans*, a vitamin B_{12}-producing bacterium. *J. Bacteriol.*, **99** (1), 347–349.

Laudert, D. and Hohmann, H.-P. (2011) in *Comprehensive Biotechnology*, 2nd edn, vol. **2011** (ed. M. Moo-Young), Academic Press, Burlington, MA, pp. 583–602.

Lee, K.M., Go, J., Yoon, M.Y., Park, Y., Kim, S.C., Yong, D.E., and Yoon, S.S. (2012) Vitamin B_{12}-mediated restoration of defective anaerobic growth leads to reduced biofilm formation in *Pseudomonas aeruginosa*. *Infect. Immun.*, **80** (5), 1639–1649.

Lenhert, P.G. and Hodgkin, D.C. (1961) Structure of the 5,6-dimethylbenzimidazolylcobamide coenzyme. *Nature*, **192**, 937–938.

Lester-Smith, E. (1948) Purification of the anti-pernicious anaemia factor from liver. *Nature*, **161**, 638–639.

Li, K.T., Liu, D.H., Li, Y.L., Chu, J., Wang, Y.H., Zhuang, Y.P., and Zhang, S.L. (2008) Improved large-scale production of vitamin B_{12} by *Pseudomonas denitrificans* with betaine feeding. *Bioresour. Technol.*, **99** (17), 8516–8520.

Li, K.-T., Peng, W.-F., Zhou, J., Wei, S.-J., and Cheng, X. (2013) Establishment of beet molasses as the fermentation substrate for industrial vitamin B_{12} production by *Pseudomonas denitrificans*. *J. Chem. Technol. Biotechnol.*, **88** (9), 1730–1735.

Li, K.-T., Zhou, J., Cheng, X., and Wei, S.-J. (2012) Study on the dissolved oxygen control strategy in large-scale vitamin B_{12} fermentation by *Pseudomonas denitrificans*. *J. Chem. Technol. Biotechnol.*, **87** (12), 1648–1653.

Lodowski, P., Jaworska, M., Kornobis, K., Andruniów, T., and Kozlowski, P.M. (2011) Electronic and structural properties of low-lying excited states of vitamin B_{12}. *J. Phys. Chem. B*, **115** (15), 13304–13319.

Long, R.A. and Parlin, N.J. (1962) Production of vitamin B_{12}. U. S. Patent Office USA 3018225.

LPSN (1998) LPSN: List of Prokaryotic names with Standing in Nomenclature, www.bacterio.cict.fr (accessed 9 July 2015).

Magnusdottir, S., Ravcheev, D., de Crecy-Lagard, V., and Thiele, I. (2015) Systematic genome assessment of B-vitamin biosynthesis suggests co-operation among gut microbes. *Front. Genet.*, **6**, 148.

Martens, J.H., Barg, H., Warren, M.J., and Jahn, D. (2002) Microbial production of vitamin B_{12}. *Appl. Microbiol. Biotechnol.*, **58** (3), 275–285.

Marwaha, S.S. and Sethi, R.P. (1984) Utilization of dairy waste for vitamin B_{12} fermentation. *Agric. Waste.*, **9**, 111–130.

Marwaha, S.S., Sethi, R.P., and Kennedy, J.F. (1983a) Influence of 5,6-dimethylbenzimidazole (DMB) on vitamin B_{12} biosynthesis by strains of *Propionibacterium*. *Enzyme Microb. Technol.*, **5**, 361–364.

Marwaha, S.S., Sethi, R.P., and Kennedy, J.F. (1983b) Role of amino acids, betaine and choline in vitamin B_{12} biosynthesis by strains of *Propionibacterium*. *Enzyme Microb. Technol.*, **5**, 454–456.

Marwaha, S.S., Sethi, R.P., Kennedy, J.F., and Rakesh, K. (1983c) Simulation of fermentation conditions for vitamin B_{12} biosynthesis from whey. *Biotechnol. Adv.*, **11**, 481–493.

Miyamoto, E., Yabuta, Y., Kwak, C.S., Enomoto, T., and Watanabe, F. (2009) Characterization of vitamin B_{12} compounds from Korean purple laver (Porphyra sp.) products. *J. Agric. Food. Chem.*, **57** (7), 2793–2796.

Miyano, K., Ye, K., and Shimizu, K. (2000) Improvement of vitamin B_{12} fermentation by reducing the inhibitory metabolites by cell recycle system and a mixed culture. *Biochem. Eng. J.*, **6** (3), 207–214.

Moine, G., Hohmann, H., Kurth, R., Paust, J., Hähnlein, W., Pauling, H., Weimann, B., and Kaesler, B. (2012) Vitamins, 6. B Vitamins, in *Ullmann's Encyclopedia of Industrial Chemistry*, John Wiley & Sons, Inc..

Molina, V.C., Medici, M., Taranto, M.P., and Font de Valdez, G. (2009) *Lactobacillus reuteri* CRL 1098 prevents side effects produced by a nutritional vitamin B deficiency. *J. Appl. Microbiol.*, **106** (2), 467–473.

Moore, S.J., Mayer, M.J., Biedendieck, R., Deery, E., and Warren, M.J. (2014) Towards a cell factory for vitamin B12 production in *Bacillus megaterium*: bypassing of the cobalamin riboswitch control elements. *N. Biotechnol.*, **31** (6), 553–561.

Moore, S.J. and Warren, M.J. (2012) The anaerobic biosynthesis of vitamin B_{12}. *Biochem. Soc. Trans.*, **40** (3), 581–586.

Moreno, P. and Salvado, V. (2000) Determination of eight water- and fat-soluble vitamins in multi-vitamin pharmaceutical formulations by high-performance liquid chromatography. *J. Chromatogr. A*, **870** (1-2), 207–215.

Muhammad, K., Briggs, D., and Jones, G. (1993) Comparison of a competitive binding assay with *Lactobacillus leichmannii* ATCC 7830 assay for the determination of vitamin B_{12} in foods. *Food Chem.*, **48** (4), 431–434.

Nielsen, M.J., Rasmussen, M.R., Andersen, C.B., Nexo, E., and Moestrup, S.K. (2012) Vitamin B_{12} transport from food to the body's cells–a sophisticated, multistep pathway. *Nat. Rev. Gastroenterol. Hepatol.*, **9** (6), 345–354.

Okuda, K. (1999) Discovery of vitamin B_{12} in the liver and its absorption factor in the stomach: a historical review. *J. Gastroenterol. Hepatol.*, **14** (4), 301–308.

Pakin, C., Bergaentzle, M., Aoude-Werner, D., and Hasselmanna, C. (2005) Alpha-ribazole, a fluorescent marker for the liquid chromatographic determination of vitamin B_{12} in foodstuffs. *J. Chromatogr. A*, **1081** (2), 182–189.

Palleroni, N.J. and Genus, I. (2005) Pseudomonas, in *Bergey's Manual of Systematic Bacteriology*, The Proteobacteria, Part B: The Gammaproteobacteria, vol. **2** (eds G. Garrity, D.J. Brenner, N.R. Krieg, and J.R. Staley), Springer.

Parizzi, L.P., Grassi, M.C., Llerena, L.A., Carazzolle, M.F., Queiroz, V.L., Lunardi, I., Zeidler, A.F., Teixeira, P.J., Mieczkowski, P., Rincones, J., and Pereira, G.A. (2012) The genome sequence of *Propionibacterium acidipropionici* provides insights into its biotechnological and industrial potential. *BMC Genomics*, **13**, 562.

Patrick, S., McDowell, A., and Genus, I. (2012) in *Bergey's Manual of Systematic Bacteriology*, The Actinobacteria, vol. **5** (eds M. Goodfellow, P. Kämpfer, H. Busse, *et al.*), Springer-Verlag New York pp. 1138–1155.

Peng, W., Cheng, X., Zhang, H.-y., and Li, K.-T. (2014) The metabolic characteristics of high-production vitamin B_{12} by *Pseudomonas denitrificans* under dissolved oxygen step-wise reduction. *J. Chem. Technol. Biotechnol*, **89** (9), 1396–1401.

Piao, Y., Yamashita, M., Kawaraichi, N., Asegawa, R., Ono, H., and Murooka, Y. (2004) Production of vitamin B_{12} in genetically engineered *Propionibacterium freudenreichii*. *J. Biosci. Bioeng.*, **98** (3), 167–173.

Poblete-Castro, I., Becker, J., Dohnt, K., dos Santos, V.M., and Wittmann, C. (2012) Industrial biotechnology of *Pseudomonas putida* and related species. *Appl. Microbiol. Biotechnol.*, **93** (6), 2279–2290.

Qin, W., Zhang, Z., and Liu, H. (1997) Chemiluminescence flow sensor for the determination of vitamin B_{12}. *Anal. Chim. Acta*, **357**, 127–132.

Quadros, E.V. (2010) Advances in the understanding of cobalamin assimilation and metabolism. *Br. J. Haematol.*, **148** (2), 195–204.

Randaccio, L., Geremia, S., Demitri, N., and Wuerges, J. (2010) Vitamin B_{12}: unique metalorganic compounds and the most complex vitamins. *Molecules*, **15** (5), 3228–3259.

Rickes, E.L., Brink, N.G., Koniuszy, F.R., Wood, T.R., and Folkers, K. (1948) Crystalline vitamin B_{12}. *Science*, **107**, 396–397.

Roche (2008) *Roche E-170 Vitamin B12 "ECLIA" Manual.*

Roth, J.R., Lawrence, J.G., and Bobik, T.A. (1996) Cobalamin (coenzyme B_{12}): synthesis and biological significance. *Annu. Rev. Microbiol.*, **50**, 137–181.

Rothenberg, S.P. (1963) Radioassay of serum vitamin B_{12} by quantitating the competition between Co57b12 and unlabeled B_{12} for the binding sites of intrinsic factor. *J. Clin. Invest.*, **42**, 1391–1398.

Rush, E.C., Katre, P., and Yajnik, C.S. (2014) Vitamin B12: one carbon metabolism, fetal growth and programming for chronic disease. *Eur. J. Clin. Nutr.*, **68** (1), 2–7.

Said, H. M. and E. Nexo (2012). Mechanisms and regulation of intestinal absorption of water-soluble vitamins: cellular and molecular aspects. In (eds Johnson LR, Ghishan FK, Merchand JL, Said HM, Wood JD) *Physiology of the Gastrointestinal Tract*, Chapter 64 (5th edn). Academic Press.

Santos, F. (2008) Vitamin B_{12} synthesis in *Lactobacillus reuteri*. PhD thesis. Microbiology Wageningen University.

Santos, F., Teusink, B., de Vos, W.M., and Hugenholtz, J. (2009a) The evidence that pseudovitamin B(12) is biologically active in mammals is still lacking – a comment on Molina *et al.*'s (2009) experimental design. *J. Appl. Microbiol.*, **107** (5), 1763; author reply 1764.

Santos, F., Teusink, B., Molenaar, D., van Heck, M., Wels, M., Sieuwerts, S., de Vos, W.M., and Hugenholtz, J. (2009b) Effect of amino acid availability on vitamin B_{12} production in *Lactobacillus reuteri*. *Appl. Environ. Microbiol.*, **75** (12), 3930–3936.

Santos, F., Vera, J.L., Lamosa, P., de Valdez, G.F., de Vos, W.M., Santos, H., Sesma, F., and Hugenholtz, J. (2007) Pseudovitamin $B_{(12)}$ is the corrinoid produced by *Lactobacillus reuteri* CRL1098 under anaerobic conditions. *FEBS Lett.*, **581** (25), 4865–4870.

Schneider, S. and Stroinski, A. (1987) *Comprehensive B12*, Walter de Gruzter, Berlin.

Seetharam, B. and Yammani, R.R. (2003) Cobalamin transport proteins and their cell-surface receptors. *Expert Rev. Mol. Med.*, **5** (18), 1–18.

Selhub, J., Jacques, P.F., Dallal, G., Choumenkovitch, S., and Rogers, G. (2008) The use of blood concentrations of vitamins and their respective functional indicators to define folate and vitamin B_{12} status. *Food Nutr. Bull.*, **29** (Suppl. 2), S67–S73.

Selvakumar, L.S. and Thakur, M.S. (2012) Dipstick based immunochemiluminescence biosensor for the analysis of vitamin B_{12} in energy drinks: a novel approach. *Anal. Chim. Acta*, **722**, 107–113.

Shorb, M.S. (1947) Unidentified essential growth factors for *Lactobacillus lactis* found in refined liver extracts and in certain natural materials. *J. Bacteriol.*, **53** (5), 669.

Shorb, M.S. (1948) Activity of vitamin B12 for the growth of *Lactobacillus lactis*. *Science*, **107**, 297–298.

Siemens Healthcare(2014) Immulite 1000 Immunoassay-System, http://www.healthcare.siemens.de/immunoassay/systems/immulite-1000-immunoassay-system (accessed 10 May 2015).

Smid, E.J. and Lacroix, C. (2013) Microbe-microbe interactions in mixed culture food fermentations. *Curr. Opin. Biotechnol.*, **24** (2), 148–154.

Snow, C.F. (1999) Laboratory diagnosis of vitamin B_{12} and folate deficiency: a guide for the primary care physician. *Arch. Int. Med.*, **159** (12), 1289–1298.

Souci, W., Fachmann, S.W., and Kraut, H. (2008) *Food Composition and Nutrition Tables*, 7th edn, Medform Scientific Publishers, Stuttgart.

Spalla, C., Grein, A., Garofano, L., and Ferni, G. (1989) Microbial production of vitamin B_{12}, in *Biotechnology of Vitamins, Pigments and Growth Factors* (ed. E.J. Vandamme), Elsevier.

Stupperich, E. and Nexo, E. (1991) Effect of the cobalt-N coordination on the cobamide recognition by the human vitamin B_{12} binding proteins intrinsic factor, transcobalamin and haptocorrin. *Eur. J. Biochem.*, **199** (2), 299–303.

Survase, S.A., Bajaj, I.B., and Singhal, R.S. (2006) Biotechnological production of vitamins. *Food Technol. Biotechnol.*, **44** (3), 381–396.

Swartz, M.E. (2005) UPLC™: an introduction and review. *J. Liq. Chromatogr. Related Technol.*, **28** (7-8), 1253–1263.

Takahashi-Iniguez, T., Garcia-Hernandez, E., Arreguin-Espinosa, R., and Flores, M.E. (2012) Role of vitamin B_{12} on methylmalonyl-CoA mutase activity. *J. Zhejiang Univ. Sci. B*, **13** (6), 423–437.

Thierry, A., Deutsch, S.M., Falentin, H., Dalmasso, M., Cousin, F.J., and Jan, G. (2011) New insights into physiology and metabolism of *Propionibacterium freudenreichii. Int. J. Food Microbiol.*, **149** (1), 19–27.

Toporok, M. (1960) The relation of binding power to intrinsic factor activity. Effect of pseudovitamin B12 on absorption of vitamin B12. *Am. J. Clin. Nutr.*, **8**, 297–300.

UPC (2008) *The United States Pharmacopeia 31/The National Formulary 26*, Supplement 1, 8 January 2008, United States Pharmacopeial Convention, Inc., Rockville, MD.

Vitafast http://www.r-biopharm.com/items/aoac-ri-certified-vitafast-vitamin-b2-riboflavin (accessed 9 May 2015).

Vorobjeva, L.I. (1999) *Propionibacteria*, Kluwer Academic Publishers, Dordrecht.

Vyas, P. and O'Kane, A.A. (2011) Determination of vitamin B12 in fortified bovine milk-based infant formula powder, fortified soya-based infant formula powder, vitamin premix, and dietary supplements by surface plasmon resonance: collaborative study. *J. AOAC Int.*, **94** (4), 1217–1226.

Vyas, P., O'Kane, A.A., and Dowell, D. (2012) Determination of vitamin B12 in infant formula and adult nutritionals by surface plasmon resonance: First Action 2011.16 (test kit method). *J. AOAC Int.*, **95** (2), 329–334.

Wang, P., Wang, Y., Liu, Y., Shi, H., and Su, Z. (2012) Novel in situ product removal technique for simultaneous production of propionic acid and vitamin B12 by expanded bed adsorption bioreactor. *Bioresour. Technol.*, **104**, 652–659.

Wang, Z.-J., Wang, H.-Y., Wang, P., Zhang, Y.-M., Chu, J., Ying-Ping Zhuang, Y.-P., and Zhang, S.-L. (2014) Enhance vitamin B_{12} production by online CO_2 concentration control optimization in 120 m^3 fermentation. *J. Bioprocess. Biotech.*, **4**, 159.

Watanabe, F., Abe, K., Fujita, T., Goto, M., Hiemori, M., and Nakano, Y. (1998) Effects of microwave heating on the loss of vitamin B12 in foods. *J. Agric. Food. Chem.*, **46**, 206–210.

Watanabe, F., Yabuta, Y., Bito, T., and Teng, F. (2014) Vitamin B_{12}-containing plant food sources for vegetarians. *Nutrients*, **6** (5), 1861–1873.

Watanabe, F., Yabuta, Y., Tanioka, Y., and Bito, T. (2013) Biologically active vitamin B12 compounds in foods for preventing deficiency among vegetarians and elderly subjects. *J. Agric. Food. Chem.*, **61** (28), 6769–6775.

WebMD (2015) Find a Vitamin or Supplement, http://www.webmd.com/vitamins-supplements/ingredientmono-926-vitamin%20b12.aspx?activeingredientid=926&activeingredientname=vitamin%20b12 (accessed 26 May 2015).

Wostmann, B.S. (1981) The germfree animal in nutritional studies. *Annu. Rev. Nutr.*, **1**, 257–279.

Xia, W., Chen, W., Peng, W.F., and Li, K.T. (2015) Industrial vitamin B_{12} production by *Pseudomonas denitrificans* using maltose syrup and corn steep liquor as the cost-effective fermentation substrates. *Bioprocess. Biosyst. Eng.*, **38** (6), 1065–1073.

Ye, K., Shijo, M., Miyano, K., and Shimizu, K. (1999) Metabolic pathway of *Propionibacterium* growing with oxygen: enzymes, 13C NMR analysis, and its application for vitamin B_{12} production with periodic fermentation. *Biotechnol. Prog.*, **15** (2), 201–207.

Yongsmith, B., Sonomoto, K., Tanaka, A., and Fukui, S. (1982) Production of vitamin

B_{12} by immobilized cells of a propionic acid bacterium. *Eur. J. Appl. Microbiol. Biotechnol.*, **16**, 70–74.

Zeuschner, C.L., Hokin, B.D., Marsh, K.A., Saunders, A.V., Reid, M.A., and Ramsay, M.R. (2013) Vitamin B_{12} and vegetarian diets. *Med. J. Aust.*, **199** (Suppl. 4), S27–S32.

Zhang, Y., Liu, J.Z., Huang, J.S., and Mao, Z.W. (2010) Genome shuffling of *Propionibacterium shermanii* for improving vitamin B_{12} production and comparative proteome analysis. *J. Biotechnol.*, **148** (2-3), 139–143.

Zhuge, X., Liu, L., Shin, H.D., Chen, R.R., Li, J., Du, G., and Chen, J. (2013) Development of a *Propionibacterium-Escherichia coli* shuttle vector for metabolic engineering of *Propionibacterium jensenii*, an efficient producer of propionic acid. *Appl. Environ. Microbiol.*, **79** (15), 4595–4602.

Zironi, E., Gazzotti, T., Barbarossa, A., Devicienti, C., Scardilli, M., and Pagliuca, G. (2013) Technical note: development and validation of a method using ultra performance liquid chromatography coupled with tandem mass spectrometry for determination of vitamin B_{12} concentrations in milk and dairy products. *J. Dairy Sci.*, **96** (5), 2832–2836.

7
Industrial Fermentation of Vitamin C

Weichao Yang and Hui Xu

7.1
Introduction and Historical Outline

Vitamin C, also named as L-ascorbic acid (Asc), is a water-soluble vitamin that is essential for humans, non-human primates and a few other mammals (Sauberlich, 1994; Padh, 2009). The discovery of Asc is related with the disease of scurvy (Sauberlich, 1997). Scurvy was a common disease in the world's navies and sailors until the beginning of the nineteenth century, with serious symptoms such as bleeding of mucous membranes, anaemia and eventually death (Hoffer, 1989; Sauberlich, 1997; Eggersdorfer *et al.*, 2012). Medical studies by doctors Lind and Blane showed that scurvy results from the lack of a nutritional factor in the human diet. This was originally designated as the antiscorbutic factor (Hoffer, 1989; Carpenter, 2012). In 1928, Albert Szent-Gyørgyi first isolated the Asc from adrenal glands and called it hexuronic acid (Svirbely and Szent-Gyorgyi, 1932). Four years later, Charles Glen King isolated Asc in his laboratory and concluded that it was the same as hexuronic acid. In 1933, Norman Haworth deduced the chemical structure of Asc (Carpenter, 2012) (Figure 7.1).

After the discovery of Asc, a demand for pure Asc began to be seen, which triggered the development of industrial production processes in the early 1930s (Pappenberger and Hohmann, 2014). Between 1933 and 1934, not only Haworth and fellow British chemist Edmund Hirsthad synthesised Asc, but also, independently, the Polish chemist Tadeus Reichstein (Figure 7.2), succeeded in synthesising the vitamin in bulk, making it the first vitamin to be artificially produced (Stacey and Manners, 1978). The latter process made possible the cheap mass-production of semi-synthetic Asc, which was quickly marketed. Haworth was awarded the 1937 Nobel Prize in Chemistry in part for this work, but the Reichstein process, a combined chemical and bacterial fermentation sequence still used today to produce vitamin C, retained Reichstein's name (Boudrant, 1990; Bremus *et al.*, 2006). In 1934, Hoffmann–La Roche, which bought the Reichstein process patent, became the first pharmaceutical company to mass-produce and market synthetic vitamin C, under the brand name Redoxon (Bächi, 2008).

Industrial Biotechnology of Vitamins, Biopigments, and Antioxidants, First Edition.
Edited by Erick J. Vandamme and José L. Revuelta.

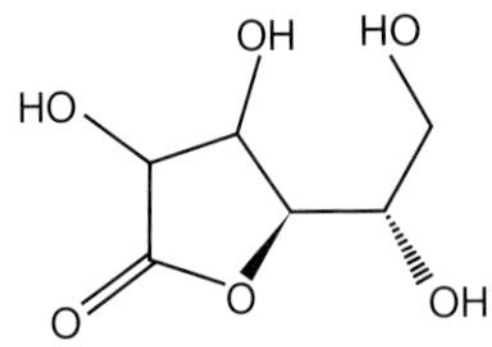

Figure 7.1 Vitamin C (L-ascorbic acid).

Figure 7.2 Tadeus Reichstein. (Source: Roche Historical Archive.)

In the late 1960s and early 1970s, a two-step microbial fermentation process was developed in China (Yin *et al.*, 1980). Compared to the Reichstein process, the new fermentation process provides a clear cost benefit: it requires not only less chemicals and energy but also significantly low investment in production equipment (Xu *et al.*, 2004). Therefore, the two-step fermentation process substituted the classical Reichstein process and was widely applied by Chinese manufacturers that produced more than 80% of vitamin C in the world market (Pappenberger and Hohmann, 2014). Currently, more than 100 000 t vitamin C are produced every year in the world and have been widely used in the food, beverage, animal feed and pharmaceutical industries (Bremus *et al.*, 2006; Mandlaa, 2014).

7.2 Occurrence in Natural/Food Sources

7.2.1 Occurrence of Asc in Foods

The majority of species of animals (but not humans or guinea pigs) and plants can synthesise their own Asc (Chatterjee *et al.*, 1975; Wheeler, Jones and Smirnoff, 1998; Linster and Van Schaftingen, 2007). Hence, the Asc are widely distributed in most foods of plant and animal origin. In foods of plant origin, the Asc amount depends on the variety of the plant, soil condition, climate where it grew, storage conditions and method of preparation (Lee and Kader, 2000; Kosheleva and Kodentsova, 2012). Generally, the richest natural sources of Asc are fruits and vegetables (Kaur and Kapoor, 2002; Proteggente *et al.*, 2002), and of those, the Kakadu plum and the camu camu fruit contain the highest concentration of the Asc (Simion *et al.*, 2008). In foods of animal origin, Asc is most present in the liver and least present in the muscle. Therefore, the animal liver can be used

Table 7.1 Vitamin C content in different fruits and vegetables.[a)]

Plant sources	Amount (mg/100 g)	Plant sources	Amount (mg/100 g)	Plant sources	Amount (mg/100 g)
Kakadu plum	1000–5300	Pineapple	48	Tomato	10
Camu camu	2800	Cauliflower	48	Pawpaw	10
Acerola	1677	Kale	41	Grape	10
Chili pepper	244	Melon	40	Watermelon	10
Guava	228.3	Garlic	31	Banana	9
Red pepper	190	Grapefruit	30	Onion	7.4
Kiwifruit	144	Raspberry	30	Cherry	7
Broccoli	90	Tangerine	30	Peach	7
Loganberry	80	Mandarin orange	30	Carrot	6
Redcurrant	80	Passion fruit	30	Apple	6
Wolfberry	73	Spinach	30	Asparagus	6
Lychee	70	Lime	30	Horned melon	5.3
strawberry	60	Mango	28	Chokecherry	5
Orange	53	Blackberry	21	Pear	4
Lemon	53	Potato	20	Cucumber	3

a) Data is from Wikipedia, http://en.wikipedia.org/wiki/Vitamin_C#cite_note-109.

as a source of dietary Asc. However, the animal muscle, not liver, provides the majority of meat consumed in human diet, which indicated that animal products are not a reliable source of the dietary Asc for people. Hence, only the food of plant origin, especially from the fruits and vegetables, supplied rich Asc to meet the needs of human body (Proteggente *et al.*, 2002; Kaur and Kapoor, 2002; Kosheleva and Kodentsova, 2012).

Table 7.1 summarises the relative abundance in different fruits or vegetables (from Wikipedia, http://en.wikipedia.org/wiki/Vitamin_C#cite_note-109). The citrus fruits (orange, grapefruit, lime and lemon) are excellent sources of Asc. Many non-citrus fruits are highly rated sources as well. Strawberries, pineapple and kiwifruit are also excellent Asc sources. In addition, blueberries, watermelons, apples, pears and bananas are examples of very good Asc sources. Many vegetables, such as broccoli, kale, spinach and carrot, are also excellent sources of Asc.

Asc concentrations in various food substances decrease with time in proportion to the temperature which they are stored at (Roig, Rivera and Kennedy, 1995). Cooking can reduce the Asc content of vegetables by around 60% possibly partly due to increased enzymatic destruction as it may be more significant at sub-boiling temperatures (Allen and Burgess, 1950). Longer cooking times also add to this effect, as will copper food vessels, which catalyse the decomposition. Another cause of Asc being lost from food is leaching, where the water-soluble vitamin dissolves into the cooking water, which is later poured away and not consumed. However, Asc does not leach in all vegetables at the same rate; research shows that broccoli seems to retain more than any other (Combs, 2001).

7.2.2
Biosynthesis of Asc in Plants and Mammals

There are different pathways for Asc biosynthesis in animals and plants (Linster and Van Schaftingen, 2007). In animals, firstly, the hydroxy function at C_6 of UDP-D-glucose is oxidised to form the UDP-D-glucuronic acid. The UDP is then removed by hydrolysis and the aldehyde function at C_1 is reduced, which leads to the inversion of the numbering of the carbon skeleton (Pappenberger and Hohmann, 2014). The new chemical product is called L-gulonic acid which is converted to its lactone and finally oxidised to Asc (Linster and Van Schaftingen, 2007). In plants, however, the carbon numbering is not inverted during the formation of Asc. After two isomerisation reactions for converting D-glucose to D-mannose, the GTP-activated GDP-D-mannose is converted to L-galactonolactone, which is eventually oxidised to Asc. Smirnoff and Wheeler, 2000.

7.3
Physiological Role of Asc

In humans, Asc performs numerous physiological functions, such as the synthesis of collagen, carnitine and neurotransmitters; the synthesis and catabolism of tyrosine and the metabolism of microsome (Chatterjee *et al.*, 1975; Gropper, Smith and Grodd, 2005). During these biosynthesis and metabolism, Asc acts as a reducing agent, donating electrons and preventing oxidation to keep iron or copper atoms in their reduced states (Sies, Stahl and Sundquist, 1992; Gropper, Smith and Grodd, 2005). In addition, Asc acts as a scavenger of many reactive oxygen species, such as singlet oxygen and superoxide anions (Linetsky, Ranson and Ortwerth, 1998; Wenzel *et al.*, 2004).

Research results have identified that Asc acts as an electron donor for several different enzymes. Three enzymes (prolyl-3-hydroxylase, prolyl-4-hydroxylase and lysyl hydroxylase) are required for the hydroxylation of proline and lysine in the synthesis of collagen (Kivirikko and Myllylä, 1985; Peterkofsky, 1991; Prockop and Kivirikko, 1995). Two enzymes (ε-N-trimethyl-L-lysine hydroxylase and γ-butyrobetaine hydroxylase) are necessary for the synthesis of carnitine (Dunn *et al.*, 1984; Rebouche, 1991), which is essential for the transport of fatty acids into mitochondria for ATP generation. In addition, dopamine beta-hydroxylase (Kaufman, 1974), Peptidylglycine alpha-amidating monooxygenase (Eipper *et al.*, 1993) and 4-hydroxyphenylpyruvate dioxygenase (Lindblad, Lindstedt and Lindstedt, 1970) also need the electron donated by Asc. Furthermore, recent findings on the specific requirement of Asc for the activity of several 2-oxoacid-dependent dioxygenases involved in cell signalling and the activation of transcription factors open new fascinating perspectives for further research (De Tullio and Arrigoni, 2004).

Asc has side effects in humans (Briggs, 1973). Relatively large doses of Asc may cause indigestion and diarrhoea, particularly when taken on an empty stomach (Pauling, 1976). In addition, some research results also showed that overdose of Asc resulted in the iron overload disorder and kidney stones (Cook and Reddy, 2001; Massey, Liebman and Kynast-Gales, 2005; Thomas *et al.*, 2013). However, it is estimated that the potential benefits of high doses of Asc outweigh the possible side effects (Lewin, 1976).

In plants, Asc is associated with chloroplasts and apparently plays a role in ameliorating the oxidative stress of photosynthesis. In addition, it has a number of other roles in cell division and protein modification (Smirnoff, 1996).

7.4 Chemical and Physical Properties

The chemical name of Asc is 2-oxo-L-threo-hexono-1,4-lactone-2,3-enediol. Its molecular formula was $C_6H_8O_6$ with a molecular weight of 176.13. Asc has two chiral carbon atoms in its molecule, so there are four kinds of optical isomers. Because of the possibility to form hydrogen bonds in the five-membered ring, the enol isomer is highly preferred to the keto-isomer (Pappenberger and Hohmann, 2014). The chemical structure of Asc is unique and the hydrogen on enolic hydroxyl groups (between C_2 and C_3) can be easily freed, which make the Asc show acidity and strong reduction. This reductive characteristic of Asc plays an important role in its biological functions (Mandl, Szarka and Bánhegyi, 2009).

Asc is a white crystal at room temperature, odourless, has a melting point of 190–192 °C, exhibits acidity, is soluble in water, slightly soluble in alcohol and glycerine, insoluble in chloroform and ethyl ether. Asc is readily oxidised by, O_2, Fe^{3+} and Cu^{2+} under light and high temperature.

7.5 Assay Methods

Many analytical methods have been reported in the literature for the determination of the ascorbic acid contents in different pharmaceutical products, fruits, vegetables and biological fluids. These include titrimetric (Verma, 1982; Arya, Mahajan and Jain, 1998; Kabasakalis, Siopidou and Moshatou, 2000), electrochemical (Li *et al.*, 2006), fluorimetric (Wu *et al.*, 2003), chemiluminescent (Kato *et al.*, 2005), gas chromatographic (GC) (Silva, 2005), high-performance liquid chromatographic (HPLC) (Lykkesfeldt, 2000; Iwase, 2003) and spectrophotometric (Jaselskis and Nelapaty, 1972; Arya, Mahajan and Jain, 2001; Fujita *et al.*, 2001; Janghel *et al.*, 2007) methods.

Of all these methods, spectrophotometric methods are, perhaps, the most commonly used. Salkić, Keran and Jašić (2009) developed a new, selective and accurate direct ultraviolet spectrophotometric method for the determination of L-ascorbic acid in pharmaceuticals. In this method, the oxidation of L-ascorbic acid by iodate in an acidic medium was used as a means of correcting for background absorption. The molar absorptivity was found to be 8.71×10^3 dm^3 /mol/cm at 250 nm. Beer's law was obeyed in the concentration range of 0.46 – 16.00 μg/cm^{-3} for L-ascorbic acid. The ingredients commonly found in vitamin C and multivitamin products did not interfere with the determination of Asc. Another UV spectrophotometric method was proposed for vitamin C determination in various fruits and vegetables (Rahman, Khan and Hosain, 2007). In this method, bromine water oxidises ascorbic acid to dehydroascorbic acid in the presence of acetic acid. After coupling with 2,4-dinitrophenyl hydrazine at 37 °C for 3 h, the solution is treated with 85% H_2SO_4 to produce a red colour complex and the absorbance was spectrophotometrically measured at 521 nm (Rahman, Khan and Hosain, 2007).

In recent years, HPLC methods are considered to be the best choice for Asc determination in fruits and vegetables. Compared with the traditional methods, HPLC methods are easier, cheaper and more efficient to perform (Odriozola-Serrano, Hernández-Jover and Martín-Belloso, 2007). In addition, many separation techniques, such as ion-pair (Ke *et al.*, 1994), NH_2 bonded-phase (Zerdin, Rooney and Vermue, 2003) and reverse-phase (Franke *et al.*, 2004) techniques, have been reported. These new methods and techniques avoid the problems of non-specific interference and make the results of determination of Asc content more accurate in different kinds of samples (Hernández, Lobo and González, 2006). Odriozola-Serrano, Hernández-Jover and Martín-Belloso (2007) validated and compared two UV-HPLC methods for the determination of ascorbic acid in strawberries, tomatoes and apples. Two different reducing agents (DL-1,4-dithiothreitol (DTT) or 2,3-dimercapto-1-propanol (BAL)) were used to differentiate dehydroascorbic acid and determine Asc. The reliability results showed that the UV-HPLC methods are useful for the routine analysis of Asc in fruits, and the best reliability was achieved when using a C_{18} column and DTT as reducing agent. Some other similar HPLC methods were also reported for Asc determination in different samples, indicating a wide application of HPLC in Asc assays.

7.6
Industrial Fermentation of Asc

Although many routes for Asc or 2-KLG (2-keto-L-gulonic acid, the precursor of Asc) biosynthesis, such as L-sorbose pathway (Yin *et al.*, 1980), D-sorbitol pathway (Motizuki, 1966; Sugisawa *et al.*, 1990), 2-keto-D-gluconic acid pathway (Shinagawa *et al.*, 1976), 2,5-diketo-D-gluconic acid pathway (Sonoyama *et al.*, 1987), D-gluconic pathway (Anderson *et al.*, 1985), have been reported and investigated for years, there are only two fermentation processes that are industrially applied, that is, the Reichstein process and the two-step fermentation process.

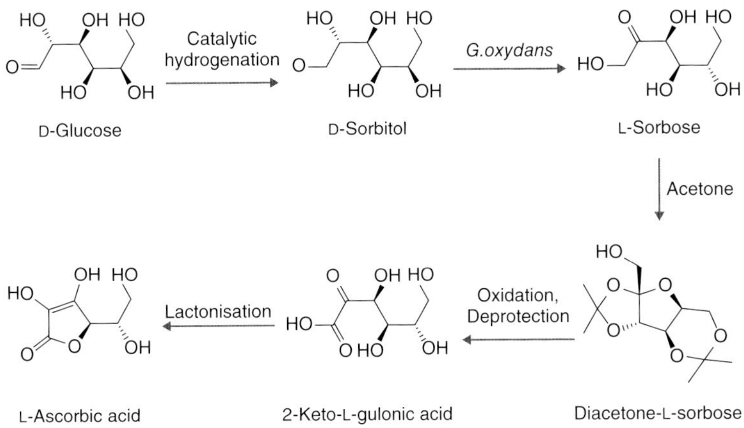

Figure 7.3 The Reichstein process for vitamin C synthesis.

7.6.1
The Reichstein Process: The Major Industrial Asc Process until the Late 1990s

Since the discovery of Asc, the growing market demand for Asc has accelerated the pace of industrial production of Asc in the 1930s (Yang, 2012). Although various efforts had been put on obtaining synthetic Asc, the process conceived by Reichstein and Grüssner is the only one applied on an industrial scale and prevailed the industrial production of Asc until the late 1990s (Crawford and Crawford, 1980; Mandlaa, 2014) (Figure 7.3).

7.6.1.1
The Establishment of the Reichstein Process

In the early 1930s, Reichstein and Grüssner had tried to develop a method for Asc synthesis. According to their hypothesis and a series of experimental results, a novel process for Asc synthesis was established in the 1930s (Reichstein and Grüssner, 1934). In this process (Figure 7.3), as an initial material, D-glucose is converted to D-sorbitol by catalytic hydrogenation. D-sorbitol is then bioconverted to L-sorbose using *Gluconobacter* spp. Thirdly, L-sorbose is oxidised to 2-KLG after several chemical steps. Finally, 2-KLG is rearranged to Asc by lactonisation.

7.6.1.2
Bioconversion of D-Sorbitol to L-Sorbose by *Gluconobacter*

In the Reichstein process, the conversion of D-sorbitol to L-sorbose is the only biological conversion step using microbes, which plays an important role for a cost-efficient synthesis of Asc. In this step, the inversion of the numbering of the carbon skeleton of D-glucose occurred because of the exclusively oxidation at C_5 of the substrate D-sorbitol (Reichstein and Grüssner, 1934). This exquisite regioselectivity was achieved by acetic acid bacteria (Gupta *et al.*, 2001). Acetic acid bacteria are well known for their ability to partially oxidise several sugars and sugar alcohols (Deppenmeier and Ehrenreich, 2009). *Gluconobacter* (previously named as *Acetobacter suboxydans*), one of the acetic acid bacteria, shows a highest effectiveness in partially oxidising sugar and sugar alcohols (Gupta *et al.*, 2001; Deppenmeier, Hoffmeister and Prust, 2002). Currently, D-sorbitol is oxidised to L-sorbose by *Gluconobacter oxydans* with an almost 100% conversion rate on the industrial scale (Giridhar and Srivastava, 2000a; Jing *et al.*, 2009).

7.6.1.3
The Key Enzyme of *Gluconobacter* for L-Sorbose Production

In 2002, Sugisawa and Hoshino purified an 80-kDa dehydrogenase from *G. oxydans* IFO3255 and designated as sorbitol dehydrogenase (SLDH). This enzyme employed the pyrroloquinoline quinone (PQQ) as a redox cofactor (Sugisawa and Hoshino, 2002). D-sorbitol, D-mannitol, glycerol and D-gluconic acid can be oxidised by SLDH (Sugisawa and Hoshino, 2002; Matsushita *et al.*, 2003; Salusjävi *et al.*, 2004). Based on the peptide sequences obtained from purified Sldh, the encoding gene (*sld*A) was sequenced and identified (Miyazaki *et al.*, 2002). The gene shows a significant sequence homology to the membrane-bound quinoprotein glucose dehydrogenases from *E. coli*, *G. oxydans* and *Acinetobacter calcoaceticus* (Pappenberger and Hohmann, 2014). Although other D-sorbitol oxidising enzymes from other *Gluconobacter* strains were also purified and characterised (Shinagawa *et al.*, 1982; Choi, Lee and Rhee, 1995), Sldh is generally considered to be the major polyol dehydrogenase of *G. oxydans* (Matsushita *et al.*, 2003).

7.6.1.4
Oxidation of L-Sorbose to 2-KLG and Rearrangement to Asc

Before L-sorbose is oxidised, the hydroxyl groups at C_2 and C_3 and C_4 and C_6 are protected by acetone. The resulting diacetone-L-sorbose is then oxidised at C_1 to the carboxy group by potassium permanganate, which produces diacetone-2-KLG. After a hydrolysis reaction, 2-KLG is obtained and eventually lactonised to Asc (Pappenberger and Hohmann, 2014).

7.6.2
The Two-Step Fermentation Process for Asc Production

Reichstein's process is the classical one via a single biocatalysis step within a series of chemically based unit operations (Reichstein and Grüssner, 1934). However, this method is highly energy consuming and relies on the use of a number of environmentally hazardous chemicals (Xu *et al.*, 2004). Compared to Reichstein's process, the application of microbial process has become more attractive for its lower cost and much less ecological problems (Boudrant, 1990; Chotani *et al.*, 2000).

In the 1970s, by co-operating with Beijing pharmaceutical factory, the researchers from Institute of Microbiology, Chinese Academy of Sciences, intended to develop a fermentation process for Asc production. After screening 5327 strains of bacteria, they obtained a strain N1197A that could produce 2-KLG from L-sorbose (Yin *et al.*, 1980). Further study showed that the N1197A contained two different strains. According to their physiological and biochemical characteristics, the strain with large colonies and the strain with small colonies were identified as *Pseudomonas striata* and *G. oxydans* (now renamed as *Ketogulonicigenium vulgare*), respectively. Yin *et al.* (1980) found that only *G. oxydans* could convert L-sorbose to 2-KLG. However, if there was no *P. striata* during the fermentation, the growth of *G. oxydans* was very poor and the yields of 2-KLG was fairly low. These results suggested that the fermentation with mixture of two strains was necessary for enhanced 2-KLG production. After study for years, several accompanying strains was found and applied in Asc fermentation instead of *P. striata*, such as *Bacillus megaterium* (Feng *et al.*, 2000), *Bacillus cereus* (Jiao *et al.*, 2002) and *Bacillus thuringiensis* (Yang *et al.*, 2013). By optimisation, the conversion rate of L-sorbose to 2-KLG of this co-culture system has been enhanced to 90–92% (Li and Zhang, 1997; Zhang *et al.*, 1998; Lu *et al.*, 2001; Xu *et al.*, 2004), which further promotes the establishment and industrial application of two-step fermentation process of Asc. Currently, the two-step fermentation process is applied by all the Chinese manufacturers for industrial production of Asc.

The two-step fermentation process of Asc, in a way, can be considered as improved Reichstein's process (Figures 7.3 and 7.5). As its name suggests, there are two fermentation steps in this fermentation process, that is, L-sorbose fermentation in the first step and 2-KLG fermentation in the second step. The biggest difference between the two processes is that the 2-KLG is produced by bioconversion in the two-step fermentation process, instead of chemical conversion in Reichstein's process.

7.6.2.1
The First Step of Fermentation: Conversion of D-Sorbitol to L-Sorbose

Producing Microorganisms Many strains show the ability to convert D-sorbitol to L-sorbose, such as *G. suboxydans* (Sugisawa and Hoshino, 2002), *A. suboxydans* (Giridhar and Srivastava, 2000a) and *G. oxydans* (Bremus *et al.*, 2006; Ge *et al.*, 2013). However, in the current industrial process of Asc fermentation, *G. oxydans* (previously named as *Acetobacter melanogenus* or *Gluconobacter melanogenus* in China) (Jing *et al.*, 2011) is widely applied in Asc production by Chinese manufacturers because of its efficient conversion of D-sorbitol to L-sorbose (Ge *et al.*, 2013; Wang *et al.*, 2013).

G. oxydans is a Gram-negative bacterium belonging to the family *Acetobacteraceae* (De Ley and Swings, 1994). *G. oxydans* is an obligate aerobe, having a respiratory type of metabolism using oxygen as the terminal electron acceptor (Gupta *et al.*, 2001). The strain has a number of membrane-bound dehydrogenases involved in many oxidation reactions for incomplete oxidation of sugars, alcohols and acids (De Ley and Swings, 1994). Incomplete oxidation leads to nearly quantitative yields of the oxidation products, making *G. oxydans* important for industrial use. *Gluconobacter* strains can be used industrially to produce L-sorbose from D-sorbitol; D-gluconic acid, 5-keto- and 2-keto-gluconic acids from D-glucose; and dihydroxyacetone from glycerol (Gupta *et al.*, 2001).

The genome of *G. oxydans* H24, an industrial strain used by Chinese Asc producer, has been sequenced and reported by Ge *et al.* (2013). The complete genome consists of a circular chromosome and a plasmid. The chromosome is composed of 3 602 424 bp, with a G + C content of 56.25%. The plasmid contains 213 808 bp, with a G + C content of 56.14%. There are a total of 3732 putative open reading frames (ORFs) (3469 in the chromosome and 263 in the plasmid), yielding a coding intensity of 89.86%. A total of 59 tRNA-encoding genes and 5 16S-23S-5S rRNA-encoding operons were identified. In addition, three kinds of Sldhs and the gene cluster responsible for the synthesis of the cofactor PQQ (*pqq*ABCDE, 3137 bp) were also found (Ge *et al.*, 2013).

Sorbitol Dehydrogenase: The Key Enzyme for Conversion of D-Sorbitol to L-Sorbose From genome information of industrial strain *G. oxydans* H24, Ge *et al.* (2013) found two different membrane-bound and one cytoplasmic Sldhs. They are pyrroloquinoline quinone–dependent D-sorbitol dehydrogenase (PQQ-SLDH), flavin adenine dinucleotide–dependent D-sorbitol dehydrogenase (FAD-SLDH) and NADP-dependent D-sorbitol dehydrogenase (NADP-SLDH), respectively. Among them, PQQ-SLDH is now generally believed to play an important role in converting D-sorbitol to L-sorbose (Matsushita *et al.*, 2003). In 2002, an 80-kDa PQQ-SLDH was purified from *G. oxydans* IFO3255 (Sugisawa *et al.*, 2002) and the encoded gene *sldA* was cloned by reverse genetics and sequenced (Miyazaki *et al.*, 2002).

Metabolic Regulation of the Bioconversion of D-Sorbitol to L-Sorbose There are few reports on regulation of bioconversion of D-sorbitol to L-sorbose because of its highly efficient conversion rate on industrial scales. More studies on optimisation of fermentation process and its conditions are described in 'fermentation process' part of this chapter.

7.6.2.2
The Second Step of Fermentation: Conversion of L-Sorbose to 2-Keto-L-Gulonic acid

Producing Microorganisms There are two kinds of strains, a companion strain (or named as helper strain) and a conversion strain, involved in the second step of Asc fermentation, that is, conversion of L-sorbose to 2-KLG (Yin *et al.*, 1980).

The Companion Strain: Stimulating Growth and 2-KLG Production of *K. vulgare* The companion strain could not convert L-sorbose to 2-KLG, but it could secret activators that stimulate the growth of conversion strain and improve greatly the yield of 2-KLG (Feng *et al.*, 2000; Jiao *et al.*, 2002). In the past 30 years, many strains had been isolated and were considered to be the companion strains, such as *P. striata* (Yin *et al.*, 1980), *B. megaterium* (Feng *et al.*, 2000), *B. thuringiensis* (Song *et al.*, 1997; Yang *et al.*, 2013), *B. cereus* (Jiao *et al.*, 2002) and *Xanthomonas maltophilia* (Takagi, Sugisawa and Hoshino, 2010). Furthermore, it was reported that *Sporobolomyces roseus* (Zhong *et al.*, 2004), a eukaryote, also can be an effective companion strain. In our opinion, it seems that all the spores-forming strains could be companion strains when they are cultured under optimal conditions.

The companion strain, such as *B. megaterium* (Feng *et al.*, 2000), *B. cereus* (Jiao *et al.*, 2002) and *B. thuringiensis* (Yang *et al.*, 2013), belongs to genera *Bacillus*. They are Gram-positive, mainly aerobic. Spore formation during the cultivation is one of the basic characteristics of a companion strain. It has been found that the companion strain could not utilise L-sorbose and 2-KLG (Feng *et al.*, 1998). The companion strain is generally considered to stimulate *K. vulgare* propagation and 2-KLG accumulation by generating and releasing some metabolites into the fermentation broth (Lu *et al.*, 2001; Zhang *et al.*, 2010b; Ma *et al.*, 2011).

Currently, *B. megaterium* and *B. cereus* were the two main companion strains applied in industrial Asc fermentation (Feng *et al.*, 2000; Jiao *et al.*, 2002). *B. megaterium* WSH-002, one of the industrial companion strains, has been sequenced (Liu *et al.*, 2011b). The 4.14-Mb genome of *B. megaterium* WSH-002 contains four replicons, a circular chromosome (4.04 Mb) encoding 5186 predicted ORFs and three circular plasmids, named pBME_100 (0.074 Mb), pBME_200 (9699 bp) and pBME_300 (7006 bp), with mean GC contents of 39.1%, 36%, 32.2% and 33.2%, respectively. There are 5482 protein-encoding genes, 99 tRNAs and 10 rRNA operons. Among them, 2460 functional descriptions, 1327 gene abbreviations and 856 EC numbers were assigned to the WSH-002 genome by function annotation. Furthermore, 782 genes were assigned according to the Kyoto Encyclopedia of Genes and Genomes metabolic pathways (Liu *et al.*, 2011b).

***K. vulgare*: The 2-KLG-Producing Strain** In contrast to numerous companion strains, only one 2-KLG-producing bacterium has been used in industrial fermentation so far. It is *K. vulgare* (previously named as *G. oxydans*, especially in China) (Urbance *et al.*, 2001). *K. vulgare* contains the whole enzymes for conversion of L-sorbose to 2-KLG, but its growth is very poor when it is cultured alone (Liu *et al.*, 2011c). Hence, the two-strain co-culture system (TSCS) of *K. vulgare* and a companion strain is necessary for efficient production of 2-KLG.

K. vulgare is a Gram-negative, facultative anaerobic, chemoheterotrophic soil microorganism with an optimum temperature range of 27–31 °C and pH range of 7.2–8.5 (Urbance *et al.*, 2001). In the second step of Asc fermentation, *K. vulgare* contains the complete enzyme system for synthesising 2-KLG from L-sorbose (Liu *et al.*, 2011a), but its growth is usually poor and the yield of 2-KLG is very low when cultured by itself (Lu *et al.*, 2001, 2003). However, the growth of *K. vulgare* can be enhanced by addition of certain nutrients (Leduc, Troostembergh and Lebeault, 2004) or by co-culturing with a companion strain (Song *et al.*, 1997; Feng *et al.*, 2000), indicating that some key metabolites involved in the metabolic pathways may be lost. Leduc, Troostembergh and Lebeault (2004) found that addition of adenine, guanine, thymine and low-molecular-weight reduced folates could improve the growth of *K. vulgare*, suggesting that the purine nucleotides and deoxythymidylate biosynthesis pathways are probably insufficient in *K. vulgare*.

Most studies on *K. vulgare* have focussed on elucidating the pathway of L-sorbose metabolism (Asakura and Hoshino, 1999; Sugisawa, Miyazaki and Hoshino, 2005; Miyazaki, Sugisawa and Hoshino, 2006). Two enzymes, L-sorbose/L-sorbosone dehydrogenase (SSDH) and L-sorbosone dehydrogenase (SNDH), have been identified as the key enzymes catalysing the oxidation of L-sorbose to 2-KLG (Asakura and Hoshino, 1999; Miyazaki, Sugisawa and Hoshino, 2006). In addition, *K. vulgare* can synthesize Asc via different substrates, such as D-sorbitol, L-sorbose, L-gulose, L-sorbosone, L-galactose, L-idose, L-talose, L-gulono-1,4-lactone and L-galactono-1,4-lactone (Sugisawa, Miyazaki and Hoshino, 2005).

The genomes of two industrial strains of *K. vulgare* (*K. vulgare* Y25 and *K. vulgare* WSH-001) have been sequenced, respectively (Liu *et al.*, 2011a; Xiong *et al.*, 2011). The genome of *K. vulgare* Y25 consists of a circular chromosome and two plasmids. The chromosome is composed of 2 776 084 bp, with a GC content of 61.72%. One plasmid contains 268 675 bp, with a GC content of 61.35%, and the other contains 243 645 bp, with a GC content of 62.63%. There are a total of 3290 putative ORFs (2807 [chromosome], 256 [pYP1] and 227 [pYP2]) using Glimmer, giving a coding intensity of 91.05% (Xiong *et al.*, 2011). The complete genome sequence of *K. vulgare* WSH-001 is composed of a circular, 2 766 400 bp chromosome and two circular plasmids named pKVU_100 (267 986 bp) and pKVU_200 (242 715 bp) with mean GC contents of 61.69%, 61.33% and 62.58%, respectively. There are 2604 protein-encoding genes, 3 rRNA operons and 51 tRNA-encoding genes in the chromosome and 246 and 215 protein-encoding genes in plasmids pKVU_100 and pKVU_200, respectively. A total of 2497 functional descriptions, 1279 gene abbreviations and 820 EC numbers were assigned in the WSH-001 genome by function annotation. Among them, the genes for the 2-KLG synthesis pathway from L-sorbose were annotated; four genes encoding SSDH, responsible for converting L-sorbose to L-sorbosone, are highly homologous to ssdA1 (AB092515), ssdA2 (AB092516), ssdA3 (AB092517) and ssdB (AB092518) of *K. vulgare* DSM 4025. The gene of SNDH that is responsible for the conversion of L-sorbosone to 2-KLG was located in plasmid pKVU_200 (Liu *et al.*, 2011a). Function annotation indicated that the *K. vulgare* lacks most of the genes or gene clusters for the biosynthesis of many kinds of amino acids, nucleotides and cofactors.

Sorbose/Sorbosone Dehydrogenase: The Key Enzyme for Conversion of L-Sorbose to 2-KLG The enzyme sorbose/sorbosone dehydrogenase (SSDH), converting L-sorbose to 2-KLG, was purified by Asakura and Hoshino from a soluble fraction of *K. vulgare* DSM 4025 in 1999 (Asakura and Hoshino, 1999). The enzyme is a unique quinoprotein dehydrogenase catalysing not only the conversion of L-sorbose to L-sorbosone but also that of L-sorbosone to 2-KLG. The molecular weight of the enzyme was about 135 kDa, consisting of two subunits with molecular weights of 64.5 and 62.5 kDa (Asakura and Hoshino, 1999). The PQQ is employed as prosthetic group for the enzyme while oxidising L-sorbose to 2-KLG (Pappenberger and Hohmann, 2014). The cytochrome c from *K. vulgare* was found to act as a physiological electron acceptor of the enzyme. The optimum enzyme activity occurred in the pH range of 7.0–9.0 (Asakura and Hoshino, 1999). It showed extremely broad substrate specificity for primary and secondary alcohols, aldehydes, aldoses, ketoses and other sugar alcohols (Asakura and Hoshino, 1999). The genes encoding SSDH had been isolated and described by Roche in 1996 (Asakura *et al.*, 1996). Based on the recent genome sequences of *K. vulgare* strains, the related genes have been well identified (Liu *et al.*, 2011a; Xiong *et al.*, 2011).

Interaction Mechanisms between *K. vulgare* and the Companion Strain The interaction relationship between the two bacteria, *K. vulgare* and the companion strain, has been a hot topic in Vc fermentation research in last 30 years. Elucidation of the mechanism of interaction is an important target for scientists in this field.

Although it is clear that the companion strain can stimulate the growth and 2-KLG production of *K. vulgare* by secreting some metabolites during the fermentation, the detailed mechanisms underlying the interactions between *K. vulgare* and the companion strain remain largely undefined (Zou, Liu and Chen, 2013). The metabolites may be the proteins (Lu *et al.*, 2001), amino acids (Zhang *et al.*, 2011) or other substances (Zhou *et al.*, 2011). In fermentation, both the cytosol and extracellular metabolites which are released during the lysis of the companion strain were shown to promote the proliferation of *K. vulgare* and 2-KLG productivity (Feng *et al.*, 2000). Meanwhile, 2-KLG secreted by *K. vulgare* can also inhibit the growth of the companion strain and shorten its growth cycle, by accelerating the lysis of the companion strain (Li and Zhang, 1997; Mandlaa, 2014). Hence, it was believed that both mutualism and antagonism exist in this artificial microbial ecosystem (Zhou *et al.*, 2011).

With the release of genome sequences for *K. vulgare* and *B. megaterium* and the development of other high-throughput techniques, several systems biology approaches have been used to achieve a more comprehensive understanding of the interaction mechanisms between *K. vulgare* and the companion strain (Figure 7.4). Based on the global pathway analysis, Zhang *et al.* (2011) reported that most of the essential amino acids were deficient during the growth of *K. vulgare*, such as L-histidine, L-glycine, L-lysine, L-proline, L-threonine, L-methionine, L-leucine and L-isoleucine. Among them, L-glycine, L-proline, L-threonine and L-isoleucine were considered to play vital roles in *K. vulgare* growth and 2-KLG

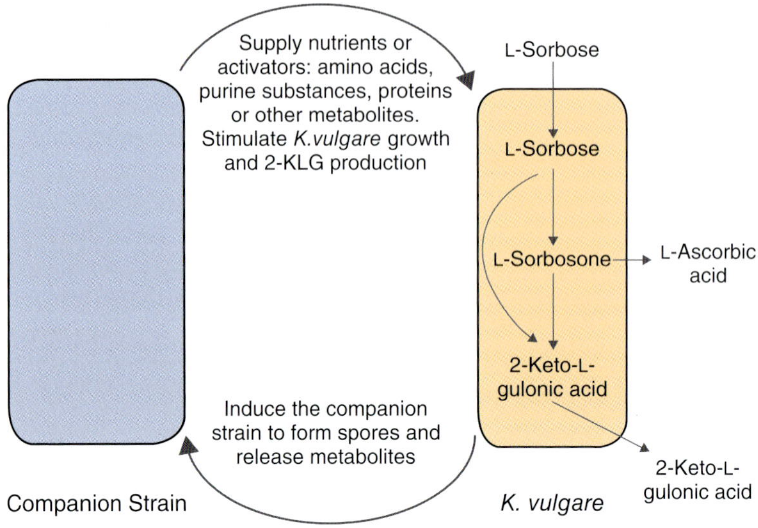

Figure 7.4 The interaction relationship between *K. vulgare* and the companion strain.

production (Liu *et al.*, 2011c). By applying an integrated time-series proteomic and metabolomic analysis, Ma *et al.* (2011) investigated interaction between the two species. They quantitatively identified approximately 100 metabolites and 258 proteins. Principal component analysis of all the metabolites showed that glutamic acid, 5-oxo-proline, L-sorbose, 2-KLG, 2,6-dipicolinic acid and tyrosine were potential biomarkers to distinguish the different time-series samples. Most of these metabolites were closely correlated with the sporulation of *B. megaterium*, indicating that sporulation process might play an important role in this microbial interaction. The proteomic results showed that proteins combating against intracellular reactive oxygen stress and proteins involved in pentose phosphate pathway, L-sorbose pathway, tricarboxylic acid cycle and amino acids metabolism were up-regulated when the cell lysis of *B. megaterium* occurred. These discoveries showed the companion strain provided key elements for the growth and 2-KLG production of *K. vulgare*. In addition, by employing metabolomics based on gas chromatography coupled with time-of-flight mass spectrometry (GC-TOF-MS), Zhou *et al.* (2011) found that the microorganisms interact through exchanging a number of metabolites. Both intracellular metabolism and cell–cell communication via metabolic cooperation were essential in determining the population dynamics in the co-culture ecosystem. The contents of amino acids and other nutritional compounds in *K. vulgare* were rather lower in comparison to those in *B. megaterium*, but the levels of these compounds in the medium surrounding *K. vulgare* were fairly high, even higher than that in fresh medium. Erythrose, erythritol, guanine and inositol accumulated around *B. megaterium* were consumed by *K. vulgare* upon its migration. Then, the oxidisation products of *K. vulgare*, including 2-KLG, were sharply increased. Upon co-culturing of *B. megaterium* and *K. vulgare*, 2,6-dipicolinic acid (the biomarker of sporulation of *B. megaterium*), was remarkably increased compared with those in the monocultures. Therefore, the interactions between *B. megaterium* and *K. vulgare* were a synergistic combination of mutualism and antagonism (Zhou *et al.*, 2011).

Regulation of the Bioconversion of L-Sorbose to 2-KLG According to the observed interaction between *K. vulgare* and the companion strain in the artificial ecosystem, special strategies can be designed to regulate this microbial process to enhance 2-KLG production (Zou, Liu and Chen, 2013). Previous work has shown that the starting inoculums, medium, pH and other environmental factors can affect the final accumulation of 2-KLG (Yan *et al.*, 1981; Zhang *et al.*, 1998; Yang *et al.*, 2008; Mandlaa *et al.*, 2011). Currently two measures have been taken for regulating this mixed fermentation: (i) regulating release of metabolites from companion strain and (ii) the genome-based methods for finding new strategies.

Regulating release of metabolites from companion strain could be a useful strategy for improving the productivity of 2-KLG. A new biochemical strategy was introduced by adding lysozyme to specifically damage the cell wall structure of *B. megaterium* with a result that its intracellular components were released. When 10 000 U/ml lysozyme were added after 12 h of co-culture in a 7 l jar fermenter, the growth rate of *K. vulgare*, L-sorbose consumption rate and 2-KLG productivity increased 27.4%, 37.1% and 28.2%, respectively. The fermentation time was decreased to 56 h, shorter by 20.6% (Zhang *et al.*, 2010b).

Application of genome information may be a new method for improving the 2-KLG production. From the genomics and proteomics study, it showed that amino acid transport and metabolism may play critical roles in the growth of *K. vulgare*. By reconstructing the amino acid metabolic pathways on the basis of the genome annotation, Liu *et al.* (2011c) found that the *de novo* biosynthesis pathways of eight different amino acids (L-histidine, glycine, L-lysine, L-proline, L-threonine, L-methionine, L-leucine and L-isoleucine) in *K. vulgare* were deficient. Among them, glycine, L-proline, L-threonine and L-isoleucine were experimentally proved to play vital roles in *K. vulgare* growth and 2-KLG production. With the aim of increasing the 2-KLG production efficiency and decreasing the cost of amino acids, a food-grade gelatin containing a high concentration of the non-essential amino acids, such as glycine and L-proline, was developed as a substitute. In the 7 l, 1 and 200 m^3 fermenters, the addition of gelatin into the ecosystem shortened the fermentation time by 17.3%, 15.6% and 8%, respectively (Liu *et al.*, 2011c).

7.6.2.3
Strain Development: Genetic Modification, Molecular Genetics and Metabolic Engineering

Strain Improvement in the First Step of Asc Fermentation Since the conversion rate of D-sorbitol of L-sorbose by *G. oxydans* had been reached to and more than 98% in industrial fermentation for decades, a few studies on *G. oxydans* breeding were reported (Jing *et al.*, 2011). On the contrary, more effort was focussed on the recombination of bacteria in the first-step fermentation (from D-sorbitol to 2-KLG) (Manning and Kahn, 1992; Hosshino *et al.*, 2012).

Strain Improvement in the Second Step of Asc Fermentation After establishment of the two-step fermentation process, many reports have been published on strain improvement by random mutagenesis and selection. By using the traditional mutagenesis method, such as UV irradiation and NTG (nitrosoguanidine), several highly efficient mutants of *K. vulgare* and companion strain were obtained. By mutation with ultraviolet light, Zhang *et al.* (1999) obtained two mutant strains of *B. megaterium* (Bn, B5) with a higher tolerance of low pH and high KLG concentration. The conversion rates of L-sorbose to 2-KLG by co-culture of Bn-K.v and co-culture of B5-K.v were increased by 3.5% and 3.3%, respectively. Lu, Guo and Li (2005) obtained a G5 strain, a mutant from *G. oxydans* (presumably a *Ketogulonicigenium* species) after UV mutation, showing an increased conversion rate (13.49%) and enhanced 2-KLG production (83.6 mg/ml) when co-cultured with *B. cereus*. By series of mutagenesis by UV and NTG, a mutant strain UN-366 with a high efficiency of association ability was screened from *B. thuringiensis* (Guo *et al.*, 2006). The conversion rate of L-sorbose to 2-KLG was increased by 6.32% when UN-366 was co-cultured with *K. vulgare*.

In recent years, two new mutagenesis technologies, ion beam implantation and spaceflight mutation, has been applied in Asc-strain screening by Chinese producers. Ion beam mutation causes a higher mutation rate and a wider mutational spectrum with lower damage (Yu, 1998). *G. oxydans* (GO29) and *B. megaterium* (BM80) were bred by ion beam implantation and mutants GO112 and BM302 were obtained (Xu *et al.*, 2004). When compared with the original mixture GO29-BM80, the averaged 2-KLG transformation rate of mutants GO112-BM302 was increased from 79.3% to 94.5% after eight passages in shake flasks and became stable at 92.0% in 180-t fermenters, indicating increased efficiency of mutants in 2-KLG accumulation. Spaceflight breeding technology has been shown to be more efficient than other radiation methods for microbes (Ermolenko *et al.*, 2000; Fang, Zhao and Gu, 2005). The biological response of microorganisms to space conditions, such as microgravity, cosmic radiation and vacuum, has led to the development of a range of new mutation-breeding techniques for industrial fermentations (Ermolenko *et al.*, 2000; Gao *et al.*, 2010). Guo *et al.* (2004) studied the effects of spaceflight on the *B. megaterium* and obtained four efficient mutants of *K. vulgare*. Yang *et al.* (2013) screened out an optimal combination of mutants *B. thuringiensis* 320 and *K. vulgare* 2194 after spaceflight of a Chinese spacecraft Shenzhou VII. When compared to the co-culture of their parent strains, the conversion rate of L-sorbose to 2-KLG by KB2194-320 in shake-flask fermentation was increased from 82.7% to 95.0%. Furthermore, a conversion rate of 94.5% and 2-KLG productivity of 1.88 g/l/h were achieved with KB2194-320 in industrial-scale fermentation (Yang *et al.*, 2013).

Many scholars have also tried to improve the capability of strains by means of gene engineering techniques. For 2-KLG producing strain, Manning and Kahn (1992) invented a method for constructing recombinant bacteria by using of transposon mutagenesis. A recombinant mutant *G. oxydans* M23-15 with a low activity of 2-KLG reductase was obtained. Hosshino *et al.* (2012) patented an overexpression system by introducing one or more copies of a polynucleotide encoding SDH into the genome of the host strain *G. oxydans* to enhance 2-KLG yields. The shuttle vectors for *K. vulgare* and *Escherichia coli* have been constructed to facilitate the process of genetic manipulation of *K. vulgare* (D'elia, 2006). Cai *et al.* (2012) successfully inserted five genes (for folate biosynthesis from *Lactococcus lactis* MG1363) (Leduc, Troostembergh and Lebeault, 2004) into *K. vulgare* and regulated the overexpression of these genes by L-sorbose dehydrogenase gene promoter Psdh from *K. vulgare* (Fu *et al.*, 2007). Compared with the parent strain *K. vulgare* DSM4025, the folate concentration in the recombinant *K. vulgare* Rif (pMCS2PsdhfolBC) was enhanced by more than eightfold, and the cell density and 2-KLG productivity were increased by 25% and 35%, respectively (Cai *et al.*, 2012). For companion strain, genetic manipulation is used to change its physiological behaviour. Zhu *et al.* (2012) constructed the spo0A and spoVFA deletion mutants of *B. megaterium* WSH002. Although the L-sorbose conversion rates of spo0A and spoVFA mutant co-culture systems were significantly reduced, the results showed that the sporulation and spore stability of *B. megaterium* play key roles in the enhancement of 2-KLG biosynthesis. Although many studies on recombination of bacteria were reported, there was no recombined strain was applied on industrial scale because of their low efficiency in 2-KLG production.

7.6.2.4
Fermentation Process

Figure 7.5 illustrates the two-step fermentation process of Asc. In the first step, as in the Reichstein process, D-sorbitol is oxidised to L-sorbose by *G. oxydans*. 2-KLG is then produced from L-sorbose by a mixed culture of *K. vulgare* (previously identified as *G. oxydans*) (Urbance *et al.*, 2001; Yang *et al.*, 2006) and *Bacillus* spp. in the second step (Yin *et al.*, 1980). 2-KLG is finally converted to Asc after several chemical steps. In this process, the first-step fermentation takes about 24 h and results in high concentration of L-sorbose (25–28%, w/v). However, in contrast to the first step of fermentation, the mixed-culture fermentation in the second step takes much more time (40–70 h) and reaches a lower product concentration (8–10%, w/v), indicating a low productivity and conversion rate. Hence, more efforts for an improved fermentation efficiency were mainly focussed on the second step of Asc fermentation (Xu *et al.*, 2004; Lv *et al.*, 2011b).

The Process Improvement in the First Step of Asc Fermentation The biotransformation of D-sorbitol to L-sorbose by *G. oxydans* is a key step in industrial production of Asc. Since L-sorbose fermentation is a typical substrate inhibition fermentation, the batch fermentation process is mainly used for the commercial production (Giridhar and Srivastava, 2000a; Wulf, Soetaert and Vandamme, 2000; Jing *et al.*,

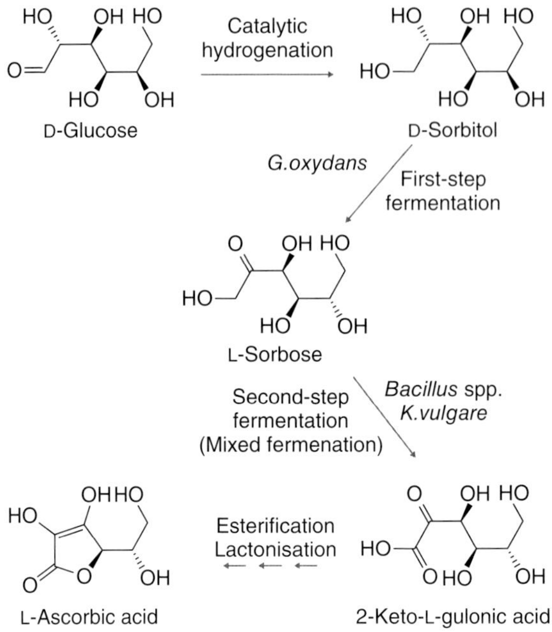

Figure 7.5 The two-step fermentation process for Asc production.

2011). In this process, the substrate concentration remains at 20–25% and the fermentation periods are 18–24 h with the batch fermentation productivity of 12–13 g/l/h. The fermentation is conducted at 32 °C with natural pH and oxygen transfer rate of 300–500 mmol/l/h by using an air-lift fermenter. Although the batch fermentation is easy to control, the disadvantage is visible. The batch process takes too much unproductive time, particularly for cleaning, sterilisation and inoculation procedures that must be conducted after each batch cycle, which significantly reduced the fermentation efficiency (Hekmat, Bauer and Neff, 2007). Optimisation of this process is of significant economic importance (Wulf, Soetaert and Vandamme, 2000; Sefcovicová *et al.*, 2009).

In order to reduce the inhibiting effects of high concentration D-sorbitol on *G. oxydans* and enhance the fermentation efficiency, several optimised fermentation processes were studied. These processes include the fed-batch fermentation and the repeated fed-batch fermentation. Giridhar and Srivastava (2000a) proposed a fed-batch fermentation to eliminate the inherent substrate inhibition present in batch fermentation. The fed-batch fermentation conducted by feeding nutrients containing 600 g/l of sorbitol at a constant feed rate of 0.36 l/h, and yielded a productivity of 17.7 g/l/h with a final sorbose concentration of 320 g/l. In fed-batch fermentation, the multiple nutrients feeding strategy can be better in improving L-sorbose productivity than the pulse nutrients feeding strategy (Giridhar and Srivastava, 2000b). In our studies, however, the additional feeding of nutrients (except for substrate of D-sorbitol) was not necessary during the fed-batch fermentation runs. By using a mutant strain G757 (Jing *et al.*, 2011), a simplified fed-batch fermentation process was proposed (Yang, 2012). In this process, batch fermentation was initiated with a substrate concentration of 23–25% (w/v). After incubation for 12–16 h, 70–75% (w/v) of the sterilised sorbitol solution was added to fermenter and the fed-batch fermentation started. 33–36% of L-sorbose was finally obtained after another 12–16 h fermentation. This process had been conducted in 110 m^3 fermenters for five batches with an average final sorbose concentration of 330 g/l and an average fermentation period of 29 h (Yang, 2012). Giridhar and Srivastava (2001) also conducted the repeated fed-batch sorbose fermentation in L-sorbose production. A batch culture with an initial concentration of 200 kg/m^3 D-sorbitol was converted to a repeated fed-batch by harvesting one-third of fermenter volume and recharging with fresh nutrient medium having the same sorbitol concentration. The average L-sorbose productivity of 19.31 kg/m^3/h was obtained after four cycles of harvesting and recharging. These results suggest that fed-batch or repeated fed-batch can enhance the substrate concentration and improve the L-sorbose productivity, indicating a better fermentation process than the traditional batch fermentation process.

The Process Improvement in the Second Step of Asc Fermentation The batch fermentation in the second step of Asc fermentation is widely applied in the current industrial production of 2-KLG, the precursor of Asc. After three-grade expanding incubation for seed culture in 40–44 h, the air-lift fermenters are then inoculated and the second-step fermentation is initiated. During the process, the fermentation is conducted at 29 °C with a constant pH of 7.0 (adjusted by sodium hydroxide). The initial substrate concentration is 10 mg/ml and a consistent addition of the pasteurised L-sorbose broth from the first-step fermentation is conducted at the fermentation stage of 10–30 h. The oxygen transmission rate is kept at 100 mmol/l/h in order to meet the needs for bacterial growth and L-sorbose conversion. The fermentation period usually varies between 40 and 65 h with a final 2-KLG concentration of 90–110 g/l.

In order to improve the fermentation efficiency, the optimisation of the fermentation process has been continuing for decades. On the basis of the traditional batch fermentation process, tremendous work has been done to improve the process stability and 2-KLG yield.

Nutrition is the basis for the growth and metabolism of microorganism. In the mixed fermentation for 2-KLG production, the medium components and their concentrations have crucial effect on both growth of bacteria and 2-KLG productivity. Yang *et al.* (2008) found that glucose supplemented with a proper concentration can effectively improve the conversion rate of 2-KLG and shorten the fermentation period. Other studies showed that folate, glutathione, amino acids (such as L-glycine, L-proline, L-threonine and L-isoleucine), adenosine 5′-triphosphate (ATP), nitrogenous bases, gelatin can also significantly enhance the fermentation efficiency (Leduc, Troostembergh and Lebeault, 2004; Lv *et al.*, 2011a; Liu *et al.*, 2011c; Zhang *et al.*, 2011; Cai *et al.*, 2012; Huang *et al.*, 2013). Inorganic salts, such as Fe^{3+}, Mg^{2+} and Mn^{2+}, can stimulate the 2-KLG production by enhancing the key enzymes, (i.e. SDH and SNDH) activities of *K. vulgare*

(Ji *et al.*, 2010). Some light rare earth elements (REEs), such as lanthanum, cerium, neodymium and samarium, showed promoting effect on 2-KLG production at concentration up to 5 mM (Lyu *et al.*, 2014). Mandlaa *et al.* (2011) optimised the fermentation medium by using the response surface method in flask fermentation. The components of optimised medium are as follows: L-sorbose 9%, corn steep liquor 1.95%, urea 1.0%, KH_2PO4 0.03%, $MgSO_4$ 0.01%, $CaCO_3$ 0.05%. By using this medium, the conversion rate of L-sorbose to 2-KLG was increased by 5%.

Another way for enhancing the 2-KLG production is to adjust the relationship and interaction between the companion strain and the 2-KLG-producing strain during the fermentation. Zhang *et al.* (1998) reported a new ecological regulation technology which contained the methods of high-quality seed preparation and a regulating strategy by changing the fermentation conditions. Zhang *et al.* (2010a) and Li *et al.* (2012) developed a control strategy of three-stage fermentation condition. By adjusting the pH and oxygen concentration to meet the needs of a companion strain and a conversion strain for their growth and conversion during the fermentation runs, the 2-KLG yields were significantly enhanced. In order to release more activators from companion strain for stimulating *K. vulgare* growth and 2-KLG production, an additional lysozyme strategy was introduced (Zhang *et al.*, 2010b). When 10 000 U/ml lysozyme were added after 12 h co-culture in a 7 l jar fermenter, the growth rate of *K. vulgare*, L-sorbose consumption rate and 2-KLG productivity increased 27.4%, 37.1% and 28.2%, respectively. Meanwhile, the fermentation time decreased to 56 h, shorter by 20.6% as compared with that of the control (Zhang *et al.*, 2010b). However, the addition of lysozyme may lead to increased costs and destruct this artificial two-strain ecosystem.

In our studies on several different companion strains, obviously different 2-KLG productivities were observed in the co-culture system with the different companion strains (Lv *et al.*, 2001; Yang *et al.*, 2009). We ascribed this to the varied growth characteristics of the different companion strains. As a rapid-growth companion strain (relatively rapid than other companion strains) in the co-culture, *B. megaterium* could supply metabolites to meet the *K. vulgare* propagation mainly at the early stage of fermentation (Lv *et al.*, 2001). Meanwhile, on the contrary, as a slow-growth companion strain, *B. cereus* released metabolites to meet the *K. vulgare* growth mainly at the late stage of fermentation (Yang *et al.*, 2009). Therefore, in order to supply enough metabolites to meet the need of *K. vulgare* during the whole period of fermentation, a novel two-helper-strain co-culture system (TSCS) was developed (Mandlaa *et al.*, 2013). *B. megaterium* and *B. cereus* (with a ratio of 1 : 3 (v/v) in the seeding culture), used as the helper strains in the co-culture system with *K. vulgare*, increased the 2-KLG yield significantly compared to the conventional one-helper-strain (either *B. cereus* or *B. megaterium*) co-culture system (OSCS). After 45 h cultivation in flask, 2-KLG concentration in the TSCS increased by 7–8.9% than that of the OSCS. The results implied that TSCS is a viable method for enhancing industrial production of 2-KLG (Mandlaa *et al.*, 2013).

The second-step fermentation can also be carried out in a continuous fermentation mode. In a single culture of *K. vulgare*, with a very high content of complex components (3% corn steep liquor, 7% baker's yeast) in the culture medium, *K. vulgare* DSM4025 produced 2-KLG at a steady-state concentration of 112.2 g/l 2-KLG for 140 h (Takagi, Sugisawa and Hoshino, 2009). The dilution rate was maintained between 0.035 and 0.043 per hour resulting in a volumetric 2-KLG productivity of 3.90–4.80 g/l/h. The average molar conversion yield of 2-KLG from L-sorbose was 91.3%. In continuous mixed-culture fermentations, with *X. maltophilia* IFO 12692 as a companion strain, 2-KLG production from L-sorbose by *K. vulgare* DSM4025 could be kept in a stable, continuous mode for more than 1300 h (Takagi, Sugisawa and Hoshino, 2010). There was a dilution rate of 0.0380 per hour and a steady concentration of 113 g/l 2-KLG. The volumetric productivity was 2.15 g/l/h. The molar conversion yield was 90.1%. These results suggested that the continuous fermentation is suitable for the mixed-culture fermentation of Asc.

7.6.2.5
Upstream and Downstream Processing

D-Sorbitol Production D-sorbitol is mainly produced by the method of hydrogenation of glucose. Currently, a new continuous hydrogenation technology was widely applied in D-sorbitol production by manufacturers (Sun, Liu and Li, 2008). The glucose solution is injected into the column reactor containing solid catalyst by high-pressure pump. The catalyst is at a stationary state in the reactor, with no effect of stirring and shock. Both the glucose solution and hydrogen continuously go through the surface of catalyst and react completely. The D-sorbitol is then formed and excreted after a period of time. After ion-exchange resin purification, evaporation and concentration, the product of D-sorbitol solution with a concentration of 50–80% is finally obtained.

Purification and Extraction of 2-KLG In the fermentation broth, 2-KLG exists as the sodium 2-KLG. Moreover, there are a lot of impurities in broth, such as bacteria, proteins, amino acids, nucleic acids and inorganic salts. To get rid of these impurities, the broth is pre-treated by enhancing the temperature or adjusting the pH to make some impurities, especially proteins, to be precipitated and removed after centrifugation. The broth is then further purified by ultrafiltration. Two efficient ultrafiltration membrane separation systems, the Sun-tar# and Sun2flo, were suggested to be applied in 2-KLG purification (Guo and Hong, 2011). Finally, the purified sodium 2-KLG is transferred to 2-KLG by ion exchange.

Chemical Conversion of 2-KLG to Asc Both in the Reichstein process and in the two-step fermentation process, the 2-KLG is converted to Asc by chemical methods. Compared with the traditional acid transformation method, the alkali transformation method is widely applied by Chinese manufacturers (Yan, 2007) (Figure 7.6). 2-KLG and methanol are transformed to methyl 2-keto-L-gulonate

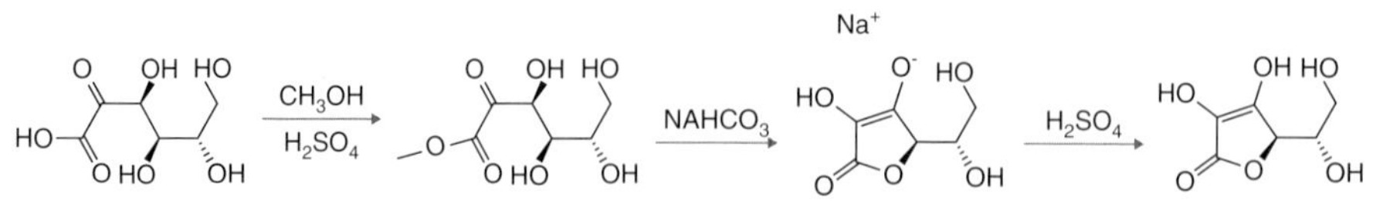

Figure 7.6 Chemical conversion of 2-keto-L-gulonic acid to vitamin C.

in the presence of concentrated sulfuric acid as a catalyst. The methyl 2-keto-L-gulonate is then converted to Asc sodium by lactonisation reaction with $NaHCO_3$. Alkali transformation method, involving simple operation process and mild reaction condition, is currently suitable for large-scale production of Asc sodium.

7.7 Application and Economics

The chemical properties of Asc make it possible to provide a wide range of applications. The applications of Asc depend on its properties as an antioxidant or on its health-related properties. Asc production has been estimated at approximately 110 000 t/year, with a worldwide market in excess of $600 million (Bremus *et al.*, 2006; Zhang *et al.*, 2011). Currently, Asc is widely used in the food, beverage, cosmetic, animal feed and pharmaceutical industries (Bremus *et al.*, 2006; Pappenberger and Hohmann, 2014) (Table 7.2).

Table 7.2 Application of Asc and its derivatives.

Fields	Usages
Pharmaceutical industry	For the treatment of scurvy, idiopathic thrombocytopenic purpura, neonatal hypoxic ischemic encephalopathy, viral myocarditis, chronic and acute viral hepatitis, ascariasis biliary colic, gastrointestinal tract ulceration, burn of cornea and conjunctiva, ocular chemical injury, thrush, virus flu, atherosclerosis, cancer, bronchial asthma
Food industry	Inhibit oxidation and browning of fruits and vegetables; change the flavour of food; maintaining the food colour; prevent lipid oxidation
Beverage industry	Antiseptic preservation function as an additive; Asc supplement as nutrition to the human body
Feed industry	Improve animal anti-stress ability; enhance immune function; accelerate growth; enhance the rate of reproduction; for the treatment of trauma or burn
Cosmetic industry	Promote collagen formation, restrain melanin formation, resist damage to the skin by body cellular oxidation, enhance skin elasticity

In the late 1950s, Asc was mainly supplied by European and Japanese manufacturers (such as Roche, BASF, Merck and Takeda Pharmaceutical). However, by the early 1990s, 26 Chinese manufacturers had already gained one-third of the world's Asc market (Pappenberger and Hohmann, 2014). The European producers conducted several rounds of price cuts to prevent Chinese companies from further entering the market (Mandlaa, 2014). In the early 2000s, because of the record low level of prices for years, many European and Japanese producers abandoned industrial production of Asc due to relatively higher production costs in the Reichstein process than two-step fermentation process. On the contrary, Chinese manufacturers produced more Asc and takes about 90% of Asc yields in the world (Yang *et al.*, 2011). Currently, five of leading Chinese producers are: (i) Weisheng Pharmaceutical Company (CSPC), Shijiazhuang, Hebei Province; (ii) Northeast Pharmaceutical Group Co., Ltd (NEPG), Shenyang, Liaoning Province; (iii) Welcome Pharmaceutical Co., Ltd (NCPC), Shijiazhuang, Hebei Province; (iv) Aland Nutraceutical Group, Jingjiang, Jiangsu Province, and a new producer (v) Shandong Luwei Pharmaceutical Co., Ltd, Zibo, Shandong Province. Today, the sole Western Asc producer is DSM Nutritional Products of Switzerland (formerly Roche Vitamins) (Mandlaa, 2014).

Approximately 80% of Asc product in China is exported to more than 150 countries and regions, while 20% is consumed in native market (Mandlaa, 2014). In the later 2000s, the price of Asc remains at record low level because of the oversupply of Asc in the world market (Pappenberger and Hohmann, 2014). Hence, two measures should be taken to make a healthy development for Asc industry: for one side, to reduce the yields of low value-added Asc products and to develop the high value-added Asc derivatives, such as Vc-crystal sodium, Vc-calcium and Vc-phosphate magnesium (Mandlaa, 2014); for the other side, to expand the new Asc consumption market, especially for Chinese domestic market. If the Asc consumption of each Chinese people is the same as that of the Western countries' people, the world demands of Asc can be increased from 120 to 220 kt every year (Mandlaa, 2014). All of these suggest that a promising future for Asc industry if the new Asc products are produced and the new markets are developed.

7.8
Outlook

The two-step fermentation process of L-ascorbic acid manufacture has been established in a commercial scale for over 40 years in China. Compared with the classical Reichstein process, which is highly energy-consuming and relies on the use of a number of environmentally hazardous chemicals, the two-step fermentation process is less costly and more environment-friendly (Xu *et al.*, 2004). However, the efficiency of the two-step fermentation still needs to be improved, especially the second step of mixed-culture fermentation for Asc production. There is still much work to be done in this area, such as regulating the artificial ecosystem to enhance 2-KLG production in the second-step fermentation, improving the fermentation efficiency by changing the current batch fermentation to fed-batch or continuous fermentation and constructing recombinant strain that directly produces 2-KLG from D-glucose or L-sorbose.

The relationship between *K. vulgare* and the companion strain, which plays an important role in 2-KLG production in the second-step fermentation, should be a research focus in the near future (Zou, Liu and Chen, 2013). Many studies have been carried out for the aim of elucidating the interaction mechanism in this artificial ecosystem (Lu *et al.*, 2001; Zhang *et al.*, 2010b; Ma *et al.*, 2011; Zhou *et al.*, 2011; Zou, Liu and Chen, 2013). Based on the genomics and other omics analyses of the two strains, the current results have shown that the relationship between the two strains is a combination of mutualism and antagonism (Zhou *et al.*, 2011). However, detailed mining and integration of the omics data will be necessary. Further, the genes involved in the biosynthesis and transportation of 2-KLG need to be elucidated, and the mechanisms of PQQ-dependent incomplete oxidation dehydrogenases should be investigated. In addition, a genome-scale metabolic ecosystem combining the two strains as well as the fermentation conditions should be constructed, which might guide us to carry on regulating the two-step fermentation more effectively and precisely.

Batch fermentation is mainly used in two-step fermentation process for industrial production of Asc (Zhang *et al.*, 1998; Jing *et al.*, 2009). In batch fermentation, the substrate concentration remains at a low level because of the substrate inhibition effect (Giridhar and Srivastava, 2000b). Moreover, a large unproductive downtime, such as the time for medium sterilisation and the time for three-grade expansion of inoculum, leads to much lower fermentation efficiency (Giridhar and Srivastava, 2001; Takagi, Sugisawa and Hoshino, 2010). Several studies on fed-batch fermentation or continuous fermentation have been reported for sorbose fermentation or 2-KLG fermentation, indicating a promising application for two-step fermentation process on industrial scale (Giridhar and Srivastava, 2000b, 2001; Jing *et al.*, 2009; Takagi, Sugisawa and Hoshino, 2010; Yang, 2012).

Many scholars have tried to improve the capability of strains by means of gene engineering techniques (Manning and Kahn, 1992; Hosshino *et al.*, 2012; Cai *et al.*, 2012; Zhu *et al.*, 2012). Based on the two-step fermentation process, recombinant strains were always inserted into a gene to express metabolites in order to improve enzyme activities or stimulate growth of strain (Cai *et al.*, 2012). A more attractive aspect would be to develop a new engineered strain which combines the metabolic traits of *K. vulgare* and the companion strain and with a higher capability of producing 2-KLG from L-sorbose or D-glucose directly. This will build a single recombinant organism that directly produces 2-KLG from D-glucose or L-sorbose. With the development of systems biology and synthetic biology, it is predictable that the engineered strain with a capability of one-step fermentation will be designed, built and applied in the future (Zou, Liu and Chen, 2013).

References

Allen, M.A. and Burgess, S.G. (1950) The losses of ascorbic acid during the large-scale cooking of green vegetables by different methods. *Br. J. Nutr.*, **4** (2-3), 95–100.

Anderson, S., Marks, C.B., Lazarus, R., Miller, J., Stafford, K., Seymour, J., Light, D., Rastetter, W., and Estell, D. (1985) Production of 2-Keto-L-gluconate, an intermediate in L-ascorbate synthesis, by a genetically modified Erwinia herbicola. *Science*, **230**, 144–149.

Arya, S.P., Mahajan, M., and Jain, P. (1998) Photometric methods for the determination of vitamin C. *Anal. Sci.*, **14**, 889–895.

Arya, S.P., Mahajan, M., and Jain, P. (2001) Spectrophotometric determination of vitamin C with iron(II)-4-(2-pyridylazo) resorcinol complex. *Anal. Chim. Acta*, **427**, 245–251.

Asakura, A. and Hoshino, T. (1999) Isolation and characterization of a new Quinoprotein dehydrogenase, l-sorbose/lsorbosone dehydrogenase. *Biosci. Biotechnol., Biochem.*, **63**, 46–53.

Asakura, A., Hoshino, T., Ojima, S., Shinjo, M. and Tomiyama, N. (1996) Alcohol/aldehyde dehydrogenase. US Patent 6730503 B1.

Bächi, B. (2008) Natural or synthetic vitamin C? A new substance's precarious status behind the scenes of World War II. *NTM*, **16** (4), 445–470 (In German).

Boudrant, J. (1990) Microbial processes for ascorbic acid biosynthesis: a review. *Enzyme Microb. Technol.*, **12** (5), 322–329.

Bremus, C., Herrmann, U., Bringer-Meyer, S., and Sahm, H. (2006) The use of microorganisms in L-ascorbic acid production. *J. Biotechnol.*, **124**, 196–205.

Briggs, M.H. (1973) Side-effects of vitamin C. *Lancet*, **302** (7843), 1439.

Cai, L., Yuan, M.Q., Li, Z.J., Chen, J.C., and Chen, G.Q. (2012) Genetic engineering of *Ketogulonigenium vulgare* for enhanced production of 2-keto-Lgulonic acid. *J. Biotechnol.*, **157**, 320–325.

Carpenter, K.J. (2012) The discovery of vitamin C. *Ann. Nutr. Metab.*, **61**, 259–264.

Chatterjee, I.B., Majumder, A.K., Nandi, B.K., and Subramanian, N. (1975) Synthesis and some major functions of vitamin C in animals. *Ann. N.Y. Acad. Sci.*, **258** (1), 24–47.

Choi, E.S., Lee, E.H., and Rhee, S.K. (1995) Purification of a membrane-bound sorbitol dehydrogenase from *Gluconobacter suboxydans. FEMS Microbiol. Lett.*, **125**, 45–49.

Chotani, G., Dodge, T., Hsu, A., Kumar, M., LaDuca, R., Trimbur, D., Weyler, W., and Sanford, K. (2000) The commercial production of chemicals using pathway engineering. *Biochim. Biophys. Acta*, **1543**, 434–455.

Combs, G.F. (2001) *The Vitamins, Fundamental Aspects in Nutrition and Health*, 2nd edn, Academic Press, San Diego, CA, pp. 245–272.

Cook, J.D. and Reddy, M.B. (2001) Effect of ascorbic acid intake on nonheme-iron absorption from a complete diet. *Am. J. Clin. Nutr.*, **73** (1), 93–98.

Crawford, T. and Crawford, S. (1980) Synthesis of L-ascorbic acid. *Adv. Carbohydr. Chem. Biochem.*, **37**, 79–155.

De Ley, J. and Swings, J. (1994) *Bergey's Manual of Systematic Bacteriology*, vol. **1**, Williams and Wilkins, Baltimore, MD, pp. 275–278.

D'elia, J. (2006) *Ketogulonigenium* shuttle vectors. US Patent 7033824.

Deppenmeier, U. and Ehrenreich, A. (2009) Physiology of acetic acid bacteria in light of the genome sequence of *Gluconobacter oxydans. J. Mol. Microbiol. Biotechnol.*, **16**, 69–80.

Deppenmeier, U., Hoffmeister, M., and Prust, C. (2002) Biochemistry and biotechnological applications of *Gluconobacter* strains. *Appl. Microbiol. Biotechnol.*, **60** (3), 233–242.

De Tullio, M.C. and Arrigoni, O. (2004) Hopes, disillusions and more hopes from vitamin C. *Cell. Mol. Life Sci.*, **61**, 209–219.

Dunn, W.A., Rettura, G., Seifter, E., and Englard, S. (1984) Carnitine biosynthesis from gamma-butyrobetaine and from exogenous protein-bound 6-N-trimethyl-L-lysine by the perfused guinea pig liver. Effect of ascorbate deficiency on the in situ activity of gamma-butyrobetaine hydroxylase. *J. Biol. Chem.*, **259** (17), 10764–10770.

Eggersdorfer, M., Laudert, D., Létinois, U., McClymont, T., Medlock, J., Netscher, T., and Bonrath, W. (2012) One hundred years of vitamins—a success story of the natural sciences. *Angew. Chem. Int. Ed.*, **51** (52), 12960–12990.

Eipper, B.A., Milgram, S.L., Husten, E.J., Yun, H.Y., and Mains, R.E. (1993) Peptidylglycine alpha-amidatingmonooxygenase: a multifunctional protein with catalytic, processing, and routing domains. *Protein Sci.*, **2** (4), 489–497.

Ermolenko, Z.M., Martovetskaya, V.A., Chugunov, V.A., and Kholodenko, V.P. (2000) Effect of space flight conditions on the properties of hydrocarbon-oxidizing bacteria. *Appl. Biochem. Microbiol.*, **36**, 559–563.

Fang, X.M., Zhao, Z.J., and Gu, H.K. (2005) A study on space mutation of *Streptomyces fradiae*. *Space Med. Med. Eng.*, **18**, 121–125 (In Chinese).

Feng, S., Sun, C.B., Zhang, Z.Z., Zhu, K.L., Zhang, H.H., and Gao, Y.T. (1998) Effects of *Bacillus megaterium* on growth and 2KGA synthesizing of *Gluconobacter oxydans* in vitamin C two-step fermentation process. *J. Microbiol.*, **18**, 6–9.

Feng, S., Zhang, Z., Zhang, C.G., and Zhang, Z.Z. (2000) Effect of *B. megaterium* on *Gluconobacter oxydans* in mixed culture. *Chin. J. Appl. Ecol.*, **11**, 119–122 (In Chinese).

Franke, A.A., Custer, L.J., Arakaki, C., and Murphy, S.P. (2004) Vitamin C and flavonoid levels of fruits and vegetables consumed in Hawaii. *J. Food Compos. Anal.*, **17**, 1–35.

Fu, S., Zhang, W., Guo, A., and Wang, J. (2007) Identification of promoters of two dehydrogenase genes in *Ketogulonicigenium vulgare* DSM 4025 and their strength comparison in *K. vulgare* and *Escherichia coli*. *Appl. Microbiol. Biotechnol.*, **75**, 1127–1132.

Fujita, Y., Mori, I., Yamaguchi, T., Hoshino, M., Shigemura, Y., and Shimano, M. (2001) Spectrophotometric determination of ascorbic acid with iron (III) and p-carboxyphenylfluorone in a cationic surfactant micellar medium. *Anal. Sci.*, **17**, 853–857.

Gao, H., Liu, M., Zhuo, Y., Zhou, X., Liu, J., Chen, D., Zhang, W., Gou, Z., Shang, P., and Zhang, L. (2010) Assessing the potential of an induced-mutation strategy for avermectin overproducers. *Appl. Environ. Microbiol.*, **76**, 4583–4586.

Ge, X., Zhao, Y., Hou, W., Zhang, W., Chen, W., Wang, J., Zhao, N., Lin, J., Wang, W., Chen, M., Wang, Q., Jiao, Y., Yuan, Z., and Xiong, X. (2013) Complete genome sequence of the industrial strain *Gluconobacter oxydans* H24. *Genome Announc.*, **1** (1), e00003-13.

Giridhar, R. and Srivastava, A.K. (2000a) Fed-batch cultivation of *Acetobacter suboxydans* for the microbial oxidation of D-sorbitol to L-sorbose. *Bioprocess. Eng.*, **23**, 575–577.

Giridhar, R. and Srivastava, A.K. (2000b) Fed-batch sorbose fermentation using pulse and multiple feeding strategies for productivity improvement. *Biotechnol. Bioprocess Eng.*, **5**, 340–344.

Giridhar, R. and Srivastava, A.K. (2001) Repeated fed-batch sorbose fermentation by *Gluconobacter oxydans*. *Chem. Biochem. Eng. Q.*, **15**, 127–129.

Gropper, S.S., Smith, J.L., and Grodd, J.L. (2005) *Advanced Nutrition and Human Metabolism*, Thomson Wadsworth, Belmont, CA, pp. 260–275.

Guo, L., Guo, L., Lu, Y., and Zhang, L. (2006) Breeding of associated bacteria *Bacillus thuringiensis* in 'Two-step fermentation' of Vc. *J. Hebei Univ. (Nat. Sci. Ed.)*, **26** (1), 108–111.

Guo, X. and Hong, S. (2011) A comparative study of extraction process of 2-keto-L-gulonic acid. *Technol. Wind*, **8**, 39 (In Chinese).

Guo, L., Liu, Q., Yang, R., Lv, Z., Zhang, L., Wu, Q., Fang, X., and Jiang, X. (2004) Screening high yield strains of vitamin C from bacterial strains retrieved after spaceflight. *J. Hebei Univ. (Nat. Sci. Ed.)*, **24**, 284–287.

Gupta, A., Singh, V.K., Qazi, G.N., and Kumar, A. (2001) *Gluconobacter oxydans*: its biotechnological applications. *J. Mol. Microbiol. Biotechnol.*, **3** (3), 445–456.

Hekmat, D., Bauer, R., and Neff, V. (2007) Optimization of the microbial synthesis of dihydroxyacetone in a semi-continuous repeated-fed-batch process by in situ immobilization of *Gluconobacter oxydans. Process Biochem.*, **42**, 71–76.

Hernández, Y., Lobo, M.G., and González, M. (2006) Determination of vitamin C in tropical fruits: a comparative evaluation of methods. *Food Chem.*, **96** (4), 654–664.

Hoffer, A. (1989) The discovery of vitamin C, Albert Szent-Gyorgyi, M.D., Ph.D., 1893–1986. *J. Orthomol. Med.*, **4** (1), 24–26.

Hosshino, T., Mouncey, N.J., Shinizu, A. and Shinjoh, M. (2012) Production of 2-keto-L-gulonic acid. US Patent 0058563.

Huang, Z., Zou, W., Liu, J., and Liu, L. (2013) Glutathione enhances 2-keto-L-gulonic acid production based on *Ketoguloni-cigenium vulgare* model iWZ663. *J. Biotechnol.*, **164**, 454–460.

Iwase, H. (2003) Routine high-performance liquid chromatographic determination of ascorbic acid in foods using L-methionine for the pre-analysis sample stabilization. *Talanta*, **60**, 1011–1021.

Janghel EK, Gupta VK, Rai MK, Rai JK. 2007. Micro determination of ascorbic acid using methyl viologen. *Talanta* **72**:1013-1016.

Jaselskis, B. and Nelapaty, S.J.J. (1972) Spectrophotometric determination of microamounts of ascorbic acid in citrus fruits. *Anal. Chem.*, **44**, 379–381.

Ji, K., Liu, J., Qing, S., Liu, L., and Chen, J. (2010) Enhancement of 2-keto-L-gulonic production in *Gluconobacter oxydans* through feeding metalions. *J. Food Sci. Biotechnol.*, **29**, 139–144 (In Chinese).

Jiao, Y.H., Zhang, W.C., Xie, L., Yuan, H.J., and Chen, M.X. (2002) Effects of *Bacillus cereus* on *Gluconobacter oxydans* in vitamin C fermentation process. *Microbiology*, **29**, 35–38 (In Chinese).

Jing, S., An, H., Li, C., Li, Y., Zhang, C., Yang, W., Jiang, M., and Xu, H. (2011) Mutation of *Gluconobacter melanogenus* in a two-step fermentation for vitamin C production by space flight. *Biotechnol. Lett.*, **21**, 76–78 (In Chinese).

Jing, S., Ji, Y., Li, C., Jiang, M., Yang, W., Chen, M., and Xu, H. (2009) Studies on the characteristics of *Gluconobacter melanogenus* with high concentration of D-sorbitol in semi-continuous fermentation. *Biotechnol. Lett.*, **19**, 77–78 (In Chinese).

Kabasakalis, V., Siopidou, D., and Moshatou, E. (2000) Ascorbic acid content of commercial fruit juices and its rate of loss upon storage. *Food Chem.*, **70**, 325–328.

Kato, T., Ohno, O., Nagoshi, T., Ichinose, Y., and Igarashi, S. (2005) Determination of small amounts of L-ascorbic acid using the chemiluminescence of an iron-chlorophyllin complex. *Anal. Sci.*, **21**, 579–581.

Kaufman, S. (1974) Dopamine-beta-hydroxylase. *J. Psychiatr. Res.*, **11**, 303–316.

Kaur, C. and Kapoor, H.C. (2002) Anti-oxidant activity and total phenolic content of some Asian vegetables. *Int. J. Food Sci. Technol.*, **37** (2), 153–161.

Ke, D., El-Wazir, F., Cole, B., Mateos, M., and Kader, A.A. (1994) Tolerance of peach and nectarine fruits to insecticidal controlled atmospheres as influenced by cultivar, maturity and size. *Postharvest Biol. Technol.*, **4**, 135–146.

Kivirikko, K.I. and Myllylä, R. (1985) Post-translational processing of procollagens. *Ann. N.Y. Acad. Sci.*, **460**, 187–201.

Kosheleva, O.V. and Kodentsova, V.M. (2012) Vitamin C in fruits and vegetables. *Vopr. Pitan.*, **82** (3), 45–52.

Leduc, S., Troostembergh, J.C., and Lebeault, J.M. (2004) Folate requirements of the 2-keto-L-gulonic acid-producing strain *Ketogulonigenium vulgare* LMP P-20356 in L-sorbose/CSL medium. *Appl. Microbiol. Biotechnol.*, **65**, 163–167.

Lee, S.K. and Kader, A.A. (2000) Preharvest and postharvest factors influencing vitamin C content of horticultural crops. *Postharvest Biol. Technol.*, **20** (3), 207–220.

Lewin, S. (1976) *Vitamin C: Its Molecular Biology and Medical Potential*, Academic Press, London.

Li, L., Li, B., Lv, S., Xu, H., Yang, W., Han, L., and Wang, Z. (2012) Enhancement of 2-keto-L-gulonic acid production using three-stage ventilation control strategy. *Chin. Brew.*, **31**, 144–147 (In Chinese).

Li, C.X., Zeng, Y.L., Liu, Y.J., and Tang, C.R. (2006) Simultaneous electrochemical determination of uric acid and ascorbic acid on a glassy carbon electrode modified with cobalt (II) tetrakisphenylporphyrin. *Anal. Sci.*, **22**, 393–397.

Li, G.C. and Zhang, Z.Z. (1997) Study on characteristics of mixed fermentation of 2-KLG producing strains and their mixed growth pattern. *J. Microbiol.*, **17**, 1–4.

Lindblad, B., Lindstedt, G., and Lindstedt, S. (1970) The mechanism of enzymic formation of homogentisate from p-hydroxyphenylpyruvate. *J. Am. Chem. Soc.*, **92** (25), 7446–7449.

Linetsky, M., Ranson, N., and Ortwerth, B.J. (1998) The aggregation in human lens proteins blocks the scavenging of UVA-generated singlet oxygen by ascorbic acid and glutathione. *Arch. Biochem. Biophys.*, **351**, 180–188.

Linster, C.L. and Van Schaftingen, E. (2007) Vitamin C: biosynthesis, recycling and degradation in mammals. *FEBS J.*, **274**, 1–22.

Liu, L.M., Li, Y., Zhang, J., Zhou, Z.M., Liu, J., Li, X.M., Zhou, J.W., Du, G.C., Wang, L., and Chen, J. (2011a) Complete genome sequence of the industrial strain *Ketogulonicigenium vulgare* WSH-001. *J. Bacteriol.*, **193**, 6108–6109.

Liu, L., Li, Y., Zhang, J., Zou, W., Zhou, Z., Liu, J., Li, X., Wang, L., and Chen, J. (2011b) Complete genome sequence of the industrial strain *Bacillus megaterium* WSH-002. *J. Bacteriol.*, **193**, 6389–6390.

Liu, L.M., Chen, K.J., Zhang, J., Liu, J., and Chen, J. (2011c) Gelatin enhances 2-keto-L-gulonic acid production based on *Ketogulonigenium vulgare* genome annotation. *J. Biotechnol.*, **156**, 182–187.

Lu, S., Feng, S., Zhang, Z., Liu, Y., Xie, Z., and An, H. (2001) The effect of *Bacillus megaterium* in vitamin C two-step fermentation. *Microbiol. China*, **28**, 10–13 (In Chinese).

Lu, X., Guo, H., and Li, B. (2005) Screening of associated bacteria *Gluconobacter oxydans* in two-step fermentation of Vc. *Biotechnology*, **15**, 23–25 (In Chinese).

Lu, S.J., Jun, W., Yao, J.M., and Yu, Z.L. (2003) Study on the effect of mutated *Bacillus megaterium* in two-stage fermentation of vitamin C. *Plasma Sci. Technol.*, **5**, 2011–2016.

Lv, S., Zhao, S., Ma, D., Lin, Y., Zhang, L., Chen, H., and Zhang, Z. (2011a) Effect of several exogenous substances on growth and 2-keto-L-gulonic acid accumulation of *Gluconobacter oxydans*. *J. Shenyang Agric. Univ.*, **42**, 184–189 (In Chinese).

Lv, S., Zhao, S., Yang, Y., Zhang, Z., and Chen, H. (2011b) Research progress on Vc precursor of 2-KGA production through mixed fermentation from L-sorbose. *Biotechnol. Bull.*, **5**, 50–54 (In Chinese).

Lv, S.X., Zhou, L.N., Feng, S., Zhang, Z.Z., Lv, Y.K., An, H.Y., and Cao, G.H. (2001) The role of Bacillus megaterium in two-step vitamin C fermentation. *J. Microbiol.*, **21** (3), 3–48 (In Chinese).

Lykkesfeldt, J. (2000) Determination of ascorbic acid and dehydroascorbic acid in biological samples by high-performance liquid chromatography using subtraction methods: reliable reduction with tris[2-carboxyethyl] phosphine hydrochloride. *Anal. Biochem.*, **282**, 89–93.

Lyu, S., Guo, Z., Pan, J., Yang, Y., Yang, W., Chen, H., and Zhang, Z. (2014) Effect of rare earth elements on vitamin C fermentation by mixed cultures. *Int. J. Agric. Biol.*, **16**, 1135–1140.

Ma, Q., Zhou, J., Zhang, W., Meng, X., Sun, J., and Yuan, Y.J. (2011) Integrated proteomic and metabolomic analysis of an artificial microbial community for two-step production of vitamin C. *PLoS One*, **6**, e26108.

Mandl, J., Szarka, A., and Bánhegyi, G. (2009) Vitamin C: update on physiology and pharmacology. *Br. J. Pharmacol.*, **157**, 1097–1110.

Mandlaa, M. (2014) *Study of Mechanism and Application of 2-Keto-L-gulonic Acid Fermentation Based on Ecological Relationships of Co-Culture*, University of Chinese Academy of Sciences, Beijing, pp. 1–5 (In Chinese).

Mandlaa, M., Yang, W., Han, L., Wang, Z., and Xu, H. (2013) Two-helper-strain co-culture system: a novel method for enhancement of 2-keto-L-gulonic acid production. *Biotechnol. Lett.*, **35**, 1853–1857.

Mandlaa, M., Yang, W., Xu, H., Li, L., and Xu, Y. (2011) Optimization of fermentation conditions for 2-keto-L-gulonic acid. *Food Ferment. Technol.*, **47** (3), 60–67 (In Chinese).

Manning, R.F. and Kahn, M.S. (1992) Biosynthesis of 2-keto-L-gulonic acid. US Patent 5082785.

Massey, L.K., Liebman, M., and Kynast-Gales, S.A. (2005) Ascorbate increases human oxaluria and kidney stone risk. *J. Nutr.*, **135** (7), 1673–1677.

Matsushita, K., Fujii, Y., Ano, Y., Toyama, H., Shinjoh, M., Tomiyama, N., Miyazaki, T., Sugisawa, T., Hoshino, T., and Adachi, O. (2003) 5-keto-D-gluconate production is catalyzed by a quinoprotein glycerol dehydrogenase, major polyol dehydrogenase, in *Gluconobacter* species. *Appl. Environ. Microbiol.*, **69**, 1959–1966.

Miyazaki, T., Sugisawa, T., and Hoshino, T. (2006) Pyrroloquinoline quinone-dependent dehydrogenases from *Ketogulonicigenium vulgare* catalyze the direct conversion of L-sorbosone to L-ascorbic acid. *Appl. Environ. Microbiol.*, **72**, 1487–1495.

Miyazaki, T., Tomiyama, N., Shinjoh, M., and Hoshino, T. (2002) Molecular cloning and functional expression of D-sorbitol dehydrogenase from *Gluconobacter suboxydans* IFO3255, which requires pyrroloquinoline quinone and hydrophobic protein SldB for activity development in *E. coli*. *Biosci. Biotechnol., Biochem.*, **66**, 262–270.

Motizuki, K. (1966) Method for producing 2-keto-L-gulonic acid. US Patent 3,234,105.

Odriozola-Serrano, I., Hernández-Jover, T., and Martín-Belloso, O. (2007) Comparative evaluation of UV-HPLC methods and reducing agents to determine vitamin C in fruits. *Food Chem.*, **105** (3), 1151–1158.

Padh, H. (2009) Vitamin C: newer insights into its biochemical functions. *Nutr. Rev.*, **49** (3), 65–70.

Pappenberger, G. and Hohmann, H. (2014) Industrial production of L-ascorbic acid (vitamin C) and D-isoascorbic acid. *Adv. Biochem. Eng./Biotechnol.*, **143**, 143–188.

Pauling, L. (1976) *Vitamin C, the Common Cold, and the Flu*, W.H. Freeman and Company, San Francisco, CA.

Peterkofsky, B. (1991) Ascorbate requirement for hydroxylation and secretion of procollagen: relationship to inhibition of collagen synthesis in scurvy. *Am. J. Clin. Nutr.*, **54** (Suppl. 6), 1135S–1140S.

Prockop, D.J. and Kivirikko, K.I. (1995) Collagens: molecular biology, diseases, and potentials for therapy. *Annu. Rev. Biochem.*, **64**, 403–434.

Proteggente, A.R., Pannala, A.S., Paganga, G., Buren, L.V., Wagner, E., Wiseman, S., and Rice-Evans, C.A. (2002) The antioxidant activity of regularly consumed fruit and vegetables reflects their phenolic and vitamin C composition. *Free Radical Res.*, **36** (2), 217–233.

Rahman, M.M., Khan, M.M.R., and Hosain, M.M. (2007) Analysis of vitamin C (ascorbic acid) contents in various fruits and vegetables by UV-spectrophotometry. *Bangladesh J. Sci. Ind. Res.*, **42** (4), 417–424.

Rebouche, C.J. (1991) Ascorbic acid and carnitine biosynthesis. *Am. J. Clin. Nutr.*, **54** (6), 1147S–1152S.

Reichstein, T. and Grüssner, A. (1934) Eine ergiebige synthese der L-ascorbinsäre (C-vitamin). *Helv. Chim. Acta*, **17**, 311–328.

Roig, M.G., Rivera, Z.S., and Kennedy, J.F. (1995) A model study on rate of degradation of L-ascorbic acid during processing using home-produced juice concentrates. *Int. J. Food Sci. Nutr.*, **46** (2), 107–115.

Salkić, M., Keran, H., and Jašić, M. (2009) Determination of L-ascorbic acid in pharmaceutical preparations using direct ultraviolet spectrophotometry. *Agric. Conspectus Sci.*, **74** (3), 263–268.

Salusjävi, T., Povelainen, M., Hvorslev, N., Eneyskaya, E.V., Kulminskaya, A.A., Shabalin, K.A., Neustroev, K.N., Kalkkinen, N., and Miasnikov, A.N. (2004) Cloning of a gluconate/polyol dehydrogenase gene from *Gluconobacter suboxydans* IFO 12528, characterisation of the enzyme and its use for the production of 5-ketogluconate in a recombinant *Escherichia coli* strain. *Appl. Microbiol. Biotechnol.*, **65**, 306–314.

Sauberlich, H.E. (1994) Pharmacology of vitamin C. *Annu. Rev. Nutr.*, **14** (1), 371–391.

Sauberlich, H.E. (1997) *Vitamin C in Health and Disease*, Antioxidants in Health and Disease Series, Marcel Decker, pp. 1–24.

Sefcovicová, J., Vikartovská, A., Pätoprstı, V., Magdolen, P., Katrlík, J., Tkac, J., and Gemeiner, P. (2009) Off-line FIA monitoring of D-sorbitol consumption during L-sorbose production using a sorbitol biosensor. *Anal. Chim. Acta*, **644**, 68–71.

Shinagawa, E., Chiyonobu, T., Adachi, O., and Ameyama, M. (1976) Distribution and solubilization of particulate gluconate dehydrogenase and particulate 2-ketogluconate dehydrogenase in acetic acid bacteria. *Agric. Biol. Chem.*, **40** (3) 475–483.

Shinagawa, E., Matsushita, K., Adachi, O., and Ameyama, M. (1982) Purifiation and characterization of D-sorbitol dehydrogenase from membrane of *Gluconobacter suboxydans* var. a. *Agric. Biol. Chem.*, **46**, 135–141.

Sies, H., Stahl, W., and Sundquist, A.R. (1992) Antioxidant functions of vitamins. *Ann. N.Y. Acad. Sci.*, **669** (1), 7–20.

Silva, F.O. (2005) Total ascorbic acid determination in fresh squeezed orange juice by gas chromatography. *Food Control*, **16** (1), 55–58.

Simion, A.I., Rusu, L., Ştefănescu, I., and Gavrilă, L. (2008) Influence of various thermal treatments over Vitamin C concentration in Lemons. *Studii şi Cercetări Ştiinţifice–Chimie şi Inginerie Chimică, Biotehnologii, Industrie Alimentară*, **9** (4) 519–530.

Smirnoff, N. (1996) Botancal briefing: the function and metabolism of ascorbic acid in plants. *Ann. Bot.*, **78** (6), 661–669.

Smirnoff, N., and Wheeler, G.L.(2000) Ascorbic acid in plants: biosynthesis and function. *Crit. Rev. Biochem. Mol.*, **35** (4), 291–314.

Song, Q., He, J., Ren, S., Ye, Q., Guo, X., Chen, C., and Yin, G. (1997) Production of vitamin C precursor-2-keto-L-gulonic acid from L-sorbose by a novel bacterial component system of SCB329-SCB933 III. The characteristics and control of 2-keto-L-gulonic acid fermentation. *Ind. Microbiol.*, **27**, 6–10 (In Chinese).

Sonoyama, T., Kageyama, B., Yagi, S., and Mitsushima, K. (1987) Biochemical aspects of 2-keto-L-gulonate accumulation from 2, 5-diketo-D-gluconate by *Corynebacterium* sp. and its mutants. *Agric. Biol. Chem.*, **51**, 3039–3047.

Stacey, M. and Manners, D.J. (1978) Edmund Langley Hirst. 1898–1975. *Adv. Carbohydr. Chem. Biochem.*, **35**, 1–29.

Sugisawa, T. and Hoshino, T. (2002) Purification and properties of membrane-bound D-sorbitol dehydrogenase from *Gluconobacter suboxydans* IFO 3255. *Biosci. Biotechnol., Biochem.*, **66**, 57–64.

Sugisawa, T., Hoshino, T., Masida, S., Momura, S., Setoguchi, Y., Tazoe, M., Shinjoh, M., Someha, S., and Fujiwara, A. (1990) Microbial production of 2-Keto-L-gulonic acid from L-sorbose and D-sorbitol by *Gluconobacter melanogenues. Agric. Biol. Chem.*, **54**, 1201–1209.

Sugisawa, T., Miyazaki, T., and Hoshino, T. (2005) Microbial production of L-ascorbic acid from D-sorbitol, L-sorbose, L-gulose, and L-sorbosone by *Ketogulonicigenium vulgare* DSM 4025. *Biosci. Biotechnol., Biochem.*, **69**, 659–662.

Sun, R., Liu, C., and Li, H. (2008) Analysis of the main application and production process of sorbitol. *China High Technol. Enterprises*, **9**, 99–100 (In Chinese).

Svirbely, J.L. and Szent-Györgyi, A. (1932) The chemical nature of vitamin C. *Biochem. J.*, **26** (3), 865.

Takagi, Y., Sugisawa, T., and Hoshino, T. (2009) Continuous 2-keto-L-gulonic acid fermentation from L-sorbose by *Ketogulonigenium vulgare* DSM 4025. *Appl. Microbiol. Biotechnol.*, **82**, 1049–1056.

Takagi, Y., Sugisawa, T., and Hoshino, T. (2010) Continuous 2-keto-L-gulonic acid fermentation by mixed culture of *Ketogulonicigenium vulgare* DSM 4025 and *Bacillus megaterium* or *Xanthomonas maltophilia. Appl. Microbiol. Biotechnol.*, **86**, 469–480.

Thomas, L.D., Elinder, C.G., Tiselius, H.G., Wolk, A., and Akesson, A. (2013) Ascorbic acid supplements and kidney stone incidence among men: a prospective study. *JAMA Intern. Med.*, **173** (5), 1–2.

Urbance, J., Bratina, B., Stoddard, S., and Schmidt, T. (2001) Taxonomic characterization of *Ketogulonigenium vulgare* gen. nov., sp. nov. and *Ketogulonigenium robustum* sp. nov., which oxidize L-sorbose to 2-keto-L-gulonic acid. *Int. J. Syst. Evol. Microbiol.*, **51**, 1059–1070.

Verma, K.K. (1982) Determination of ascorbic acid with o-iodosobenzoate. *Talanta*, **29**, 41–45.

Wang, X.B., Liu, J., Du, G.C., Zhou, J.W., and Chen, J. (2013) Efficient production of L-sorbose from D-sorbitol by whole cell immobilization of *Gluconobacter oxydans* WSH-003. *Biochem. Eng. J.*, **77**, 171–176.

Yang, W.C., Ji, Y., Jing, S.H., Zhang, C.Y., Jiang, M.Y., Chen, M., and Xu, H. (2009)

A preliminary screening method of associated bacteria in two-stage fermentation of vitamin C. *Biotechnol.*, **19** (6), 51–53 (In Chinese).

Wenzel, U., Nickel, A., Kuntz, S., and Daniel, H. (2004) Ascorbic acid suppresses drug-induced apoptosis in human colon cancer cells by scavenging mitochondrial superoxide anions. *Carcinogenesis*, **25**, 703–712.

Wheeler, G.L., Jones, M.A., and Smirnoff, N. (1998) The biosynthetic pathway of vitamin C in higher plants. *Nature*, **393** (6683), 365–369.

Wu, X., Diao, Y., Sun, C., Yang, J., Wang, Y., and Sun, S. (2003) Fluorimetric determination of ascorbic acid with o-phenylenediamine. *Talanta*, **59**, 95–99.

Wulf, P.D., Soetaert, W., and Vandamme, E.J. (2000) Optimized synthesis of L-sorbose by C5-dehydrogenation of D-sorbitol with *Gluconobacter oxydans. Biotechnol. Bioeng.*, **69**, 339–343.

Xiong, X.H., Han, S., Wang, J.H., Jiang, Z.H., Chen, W., Jia, N., Wei, H.L., Cheng, H., Yang, Y.X., Zhu, B., You, S., He, J.Y., Hou, W., Chen, M.X., Yu, C.J., Jiao, Y.H., and Zhang, W.C. (2011) Complete genome sequence of the bacterium *Ketogulonicigenium vulgare*Y25. *J. Bacteriol.*, **193**, 315–316.

Xu, A., Yao, J., Yu, L., Lv, S., Wang, J., Yan, B., and Yu, Z. (2004) Mutation of *Gluconobacter oxydans* and *Bacillus megaterium* in a two-step process of L-ascorbic acid manufacture by ion beam. *J. Appl. Microbiol.*, **96**, 1317–1323.

Yan, F. (2007) Advances in the production process of vitamin C. *Shanghai Med. Pharm. J.*, **28**, 559–562 (In Chinese).

Yan, Z., Tao, Z., Yu, L., Yin, G., Ning, W., Wang, C., Wang, S., Jiang, H., Yu, J., Wang, M., and Yu, X. (1981) Studies on production of vitamin C precursor 2-keto-L-gulonic acid from L-sorbose by fermentation. II. Conditions for submerged fermentation of 2-keto-L-gulonic acid. *Acta Microbiol. Sin.*, **21**, 185–191 (In Chinese).

Yang, W. (2012) *Mutation Breeding and the Regulation of Metabolic Activity for Strains in Vitamin C Fermentation*, University of Chinese Academy of Sciences, Beijing, pp. 31–57 (In Chinese).

Yang, W., Han, L., Mandlaa, M., Chen, H., Jiang, M., and Xu, H. (2013) Spaceflight-induced enhancement of 2-keto-L-gulonic acid production by a mixed culture of *Ketogulonigenium vulgare* and *Bacillus thuringiensis. Lett. Appl. Microbiol.*, **57**, 54–62.

Yang, F., Jia, Q., Xiong, Z., Zhang, X., Wu, H., Zhao, Y., Yang, J., Zhu, J., Dong, J., and Xue, Y. (2006) Complete genome analysis of Ketogulonigenium sp. WB0104. *Chin. Sci. Bull.*, **51**, 941–945.

Yang, W., Kang, Y., Zhao, J., Jin, S., Liu, J., and Xu, H. (2008) Effect of addition of glucose on yields of 2-keto-L-gulonic acid in two-step vitamin C fermentation. *Biotechnology*, **18**, 85–86 (In Chinese).

Yang, L., Zhu, X., Xiu, J., and Li, J. (2011) Ketogulonigenium vulgare and research progress on its molecular biology. *Hebei Chem. Ind.*, **34**, 11–15. (In Chinese).

Yin, G., Tao, Z., Yu, L., Yan, Z., Ning, W., Wang, C., Wang, S., Jiang, H., Zhang, X., Feng, X., Zhao, Q., and Wei, W. (1980) Studies on the production of vitamin C precursor-2-keto-L-gulonic acid from L-sorbose by fermentation. I. Isolation, screening and identification of 2-keto-L-gulonic acid producing bacteria. *Acta Microbiol. Sin.*, **20**, 246–251 (In Chinese).

Yu, Z.L. (1998) *Introduction of Ion Beam Biotechnology*, Anhui Science and Technology Press, Anhui, pp. 223–250.

Zerdin, K., Rooney, M.L., and Vermue, J. (2003) The vitamin C content of orange juice packed in an oxygen scavenger material. *Food Chem.*, **82**, 387–395.

Zhang, Z., Feng, S., Jiang, J., Wang, Y., and Zhang, Z. (1999) Breeding of associated bacteria *Bacillus megaterium* in 'Two-step fermentation' of vitamin C. *J. Microbiol.*, **19**, 8–10 (In Chinese).

Zhang, Z., Zhang, C., Sun, C., Zhang, Z., Feng, S., Zhang, H., Zhu, K., Li, G., Zhang, H. and An, H. (1998) A mini-ecological regulation technology in two-step fermentation of Vc. CN Patent 98114478.0.

Zhang, J., Zhou, J., Liu, L., Liu, J., Chen, K., Du, G., and Chen, J. (2010a) Enhancement of 2-keto-L-gulonic acid production using three-stage pH control strategy. *Chin. J. Biotechnol.*, **26**, 1263–1268 (In Chinese).

Zhang, J., Liu, J., Shi, Z.P., Liu, L.M., and Chen, J. (2010b) Manipulation of *B.*

megaterium growth for efficient 2-KLG production by *K. vulgare*. *Process Biochem.*, **45**, 602–606.

Zhang, J., Zhou, J.W., Liu, J., Chen, K.J., Liu, L.M., and Chen, J. (2011) Development of chemically defined media supporting high cell density growth of *Ketogulonicigenium vulgare* and *Bacillus megaterium*. *Bioresour. Technol.*, **102**, 4807–4814.

Zhong, C., Zhang, Z., Zhang, W., and Li, C. (2004) Study of *Sporobolomyces roseus* on production of vitamin C precursor-2-keto-L-gulonic acid. *Biotechnology*, **14**, 45–47 (In Chinese).

Zhou, J., Ma, Q., Yi, H., Wang, L.L., Hao, S., and Yuan, Y.J. (2011) Metabolome profiling reveals metabolic cooperation between *Bacillus megaterium* and *Ketogulonicigenium vulgare* during induced swarm motility. *Appl. Environ. Microbiol.*, **77**, 7023–7030.

Zhu, Y., Liu, J., Du, G., Zhou, J., and Chen, J. (2012) Sporulation and spore stability of *Bacillus megaterium* enhance *Ketogulonigenium vulgare* propagation and 2-keto-L-gulonic acid biosynthesis. *Bioresour. Technol.*, **107**, 399–404.

Zou, W., Liu, L.M., and Chen, J. (2013) Structure, mechanism and regulation of an artificial microbial ecosystem for vitamin C production. *Crit. Rev. Microbiol.*, **39**, 247–255.

8
Direct Microbial Routes to Vitamin C Production

Günter Pappenberger and Hans-Peter Hohmann

8.1
Introduction and Scope

The industrial production of vitamin C (L-ascorbic acid, L-Asc) has a prominent role in the industrial biotechnology. L-Asc is the vitamin with the biggest production volume, with well over 100 000 metric tons produced annually worldwide. It has a long history of industrial production, the first commercial implementation of the Reichstein-Grüssner L-Asc process dating back to 1934. It has been spearheading the industrial application of biotechnology for vitamin production. A fermentative step using *G. oxydans* had been central to the industrial process from its beginning, and the most relevant advancement of this process was the implementation of additional fermentative steps towards the key intermediate 2-keto-L-gulonic acid (2-KGA; alternatively abbreviated in the scientific and patent literature as 2-KLGA or 2-KLG) since the 1990s.

The focus of this chapter is directed towards novel and disruptive technologies for industrial L-Asc production, which, despite still being at their infancy now, have the potential of becoming a clear technology and cost leader and may thus become the next generation of L-Asc production. The 2-KGA fermentation technology can, nowadays, be considered as mature, being implemented as industrial standard by all major vitamin C producers and nearing the end of its learning curve with respect to further improvements. It will, therefore, not be a future key differentiating technology. Nevertheless, achieving further incremental improvements remains critical for maximising competitiveness in a highly contested market environment with all major players rooted in the same technology. The 2-KGA fermentation technology, its history, development, status and future improvement potential have been discussed in several recent reviews (Bremus *et al.*, 2006; Zhou, Du and Chen, 2012; Pappenberger and Hohmann, 2014), including this book (Yang and Xu, 2016), and will not be part of this review. The historical, scientific and commercial background on L-Asc production can also be found there.

Industrial Biotechnology of Vitamins, Biopigments, and Antioxidants, First Edition.
Edited by Erick J. Vandamme and José L. Revuelta.

With the high efficiency of the D-sorbitol to 2-KGA conversion established in the current 2-KGA fermentation process, there is little room for improvement by alternative routes to 2-KGA such as via 2,5-diketo-D-gluconic acid (Sonoyama *et al.*, 1982; Anderson *et al.*, 1985; Grindley *et al.*, 1988). These approaches have been summarised elsewhere (Hancock and Viola, 2001; Bremus *et al.*, 2006) and will not be covered here. Similarly, alternative feedstocks such as second-generation carbon sources are not generally considered here. The biotechnological conversion steps towards L-Asc build on the specific configuration of D-glucose, where much of the carbon skeleton and functionalities of L-Asc are already preformed. Rather than a classical fermentation of feedstock towards a product via central metabolism, the 2-KGA 'fermentation' is indeed a biocatalytic functionalisation of a specific substrate. The choice of carbon source is therefore limited to those few with the appropriate stereochemistry and number of carbon atoms.

The most promising approach towards significant process improvement is to develop a microbial fermentation directly proceeding to L-Asc. This would make the chemical conversion step of 2-KGA to L-Asc of the industrially established L-Asc process obsolete. It will save not only a conversion step, but also the associated down-streaming steps required to work up fermentatively produced 2-KGA as well as the recycling of the organic solvents required in the subsequent chemical steps. The concepts and realisations of such direct microbial production of L-Asc are the main topics of this chapter.

An obvious biotechnological route towards L-Asc would be an extension of the current 2-KGA fermentation by a biocatalytic conversion of 2-KGA to L-Asc in aqueous solution, thus simply replacing the current chemical step. In fact, numerous attempts have been made to identify enzymes catalysing this step and proof of principle has been shown several times (Hubbs, 1997; Kumar, 1998; Asakura, Hoshino and Shinjoh, 2001; Hoshino and Kiyasu, 2002). In all cases, however, conversion rates and yields have been very low and no improvements have been obtained so far. 2-KGA has proven extremely refractory for rearrangement to L-Asc in neutral or weakly acidic aqueous environment. This approach will, therefore, be not discussed further here.

Microbes (bacteria and yeasts) do not have natural biosynthetic capabilities for L-Asc production. A self-evident direct source of L-Asc is the utilisation of the natural biosynthetic routes in plants. Attempts to increase pathway flux and L-Asc titres in genetically modified plants have demonstrated proof of principle for such an approach, but achieved titres remain orders of magnitude below commercial requirements for the bulk production of pure L-Asc substance. It seems unlikely that these approaches will reach competitiveness with L-Asc derived from microbial 2-KGA fermentation. Markets for higher priced, plant-derived L-Asc do exist, but will remain small. These routes based on L-Asc biosynthesis in plants will therefore not be discussed here in further detail, and the reader is referred to recent reviews on this topic (Hancock and Viola, 2005; Zhang *et al.*, 2007; Cruz-Rus, Amaya and Valpuesta, 2012; Gallie, 2013; Zhu *et al.*, 2013).

8.2
Principles of Direct L-Ascorbic Acid Formation: The Major Challenges

With respect to its chemical structure, L-Asc is the vitamin which is closest to abundant feedstocks. It bears resemblance to common hexose carbohydrates such as D-glucose (Figure 8.1), and it may seem obvious to use this as starting material for commercial L-Asc production. Still, the industrial synthesis of L-Asc from common hexose carbohydrates faces several challenges.

8.2.1
Stereochemistry of L-Ascorbic Acid

A first challenge is caused by the stereocenters at C_4 and C_5 of L-Asc, in particular the stereocenter at C_5, which is responsible for the assignment to the L-series of sugars. All naturally abundant hexose sugars fall into the D-series. To convert D-glucose to L-Asc, nature uses two different approaches: the plant pathway comprising the epimerisation of D-sugars at C_5 or the animal pathway characterised by carbon skeleton inversion (Smirnoff and Wheeler, 2000; Valpuesta and Botella, 2004; Kondo *et al.*, 2006; Linster and Van Schaftingen, 2007). Conceptually, this carbon skeleton inversion is achieved by oxidation of the primary hydroxyl group at C_6 and reduction of the carbonyl group at C_1 to the primary alcohol. As a consequence, the structure in the Fischer projection is turned upside down (the numbering of carbon atoms is inverted) and the newly assigned C_4 and C_5 (former

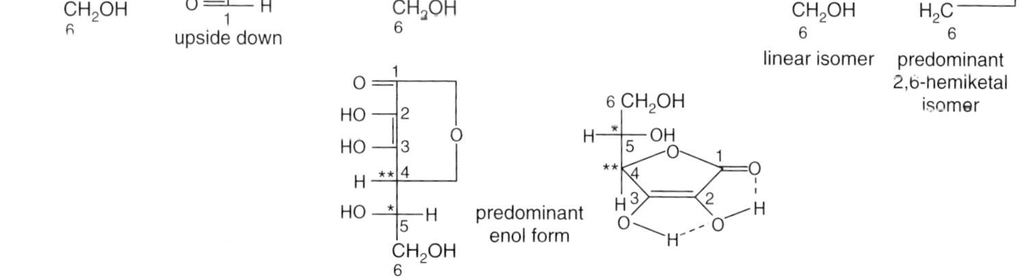

Figure 8.1 Structures of D-glucose, L-ascorbic acid and 2-keto-L-gulonic acid. The stereocenters at C_5 and C_4 of L-ascorbic acid and 2-keto-L-gulonic acid are indicated by one or two asterisks, respectively. The corresponding C_2 and C_3 positions in D-glucose, relevant for L-Asc synthesis routes involving 'carbon inversion' (natural pathway in animals, industrial synthesis route) are similarly indicated.

C_3 and C_2 of D-glucose, respectively) are then of correct stereochemistry for further conversion to L-Asc. Any industrial synthesis towards L-Asc will need to draw on common hexose sugars from the D-series and thus find an efficient way for adjusting the stereochemistry.

8.2.2
Enzymes Producing L-Ascorbic Acid and Their By-Product Spectrum

Once the stereochemistry is settled, the further conversion steps towards Asc are evident (Figure 8.2). By oxidation reactions at C_1 and C_2 and the formation of the 1,4-lactone, the hexose sugar of appropriate stereochemistry is converted into L-Asc. Here, a second challenge arises specifically for direct routes towards L-Asc. For directly generating L-Asc (the 1,4-lactone of 2-keto-L-gulonic acid), it is critical that the 1,4 ring structure of L-Asc is already formed before the final oxidation step is completed. Otherwise, the linear molecule, 2-KGA (Figure 8.1) would form, which, after its immediate deprotonation in a neutral or weakly acidic fermentation broth, will not further convert into L-Asc. In the context of direct L-Asc formation, 2-KGA, the main intermediate of the industry standard 2-KGA

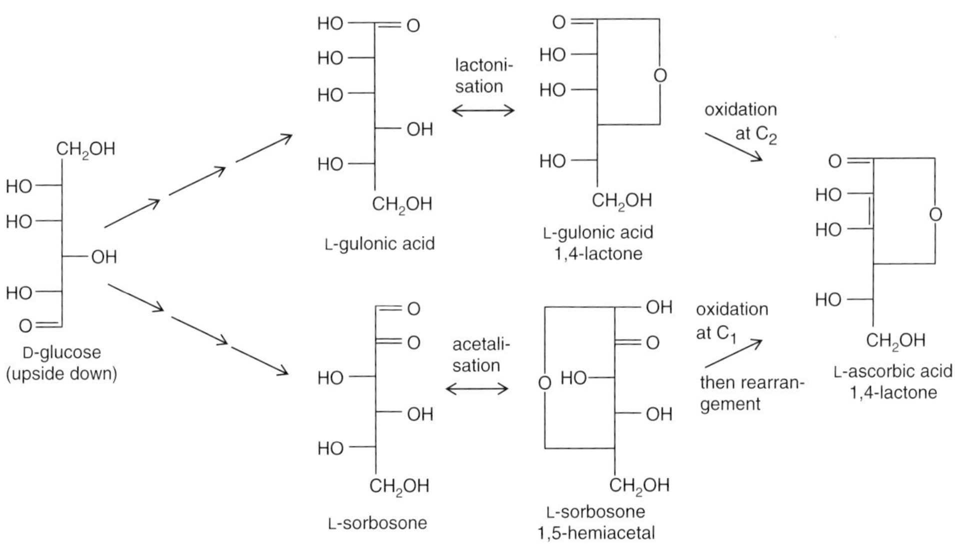

Figure 8.2 Conceptual outline of direct routes from D-glucose to L-Asc. The final oxidation step leading to L-Asc will be at C_2 for 1,4-lactone intermediates (such as L-gulonic acid 1,4-lactone), or C_1 for 2-keto aldose intermediates (such as L-sorbosone). Both types of intermediates have different propensities for different types of ring structures. 1,4-lactones are prevalent, while for the 2-keto aldoses, the corresponding 1,4-hemiacetals are not formed to appreciable extend. L-Asc formation from 2-keto aldoses may proceed via 1,5-hemiacetals, but an additional re-arrangement step to L-Asc is required.

fermentation process towards L-Asc, represents a dead-end compound and has to be avoided.

Which of the respective positions (C_1 or C_2) is oxidised in the final step towards L-Asc has significant impact on the propensity towards L-Asc of 2-KGA formation. This is because of the different predisposition of the respective intermediates, sugar acid or 2-keto aldoses, towards ring formation. The natural biosynthesis routes for L-Asc, both the plant and the animal pathway, proceed via sugar acids, which spontaneously or with enzymatic assistance (Kondo *et al.*, 2006) convert to their corresponding 1,4-lactones. The 1,4-lactone is the most prevalent form of most sugar acids (Levene and Simms, 1925; Xavier, Rauter and Queneau, 2010). It is favoured by the sp2 hybridisation of the carbonyl group as part of the ring, which can be better accommodated in the five-membered furanose ring compared to the six-membered pyranose ring (Brown, Brewster and Shechter, 1954). With this, the 1,4-lactone ring of L-Asc is already pre-formed before the second oxidation by oxidoreductases specific for C_2 of 1,4-lactones completes the synthesis. By-product formation of 2-KGA is negligible.

An alternative conceptual route first oxidises at C_2 to generate a 2-keto aldose with appropriate stereochemistry, for example, L-sorbosone. With its two carbonyl and four hydroxyl functionalities, such 2-keto aldoses can adapt numerous different ring forms (Köpper and Freimund, 2003). The 1,4-furanose form has the ring structure of L-Asc pre-formed and would be the obvious substrate to directly yield L-Asc upon oxidation at C_1, but is strongly disfavoured compared to the other isomers and present only in traces (see Section 8.4.2).

On the other hand, enzymes (further described in Section 8.4.1) which convert L-sorbosone directly into L-Asc are known to exist. Due to the virtual absence of 1,4-furanose structures for L-sorbosone, this reaction has to start from a different L-sorbosone isomer. Most plausible substrate for L-Asc formation is the L-sorbosone 1,5-hemiacetal (see Section 8.4.2), but additional rearrangement is required to arrive at the 1,4-lactone structure of L-Asc, which carries the risk of 2-KGA by-product formation upon being trapped as deprotonised free acid.

8.3 Direct L-Ascorbic Acid Formation via 1,4-Lactones

The formation of L-Asc from L-gulono-1,4-lactone (Figure 8.2) or L-galactono-1,4-lactone (the C_3 epimer of L-gulono-1,4-lactone) corresponds to the natural biosynthetic routes. A diverse range of approaches via 1,4-lactones has been explored for L-Asc production in microorganisms, either by exploiting the natural L-Asc biosynthesis in microalgae or by introducing routes towards 1,4-lactones and L-Asc in fungi, yeasts or bacteria. This review focusses on the new developments in this route over the past 10 years since 2005. Excellent background on older work can be found in a range of reviews (Hancock and Viola, 2001, 2002; Running, Peng and Rosson, 2004; Bremus *et al.*, 2006).

8.3.1
L-Ascorbic Acid Forming Enzymes: 1,4-Lactone Oxidoreductases

Nature offers a range of different L-Asc-forming 1,4-lactone oxidoreductases from animals, plants, fungi (such as yeast) and even bacteria, which have recently been reviewed by Leferink and van Berkel (2014). Their different properties allow choosing the most suitable enzyme in a given strain or process set-up. Important differences are not only substrate specificity and catalytic performance, but also subcellular localisation and nature of the electron acceptor.

The two different natural L-Asc biosynthesis routes in plants and animals have a match in two distinct groups of L-Asc-forming enzymes: L-galactono-1,4-lactone dehydrogenases (GALDH) in plants and L-gulono-1,4-lactone oxidases (GULO) in animals. GALDH (plants) are specific to their namesake substrate with a K_M in the range of 0.1–3 mM and k_{cat} of 10–135 s^{-1}, depending on the study and enzyme source (Oba *et al.*, 1995; Ostergaard *et al.*, 1997; Leferink, van den Berg and van Berkel, 2008). GALDH are located on the inner mitochondrial membrane and feed the electrons from the oxidised substrate via cytochrome c into the respiratory chain, ultimately reducing molecular oxygen to water (Schertl *et al.*, 2012; Hervas *et al.*, 2013). GULO (animals) have a wider substrate range. They also accept, besides their physiological substrate L-gulono-1,4-lactone, L-galactono-1,4-lactone and several (but not all) additional 1,4-lactones (Linster and Van Schaftingen, 2007). Affinity to L-gulono-1,4-lactone is very high with (depending on the species and study) K_M reported in the range of 0.01–0.1 mM, while the specific activity is rather low with k_{cat} around 0.5–3.5 s^{-1} (Nishikimi, Tolbert and Udenfriend, 1976; Kiuchi, Nishikimi and Yagi, 1982). GULO is attached to the membrane of the endoplasmic reticulum (ER), with its active site presumably facing the ER lumen. Electrons are transferred directly onto molecular oxygen, resulting in stoichiometric formation of hydrogen peroxide (Linster and Van Schaftingen, 2007). Despite these differences, both enzyme families GALDH and GULO are related, both contain FAD as cofactor, belong to the same VAO (vanillyl alcohol oxidase) family of flavoproteins (Leferink *et al.* 2008) and are probably derived from the same ancestral L-Asc-forming enzyme (Wheeler *et al.*, 2015). With mutations of just single amino acids, key functional differences between GALDH and GULO could be levelled. An A113G replacement converted GALDH from *Arabidopsis* to an oxidase, enabling access of molecular oxygen to the flavin cofactor and direct transfer of electrons to molecular oxygen, just as in the case for GULO (Leferink *et al.*, 2009a). An E386D replacement in the same GALDH enzyme altered substrate specificity towards equally accepting L-galactono- and L-gulono-1,4-lactone, just as in the case for GULO (Leferink *et al.*, 2009b). Despite of this ease of interconverting key features such as substrate spectrum and mode of electron transfer of both enzyme groups, these properties are strictly separated in the wild-type enzymes according to their phylogeny (plants or animals). This suggests there is an important difference in the functional needs between animal and plant L-Asc biosynthesis, which may relate to the high level of reactive oxygen species and the increased need for water-soluble antioxidants in photosynthetic organisms (Wheeler *et al.*, 2015).

Yeast and many other fungi do not have an endogenous pathway for L-Asc biosynthesis, but rather produce L-Asc analogues such as the five-carbon erythroascorbic acid (see Section 8.3.3). Its biosynthesis proceeds by oxidation of D-arabinose at C_1 and C_2 in analogy to the final steps of the plant route for L-Asc biosynthesis. The enzyme catalysing the final step, D-arabinono-1,4-lactone oxidase (ALO) is localised at the mitochondrial membrane similarly to GALDH (Nishikimi, Noguchi and Yagi, 1978), but otherwise is closely related in function and phylogeny (Wheeler *et al.*, 2015) to the mammalian GULO enzyme. As an oxidase, it transfers electrons directly to molecular oxygen, resulting in the formation of hydrogen peroxide. Besides its physiological five-carbon substrate, it accepts L-gulono- and L-galactono-1,4-lactones equally well (Nishikimi, Noguchi and Yagi, 1978; Huh *et al.*, 1994, 1998) with comparatively high K_M of 50 mM and k_{cat} of 20 s^{-1} for both D-arabino- and L-galactono-1,4-lactone substrates.

Prokaryotic organisms are not known to produce or require L-Asc as antioxidant or enzyme co-substrate. Nevertheless, sequence homologues to eukaryotic 1,4-lactone oxidoreductases can be detected in several bacterial organisms. One such homologue from *Mycobacterium tuberculosis* was recombinantly expressed and characterised and shown to indeed be capable of converting L-gulono-1,4-lactone to L-Asc (Wolucka and Communi, 2006). Similar to the plant GALDH, this enzyme does not accept molecular oxygen, but can transfer electrons to cytochrome c. It is highly selective for L-gulono-1,4-lactone, with K_M of 3.3 mM. Turnover, however, is slow at 0.05 s^{-1}. Surprisingly, no FAD cofactor could be detected in the recombinant protein, and the low specific activity could thus be caused by only a small fraction of the enzyme being charged with FAD cofactor. Activity of this enzyme could also be detected in *Mycobacterium* cell lysates, but whether L-Asc formation is indeed its physiological role remains unclear.

A highly unusual bacterial L-gulono-1,4-lactone dehydrogenase had been identified from *Ketogulonicigenium vulgare* (Sugisawa *et al.*, 1995). This is the very organism used in the industrial fermentation of the 2-KGA intermediate towards L-Asc. This L-Asc-forming enzyme, however, has no role in this industrial process, where oxidation at C_2 already occurs at the very beginning upon converting D-sorbitol to L-sorbose (see Section 8.4). No gene for a VAO-family oxidoreductase is found in the genome sequence of *Ketogulonicigenium;* thus, this enzyme is unrelated to any of the other Asc-forming 1,4-lactone oxidoreductases. The sequence or gene for this enzyme has not been described, but the enzyme was biochemically characterised as a heterotrimeric assembly of a flavoprotein (61 kD), a cytochrome c subunit (32.5 kD) and a small subunit (16.5 kD). Several such heterotrimeric flavoproteins are known from other organisms for carbohydrate oxidation, such as FAD-linked dehydrogenases of D-sorbitol (Shinagawa *et al.*, 1982; Toyama *et al.*, 2005), D-gluconic acid (Shinagawa *et al.*, 1984; Toyama *et al.*, 2007), D-fructose (Ameyama *et al.*, 1981; Kawai *et al.*, 2013) and D-glucose (Tsuya *et al.*, 2006). They all have in common that they oxidize the hydroxyl function a C_2 or C_5, one carbon away from the terminal position in the carbohydrate substrates. This is well consistent with the *Ketogulonicigenium* enzyme oxidising C_2 of L-gulono-1,4-lactone. Typically, these heterotrimeric flavoproteins belong to the glucose-methanol-choline (GMC) oxidoreductase family, which are linked with a cytochrome c and an additional small subunit in a single operon. The genome sequence of *Ketogulonicigenium* shows one example of such operon (KVU_PB0007, KVU_PB0008 and KVU_PB0009), and the predicted sizes of the protein products (58.8, 28.2 and 15.2 kD; after cleavage of putative signal sequences, but including covalently linked cofactors) are very close to the biochemical observations. The cytochrome c subunit sequence is peculiar, being significantly smaller than its homologues, consisting only of two instead of three cytochrome c repeats. This is also in agreement with the biochemical size observation of 32.5 kD compared to the standard around 50 kD. Overall, it seems highly likely that this set of *Ketogulonicigenium* genes is indeed the L-gulono-1,4-lactone dehydrogenase described by Sugisawa *et al.* This enzyme is localised in the periplasmic space between the inner and the outer membrane of this Gram-negative bacterium. It is thus readily accessible to small hydrophilic compounds from the environment. It cannot use molecular oxygen directly, but it is membrane-associated and feeds the electrons from the substrate oxidation into the respiratory chain. Substrate affinity is low with K_M estimated at 35 mM, but specific activity is very high with a k_{cat} of 850 s^{-1}. The substrate range of this enzyme was found to also include, besides L-gulono-1,4-lactone, D-glucose and D-xylose, both also being oxidised at C_2. It is therefore plausible that standard sugars such as D-glucose or D-xylose are the actual physiological substrates of this *Ketogulonicigenium* enzyme and its Asc-forming capability from L-gulono-1,4-lactone is a mere coincidence of substrate promiscuity.

8.3.2
Direct L-Ascorbic Acid Formation in Heterotrophic Microalgae

The right choice from the range of 1,4-lactone oxidoreductases is important, but only part of a direct microbial process to L-Asc. Equally important is the good provision of the 1,4-lactone substrate from a commodity carbon source. Microalgae are the only microorganisms with natural L-Asc formation capability, having, similar to higher plants, an endogenous route from D-glucose to L-Asc via L-galactono-1,4-lactone. This makes them obvious candidates as fermentative L-Asc production hosts. While phototrophically grown microalgae are indeed known for their high vitamin C content (Brown *et al.*, 1997), this type of growth is not suitable for industrial L-Asc production. Photobioreactors are of prohibitive investment and operation costs, while in an open pond set-up, growth is too slow and dilute to enable meaningful L-Asc titres.

A number of microalgae, however, also have the ability to grow heterotrophically in fermenters on added carbon source, while retaining their capability for L-Asc formation. This concept was worked out by researchers at BTR (Biotechnology Resources) in the 1990s (reviewed in detail in Running, Peng and Rosson,

2004). Initial work in *Chlorella* was hampered by the incompatibility of conditions for Asc stability (low pH, low dissolved oxygen) and *Chlorella* growth (neutral pH, high dissolved oxygen). A way forward was given by adapting the process set-up towards stabilisation of L-Asc, such as separation of biomass growth under high aeration and L-Asc production under low aeration. Still, a large fraction of the L-Asc produced was not secreted and remained intracellular. Subsequently, acidophilic species of the related microalgae *Prototheca* were selected, where low-pH growth allowed for stable L-Asc even under high aeration and most L-Asc was secreted and found in the fermentation broth. These strains were further improved for extracellular L-Asc accumulation in a strain development programme based on random genome mutagenesis and screening. Respectable extracellular L-Asc titres of up to 5 g/l in fermentation on D-glucose have been achieved, but industrial process requirements demand further improvements by at least an order of magnitude. The lack of genetic engineering tools for *Prototheca* prevented further progress, leading to abandonment of this program (Running, Peng and Rosson, 2004).

8.3.3 Direct L-Ascorbic Acid Formation in Recombinant Yeast

Baker's yeast (*Saccharomyces cerevisiae*) and other yeast strains are workhorses of industrial biotechnology, thanks to their ease of cultivation and genetic engineering. In addition, many yeast strains are capable of growth at low pH, which in a direct fermentative L-Asc production could be critical for the stabilisation of L-Asc in the fermentation broth. Due to the substrate promiscuity of the yeast enzyme ALO (see Section 8.3.1), yeast not only produces its physiological antioxidant erythroascorbic acid, but also supports L-Asc synthesis (Figure 8.3) when provided with the appropriate precursors L-galactono-1,4-lactone or L-gulono-1,4-lactone (Nishikimi, Noguchi and Yagi, 1978; Roland *et al.*, 1983). Even the preceding step, oxidation of L-galactose to L-galactonic acid (which spontaneously rearranges to the 1,4-lactone), is supported by *S. cerevisiae* biomass, possibly because of similar substrate promiscuity of the corresponding yeast pathway enzyme D-arabinose dehydrogenase (Kim *et al.*, 1996; Hancock, Galpin and Viola, 2000). Looking at the front end of the L-Asc biosynthesis, *S. cerevisiae* has the added advantage of a high endogenous flux into GDP-D-mannose for cell wall biosynthesis and protein glycosylation (Hashimoto *et al.*, 1997; Kruszewska *et al.*, 1999). With this, only three genes are missing to establish the plant route towards L-Asc in *S. cerevisiae:* GDP-D-mannose-3,5-epimerase, GDP-L-galactose-phosphorylase and L-galactose-1-phosphatase (Figure 8.3).

This physiological framework in *S. cerevisiae* provided a clear outline for genetic engineering work to enable *S. cerevisiae* for L-Asc synthesis from D-glucose. Branduardi *et al.* established the heterologous expression of the three missing plant pathway enzymes in *S. cerevisiae* (Branduardi *et al.*, 2007). In addition, the endogenous yeast enzymes for the two final steps were complemented by overexpression of the plant enzyme L-galactose dehydrogenase and of the

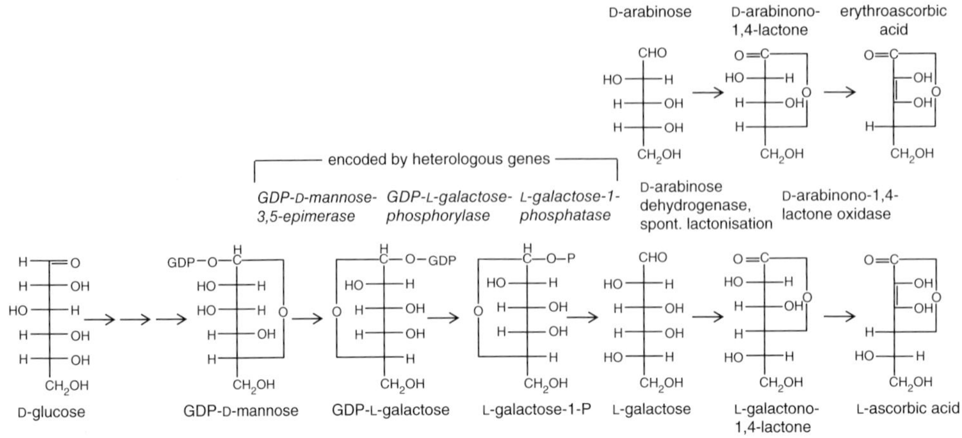

Figure 8.3 Top: Erythroascorbic acid biosynthesis in *S. cerevisiae*. Bottom: L-Ascorbic acid biosynthesis in recombinant *S. cerevisiae* provided with heterologous genes encoding (i) GDP-D-mannose-3,5-epimerase, (ii) GDP-L-galactose-phosphorylase and (iii) L-galactose-1-phosphatase. Enzymes encoded by the heterologous genes are marked in *italic*. The recombinant pathway taps into the abundant GDP-D-mannose pool of the host strain.

endogenous L-arabinono-1,4-lactone dehydrogenase to prevent bottlenecks at the final steps of the pathway. With this, production of L-Asc from D-glucose in *S. cerevisiae* could be demonstrated, albeit only intracellularly and at very low level (0.4 mg/l) (Branduardi *et al.*, 2007; Fossati *et al.*, 2011). Expression of enzymes for the recovery of the oxidised L-dehydroascorbic acid had no positive effect on the L-Asc titres (Fossati *et al.*, 2011). The same picture was observed in a similar study incorporating the missing L-Asc biosynthesis genes from plant in the yeast *Kluyveromyces lactis* (Rosa *et al.*, 2013). Intracellular L-Asc formation was shown at 14 mg/l after 48 h fermentation, but again no L-Asc accumulation in the fermentation broth was reported. Could it be that L-Asc cannot easily leave the yeast biomass? Previous work starting from L-galactose clearly showed L-Asc accumulating to 50 mg/l in 48 h fermentation broth (Sauer *et al.*, 2004). Therefore, L-Asc is well capable of leaving the *S. cerevisiae* cells and can accumulate in the medium. Instead, it is likely that the pathway flux from D-glucose towards the 1,4-lactone is too low, and the resulting low L-Asc productivity is more than offset by the rate of L-Asc degradation in the fermentation broth. Much improvement will be needed in the pathway flux from D-glucose to L-galactose to enable L-Asc accumulation in fermentation broth, even more to achieve the productivities and titres needed for commercialisation. This will require further insight into the expression level, localisation and enzymatic activity of the heterologously expressed plant enzymes to clarify if their enzymatic activities, the precursor supply, or the compartmentalisation of the reactions in the cell is causing critical limitations.

8.3.4 Direct L-Ascorbic Acid Formation from Orange Processing Waste in Recombinant *Aspergillus niger*

Direct production of L-Asc via L-galactono-1,4-lactone obtained from D-galacturonic acid was established in a recent study (Kuivanen *et al.*, 2012, 2014; Kuivanen, Penttila and Richard, 2015). Interestingly, rather than using the epimerisations of the standard plant pathway, the correct stereochemistry was achieved as in the animal pathway by inversion of the carbon skeleton, generating the L-galactonic acid intermediate from D-galacturonic acid (Figure 8.4). D-galacturonic acid is a major constituent of pectin and abundantly present in waste streams from fruit or sugar beet processing, such as ~375 000 t/year from orange processing waste only (Kuivanen *et al.*, 2014). Conceptually this builds on earlier proposals (Danehy, 1979; Roland *et al.*, 1983) to apply the inversion of the carbon skeleton on D-galactose or D-galacturonic acid containing waste streams such as whey or citrus pectin for L-Asc formation. Kuivanen *et al.* now combined the individual reaction steps of pectin hydrolysis, sugar conversion towards the correct stereochemistry and L-Asc formation in *Aspergillus niger* as single microorganism (Figure 8.4). The filamentous fungus *A. niger* is capable of growth on pectin or on orange processing waste by hydrolysing pectin to yield the D-galacturonic acid monomers. It naturally catabolises the monomers via an NADPH-linked reductase first to L-galactonic acid (the desired intermediate towards L-Asc synthesis) and further on via dehydratase and aldolase to pyruvate

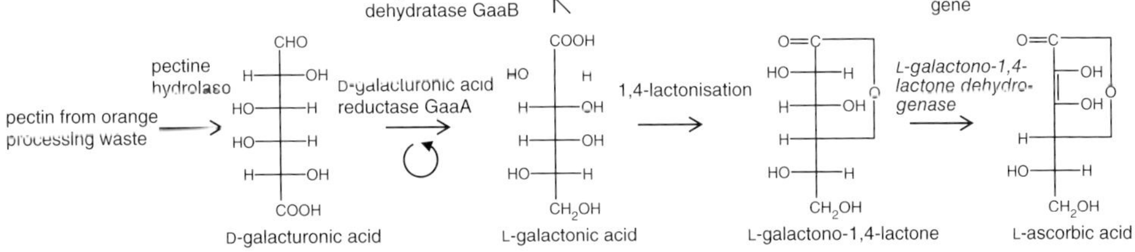

Figure 8.4 Direct production of L-Asc via L-galactono-1,4-lactone obtained from pectin in *Aspergillus niger* as single microorganism. Knockout of *gaaB* encoding L-galactonic acid dehydratase and overexpression of *gaaA* encoding D-galacturonic acid reductase re-directs the flux towards L-galactono-1,4-lactone. The synthesis route to L-Asc is completed by heterologous expression of L-galactono-1,4-lactone dehydrogenase from Acerola (marked by *italics*).

and glycerol (Kuivanen *et al.*, 2014). Upon knockout of the L-galactonic acid dehydratase (*gaaB*) and additional overexpression of the D-galacturonic acid reductase (*gaaA*), this assimilation route is blocked and L-galactonic acid was shown to accumulate to up to 8 g/l in the fermentation broth and to even much higher levels in the biomass (Kuivanen *et al.*, 2012, 2014). Molar yields of L-galactonic acid on D-galacturonic acid were at 60% in submerged fermentation and close to 90% in solid-state fermentation. The volumetric productivity, however, was low at 0.07 g/l/h. Up to this point, the pathway stops at L-galactono-1,4-lactone, since the *Aspergillus* host itself does not express 1,4-lactone dehydrogenase activity, such as an arabinono-1,4-lactone dehydrogenase activity as in yeast, for conversion of L-galactono-1,4-lactone to L-Asc. Setting up *Aspergillus* towards production of L-Asc was completed by heterologous expression of the GALDH from Acerola (*Malpighia glabra*), a plant known for the high vitamin C content of its fruits. Titres of 0.17 g/l L-Asc were achieved in the fermentation broth after 96 h fermentation on orange processing waste (Kuivanen, Penttila and Richard, 2015), which compares favourably to the recombinant yeast approach that only yielded intracellular L-Asc. Still, more than 5 g/l L-galactonic acid was found in the supernatant (Kuivanen, Penttila and Richard, 2015). Clearly, the conversion of L-galactonic acid to L-Asc is limiting, despite high GALDH enzymatic activity reported in a cell-free extract of the recombinant *Aspergillus* strain with externally added cytochrome c (0.1 U/mg total protein, corresponding to ~1 g/l/h volumetric productivity for L-Asc at 5 g/l cell dry weight biomass concentration). The reaction step of 1,4-lactone formation from L-galactonic acid was not engineered in this set-up and could form a limitation as well. Additionally, proper targeting and localisation of GALDH to the inner mitochondrial membrane are essential to enable the link to the respiratory chain to regenerate its cofactor. Problems of such targeting may arise upon heterologous expression of plant genes in fungi and would not show in the enzymatic assay which used externally added electron acceptor. Further thorough investigations will be required to elucidate the molecular cause of the current limitation and the full potential of this approach.

8.3.5
Overall Conclusion on 1,4-Lactone Routes

With the elucidation of the plant pathway for L-Asc biosynthesis in 1998 (Wheeler, Jones and Smirnoff, 1998), new possibilities arose for engineering this pathway into microorganisms for fermentative production of L-Asc. This applies in particular to yeast since this organism is capable of many of the required conversions, leaving the need for only three plant enzymes to close the gap in the pathway. Based on these expectations, it is disillusioning how poorly this engineered pathway performs. There clearly is potential for better performance and need for better understanding of the current limitations, but it seems questionable if the ambitious targets for industrial production will ever be within reach. One conceptual limitation is in the complexity of the eight reaction steps required for converting D-glucose to L-galactose to provide the correct stereochemistry.

Rather than emulating the natural L-Asc biosynthesis in microbial hosts, there may therefore be more promise in selecting an appropriate abundant raw material which can be efficiently converted to an L-aldono-1,4-lactone precursor. A good example is given by the conversion of D-galacturonic acid to L-galactonic acid and L-Asc by genetically engineered *Aspergillus*, in just three steps. The orange processing waste as source of D-galacturonic acid is sufficiently abundant to also serve L-Asc production. While the proof of principle could be shown, significant breakthroughs will be required to improve performance parameters towards industrial relevance. Even if this can be achieved, however, there will be major logistical challenges related to the use of orange processing waste, based on the low content of D-galacturonic acid (3–4% of wet weight, (Kuivanen *et al.*, 2014)), the need to avoid the drying costs, the perishable nature of this wet waste stream and the seasonal fluctuations of its supply.

A related and possibly even more attractive route towards L-Asc would be based on the conversion of starch (D-glucose) to L-gulonic acid, which could be converted to L-Asc by any of the known L-gulono-1,4-lactone oxidoreductases (Figure 8.5) (Hearon and Witte, 1981). Here, the rooting in the most universal feedstock D-glucose would form a key advantage. Interesting concepts have been worked out for chemical steps towards L-gulonic acid, starting with methanolysis of starch. The methyl-glucoside can be selectively oxidised at C_6 by O_2 on Pd/C catalyst to yield methyl-glucuronic acid up to 70% yield. By reductive demethylation with H_2 on Ru catalyst, L-gulono-1,4-lactone is achieved up to 90% yield. The performance of this process is clearly more advanced than that of any biotechnological route to 1,4-lactone intermediates. Still, further optimisation of catalysts and reaction conditions is needed to meet industrial process requirements.

For the final step of oxidation of the L-gulono-1,4-lactone to L-Asc, there is the choice between several families of enzymes from mammals, yeasts and bacteria (see Section 8.3.1). Here, the *Ketogulonicigenium* enzyme may offer

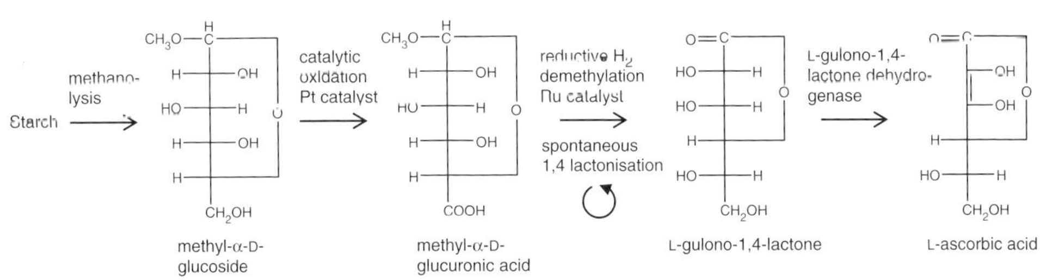

Figure 8.5 Direct production of L-Asc via L-gulono-1,4-lactone from starch by combinations of chemical and biocatalytic reaction steps. Chemical conversion of starch to L-gulono-1,4-lactone is followed by biocatalytic oxidation to L-Asc by any of the known L-gulono-1,4-lactone oxidoreductases.

unique advantages, not only because of its reported high specific activity. Being linked to the respiratory chain avoids the release of stoichiometric H_2O_2, which would be bound to cause havoc with a product as sensitive to oxidation as L-Asc. Unique amongst the 1,4-lactone oxidoreductase is its periplasmic localisation, which provides direct access to the substrate in the production medium without any transport limitations. This could provide critical advantage over the existing approaches in microalgae, yeast or *Aspergillus*, which all require transfer of the 1,4-lactone into the mitochondrion before the L-Asc product needs to become released to the fermentation medium. This set-up is prone to yield losses, by-product formation and transport limitations. Such fresh approaches towards direct L-Asc fermentation have the potential for industrial realisation by tapping into an abundant feedstock from which in an efficient way the substrate for a highly active and product specific 1,4-lactone oxidoreductase can be obtained.

8.4
Direct L-Ascorbic Acid Formation via 2-Keto Aldoses

All approaches so far to synthesise L-Asc via 1,4-lactones struggle with efficiently providing the 1,4-lactone intermediate with the appropriate stereochemistry for L-Asc. More than 80 years ago, Tadeus Reichstein and Andreas Grüssner invented an efficient approach for the synthesis of L-Asc from D-glucose (Reichstein and Grüssner, 1934) and solved the problem of making the stereochemistry of D-glucose suitable for L-Asc. This 'Reichstein synthesis' route towards L-Asc has been the industry standard for 60 years. Even the current standard, the 2-KGA fermentation process, still follows the same concept and shares most reaction steps with the Reichstein synthesis, including those critical for the stereochemistry (Pappenberger and Hohmann, 2014). In contrast to the natural biosynthetic routes, the precursor for the final oxidation step is not a 1,4-lactone (to be oxidised at C_2), but L-sorbosone, a 2-keto aldose (to be oxidised at C_1) (Figure 8.2).

The stereochemical trick in the approach by Reichstein and Grüssner is based on inversion of the carbon skeleton of D-glucose (see Section 8.2.1), similar to the biosynthetic route in animals (which was then unknown). D-glucose is first reduced to its corresponding sugar alcohol D-sorbitol, for example, by catalytic hydrogenation, a highly efficient chemical process step (Figure 8.6). To invert the numbering of the carbon atoms, D-sorbitol is next oxidised at the 'opposite end'. Exquisite regioselectivity is required for selective oxidation of only one of the six hydroxyl functions. For this, already Reichstein and Grüssner used enzymatic catalysis. No enzyme, however, is known which is selective for oxidation of the hydroxyl at C_6 of D-sorbitol (which would result in L-gulose as product). There is therefore no route from D-sorbitol via C_1 oxidation to a 1,4-lactone intermediate that would then allow direct L-Asc formation using the set of available 1,4-lactone oxidoreductases for the final oxidation step at C_2.

The only enzymes available that achieve quantitative inversion of the carbon skeleton of D-sorbitol are dehydrogenases specific for the C_5 position of D-sorbitol

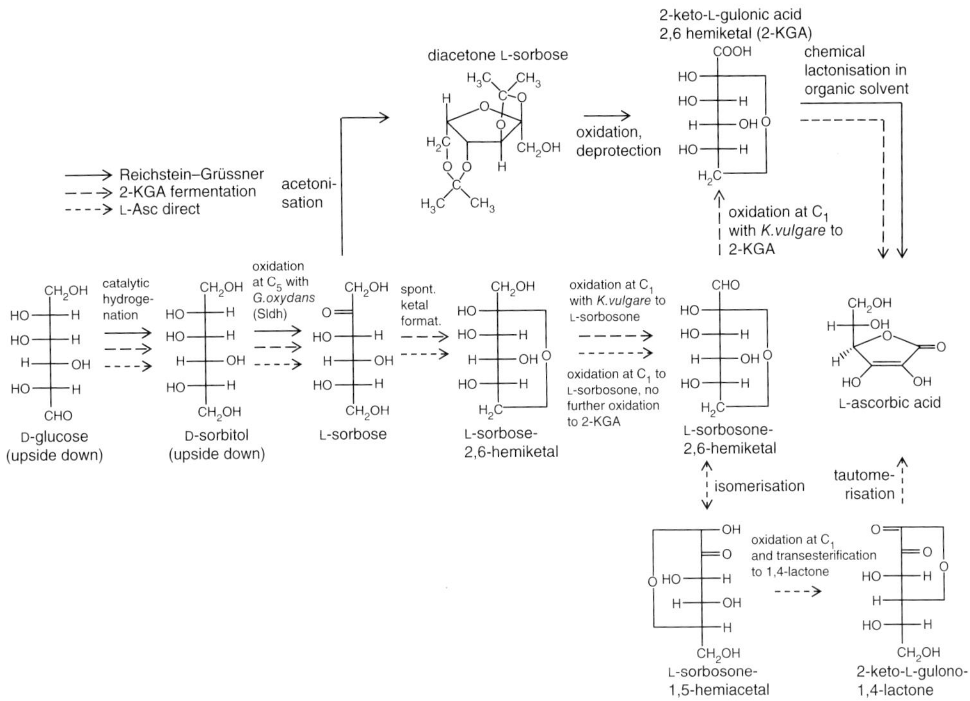

Figure 8.6 A possible novel biocatalytic route from D-sorbitol directly to L-Asc. Correct stereo conformation is obtained as in the Reichstein-Grüssner and the 2-KGA fermentation process by inversion of the carbon skeleton of D-glucose to L-sorbose via D-sorbitol. The key enabling elements of the direct route to L-Asc are (i) oxidation of L-sorbose to L-sorbosone at C_1 without further oxidation to 2-KGA and (ii) oxidation of L-sorbosone at C_1 by specific L-Asc-forming dehydrogenase that provides 2-keto-L-gulonic acid as 1,4-lactone instead of the free acid. This reaction presumably proceeds from the 1,5-hemiacetal isomer of L-sorbosone as substrate and entails an additional re-arrangement step to the 1,4-lactone.

(which corresponds to C_2 in L-Asc), yielding the 2-keto compound L-sorbose. The oxidation at C_2, which is the final step in L-Asc biosynthesis, is thus already done in the first steps of industrial L-Asc synthesis. Examples for such C_5-specific enzymes are the FAD-linked sorbitol dehydrogenase (FAD-Sldh) and the pyrroloquinoline quinone (PQQ)-linked glycerol/sorbitol dehydrogenase (PQQ)-Gldh/PQQ-Sldh) of *G. oxydans* (Shinagawa *et al.*, 1982; Shinjoh *et al.*, 2002; Sugisawa and Hoshino, 2002; Soemphol *et al.*, 2008) or the SsdB enzyme of *K. vulgare* (Pappenberger and Hohmann, 2014). In all the past and present industrial synthetic routes, this conversion occurs in a fermentation step using *G. oxydans* (original nomenclature: *Acetobacter suboxydans*).

Conversion of L-sorbose to L-Asc requires two further oxidation steps at C_1. The original Reichstein-Grüssner process used chemical oxidation of a protected L-sorbose substrate (diacetone L-sorbose), while in the current 2-KGA fermentation process, both oxidation steps are done biocatalytically with the Gram-negative bacterium *K. vulgare.* In contrast to D-sorbitol, a selective oxidation at C_1 of L-sorbose is possible. The key difference to D-sorbitol is the quantitative formation of a 2,6-pyranose ring in L-sorbose, where the second primary hydroxyl group at C_6 is protected and thus not reactive (Figure 8.6). On D-sorbitol, both 'ends' (C_1 and C_6) are subject to oxidation by the *K. vulgare* SsdA-type enzymes (see Section 8.4.3.1), leading to a mixture of D-glucose and L-gulose (Pappenberger and Hohmann, 2014).

In both, the original Reichstein-Grüssner process and in the current 2-KGA fermentation process, 2-KGA is the immediate product, which is re-arranged to L-Asc in a chemical reaction step in the absence of water. For direct access to L-Asc, formation of 2-KGA as free acid has to be avoided in the second oxidation step, since after its immediate deprotonation in fermentation broth, no rearrangement to L-Asc is possible any more in this aqueous environment. Instead, this second oxidation at C_1, starting from L-sorbosone, has to yield the 1,4-lactone of 2-keto-L-gulonic acid, which rapidly and quantitatively isomerises to its enol tautomer L-Asc.

8.4.1
L-Ascorbic Acid Forming Enzymes: L-Sorbosone Dehydrogenases

Several enzymes have been described as oxidising the aldehyde functionality at C_1 of L-sorbosone to the acid. Either product, 2-KGA or L-Asc (the enol tautomer of the 1,4-lactone of 2-KGA), can indeed be observed, depending on the enzyme.

Well-characterised 2-KGA forming sorbosone dehydrogenases are the sorbose / sorbosone dehydrogenases (Ssdhs) from *K. vulgare* (Asakura and Hoshino, 1996, 1999; Gao *et al.*, 2013, 2014) (see also Section 8.4.3.1), the membrane-bound sorbosone dehydrogenases (mSndhs) from *K. vulgare* (Gao *et al.*, 2013, 2014) or *Gluconacetobacter liquefaciens* (Shinjoh *et al.*, 1995) and the cytosolic sorbosone dehydrogenase (cSndh) from *G. oxydans* IFO3293 (Hoshino, Sugisawa and Fujiwara, 1991; Saito *et al.*, 1997). These three enzyme types are unrelated on amino acid sequence level. Ssdh and mSndh are both PQQ enzymes with periplasmic localisation, while the cytosolic cSndh uses NAD as redox cofactor.

L-Asc-forming activity from L-sorbosone was first described by Loewus *et al.* (1990) in leaves from beans and spinach. Based on a partial purification of this activity, the enzyme was characterised as an NADP-linked dehydrogenase with molecular weight of 21–29 kD and a K_M towards L-sorbosone of 12–18 mM, but with a low specific activity towards L-sorbosone oxidation. No amino acid or gene sequence has been described for this enzyme, and no further work has been published towards exploiting it for industrial application. It seems unlikely that this enzymatic activity plays a role towards L-Asc biosynthesis in plants (Davey *et al.*, 1999). The described activity may rather reflect a cross-reactivity with the non-physiological substrate L-sorbosone, which, remarkably, results in (at least some) L-Asc formation rather than (exclusive) 2-KGA formation. 2-KGA formation as by-product next to L-Asc was not analysed and cannot be excluded.

8.4.1.1 Sndhak

The next report on L-Asc-forming activity from L-sorbosone was from *K. vulgare*, one key organism of the industrial 2-KGA fermentation process. This bacterium had already been known to convert L-gulono-1,4-lactone to L-Asc (Sugisawa *et al.*, 1995) (see Section 8.3.1), but it came as surprise that even L-sorbosone, the intermediate of the well-established 2-KGA fermentation, results in L-Asc formation in *K. vulgare* (Sugisawa, Miyazaki and Hoshino, 2005). This had gone unnoticed, despite its industrial application, presumably due to the rapid degradation of L-Asc in fermentation broth. For 2-KGA fermentation, this L-Asc formation is undesirable since its degradation significantly contributes to discoloration and yield loss during 2-KGA fermentation (Hoshino, Miyazaki and Sugisawa, 2002a). The enzyme responsible for L-Asc formation from L-sorbosone was purified and found to be a soluble periplasmic protein with PQQ as cofactor (Hoshino, Miyazaki and Sugisawa, 2002b,c; Miyazaki, Sugisawa and Hoshino, 2006). Despite having the same cofactor and subcellular localisation, it bears no sequence similarity to the Ssdh enzymes of *Ketogulonicigenium* (see Section 8.4.3.1). Originally, this enzyme was named Sndh, but was subsequently renamed as Sndhak (sorbosone dehydrogenase, L-Asc forming, from *Ketogulonicigenium* (Pappenberger and Hohmann, 2014)).

Sndhak falls in the superfamily of 'quinoprotein glucose dehydrogenase B', several members of which have known crystal structure (Figure 8.7). These enzymes form a β-propeller of six β-sheets ('blades'), with the PQQ cofactor located on one face of the propeller and the substrate binding on top of the cofactor, surrounded by loops. Besides this common architecture, there are structural differences between the members of this superfamily. Certain members such as the soluble glucose dehydrogenase (sGdh) from *Acinetobacter* (Oubrie, Rozeboom and Dijkstra, 1999; Oubrie *et al.*, 1999a,b) and, by sequence homology, also Sndhak have long loop extensions around the active site, while others such as the aldose sugar dehydrogenase (Asd) from *Escherichia coli* (Southall *et al.*, 2006) have most of these loops sized to the bare minimum (Figure 8.7). Further differences are found in the fusion of additional domains to this structure core. Several members of this superfamily, including Sndhak, have a cytochrome c domain fused either N- or C-terminal, presumably involved in transferring the electrons from the cofactor to the respiratory chain. Most others, including sGdh and Asd, however, have no such additional domain.

Sndhak was shown to produce both L-Asc and 2-KGA from L-sorbosone at a ratio of ~5 : 1, with a k_{cat} for L-Asc formation around $4\,s^{-1}$ (Miyazaki, Sugisawa and Hoshino, 2006). The L-Asc/2-KGA product ratio and the L-Asc productivity may be underestimated due to the need for artificial electron acceptors in the enzymatic assay, which accelerate L-Asc, but not 2-KGA degradation. K_M

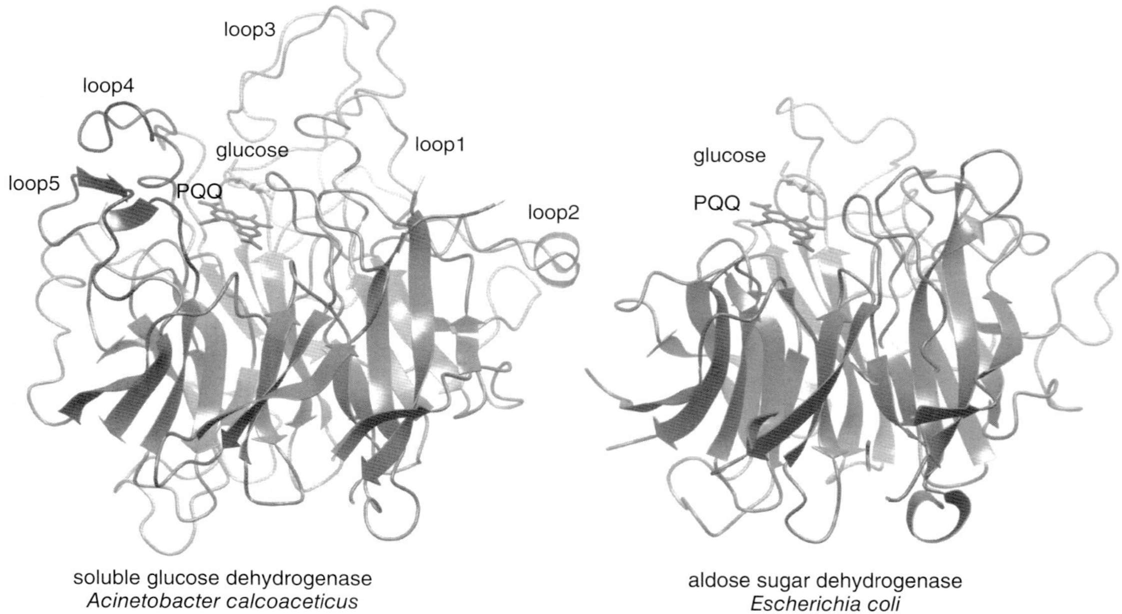

Figure 8.7 Structure of the Sndhak homologous enzymes soluble glucose dehydrogenase (sGdh) from *Acinetobacter calcoaceticus* and aldose sugar dehydrogenase (Asd) from *Escherichia coli*. The course of the polypeptide backbone is indicated in cartoon representation. The bound cofactor PQQ and substrate β-D-glucose are shown in stick representation. The overall architecture of both enzymes is highly similar, but sGdh is distinguished by a set of five strongly extended loops (indicated), which form a rim around the substrate binding site. The drawings were prepared with the Maestro molecular graphics software (Schrödinger, LLC, New York, NY, 2014) using the coordinates with the pdb access codes 1CRU for sGdh and 2G8S for Asd. The glucose in the Asd structure is added for illustration based on the position of glucose in the sGdh structure.

for L-sorbosone is rather high, around 50 mM (our own unpublished results). D-glucosone, the C_5 epimer of L-sorbosone, is similarly converted by Sndhak to both D-iso-ascorbic acid and 2-keto-D-gluconic acid, the C_5 epimers of L-Asc and 2-KGA, respectively. The specific activity of Sndhak against D-glucosone is even an order of magnitude higher that of L-sorbosone (Hoshino, Miyazaki and Sugisawa, 2002b). Also, D-glucose and D-xylose, but not myo-inositol (the natural substrate for the Sndhai enzyme, see Section 8.4.1.2), are rapidly oxidised at C_1 by Sndhak with a low K_M of 5 mM (our own unpublished results), yielding D-gluconic acid and D-xylonic acid. The physiological role of Sndhak may therefore be the oxidation of the common sugars D-glucose and D-xylose. The oxidation of L-sorbosone and moreover the production of L-Asc by Sndhak seem to be an evolutionarily unintended side effect.

8.4.1.2
Sndhai

Shortly after the discovery of Sndhak, another L-sorbosone to L-Asc converting enzyme could be identified (Berry *et al.*, 2003), this time from *G. oxydans*, the other key organism of the 2-KGA fermentation process, responsible for the D-sorbitol to L-sorbose conversion. This enzyme was named Sndhai (sorbosone dehydrogenase, L-Asc forming, from *G. oxydans* IFO3293 (Pappenberger and Hohmann, 2014)) and again found to be a periplasmic PQQ enzyme, but membrane-associated and with no sequence similarity to any of the other PQQ-linked sorbosone dehydrogenases (Ssdhs, mSndh, Sndhak).

Sndhai and its homologues belong to the family of 'membrane-bound PQQ-dependent glucose dehydrogenase'. There is no structure available so far for any member of this family, but in analogy to structures from distantly related PQQ enzymes and similar to the (unrelated) Sndhak, Sndhai is formed of an β-propeller, here composed of eight β-sheets, with PQQ and substrate binding on one side of the propeller, surrounded by loops. The distinguishing mark of this family is an additional N-terminal domain of five transmembrane helices which anchors this enzyme to the outside of the inner membrane, with the β-propeller domain facing the periplasmic space. Enzyme constructs lacking this N-terminal anchor are still functional, indicating that all key features relevant for dehydrogenase activity reside in the β-propeller domain (Elias *et al.*, 2001).

Similarly to Sndhak, it converts L-sorbosone to both L-Asc and 2-KGA at a ratio of ~5 : 1. K_M towards L-sorbosone is even higher than for Sndhak, around or above 100 mM (our own unpublished results). Again, this low affinity points towards the non-physiological nature of the L-sorbosone to L-Asc conversion by Sndhai. The presumptive natural substrate, by which this enzyme has indeed been originally identified (Kluyver and Boezaardt, 1939; Chargaff and Magasanik, 1946; Wissler, Freivogel and Wiesner, 1995; Hölscher, Weinert-Sepalage and Görisch, 2007) is myo-inositol with a K_M of 5 mM and a 10-fold higher specific activity (our own unpublished results). It also readily oxidises D-xylose, but, in contrast to Sndhak, not D-glucose.

8.4.1.3
Prevalence of L-Asc Forming Sorbosone Dehydrogenases in Nature

It is remarkable that both L-Asc-forming enzymes Sndhak and Sndhai have been found in organisms involved in the industrial 2-KGA fermentation process. In this industrial process, L-Asc is only formed subsequently by chemical conversion, and there is no reason to assume that such industrial application towards 2-KGA would correlate with the presence of L-Asc-forming enzymes. It does correlate, however, with formation and exposure towards L-sorbosone. Only in the context of the 2-KGA fermentation, one would routinely expose organisms to L-sorbosone and the chance of L-Asc formation from L-sorbosone would become apparent to those researchers with a prepared mind. One must therefore conclude that enzymes converting L-sorbosone to L-Asc should not be uncommon amongst microbes. Further screening of natural diversity with L-sorbosone as substrate is therefore likely to uncover further L-Asc-forming enzymes. Both known L-Asc-forming L-sorbosone dehydrogenases Sndhak and Sndhai bear no detectable sequence similarity, and it will be interesting if additional L-Asc-forming enzymes are found from yet further distinct families of dehydrogenases. In addition, homologues to Sndhak and Sndhai at around 40% sequence identity level are widespread, but it remains to be seen if this level of similarity is sufficient to enable L-Asc formation.

8.4.2
L-Asc or 2-KGA from L-Sorbosone: One Substrate, Several Isomers, Two Products

Overall, six different classes of L-sorbosone dehydrogenases have been described in the literature, four even having the same PQQ redox cofactor. There is no obvious differentiator to explain why some of those enzymes only form 2-KGA as reaction product, while others result in L-Asc formation. In addition, it needs to be understood why L-Asc formation is generally accompanied by 2-KGA formation and how this 2-KGA by-product can be minimised.

L-Sorbosone does form a variety of isomeric ring structures in aqueous solution. Some of the different ring isomers are predestined to yield a single specific product, either L-Asc or 2-KGA, upon oxidation of C_1 (Figure 8.8). Oxidation of the 2,6-pyranose isomer of L-sorbosone leads directly to 2-KGA. This isomer has an abundance (predicted by quantum mechanical calculations, our own unpublished results) of 88%, almost exclusively as α-anomer with the bulky $C_1(OH)_2$ substituent in equatorial position. This is the presumptive substrate for the unspecific alcohol/aldehyde dehydrogenase Ssdh (see Section 8.4.3.1) during the industrial 2-KGA fermentation, as well as for the other 2-KGA-forming enzymes mSndh and cSndh (see Section 8.4.1). Oxidation of the 1,4-furanose would lead directly to L-Asc. This isomer, however, should be strongly disfavoured compared to 1,5- or 2,6-pyranose structures, particularly in the case where the substituents at C_3 and C_4 are both at the same face of the ring. This is the case for D-glucosone, where no 1,4-furanose structure could be detected by NMR (Köpper and Freimund, 2003), and this is also expected for L-sorbosone with the

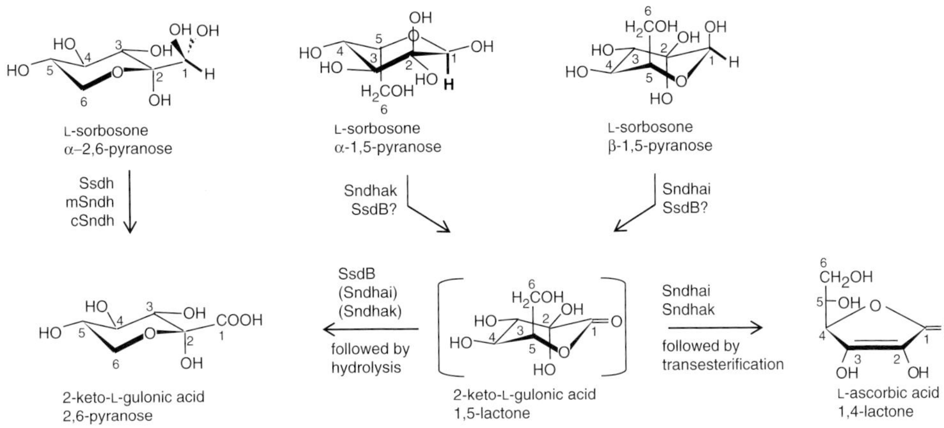

Figure 8.8 Routes to L-Asc and 2-KGA from the prevalent L-sorbosone isomers. Oxidation of L-sorbosone to 2-KGA generally proceeds from the 2,6-pyranose form (Ssdh, mSndh, cSndh). The 1,4 furanose form would be predestined for resulting in L-Asc upon oxidation, but has only very minor abundance and is therefore not shown. Oxidation of L-sorbosone to L-Asc presumably proceeds via the 1,5-pyranose form with subsequent transesterification (Sndhai, Sndhak), but it may also yield 2-KGA by-product upon hydrolysis. The specificity of Sndhai and Sndhak to the anomers of L-sorbosone 1,5-pyranose is discussed in Figure 8.9. SsdB presumably uses the same 1,5-pyranose form of L-sorbosone as substrate, but yields exclusively 2-KGA upon hydrolysis of the intermediate.

same configuration at C_3 and C_4, based on quantum mechanical calculations (our own unpublished results). It is therefore excluded as relevant substrate for L-Asc formation and not shown in Figure 8.8.

The one other isomer with significant abundance is the 1,5-pyranose, with 2% abundance for the α-anomer and 9% for the β-anomer, based on quantum mechanical calculations. Indeed, there are indications that support the 1,5-pyranose anomers of L-sorbosone as substrate for Sndhai and Sndhak, based on the similarity to the natural substrates of these two enzymes (Figure 8.9). D-glucose and D-xylose (but not myo-inositol) are substrates for Sndhak. For D-glucose (and, by analogy, presumably also for D-xylose), only the β-anomer is bound at the active site of sGdh, the Sndhak homologue from *Acinetobacter* (Oubrie *et al.*, 1999b). Myo-inositol and D xylose (but not D-glucose) are substrates for Sndhai. Only the single-axial hydroxyl of myo-inositol is being oxidised by Sndhai. To provide for an equivalent axial hydroxyl group, it has to be the α-anomer of D-xylose which is the substrate for Sndhai. Even though both, Sndhai and Sndhak, oxidise D-xylose, they may act on the different anomers of this sugar, in line with the distinct substrate specificities of both enzymes.

The conformation and substitution pattern of the natural substrates have close counterparts in the L-sorbosone isomer structures. In accordance with the different substrate specificity, these L-sorbosone counterparts of Sndhak and Sndhai are different, but both are anomers of the 1,5-pyranose: the α-anomer for

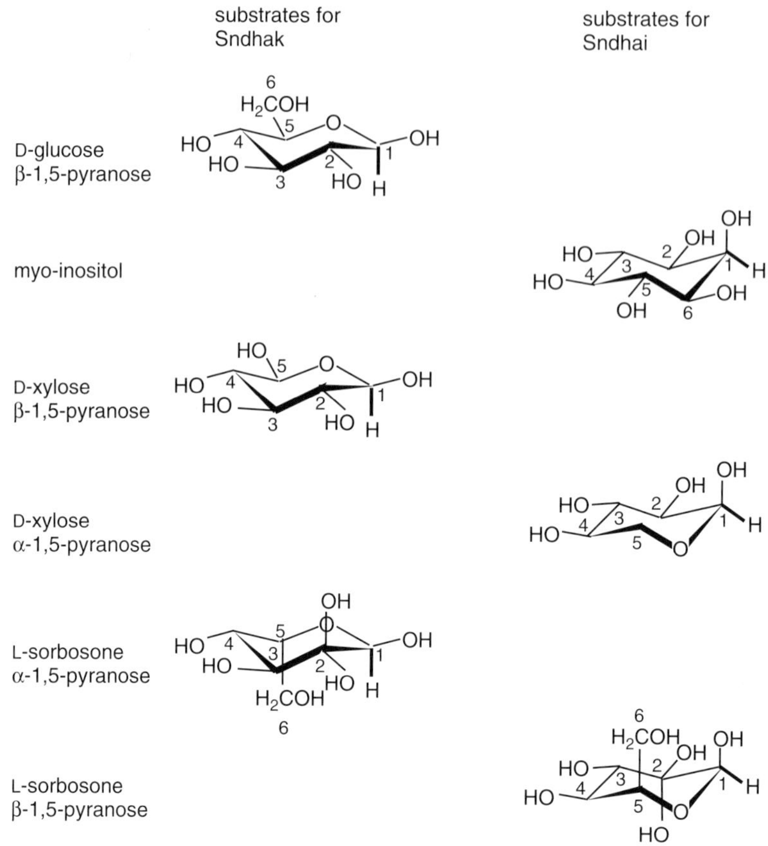

Figure 8.9 Structural match of the putative natural substrates of Sndhak (β-D-glucose, β-D-xylose) and Sndhai (myo-inositol, α-D-xylose) to the 1,5-pyranose isomers of L-sorbosone. The numbering of the carbon atoms is indicated, as is the hydrogen at C_1 which is abstracted as hydride upon oxidation (Oubrie *et al.*, 1999b). The structures are drawn to result in the same orientation of this catalytically critical hydrogen at C_1. Note that the assignment of the anomers as α and β is inverted in L-sorbosone compared to D-glucose and D-xylose due to the change of the stereochemistry of the anomeric reference atom C_5.

Sndhak and the corresponding β-anomer for Sndhai. While some differences in the natural substrates exist, such as the additional axial hydroxyl at C_2 and the different stereochemistry at C_5, biochemical studies of Sndhai substrate specificity show that they can likely be accommodated (our own unpublished results).

The two different anomers of L-sorbosone 1,5-pyranose yield the same product upon oxidation at C_1, the 1,5-lactone of 2-KGA (Figure 8.8). This presumed immediate oxidation product has not been observed so far and presumably rapidly reacts by re-arranging its ring structure. This could happen by lactone hydrolysis and subsequent formation of the 2,6-pyranose of 2-KGA. Alternatively, intramolecular attack of the C_4-hydroxyl on the lactone could lead to transesterification and L-Asc formation (Pappenberger and Hohmann, 2014). Such a transesterification mechanism has been discussed previously for the interconversion of the D-glucono lactones (Jermyn, 1960; Takahashi and Mitsumoto, 1963). For Sndhai and Sndhak, both possible reaction paths, transesterification and hydrolysis, seem to occur, with bias towards transesterification and L-Asc formation. Other sorbosone dehydrogenases, such as the SsdA-homologue SsdB (Asakura and Hoshino, 1996; Pappenberger and Hohmann, 2014), which presumably also uses the same 1,5-pyranose form of L-sorbosone as substrate, yield exclusively 2-KGA. What factors play a role in determining the fate of the primary oxidation product at this bifurcation is unknown. But it is obvious that access of water to the immediate oxidation product is a prerequisite for hydrolysis to 2-KGA and likely is a critical factor. Further improvement of Sndhak and Sndhai with respect to minimising 2-KGA by-product formation will be required for industrial application.

8.4.3
L-Sorbose Dehydrogenase, Accumulating L-Sorbosone

With this set of useful L-Asc-forming L-sorbosone dehydrogenases at hand, one requires efficient provision of the precursor L-sorbosone for their industrial application. The 2-KGA fermentation in *K. vulgare* proceeds via L-sorbosone and seems to be an obvious candidate, but it is not trivial to defy the strong tendency of this microorganism to form 2-KGA and make it stop at the level of L-sorbosone.

8.4.3.1
Ssdh from *K. vulgare*

In the industry standard 2-KGA fermentation, the conversion of L-sorbose to 2-KGA by *K. vulgare* is a rapid and highly efficient step. *K. vulgare* expresses a set of closely related dimeric enzymes consisting of monomers encoded by the genes *ssda1, ssda2, ssda3* or *ssdb*, which are collectively called sorbose/sorbosone dehydrogenases (Ssdh) (Asakura and Hoshino, 1999; Pappenberger and Hohmann, 2014). In the annotated genome of *K. vulgare* WSH-001, these correspond to the genes KVU_2159, KVU_2142, KVU_0203, KVU_1366 (Gao *et al.*, 2013). A fifth *ssd* gene, slightly more distantly related compared to the others, has recently been reported (KVU_pmdA_0245, (Gao *et al.*, 2013)). The Ssdh enzymes are soluble periplasmic dehydrogenases using PQQ as redox cofactor and cytochrome c551 as primary electron acceptor, feeding the electrons into the respiratory chain (Asakura and Hoshino, 1999). To characterise the substrate specificity, *ssda1, ssda2, ssda3* and *ssdb* were individually expressed in *Pseudomonas putida*, purified and their activities measured with various substrates, using PMS (phenazine methosulfate) as artificial electron acceptor (Asakura and Hoshino, 1996). The three SsdA enzymes showed activity against primary hydroxyl groups in a broad substrate range of sugars or polyols, including C_1 of both L-sorbose and L-sorbosone. Both substrates feature accessible primary hydroxyl groups only at C_1 since both L-sorbose (Angyal, 1984) and L-sorbosone (our own unpublished results, see Section 8.4.1) are predominantly present as 2,6-pyranose rings in aqueous solution, leaving the C_1 position exocyclic and sequestering the hydroxyl moiety at C_6 (Figure 8.6). Furthermore, the carbonyl moieties of 2-keto aldoses are fully hydrated in aqueous solution (Köpper and Freimund, 2003), resulting in a geminal diol (two hydroxyl groups) at C_1 of L-sorbosone. This explains why also the aldehyde function at C_1 of L-sorbosone is efficiently oxidised by SsdA-type alcohol dehydrogenases. In an *in vitro* assay set-up, using the physiological electron acceptor cytochrome c551 from *K. vulgare*, only small amounts in L-sorbosone were found to accumulate besides major 2-KGA formation (Asakura and Hoshino, 1999). In the natural host *K. vulgare*, expressing additional enzymes with L-sorbosone dehydrogenase activity (such as mSndh, see Section 8.4.1), no L-sorbosone accumulation is detected during 2-KGA fermentation, indicating that all transiently formed L-sorbosone is immediately oxidised further to 2-KGA.

For a direct conversion to L-Asc, this efficient conversion of L-sorbosone to 2-KGA in *K. vulgare* poses a problem, since 2-KGA cannot be converted to L-Asc in aqueous solution and therefore has to be avoided. Since the same enzymes are responsible for both oxidations of L-sorbose and L-sorbosone, one cannot simply engineer *K. vulgare* for L-sorbosone accumulation by gene knockout. Also, enzyme engineering of SsdA-type enzymes towards better distinguishing between L-sorbose and L-sorbosone is not a promising approach, given the high similarity of both substrates. The efficient oxidation of L-sorbose by *K. vulgare* therefore cannot be exploited for a direct approach to L-Asc.

Recently, cloning and expression of the genes coding for Ssdh in *E. coli* and *G. oxydans* were described (Gao *et al.*, 2013, 2014). Surprisingly, *in vivo* activity of L-sorbose oxidation by the individual Ssdh homodimers was detected in *G. oxydans*, despite the absence of the native electron acceptor cytochrome c551 or near homologues thereof (Gao *et al.*, 2014). This is in contrast to the results for the expression of these genes in *P. putida*, where activity is only observed upon addition of an artificial electron acceptor such as PMS (Asakura and Hoshino, 1996; our own unpublished results). Furthermore, the Ssdh enzymes expressed in *G. oxydans* are described to be membrane-associated (Gao *et al.*, 2014), as opposed to the situation in the native host *K. vulgare* or in *P. putida* (Asakura and Hoshino, 1996, 1999). Most surprisingly, high accumulation of L-sorbosone from oxidation of L-sorbose is described in the *Gluconobacter* expression system, of up to 40 g/l (Gao *et al.*, 2014) with SsdA3 (KVU_0203) or even up to 72 g/l (Chen *et al.*, 2012) with SsdA1 (KVU_2159) in the fermentation broth. Gao *et al.* conclude that the Ssdh enzymes primarily act as sorbose dehydrogenases (Sdhs) and attribute the efficient 2-KGA formation in *K. vulgare* exclusively to the separate mSndh enzyme activity. Here is need for consolidating the contradicting results from *Pseudomonas* and *Gluconobacter* expression of SsdA-type enzymes. Could the unexpected membrane association in *Gluconobacter* enable the connection of the Ssdh enzymes to the respiratory chain? Does it affect the substrate preference of the Ssdh enzymes? Is the accumulation of L-sorbosone specific for the *Gluconobacter* expression system only, or a general feature of these enzymes, but not apparent in the *Pseudomonas* expression? The applicability of these enzymes for L-sorbosone accumulation and thus L-Asc formation needs further clarification.

8.4.3.2
Sorbose Dehydrogenase Sdh from *G. oxydans*

Apart from Ssdh, another L-sorbose to L-sorbosone converting enzyme had been identified, and this one had been clearly demonstrated to differentiate well between L-sorbose and L-sorbosone molecules. This sorbose dehydrogenase (Sdh), originating from *G. oxydans* has been first described by Sugisawa *et al.* (1991) and was later cloned, sequenced and heterologously expressed (Saito *et al.*, 1997; Shibata *et al.*, 2000). It is a membrane-bound GMC-family flavoprotein of 58 kD size and with exquisite selectivity towards L-sorbose. Activity against L-sorbosone was later shown to be below 10% of that against L-sorbose (our own unpublished results).

In the *Gluconobacter* genome, this enzyme forms an operon together with the cytosolic NAD-linked L-sorbosone dehydrogenase (cSndh, see Section 8.4.1), converting L-sorbosone to 2-KGA. Intracellular accumulation of the L-sorbosone intermediate, which presumably would be toxic due to chemical reactivity of the 2-keto aldehyde function, is avoided here by the concerted expression of two enzymes functionalities. This set-up of Sdh and cSndh was initially explored in *Gluconobacter* and *Pseudomonas* for an improved 2-KGA fermentation process, combining the already industrially applied D-sorbitol to L-sorbose conversion capability of *Gluconobacter* with further oxidation of L-sorbose to 2-KGA in the same strain. With this, an impressive conversion of 130 g/l 2-KGA from 150 g/l D-sorbitol was achieved over 72 h, with no accumulation of L-sorbosone due to concomitant overexpression of cSndh (Saito *et al.*, 1997).

The separation of the two oxidation steps in two different enzymes in *Gluconobacter*, however, now allows to exploit Sdh beyond 2-KGA fermentation. Based on the proven high activity of this system towards 2-KGA formation from D-sorbitol, it opens the possibility to metabolically engineer L-Asc formation in *Gluconobacter* or other hosts by combining this FAD-linked L-Sdh with a L-Asc-forming L-sorbosone dehydrogenase and removing all host-mediated activities converting L-sorbosone to 2-KGA.

8.4.4
Gluconobacter as Host for Direct L-Ascorbic Acid Formation

Since the inception of the Reichstein-Grüssner process in 1934, *G. oxydans* has played a key role in industrial L-Asc synthesis. The fermentative conversion of D-sorbitol to L-sorbose is the critical step to achieve the stereochemistry of L-Asc by inverting the carbon skeleton of D-glucose. In this application, *Gluconobacter* has proven as a robust industrial host, distinguished by very high specific activity (14 g product per gram cell dry weight and hour), volumetric activity (exceeding 10 g product per litre reaction volume and hour) and yields (exceeding 95%) (De Wulf, Soetaert and Vandamme, 2000; Hu *et al.*, 2015). This exemplifies the catalytic potency of the Sldh enzymes and of its cofactor recycling by the respiratory chain of *Gluconobacter*. It also demonstrates the power of biocatalytic conversion in the periplasm of Gram-negative bacteria, where no uptake in the cell is required, transport limitations are avoided and by-product formation by the metabolic diversity within the cell is minimised. These benefits of *Gluconobacter* have been exploited in several other biosyntheses, often building on the broad substrate range of the same Sldh enzyme towards stereoselective oxidation of diverse compounds (Deppenmeier, Hoffmeister and Prust, 2002; De Muynck *et al.*, 2007). Several approaches have been directed towards expanding the fermentative capabilities of *Gluconobacter* to include the further oxidation steps to 2-KGA (Tsukada and Perlman, 1972a,1972b; Sugisawa *et al.*, 1990; Saito *et al.*, 1998; Gao *et al.*, 2014). These attempts had so far been outperformed by the even more efficient conversion of L-sorbose to 2-KGA with *K. vulgare* (Pappenberger and Hohmann, 2014; Yang and Xu, 2016).

With the different approach of direct microbial production of L-Asc, however, *Gluconobacter* is back at the centre of attention. Three periplasmic oxidation steps are sufficient to convert D-sorbitol to L-Asc (Figure 8.6). *Gluconobacter* has already proven its potency for the first reaction of D-sorbitol to L-sorbose conversion by D-sorbitol dehydrogenase (Sldh). Also, the other two enzymatic activities required, L-sorbose dehydrogenase (Sdh) and Asc-forming L-sorbosone dehydrogenase (Sndhai), are endogenous *Gluconobacter* enzymes. They therefore link well to the *Gluconobacter* respiratory chain for efficient cofactor recycling, which is not a given for heterologous expression in different organisms. *Ketogulonicigenium* is also an established host for periplasmic oxidations as exemplified in the 2-KGA fermentation process, but its key Ssdh enzymes are not suitable for a direct L-Asc process due to their strong tendency for 2-KGA formation (except, possibly, when expressed in *Gluconobacter*, see Section 8.4.3.1).

Next to *Gluconobacter*'s potential for high productivities in periplasmic oxidations, a direct microbial L-Asc process has to meet particular challenges with respect to yield. Any 2-KGA by-product will amount to yield loss, as it cannot be converted to L-Asc in aqueous set-up. Besides the 2-KGA generated as by-product by the Asc-forming sorbosone dehydrogenases, additional attention needs to be given to host-mediated L-sorbosone oxidation to 2-KGA. Many organisms feature enzymatic activities for rapid conversion of L-sorbosone to 2-KGA (Isono *et al.*, 1968), presumably for detoxification of this rather reactive aldehyde compound. In *G. oxydans* IFO3293, only the cytosolic cSndh enzyme (see Section 8.4.1) is described for 2-KGA formation. Knockout of the gene encoding cSndh results in a mutant strain with improved L-Asc and very low background 2-KGA formation (Hoshino *et al.*, 2006) (our own unpublished results).

Especially critical for process yield is the stability of the product L-Asc in the fermentation broth. Enzymatic activities for degradation and assimilation of L-Asc have been described for *E. coli* (Yew and Gerlt, 2002; Campos *et al.*, 2007). Accordingly, incubation of an L-Asc solution with biomass leads to rapid loss of L-Asc titre, not only in the case of *E. coli* but also, for example, for *Pseudomonas*. In contrast, biomass of *Gluconobacter* has hardly any impact on L-Asc titre (our own unpublished results) and thus seems largely devoid of L-Asc degrading activities. The biggest challenge for L-Asc stability, however, is its distinguished reductive properties and readiness to donate electrons. This is key to many physiological roles of L-Asc and *in vivo* compensated by a recycling system which transfers electrons back on the transient oxidation product L-dehydroascorbic acid to recover L-Asc (Linster and Van Schaftingen, 2007). If not recovered, L-dehydroascorbic acid spontaneously and irreversibly degrades with a half life of minutes. This is happening to significant extent during microbial fermentation processes yielding L-Asc, which lack such recycling system. L-Asc is oxidised by molecular oxygen, and this reaction is strongly accelerated at above-neutral pH, increased temperature or the presence of even traces of transition metal ions (Cu^{2+}, Fe^{3+}), which are required in the fermentation broth to support microbial growth. This combination of L-Asc, molecular oxygen and transition metal ions is therefore a major challenge in direct L-Asc fermentation. Since L-Asc is formed from D-sorbitol in oxidation reactions, the presence of molecular oxygen in such process cannot be avoided.

One possible way forward to minimise L-Asc degradation is to separate biomass growth (requiring transition metal ions) from L-Asc production in a subsequent biotransformation step with resting cells in the absence of detrimental medium components. Using such set-up, conversion of D-sorbitol to titres of 1.8 g/l L-Asc in the medium were reported after 48 h incubation with OD 10 biomass (Berry *et al.*, 2003), using *G. oxydans* N44-1, a derivative of *G. oxydans* IFO3293 with deregulated expression of Sdh. Further optimisation of strain and process set-up, including vector-based overexpression of Sdh and Sndhai and knockout of cSndh, resulted in L-Asc titres well above 10 g/l with yields on consumed substrate near 90% (our own unpublished results).

8.5 Outlook

The industrial production of L-Asc via 2-KGA fermentation is a well-established technology with several decades of experience towards optimisation of this process. It is a challenge for any new technology to build up to a level where it will be able to compete with such an established process. This is also true for technologies which clearly have the potential to supersede the 2-KGA fermentation process, such as direct microbial fermentation towards L-Asc. Of the two principal approaches shown here, the route via L-sorbosone has seen further progress compared to the route via 1,4-lactone intermediates. This advantage is based on the efficient pathway providing the L-sorbosone intermediate, which builds on synergies with the established 2-KGA fermentation. L-Asc productivities and titers on this route show encouraging progress towards the right order of magnitude for commercial implementation. Several obstacles, however, remain, one being the formation of 2-KGA by-product during the final step of enzymatic L-Asc formation. This needs to be significantly improved to fall within the commercial yield requirements. The paramount challenge, however, is the instability of the L-Asc product in the fermentation broth. Concepts such as separation of biomass growth and L-Asc production show the way forward, but further improvements need to be achieved by means that are compatible with a low-cost production set-up. Direct microbial fermentation of L-Asc via L-sorbosone is closely related to the 2-KGA fermentation process and can build on this established and efficient process in many aspects. The chemical instability of the product L-Asc, however, is a key difference, and specific novel approaches are needed to overcome this challenge.

Acknowledgement

The contributions by Christian Manhart, Fabian Hentschel, Manuela Dallügge and Betty Coussens to the unpublished results reported here are gratefully acknowledged.

References

Ameyama, M., Shinagawa, E., Matsushita, K., and Adachi, O. (1981) D-fructose dehydrogenase of *Gluconobacter industrius:* purification, characterization, and application to enzymatic microdetermination of D-fructose. *J. Bacteriol.*, **145** (2), 814–823.

Anderson, S., Marks, C.B., Lazarus, R., Miller, J., Stafford, K., Seymour, J., Light, D., Rastetter, W., and Estell, D. (1985) Production of 2-keto-L-gulonate, an intermediate in L-ascorbate synthesis, by a genetically modified *Erwinia herbicola. Science*, **230** (4722), 144–149.

Angyal, S.J. (1984) The composition of reducing sugars in solution. in *Advances in Carbohydrate Chemistry and Biochemistry*, vol. **42** (eds R.S. Tipson and H. Derek), Academic Press, pp. 15–68.

Asakura, A., Hoshino T., Ojima, S., Shinjoh, M. and Tomiyama, N. (1996) Alcohol/aldehyde dehydrogenase. Patent US6730503 B1.

Asakura, A. and Hoshino, T. (1999) Isolation and characterization of a new quinoprotein dehydrogenase, L-sorbose/L-sorbosone dehydrogenase. *Biosci. Biotechnol., Biochem.*, **63** (1), 46–53.

Asakura, A., Hoshino, T. and Shinjoh, M. (2001) Microbial process for producing L-ascorbic acid, D-erythorbic acid, and salts thereof. Patent US6777212.

Berry, A., Lee, C., Mayer, A. and Shinjoh, M. (2003) Microbial production of L-ascorbic acid. Patent EP2348113.

Branduardi, P., Fossati, T., Sauer, M., Pagani, R., Mattanovich, D., and Porro, D. (2007) Biosynthesis of vitamin C by yeast leads to increased stress resistance. *PLoS One*, **2** (10), e1092.

Bremus, C., Herrmann, U., Bringer-Meyer, S., and Sahm, H. (2006) The use of microorganisms in L-ascorbic acid production. *J. Biotechnol.*, **124** (1), 196–205.

Brown, H.C., Brewster, J.H., and Shechter, H. (1954) An interpretation of the chemical behavior of five- and six-membered ring compounds. *J. Am. Chem. Soc.*, **76** (2), 467–474.

Brown, M.R., Jeffrey, S.W., Volkman, J.K., and Dunstan, G.A. (1997) Nutritional properties of microalgae for mariculture. *Aquaculture*, **151**, 315–331.

Campos, E., Montella, C., Garces, F., Baldoma, L., Aguilar, J., and Badia, J. (2007) Aerobic L-ascorbate metabolism and associated oxidative stress in *Escherichia coli. Microbiology*, **153** (10), 3399–3408.

Chargaff, E. and Magasanik, B. (1946) Oxidation of stereoisomers of the inositol group by *Acetobacter suboxydans. J. Biol. Chem.*, **165** (1), 379–380.

Chen, J., Zhou, J., Gao, L. and Du, G. (2012), *Gluconobacter oxydans* engineering bacterium for producing sorbic ketone in high yield mode and construction method thereof. Patent CN102851252.

Cruz-Rus, E., Amaya, I., and Valpuesta, V. (2012) The challenge of increasing vitamin C content in plant foods. *Biotechnol. J.*, **7** (9), 1110–1121.

Danehy, P. (1979) Synthesis of ascorbic acid from lactose. Patent US4259443.

Davey, M.W., Gilot, C., Persiau, G., Ostergaard, J., Han, Y., Bauw, G.C., and Van Montagu, M.C. (1999) Ascorbate biosynthesis in *Arabidopsis* cell suspension culture. *Plant Physiol.*, **121** (2), 535–543.

De Muynck, C., Pereira, C.S., Naessens, M., Parmentier, S., Soetaert, W., and Vandamme, E.J. (2007) The genus *Gluconobacter oxydans*: comprehensive overview of biochemistry and biotechnological applications. *Crit. Rev. Biotechnol.*, **27** (3), 147–171.

De Wulf, P., Soetaert, W., and Vandamme, E.J. (2000) Optimized synthesis of L-sorbose by C(5)-dehydrogenation of D-sorbitol with *Gluconobacter oxydans*. *Biotechnol. Bioeng.*, **69** (3), 339–343.

Deppenmeier, U., Hoffmeister, M., and Prust, C. (2002) Biochemistry and biotechnological applications of *Gluconobacter* strains. *Appl. Microbiol. Biotechnol.*, **60** (3), 233–242.

Elias, M., Tanaka, M., Sakai, M., Toyama, H., Matsushita, K., Adachi, O., and Yamada, M. (2001) C-terminal periplasmic domain of *Escherichia coli* quinoprotein glucose dehydrogenase transfers electrons to ubiquinone. *J. Biol. Chem.*, **276** (51), 48356–48361.

Fossati, T., Solinas, N., Porro, D., and Branduardi, P. (2011) L-ascorbic acid producing yeasts learn from plants how to recycle it. *Metab. Eng.*, **13** (2), 177–185.

Gallie, D.R. (2013) Increasing vitamin C content in plant foods to improve their nutritional value-successes and challenges. *Nutrients*, **5** (9), 3424–3446.

Gao, L., Du, G., Zhou, J., Chen, J., and Liu, J. (2013) Characterization of a group of pyrroloquinoline quinone-dependent dehydrogenases that are involved in the conversion of L-sorbose to 2-Keto-L-gulonic acid in *Ketogulonicigenium vulgare* WSH-001. *Biotechnol. Prog.*, **29** (6), 1398–1404.

Gao, L., Hu, Y., Liu, J., Du, G., Zhou, J., and Chen, J. (2014) Stepwise metabolic engineering of *Gluconobacter oxydans* WSH-003 for the direct production of 2-keto-L-gulonic acid from D-sorbitol. *Metab. Eng.*, **24**, 30–37.

Grindley, J.F., Payton, M.A., van de Pol, H., and Hardy, K.G. (1988) Conversion of glucose to 2-keto-L-gulonate, an intermediate in L-ascorbate synthesis, by a recombinant strain of *Erwinia citreus*. *Appl. Environ. Microbiol.*, **54** (7), 1770–1775.

Hancock, R.D., Galpin, J.R., and Viola, R. (2000) Biosynthesis of L-ascorbic acid (vitamin C) by *Saccharomyces cerevisiae*. *FEMS Microbiol. Lett.*, **186** (2), 245–250.

Hancock, R.D. and Viola, R. (2001) The use of micro-organisms for L-ascorbic acid production: current status and future perspectives. *Appl. Microbiol. Biotechnol.*, **56** (5-6), 567–576.

Hancock, R.D. and Viola, R. (2002) Biotechnological approaches for L-ascorbic acid production. *Trends Biotechnol.*, **20** (7), 299–305.

Hancock, R.D. and Viola, R. (2005) Improving the nutritional value of crops through enhancement of L-ascorbic acid (vitamin C) content: rationale and biotechnological opportunities. *J. Agric. Food. Chem.*, **53** (13), 5248–5257.

Hashimoto, H., Sakakibara, A., Yamasaki, M., and Yoda, K. (1997) *Saccharomyces cerevisiae* VIG9 encodes GDP-mannose pyrophosphorylase, which is essential for protein glycosylation. *J. Biol. Chem.*, **272** (26), 16308–16314.

Hearon, W.M. and Witte, J.F. (1981) Process of making L-gulono gamma lactone. Patent US4337202.

Hervas, M., Bashir, Q., Leferink, N.G., Ferreira, P., Moreno-Beltran, B., Westphal, A.H., Diaz-Moreno, I., Medina, M., de la Rosa, M.A., Ubbink, M., Navarro, J.A., and van Berkel, W.J. (2013) Communication between (L)-galactono-1,4-lactone dehydrogenase and cytochrome c. *FEBS J.*, **280** (8), 1830–1840.

Hölscher, T., Weinert-Sepalage, D., and Görisch, H. (2007) Identification of membrane-bound quinoprotein inositol dehydrogenase in *Gluconobacter oxydans* ATCC 621H. *Microbiology*, **153** (2), 499–506.

Hoshino, T., Kiyasu, T. and Shinjoh, M. (2002) Enzymatic process for the manufacture of L-ascorbic acid and D-erythorbic acid. Patent US7465563.

Hoshino, T., Miyazaki, T. and Sugisawa, T. (2002a) Aldehyde dehydrogenase gene. Patent US7135315.

Hoshino, T., Miyazaki, T. and Sugisawa, T. (2002b) Aldehyde dehydrogenase II. Patent US2005153412.

Hoshino, T., Miyazaki, T. and Sugisawa, T. (2002c) Vitamin C from sorbosone. Patent US7544494.

Hoshino, T., Shinjoh, M., Toepfer, C. and Tomiyama, N. (2006) Gene SMS 05. Patent US2009/0142815.

Hoshino, T., Sugisawa, T., and Fujiwara, A. (1991) Isolation and characterization of NAD(P)-dependent L-sorbosone dehydrogenase from Gluconobacter melanogenus UV10. *Agric. Biol. Chem.*, **55** (3), 665–670.

Hu, Y., Wan, H., Li, J., and Zhou, J. (2015) Enhanced production of L-sorbose in an industrial *Gluconobacter oxydans* strain by identification of a strong promoter based on proteomics analysis. *J. Ind. Microbiol. Biotechnol.*, **42**, 1039–1047.

Hubbs, J.C. (1997) Enzymatic process for the manufacture of ascorbic acid 2-keto-L-gulonic acid and esters of 2-keto-L-gulonic acid. Patent US5817490.

Huh, W.K., Kim, S.T., Yang, K.S., Seok, Y.J., Hah, Y.C., and Kang, S.O. (1994) Characterisation of D-arabinono-1,4-lactone oxidase from *Candida albicans* ATCC 10231. *Eur. J. Biochem.*, **225** (3), 1073–1079.

Huh, W.K., Lee, B.H., Kim, S.T., Kim, Y.R., Rhie, G.E., Baek, Y.W., Hwang, C.S., Lee, J.S., and Kang, S.O. (1998) D-Erythroascorbic acid is an important antioxidant molecule in *Saccharomyces cerevisiae. Mol. Microbiol.*, **30** (4), 895–903.

Isono, M., Nakanishi, I., Sasajima, K., Motizuki, K., Kanzaki, T., Okazaki, H., and Yoshino, H. (1968) 2-keto-L-gulonic acid fermentation part I. Paper chromatographic characterization of metabolic products from sorbitol and L-sorbose by various bacteria. *Agric. Biol. Chem.*, **32** (4), 424–431.

Jermyn, M.A. (1960) Studies on the glucono-delta-lactonase of *Pseudomonas fluorescens. Biochim. Biophys. Acta*, **37**, 78–92.

Kawai, S., Goda-Tsutsumi, M., Yakushi, T., Kano, K., and Matsushita, K. (2013) Heterologous overexpression and characterization of a flavoprotein-cytochrome c complex fructose dehydrogenase of *Gluconobacter japonicus* NBRC3260. *Appl. Environ. Microbiol.*, **79** (5), 1654–1660.

Kim, S.T., Huh, W.K., Kim, J.Y., Hwang, S.W., and Kang, S.O. (1996) D-arabinose dehydrogenase and biosynthesis of erythroascorbic acid in *Candida albicans. Biochim. Biophys. Acta*, **1297** (1), 1–8.

Kiuchi, K., Nishikimi, M., and Yagi, K. (1982) Purification and characterization of L-gulonolactone oxidase from chicken kidney microsomes. *Biochemistry*, **21** (20), 5076–5082.

Kluyver, A. and Boezaardt, A. (1939) Note on the biochemical preparation of inosose. *Rec. Trav. Chim. Pays-Bas*, **58**, 956–958.

Kondo, Y., Inai, Y., Sato, Y., Handa, S., Kubo, S., Shimokado, K., Goto, S., Nishikimi, M., Maruyama, N., and Ishigami, A. (2006) Senescence marker protein 30 functions as gluconolactonase in L-ascorbic acid biosynthesis, and its knockout mice are prone to scurvy. *Proc. Natl. Acad. Sci. U.S.A.*, **103** (15), 5723–5728.

Köpper, S. and Freimund, S. (2003) The composition of keto aldoses in aqueous solution as determined by NMR spectroscopy. *Helv. Chim. Acta*, **86** (3), 827–843.

Kruszewska, J., Janik, A., Lenart, U., and Palamarczyk, G. (1999) Glycosylation defects corrected by the changes in GDP-mannose level. *Acta Biochim. Pol.*, **46** (2), 315–324.

Kuivanen, J., Dantas, H., Mojzita, D., Mallmann, E., Biz, A., Krieger, N., Mitchell, D., and Richard, P. (2014) Conversion of orange peel to L-galactonic acid in a consolidated process using engineered strains of *Aspergillus niger. AMB Express*, **4**, 33.

Kuivanen, J., Mojzita, D., Wang, Y., Hilditch, S., Penttila, M., Richard, P., and Wiebe, M.G. (2012) Engineering filamentous fungi for conversion of D-galacturonic acid to L-galactonic acid. *Appl. Environ. Microbiol.*, **78** (24), 8676–8683.

Kuivanen, J., Penttila, M., and Richard, P. (2015) Metabolic engineering of the fungal D-galacturonate pathway for L-ascorbic acid production. *Microb. Cell Fact.*, **14** (1), 2.

Kumar, M. (1998) Production of ascorbic acid. Patent US6358715.

Leferink, N.G., van den Berg, W.A., and van Berkel, W.J. (2008a) L-Galactono-gamma-lactone dehydrogenase from *Arabidopsis thaliana*, a flavoprotein involved in vitamin C biosynthesis. *FEBS J.*, **275** (4), 713–726.

Leferink, N.G., Heuts, D.P., Fraaije, M.W., and van Berkel, W.J. (2008b) The growing VAO flavoprotein family. *Arch. Biochem. Biophys.*, **474** (2), 292–301.

Leferink, N.G. and van Berkel, W.J. (2014) Aldonolactone oxidoreductases. *Methods Mol. Biol.*, **1146**, 95–111.

Leferink, N.G., Fraaije, M.W., Joosten, H.J., Schaap, P.J., Mattevi, A., and van Berkel, W.J. (2009a) Identification of a gatekeeper residue that prevents dehydrogenases from acting as oxidases. *J. Biol. Chem.*, **284** (7), 4392–4397.

Leferink, N.G., Jose, M.D., van den Berg, W.A., and van Berkel, W.J. (2009b) Functional assignment of Glu386 and Arg388 in the active site of L-galactono-gamma-lactone dehydrogenase. *FEBS Lett.*, **583** (19), 3199–3203.

Levene, P.A. and Simms, H.S. (1925) Lactone formation from mono- and dicarboxylic sugar acids. *J. Biol. Chem.*, **65**, 31–47.

Linster, C.L. and Van Schaftingen, E. (2007) Vitamin C. Biosynthesis, recycling and degradation in mammals. *FEBS J.*, **274** (1), 1–22.

Loewus, M., Bedgar, D., Saito, K., and Loewus, F. (1990) Conversion of L-sorbosone to L-ascorbic acid by a NADP-dependent dehydrogenase in bean and spinach; leaf. *Plant Physiol.*, **94** (3), 1492–1495.

Miyazaki, T., Sugisawa, T., and Hoshino, T. (2006) Pyrroloquinoline quinone-dependent dehydrogenases from *Ketogulonicigenium vulgare* catalyze the direct conversion of L-sorbosone to L-ascorbic acid. *Appl. Environ. Microbiol.*, **72** (2), 1487–1495.

Nishikimi, M., Noguchi, E., and Yagi, K. (1978) Occurrence in yeast of L-galactonolactone oxidase which is similar to a key enzyme for ascorbic acid biosynthesis in animals, L-gulonolactone oxidase. *Arch. Biochem. Biophys.*, **191** (2), 479–486.

Nishikimi, M., Tolbert, B.M., and Udenfriend, S. (1976) Purification and characterization of L-gulono-gamma-lactone oxidase from rat and goat liver. *Arch. Biochem. Biophys.*, **175** (2), 427–435.

Oba, K., Ishikawa, S., Nishikawa, M., Mizuno, H., and Yamamoto, T. (1995) Purification and properties of L-galactono-gamma-lactone dehydrogenase, a key enzyme for ascorbic acid biosynthesis, from sweet potato roots. *J. Biochem.*, **117** (1), 120–124.

Ostergaard, J., Persiau, G., Davey, M.W., Bauw, G., and Van Montagu, M. (1997) Isolation of a cDNA coding for L-galactono-gamma-lactone dehydrogenase, an enzyme involved in the biosynthesis of ascorbic acid in plants. Purification, characterization, cDNA cloning, and expression in yeast. *J. Biol. Chem.*, **272** (48), 30009–30016.

Oubrie, A., Rozeboom, H.J., and Dijkstra, B.W. (1999) Active-site structure of the soluble quinoprotein glucose dehydrogenase complexed with methylhydrazine: a covalent cofactor-inhibitor complex. *Proc. Natl. Acad. Sci. U.S.A*, **96** (21), 11787–11791.

Oubrie, A., Rozeboom, H.J., Kalk, K.H., Duine, J.A., and Dijkstra, B.W. (1999a) The 1.7 A crystal structure of the apo form of the soluble quinoprotein glucose dehydrogenase from *Acinetobacter calcoaceticus* reveals a novel internal conserved sequence repeat. *J. Mol. Biol.*, **289** (2), 319–333.

Oubrie, A., Rozeboom, H.J., Kalk, K.H., Olsthoorn, A.J., Duine, J.A., and Dijkstra, B.W. (1999b) Structure and mechanism of soluble quinoprotein glucose dehydrogenase. *EMBO J.*, **18** (19), 5187–5194.

Pappenberger, G. and Hohmann, H.P. (2014) Industrial production of L-ascorbic acid (vitamin C) and D-isoascorbic acid. *Adv. Biochem. Eng. Biotechnol.*, **143**, 143–188.

Reichstein, T. and Grüssner, A. (1934) Eine ergiebige Synthese der L-Ascorbinsäure (C-Vitamin). *Helv. Chim. Acta*, **17** (1), 311–328.

Roland, J.F., Cayle, T., Dinwoodie, R.C. and Mahnert, D.W. (1983) Fermentation production of ascorbic acid from L-galactonic substrate. Patent US4595659.

Rosa, J.C., Colombo, L.T., Alvim, M.C., Avonce, N., Van Dijck, P., and Passos, F.M. (2013) Metabolic engineering of *Kluyveromyces lactis* for L-ascorbic acid (vitamin C) biosynthesis. *Microb. Cell Fact.*, **12**, 59.

Running, J.A., Peng, S., and Rosson, R.A. (2004) The biotechnology of ascorbic acid manufacture. in *Vitamin C* (eds H. Asard,

J.M. May, and N. Smirnof), Scientific Publishers Ltd., Oxford, pp. 49–64.

Saito, Y., Ishii, Y., Hayashi, H., Imao, Y., Akashi, T., Yoshikawa, K., Noguchi, Y., Soeda, S., Yoshida, M., Niwa, M., Hosoda, J., and Shimomura, K. (1997) Cloning of genes coding for L-sorbose and L-sorbosone dehydrogenases from *Gluconobacter oxydans* and microbial production of 2-keto-L-gulonate, a precursor of L-ascorbic acid, in a recombinant *G. oxydans* strain. *Appl. Environ. Microbiol.*, **63** (2), 454–460.

Saito, Y., Ishii, Y., Hayashi, H., Yoshikawa, K., Noguchi, Y., Yoshida, S., Soeda, S., and Yoshida, M. (1998) Direct fermentation of 2-keto-L-gulonic acid in recombinant *Gluconobacter oxydans*. *Biotechnol. Bioeng.*, **58** (2-3), 309–315.

Sauer, M., Branduardi, P., Valli, M., and Porro, D. (2004) Production of L-ascorbic acid by metabolically engineered *Saccharomyces cerevisiae* and *Zygosaccharomyces bailii*. *Appl. Environ. Microbiol.*, **70** (10), 6086–6091.

Schertl, P., Sunderhaus, S., Klodmann, J., Grozeff, G.E., Bartoli, C.G., and Braun, H.P. (2012) L-galactono-1,4-lactone dehydrogenase (GLDH) forms part of three subcomplexes of mitochondrial complex I in *Arabidopsis thaliana*. *J. Biol. Chem.*, **287** (18), 14412–14419.

Shibata, T., Ichikawa, C., Matsuura, M., Takata, Y., Noguchi, Y., Saito, Y., and Yamashita, M. (2000) Metabolic engineering study on the direct fermentation of 2-keto-L-gulonic acid, a key intermediate of L-ascorbic acid in *Pseudomonas putida* IFO3738. *J. Biosci. Bioeng.*, **90** (2), 223–225.

Shinagawa, E., Matsushita, K., Adachi, O., and Ameyama, M. (1982) Purification and characterization of D-sorbitol dehydrogenase from membrane of *Gluconobacter suboxydans* var alpha. *Agric. Biol. Chem.*, **46**, 135–141.

Shinagawa, E., Matsushita, K., Adachi, O., and Ameyama, M. (1984) D-Gluconate dehydrogenase, 2-keto-D-gluconate yielding, from *Gluconobacter dioxyacetonicus:* purification and characterization. *Agric. Biol. Chem.*, **48**, 1517–1522.

Shinjoh, M., Tomiyama, N., Asakura, A., and Hoshino, T. (1995) Cloning and nucleotide sequencing of the membrane-bound L-sorbosone dehydrogenase gene of *Acetobacter liquefaciens* IFO 12258 and its expression in *Gluconobacter oxydans*. *Appl. Environ. Microbiol.*, **61** (2), 413–420.

Shinjoh, M., Tomiyama, N., Miyazaki, T., and Hoshino, T. (2002) Main polyol dehydrogenase of *Gluconobacter suboxydans* IFO 3255, membrane-bound D-sorbitol dehydrogenase, that needs product of upstream gene, *sldB*, for activity. *Biosci. Biotechnol., Biochem.*, **66** (11), 2314–2322.

Smirnoff, N. and Wheeler, G. (2000) Ascorbic acid in plants: biosynthesis and function. *Crit. Rev. Biochem. Mol. Biol.*, **35** (4), 291–314.

Soemphol, W., Adachi, O., Matsushita, K., and Toyama, H. (2008) Distinct physiological roles of two membrane-bound dehydrogenases responsible for D-sorbitol oxidation in *Gluconobacter frateurii*. *Biosci. Biotechnol., Biochem.*, **72** (3), 842–850.

Sonoyama, T., Tani, H., Matsuda, K., Kageyama, B., Tanimoto, M., Kobayashi, K., Yagi, S., Kyotani, H., and Mitsushima, K. (1982) Production of 2-keto-L-gulonic acid from D-glucose by two-stage fermentation. *Appl. Environ. Microbiol.*, **43** (5), 1064–1069.

Southall, S.M., Doel, J.J., Richardson, D.J., and Oubrie, A. (2006) Soluble aldose sugar dehydrogenase from *Escherichia coli*: a highly exposed active site conferring broad substrate specificity. *J. Biol. Chem.*, **281** (41), 30650–30659.

Sugisawa, T. and Hoshino, T. (2002) Purification and properties of membrane-bound D-sorbitol dehydrogenase from *Gluconobacter suboxydans* IFO 3255. *Biosci. Biotechnol., Biochem.*, **66** (1), 57–64.

Sugisawa, T., Hoshino, T., Masuda, S., Nomura, S., Setoguchi, Y., Tazoe, M., Shinjoh, M., Someha, S., and Fujiwara, A. (1990) Microbial production of 2-keto-L-gulonic acid from L-sorbose and D-sorbitol by *Gluconobacter melanogenus*. *Agric. Biol. Chem.*, **54** (5), 1201–1209.

Sugisawa, T., Hoshino, T., Nomura, S., and Fujiwara, A. (1991) Isolation and characterization of membrane-bound L-sorbose dehydrogenase from *Gluconobacter melanogenus* UV10. *Agric. Biol. Chem.*, **55** (2), 363–370.

Sugisawa, T., Miyazaki, T., and Hoshino, T. (2005) Microbial production of L-ascorbic acid from D-sorbitol, L-sorbose, L-gulose, and L-sorbosone by *Ketogulonicigenium vulgare* DSM 4025. *Microbiol. Ferment. Technol. Commun.*, **69** (3), 659–662.

Sugisawa, T., Ojima, S., Matzinger, P., and Hoshino, T. (1995) Isolation and characterization of a new vitamin C producing enzyme (L-gulono-γ-lactone dehydrogenase) of bacterial origin. *Biosci. Biotechnol., Biochem.*, **59**, 190–196.

Takahashi, T. and Mitsumoto, M. (1963) Transformation and hydrolysis of D-gulono-gamma and delta-lactone. *Nature*, **199**, 765–767.

Toyama, H., Furuya, N., Saichana, I., Ano, Y., Adachi, O., and Matsushita, K. (2007) Membrane-bound, 2-keto-D-gluconate-yielding D-gluconate dehydrogenase from *Gluconobacter dioxyacetonicus* IFO 3271: molecular properties and gene disruption. *Appl. Environ. Microbiol.*, **73** (20), 6551–6556.

Toyama, H., Soemphol, W., Moonmangmee, D., Adachi, O., and Matsushita, K. (2005) Molecular properties of membrane-bound FAD-containing D-sorbitol dehydrogenase from thermotolerant *Gluconobacter frateurii* isolated from Thailand. *Biosci. Biotechnol., Biochem.*, **69** (6), 1120–1129.

Tsukada, Y. and Perlman, D. (1972a) The fermentation of L-sorbose by *Gluconobacter melanogenus*. I. General characteristics of the fermentation. *Biotechnol. Bioeng.*, **14** (5), 799–810.

Tsukada, Y. and Perlman, D. (1972b) The fermentation of L-sorbose by *Gluconobacter melanogenus*. II. Inducible formation of enzyme catalyzing conversion of L-sorbose to 2-keto-L-gulonic acid. *Biotechnol. Bioeng.*, **14** (5), 811–818.

Tsuya, T., Ferri, S., Fujikawa, M., Yamaoka, H., and Sode, K. (2006) Cloning and functional expression of glucose dehydrogenase complex of *Burkholderia cepacia* in *Escherichia coli*. *J. Biotechnol.*, **123** (2), 127–136.

Valpuesta, V. and Botella, M.A. (2004) Biosynthesis of L-ascorbic acid in plants: new pathways for an old antioxidant. *Trends Plant Sci.*, **9** (12), 573–577.

Wheeler, G., Ishikawa, T., Pornsaksit, V., and Smirnoff, N. (2015) Evolution of alternative biosynthetic pathways for vitamin C following plastid acquisition in photosynthetic eukaryotes. *Elife*, **4**, e06369.

Wheeler, G.L., Jones, M.A., and Smirnoff, N. (1998) The biosynthetic pathway of vitamin C in higher plants. *Nature*, **393** (6683), 365–369.

Wissler, J., Freivogel, K. and Wiesner, W. (1995) Cyclitol. Patent WO9704101.

Wolucka, B.A. and Communi, D. (2006) *Mycobacterium tuberculosis* possesses a functional enzyme for the synthesis of vitamin C, L-gulono-1,4-lactone dehydrogenase. *FEBS J.*, **273** (19), 4435–4445.

Xavier, N.M., Rauter, A.P., and Queneau, Y. (2010) Carbohydrate-based lactones: synthesis and applications. *Top. Curr. Chem.*, **295**, 19–62.

Yang, W. and Xu, H. (2016) Industrial fermentation of vitamin C, in *Industrial Biotechnology of Vitamins, Biopigments, and Antioxidants*, 1st edn (ed E.J. Vandamme), Wiley-VCH Verlag GmbH & Co. KGaA.

Yew, W.S. and Gerlt, J.A. (2002) Utilization of L-ascorbate by *Escherichia coli* K-12: assignments of functions to products of the *yjf-sga* and *yia-sgb* operons. *J. Bacteriol.*, **184** (1), 302–306.

Zhang, L., Wang, Z., Xia, Y., Kai, G., Chen, W., and Tang, K. (2007) Metabolic engineering of plant L-ascorbic acid biosynthesis: recent trends and applications. *Crit. Rev. Biotechnol.*, **27** (3), 173–182.

Zhou, J., Du, G., and Chen, J. (2012) Metabolic engineering of microorganisms for vitamin C production, in *Reprogramming Microbial Metabolic Pathways*, vol. **64** (eds X. Wang, J. Chen, and P. Quinn), Springer, pp. 241–259.

Zhu, C., Sanahuja, G., Yuan, D., Farre, G., Arjo, G., Berman, J., Zorrilla-Lopez, U., Banakar, R., Bai, C., Perez-Massot, E., Bassie, L., Capell, T., and Christou, P. (2013) Biofortification of plants with altered antioxidant content and composition: genetic engineering strategies. *Plant Biotechnol. J.*, **11** (2), 129–141.

Part II
Fat Soluble Vitamins

Industrial Biotechnology of Vitamins, Biopigments, and Antioxidants, First Edition.
Edited by Erick J. Vandamme and José L. Revuelta.

9
Synthesis of β-Carotene and Other Important Carotenoids with Bacteria

Christoph Albermann and Holger Beuttler

9.1
Introduction

Carotenoids are a large group of lipophilic isoprenoid compounds, which are found in many eukaryotic and prokaryotic organisms. Currently, more than 700 structurally different carotenoids have been identified in nature (Britton, Liaaen-Jensen and Pfander, 2004). The most important function for photosynthetic and non-photosynthetic organisms is the antioxidative property of these molecules that protects the cell against reactive oxygen species, particularly the peroxidation of lipids (Kiokias, Varzakas and Oreopoulou, 2008). In photosynthetic organisms, carotenoids are also part of the light harvesting complexes to absorb light in a broader range than chlorophyll (Croce and van Amerongen, 2014).

Carotenoids have received great attention due to their beneficial effects on human health (Kaulmann and Bohn, 2014) and their application as colourant in food, feed and cosmetic industries (Jaswir *et al.*, 2011; Aberoumand, 2011). To meet the industrial demand for carotenoids, chemical methods for their large-scale production have been developed (Pfander, Traber and Lanz, 1997; Alvarez *et al.*, 2014). However, the interest in carotenoids from natural sources has been increasing during recent years. Therefore, carotenoid-producing microorganisms, such as fungi (e.g. *Blakeslea trispora*, *Xanthophyllomyces dendrorhous*) and microalgae (e.g. *Haematococcus pluvialis*), have been considered for the large-scale production of carotenoids by fermentation (Olaizola, 2000; Mehta, Obraztsova and Cerda-Olmedo, 2003; Raja *et al.*, 2007). Microbial production allows a more sustainable and environmental-friendly production of carotenoids compared to chemical synthesis methods.

Besides fungi and microalgae, carotenogenic bacteria are another microbial source for the extraction of carotenoids. But in contrast to fungi and microalga, which are used for the production of, for example, lycopene, β-carotene or astaxanthin, no effective production of these carotenoids by native carotenogenic bacteria is described. The great potential of bacteria for the production of carotenoids lies in the use of metabolically engineered bacterial strains. The successful production of industrially important carotenoids has been demonstrated

Industrial Biotechnology of Vitamins, Biopigments, and Antioxidants, First Edition.
Edited by Erick J. Vandamme and José L. Revuelta.

by recombinant carotenogenic and non-carotenogenic bacteria, for example, *Corynebacterium glutamicum, Escherichia coli, Pseudomonas putida, Rhodospirillum rubrum* and *Zymomonas mobilis.* Because of the available tools for genetic modification and the scalability of cultivation, industrially important bacteria, such as *C. glutamicum* and *E. coli*, are suitable host organisms for the production of carotenoids.

In this chapter, we give an overview on the properties, analytics and bacterial syntheses of carotenoids and focus especially on the production of lycopene, β-carotene, zeaxanthin and astaxanthin by metabolically engineered bacteria.

9.2
Carotenoids: Chemical Properties, Nomenclature and Analytics

Most carotenoids are tetraterpenes, meaning that they are built up of eight C_5 isoprene units. Carotenoids formed of more than eight units, such as decaprenoxanthin, are called *homocarotenoids* (Barua *et al.*, 2000). Homocarotenoids are exclusively synthesised by some bacterial organisms. Carotenoids occur as two major types: carotenes, consisting of only hydrocarbons and xanthophylls, which are oxygen-substituted hydrocarbons. Xanthophylls contain various functional groups, as hydroxy groups (e.g. zeaxanthin or lutein), epoxy groups (e.g. violaxanthin) and oxo groups (e.g. canthaxanthin) found in carotenogenic bacteria. By the hydroxy group, xanthophylls can be further modified, for example, by formation of esters or glycosides. Furthermore, some carotenoids contain olefinic or aromatic groups.

Besides α-, β-, γ-, δ-, ε-carotene and the open-chain lycopene, the not completely conjugated natural occurring precursors such as neurosporene and ζ-carotene belong to carotenes. Since there are no substituents, the only difference between the completely conjugated carotenes is the presence of the terminal groups. α-, β- and ε-carotene are regioisomers as well as γ- and δ-carotene. The difference between these isomers is the position of the double bond in the ring. In β-carotene, two β-rings can be found while α-carotene contains one β- and one ε-ring and ε-carotene carries two ε-rings. In γ-carotene, one β-ring on one side and an open chain on the other side exist. The difference in δ-carotene is the ε-ring. Lycopene has just open chains on both sides. As this configuration is called ψ, another name for this substance is ψ,ψ-carotene.

Xanthophylls are a more heterogenic group of compounds. The skeletal structure is substituted at different positions with hydroxyl groups, epoxides, oxo groups or, in rare cases, with carboxyl groups. Hydroxyl groups are the most important substituents, found in lutein or zeaxanthin, for example. An example for xanthophyll-carrying epoxygroups is violaxanthin, and oxo groups can be found in canthaxanthin. The rare carboxyl groups can be found in crocetin which is a derivate of the yellowish colourant crocin of saffron. In crocetin, the carboxyl groups are connected to gentiobiose. Combinations of all the substituents lead to a very heterogenic groups of compounds.

9.2.1 Nomenclature

The usage of trivial names is common for carotenoids because systematic names are very complicated or confusing. But even for trivial names, nomenclature rules are set up. A short overview is given in the next paragraph.

The root name is carotene. The atoms of the main chain will not be numbered continuously, but the first lineage will get number 1 until the middle of the molecule which will be number 15. The methyl groups shall be numbered from 16 to 20. The other parts of the molecule shall get the numbers $1'-20'$.

Acyclic and cyclic ends are marked with Greek characters depending on their form. An open chain, as in lycopene, is called ψ. Cyclohexylic or methylenecyclohexylic rings are called, β, ε and γ, depending on the position of the double bond. A cyclopentyl ring is called κ, and aromatic rings are named as φ or χ. The characters shall be set in the order of the Greek alphabet prior to carotene. Substituents are marked with type and position. For example, zeaxanthin is β,β-carotene-3,3′-diol.

If the molecule is shortened, it is named 'apo' with the positions as prefix (e.g. apo-8′-carotenal) (Weedon and Moss, 1995) (Table 9.1).

9.2.2 Analysis of Carotenoids

A first step in investigating the production of carotenoids in non-photosynthetic bacteria is visual inspection of the colony or the centrifuged pellet, respectively. Due to their intensive colouring of the cells, carotenoid-producing cells can be easily identified by the red, yellow or orange colour of the colonies. But this could only be the indication that the strain produces carotenoids, because intermediates of the carotenoids may have the same colour as the final product. A better insight into the production could be received after the extractions using an organic solvent and the separation of the obtained solution using a chromatographic method.

9.2.2.1 Handling Precautions

Since many carotenoids are sensitive to light or oxygen, some precautions have to be undertaken in order to prevent them from degradation. Light and molecular oxygen can cause *cis*/*trans*-isomerisation (Feltl *et al.*, 2005).

If the bacterial carotenoid formation is not induced by light, it is advised to perform cultivation in the dark. In particularly, the sensitivity of carotenoids to light and oxygen has to be considered during the extraction of carotenoids. For that, rapid working with as short as possible steps for the extraction, the minimisation of direct exposure to day light is usually adequate. Small amounts of the carotenoids can isomerise, but the huge amount of the products should stay in the produced form. The addition of small amounts of butylated hydroxytoluene (BHT), ascorbic acid or other antioxidants is a standard procedure in carotenoid analytics and might be also helpful to minimise oxidation processes in the solution

Table 9.1 Overview on prefixes and their structures for carotenoid nomenclature.

Structure	Description	Prefix
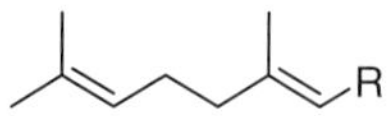	Open chain	Ψ
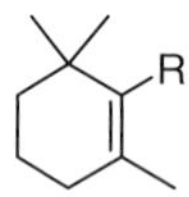	Cyclohexylic	*B*
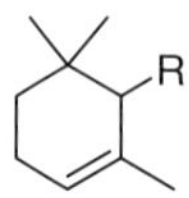	Cyclohexylic	*E*
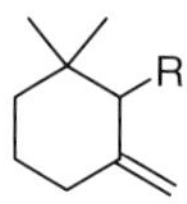	Methylencyclohexylic	Γ
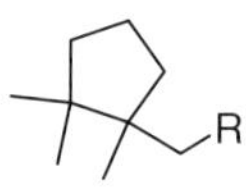	Cyclopentylic	*K*
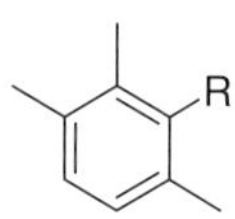	Arylic	Φ
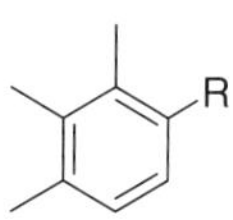	Arylic	*χ*

(Breithaupt and Bamedi, 2002; Breithaupt and Schwack, 2000; Feltl *et al.*, 2005). It might be also helpful to use tinted glass or plastic labware and to store solution in a cold place. It has been also been described that the use of plastic lab ware results in a significantly lower recovery of carotenoids as compared to glassware (5.4% loss to 1.1% in tinted glass vials) (Kaiser, 2009). For longer storage, carotenoids should be stored in a freezer (−20 °C), and, if possible, the solvent should be removed to avoid conformational rearrangement (De Ritter, Purcell and Bauernfeind, 1981).

9.2.2.2
Extraction

An important point for accurate and reliable results is an extraction method that led to a complete extract of all carotenoids. Most carotenoids have a low solubility in polar solvents but could be solved in small amounts in mid-polar or completely in non-polar organic solvents. The use of non-polar solvents, such as hexane, for the extraction from bacterial cells is inapplicable, because these solvents are not able to break up the cells. There are many methods described in the literature, some use hot methanol (Breitenbach *et al.*, 1996), acetone (Beuttler, 2010), acetone followed by a re-extraction with hexane (Blasco, Kauffmann and Schmid, 2004), mixtures of solvents (Beuttler *et al.*, 2011) or methanolic potassium hydroxide solution followed by ethyl ether (Kim *et al.*, 2003). The addition of tin acetate or zinc acetate might increase the recovery rate of hydrophilic substituted carotenoids (Kaiser, 2009).

An easy method for extraction is the use of acetone (Beuttler, 2010). The use of more unipolar solvents was very ineffective because the cells could not be broken up by these solvents, and even the addition of glass beads or ultrasound was not really effective. The use of acetone is a very environmental-friendly, cost-effective and easy method. With this method, a recovery rate of 98.1% for lycopene, 97.3% for zeaxanthin and 98.0% for apo-8-carotenal was received. The rate was in the same range as the recovery rates for other carotenoids, solvents and matrices (Breithaupt and Bamedi, 2001, 2002; Breithaupt, 2004; Thürmann *et al.*, 2005).

The main disadvantage is the miscibility of acetone and water, which resulted in long drying times and makes this method unfeasible for the extraction in larger scale, because up to 25% of the solution was water. This method is preferred for laboratory-scale extraction. A re-extraction using a more unipolar solvent such as hexane (Blasco, Kauffmann and Schmid, 2004) or chloroform (Kaiser, 2009) could help to minimise this problem. The extraction of carotenoids by acetone was successfully used by several studies (e.g., Blasco, Kauffmann and Schmid, 2004; Beuttler, 2010; Hoffmann *et al.*, 2012). However, some studies observed the formation of reaction products of hydroxyl or keto carotenoids and acetone, which could complicate the analytical procedure (Britton, Liaaen-Jensen and Pfander, 2004; Lea, 1988).

For a quantitative analysis of carotenoids from bacterial cells, it is necessary to add an internal standard to the cells before extraction. Apo-8′-carotenal is widely used in all fields of carotenoid analytics for this purpose, whereby the concentration of the internal standard should be in the same range as the substances to be analysed.

9.2.2.3
Chromatography Methods for Analysis of Carotenoids

The High-Performance Liquid Chromatography (HPLC) is the method of choice to analyse carotenoids. Because of the importance of carotenoids, many methods for the separation of carotenoids were developed (Oliver and Palou, 2000). The following overview on this field is limited to the most important methods.

Today, C_{30} columns, which are specifically developed for the separation of carotenoids (Sander and Wise, 1987; Sander *et al.*, 1994), are preferentially used instead of the standard C_{18} columns in HPLC (Breithaupt and Schlatterer, 2005; Breithaupt and Schwack, 2000; Breithaupt, 2000, 2004). The longer alkane chains on C_{30} phase columns, compared to the C_{18} phases, lead to a stronger interaction of the carotenoids with the matrix and, as a result, to better separation. Disadvantage of these columns are problems with peak geometry, especially tailing (Emenhiser *et al.*, 1996).

An easy method for C_{18} columns was developed using water and acetonitrile as mobile phase (Steel and Keller, 2000). With this method, apo-8′-carotenal, lutein and β-carotene could be separated. The use of this column type has also been reported in other study for the separation of the most important carotenoids generated by recombinant bacterial strain, such as lycopene, β-carotene and zeaxanthin (Barba *et al.*, 2006; Blasco, Kauffmann and Schmid, 2004; Cortés *et al.*, 2004).

Sometimes, special columns are used, such as a nitrile phase, which can perform the separation of geometric isomers of lutein and zeaxanthin (Khachik *et al.*, 1992).

Many other published methods deal with the separation of carotenoids used as food colourants, such as the methods developed by Breithaupt, who used a C_{30} column both for LC-MS and LC/UV-Vis and the separation of 12 carotenoids (Breithaupt and Bamedi, 2001, 2002; Breithaupt and Schlatterer, 2005; Breithaupt, 2004). As mobile phase, a mixture of methanol, water and methyl tert-butyl ester (MTBE) was used (Breithaupt, 2000).

Besides HPLC, thin-layer chromatography (TLC) is an easy, fast and cheap way to investigate carotenoids from cell extracts. Because of their strong colouring, carotenoids are easy to detect without any staining. An overview of methods for the TLC separation of carotenoids including vitamin A, also coupled with flame ionization detector (FID) detection, are reviewed by Cimpoiu and Hosu (2007). TLC separation could be done in normal phase (Deli, 1998) using two different binary mobile phases and one ternary mobile phase or reversed-phase TLC sheets using a ternary mobile phase (Isaksen and Francis, 1986).

9.3
Natural Occurrence in Bacteria

In nature, approximately 100 million tons of carotenoids are synthesised each year (Britton, Liaaen-Jensen and Pfander, 1995). Carotenoids occur in nature mostly in plants where they are used as light-collecting pigments for photosynthesis, as photo-protector or due to their anti-oxidative effect.

But there are also several photosynthetic and non-photosynthetic bacteria known, which produce carotenoids. For *Brevibacterium* sp. KY-4313 canthaxanthin, echinenone and β-carotene was reported (Nelis and De Leenheer, 1989), while *Rhodobacter capsulatus* produces spheroidenone and some unidentified carotenoids (Nelis and De Leenheer, 1989). A very widespread range of carotenoids has *Rhodomicrobium vannielii* which can synthesise rhodovibrin, rhodopin, spirilloxanthin, anhydrorhodovibrin, lycopene and neurosporene in different amounts (Nelis and De Leenheer, 1989). The wild type of *Pantoea ananatis* produces a range of different carotenoids, mainly zeaxanthin and glycosylated derivatives, as well as the regioisomer lutein (Beuttler, 2010), while a zeaxanthin dirhamnoside can be found in *Corynebacterium autotrophicum* (Hertzberg, Borch and Liaaen-Jensen, 1976). In costal bacteria, diadinoxanthin has been found (Nugraheni *et al.*, 2010). Different carotenoids were also found in *Microbacterium arborescens* (Godinho and Bhosle, 2008) and some Antarctic bacteria (Dieser, Greenwood and Foreman, 2010).

Despite the intensive colour of carotenoid-producing organisms, the concentration of carotenoids in a natural matrix is low and in particular in bacterial cells quite diverse. Some examples of bacteria that produce industrially important carotenoids are given in Table 9.2.

Table 9.2 Examples of native bacterial strains that produce industrially important carotenoids.

Carotenoid	Bacterial organism	Carotenoid content or concentration	References
β-Cryptoxanthin	*Brevibacterium linens*	0.3 mg/l culture	Guyomarch, Binet and Dufosse (2000)
Canthaxanthin	*Bradyrhizobium* sp.	1.3 mg/g biomass	Lorquin, Molouba and Dreyfus (1997)
Canthaxanthin	*Gordonia jacobaea* MV-1	13 mg/l culture	De Miguel *et al.* (2001) and Veiga-Crespo *et al.* (2005)
Canthaxanthin	*Micrococcus roseus*	1.7 mg/l culture	Cooney *et al.* (1966)
Zeaxanthin	*Paracoccus zeaxanthinifaciens*	—	McDermott, Britton and Goodwin (1973)
Zeaxanthin	*Erwinia herbicola*	—	Hundle *et al.* (1993)
Zeaxanthin	*Synechocystis* sp. PCC 6803	0.9 mg/l culture	Lagarde, Beuf and Vermaas (2000)
Zeaxanthin	*Sphingobacterium multivorum*	10.6 mg/l culture	Bhosale, Larson and Bernstein (2004)
Astaxanthin-glycoside	*Sphingomonas* sp. PB304	—	Kim *et al.* (2014)
Astaxanthin	*Brevundimonas* spp. (wild type)	0.36 mg/g biomass	Asker *et al.* (2012)
Astaxanthin	*Brevundimonas* sp. M7 (mutant)	1.3 mg/g biomass	Asker *et al.* (2012)
Astaxanthin	*Paracoccus* sp. N81106	0.05 mg/l culture	Yokoyama, Izumida and Miki (1994)
Astaxanthin	*Paracoccus carotinifaciens* E-396	—	Tsubokura, Yoneda and Mizuta (1999)
Astaxanthin	*Brevundimonas* sp. SD212	0.05 mg/l culture	Yokoyama *et al.* (1996)
Astaxanthin	*Paracoccus* sp. PC1	—	Yokoyama, Izumida and Miki (1994)
Astaxanthin	*Paracoccus bogoriensis*	0.4 mg/g biomass	Osanjo *et al.* (2009)

9.4 Biosynthesis of Carotenoids in Bacteria

As for all isoprenoid compounds, the biosynthesis of carotenoids depends on the precursor metabolites isopentenyl-pyrophosphate (IPP) and dimethylallyl-pyrophosphate (DMAPP). Two distinct pathways of the synthesis of these isoprenoid precursors were identified in nature. The mevalonate pathway produces IPP from acetyl-CoA in six enzymatic steps via mevalonate (Maury *et al.*, 2005). This pathway mostly exists in eukaryotic organisms, but also in some bacteria (Kuzuyama and Seto, 2012). The second pathway to the isoprenoid precursors is called *deoxy-xylulose-phosphate* (DXP) pathway or methyl-erythritol-phosphate (MEP) pathway (Rohmer *et al.*, 1993). This pathway starts from the glycolysis intermediates pyruvate and glyceraldehyde-3-phosphate and proceeds via seven enzymatic steps to IPP and DMAPP (Figure 9.1). In contrast to the mevalonate pathway, both IPP and DMAPP are formed simultaneously by the activity of IspH (Adam *et al.*, 2002; Gräwert *et al.*, 2004).

Besides eukaryotes, all archaea and some bacteria, such as *Staphylococcus, Streptococcus, Enterococcus, Lactococcus, Lactobacillus, Leuconostoc* and *Myxobacteria*, use the mevalonate pathway. Most bacteria including cyanobacteria, however, rely on the MEP pathway for the synthesis of isoprenoids. Only a few

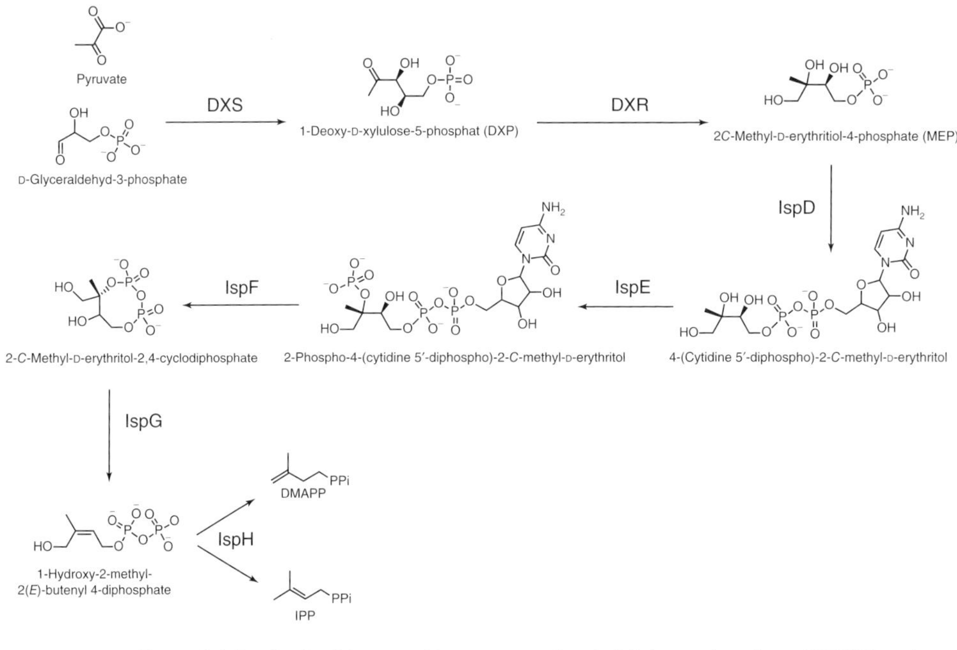

Figure 9.1 Synthesis of isoprenoid precursors dimethylallyl-pyrophosphate (DMAPP) and isopentenyl-pyrophosphate (IPP) via the methyl-D-erythritol-4-phosphate (MEP) pathway.

bacteria, such as some *Streptomyces* strains, possess both pathways (Kuzuyama and Seto, 2012).

After the synthesis of IPP and DMAPP, either by mevalonate or MEP pathway, the elongation to oligoprenyl-pyrophosphate takes place by a head-to-tail condensation. These reactions are catalysed by short-chain prenyl-pyrophosphate synthases forming geranyl-pyrophosphate (C_{10}), farnesyl-pyrophosphate (FPP) (C_{15}) and geranylgeranyl-pyrophosphate (GGPP) (C_{20}), which are precursors of various isoprenoid compounds, including carotenoids. In *E. coli*, the synthesis of geranyl-pyrophosphate and FPP is catalysed by IspA (Fujisaki, Nishino and Katsuki, 1986). Some prenyl-pyrophosphate synthases from other organisms can synthesise the condensation of four C_5-unit to form GGPP, such as GGPP-synthase (Gps) from *Archaeoglobus fulgidus* (Payandeh *et al.*, 2006).

The genes responsible for the synthesis of lycopene and β-carotene, starting from FPP, have been isolated from the epiphytic bacterium *Erwinia* and plants, and the functions of the carotenogenic genes have been identified (Misawa *et al.*, 1990; Hundle *et al.*, 1991; To *et al.*, 1994; Albrecht *et al.*, 1995; Sandmann, 1994).

The specific pathway that leads to C_{40} carotenoids (Figure 9.2) starts from two molecules of GGPP that react to phytoene, catalysed by the phytoene-synthase (CrtB). Phytoene is the first carotenoid in the pathway, but is still colourless (mainly maximum at 286 nm in hexane). Most bacteria synthesise (15*Z*)-phytoene. This molecule is converted by the phytoene-desaturase (CrtI) via phytofluene, ζ-carotene and neurosporene to lycopene. In non-photosynthetic bacteria, the desaturation is catalysed by one enzyme (CrtI), whereas in cyanobacteria, the reaction is carried out by two desaturases and a carotenoid-isomerase (Umeno *et al.*, 2005). The synthesis of cyclic carotenoids occurs by the cyclisation at one or both ends of lycopene. The formation of β-carotene is catalysed by the lycopene-β-cyclase (CrtY) (Figure 9.2).

The majority of bacteria that had been investigate concerning the formation of carotenoids showed that most bacteria produce carotenoids with oxygen-containing functional groups (xanthophyll), such as hydroxyl, keto or epoxy groups. The synthesis of the most important xanthophyll astaxanthin requires the introduction of two keto groups and two hydroxyl groups into β-carotene (Figure 9.4). The addition of the keto groups is catalysed by the β-carotene ketolase, which is encoded by *crtO* or by *crtW* genes (Misawa *et al.*, 1995; Tao and Cheng, 2004). The introduction of the hydroxyl groups is catalysed by the β-carotene hydroxylase, for which three known isoforms of this enzyme exist: CrtZ, CrtR and cytochrome-P450 hydroxylase (Blasco, Kauffmann and Schmid, 2004; Masamoto *et al.*, 1998; Misawa *et al.*, 1990; Alvarez *et al.*, 2006).

The majority of carotenoids produced by eukaryotes and prokaryotes consist of eight isoprenoid units with 40 carbon atoms. However, a few non-photosynthetic bacteria strains are able to synthesis carotenoids with 30, 45 or 50 carbon atoms. C_{30} carotenoids, for example, are produced by *Staphylococcus aureus, Heliobacillus mobilis* and *Streptococcus faecium* (Marshall and Wilmoth, 1981; Takaichi *et al.*, 1997). The biosynthesis of the C_{30} carotenoids proceeds similarly

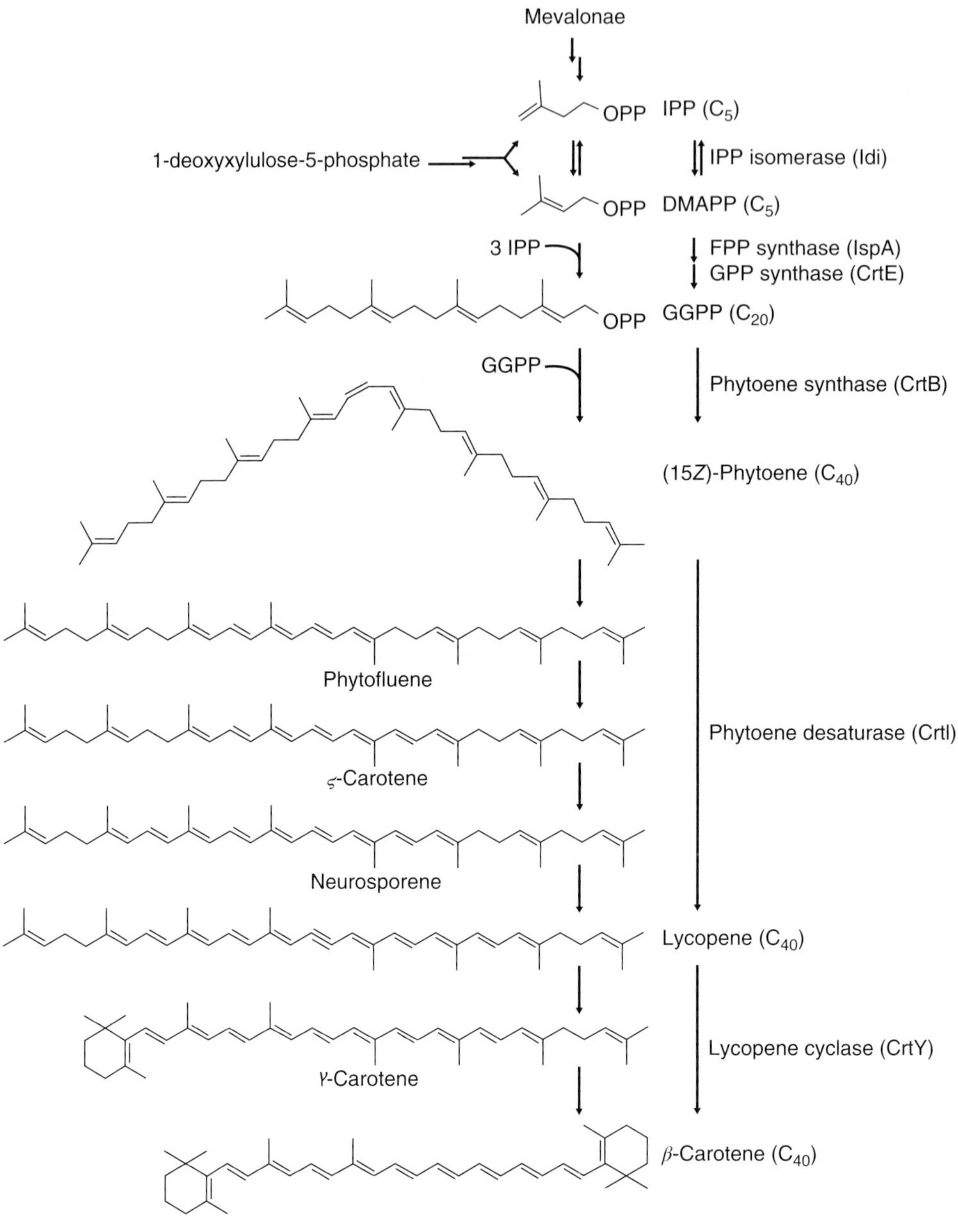

Figure 9.2 Biosynthesis pathway of β-carotene. Dimethylallyl-pyrophosphate (DMAPP); isopentenyl-pyrophosphate (IPP), geranyl-pyrophosphate (GPP); farnesyl-pyrophosphate (FPP) and geranylgeranyl-pyrophosphate (GGPP).

to the formation of C_{40} carotenoids by the condensation of two molecules of farnesyl-diphosphate forming diapophytoene, followed by a desaturase reaction (Maeda, 2012).

The formation of C_{45} and C_{50} homocarotenoids has been reported for several Gram-positive and Gram-negative organisms (Heider *et al.*, 2014a), but only a few biosynthesis pathways have been reported. So far, the synthesis of decaprenoxanthin in *Corynebacterium glutamincum* (Krubasik and Sandmann, 2000; Krubasik *et al.*, 2001) of the β-cyclic C_{50} carotenoid C.p. 450 in Dietz asp. (Tao, Yao and Cheng, 2007) and of the γ-cyclic C_{50} carotenoid sarcinaxanthin in *Micrococcus luteus* (Netzer *et al.*, 2010) have been investigated. The biosynthetic pathway of these homocarotenoids follows in the first step of the synthesis of C_{40} carotenoids via GGDP and phytoene to the formation of lycopene. Lycopene is then elongated by C_5 isoprenyl units at one or both ends of the molecule, respectively. Subsequently, this molecule can be further modified by cyclisation, hydroxylation and glycosylation (Heider *et al.*, 2014a).

9.5 Biotechnological Synthesis of Carotenoids by Carotenogenic and Non-Carotenogenic Bacteria

Due to the importance of carotenoids, both in the diet of humans and animals and as colourant, an efficient way for the production of carotenoids is required. In order to meet this demand, the development of a chemical synthesis of carotenoids had started more than 60 year ago (Pfander, Traber and Lanz, 1997). Currently, the large-scale production of most commercially important carotenoids is conducted to about 90% by chemical syntheses (Alvarez *et al.*, 2014). Based on safety concerns about the chemical synthesis and by the ambition for a sustainable production of chemical materials, an increasing demand for carotenoids, such as β-carotene, lycopene, lutein, zeaxanthin and astaxanthin from natural sources carotenoids is raising. However, the amounts of these carotenoids in most of the natural producing organisms are very low, and so, it appears unrealistic to produce large quantities of carotenoids by isolation from natural sources in an economical manner. But only a few carotenogenic organisms are currently used for the large-scale production of carotenoids, such as the fungus *B. trispora* and the microalga *Dunaliella salina* for the production of β-carotene (Mantzouridou, Roukas and Kotzekidou, 2004; Mantzouridou, Naziri and Tsimidou, 2008; Hejazi and Wijffels, 2004; Prieto, Canavate and García-González, 2011) or microalga *H. pluvialis* and yeast *X. dendrorhous* for the production of astaxanthin (Rodríguez-Sáiz, de la Fuente and Barredo, 2010; Li *et al.*, 2011).

A classical way to enhance the productivity of carotenoids in the native organism can be achieved, for example, by breeding or mutagenesis experiments. These methods were successfully used to increase the amount of carotenoids in tomatoes and manioc roots (Kinkade and Foolad, 2013; Ceballos *et al.*, 2013) as well as in *H. pluvialis*, *X. dendrorhous* and *B. trispora* (Gómez *et al.*, 2013; Schmidt *et al.*, 2010; Mehta, Obraztsova and Cerda-Olmedo, 2003).

Due to the availability and knowledge of biosynthesis genes and enzymes involved in the biosynthetic pathways of carotenoids from microorganisms, it is possible to create new carotenoid biosynthesis pathways in carotenogenic or non-carotenogenic organisms or to modify producing organisms to increase their productivity by means of pathway or metabolic engineering, which has been demonstrated in several distinguished publications that are reviewed here.

In this chapter, we give a summary of the approaches that have been undertaken to produce β-carotene as well as other industrially important carotenoids and xanthophylls in a bacterial strain, both by modifying natural bacterial carotenoid-producing strain and by establishing synthetic carotenoid pathways in a non-carotenogenic bacterial host.

9.5.1 Heterologous Expression of Carotenoid Biosynthesis Genes

The vast majority of studies on the heterologous biosynthesis of carotenoids were conducted in *E. coli*. Since multiple tools for the genetic modification are available, this Gram-negative bacterium is by far the most suitable host for cloning and expression of foreign genes.

The first successful attempt of cloning and expression of carotenoid genes in *E. coli* was shown for the gene cluster from *R. capsulatus* and *Pantoea agglomerans*, formerly called *Erwinia herbicola* (Marrs, 1981; Perry *et al.*, 1986). The first elucidation of a carotenoid biosynthesis gene cluster and pathway by heterologous expression in *E. coli* was demonstrated by Misawa *et al.* (1990). In this study, the expression of a gene cluster, containing six open-reading frames, from the phytopathogenic bacterium *P. ananatis*, formerly called *Erwinia uredovora*, led to the conversion of the *E. coli* derived precursor farnesyl-diphosphate to zeaxanthin-diglucoside by the step-wise action of the recombinant enzymes CrtE, CrtB, CrtI, CrtY, CrtZ and CrtX. By truncation or modification of the gene cluster, the formation of the pathway intermediates phytoene, lycopene, β-carotene and zeaxanthin was demonstrated. In the years after this pioneering work by Misawa and co-workers, several studies followed that showed the cloning and expression of further biosynthesis genes and gene clusters form other carotenogenic organisms as well as the formation of other important carotenoids, for example, astaxanthin and canthaxanthin (reviewed by Misawa and Shimada, 1997).

The amount of carotenoids produced by a simple plasmid-based overexpression of heterologous carotenoid genes in *E. coli* cells varies by the genes or gene clusters employed from different sources (Yoon *et al.*, 2007b). However, the content is relatively low (<2 mg/g cell dry weight (CDW) (Table 9.3)) compared to several times higher content of carotenoids found in organisms such as *Blakeslea*, *Dunaliella* or *Haematococcus* (Johnson and Schroeder, 1996). In order to increase the carotenoid yield in the recombinant bacterial host strains, several limiting

Table 9.3 Synthesis of industrially important carotenoids by recombinant bacteria expressing a heterologous and/or modified native carotenoid-biosynthesis pathway.

Carotenoid product	Modification to improve carotenoid formation	Bacterial host	Product yield	References
β-Carotene	—	*A. tumefaciens*	0.35 mg/g CDW	Misawa, Yamano and Ikenaga (1991)
β-Carotene	—	*H. elongate*	0.56 mg/g CDW	Rodríguez-Sáiz *et al.* (2007)
β-Carotene	—	*Z. mobilis*	0.22 mg/g CDW	Misawa, Yamano and Ikenaga (1991)
β-Carotene	—	*E. coli*	2 mg/g CDW	Misawa *et al.* (1990)
β-Carotene	Low-copy expression plasmid	*E. coli*	390 g/l culture[a]	Kim *et al.* (2006)
β-Carotene	MEP pathway	*E. coli*	6 mg/g CDW	Yuan *et al.* (2006)
β-Carotene	Chromosomal insertion MEP pathway	*E. coli*	6.2 mg/g CDW	Lemuth, Steuer and Albermann (2011)
β-Carotene	Mevalonate pathway	*E. coli*	49.3.2 mg/g CDW	Yoon *et al.* (2007a)
β-Carotene	Engineering central metabolism MEP pathway	*E. coli*	30.2 mg/g CDW 59.9 mg/g CDW[a] 2100 mg/l culture[a]	Zhao *et al.* (2013)
β-Carotene	MEP pathway Mevalonate pathway	*E. coli*	3200 mg/l culture[a]	Yang and Guo (2014)
Lycopene	—	*C. glutamicum*	2.4 mg/g CDW	Heider, Peters-Wendisch and Wendisch (2012)
Lycopene	—	*R. rubrum*	2 mg/g CDW	Wang *et al.* (2012)
Lycopene	Genome-wide modification	*E. coli*	6.6 mg/g CDW	Alper *et al.* (2005)
Lycopene	Genome-wide modification	*E. coli*	18 mg/g CDW	Alper, Miyaoku and Stephanopoulos (2005)
Lycopene	Genome-wide modification	*E. coli*	16 mg/g CDW	Jin and Stephanopoulos (2007)

(*continued overleaf*)

Table 9.3 *(Continued)*

Carotenoid product	Modification to improve carotenoid formation	Bacterial host	Product yield	References
Lycopene	Chromosomal insertion Balancing gene expression	*E. coli*	11 mg/g CDW	Tyo, Ajikumar and Stephanopoulos (2009)
Lycopene	Mevalonate pathway	*E. coli*	22 mg/g CDW	Yoon *et al.* (2006)
Lycopene	Chromosomal insertion MEP pathway Balancing gene expression	*E. coli*	33.4 mg/g CDW	Chen *et al.* (2013)
Lycopene	Engineering central metabolism MEP pathway	*E. coli*	50.6 mg/g CDW[a)] 3520 mg/l culture[a)]	Sun *et al.* (2014)
Lycopene	MEP pathway Gene disruption	*E. coli*	7.55 mg/g CDW	Zhou *et al.* (2013)
Lycopene	Engineering central metabolism MEP pathway	*E. coli*	14 mg/g CDW	Farmer and Liao (2001)
Canthaxanthin	Pathway engineering	*Bradyrhizobium sp.*	1.5 mg/g CDW	Giraud and Verméglio (2012)
Canthaxanthin	—	*E. coli*	>90% of total carotenoid	Cheng and Tao (2012)
Zeaxanthin	Balancing gene expression	*E. coli*	0.82 mg/g CDW	Nishizaki *et al.* (2007)
Zeaxanthin	—	*E. coli*	2.2 mg/g CDW	Misawa *et al.* (1990)
Zeaxanthin	Addition of surfactants to culture	*P. putida*	7 mg/g CDW	Beuttler *et al.* (2011)
Astaxanthin	—	*C. glutamicum*	0.14 mg/g CDW	Heider *et al.* (2014b)
Astaxanthin	—	*Methylomonas* sp.	>95% of total carotenoid	Ye *et al.* (2007)
Astaxanthin	Chromosomal insertion MEP pathway Balancing gene expression	*E. coli*	1.4 mg/g CDW	Lemuth, Steuer and Albermann (2011)

a) Results from fed-batch bioreactor cultivation.

factors have to be considered: first, the availability of the carotenoid precursor substrates IPP/DMAPP and FPP; second, the activity and expression of biosynthesis enzymes should be balanced for a efficient conversion of pathway intermediates and to avoid an effect of metabolic burden (Jones *et al.*, 2000), due to the overexpression of recombinant proteins; and third, an increased storage capacity or an *in situ* extraction is required to avoid high concentration of the lipophilic carotenoids in the cytoplasmic membrane.

9.5.2 Increased Isoprenoid Precursor Supply

The formation of isoprenoid precursor in bacteria, such as *E. coli*, is required for the synthesis of ubiquinone, menaquinone and undecaprenyl-phosphate (Fujisaki, Nishino and Katsuki, 1986). These compounds occur only in small quantities in *E. coli*. A heterologous carotenoid pathway competes for this small precursor pool, and therefore, an increase in precursor supply should result in a higher carotenoid yield. Several studies have shown that the increased expression of the 1-deoxy-xylose-5-phosphate synthase (DXS, Figure 9.1), which catalysed the first step of the MEP-pathway, led to about threefold increase in the formation of lycopene (Harker and Bramley, 1999; Matthews and Wurtzel, 2000) or β-carotene (Yuan *et al.*, 2006; Lemuth, Steuer and Albermann, 2011) by *E. coli*. A comparable strong effect was observed by the enhanced expression of the isoprenyl-diphosphate isomerase (Idi) (Lee and Schmidt-Dannert, 2002). Further increase in carotenoid formation, but to less extent than by change of the Dxs and Idi activities, was observed by the overexpression of the MEP-pathway genes *dxr*, *ispD*, *ispG* or *ispF* (Kim and Keasling, 2001; Yuan *et al.*, 2006). Zhou *et al.* (2012) showed that the overexpression of *ispG* led to reduced efflux of the MEP-pathway intermediate methylerythritol cyclodiphosphate.

By the combination of the overexpression of the genes *dxs*, *idi*, *ispD* and *ispF* in an *E. coli* strain carrying a plasmid-based *crtEBIY* operon led to the formation of 6 mg/g CDW of β-carotene (Yuan *et al.*, 2006). This β-carotene content is comparable to the amount of carotenoids that is formed in natural producer organisms, such as *H. pluvialis* or *X. dendrorhous* (Das *et al.*, 2007).

Another bottleneck in the carotenoid formation by *E. coli* was identified in the conversion of the C_5 isoprenoids to the C_{15} FPP by IspA. The use of a recombinant GpS from *Archaeoglobus fulgidus* that converts IPP directly to GGPP increased the yield of, for example, astaxanthin in *E. coli* by about two-times (Wang, Oh and Liao, 1999). By using *in vitro* evolution methods, a gene variant of *gps* improved the carotenoid formation even further (Wang, Oh and Liao, 2000).

This shows that the activities of the biosynthesis enzymes that catalyse the conversion of the glycolysis intermediates glyceraldehyde-3-phosphate and pyruvate to FPP are limiting for the formation of carotenoid by *E. coli*. The flux through the MEP pathway was increased by the enhanced expression of single or multiple biosynthesis genes, as shown earlier.

The alternative mevalonate pathway for the synthesis of IPP does not exist in *E. coli* and most eubacteria, but this pathway can be functionally expressed in *E. coli* strains, as it was demonstrated for the first time by the introduction of the mevalonate pathway from *Saccharomyces cerevisiae* into *E. coli* for the synthesis of the bio-active terpenoid artemisinin (Martin *et al.*, 2003). By using a recombinant mevalonate pathway from *Streptococcus pneumoniae* for the synthesis of β-carotene in *E. coli*, the product yield, compared to studies with only engineered MEP pathway, was significantly higher and reached a content of 22 mg/g CDW of lycopene (Yoon *et al.*, 2006) and 49 mg/g CDW of β-carotene (Yoon *et al.*, 2007a), respectively. On the other hand, the recombinant expression of a mevalonate pathway from *Streptomyces* sp. strain CL190 in *E. coli* resulted only in the formation of 4.3 mg/g CDW of lycopene (Vadali *et al.*, 2005), demonstrating the variable activity of recombinant enzyme from different organisms.

9.5.3
Genome-Wide Modification of *E. coli* to Increase Carotenoid Formation

Besides the optimisation of the specific pathways towards IPP/DMAPP supply and carotenoid formation, it is of importance in metabolic engineering approaches to consider the whole cell metabolism, which can be modulated by rational engineering or by evolutionary engineering approaches (Sauer, 2001). Besides the direct isoprenoid synthesis, another important issue that has to be deal with is the availability of precursors for the MEP-(glyceraldehyde-3-phosphate, pyruvate) and mevalonate pathway (acetyl-CoA) (Figure 9.3), respectively, as well as of cofactors, such as ATP and NADPH.

The precursors of the MEP pathway, glyceraldehyde-3-phosphate and pyruvate, are needed in stoichiometric amount for the Dxs catalysed reaction and the formation of one β-carotene molecule via the MEP pathway requires 8 ATP, 8 CTP and 16 NADPH. The flux distribution of glyceraldehyde-3-phosphate and pyruvate could not be varied by exogenous supply of pyruvate or glycerol, showing the strict regulation of the intracellular metabolic fluxes (Farmer and Liao, 2001). Attempts to change the distribution by inactivation or overexpression of flux regulating enzymes, however, showed that the lycopene formation can be increased three times by a redirection of the lower glycolysis pathway towards glyceraldehyde-3-phosphate (Farmer and Liao, 2000, 2001).

During growth on glucose, the uptake of glucose via the phosphotransferase-system (PTS) is a strong shunt of phosphoenol-pyruvate in *E. coli*. The deletion of the glucose-PTS leads to an increase in the intracellular phosphoenol-pyruvate

Figure 9.3 Pathway of the synthesis of C_{40} carotenoids by recombinant *Escherichia coli*. The central precursor substrates IPP and DMAPP can be formed via MEP and/or mevalonate pathway. G3P, glyceraldehyde-3-phosphate; DXP, 1-deoxyxylulose-5-phosphate; MEP, 2-*C*-methyl-erythritol-4-phosphate; IPP, isopentenyl-pyrophosphate; DMAPP, dimethylallyl-pyrophosphate; GPP, geranyl-pyrophosphate; FPP, farnesyl-pyrophosphate and GGPP, geranylgeranyl-pyrophosphate.

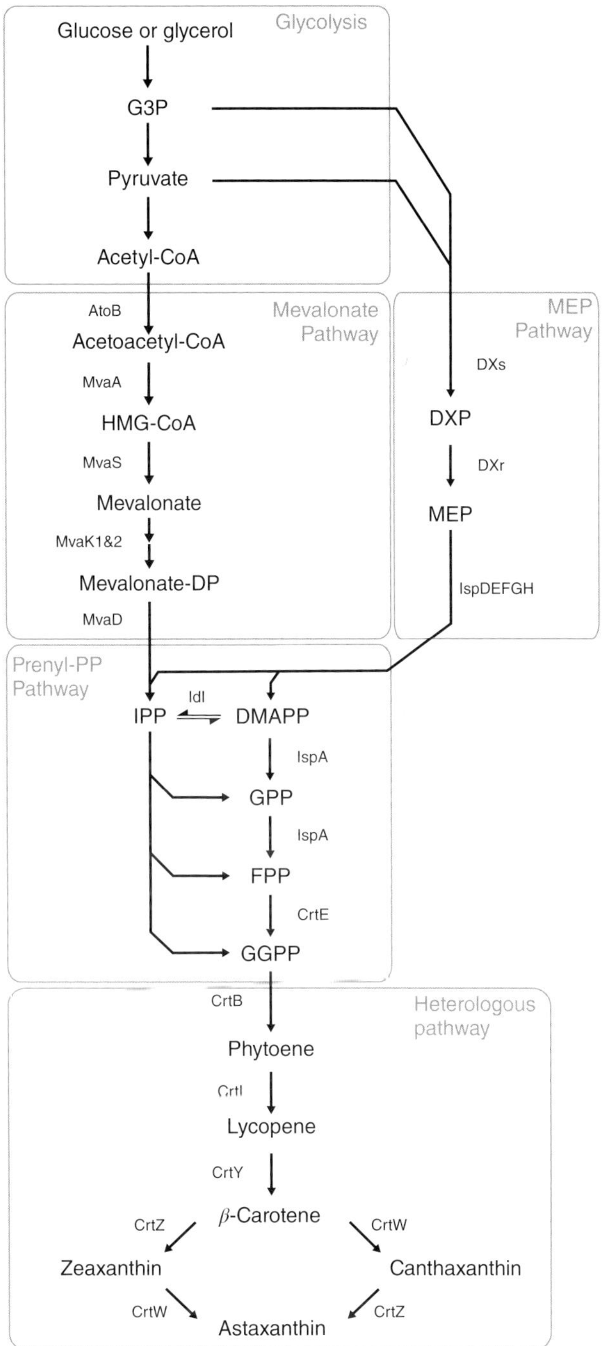
Glycolysis
Glucose or glycerol
G3P
Pyruvate
Acetyl-CoA
Mevalonate Pathway
AtoB
Acetoacetyl-CoA
MvaA
HMG-CoA
MvaS
Mevalonate
MvaK1&2
Mevalonate-DP
MvaD
MEP Pathway
DXs
DXP
DXr
MEP
IspDEFGH
Prenyl-PP Pathway
IPP
Idl
DMAPP
IspA
GPP
IspA
FPP
CrtE
GGPP
Heterologous pathway
CrtB
Phytoene
CrtI
Lycopene
CrtY
β-Carotene
CrtZ
CrtW
Zeaxanthin
Canthaxanthin
CrtW
CrtZ
Astaxanthin

concentration (Gabor *et al.*, 2011). Zhang *et al.* (2013) showed that lycopene formation in a PTS-mutant increased seven times over the parental strain and reached a content of 20 mg/g CDW. The increase in lycopene formation was thereby observed on glucose as well as on glycerol as carbon source.

The use of an additional recombinant mevalonate pathway in *E. coli* requires pyruvate, glyceraldehyde-3-phosphate and acetyl-CoA as precursors for the formation of IPP. By heterologous expression of the mevalonate pathway from *Streptomyces* strain and the knockout of genes encoding for enzymes or pathways competing at the nodes of pyruvate and acetyl-CoA (e.g. acetate formation), the formation of lycopene increased by three times compared with the parental strain (Vadali *et al.*, 2005).

Zhao *et al.* (2013) described the engineering of several *E. coli* pathways of the central metabolism by the insertion of regulatory element to modulate the formation of ATP and NADPH. Gene operons coding for enzymes of the ATP synthesis as well as of the electron transport chain were modulated. With the variation of the expression of the genes *nuo* (NADH: ubiquinone oxidoreductase I genes), *cyd* (Cytochrome bd-I oxidase gene), *cyo* (cytochrome bo oxidase gene) or *atp* (ATP-synthase gene), an increase in the β-carotene content by 20%, 16%, 5% or 21%, respectively, was achieved. An identical method was used for changing the NADPH formation by tuning the tricarbon acid cycle as well as the pentose-phosphate pathway. Modulation of the expression of the tricarbon acid cycle genes *sucAB* (α-ketoglutarate dehydrogenase genes), *gltA* (citrate synthase gene) and *sdhABCD* (succinate dehydrogenase genes) led to an increase of 39%, 35% and 25%, respectively, in β-carotene synthesis. In the case of the pentose-phosphate pathway, tuning of the genes *tktA* (transketolase I gene) and *talB* (transaldolase B gene) resulted in a 16% and 17% higher β-carotene content, respectively. By combining the modulations of *sucAB*, *sdhABCD* and *talB* in one *E. coli* strain along with an increased expression of the IPP isomerase (Idi) and Dxs, the β-carotene content reached 30.17 mg/g CDW in shake-flask cultivations. Under optimised growth conditions during fed-batch cultivation in a bioreactor, the β-carotene yield could be further improved to a total concentration of 2.1 g/l in the culture and a content of 59.88 mg/g CDW (Zhao *et al.*, 2013). This value presents the highest content of β-carotene in *E. coli* that has been reported so far (Table 9.3).

In a comparable study, the modulation of tricarbon acid cycle, pentose-phosphate-, MEP-pathway was also applied to a lycopene-producing *E. coli* strain. During fed-batch cultivation, the optimised strain reached a concentration of 3.52 g/l and 50.6 mg/g CDW of lycopene after a 100 h fermentation process (Sun *et al.*, 2014).

Other approaches to increase the carotenoid production by *E. coli* used random mutagenesis or used a systematic computational search based on a stoichiometric *E. coli* model to target possible candidates for gene disruption (Table 9.4).

Alper, Miyaoku and Stephanopoulos (2005) identified several knockout strains based on stoichiometric analysis which could enhance the formation of lycopene by increasing precursor supply for the MEP pathway. After construction of several

Table 9.4 *E. coli* genes, which showed increased formation of carotenoids upon overexpression or disruption.

Overexpressed genes	Known or putative function of protein
appY	Regulatory protein affecting appA and other genes
Crl	RNA polymerase holoenzyme assembly factor
dxs	1-Deoxy-D-xylulose 5-phosphate (DXP) synthase
elbB (yhbL)	Isoprenoid biosynthesis protein with amidotransferase-like domain
idi	Isopentenyl diphosphate isomerase
iraD (yjiD)	Inhibitor of σ^S proteolysis
iraM (ycgW)	Inhibitor of σ^S proteolysis
preA (yeiA)	NADH-dependent dihydropyrimidine dehydrogenase subunit
purDH	Phosphoribosylamine-glycine ligase and AICAR transformylase/IMP cyclohydrolase
rnlA (yfjN)	CP4-57 prophage; RNase LS, toxin of the RnlAB toxin-antitoxin system
rpoS	RNA polymerase sigma factor 38 (σ^S)
torC	Trimethylamine N-oxide reductase, cytochrome c-type subunit
torT	Inducer-binding protein in TorSR two-component signal transduction system
ydgk	Conserved inner membrane protein
yedR	Predicted inner membrane protein
yggT	Hypothetical integral membrane protein, putative resistance protein

Disrupted genes	Known or putative function of protein
ackA	Acetate kinase
appYp	Acid (poly)phosphatase, starvation response
aspC	Aspartate aminotransferase
aspC	Aspartate aminotransferase
clpP	ATP-dependent protease, (one target: σS)
cyaA	Adenylate cyclase
fdhA	(selB) selenocysteine incorporation (into fdhF)
fdhD	fdhF formation protein
fdhF	Formate dehydrogenase
glnE	Protein adenylyl transferase, modifies glutamine synthase
glxR	Tartronate semialdehyde reductase
gntK	Gluconokinase
hnr	σS degradation
lipB	Lipoate biosynthesis (related with aceE activity)
lysU	Lysine-tRNA ligase
modA	Periplasmic molybdate-binding protein
moeA	Molybdopterin biosynthesis
nadA	Quinolinate synthetase
pitA	Low-affinity phosphate transport
pst	High-affinity phosphate transport (membrane proteins)
pstC	High-affinity phosphate transport

(*continued overleaf*)

Table 9.4 (*Continued*)

Disrupted genes	Known or putative function of protein
pstC	High-affinity phosphate transport
sohA	Putative protease
stpA	Putative regulator – chaperone
yagR	Putative molybdenum cofactor-binding oxidoreductase
ybaS	Putative glutaminase
ycfZ	Putative factor
ydeN	Putative enzyme (possible sulfur metabolism)
yebB	Hypothetical protein
yedN	Hypothetical protein
yfcC	Putative integral membrane protein
ygjP	Putative transcriptional regulator
yibD	Putative glycosyl transferase
yjfP	Hypothetical protein
yjhH	Putative enzyme
yliE	Hypothetical protein
zwf	Glucose-6-phosphate dehydrogenase

Kang *et al.* (2005), Alper *et al.* (2005), Alper, Miyaoku and Stephanopoulos (2005), Jin and Stephanoupolos (2007), Alper and Stephanopoulos (2008) and Zhou *et al.* (2013).

strains with single or multiple gene deletions, a strain that carries the gene disruption of *gdhA* (glutamate dehydrogenase), *aceE* (part of pyruvate dehydrogenase complex) and *fdhF* (part of formiate dehydrogenase complex) showed the most significant increase in lycopene yield (6.6 mg/g CDW). A stoichiometric model, however, is unable to predict the flux balance, which is mediated by multiple factors. In order to overcome this disadvantage, random transposon mutagenesis was conducted (Alper *et al.*, 2005; Alper and Stephanopoulos, 2008). These studies revealed several new gene deletions, including biosynthesis and regulatory genes, that effect the formation of lycopene (Table 9.4). The combination of gene knockout targets gained by the systematic and combinatorial approach resulted in a triple mutant strain (Δ*gdhA*, Δ*aceE* and Δ*yjiD*) that produced 18 mg/g CDW of lycopene (Das *et al.*, 2007).

Besides the identification of deletion strains based on random mutagenesis or on model-predicted gene targets that enhance the synthesis of carotenoids, shotgun libraries of *E. coli* DNA have been constructed and introduced into lycopene-producing *E. coli* strains (Kang *et al.*, 2005; Jin and Stephanopoulos, 2007). Through this random approach, an increased expression of single or multiple *E. coli* genes in a carotenoid-producing strain is generated. By visual screening of the gained strain libraries, several overexpressed genes were identified, which positively influence the lycopene formation (Table 9.4). Among them are *dxs* and *idi*, which are already known to promote carotenoid synthesis by their overexpression (Harker and Bramley, 1999; Matthews and Wurtzel, 2000). Furthermore, genes that encode for regulator proteins, such as AppY, Crl and RpoS, as well as for putative regulatory proteins, such as YjiD and YcgW, were identified to increase lycopene formation. Among the different regulatory proteins, the overexpression of AppY had the strongest effect on carotenoid formation (Kang *et al.*, 2005). The transcriptional regulator AppY induces the expression of genes involved in energy metabolism in the stationary phase and under anaerobic conditions (Brøndsted and Atlung, 1996; Yang *et al.*, 2004). Because of the multitude of genes that are regulated by AppY, it is difficult to explain the link between *appY* expression and the increased synthesis of lycopene. Although the mechanism for the positive lycopene effect induced by the different regulator proteins is not solved yet, the various knockout and overexpression targets were used for the construction of lycopene-producing strains with improved performance (Jin and Stephanopoulos, 2007). Among all the tested strains, the enhanced expression of *dxs, idi, yjiD* and *ycgW* in a lycopene-producing *E. coli* strain with *aceE, fdhF* and *gdhA* disruption resulted in the highest accumulation of lycopene (16 mg/g CDW) (Jin and Stephanopoulos, 2007) (Table 9.3).

9.5.4
Balancing Recombinant Enzyme Activities for an Improved Synthesis of Carotenoids by *E. coli*

The enzymatic activities of a recombinant, heterologous biosynthesis pathway need to be adapted to host organism to enable an efficient transformation of available precursors to the desired product (Jones *et al.*, 2000). A low enzyme activity would lead to an incomplete conversion and thereby to the accumulation of pathway intermediates (Nishizaki *et al.*, 2007). Too high activity and expression of a recombinant protein can lead to unwanted side reactions (Albermann, 2011) or to a metabolic burden effect (Jones *et al.*, 2000).

The vast majority of studies about the heterologous synthesis of carotenoids, as described in the previous sections, use recombinant high-copy plasmids with strong artificial promoter for the expression of the carotenoid biosynthesis enzymes (Crt). However, a couple of publications have demonstrated that a lower protein expression, mediated by low-copy plasmids or by chromosomal integration of biosynthesis genes, can result in a higher carotenoid yield and in genetically more stable production hosts (Jones *et al.*, 2000; Kim *et al.*, 2006; Chiang, Chen and Chao, 2008; Tyo, Ajikumar and Stephanopoulos, 2009; Albermann *et al.*, 2010; Albermann, 2011; Lemuth, Steuer and Albermann, 2011; Chen *et al.*, 2013; Zhao *et al.*, 2013). The first chromosomal integration of the lycopene biosynthesis genes in *E. coli* was achieved by the site-specific insertion of a T_7-promoter-gps-*crtBI*-operon via the phage attachment sites *attP/attB* (Chiang, Chen and Chao, 2008). The insertion of individual expression cassettes of each biosynthesis gene (*crtE, crtB, crtI* from *P. ananatis*) was realised by the λ-Red recombination method (Datsenko and Wanner, 2000), by which each individual cassette was site-specifically integrated into a sugar-degradation locus (*malEFG*, *fucPIK*, *xylAB*) on the *E. coli* chromosome (Albermann *et al.*, 2010). In order to investigated the effect of single- and multi-copy expression of *crtE*, *crtB* and *crtI* on the formation of lycopene, strains with all different combinations of two chromosomal and one plasmid-encoded *crt*-gene were constructed. Comparison of the different strains concerning lycopene and biomass formation revealed that the highest lycopene and biomass yields were gained by the strain in which all genes are integrated in the chromosome. A high expression of *crtB* had a strong growth-inhibiting effect, whereas the high expression of *crtI* led to the formation of the by-product tetradehydrolycopene (Albermann, 2011).

The expression of the chromosomal integrated genes led to an enzyme activity that is high enough to convert the available precursor completely to lycopene or β-carotene, even if the IPP/DMAPP supply is increased, for example, when *dxs* and *idi* are additionally expressed (Chiang *et al.*, 2008; Albermann *et al.*, 2010; Albermann, 2011; Lemuth, Steuer and Albermann, 2011).

If an increased activity of recombinant enzyme that derived from chromosomally integrated genes is needed, DNA fragments can be inserted twice or multiple times at different sites on the bacterial chromosome (Albermann *et al.*, 2010; Ye and Kelly, 2012). It can also be achieved by the so-called *chemical inducible chromosomal evolution* (CIChE), by which inserted expression cassettes have been amplified through an increased selection pressure (e.g. antibiotics) in a *recA*-positive *E. coli* strain (Tyo, Ajikumar and Stephanopoulos, 2009). This method was shown to be successful to increase the formation of lycopene by *E. coli* (Tyo, Ajikumar and Stephanopoulos, 2009; Chen *et al.*, 2013).

For the *E. coli* synthesis of zeaxanthin, which requires five enzymatic steps from FPP, it was shown that no accumulation of pathway intermediates (phytoene, lycopene, β-carotene) occurs when the five genes (*crtEBIYZ*) were expressed by individual chromosomal cassettes, respectively (Albermann *et al.*, 2010). In contrast, the plasmid-based expression of *crtEBIYZ* in one operon can result in an incomplete conversion, depending on the order for genes in the operon (Nishizaki *et al.*, 2007). By assembling zeaxanthin operons with all possible arrangement of the five genes, Nishizaki and co-workers could demonstrate that the quantity of the mRNA of each individual gene, and thereby the amount of protein, decreases by an increasing distance to the promoter. The gene order *crtEBIYZ* in the operon led to the highest zeaxanthin content (0.82 mg/g CDW), and the CrtE catalysed reaction was identified as the most limiting step in the formation zeaxanthin.

The synthesis of the industrially important xanthophyll astaxanthin by recombinant organisms is more challenging than the synthesis of lycopene or β-carotene, because of an inefficient conversion of β-carotene to astaxanthin (Misawa *et al.*, 1995; Fraser, Miura and Misawa, 1997; Hasunuma *et al.*, 2008). The conversion of β-carotene to astaxanthin requires the introduction of keto groups at 4 and 4′ as well as hydroxyl groups at 3 and 3′ positions of the β-ionone rings (Figure 9.4). The introduction of the keto groups is catalysed by the β-carotene ketolase, encoded by *crtO* or by *crtW* (Misawa *et al.*, 1995; Tao and Cheng, 2004). The hydroxylation is catalysed by the β-carotene hydroxylase CrtZ, CrtR or cytochrome-P450

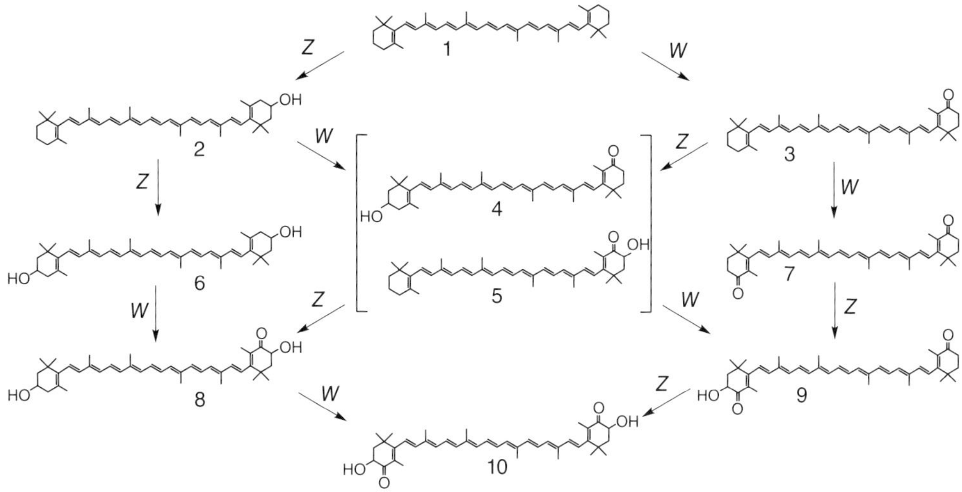

Figure 9.4 Scheme of the astaxanthin biosynthesis pathway in recombinant *E. coli*, proceeding from β-carotene (**1**) to astaxanthin (**10**). *W* β-carotene ketolase (CrtW), *Z* β-carotene hydroxylase (CrtZ), (**2**) β-cryptoxanthin, (**3**) echinenone, (**4**) 3′-hydroxyechinenone, (**5**) 3-hydroxyechinenone, (**6**) zeaxanthin, (**7**) canthaxanthin, (**8**) adonixanthin and (**9**) adonirubin.

hydroxylase (Misawa *et al.*, 1990; Masamoto *et al.*, 1998; Blasco, Kauffmann and Schmid, 2004; Alvarez *et al.*, 2006). The different iso-enzymes of the ketolase and hydroxylase as well as their variants from different organisms show differences in their substrate specificity concerning the 3-hydroxylation or 4-ketolation status of the β-ionone rings (Fraser, Miura and Misawa, 1997; Steiger and Sandmann, 2004; Makino *et al.*, 2008). As a result, the expression of ketolase and hydroxylase in naturally producing organisms as well as in a heterologous host leads to the formation of up to eight intermediates besides astaxanthin (Figure 9.4).

In order to overcome the incomplete conversion of β-carotene to astaxanthin, protein variants of the β-carotene ketolase were generated. By the increased enzyme activity of some muteins towards the hydroxylated substrate, astaxanthin was obtained as major product, up to 90% of the total carotenoid content, by *in vivo* synthesis (Ye *et al.*, 2006; Tao *et al.*, 2006). Another work to overcome the insufficient conversion of β-carotene to astaxanthin showed that, by balancing the gene expression of the native *crtZ* (*P. ananatis*) and *crtW* (*Nostoc punctiforme*) in a β-carotene-producing *E. coli* strain, astaxanthin was product as the sole carotenoid with a content of 1.4 mg/g CDW (Lemuth, Steuer and Albermann, 2011). By the high expression of *crtW* and low expression of *crtZ*, achieved by the used of different inducible promoters, the conversion of β-carotene proceeds mainly via canthaxanthin to astaxanthin. In contrast, an equal expression of both genes leads mainly to the accumulation of zeaxanthin (>70%) (Lemuth, Steuer and Albermann, 2011).

9.5.5
Production of Industrially Important Carotenoids by Other Recombinant Bacteria

The vast majority of studies on the biosynthesis of carotenoids by recombinant bacteria have been conducted with the non-carotenogenic bacterium *E. coli*. This is mainly because of the knowledge on the molecular biology and physiology of this organism that allows extensive studies on metabolic engineering. However, a couple of other bacterial organisms have also been considered for whole-cell synthesis of industrially important carotenoids, such as lycopene, β-carotene, zeaxanthin, canthaxanthin and astaxanthin.

C. glutamicum is an industrially important organism of the production of amino acids and other valuable chemicals. Naturally this organism produces C_{50} homocarotenoids, for example, decaprenoxanthin (Krubasik, Kobayashi and Sandmann, 2001). It was demonstrated that by overexpression of the lycopene and astaxanthin biosynthesis genes and deletion of the genes responsible for C_{50} carotenoid-biosynthesis, the synthesis of the respective carotenoids in *C. glutamicum* can be achieved (Heider, Peters-Wendisch and Wendisch, 2012; Heider *et al.*, 2014b; Matano *et al.*, 2014; Peters-Wendisch *et al.*, 2014). Comparable procedures were also used for the pathway engineering in other carotenogenic bacteria to allow, for example, the lycopene formation by *R. rubrum* (Wang *et al.*, 2012), β-carotene formation by the halophilic bacterium *Halomonas elongate* (Rodríguez-Sáiz *et al.*, 2007; De Lourdes Moreno *et al.*, 2012) or the improved canthaxanthin synthesis by *Bradyrhizobium* sp. (Giraud and Verméglio, 2012) (Table 9.3).

Besides *E. coli*, the formation of carotenoids was also achieved in other non-carotenogenic upon expression of *crt*-genes from carogenogenic microorganisms (Table 9.4). By this way, β-carotene was synthesised by *Z. mobilis* and *Agrobacterium tumefaciens*, respectively (Misawa, Yamano and Ikenaga, 1991). Canthaxanthin and astaxanthin biosynthesis pathways were introduced into *Methylmonas* sp. strain 16a (Sharpe *et al.*, 2007; Tang *et al.*, 2007; Ye *et al.*, 2007; Tao *et al.*, 2007a; Ye and Kelly, 2012). In order to have a stable production strain, for large-scale fermentation processes, the recombinant genes were integrated into the chromosome of *Methylmonas* sp. (Tao *et al.*, 2007b). For the production of astaxanthin by *Methylmonas* sp., a high content of 95% could be achieved (Ye *et al.*, 2007). A further bacterium that was applied for carotenoid production is *P. putida*. By introducing an expression plasmid containing the zeaxanthin biosynthesis genes (*crtEIBYZ*) along with the *dxs*, *idi* and *ispA* into *P. putida*, the resulting strain was able to produce 7 mg/g CDW of zeaxanthin (Beuttler *et al.*, 2011).

9.5.6
Culture Conditions of Improved Formation of Carotenoids by Recombinant Bacteria

Due to relevance of carotenoid for many industrial applications, an efficient way to synthesise carotenoid compounds is required. The cultivation of carotenoid-producing organisms is a promising approach for a sustainable production of these compounds. As mentioned here, the metabolic engineering of bacteria, especially

E. coli, has led to significant progress in the development of strains for the production of carotenoids. But so far, only a few studies have dealt with the large-scale fermentation and the optimisation of culture and extraction conditions of carotenoid-producing bacteria.

Several studies have shown that the cultivation of *E. coli* strain on glycerol leads to higher carotenoid formation than on culture media with glucose (Lee, Mijts and Schmidt-Dannert, 2004; Zhang *et al.*, 2013; Yang and Guo, 2014). The growth on glucose can lead to the formation of high concentration of acetate, which might be an inhibitor metabolite and so lowers the formation of carotenoids (Yang and Guo, 2014). Zhang *et al.* (2013) investigated different growth media and cultivation condition in a comprehensive study and demonstrated that temperature and shaking speed of a shake-flask cultivation influence lycopene synthesis by *E. coli*. The strongest effect was observed by different growth media, an optimised minimal medium with adapted concentration for the C-source (glycerol), ammonium phosphate and potassium phosphate resulted in a 4–10 times higher yield of lycopene than standard media, such as M9-minimal medium or 2xYT complex media.

By examining the results from shake-flask cultivation and fed-batch bioreactor cultivation, it becomes evident that under controlled conditions in bioreactor experiments (controlled pH, pO2, feeding rates, etc.), the product formation is significantly high. Zhao *et al.* (2013), for example, reported a yield of 39 mg/g CDW β-carotene by shake-flask cultivation and 60 mg/g CDW by fed-batch bioreactor cultivation.

In some studies, the addition of surfactants had an increasing effect on growth and on product formation. The addition of Tween 80, lecithin and oleic acid, respectively, was reported to be beneficial for carotenoid formation and prevented the formation of cell clumps (Yoon *et al.*, 2006; Beuttler *et al.*, 2011). The emergence of cell clumps is obviously a result of the carotenoid formation. The carotenoids that are produced by the bacteria accumulate in the cells associated with the lipophilic membrane or proteins. In order to avoid harmful effects by the strong accumulation of carotenoids, some studies (Yoon *et al.*, 2008; Beuttler *et al.*, 2011; Jang *et al.*, 2014) tried to remove the carotenoids during the cultivation process by *in situ* extraction. Yoon *et al.*, 2008 reported the use of an organic/aqueous two-phase system to extract lycopene during bacterial growth. The use of octane as organic solvent had a strong negative effect on growth, whereas the use of decane allowed a sufficient growth. But only a small amount of lycopene could be extracted from the cells by this approach. However, by the addition of lysozyme, which degrades the cell wall, more than 40% of the synthesised lycopene was recovered from the organic phase.

9.6 Conclusion

Carotenoids, especially lycopene, β-carotene and astaxanthin, are widely used in the industry as nutraceuticals, animal feed additives, food colourants and functional cosmetics. Carotenogenic bacteria are a source for a variety of carotenoids, especially xanthophylls. However, at present, no native bacterial organism is considered for the industrial production of β-carotene or lycopene, which is mainly due to a lower content and a higher diversity of carotenoids in most bacterial cells, as compared to fungi or microalgae, for example, *B. trispora* or *D. salina*.

Over the past 20 years, pathway and metabolic engineering have enabled the production and improvement of carotenoids by bacteria, especially by *E. coli*. The key issue for the increased formation of carotenoids by recombinant bacteria is the improved supply of the precursors IPP and DMAPP. The synthesis of these central intermediates was successfully enhanced by rational and evolutionary metabolic engineering approaches. Improved *E. coli* strains now exceed the productivity of lycopene or β-carotene by native carotenogenic microorganisms. But despite the significant progress made in metabolic engineering, only little efforts have been made in terms of process engineering. To meet the industrial demands, further optimisation of culturing conditions and media and product recovery are needed. But also, a further generation of improved *E. coli* or other bacterial strains by metabolic engineering is required to compete with the chemical synthesis of carotenoids.

As shown for photosynthetic organisms, carotenoids and especially xanthophylls span the lipid bilayer membrane (Gruszecki and Strzałka, 2005). The capacity of a bacterial cell to accumulated carotenoids in the cytoplasmic membrane might therefore be a limiting factor for the production of carotenoids. To overcome this possible limitation, *in situ* extraction with organic solvent was demonstrated (Yoon *et al.*, 2008), but the toxicity of organic solvent reduces or stops growth of *E. coli*. Further development of recombinant bacteria with higher tolerance to organic solvents, as, for example, *Pseudomonas* strains, might be a promising approach.

Besides the improved synthesis of important and well-known carotenoids by *E. coli*, the heterologous *in vivo* synthesis also allows the generation of novel carotenoid compounds by combinatorial biosynthesis and enzyme engineering (Schmidt-Dannert, Umeno and Arnold, 2000; Sandmann, 2002; Furubayashi and Umeno, 2012). Such novel carotenoids could lead to new industrial applications.

References

Aberoumand, A. (2011) A review article on edible pigments properties and sources as natural biocolorants in foodstuff and food industry. *World J. Dairy Food Sci.*, **6** (1), 71–78.

Adam, P., Hecht, S., Eisenreich, W., Kaiser, J., Grawert, T., Arigoni, D., Bacher, A., and Rohdich, F. (2002) Biosynthesis of terpenes: studies on 1-hydroxy-2-methyl-2-(E)-butenyl 4-diphosphate reductase. *Proc. Natl. Acad. Sci. U.S.A.*, **99** (19), 12108–12113.

Albermann, C., Trachtmann, N., Sprenger, G.A. (2010) A simple and reliable method to conduct and monitor expression cassette integration into the *Escherichia coli* chromosome. *Biotechnol. J.*, **5** (1), 32–38.

Albermann, C. (2011) High versus low level expression of the lycopene biosynthesis genes from *Pantoea ananatis* in

Escherichia coli. Biotechnol. Lett., **33** (2), 313–319.

Albrecht, M., Klein, A., Hugueney, P., Sandmann, G., Kuntz, M. (1995) Molecular cloning and functional expression in *E. coli* of a novel plant enzyme mediating zeta-carotene desaturation. *FEBS Lett.*, **372** (2–3), 199–202.

Alper, H., Jin, Y.S., Moxley, J.F., and Stephanopoulos, G. (2005) Identifying gene targets for the metabolic engineering of lycopene biosynthesis in *Escherichia coli. Metab. Eng.*, **7** (3), 155–164.

Alper, H., Miyaoku, K., and Stephanopoulos, G. (2005) Construction of lycopene-overproducing E. coli strains by combining systematic and combinatorial gene knock-out targets. *Nat. Biotechnol.*, **23**, 612–616.

Alper, H. and Stephanopoulos, G. (2008) Uncovering the gene knockout landscape for improved lycopene production in E. coli. *Appl. Microbiol. Biotechnol.*, **78** (5), 801–810.

Alvarez, V., Rodriguez-Saiz, M., de la Fuente, J.L., Gudina, E.J., Godio, R.P., Martin, J.F., and Barredo, J.L. (2006) The crtS gene of *Xanthophyllomyces dendrorhous* encodes a novel cytochrome-P450 hydroxylase involved in the conversion of betacarotene into astaxanthin and other xanthophylls. *Fungal Genet. Biol.*, **43**, 261–272.

Alvarez, R., Vaz, B., Gronemeyer, H., and de Lera, A.R. (2014) Functions, therapeutic applications, and synthesis of retinoids and carotenoids. *Chem. Rev.*, **114** (1), 1–125.

Asker, D., Awad, T.S., Beppu, T., and Ueda, K. (2012) Isolation, characterization, and diversity of novel radiotolerant carotenoid-producing bacteria. *Methods Mol. Biol.*, **892**, 21–60.

Barba, A.I.O., Hurtado, M.C., Mata, M.C.S., Ruiz, V.F., and Tejada, M.L.S. (2006) Application of a UV–vis detection-HPLC method for a rapid determination of lycopene and β-carotene in vegetables. *Food Chem.*, **95**, 328–336.

Barua, A.B., Olson, J.A., Van Breemen, R.B. (2000) Modern Chromatographic Analysis of Vitamins, in *Modern Chromatographic Analysis of Vitamins* (eds De Leenheer, A.P., Lambert, W.E., Van Bocxlaer, J.F.), Marcel Dekker, New York.

Beuttler, H. (2010) Biosynthese von Carotinoiden mittels rekombinanten Mikroorganismen. Doctoral thesis. Universität Stuttgart, Stuttgart.

Beuttler, H., Hoffmann, J., Jeske, M., Hauer, B., Schmid, R.D., Altenbuchner, J., and Urlacher, V.B. (2011) Biosynthesis of zeaxanthin in recombinant *Pseudomonas putida. Appl. Microbiol. Biotechnol.*, **89** (4), 1137–1147.

Bhosale, P., Larson, A.J., and Bernstein, P.S. (2004) Factorial analysis of tricarboxylic acid cycle intermediates for optimization of zeaxanthin production from Flavobacterium multivorum. *J. Appl. Microbiol.*, **96**, 623–629.

Blasco, F., Kauffmann, I., and Schmid, R.D. (2004) CYP175A1 from Thermus thermophilus HB27, the first beta-carotene hydroxylase of the P450 superfamily. *Appl. Microbiol. Biotechnol.*, **64**, 671–674.

Breitenbach, J., Misawa, N., Kajiwara, S., and Sandmann, G. (1996) Expression in *Escherichia coli* and properties of the carotene ketolase from *Haematococcus pluvialis. FEMS Microbiol. Lett.*, **140** (2-3), 241–246.

Breithaupt, D.E. (2000) Enzymatic hydrolysis of carotenoid fatty acid esters of red pepper (Capsicum annuum L.) by a lipase from Candida rugosa. *Z. Naturforsch., C*, **55**, 971–975.

Breithaupt, D.E. (2004) Simultaneous HPLC determination of carotenoids used as food coloring additives: applicability of accelerated solvent extraction. *Food Chem.*, **86**, 449–456.

Breithaupt, D.E. and Bamedi, A. (2001) Carotenoid esters in vegetables and fruits: a screening with emphasis on β-cryptoxanthin esters. *J. Agric. Food. Chem.*, **49**, 2064–2070. doi: 10.1021/jf001276t

Breithaupt, D.E. and Bamedi, A. (2002) Carotenoids and carotenoid esters in potatoes (Solanum tuberosum L.): new insights into an ancient vegetable. *J. Agric. Food. Chem.*, **50**, 7175–7181.

Breithaupt, D.E. and Schlatterer, J. (2005) Lutein and zeaxanthin in new dietary supplements—analysis and quantification. *Eur. Food Res. Technol.*, **220**, 648–652. doi: 10.1007/s00217-004-1075-2

Breithaupt, D.E. and Schwack, W. (2000) Determination of free and bound carotenoids in paprika (Capsicum annuum L.) by LC/MS. *Eur.*

Food Res. Technol., **211**, 52–55. doi: 10.1007/s002170050588

Britton, G., Liaaen-Jensen, S., and Pfander, H. (1995) Carotenoids: isolation and analysis, in *Carotenoids* (eds G. Britton, S. Liaaen-Jensen, and H. Pfander), Birkhäuser Verlag, Basel, p. 13.

Britton, G., Liaaen-Jensen, S., and Pfander, H. (2004) *Carotenoids Handbook*, Birkhäuser Verlag, Basel. ISBN: 3-7643-6180-8.

Brøndsted, L. and Atlung, T. (1996) Effect of growth conditions on expression of the acid phosphatase (cyx-appA) operon and the appY gene, which encodes a transcriptional activator of *Escherichia coli*. *J. Bacteriol.*, **178** (6), 1556–1564.

Ceballos, H., Morante, N., Sánchez, T., Ortiz, D., Aragón, I., Chávez, A.L., Pizarro, M., Calle, F., and Dufour, D. (2013) Rapid cycling recurrent selection for increased carotenoids content in cassava roots. *Crop Sci.*, **53**, 1–10.

Chen, Y.Y., Shen, H.J., Cui, Y.Y., Chen, S.G., Weng, Z.M., Zhao, M., and Liu, J.Z. (2013) Chromosomal evolution of *Escherichia coli* for the efficient production of lycopene. *BMC Biotech.*, **13**, 6.

Cheng, Q., Tao, L. (2012) Engineering *Escherichia coli* for canthaxanthin and astaxanthin biosynthesis. *Methods Mol. Biol.*, **892**, 143–158.

Chiang, C.J., Chen, P.T., and Chao, Y.P. (2008) Replicon-free and markerless methods for genomic insertion of DNAs in phage attachment sites and controlled expression of chromosomal genes in *Escherichia coli*. *Biotechnol. Bioeng.*, **101** (5), 985–995.

Cimpoiu, C. and Hosu, A. (2007) Thin layer chromatography for the analysis of vitamins and their derivatives. *J. Liq. Chromatogr. Related Technol.*, **30**, 701–728. doi: 10.1080/10826070701191011

Cooney, J.J., Marks, H.W., Smith, A.M. (1966) Isolation and Identification of Canthaxanthin from Micrococcus roseus. *J. Bacteriol.*, **92** (2), 342–345.

Cortés, C., Esteve, M.J., Frígola, A., and Torregrosa, F. (2004) Identification and quantification of carotenoids including geometrical isomers in fruit and vegetable juices by liquid chromatography with ultraviolet–diode array detection. *J. Agric. Food. Chem.*, **52**, 2203–2212.

Croce, R. and van Amerongen, H. (2014) Natural strategies for photosynthetic light harvesting. *Nat. Chem. Biol.*, **10** (7), 492–501.

Das, A., Yoon, S.H., Lee, S.H., Kim, J.Y., Oh, D.K., and Kim, S.W. (2007) An update on microbial carotenoid production: application of recent metabolic engineering tools. *Appl. Microbiol. Biotechnol.*, **77** (3), 505–512.

Datsenko, K.A., Wanner, B.L. (2000) One-step inactivation of chromosomal genes in *Escherichia coli* K-12 using PCR products. *Proc. Natl. Acad. Sci. USA*, **97** (12), 6640–6645.

Deli, J. (1998) Thin-layer chromatography of carotenoids. *J. Planar. Chromatogr. - Mod. TLC*, **11**, 311–312.

De Lourdes Moreno, M., Sánchez-Porro, C., García, M.T., and Mellado, E. (2012) Carotenoids' production from halophilic bacteria. *Methods Mol. Biol.*, **892**, 207–217.

De Miguel, T., Sieiro, C., Poza, M., and Villa, T.G. (2001) Analysis of canthaxanthin and related pigments from Gordonia jacobaea mutants. *J. Agric. Food. Chem.*, **49**, 1200–1202.

De Ritter, E., Purcell, A.E., and Bauernfeind, J.C. (1981) *Carotenoids as Colorants and Vitamin A Precursors*, Academic Press, London.

Dieser, M., Greenwood, M., and Foreman, C.M. (2010) Carotenoid pigmentation in antarctic heterotrophic bacteria as a strategy to withstand environmental stresses. *Arct. Antarct. Alp. Res.*, **42**, 396–405.

Emenhiser, C., Simunovic, N., Sander, L.C., and Schwartz, S.J. (1996) Separation of geometrical carotenoid isomers in biological extracts using a polymeric C 30 column in reversed-phase liquid chromatography. *J. Agric. Food. Chem.*, **44**, 3887–3893. doi: 10.1021/jf960104m

Farmer, W.R. and Liao, J.C. (2000) Improving lycopene production in *Escherichia coli* by engineering metabolic control. *Nat. Biotechnol.*, **18** (5), 533–537.

Farmer, W.R. and Liao, J.C. (2001) Precursor balancing for metabolic engineering of lycopene production in *Escherichia coli*. *Biotechnol. Progr.*, **17** (1), 57–61.

Feltl, L., Pacakova, V., Stulik, K., and Volka, K. (2005) Reliability of carotenoid analyses: a review. *Curr. Anal. Chem.*, **1**, 93–102.

Fraser, P.D., Miura, Y., and Misawa, N. (1997) In vitro characterization of astaxanthin biosynthetic enzymes. *J. Biol. Chem.*, **272**, 6128–6135.

Fujisaki, S., Nishino, T., and Katsuki, H. (1986) Isoprenoid synthesis in *Escherichia coli*. Separation and partial purification of four enzymes involved in the synthesis. *J. Biochem.*, **99** (5), 1327–1337.

Furubayashi, M. and Umeno, D. (2012) Directed evolution of carotenoid synthases for the production of unnatural carotenoids. *Methods Mol. Biol.*, **892**, 245–253.

Gabor, E., Göhler, A.K., Kosfeld, A., Staab, A., Kremling, A., and Jahreis, K. (2011) The phosphoenolpyruvate-dependent glucose- phosphotransferase system from *Escherichia coli* K-12 as the center of a network regulating carbohydrate flux in the cell. *Eur. J. Cell Biol.*, **90** (9), 711–720.

Giraud, E. and Verméglio, A. (2012) Isolation and light-stimulated expression of canthaxanthin and spirilloxanthin biosynthesis genes from the photosynthetic bacterium Bradyrhizobium sp. strain ORS278. *Methods Mol. Biol.*, **892**, 173–183.

Godinho, A. and Bhosle, S. (2008) Carotenes produced by alkaliphilic orange-pigmented strain of *Microbacterium arborescens* – AGSB isolated from coastal sand dunes. *Indian J. Geo-Mar. Sci.*, **37**, 307–312.

Gómez, P.I., Inostroza, I., Pizarro, M., and Pérez, J. (2013) From genetic improvement to commercial-scale mass culture of a Chilean strain of the green microalga *Haematococcus pluvialis* with enhanced productivity of the red ketocarotenoid astaxanthin. *AoB Plants*, **5**, plt026.

Gräwert, T., Kaiser, J., Zepeck, F., Laupitz, R., Hecht, S., Amslinger, S., Schramek, N., Schleicher, E., Weber, S., Haslbeck, M., Buchner, J., Rieder, C., Arigoni, D., Bacher, A., Eisenreich, W., and Rohdich, F. (2004) IspH protein of *Escherichia coli*: studies on iron-sulfur cluster implementation and catalysis. *J. Am. Chem. Soc.*, **126** (40), 12847–12855.

Gruszecki, W.I. and Strzałka, K. (2005) Carotenoids as modulators of lipid membrane physical properties. *Biochim. Biophys. Acta*, **1740** (2), 108–115.

Guyomarch, F., Binet, A., and Dufosse, L. (2000) Production of carotenoids by Brevibacterium lines: variation among strains, kinetic aspects and HPLC profiles. *J. Ind. Microbiol. Biotechnol.*, **24**, 64–70.

Harker, M. and Bramley, P.M. (1999) Expression of prokaryotic 1-deoxy-D-xylulose-5-phosphatases in *Escherichia coli* increases carotenoid and ubiquinone biosynthesis. *FEBS Lett.*, **448**, 115–119.

Hasunuma, T., Miyazawa, S., Yoshimura, S., Shinzaki, Y., Tomizawa, K., Shindo, K., Choi, S.K., Misawa, N., and Miyake, C. (2008) Biosynthesis of astaxanthin in tobacco leaves by transplastomic engineering. *Plant J.*, **55**, 857–868.

Heider, S.A., Peters-Wendisch, P., and Wendisch, V.F. (2012) Carotenoid biosynthesis and overproduction in *Corynebacterium glutamicum*. *BMC Microbiol.*, **12**, 198.

Heider, S.A., Peters-Wendisch, P., Wendisch, V.F., Beekwilder, J., and Brautaset, T. (2014a) Metabolic engineering for the microbial production of carotenoids and related products with a focus on the rare C50 carotenoids. *Appl. Microbiol. Biotechnol.*, **98** (10), 4355–4368.

Heider, S.A., Wolf, N., Hofemeier, A., Peters-Wendisch, P., and Wendisch, V.F. (2014b) Optimization of the IPP precursor supply for the production of lycopene, decaprenoxanthin and astaxanthin by *Corynebacterium glutamicum*. *Front. Bioeng. Biotechnol.*, **2**, 28.

Hejazi, M.A. and Wijffels, R.H. (2004) Milking of microalgae. *Trends Biotechnol.*, **22** (4), 189–194.

Hertzberg, S., Borch, G., and Liaaen-Jensen, S. (1976) Bacterial carotenoids: L. Absolute configuration of zeaxanthin dirhamnoside. *Arch. Microbiol.*, **110**, 95–99.

Hoffmann, J., Bóna-Lovász, J., Beuttler, H., and Altenbuchner, J. (2012) *In vivo* and in vitro studies on the carotenoid cleavage oxygenases from Sphingopyxis alaskensis RB2256 and Plesiocystis pacifica SIR-1 revealed their substrate specificities and non-retinal-forming cleavage activities. *FEBS J.*, **279**, 3911–3924.

Hundle, B.S., Beyer, P., Kleinig, H., Englert, G., Hearst, J.E. (1991) Carotenoids of

Erwinia herbicola and an *Escherichia coli* HB101 strain carrying the Erwinia herbicola carotenoid gene cluster. *Photochem. Photobiol.*, **54** (1), 89–93.

Hundle, B.S., O'Brien, D.A., Beyer, P., Kleinig, H., and Hearst, J.E. (1993) In vitro expression and activity of lycopene cyclase and beta- carotene hydroxylase from *Erwinia herbicola. FEBS Lett.*, **315**, 329–334.

Isaksen, M. and Francis, G.W. (1986) Reversed-phase thin-layer chromatography of carotenoids. *J. Chromatogr. A*, **355**, 358–362.

Jang, H.J., Ha, B.K., Kim, J.W., Jung, K.H., Ahn, J., Yoon, S.H., Kim, S.W. (2014) Comparison of extraction phases for a two-phase culture of a recombinant *E. coli* producing retinoids. *Biotechnol. Lett.*, **36** (3), 497–505.

Jaswir, I., Noviendri, D., Hasrini, R.F., and Octavianti, F. (2011) Carotenoids: sources, medicinal properties and their application in food and nutraceutical industry. *J. Med. Plants Res.*, **5** (33), 7119–7131.

Jin, S., Stephanopoulos, G. (2007) Multi-dimensional gene target search for improving lycopene biosynthesis in *Escherichia coli. Metab. Eng.*, **9**, 337–347.

Johnson, E.A. and Schroeder, W.A. (1996) Microbial carotenoids. *Adv. Biochem. Eng./Biotechnol.*, **53**, 119–178.

Jones, K.L., Kim, S.W., Keasling, J.D. (2000) Low-copy plasmids can perform as well as or better than high-copy plasmids for metabolic engineering of bacteria. *Metab. Eng.*, **2** (4), 328–338.

Kaiser, P. (2009) Entwicklung, Optimierung und Anwendung einer effizienten analytischen Methode zur Bestimmung von Carotinoiden aus Bakterien und Hefen. Doctoral thesis. Freie Universität, Berlin.

Kang, M.J., Lee, Y.M., Yoon, S.H., Kim, J.H., Ock, S.W., Jung, K.H., Shin, Y.C., Keasling, J.D., and Kim, S.W. (2005) Identification of genes affecting lycopene accumulation in *Escherichia coli* using a shot-gun method. *Biotechnol. Bioeng.*, **91** (5), 636–642.

Kaulmann, A. and Bohn, T. (2014) Carotenoids, inflammation, and oxidative stress-implications of cellular signaling pathways and relation to chronic disease prevention. *Nutr. Res.*, **34** (11), 907–929.

Khachik, F., Englert, G., Daitch, C.E., Beecher, G.R., Tonucci, L.H., and Lusby, W.R. (1992) Isolation and structural elucidation of the geometrical isomers of lutein and zeaxanthin in extracts from human plasma. *J. Chromatogr.*, **582**, 153–166.

Kim, S.W. and Keasling, J.D. (2001) Metabolic engineering of the nonmevalonate isopentenyl diphosphate synthesis pathway in *Escherichia coli* enhances lycopene production. *Biotechnol. Bioeng.*, **72** (4), 408–415.

Kim, S.W., Kim, J.B., Jung, W.H., Kim, J.H., and Jung, J.K. (2006) Over-production of beta-carotene from metabolically engineered *Escherichia coli. Biotechnol. Lett*, **28** (12), 897–904.

Kim, S.H., Kim, J.H., Lee, B.Y., and Lee, P.C. (2014) The astaxanthin dideoxyglycoside biosynthesis pathway in Sphingomonas sp. PB304. *Appl. Microbiol. Biotechnol.*, **98** (24), 9993–10003.

Kim, I.J., Ko, K.C., Nam, T.S., Kim, Y.W., Chung, W.I., and Kim, C.S. (2003) Expression and activity of citrus phytoene synthase and b-carotene hydroxylase in *Escherichia coli. J. Microbiol.*, **41**, 212–218.

Kinkade, M.P. and Foolad, M.R. (2013) Validation and fine mapping of lyc12.1, a QTL for increased tomato fruit lycopene content. *Theor. Appl. Genet.*, **126** (8), 2163–2175.

Kiokias, S., Varzakas, T., and Oreopoulou, V. (2008) In vitro activity of vitamins, flavonoids, and natural phenolic antioxidants against the oxidative deterioration of oil-based systems. *Crit. Rev. Food Sci. Nutr.*, **48** (1), 78–93.

Krubasik, P., Kobayashi, M., and Sandmann, G. (2001) Expression and functional analysis of a gene cluster involved in the synthesis of decaprenoxanthin reveals the mechanisms for C50 carotenoid formation. *Eur. J. Biochem.*, **268**, 3702–3708.

Krubasik, P. and Sandmann, G. (2000) A carotenogenic gene cluster from Brevibacterium linens with novel lycopene cyclase genes involved in the synthesis of aromatic carotenoids. *Mol. Gen. Genet.*, **263**, 423–432.

Krubasik, P., Takaichi, S., Maoka, T., Kobayashi, M., Masamoto, K., and Sandmann, G. (2001) Detailed biosynthetic pathway to decaprenoxanthin diglucoside in *Corynebacterium glutamicum* and identification of novel intermediates. *Arch. Microbiol.*, **176** (3), 217–223.

Kuzuyama, T. and Seto, H. (2012) Two distinct pathways for essential metabolic precursors for isoprenoid biosynthesis. *Proc. Jpn. Acad., Ser. B, Phys. Biol. Sci.*, **88** (3), 41–52.

Lagarde, D., Beuf, L., and Vermaas, W. (2000) Increased production of zeaxanthin and other pigments by application of genetic engineering techniques to Synechocystis sp. strain PCC 6803. *Appl. Environ. Microbiol.*, **66**, 64–72.

Lea, A.G.H. (1988) in *HPLC in Food Analysis* (ed R. Macrae), Academic Press, London, pp. 298–314.

Lee, P.C., Mijts, B.N., and Schmidt-Dannert, C. (2004) Investigation of factors influencing production of the monocyclic carotenoid torulene in metabolically engineered *Escherichia coli*. *Appl. Microbiol. Biotechnol.*, **65** (5), 538–546.

Lee, P.C. and Schmidt-Dannert, C. (2002) Metabolic engineering towards biotechnological production of carotenoids in microorganisms. *Appl. Microbiol. Biotechnol.*, **60**, 1–11.

Lemuth, K., Steuer, K., and Albermann, C. (2011) Engineering of a plasmid-free *Escherichia coli* strain for improved *in vivo* biosynthesis of astaxanthin. *Microb. Cell Fact.*, **10**, 29.

Li, J., Zhu, D., Niu, J., Shen, S., and Wang, G. (2011) An economic assessment of astaxanthin production by large scale cultivation of *Haematococcus pluvialis*. *Biotechnol. Adv.*, **29**, 568–574.

Lorquin, J., Molouba, F., and Dreyfus, B.L. (1997) Identification of the carotenoid pigment canthaxanthin from photosynthetic bradyrhizobium strains. *Appl. Environ. Microbiol.*, **63**, 1151–1154.

Maeda, I. (2012) Genetic modification in Bacillus subtilis for production of C30 carotenoids. *Methods Mol. Biol.*, **892**, 197–205.

Makino, T., Harada, H., Ikenaga, H., Matsuda, S., Takaichi, S., Shindo, K., Sandmann, G., Ogata, T., and Misawa, N. (2008) Characterization of cyanobacterial carotenoid ketolase CrtW and hydroxylase CrtR by complementation analysis in *Escherichia coli*. *Plant Cell Physiol.*, **49**, 1867–1878.

Mantzouridou, F., Naziri, E., and Tsimidou, M.Z. (2008) Industrial glycerol as a supplementary carbon source in the production of β-carotene by *Blakeslea trispora*. *J. Agric. Food. Chem.*, **56**, 2668–2675.

Mantzouridou, F., Roukas, T., and Kotzekidou, P. (2004) Production of beta-carotene from synthetic medium by *Blakeslea trispora* in fed-batch culture. *Food Biotechnol.*, **18** (3), 343–361.

Marrs, B. (1981) Mobilization of the genes for photosynthesis from Rhodopseudomonas capsulata by a promiscuous plasmid. *J. Bacteriol.*, **146**, 1003–1012.

Marshall, J.H. and Wilmoth, G.J. (1981) Pigments of *Staphylococcus aureus*, a series of triterpenoid carotenoids. *J. Bacteriol.*, **147**, 900–913.

Martin, V.J., Pitera, D.J., Withers, S.T., Newman, J.D., and Keasling, J.D. (2003) Engineering a mevalonate pathway in *Escherichia coli* for production of terpenoids. *Nat. Biotechnol.*, **21** (7), 796–802.

Masamoto, K., Misawa, N., Kaneko, T., Kikuno, R., and Toh, H. (1998) Beta-carotene hydroxylase gene from the cyanobacterium Synechocystis sp. PCC6803. *Plant Cell Physiol.*, **39**, 560–564.

Matano, C., Uhde, A., Youn, J.W., Maeda, T., Clermont, L., Marin, K., Krämer, R., Wendisch, V.F., and Seibold, G.M. (2014) Engineering of *Corynebacterium glutamicum* for growth and L-lysine and lycopene production from N-acetyl-glucosamine. *Appl. Microbiol. Biotechnol.*, **98** (12), 5633–5643.

Matthews, P.D. and Wurtzel, E.T. (2000) Metabolic engineering of carotenoid accumulation in *Escherichia coli* by modulation of the isoprenoid precursor pool with expression of deoxyxylulose phosphate synthase. *Appl. Microbiol. Biotechnol.*, **53**, 396–400.

Maury, J., Asadollahi, M.A., Møller, K., Clark, A., and Nielsen, J. (2005) Microbial isoprenoid production: an example of green chemistry through metabolic engineering. *Adv. Biochem. Eng./Biotechnol.*, **100**, 19–51.

McDermott, J.C., Britton, G., and Goodwin, T.W. (1973) Carotenoid biosynthesis in a Flavobacterium sp.: stereochemistry of hydrogen elimination in the desaturation of phytoene to lycopene, rubixanthin and zeaxanthin. *Biochem. J.*, **134**, 1115–1117.

Mehta, B.J., Obraztsova, I.N., and Cerda-Olmedo, E. (2003) Mutants and intersexual heterokaryons of *Blakeslea trispora* for production of β-carotene and lycopene. *Appl. Environ. Microbiol.*, **69**, 4043–4048.

Misawa, N., Kajiwara, S., Kondo, K., Yokoyama, A., Satomi, Y., Saito, T., Miki, W., and Ohtani, T. (1995) Canthaxanthin biosynthesis by the conversion of methylene to keto groups in a hydrocarbon beta-carotene by a single gene. *Biochem. Biophys. Res. Commun.*, **209**, 867–876.

Misawa, N., Nakagawa, M., Kobayashi, K., Yamano, S., Izawa, Y., Nakamura, K., and Harashima, K. (1990) Elucidation of the *Erwinia uredovora* carotenoid biosynthetic pathway by functional analysis of gene products expressed in *Escherichia coli*. *J. Bacteriol.*, **172** (12), 6704–6712.

Misawa, N., Shimada, H. (1997) Metabolic engineering for the production of carotenoids in non-carotenogenic bacteria and yeasts. *J. Biotechnol.*, **59** (3), 169–181.

Misawa, N., Yamano, S., and Ikenaga, H. (1991) Production of beta-carotene in *Zymomonas mobilis* and *Agrobacterium tumefaciens* by introduction of the biosynthesis genes from *Erwinia uredovora*. *Appl. Environ. Microbiol.*, **57** (6), 1847–1849.

Nelis, H.J. and De Leenheer, A.P. (1989) Profiling and quantitation of bacterial carotenoids by liquid chromatography and photodiode array detection. *Appl. Environ. Microbiol.*, **55**, 3065–3071.

Netzer, R., Stafsnes, M.H., Andreassen, T., Goksoyr, A., Bruheim, P., and Brautaset, T. (2010) Biosynthetic pathway for gamma-cyclic sarcinaxanthin in *Micrococcus luteus*: heterologous expression and evidence for diverse and multiple catalytic functions of C(50) carotenoid cyclases. *J. Bacteriol.*, **192**, 5688–5699.

Nishizaki, T., Tsuge, K., Itaya, M., Doi, N., and Yanagawa, H. (2007) Metabolic engineering of carotenoid biosynthesis in *Escherichia coli* by ordered gene assembly in Bacillus subtilis. *Appl. Environ. Microbiol.*, **73** (4), 1355–1361.

Nugraheni, S.A., Khoeri, M.M., Kusmita, L., Widyastuti, Y., and Radjasa, O.K. (2010) Characterization of carotenoids pigments from bacterial symbionts of seagrass Thalassia hemprichii. *J. Coastal Dev.*, **14**, 51–60.

Olaizola, M. (2003) Commercial development of microalgal biotechnology: from the test tube to the marketplace. *Biomol. Eng.*, **20** (4–6), 459–466.

Oliver, J. and Palou, A. (2000) Chromatographic determination of carotenoids in foods. *J. Chromatogr. A*, **881**, 543–555.

Osanjo, G.O., Muthike, E.W., Tsuma, L., Okoth, M.W., Bulimo, W.D., Lünsdorf, H., Abraham, W.-R., Dion, M., Timmis, K.N., Golyshin, P.N., and Mulaa, F.J. (2009) A salt lake extremophile, Paracoccus bogoriensis sp.nov., efficiently produces xanthophyll carotenoids. *Afr. J. Microbiol. Res.*, **3**, 426–433.

Payandeh, J., Fujihashi, M., Gillon, W., and Pai, E.F. (2006) The crystal structure of (S)-3-O-geranylgeranylglyceryl phosphate synthase reveals an ancient fold for an ancient enzyme. *J. Biol. Chem.*, **281** (9), 6070–6078 (Epub 2005 Dec 23).

Perry, K.L., Simonitch, T.A., Harrison-Lavoie, K.J., and Liu, S.T. (1986) Cloning and regulation of *Erwinia herbicola* pigment genes. *J. Bacteriol.*, **168**, 607–612.

Peters-Wendisch, P., Götker, S., Heider, S.A., Komati Reddy, G., Nguyen, A.Q., Stansen, K.C., and Wendisch, V.F. (2014) Engineering biotin prototrophic *Corynebacterium glutamicum* strains for amino acid, diamine and carotenoid production. *J. Biotechnol.*, **192** (Part B), 346–354.

Pfander, H., Traber, B., and Lanz, M. (1997) Carotenoid synthesis: a progress report. *Pure Appl. Chem.*, **69** (10), 2047–2060.

Prieto, A., Canavate, P.J., and García-González, M. (2011) Assessment of carotenoid production by *Dunaliella salina* in different culture systems and operation regimes. *J. Biotechnol.*, **151**, 180–185.

Raja, R., Hemaiswarya, S., Rengasamy, R. (2007) Exploitation of Dunaliella for beta-carotene production. *Appl. Microbiol. Biotechnol.*, **74** (3), 517–523.

Rodríguez-Sáiz, M., de la Fuente, J., and Barredo, J. (2010) *Xanthophyllomyces dendrorhous* for the industrial production of astaxanthin. *Appl. Microbiol. Biotechnol.*, **88**, 645–658.

Rodríguez-Sáiz, M., Sánchez-Porro, C., De La Fuente, J.L., Mellado, E., and Barredo, J.L. (2007) Engineering the halophilic bacterium Halomonas elongata to produce beta-carotene. *Appl. Microbiol. Biotechnol.*, **77** (3), 637–643.

Rohmer, M., Knani, M., Simonin, P., Sutter, B., and Sahm, H. (1993) Isoprenoid biosynthesis in bacteria: a novel pathway for the early steps leading to isopentenyl diphosphate. *Biochem. J*, **295** (Pt. 2), 517–524.

Sander, L.C., Sharpless, K.E., Craft, N.E., and Wise, S.A. (1994) Development of Engineered Stationary Phases for the Separation of Carotenoid Isomers. *Anal. Chem.*, **66**, 1667–1674.

Sander, L.C. and Wise, S.A. (1987) Effect of phase length on column selectivity for the separation of polycyclic aromatic hydrocarbons by reversed-phase liquid chromatography. *Anal. Chem.*, **59**, 2309–2313.

Sandmann, G. (1994) Carotenoid biosynthesis in microorganisms and plants. *Eur. J. Biochem.*, **223**, 7–24.

Sandmann, G. (2002) Combinatorial biosynthesis of carotenoids in a heterologous host: a powerful approach for the biosynthesis of novel structures. *ChemBioChem*, **3** (7), 629–635.

Sauer, U. (2001) Evolutionary engineering of industrially important microbial phenotypes. *Adv. Biochem. Eng./Biotechnol.*, **73**, 129–169.

Schmidt, I., Schewe, H., Gassel, S., Jin, C., Buckingham, J., Hümbelin, M., Sandmann, G., and Schrader, J. (2010) Biotechnological production of astaxanthin with Phaffia rhodozyma/*Xanthophyllomyces dendrorhous*. *Appl. Microbiol. Biotechnol.*, **89**, 555–571.

Schmidt-Dannert, C., Umeno, D., and Arnold, F.H. (2000) Molecular breeding of carotenoid biosynthetic pathways. *Nat. Biotechnol.*, **18** (7), 750–753.

Sharpe, P.L., Dicosimo, D., Bosak, M.D., Knoke, K., Tao, L., Cheng, Q., and Ye, R.W. (2007) Use of transposon promoter-probe vectors in the metabolic engineering of the obligate methanotroph Methylomonas sp. strain 16a for enhanced C40 carotenoid synthesis. *Appl. Environ. Microbiol.*, **73** (6), 1721–1728.

Steel, C.C. and Keller, M. (2000) Influence of UV-B irradiation on the carotenoid content of Vitis vinifera tissues. *Biochem. Soc. Trans.*, **28**, 883.

Steiger, S. and Sandmann, G. (2004) Cloning of two carotenoid ketolase genes from Nostoc punctiforme for the heterologous production of canthaxanthin and astaxanthin. *Biotechnol. Lett*, **26**, 813–817.

Sun, T., Miao, L., Li, Q., Dai, G., Lu, F., Liu, T., Zhang, X., and Ma, Y. (2014) Production of lycopene by metabolically-engineered *Escherichia coli*. *Biotechnol. Lett*, **36** (7), 1515–1522.

Takaichi, S., Inoue, K., Akaike, M., Kobayashi, M., Oh-oka, H., and Madigan, M.T. (1997) The major carotenoid in all known species of heliobacteria is the C30 carotenoid 4,4 ¢ -diaponeurosporene, not neurosporene. *Arch. Microbiol.*, **168**, 277–281.

Tang, X.S., Shyr, J., Tao, L., Sedkova, N., and Cheng, Q. (2007) Improvement of a CrtO-type of beta-carotene ketolase for canthaxanthin production in Methylomonas sp. *Metab. Eng.*, **9** (4), 348–354.

Tao, L. and Cheng, Q. (2004) Novel beta-carotene ketolases from non-photosynthetic bacteria for canthaxanthin synthesis. *Mol. Genet. Genomics*, **272**, 530–537.

Tao, L., Yao, H., and Cheng, Q. (2007a) Genes from a Dietzia sp. for synthesis of C40 and C50 beta-cyclic carotenoids. *Gene*, **386**, 90–97.

Tao, L., Sedkova, N., Yao, H., Ye, R.W., Sharpe, P.L., and Cheng, Q. (2007b) Expression of bacterial hemoglobin genes to improve astaxanthin production in a methanotrophic bacterium Methylomonas sp. *Appl. Microbiol. Biotechnol.*, **74** (3), 625–633.

Tao, L., Wilczek, J., Odom, J.M., and Cheng, Q. (2006) Engineering a beta-carotene ketolase for astaxanthin production. *Metab. Eng.*, **8**, 523–531.

Thürmann, P.A., Schalch, W., Aebischer, J.C., Tenter, U., and Cohn, W. (2005) Plasma kinetics of lutein, zeaxanthin, and 3-dehydro- lutein after multiple oral doses of a lutein supplement. *Am. J. Clin. Nutr.*, **82**, 88–97.

To, K.Y., Lai, E.M., Lee, L.Y., Lin, T.P., Hung, C.H., Chen, C.L., Chang, Y.S.,

Liu, S.T. (1994) Analysis of the gene cluster encoding carotenoid biosynthesis in Erwinia herbicola Eho13. *Microbiol.*, **140**, 331–339.

Tsubokura, A., Yoneda, H., and Mizuta, H. (1999) Paracoccus carotinifaciens sp. nov., a new aerobic gram-negative astaxanthin-producing bacterium. *Int. J. Syst. Bacteriol.*, **49** (Pt. 1), 277–282.

Tyo, K.E., Ajikumar, P.K., and Stephanopoulos, G. (2009) Stabilized gene duplication enables long-term selection-free heterologous pathway expression. *Nat. Biotechnol.*, **27** (8), 760–765.

Umeno, D., Tobias, A.V., Arnold, F.H. (2005) Diversifying carotenoid biosynthetic pathways by directed evolution. *Microbiol. Mol. Biol. Rev.*, **69** (1), 51–78.

Vadali, R.V., Fu, Y., Bennett, G.N., and San, K.Y. (2005) Enhanced lycopene productivity by manipulation of carbon flow to isopentenyl diphosphate in *Escherichia coli*. *Biotechnol. Progr.*, **21**, 1558–1561.

Veiga-Crespo, P., Blasco, L., Rosa-Dos-Santos, F., Poza, M., and Villa, T.G. (2005) In fl uence of culture conditions of Gordonia jacobaea MV-26 on canthaxanthin production. *Int. Microbiol.*, **8**, 55–58.

Wang, G.S., Grammel, H., Abou-Aisha, K., Sägesser, R., and Ghosh, R. (2012) High-level production of the industrial product lycopene by the photosynthetic bacterium *Rhodospirillum rubrum*. *Appl. Environ. Microbiol.*, **78** (20), 7205–7215.

Wang, C., Oh, M., and Liao, J. (1999) Engineered isoprenoid pathway enhances astaxanthin production in *Escherichia coli*. *Biotechnol. Bioeng.*, **62**, 235–241.

Wang, C., Oh, M., and Liao, J. (2000) Directed evolution of metabolically engineered *Escherichia coli* for carotenoid production. *Biotechnol. Progr.*, **16**, 922–926.

Weedon, B.C.L., Moss, G.P. (1995) Structure and Nomenclature, in *Carotenoids 1A* (eds Britton, G., Liaaen-Jensen, S., Pfander, H.), Birkhäuser Verlag, Basel.

Yang, J. and Guo, L. (2014) Biosynthesis of ß-carotene in engineered E. coli using the MEP and MVA pathways. *Microb. Cell Fact.*, **13** (1), 160.

Yang, H., Wolff, E., Kim, M., Diep, A., and Miller, J.H. (2004) Identification of mutator genes and mutational pathways in *Escherichia coli* using a multicopy cloning approach. *Mol. Microbiol.*, **53** (1), 283–295.

Ye, R.W. and Kelly, K. (2012) Construction of carotenoid biosynthetic pathways through chromosomal integration in methane-utilizing bacterium Methylomonas sp. strain 16a. *Methods Mol. Biol.*, **892**, 185–195.

Ye, R.W., Stead, K.J., Yao, H., and He, H. (2006) Mutational and functional analysis of the beta-carotene ketolase involved in the production of canthaxanthin and astaxanthin. *Appl. Environ. Microbiol.*, **72**, 5829–5837.

Ye, R.W., Yao, H., Stead, K., Wang, T., Tao, L., Cheng, Q., Sharpe, P.L., Suh, W., Nagel, E., Arcilla, D., Dragotta, D., and Miller, E.S. (2007) Construction of the astaxanthin biosynthetic pathway in a methanotrophic bacterium Methylomonas sp. strain 16a. *J. Ind. Microbiol. Biotechnol.*, **34** (4), 289–299.

Yokoyama, A., Izumida, H., and Miki, W. (1994) Production of astaxanthin and 4-ketozeaxanthin by the marine bacterium, Agrobacterium aurantiacum. *Biosci. Biotechnol., Biochem.*, **58**, 1842–1844.

Yokoyama, A., Miki, W., Izumida, H., and Shizuri, Y. (1996) New trihydroxy-keto-carotenoids isolated from an astaxanthin-producing marine bacterium. *Biosci. Biotechnol., Biochem.*, **60**, 200–203.

Yoon, K.W., Doo, E.H., Kim, S.W., and Park, J.B. (2008) *In situ* recovery of lycopene during biosynthesis with recombinant *Escherichia coli*. *J. Biotechnol.*, **135** (3), 291–294.

Yoon, S.H., Lee, Y.M., Kim, J.E., Lee, S.H., Lee, J.H., Kim, J.Y., Jung, K.H., Shin, Y.C., Keasling, J.D., and Kim, S.W. (2006) Enhanced lycopene production in *Escherichia coli* engineered to synthesize isopentenyl diphosphate and dimethylallyl diphosphate from mevalonate. *Biotechnol. Bioeng.*, **94** (6), 1025–1032.

Yoon, S.H., Park, H.M., Kim, J.E., Lee, S.H., Choi, M.S., Kim, J.Y., Oh, D.K., Keasling, J.D., and Kim, S.W. (2007a) Increased beta-carotene production in recombinant *Escherichia coli* harboring an engineered isoprenoid precursor pathway with mevalonate addition. *Biotechnol. Progr.*, **23** (3), 599–605.

Yoon, S.H., Kim, J.E., Lee, S.H., Park, H.M., Choi, M.S., Kim, J.Y., Lee, S.H., Shin, Y.C., Keasling, J.D., and Kim, S.W. (2007b) Engineering the lycopene synthetic pathway in E. coli by comparison of the carotenoid genes of *Pantoea agglomerans* and *Pantoea ananatis*. *Appl. Microbiol. Biotechnol.*, **74** (1), 131–139.

Yuan, L.Z., Rouvière, P.E., Larossa, R.A., Suh, W. (2006) Chromosomal promoter replacement of the isoprenoid pathway for enhancing carotenoid production in *E. coli*. *Metab. Eng.*, **8** (1), 79–90.

Zhang, C., Chen, X., Zou, R., Zhou, K., Stephanopoulos, G., and Too, H.P. (2013) Combining genotype improvement and statistical media optimization for isoprenoid production in E. coli. *PLoS One*, **8** (10), e75164.

Zhao, J., Li, Q., Sun, T., Zhu, X., Xu, H., Tang, J., Zhang, X., and Ma, Y. (2013) Engineering central metabolic modules of *Escherichia coli* for improving β-carotene production. *Metab. Eng.*, **17**, 42–50.

Zhou, Y., Nambou, K., Wei, L., Cao, J., Imanaka, T., and Hua, Q. (2013) Lycopene production in recombinant strains of *Escherichia coli* is improved by knockout of the central carbon metabolism gene coding for glucose-6-phosphate dehydrogenase. *Biotechnol. Lett.*, **35** (12), 2137–2145.

Zhou, K., Zou, R., Stephanopoulos, G., and Too, H.P. (2012) Metabolite profiling identified methylerythritol cyclodiphosphate efflux as a limiting step in microbial isoprenoid production. *PLoS One*, **7** (11), e47513.

10
β-Carotene and Other Carotenoids and Pigments from Microalgae

Borhane Samir Grama, Antoine Delhaye, Spiros N. Agathos, and Clayton Jeffryes

10.1
Introduction and Historical Outline

Humans have used microalgae for hundreds of years as food, fodder, medicine and fertiliser, but microalgal biotechnology is currently attracting an unparalleled interest and investment worldwide (Cadoret, Garnier and Saint-Jean, 2012). The problem is that microalgae are not a well-studied group from a biotechnological standpoint. Of the tens of thousands of microalgal species that are thought to exist, only a few thousand strains are kept in collections around the world, only a few hundred have been explored for chemical content and only a handful have been cultivated in industrial (tons per year) quantities (Olaizola, 2003).

During the twentieth century, researchers and industries had developed various cultivation technologies that are suitable for microalgal biomass production. Along with the conventional fermentation reactors, there are two frequently used techniques, the production in open ponds and in closed photobioreactors (PBRs) (Borowitzka, 1999). It is thus crucial to understand the diverse parameters of microalgal cultivation to optimise yields and production costs (Harun *et al.*, 2010). Growth rates significantly fluctuate between species and greatly depend on cultivation methods.

The biodiversity of microalgae offers a wide potential of applications as a food or feed, source of biomaterials and as an original source of biotechnologically produced molecules (Cadoret, Garnier and Saint-Jean, 2012; Jeffryes, Agathos and Rorrer, 2015; Murray *et al.*, 2013). This diverse phylogeny is also reflected in a large biochemical diversity of pigments such as carotenoids, photosynthetic storage products, cell wall and mucilage materials, fatty acids and lipids, oils and hydrocarbons, sterols and bioactive secondary metabolite compounds (Metting, 1996).

Currently, the principal industrial products from green microalgae are carotenoids and biomass for food, health and aquaculture (Skjånes, Rebours and Lindblad, 2013), with production from a limited number of species. Cyanobacteria from the genus *Arthrospira* constitute 50% of the global biomass production, followed by green microalgae from the genera *Chlorella*, *Dunaliella*,

Industrial Biotechnology of Vitamins, Biopigments, and Antioxidants, First Edition.
Edited by Erick J. Vandamme and José L. Revuelta.

Haematococcus, *Nannochloropsis* and the diatom *Odontella* (Person, 2011). Microalgae are usually selected on the basis of growth rate and their capacity to produce considerable amounts of specific metabolites. The efficiency of biomass production is a pivotal element to financial success in most of today's commercial systems (Skjånes, Rebours and Lindblad, 2013).

10.2 Occurrence in Nature and Food Sources

More than 750 structurally defined carotenoids are found in nature within land plants, algae, cyanobacteria, photosynthetic bacteria, archaea, fungi and animals (Britton, 1998). Their characteristic yellow, orange and red colours are related to the presence of a number of conjugated double bonds in a polyene chain that operates as a chromophore. The hundreds of carotenoid structures known today can be partitioned into two dominant groups: carotenes (non-oxygenated molecules) and xanthophylls (oxygenated carotenoids) (Rodriguez-Concepcion and Stange, 2013).

Very often, a distinction is made between primary and secondary carotenoids: primary carotenoids are structural and functional components of the photosynthetic apparatus and are crucial for survival. Secondary carotenoids are also produced by microalgae at significant levels, but only after exposure to particular environmental stimuli (Jin *et al.*, 2003). Unlike primary carotenoids, which act directly in photosynthesis, secondary carotenoids are synthesised in cells as a protective response to diverse environmental factors such as exposure to high-intensity light, nutrient deficiency, temperature changes, high or low pH, high salinity and oxidative stress (Solovchenko, 2013). Among primary carotenoids are α-carotene, β-carotene, lutein, violaxanthin, zeaxanthin and neoxanthin, while secondary carotenoids include bioactives such as astaxanthin, canthaxanthin and echinenone (Leya *et al.*, 2009).

Carotenoids are hydrophobic and are therefore enriched in membranes and other lipophilic cell structures (Stahl and Sies, 2005). The xanthophylls are generally localised in the thylakoid membrane and can also be included in non-covalent bonds to particular proteins. Secondary carotenoids are primarily located in lipid vesicles in the plastid stroma or cytosol (Guedes, Amaro and Malcata, 2011).

Chlorophylls, carotenoids and phycobilins constitute the most common pigments produced by photosynthetic microalgae. Chlorophyll is present in all algae and land plants, including the prokaryotic cyanobacteria and prochlorophytes. Other chlorophylls (*b*, *c*1, *c*2, *d*) are accessory light-harvesting molecules whose distribution among algal groups is used partly for taxonomic purposes at the division level (Metting, 1996).

Carotenes and xanthophylls are usually synthesised and stored within plastids. However, in some green microalgae – for example, astaxanthin in *Haematococcus* sp., they accumulate in the cytoplasm. This observation raises the possibility of an extra-plastidic site of carotenoid biosynthesis. Xanthophylls synthesised in the chloroplast may also be exported and, as a result, accumulate in the cytoplasm (Rabbani *et al.*, 1998; Tardy and Havaux, 1996).

Many carotenes and xanthophylls are usually confined to one or a few algal groups. For instance, lutein is found in chlorophytes and land plants, myxoxanthin and myxoxanthophyll are characteristic of cyanobacteria, as are peridinin for dinoflagellates and fucoxanthin for brown algae and diatoms (Metting, 1996). However, the carotenoid composition of cyanobacteria is extremely dissimilar from those of chloroplasts in algae (Takaichi and Mochimaru, 2007).

10.3 Physiological Role as a Vitamin or as a Coenzyme

Reactive oxygen species (ROS), such as O_2^-, H_2O_2 or singlet-state oxygen, which are harmful to pigments, proteins and lipids, are produced in algal cells in stressful conditions but are found in most living cells, including human cells. Carotenoids can provide protection against ROS, either by preventing their formation or by acting as an antioxidant that inactivates them. The quenching effect of carotenoids is due to their polyene structure of conjugated double bonds. It also implies that the photoprotective properties of carotenoids strongly depend on their chemical characteristics (Domonkos *et al.*, 2013).

Secondary carotenoids are distinct from primary carotenoids. They are not structurally bound to the photosynthetic apparatus, and their functions are still under debate, although they are known to be involved in the screening of excessive photosynthetically active radiation (PAR), the inhibition of ROS generation and the remediation of already produced ROS (Solovchenko, 2012). These compounds have been shown to accumulate in high amounts in several algal species, usually under unfavourable conditions (Solovchenko, 2012).

In humans, carotenoids have been shown to directly provide photoprotection against UV light in the skin, effectively diminishing UV-induced erythema (Sies and Stahl, 2004; Stahl and Sies, 2007) and inhibit photooxidative damage in the eye (Beatty *et al.*, 2000). Lutein and zeaxanthin, carotenoids that accumulate in the human retina (Fernandez-Sevilla, Fernandez and Grima, 2010), have been associated with a reduced risk of developing age-related macular degeneration and age-related cataracts (Guerin, Huntley and Olaizola, 2003). Carotenoids are a key factor in the prevention of many diseases, especially those associated with light, due to their antioxidant properties (Cardozo *et al.*, 2007; Sies and Stahl, 2004). Furthermore, carotenoids have been associated with inhibition of angiogenesis and exhibit anti-tumour-promoting and anti-carcinogenic activities (Ganesan *et al.*, 2013; Maoka *et al.*, 2013). A carotenoid-rich diet could be helpful in the prevention of many cancers and chronic illnesses (Berman *et al.*, 2015).

Traditionally, nutritional supplements derived from terrestrial plants have been most prevalent in the marketplace. Nonetheless, the health benefits of aquatic microorganisms such as algae are being explored and more acknowledged and appreciated. This is particularly the case within the last three to four decades with the introduction of probiotic supplements. In addition, the U.S.FDA (Food and Drug Administration) for marketing has cleared *Haematococcus pluvialis* as a dietary supplement, and it has also been approved in different European countries for human consumption (Mata, Martins and Caetano, 2010).

10.4 Chemical and Physical Properties; Technical Functions

Carotenoids form a class of terpenoid pigments, derived from a 40-carbon polyene chain, which serves as their molecular backbone. The polyene system gives carotenoids both their special molecular structures and the related chemical properties involving light-absorption characteristics (Del Campo, Garcia-Gonzalez and Guerrero, 2007). The majority of the carotenoids consist of a central carbon chain of alternating single and double bonds with various cyclic or acyclic end groups (Stahl and Sies, 2005). Carotenoids are moreover depicted as being acyclic, monocyclic or bicyclic, depending on the structure of the hydrocarbon backbone's end groups (Armstrong, 1997). Disparate carotenoids derive primarily from alteration in the base structure by cyclisation of the end groups and by introduction of oxygen functions, giving them their characteristic colours and antioxidant properties (Rao and Rao, 2007), while the polyene chains, made up of conjugated double bonds, are responsible for the pigmentation of carotenoids because of their absorbance in the visible-light spectrum (Jin *et al.*, 2003). The term *carotenoid* encloses both hydrocarbon carotenes and xanthophylls: carotene derivatives that contain one or more oxygen atoms assimilated into hydroxy-, methoxy-, oxo-, epoxy-, carboxy-aldehydic and glycosidic functional groups (Armstrong, 1997). Carotenes are non-polar molecules, and xanthophylls are more polar (Arnold, Schwarzenbolz and Böhm, 2014; Skjånes, Rebours and Lindblad, 2013). β-Carotene, α-carotene and lycopene are distinguished members of the carotene group which involves carotenoids consisting of only carbon and hydrogen atoms, whereas xanthophylls carry at least one oxygen atom. Zeaxanthin, lutein, α- and β-cryptoxanthin, canthaxanthin and astaxanthin are essential xanthophylls with hydroxy and keto groups as structural elements (Stahl and Sies, 2005).

All xanthophylls produced by higher plants, for instance, violaxanthin, antheraxanthin, zeaxanthin, neoxanthin and lutein, are also synthesised by green algae (Jin *et al.*, 2003). Nevertheless, they have supplementary xanthophylls, for example, loroxanthin, astaxanthin and canthaxanthin. Diatoxanthin, diadinoxanthin and fucoxanthin can also be produced by brown algae or diatoms (Jin *et al.*, 2003).

As hydrophobic antioxidants, the main mechanisms of carotenoid activity are located within biological membranes. Carotenoids are rarely found in microalgae as free monomeric molecules. Very often, they are combined with proteins or lipoprotein structures (Jomova and Valko, 2013). For photosynthesis, both carotenoids and chlorophylls are essentially connected to peptides in order to produce pigment–protein complexes in the thylakoid membrane (Takaichi, 2011). Usually, xanthophylls such as zeaxanthin and astaxanthin span the membrane and strengthen the membrane, but diminish oxygen penetration into the membranes. The interaction of carotenoids with the membrane lipids affects the membrane fluidity and thermostability in addition to affecting the signal transduction pathways (Jomova and Valko, 2013).

Carotenoids are fundamental for the assembly and preservation of photosystem II (PSII) and may contribute to the electron transfer reactions in this system. Contrarily, the assembly of the reaction centre in photosystem I (PSI) does not necessitate the presence of carotenoids (Jomova and Valko, 2013). β-Carotene is present in the reaction-centre complexes (RC) and the light-harvesting complexes (LHCs) of (PSI) as well as the RC and the core LHC of PSII; exceptionally, zeaxanthin is present in some red algae of the LHC of PSI and on the other hand, in the peripheral LHC of PSII (Takaichi, 2011). In their supplementary light-harvesting function, carotenoids augment the cross section for the absorption of radiant energy, which is eventually transferred via chlorophyll molecules to the photosynthetic reaction centre, the site of primary charge separation (Armstrong, 1997). The ratio of chlorophylls to carotenoids is a major factor in preserving the integrity of the photosynthesis system. As a result, a metabolic equilibrium between biosynthesis and catabolism of carotenoids is important to preserve carotenoids at physiological levels, particularly in photosynthetic tissues (Beisel *et al.*, 2010).

It is commonly thought that the secondary carotenoids operate as passive photoprotectants that diminish the quantity of light which can reach the light-harvesting pigment complexes of PSII (Hagen, Braune and Bjorn, 1994). However, the ability of secondary carotenoids to protect against excessive PAR is limited to wavelengths between 390 and 600 nm. As a consequence, most carotenoids cannot function as UV-screening compounds (Leya *et al.*, 2009). The antioxidant behaviour of a carotenoid molecule (whether mediated by direct or indirect means) depends not only on its structure but also on the nature of the oxidising species(Young and Lowe, 2001). Commonly, there are three major features determining the rate and type of mechanism for the reactions of carotenoids with free radicals: structure and redox potential of a carotenoid and the polarity of the medium (Jomova and Valko, 2013). This protective effect against free radicals is corroborated by the coordinated actions of both hydrophobic and hydrophilic antioxidants (Jomova and Valko, 2013). Among the different defence strategies, carotenoids are most likely included in the scavenging of two of the ROS; singlet molecular oxygen (O_2^*) and peroxyl radicals (Stahl and Sies, 2005). Singlet oxygen quenching by carotenoids occurs via physical or chemical quenching; physical quenching includes the transmission of excitation energy from O_2^* to the carotenoid, resulting in ground-state oxygen and an excited triplet-state carotenoid. The energy is dispersed between the excited carotenoid and the encircling solvent to yield the ground-state carotenoid and thermal energy. In the process of physical quenching, the carotenoid is kept in a perfect state and can go through further cycles of singlet oxygen quenching (Stahl and Sies, 2005).

On the other hand, other organisms that started developing antioxidant defence systems to protect against O_2 toxicity existed. It is considered as a more fruitful path in retrospect, because organisms that tolerated O_2 could also evolve to use it for metabolic transformations (e.g. oxidase and hydroxylase enzymes, such as cytochromes P450) and for efficient energy production using electron transport chains with O_2 as the terminal electron acceptor (Halliwell, 1996).

10.5 Assay Methods and Units

Simple, cheap and reliable assay methods are critical in research and industry where the determination of the quantity and quality of pigments as well as other algal components is of the utmost importance for the design and optimisation of bioprocesses. The main issue to overcome while trying to measure the algal carotenoid content is an effective method of lysing algal cell walls in order to extract the intracellular components. While some methods are non-destructive and allow the monitoring of pigments by reading the optical density of intact, living whole cells, they only allow the quantification of the total chlorophyll/carotenoid content and not the identification and quantification of the different compounds that are present (Solovchenko *et al.*, 2013). Flow cytometry methods could be used for counting, examine the properties and accordingly sort microalgae for the high-throughput screening of microalgae with bioindustrial interest. However, for many applications, the cells have to be permeabilised and stained, which is time-consuming (Hyka *et al.*, 2012). Voltammetric methods can also be used to rapidly measure the antioxidant power of algal biomass (Goiris *et al.*, 2012). These methods are easy and fast to carry out, but they also usually require a considerable amount of calibration. More precise assays must rely on a mechanical disruption of the cell wall and extraction of the compounds of interest with organic solvents. This disruption can be achieved by various means such as sonication, bead-beating or freeze-drying and then grinding of the sample.

The extracted carotenoids can be readily quantified by RP-HPLC analysis, which provides excellent separation efficiency (Hu *et al.*, 2008), and the peaks can be identified using standards or by comparing the UV–visible spectrum with those available in the literature (Plaza *et al.*, 2012).

10.6 Biotechnological Synthesis

10.6.1 Producing Organisms

The commercial production of high-value carotenoids, such as β-carotene and astaxanthin, has historically been by chemical synthesis because production from natural sources has been cost-prohibitive (Milledge, 2011; Olaizola, 2003). However, recent economic analyses have shown that the biotechnological synthesis of astaxanthin in raceway cultivation systems has the potential to be more economical than synthetic sources (Li *et al.*, 2011). Indeed, the biotechnological synthesis of β-carotene from *Dunaliella salina* (Borowitzka and Borowitzka, 1989) and astaxanthin from *H. pluvialis* (Lorenz and Cysewski, 2000) has already achieved some commercial success, despite being cost-prohibitive, because of consumer desire for naturally sourced antioxidants.

The principal organism for β-carotene production is *D. salina*, sometimes known as *Dunaliella bardawil*. This species has been cultivated for the industrial production of β-carotene since the 1980s in Israel and Australia (Ben Amotz and Avron, 1989; Borowitzka and Borowitzka, 1989) using a two-stage production technique. Currently, the largest producer of *D. salina* is BASF, mainly through its installations in Western Australia that consist of natural salt lakes with an area of 400 ha. *Dunaliella* powder is sold as a dietary supplement both for human nutrition and for animal feed. Extracts from the microalgae, a mix of several carotenoids, are manufactured under the product name Betatene® (Rosello Sastre, 2012).

Industrial-scale carotenoid production is carried out in two stages. The first stage is for biomass production followed by a phase which imposes high salinity under an increased mean cell-culture light intensity, which initiates the production of β-carotene. Using this method, the production in pond cultures has been reported at 450 mg β-carotene m^{-2}/day in pond volumes up to 4000 m^3 (Ben-Amotz, 1995; Ben Amotz and Avron, 1989). The productivity of *D. salina* biomass can be increased to 2 g DW m^{-2}/day (DW, dry weight) aerial productivity, equivalent to 80 g DW m^{-3}/day on a basis of reactor volume, in a closed 55 l tubular PBR with a tube diameter of 2.4 cm and a 90 m long solar receiver. The β-carotene content of the biomass in this process was up to 10 wt% (Garcia-Gonzalez *et al.*, 2005). A method for the continuous production of β-carotene from *D. salina* has also been demonstrated in a lab-scale helical PBR with production rates up to 26 mg/l/h (Zhu and Jiang, 2008).

The production of astaxanthin has been achieved in numerous organisms, such as *Chlorella zofingiensis*, *Chlorococcum* sp., *Neochloris wimmeri* and *Scenedesmus obliquus*, although the highest biomass concentrations are achieved with *H. pluvialis*. Astaxanthin produced by microorganisms, in contrast to chemosynthetic routes, possesses high enantiomeric purity. Specifically, astaxanthin from *H. pluvialis* is exclusively in the (3*S*, 3′*S*) form, while the yeast *Xanthophyllomyces dendrorhous* synthesises the (3*R*, 3′*R*) form, and chemosynthetic astaxanthin is a stereoisomeric mixture of (3*S*, 3′*S*), (3*R*, 3′*S*) and (3*R*, 3′*R*) forms in the ratio 1 : 2 : 1 (Grewe, Menge and Griehl, 2007). Indeed, increased bioavailability, as well as increased health benefits of astaxanthin from (3*S*, 3′*S*) have been reviewed elsewhere (Fassett and Coombes, 2011).

As with β-carotene, the production of astaxanthin follows the two-stage process of cellular growth to increase biomass, followed by the induction of carotenogenesis by a stress phase, with nitrogen starvation and high light intensities being the most common, which increases the intracellular content of astaxanthin. During the growth phase, the highest cell densities have been reported for *H. pluvialis* in a growth medium that allowed the luxury uptake of phosphorous (Tocquin, Fratamico and Franck, 2012). The carotenogenesis phase has been induced in both pond and large PBR (25 m^3) cultures, achieving between 2.5 and 4 wt% astaxanthin in the biomass (Aflalo *et al.*, 2007; Olaizola, 2000). However, while accumulating the highest intracellular content of astaxanthin, *H. pluvialis* grows slowly ($\mu = 0.02\,h^{-1}$) (Tocquin, Fratamico and Franck, 2012), so it is biotechnologically interesting to examine production using faster growing strains, such as those from the *Chlorella* genus.

C. zofingiensis is a known producer of astaxanthin with observed growth rates up to 0.04 h^{-1} and can achieve cell densities up to 7 g/l while attaining up to 0.15 wt% astaxanthin and 0.4 wt% lutein, which was induced by the addition of NaCl (50–400 mM) and an incident light intensity of 460 $\mu E/m^2/s$ (Del Campo *et al.*, 2004). Another study using *C. zofingiensis* stressed with 300 mM NaCl and nitrate deficiency under illumination intensities between 50 and 300 $\mu E/m^2/s$ light accumulated up to 1.2 wt% total carotenoids consisting primarily of astaxanthin and canthaxanthin (Pelah, Sintov and Cohen, 2004). In addition to stress induced by nitrate starvation, NaCl and high light, it has also been shown that the addition of reactive oxygen species, peroxynitrite or nitryl chloride can induce astaxanthin production in *C. zofingiensis* (Ip and Chen, 2005a,b). High cell densities of *C. zofingiensis* have been achieved in the dark using mannose and glucose as carbon substrates with cell densities up to 10 g/l and astaxanthin contents of 0.1 wt% (Sun *et al.*, 2008).

The production of astaxanthin from *Chlorococcum* sp. in a tubular PBR has also been investigated with growth rates up to 0.066 h^{-1} and secondary carotenoid accumulations of 0.5 wt%, with astaxanthin as the primary pigment (Zhang *et al.*, 1997). The optimal production conditions for cellular growth were a pH of 8.0 and a temperature of 30 °C, while astaxanthin production was optimal at a temperature of 35 °C and a pH of 6.0. Significant production of astaxanthin at the PBR scale has also been observed in *N. wimmeri* (Orosa *et al.*, 2000), *S. obliquus* (Qin, Liu and Hu, 2008) and *Dactylococcus dissociatus* (Grama *et al.*, 2014b).

The production of lutein has been primarily investigated in the *Scenedesmus, Chlorella* and *Muriellopsis* genera. Currently, the largest PBR production facility for lutein uses *Scenedesmus almeriensis* in a 28 m^3 facility of tubular airlift reactors (Fernandez-Sevilla, Fernandez and Grima, 2010; Sanchez *et al.*, 2008a,b). This thermophilic strain grows optimally at 44 °C with a corresponding productivity of 3.8 g lutein m^{-3}/day. Among the *Chlorellas*, lutein contents up to 0.25 wt% lutein are observed in *Chlorella vulgaris* under nitrate and salinity stress, while also producing up to 0.15 wt% of other carotenoids, such as astaxanthin and β-carotene (Gouveia *et al.*, 1996). *C. sorokiniana* can produce up to 0.43 wt% lutein in flat-panel PBRs at 35 °C and 10% CO_2 (Matsukawa *et al.*, 2000), while *C. protothecoides* can produce up to 49 mg lutein l^{-1}/day under heterotrophic growth using glucose and urea as organic carbon and nitrogen substrates (Shi, Jiang and Chen, 2002).

The use of light and salinity stress has also been used to produce lutein from cultures of *Muriellopsis* sp., which achieved growth rates of up to 0.17–0.23 h^{-1} and lutein concentrations of up to 35 mg/l in outdoor tubular PBR cultivations, equivalent to an areal productivity of 180 mg lutein m^{-2}/day (Del Campo *et al.*, 2000, 2001). Another study examined the production of lutein from *Muriellopsis* sp. in open paddle-wheel agitated tanks (3 m^2 illuminated surface, 30 cm culture depth) under a salinity stress of 150 mM NaCl. Accumulations of up to 0.4–0.6 wt% free lutein in the biomass were achieved for a semi-continuous production rate of 100 mg lutein m^{-2}/day (Blanco *et al.*, 2007).

Lutein production has also been examined in *Chlamydomonas acidophila*, which autotrophically produced 57.5 ± 1.6 mg/l total carotenoids, of which 20 ± 0.5 mg/l was lutein and 8.3 ± 0.2 mg/l was β-carotene (Garbayo *et al.*, 2008). In this process, high levels of iron (5–35 mM) were found to inhibit carotenogenesis, while the addition of 1–5 mM of Cu^{2+} shifted the pigment composition towards β-carotene. However, nutrient stress conditions in *Desmodesmus* sp. F51 led to a shift in carotenoid content from β-carotene to lutein, up to a production level of 3.91 mg lutein l^{-1}/day (Xie *et al.*, 2014).

The production of canthaxanthin is also achieved in a two-stage process. Under nitrogen starvation and 65 μE/m^2/s, *Coelastrella striolata* produced canthaxanthin up to 4.8 wt% (Abe, Hattori and Hirano, 2007). Under nitrogen and salinity stress, canthaxanthin was also the primary carotenoid produced by *Chlorella protothecoides* (Campenni *et al.*, 2013), *Scenedesmus komarekii* (Hanagata and Dubinsky, 1999) and *D. dissociatus* (Grama *et al.*, 2014a).

10.6.2
Biosynthesis and Metabolic Regulation

Carotenogenesis in algae is induced by different stressors and specifically their combinations (Lemoine and Schoefs, 2010). For example, induction can be by osmotic stress or nitrogen and phosphorus insufficiency, together with high light intensities (Solovchenko, 2012).

Carotenogenesis can be induced by either light quality (wavelength, frequency) or intensity. Specifically, blue light induces greater astaxanthin production in *H. pluvialis* than red light, while red light is more effective at increasing biomass production (Steinbrenner and Linden, 2003). Production levels are also known to increase with increasing light intensities at a wavelength of 470 nm (Park and Lee, 2001), and high biomass concentrations of astaxanthin (5–6%) have been observed when using LED lighting with wavelengths as short as 380 nm (Katsuda *et al.*, 2004). Astaxanthin production efficiency has also been increased by using flashing LED lighting, rather than a continuous photonic input (Kim *et al.*, 2006; Katsuda *et al.*, 2008). Thus, a strategy of using high intensity, pulsing, LED lighting could be an appropriate astaxanthin production strategy if red wavelengths are used during the growth phase and blue wavelengths are used to induce carotenogenesis.

In conventional high light conditions, such as solar illumination, the photonic input is greater than the maximum photochemical reaction rate and the excess energy cannot be used by the photosynthetic apparatus (Bar *et al.*, 1995; Solovchenko, 2010). This results in an imbalance between reactants and products in the redox reactions and the production of excited intermediates. As a result, this superfluous energy contributes to the production of highly active oxygen molecules (Bar *et al.*, 1995; Niyogi, 1999). ROS are highly toxic and result from the partial reduction of harmless molecular oxygen. The ROS are superoxide radicals (O_2^-), the hydroxyl radical (OH^-) and hydrogen peroxide (H_2O_2) (Mallick and Mohn, 2000). The mixed biological effect of these toxic oxygen species on organisms is labelled 'oxidative stress' (Mallick and Mohn, 2000). The increased levels of ROS produce oxidative damage to proteins, nucleic acids and lipids and eventually result in the damage of various cellular organelles. Algal chloroplasts are formed by an intricate system of membranes rich in polyunsaturated fatty acids, which are potential targets for peroxidation (Halliwell and Gutteridge, 1999). These potentially damaging molecules are produced at three main sites in the photosynthetic apparatus: the PSII LHC, the PSII reaction centre and the PSI acceptor side (Niyogi, 1999).

Carotenoids can be synthesised *in vivo* through two diverging pathways: (i) mevalonic acid (MVA) pathway in the cytoplasm and (ii) non-mevalonic acid pathway (2-*C*-methyl-D-erythritol 4-phosphate pathway: MEP) in plastids that provides the precursors for carotenoids (Bolhassani, Khavari and Bathaie, 2014). The pathways of secondary carotenoids biosynthesis, the factors of their induction, in addition to the mechanisms of their synthesis regulation are less studied compared to those of primary carotenoids (Cunningham and Gantt, 1998). However, it is known that algae have common carotenogenic pathways with land plants in addition to supplementary algae-specific pathways and algae-specific enzymes (Takaichi, 2011).

Along with the secondary carotenoids production, to alleviate and restore the damage caused by ROS and to protect cellular membranes and organelles from the damaging effects of ROS, intricate antioxidant systems are activated. The well-developed cellular antioxidant defence system incorporates enzymatic antioxidants and reduced non-enzymatic components (Mallick and Mohn, 2000). Biotic and environmental stresses result in upsurges in cellular oxidants that induce increased synthesis of non-enzymic antioxidants such as the tripeptide thiol, glutathione and vitamins C (ascorbate) and E (tocopherol), in addition to upsurges in antioxidant enzymes such as superoxide dismutases (SODs), glutathione peroxidase (GPX), glutathione reductase (GR), ascorbate peroxidase (APX) and catalases (CAT) (Foyer *et al.*, 1997). SODs are the first line of defence against oxyradical-mediated injury (Alscher, Donahue and Cramer, 1997). In enzymatic antioxidants, SOD conversion of superoxide anion (O_2^-) to hydrogen peroxide (H_2O_2) and CAT Catalase involvement in direct removal of H_2O_2 are extremely important. Ascorbic acid (AsA) and reduced glutathione (GSH) are crucial non-enzymatic antioxidants for H_2O_2 removal in the ascorbate–glutathione cycle (AGC). In the AGC, GSH is frequently used as an electron donor to be oxidised into glutathione disulfide (i.e. oxidised glutathione, GSSG) to diminish dehydroascorbate (DHAsA) in order to produce AsA. The balance of reduced and oxidised non-enzymatic components is precisely regulated in normal cells (Hong *et al.*, 2008). It is reported by Park *et al.* (2008) that *H. pluvialis* exposed to excess irradiation experiences a 70% reduction of catalase activity, while GPX activity was somewhat improved. Total activity of SOD and APX also slightly diminished. H_2O_2 content increased about sixfold after high light exposure, exhibiting harsh cellular oxidative stress, while lipid peroxidation was remarkably reduced.

During the carotenogenesis process, it is believed that the formation of hydrophobic structures which are capable of secondary carotenoid accumulation and sequestration moves the chemical equilibrium towards secondary carotenoid biosynthesis (Solovchenko, 2013). Under ordinary conditions, the carotenogenic pathway is not maximally active, but may be up-regulated in the presence of the sequestering structures, thereby creating an extra-plastid localised sink for the carotenoid end products. (Rabbani *et al.*, 1998).

The formation of lipid structures serving as depots for secondary carotenoids is possible in the absence of secondary carotenoid biosynthesis. However, the induction of carotenogenesis is impossible in the absence of such structures. Thus, the chief factor of carotenogenesis in *D. bardawil* is the synthesis of triacylglycerls (TAG) (Solovchenko, 2013). Zhekisheva *et al.* (2005) used two carotenogenesis inhibitors (norflurazon, an inhibitor of phytoene desaturase, and diphenylamine (DPA), an inhibitor of β-carotene C-4 oxygenase) in cultures of *H. pluvialis* cells exposed to high light and determined that while astaxanthin accumulation was significantly inhibited, fatty acid synthesis was not proportionally decreased. The authors report that neutral lipids and specifically TAG were still produced, arguing that their synthesis was not dependent on carotenoids synthesis. In contrast to this work, (Rabbani *et al.*, 1998) used the herbicide sethoxydim for inhibition of fatty acid biosynthesis during carotenogenesis and determined that the formation of these fatty-acid-based sequestering structures and β-carotene accumulation are interdependent. When the synthesis of triacylglycerol is blocked, the overproduction of β-carotene is also inhibited. Also, the enzymes responsible for astaxanthin synthesis could also be under feedback control when fatty acid biosynthesis is inhibited by cerulenin. Astaxanthin esterification, the key point for flux control, needs to be further characterised at the gene, enzyme and subcellular levels (Han, Li and Hu, 2013). Moreover, the inhibition of fatty acid synthesis by cerulenin, an inhibitor of 3-ketoacyl-acyl carrier protein synthase, simultaneously inhibited astaxanthin accumulation in *H. pluvialis* under high light (Schoefs *et al.*, 2001).

It has been proposed that interactions between structurally diverse compounds with variable antioxidant activity enable enhanced resistance to oxidative stress. A mixture of both lipophilic antioxidants culminated in an inhibition of lipid peroxidation more than the total of the individual effects (Stahl and Sies, 2005). Mixtures had a stronger effect than single compounds, and the synergistic effect was most pronounced when lycopene or lutein was present. The higher protection of mixtures may be connected to the particular location of various carotenoids in membranes (Stahl and Sies, 2005).

10.6.3
Strain Development: Genetic Modification, Molecular Genetics and Metabolic Engineering

Algal molecular biology is a field that has gained a lot of interest these past few years. To date, 18 algal genomes have been sequenced, much fewer than the almost 3000 closed bacterial genomes and the 17 615 sequencing projects including permanent drafts. Nevertheless, more and more algal genomes are being sequenced, and our knowledge base is slowly building up. *Chlamydomonas reinhardtii* is currently the model organism for the genetic engineering of algae, with most molecular biology methods and tools working specifically with this organism. Currently, transformation methods that give rise to stable expression of transgenes are available for more than 30 strains of algae (Radakovits *et al.*, 2010). These methods include electroporation, biolistic transformation, agitation with glass beads, agitation with silicon carbon whiskers and *Agrobacterium tumefaciens*–mediated genetic transformation (Qin, Lin and Jiang, 2012). To date, most of the metabolic engineering has been carried out for the overproduction of compounds aimed for the biofuel industry, mainly biohydrogen. Molecular strategies for the optimisation of carotenoid production have been proposed but with the aim of microbial production in hosts such as *Escherichia coli* or *Saccharomyces cerevisiae* (Heider *et al.*, 2014). The development of a molecular toolkit that works efficiently with various algal species could enable the development of strategies for the enhancement of carotenoid production in algal hosts.

Algae are now also a new potential platform for industrial recombinant protein expression. Previously, the poor expression of heterologous genes from the nuclear genome was a major obstacle, but it was discovered that linking the foot-and-mouth-disease-virus 2A self-cleavage peptide to the heterologous gene (here, green fluorescent protein or the xylanase enzyme) and an appropriate selectable marker (here, bleomycin resistance gene) resulted in increased expression (about 100-fold) (Rasala *et al.*, 2012). Such a technology reduces the number of steps and selectable markers involved in the integration of multiple biosynthetic pathways, which should greatly facilitate metabolic engineering in the future (Gimpel *et al.*, 2013).

10.6.4
Downstream Processing, Purification and Formulation

The downstream processing of microalgal bioproducts is analogous to downstream bioprocess engineering of other microbial products. Carotenoids are intracellular products, so the first step towards the final product is the separation of the biomass from the liquid medium. This is usually done by first pre-concentrating the cell mass and then thickening the dewatered cell culture to a paste. The following step is to dry the biomass. The dried biomass can be the final product, but if purified carotenoids are desired, the cells must be disrupted and the carotenoids extracted from the biomass. The crude carotenoid extract can then be further purified before being formulated or stabilised, before going to market. The bioprocess steps are outlined in detail as follows.

The first step in downstream processing is to increase the solids (biomass) to liquid ratio to at least 20–70 g/l. The least expensive method to achieve this is by sedimentation in place, removal of the clarified supernatant and collection of the biomass. While simple and inexpensive, this method is also slow (Li *et al.*, 2011). The speed of sedimentation and the efficacy of separation can be increased by flocculation, which is the formation of cell aggregates. Negative ions on the cell surface normally inhibit cell aggregation because of like-charge repulsion, so cations such as $Al_2(SO_4)_3$, $FeCl_3$, $Fe_2(SO_4)$, calcium and magnesium phosphates or magnesium hydroxide can be added to the culture, which neutralise the charges on the cell surface. As a result, the van der Waals forces between the cells dominate, leading to aggregation. However, since metal salts can contaminate biomass products, cationic polymers such as chitosan or cationic starch which physically link cells together could also be used (Vandamme, Foubert and Muylaert, 2013). Another method to induce flocculation is to increase the pH by removal of the dissolved CO_2 in the cell culture by the action of the photosynthetically active organisms. This is achieved simply by halting the mixing or stopping the sparge gas. The increased pH reduces the charge repulsion between cells and permits flocculation (Christenson and Sims, 2011). It has also recently been shown that the application of low-energy electric potential (4 V) can alter cell surface charges and facilitate electrocoagulation flocculation (Shuman *et al.*, 2014).

Cell culture pre-concentration can also be achieved by floatation, which is most efficient on hydrophobic cells, and can be applied by itself or as a secondary dewatering step to compliment flocculation (Garg, Wang and Schenk, 2014). Cell floatation is achieved when the culture is sparged with micro-air bubbles from specially designed spargers or air–liquid agitators. The cells are trapped in the bubbles and float to the liquid surface, which creates a foam or froth enriched in cell mass (Sharma *et al.*, 2013). Solid–liquid separation is then achieved by skimming the foam or froth from the top of the culture.

Thickening is a secondary dewatering step which normally concentrates the dewatered culture to a paste containing up to 75% solid content. The most common forms of thickening are centrifugation and filtration. Centrifugal equipment, such as disc-stack, tubular bowl or spiral plate centrifuges, are rapid but energy-intensive (Harun *et al.*, 2010). Additionally, there is the risk of cell disruption, difficulties in scale-up and maintenance costs are often prohibitive (Harun *et al.*, 2010; Molina-Grima *et al.*, 2003).

Culture thickening can also be achieved by methods of filtration, such as dead end, micro-, ultra-, vacuum, rotary drum and tangential flow filtration, with the filter press being the most cost-effective and scalable filtration method for larger cells (Harun *et al.*, 2010; Sharma *et al.*, 2013).

The drying step removes moisture from the algal paste to attain a concentration of 90–95% dry matter. This process is energy-intensive and therefore costly. During high-temperature drying, there is also the potential for product degradation, which will reduce the final yield. Solar drying is the most economical, but is space-intensive (Ryckebosch *et al.*, 2011). Other drying methods include spray-drying and freeze-drying, which are excellent for product preservation, but also more costly (Ryckebosch *et al.*, 2011).

To recover intracellular products, such as carotenoids, the cell walls must be weakened in order to release the intracellular contents for downstream processing. The cell disruption process is more effective, particularly for algae with thick cell walls, on cells that have been pre-dried (Mäki-Arvela, Hachemi and Murzin, 2014). The disruption process can be carried out by mechanical or chemical means. Acid treatment with HCl at 70 °C has been used in several studies. After exposure to 4 N HCl for 5–10 min, astaxanthin could be acetone-extracted from encysted cells of *H. pluvialis* (Sarada *et al.*, 2006). Alkaline treatments have also been used and were found to be effective with the added advantage of freeing esterified carotenoids from their lipid moieties. Typically, a 4% (w/v) solution of KOH is added to the biomass in the dark under argon and agitation. Following this treatment, the chlorophylls and ionised lipids will remain solubilised in the aqueous phase while the carotenoids can be extracted by hexane. (Fernandez-Sevilla, Fernandez and Grima, 2010).

The most effective and scalable method of mechanical cell disruption at the process scale has been bead milling (Balasundaram, Skill and Llewellyn, 2012; Fernandez-Sevilla, Fernandez and Grima, 2010), or steam explosion (Nurra *et al.*, 2014), although other methods such as ultrasound or repetitive cycles of freezing and thawing have also been employed. However, steam explosion is prone to degrade thermosensitive products, such as carotenoids. The efficacy of bead milling can be increased by adding disintegrating agents, such as alumina, into the mill with the dried cell mass (Balasundaram, Skill and Llewellyn, 2012). Cell cracking methods, such as the use of a BeadBeater with a 0.02 bead to DW ratio, are particularly effective for disrupting cells with extremely resistant cell walls (Chan *et al.*, 2013).

Following cell disruption, carotenoids and bioactive compounds can be separated from the spent biomass by extraction. Extraction is most effective after cell disruption but can be possible with intact cells if they have no cell walls, such as with *D. salina*. Extraction is carried out primarily with organic solvents such as acetone (Fiedler *et al.*, 2007; Sarada *et al.*, 2006), diethylether (Chan *et al.*, 2013), hexane (Fernandez-Sevilla, Fernandez and Grima, 2010), ethyl acetate, diethyl-propylene glycol (Fiedler *et al.*, 2007), tetrahydrofuran (Fiedler *et al.*, 2007; Sarkar *et al.*, 2012), dichloromethane, methanol, ethanol, dimethylether, dimethylsulfoxide, toluene, 2-propanol, n-butanol, heptane or acetonitrile (Sarkar *et al.*, 2012). Carotenoid extraction has also been carried out using more environmentally friendly solvents, such as ethyl lactate (Ishida and Chapman, 2009) or safflower oil (Splinter, Pare and Kadali, 2013). One process employed corn and safflower oil which was homogenised directly with an algae paste for the extraction of β-carotene from *Dunaliella* and *Chlorococcus* (Nonomura, 1987). More recently, super critical extractions of β-carotene from *D. salina* have been carried out with supercritical solvents such as CO_2 (Mäki-Arvela, Hachemi and Murzin, 2014), ethane or ethylene (Talisic, Yumang and Salta, 2012).

Product polishing is the final step of purification and is intended to yield a product of at least 99% purity. The most commonly employed method of polishing is crystallisation by successive steps of crystallisation and crystal washing. Carotenoids have been proven to be difficult to crystallise and are in fact prone to oxidation in this form. However, precipitation and crystallisation processes have been developed for lutein, lycopene and β-carotene (Miguel *et al.*, 2008). The preferred final formulation for purified carotenoids is therefore solvation in an oil suspension, with safflower, corn or olive oil suspensions having been used to demonstrate good carotenoid stability (Fernandez-Sevilla, Fernandez and Grima, 2010).

10.7 Chemical Synthesis or Extraction

The chemical synthesis of carotenoids such as lycopene, canthaxanthin and astaxanthin are well established at the industrial scale, primarily by companies such as BASF and Hoffmann-La Roche (Ernst, 2002). Synthetic β-carotene also dominates with a market share of about 90%, mostly produced by the Dutch company DSM (Raja, Hemaiswarya and Rengasamy, 2007). The dominance of synthetic carotenoids is a result of high profitability, with an estimated production cost of \$1000 kg^{-1} and a market price above \$2000 kg^{-1} (Milledge, 2011; Olaizola, 2003).

The standard method for carotenoid synthesis is by the double Wittig olefination of a symmetrical 10 carbon dialdehyde with 2 equiv. of a C_{15}-phosphonium salt. The phosphonium salt carries the characteristic functional end groups of the carotenoid to be synthesised (Ernst, 2002). However, this industrial manufacturing process for carotenoids produces, without exception, symmetrical molecules. Thus, the synthesis of high-value, asymmetrical carotenoids such as adonixanthin is excluded. To synthesise the functional ends of the carotenoid molecule, 9-carbon building blocks, such as C_9-ketoisophorone for astaxanthin, are subject to a sequence of reactions, reviewed elsewhere, to produce the C_{15}-phosphonium salt of interest, with final carotenoid yields of up to 95% (Ernst, 2002).

10.8 Process Economics

The current global market for carotenoids was about \$1.2 billion in 2010, estimated to reach \$1.4 billion by 2018. The largest markets are for β-carotene (over \$270 million), lutein and astaxanthin (each valued at over \$200 million) (Borowitzka, 2013; Li *et al.*, 2011). An economic analysis based on pilot- and production-scale cultivations of *H. pluvialis* for the biotechnological synthesis of astaxanthin has estimated production costs as low as \$882 kg^{-1}. This analysis is comprehensive, including both fixed and variable costs, capital investment, labour, land, design, maintenance, management and all expenditures related to upstream and downstream processes as previously described in this chapter (Li *et al.*, 2011). This cost analysis indicates that it is possible for natural sources to be cost-competitive with synthetic carotenoids, which have an estimated production cost of \$1000 kg^{-1} (Milledge, 2011; Olaizola, 2003). The largest producers of natural astaxanthin are Cyanotech Inc., Mera Pharmaceuticals Inc., Biogenic Inc. and Algatechnologies Ltd. (Del Campo, Garcia-Gonzalez and Guerrero, 2007; Li *et al.*, 2011; Milledge, 2011). Recently, Atacama Bionatural Products started producing astaxanthin from *H. pluvialis* in the Atacama Desert of Chile, taking advantage of the high solar irradiation. Its processing facilities cover 80 000 m^2 with a culturing surface of 40 000 m^2 (open ponds) and a much higher growing potential. Currently, the company markets astaxanthin in powder form known as *Supreme Asta Powder* and the oleoresin known as *Supreme Asta Oil* (http://www.atacamabionatural.com/about.html).

The cost of production of β-carotene from *D. salina* in large open ponds can also compete with synthetic sources. The low capital and operating costs of large open ponds in Western Australia (Western Biotechnology Ltd and Betatene Ltd) combined with the high biomass accumulation of β-carotene in algal biomass have reduced the production cost to below \$300 kg^{-1} (Borowitzka, 2013). Betatene Ltd is currently a subsidiary of BASF and holds 80% of the natural β-carotene market.

References

Abe, K., Hattori, H., and Hirano, M. (2007) Accumulation and antioxidant activity of secondary carotenoids in the aerial microalga *Coelastrella striolata* var. multistriata. *Food Chem.*, **100**, 656–661.

Aflalo, C., Meshulam, Y., Zarka, A., and Boussiba, S. (2007) On the relative efficiency of two- vs. one-stage production of astaxanthin by the green alga *Haematococcus pluvialis. Biotechnol. Bioeng.*, **98**, 300–305.

Alscher, R.G., Donahue, J.L., and Cramer, C.L. (1997) Reactive oxygen species and antioxidants: relationships in green cells. *Physiol. Plant.*, **100**, 224–233.

Armstrong, G.A. (1997) Genetics of eubacterial carotenoid biosynthesis: a colorful tale. *Annu. Rev. Microbiol.*, **51**, 629–659.

Arnold, C., Schwarzenbolz, U., and Böhm, V. (2014) Carotenoids and chlorophylls in processed xanthophyll-rich food. *LWT Food Sci. Technol.*, **57**, 442–445.

Balasundaram, B., Skill, S.C., and Llewellyn, C.A. (2012) A low energy process for the recovery of bioproducts from cyanobacteria using a ball mill. *Biochem. Eng. J.*, **69**, 48–56.

Bar, E., Rise, M., Vishkautsan, M., and Arad, S. (1995) Pigment and structural changes in *Chlorella zofingiensis* upon light and nitrogen stress. *J. Plant Physiol.*, **146**, 527–534.

Beatty, S., Koh, H.H., Phil, M., Henson, D., and Boulton, M. (2000) The role of oxidative stress in the pathogenesis of age-related macular degeneration. *Surv. Ophthalmol.*, **45**, 115–134.

Beisel, K.G., Jahnke, S., Hofmann, D., Köppchen, S., Schurr, U., and Matsubara, S. (2010) Continuous turnover of carotenes and chlorophyll a in mature leaves of *Arabidopsis* revealed by $^{14}CO_2$ pulse-chase labeling. *Plant Physiol.*, **152**, 2188–2199.

Ben-Amotz, A. (1995) New mode of *Dunaliella* biotechnology: two-phase growth for β-carotene production. *J. Appl. Phycol.*, **7**, 65–68.

Ben Amotz, A. and Avron, M. (1989) in *Algal and Cyanobacterial Biotechnology* (eds

R.C. Cresswell, T.A.V. Rees, and N. Shah), Longman Scientific & Technical, New York, pp. 91–114.

Berman, J., Zorrilla-López, U., Farré, G., Zhu, C., Sandmann, G., Twyman, R.M., Capell, T., and Christou, P. (2015) Nutritionally important carotenoids as consumer products. *Phytochem. Rev.*, **14**, 727–743.

Blanco, A.M., Moreno, J., Del Campo, J.A., Rivas, J., and Guerrero, M.G. (2007) Outdoor cultivation of lutein-rich cells of *Muriellopsis* sp. in open ponds. *Appl. Microbiol. Biotechnol.*, **73**, 1259–1266.

Bolhassani, A., Khavari, A., and Bathaie, S.Z. (2014) Saffron and natural carotenoids: biochemical activities and anti-tumor effects. *Biochim. Biophys. Acta, Rev. Cancer*, **1845**, 20–30.

Borowitzka, M.A. (1999) Commercial production of microalgae: ponds, tanks, tubes and fermenters. *J. Biotechnol.*, **70**, 313–321.

Borowitzka, M.A. (2013) High-value products from microalgae-their development and commercialisation. *J. Appl. Phycol.*, **25**, 743–756.

Borowitzka, M.A. and Borowitzka, L.J. (1989) in *Biotechnology of Vitamins, Pigments and Growth Factors* (ed. E. Vandamme), Elsevier Applied Science, London, pp. 15–26.

Britton, G. (1998) in *Carotenoids: Biosynthesis and Metabolism* (eds G. Britton, S. Liaaen-Jensen, and H. Pfander), Burkhäuser, Basel, pp. 13–147.

Cadoret, J.P., Garnier, M., and Saint Jean, B. (2012) Microalgae, functional genomics and biotechnology. *Adv. Bot. Res.*, **64**, 285–341.

Campenni, L., Nobre, B.P., Santos, C.A., Oliveira, A.C., Aires-Barros, M.R., Palavra, A.M.F., and Gouveia, L. (2013) Carotenoid and lipid production by the autotrophic microalga *Chlorella protothecoides* under nutritional, salinity, and luminosity stress conditions. *Appl. Microbiol. Biotechnol.*, **97**, 1383–1393.

Cardozo, K.H.M., Guaratini, T., Barros, M.P., Falcao, V.R., Tonon, A.P., Lopes, N.P., Campos, S., Torres, M.A., Souza, A.O., Colepicolo, P. *et al.* (2007) Metabolites from algae with economical impact. *Comp. Biochem. Physiol. C: Toxicol. Pharmacol.*, **146**, 60–78.

Chan, M.C., Ho, S.H., Lee, D.J., Chen, C.Y., Huang, C.C., and Chang, J.S. (2013) Characterization, extraction and purification of lutein produced by an indigenous microalga *Scenedesmus obliquus* CNW-N. *Biochem. Eng. J.*, **78**, 24–31.

Christenson, L. and Sims, R. (2011) Production and harvesting of microalgae for wastewater treatment, biofuels, and bioproducts. *Biotechnol. Adv.*, **29**, 686–702.

Cunningham, F.X.J. and Gantt, E. (1998) Genes and enzymes of carotenoid biosynthesis in plants. *Annu. Rev. Plant Physiol. Plant Mol. Biol.*, **49**, 557–583.

Del Campo, J.A., Garcia-Gonzalez, M., and Guerrero, M.G. (2007) Outdoor cultivation of microalgae for carotenoid production: current state and perspectives. *Appl. Microbiol. Biotechnol.*, **74**, 1163–1174.

Del Campo, J.A., Moreno, J., Rodriguez, H., Vargas, M.A., Rivas, J., and Guerrero, M.G. (2000) Carotenoid content of chlorophycean microalgae: factors determining lutein accumulation in *Muriellopsis* sp. (Chlorophyta). *J. Biotechnol.*, **76**, 51–59.

Del Campo, J.A., Rodriguez, H., Moreno, J., Angeles Vargas, M., Rivas, J., and Guerrero, M.G. (2001) Lutein production by *Muriellopsis* sp. in an outdoor tubular photobioreactor. *J. Biotechnol.*, **85**, 289–295.

Del Campo, J.A., Rodríguez, H., Moreno, J., Vargas, M.Á., Rivas, J., and Guerrero, M.G. (2004) Accumulation of astaxanthin and lutein in *Chlorella zofingiensis* (Chlorophyta). *Appl. Microbiol. Biotechnol.*, **64**, 848–854.

Domonkos, I., Kis, M., Gombos, Z., and Ughy, B. (2013) Carotenoids, versatile components of oxygenic photosynthesis. *Prog. Lipid Res.*, **52**, 539–561.

Ernst, H. (2002) Recent advances in industrial carotenoid synthesis. *Pure Appl. Chem.*, **74**, 1369–1382.

Fassett, R.G. and Coombes, J.S. (2011) Astaxanthin: a potential therapeutic agent in cardiovascular disease. *Mar. Drugs*, **9**, 447–465.

Fernandez-Sevilla, J.M., Fernandez, F.G.A., and Grima, E.M. (2010) Biotechnological production of lutein and its applications. *Appl. Microbiol. Biotechnol.*, **86**, 27–40.

Fiedler, D., Hager, U., Franke, H., Soltmann, U., and Bottcher, H. (2007) Algae biocers: astaxanthin formation in sol-gel immobilised living microalgae. *J. Mater. Chem.*, **17**, 261–266.

Foyer, C.H., Lopez-Delgado, H., Dat, J.F., and Scott, I.M. (1997) Hydrogen peroxide- and glutathione-associated mechanisms of acclimatory stress tolerance and signalling. *Physiol. Plant.*, **100**, 241–254.

Ganesan, P., Matsubara, K., Sugawara, T., and Hirata, T. (2013) Marine algal carotenoids inhibit angiogenesis by down-regulating FGF-2-mediated intracellular signals in vascular endothelial cells. *Mol. Cell. Biochem.*, **380**, 1–9.

Garbayo, I., Cuaresma, M., Vilchez, C., and Vega, J.M. (2008) Effect of abiotic stress on the production of lutein and beta-carotene by *Chlamydomonas acidophila. Process Biochem.*, **43**, 1158–1161.

Garcia-Gonzalez, M., Moreno, J., Manzano, J.C., Florencio, F.J., and Guerrero, M.G. (2005) Production of *Dunaliella salina* biomass rich in 9-cis-beta-carotene and lutein in a closed tubular photobioreactor. *J. Biotechnol.*, **115**, 81–90.

Garg, S., Wang, L., and Schenk, P.M. (2014) Effective harvesting of low surface-hydrophobicity microalgae by froth flotation. *Bioresour. Technol.*, **159**, 437–441.

Gimpel, J.A., Specht, E.A., Georgianna, D.R., and Mayfield, S.P. (2013) Advances in microalgae engineering and synthetic biology applications for biofuel production. *Curr. Opin. Chem. Biol.*, **17**, 489–495.

Goiris, K., De Vreese, P., De Cooman, L., and Muylaert, K. (2012) Rapid screening and guided extraction of antioxidants from microalgae using voltammetric methods. *J. Agric. Food. Chem.*, **60**, 7359–7366.

Gouveia, L., Veloso, V., Reis, A., Fernandes, H., Novais, J., and Empis, J. (1996) Evolution of pigment composition in *Chlorella vulgaris. Bioresour. Technol.*, **57**, 157–163.

Grama, B.S., Chader, S., Khelifi, D., Agathos, S.N., and Jeffryes, C. (2014a) Induction of canthaxanthin production in a *Dactylococcus* microalga isolated from the Algerian Sahara. *Bioresour. Technol.*, **151**, 297–305.

Grama, B.S., Delhaye, A., Chader, S., Khelifi, D., Agathos, S.N., and Jeffryes, C. (2014b) Canthaxanthin, astaxanthin and adonixanthin production from a *Dactylococcus* microalga in a new flat plate airlift photobioreactor. *Commun. Agric. Appl. Biol. Sci.*, **79**, 65–70.

Grewe, C., Menge, S., and Griehl, C. (2007) Enantioselective separation of all-E-astaxanthin and its determination in microbial sources. *J. Chromatogr. A*, **1166**, 97–100.

Guedes, A.C., Amaro, H.M., and Malcata, F.X. (2011) Microalgae as sources of high added-value compounds-a brief review of recent work. *Biotechnol. Progr.*, **27**, 597–613.

Guerin, M., Huntley, M.E., and Olaizola, M. (2003) *Haematococcus* astaxanthin: applications for human health and nutrition. *Trends Biotechnol.*, **21**, 210–216.

Hagen, C., Braune, W., and Bjorn, L.O. (1994) Functional aspects of secondary carotenoids in *Haematococcus lacustris* (Volvocales). 3. Action as a sunshade. *J. Phycol.*, **30**, 241–248.

Halliwell, B. (1996) Antioxidants in human health and disease. *Annu. Rev. Nutr.*, **16**, 33–50.

Halliwell, B. and Gutteridge, J.M.C. (1999) *Free Radicals in Biology and Medicine*, Clarendon Press, Oxford.

Han, D., Li, Y., and Hu, Q. (2013) Astaxanthin in microalgae: pathways, functions and biotechnological implications. *Algae*, **28**, 131–147.

Hanagata, N. and Dubinsky, Z. (1999) Secondary carotenoid accumulation in *Scenedesmus komarekii* (Chlorophyceae, Chlorophyta). *J. Phycol.*, **35**, 960–966.

Harun, R., Singh, M., Forde, G.M., and Danquah, M.K. (2010) Bioprocess engineering of microalgae to produce a variety of consumer products. *Renewable Sustainable Energy Rev.*, **14**, 1037–1047.

Heider, S.A.E., Peters-Wendisch, P., Wendisch, V.F., Beekwilder, J., and Brautaset, T. (2014) Metabolic engineering for the microbial production of carotenoids and related products with a focus on the rare C50 carotenoids. *Appl. Microbiol. Biotechnol.*, **98**, 4355–4368.

Hong, Y., Hu, H.Y., Xie, X., and Li, F.M. (2008) Responses of enzymatic antioxidants and non-enzymatic antioxidants in the cyanobacterium *Microcystis aeruginosa* to the allelochemical ethyl 2-methyl acetoacetate (EMA) isolated from reed (*Phragmites communis*). *J. Plant Physiol.*, **165**, 1264–1273.

Hu, C.C., Lin, J.T., Lu, F.J., Chou, F.P., and Yang, D.J. (2008) Determination of carotenoids in *Dunaliella salina* cultivated in Taiwan and antioxidant capacity of the algal carotenoid extract. *Food Chem.*, **109**, 439–446.

Hyka, P., Lickova, S., Přibyl, P., Melzoch, K., and Kovar, K. (2013) Flow cytometry for the development of biotechnological processes with microalgae. *Biotechnol. Adv.*, **31**, 2–16.

Ip, P.F. and Chen, F. (2005a) Employment of reactive oxygen species to enhance astaxanthin formation in *Chlorella zofingiensis* in heterotrophic culture. *Process Biochem.*, **40**, 3491–3496.

Ip, P.F. and Chen, F. (2005b) Peroxynitrite and nitryl chloride enhance astaxanthin production by the green microalga *Chlorella zofingiensis* in heterotrophic culture. *Process Biochem.*, **40**, 3595–3599.

Ishida, B.K. and Chapman, M.H. (2009) Carotenoid extraction from plants using a novel, environmentally friendly solvent. *J. Agric. Food. Chem.*, **57**, 1051–1059.

Jeffryes, C., Agathos, S.N., and Rorrer, G. (2015) Biogenic nanomaterials from photosynthetic microorganisms. *Curr. Opin. Biotechnol.*, **33**, 23–31.

Jin, E., Polle, J.E.W., Lee, H.K., Hyun, S.M., and Chang, M. (2003) Xanthophylls in microalgae: from biosynthesis to biotechnological mass production and application. *J. Microbiol. Biotechnol.*, **13**, 165–174.

Jomova, K. and Valko, M. (2013) Health protective effects of carotenoids and their interactions with other biological antioxidants. *Eur. J. Med. Chem.*, **70**, 102–110.

Katsuda, T., Lababpour, A., Shimahara, K., and Katoh, S. (2004) Astaxanthin production by *Haematococcus pluvialis* under illumination with LEDs. *Enzyme Microb. Technol.*, **35**, 81–86.

Katsuda, T., Shiraishi, H., Ishizu, N., Ranjbar, R., and Katoh, S. (2008) Effect of light intensity and frequency of flashing light from blue light emitting diodes on astaxanthin production by *Haematococcus pluvialis*. *J. Biosci. Bioeng.*, **105**, 216–220.

Kim, Z.H., Kim, S.H., Lee, H.S., and Lee, C.G. (2006) Enhanced production of astaxanthin by flashing light using *Haematococcus pluvialis*. *Enzyme Microb. Technol.*, **39**, 414–419.

Lemoine, Y. and Schoefs, B. (2010) Secondary ketocarotenoid astaxanthin biosynthesis in algae: a multifunctional response to stress. *Photosynth. Res.*, **106**, 155–177.

Leya, T., Rahn, A., Lütz, C., and Remias, D. (2009) Response of arctic snow and permafrost algae to high light and nitrogen stress by changes in pigment composition and applied aspects for biotechnology. *FEMS Microbiol. Ecol.*, **67**, 432–443.

Li, J., Zhu, D., Niu, J., Shen, S., and Wang, G. (2011) An economic assessment of astaxanthin production by large scale cultivation of *Haematococcus pluvialis*. *Biotechnol. Adv.*, **29**, 568–574.

Lorenz, R.T. and Cysewski, G.R. (2000) Commercial potential for *Haematococcus* microalgae as a natural source of astaxanthin. *Trends Biotechnol.*, **18**, 160–167.

Mäki-Arvela, P., Hachemi, I., and Murzin, D.Y. (2014) Comparative study of the extraction methods for recovery of carotenoids from algae: extraction kinetics and effect of different extraction parameters. *J. Chem. Technol. Biotechnol.*, **89**, 1607–1626.

Mallick, N. and Mohn, F.H. (2000) Reactive oxygen species: response of algal cells. *J. Plant Physiol.*, **157**, 183–193.

Maoka, T., Yasui, H., Ohmori, A., Tokuda, H., Suzuki, N., Osawa, A., Shindo, K., and Ishibashi, T. (2013) Anti-oxidative, anti-tumor-promoting, and anti carcinogenic activities of adonirubin and adonixanthin. *J. Oleo Sci.*, **62**, 181–186.

Mata, T.M., Martins, A.A., and Caetano, N.S. (2010) Microalgae for biodiesel production and other applications: a review. *Renewable Sustainable Energy Rev.*, **14**, 217–232.

Matsukawa, R., Hotta, M., Masuda, Y., Chihara, M., and Karube, I. (2000) Antioxidants from carbon dioxide fixing *Chlorella sorokiniana*. *J. Appl. Phycol.*, **12**, 263–267.

Metting, F.B. (1996) Biodiversity and application of microalgae. *J. Ind. Microbiol. Biotechnol.*, **17**, 477–489.

Miguel, F., Martín, A., Mattea, F., and Cocero, M.J. (2008) Precipitation of lutein and co-precipitation of lutein and poly-lactic acid with the supercritical anti-solvent process. *Chem. Eng. Process. Process Intensif.*, **47**, 1594–1602.

Milledge, J.J. (2011) Commercial application of microalgae other than as biofuels: a brief review. *Rev. Environ. Sci. Biotechnol.*, **10**, 31–41.

Molina-Grima, E.M., Belarbi, E.H., Fernandez, F.G.A., Medina, A.R., and Chisti, Y. (2003) Recovery of microalgal biomass and metabolites: process options and economics. *Biotechnol. Adv.*, **20**, 491–515.

Murray, P.M., Moane, S., Collins, C., Beletskaya, T., Thomas, O.P., Duarte, A.W.F., Nobre, F.S., Owoyemi, I.O., Pagnocca, F.C., Sette, L.D. *et al.* (2013) Sustainable production of biologically active molecules of marine based origin. *New Biotechnol.*, **30**, 839–850.

Niyogi, K.K. (1999) Photoprotection revisited: genetic and molecular approaches. *Annu. Rev. Plant Biol.*, **50**, 333–359.

Nonomura, A.M. (1987) Process for producing a naturally-derived carotene/oil composition by direct extraction from algae. US Patent 4680314 A.

Nurra, C., Torras, C., Clavero, E., Ríos, S., Rey, M., Lorente, E., Farriol, X., and Salvadó, J. (2014) Biorefinery concept in a microalgae pilot plant. Culturing, dynamic filtration and steam explosion fractionation. *Bioresour. Technol.*, **163**, 136–142.

Olaizola, M. (2000) Commercial production of astaxanthin from *Haematococcus pluvialis* using 25,000-liter outdoor photobioreactors. *J. Appl. Phycol.*, **12**, 499–506.

Olaizola, M. (2003) Commercial development of microalgal biotechnology: from the test tube to the marketplace. *Biomol. Eng.*, **20**, 459–466.

Orosa, M., Torres, E., Fidalgo, P., and Abalde, J. (2000) Production and analysis of secondary carotenoids in green algae. *J. Appl. Phycol.*, **12**, 553–556.

Park, S.K., Jin, E., Lee, C.G., and Lee, M.Y. (2008) High light-induced changes in the activities of antioxidant enzymes and the accumulation of astaxanthin in the green alga *Haematococcus pluvialis. Mol. Cell. Toxicol.*, **4**, 300–306.

Park, E.K. and Lee, C.G. (2001) Astaxanthin production by *Haematococcus pluvialis* under various light intensities and wavelengths. *J. Microb. Biotechnol.*, **11**, 1024–1030.

Pelah, D., Sintov, A., and Cohen, E. (2004) The effect of salt stress on the production of canthaxanthin and astaxanthin by *Chlorella zofingiensis* grown under limited light intensity. *World J. Microbiol. Biotechnol.*, **20**, 483–486.

Person, J. (2011) Algues, filières du future (Romainville, France, Adebiotech), http://www.adebiotech.org/home/img/algues/LIVRE_TURQUOISE-V.screen.pdf (accessed 10 January 2015).

Plaza, M., Santoyo, S., Jaime, L., Avalo, B., Cifuentes, A., Reglero, G., García-Blairsy Reina, G., Señoráns, F.J., and Ibáñez, E. (2012) Comprehensive characterization of the functional activities of pressurized liquid and ultrasound-assisted extracts from *Chlorella vulgaris. LWT Food Sci. Technol.*, **46**, 245–253.

Qin, S., Lin, H., and Jiang, P. (2012) Advances in genetic engineering of marine algae. *Biotechnol. Adv.*, **30**, 1602–1613.

Qin, S., Liu, G.-X., and Hu, Z.-Y. (2008) The accumulation and metabolism of astaxanthin in *Scenedesmus obliquus* (Chlorophyceae). *Process Biochem.*, **43**, 795–802.

Rabbani, S., Beyer, P., Lintig, J.V., Hugueney, P., and Kleinig, H. (1998) Induced β-Carotene synthesis driven by triacylglycerol deposition in the unicellular alga *Dunaliella bardawil. Plant Physiol.*, **116**, 1239–1248.

Radakovits, R., Jinkerson, R.E., Darzins, A., and Posewitz, M.C. (2010) Genetic engineering of algae for enhanced biofuel production. *Eukaryotic Cell*, **9**, 486–501.

Raja, R., Hemaiswarya, S., and Rengasamy, R. (2007) Exploitation of *Dunaliella* for beta-carotene production. *Appl. Microbiol. Biotechnol.*, **74**, 517–523.

Rao, A.V. and Rao, L.G. (2007) Carotenoids and human health. *Pharmacol. Res.*, **55**, 207–216.

Rasala, B.A., Lee, P.A., Shen, Z., Briggs, S.P., Mendez, M., and Mayfield, S.P. (2012) Robust expression and secretion of xylanase1 in *Chlamydomonas reinhardtii* by fusion to a selection gene and processing with the FMDV 2A peptide. *PLoS One*, **7**, e43349.

Rodriguez-Concepcion, M. and Stange, C. (2013) Biosynthesis of carotenoids in carrot: an underground story comes to light. *Arch. Biochem. Biophys.*, **539**, 110–116.

Rosello Sastre, R. (2012) in *Microalgal Biotechnology: Integration and Economy* (eds C. Posten and C. Walter), De Gruyter, Berlin, pp. 13–44.

Ryckebosch, E., Muylaert, K., Eeckhout, M., Ruyssen, T., and Foubert, I. (2011) Influence of drying and storage on lipid and carotenoid stability of the microalga *Phaeodactylum tricornutum*. *J. Agric. Food. Chem.*, **59**, 11063–11069.

Sanchez, J.F., Fernandez-Sevilla, J.M., Acien, F.G., Ceron, M.C., Perez-Parra, J., and Molina-Grima, E. (2008a) Biomass and lutein productivity of *Scenedesmus almeriensis*: influence of irradiance, dilution rate and temperature. *Appl. Microbiol. Biotechnol.*, **79**, 719–729.

Sanchez, J.F., Fernandez-Sevilla, J.M., Acien, F.G., Rueda, A., Perez-Parra, J., and Molina-Grima, E. (2008b) Influence of culture conditions on the productivity and lutein content of the new strain *Scenedesmus almeriensis*. *Process Biochem.*, **43**, 398–405.

Sarada, R., Vidhyavathi, R., Usha, D., and Ravishankar, G.A. (2006) An efficient method for extraction of astaxanthin from green alga *Haematococcus pluvialis*. *J. Agric. Food. Chem.*, **54**, 7585–7588.

Sarkar, C.R., Das, L., Bhagawati, B., and Goswami, B.C. (2012) A comparative study of carotenoid extraction from algae in different solvent systems. *Asian J. Plant Sci. Res.*, **2**, 546–549.

Schoefs, B., Rmiki, N.E., Rachadi, J., and Lemoine, Y. (2001) Astaxanthin accumulation in *Haematococcus* requires a cytochrome P450 hydroxylase and an active synthesis of fatty acids. *FEBS Lett.*, **500**, 125–128.

Sharma, K.K., Garg, S., Li, Y., Malekizadeh, A., and Schenk, P.M. (2013) Critical analysis of current microalgae dewatering techniques. *Biofuels*, **4**, 397–407.

Shi, X.M., Jiang, Y., and Chen, F. (2002) High-yield production of lutein by the green microalga *Chlorella protothecoides* in heterotrophic fed-batch culture. *Biotechnol. Progr.*, **18**, 723–727.

Shuman, T.R., Mason, G., Marsolek, M.D., Lin, Y., Reeve, D., and Schacht, A. (2014) An ultra-low energy method for rapidly pre-concentrating microalgae. *Bioresour. Technol.*, **158**, 217–224.

Sies, H. and Stahl, W. (2004) Nutritional protection against skin damage from sunlight. *Annu. Rev. Nutr.*, **24**, 173–200.

Skjånes, K., Rebours, C., and Lindblad, P. (2013) Potential for green microalgae to produce hydrogen, pharmaceuticals and other high value products in a combined process. *Crit. Rev. Biotechnol.*, **33**, 172–215.

Solovchenko, A. (2010) *Photoprotection in Plants*, Springer-Verlag, Berlin.

Solovchenko, A.E. (2012) Physiological role of neutral lipid accumulation in eukaryotic microalgae under stresses. *Russ. J. Plant Physiol.*, **59**, 167–176.

Solovchenko, A.E. (2013) Physiology and adaptive significance of secondary carotenogenesis in green microalgae. *Russ. J. Plant Physiol.*, **60**, 1–13.

Solovchenko, A., Aflalo, C., Lukyanov, A., and Boussiba, S. (2013) Nondestructive monitoring of carotenogenesis in *Haematococcus pluvialis* via whole-cell optical density spectra. *Appl. Microbiol. Biotechnol.*, **97**, 4533–4541.

Splinter, S., Pare, J.J.R. and Kadali, S. (2013) Method for direct extraction and concentration of naturally-derived active compounds. US Patent 20130338234 A1.

Stahl, W. and Sies, H. (2005) Bioactivity and protective effects of natural carotenoids. *Biochim. Biophys. Acta, Mol. Basis Dis.*, **1740**, 101–107.

Stahl, W. and Sies, H. (2007) Carotenoids and flavonoids contribute to nutritional protection against skin damage from sunlight. *Mol. Biotechnol.*, **37**, 26–30.

Steinbrenner, J. and Linden, H. (2003) Light induction of carotenoid biosynthesis genes in the green alga *Haematococcus pluvialis*: regulation by photosynthetic redox control. *Plant Mol. Biol.*, **52**, 343–356.

Sun, N., Wang, Y., Li, Y.T., Huang, J.C., and Chen, F. (2008) Sugar-based growth, astaxanthin accumulation and carotenogenic transcription of heterotrophic *Chlorella zofingiensis* (Chlorophyta). *Process Biochem.*, **43**, 1288–1292.

Takaichi, S. (2011) Carotenoids in algae: distributions, biosyntheses and functions. *Mar. Drugs*, **9**, 1101–1118.

Takaichi, S. and Mochimaru, M. (2007) Carotenoids and carotenogenesis in cyanobacteria: unique ketocarotenoids and carotenoid glycosides. *Cell. Mol. Life Sci.*, **64**, 2607–2619.

Talisic, G.C., Yumang, A.N., and Salta, M.T.S. (2012) Supercritical fluid extraction of β-carotene from *D. salina* algae using C_2H_6 and C_2H_2. *Int. Proc. Chem., Biol. Environ. Eng.*, **36**, 30–34.

Tardy, F. and Havaux, M. (1996) Photosynthesis, chlorophyll fluorescence, light-harvesting system and photoinhibition resistance of a zeaxanthin-accumulating mutant of *Arabidopsis thaliana. J. Photochem. Photobiol., B*, **34**, 87–94.

Tocquin, P., Fratamico, A., and Franck, F. (2012) Screening for a low-cost *Haematococcus pluvialis* medium reveals an unexpected impact of a low N/P ratio on vegetative growth. *J. Appl. Phycol.*, **24**, 365–373.

Vandamme, D., Foubert, I., and Muylaert, K. (2013) Flocculation as a low-cost method for harvesting microalgae for bulk biomass production. *Trends Biotechnol.*, **31**, 233–239.

Xie, Y.P., Ho, S.H., Chen, C.Y., Chen, C.N.N., Liu, C.C., Ng, I.S., Jing, K.J., Yang, S.C., Chen, C.H., Chang, J.S. *et al.* (2014) Simultaneous enhancement of CO_2 fixation and lutein production with thermo-tolerant *Desmodesmus* sp. F51 using a repeated fed-batch cultivation strategy. *Biochem. Eng. J.*, **86**, 33–40.

Young, A.J. and Lowe, G.M. (2001) Antioxidant and prooxidant properties of carotenoids. *Arch. Biochem. Biophys.*, **385**, 20–27.

Zhang, D.H., Lee, Y.K., Ng, M.L., and Phang, S.M. (1997) Composition and accumulation of secondary carotenoids in *Chlorococcum* sp. *J. Appl. Phycol.*, **9**, 147–155.

Zhekisheva, M., Zarka, A., Khozin-Goldberg, I., Cohen, Z., and Boussiba, S. (2005) Inhibition of astaxanthin synthesis under high irradiance does not abolish triacylglycerol accumulation in the green alga *Haematococcus pluvialis* (Chlorophyceae). *J. Phycol.*, **41**, 819–826.

Zhu, Y.H. and Jiang, J.G. (2008) Continuous cultivation of *Dunaliella salina* in photobioreactor for the production of beta-carotene. *Eur. Food Res. Technol.*, **227**, 953–959.

11
Microbial Production of Vitamin F and Other Polyunsaturated Fatty Acids

Colin Ratledge

Lipid Nomenclature

The conventional nomenclature for fatty acids is used in this chapter but, for those readers not familiar with this system, the following notes may be helpful. Fatty acids are delineated by the number of carbon atoms in their chain. With some exceptions found in bacteria, the chain of carbon atoms is normally straight and the number of carbons is usually even and varies between 12 and 22 although shorter and longer chain lengths occur. These are the saturated fatty acids. Fatty acids frequently have one or more double bonds (–C=C–). These are termed *unsaturated* fatty acids or *polyunsaturated* if there are multiple double bonds. These are given by a standard nomenclature. Thus, oleic acid is 18 : 1, indicating that it has 18 carbon atoms and one double bond. If the position of the double bond in the chain needs to be specified, then this is given an appropriate number thus, 18 : 1(9) indicates that the bond in oleic acid is between the 9th and 10th carbon atoms starting the numbering from the carboxylic acid head (see Figure 11.1). If there are multiple double bonds then these are similarly specified; linoleic acid is thus 18 : 2(9,12). In the vast majority of cases, multiple double bonds are methylene-interrupted $-CH{=}CH{-}CH_2{-}CH{=}CH-$ where $-CH_2-$ is the methylene group. The double bonds, though, may be of either a *cis*- or a *trans*-configuration depending on the orientation of the two H atoms on the double bond. Normally, bonds of naturally occurring fatty acids are in the *cis*-form and, unless stated otherwise, this is to be assumed. Once the position of one double bond has been given and, if the others are methylene-interrupted, the positions of all the other bonds are then established. In these cases, the bond position of a polyunsaturated fatty acid (PUFA) is specified starting the counting from the terminal methyl group and is indicated usually as ω-3, ω-6 or ω-9 or, alternatively, as n-3, n-6 or n-9. The latter nomenclature is used in this chapter. Where the double bonds may not follow this pattern, the position of each must be specified individually but are then counted from the carboxylic acid end of the fatty acid. It should also be appreciated that fatty acids *per se* do not occur as natural entities in cells; they are always linked to another molecule, usually glycerol to form acylglycerols.

Industrial Biotechnology of Vitamins, Biopigments, and Antioxidants, First Edition.
Edited by Erick J. Vandamme and José L. Revuelta.

(a) $CH_3 \cdot (CH_2)_4 \cdot CH{:}CH \cdot CH_2 \cdot CH{:}CH \cdot (CH_2)_7 \cdot COOH$

Linoleic acid (*cis,cis*-9,12-octadecadienoic acid); 18:2 (*n*-6)

(originally referred to as vitamin F)

$CH_3 \cdot CH_2 \cdot CH{:}CH \cdot CH_2 \cdot CH{:}CH \cdot CH_2 \cdot CH{:}CH \cdot (CH_2)_7 \cdot COOH$

Alpha-linolenic acid, ALA (*cis,cis,cis*-9,12,15-octadecatrienoic acid); 18:3 (*n*-3)

$CH_3 \cdot (CH_2)_4 \cdot CH{:}CH \cdot CH_2 \cdot CH{:}CH \cdot CH_2 \cdot CH{:}CH \cdot (CH_2)_4 \cdot COOH$

Gamma-linolenic acid, GLA (*cis,cis,cis*-6,9,12-octadecatrienoic acid); 18:3 (*n*-6)

(also referred to as vitamin FF)

$CH_3 \cdot (CH_2)_4 \cdot CH{:}CH \cdot CH_2 \cdot CH{:}CH \cdot CH_2 \cdot CH{:}CH \cdot CH_2 \cdot CH{:}CH \cdot (CH_2)_3 \cdot COOH$

Arachidonic acid, ARA (all *cis*-5,8,11,14-eicosatetraenoic acid); 20:4 (*n*-6)

$CH_3 \cdot CH_2 \cdot CH{:}CH \cdot CH_2 \cdot CH{:}CH \cdot CH_2 \cdot CH{:}CH \cdot CH_2 \cdot CH{:}CH \cdot CH_2 \cdot CH{:}CH \cdot (CH_2)_3 \cdot COOH$

Eicosapentaenoic acid, EPA (all *cis*-5,8,11,14,17-EPA); 20:5 (*n*-3)

$CH_3 \cdot CH_2 \cdot CH{:}CH \cdot CH_2 \cdot CH \cdot CH \cdot CH_2 \cdot CH{:}CH \cdot CH_2 \cdot CH{:}CH \cdot CH_2 \cdot CH{:}CH \cdot CH_2 \cdot CH{:}CH \cdot (CH_2)_2 \cdot COOH$

Docosahexaenoic acid, DHA (all *cis*-4,7,10,13,16,19-DHA); 22:6 (*n*-3)

(b)
$CH_2—O—CO—R_1$
|
$CH—O—CO—R_2$
|
$CH_2—O—CO—R_3$

Figure 11.1 (a) Structures of the principal essential fatty acids and major polyunsaturated fatty acids. (b) Structure of a triacylglycerol (sometimes but incorrectly termed triglyceride), where R_1, R_2 and R_3 are the long carbon chains (alkyl groups) of fatty acids. These may be the same or different.

There may be one, two or three fatty acids attached to glycerol, and the structure of a triacylglycerol, which is the usual form of oils and fats, is shown in Figure 11.1.

11.1 Introduction: Essential Fatty Acids

It may be strange to consider fatty acids that are the major components of all lipids found in the human body as vitamins, but there are, in fact, two essential fatty acids that cannot be synthesised by the conventional fatty-acid biosynthetic pathway. These are linoleic acid (18 : 2 *n*-6) and alpha-linolenic acid (ALA) (18 : 3 *n*-3). Although vertebrate animals can synthesise oleic acid [18 : 1(9)] from stearic acid (18 : 0), they lack the necessary mechanism of a desaturating enzyme (i.e. a desaturase) to introduce the second and the third double bonds between the existing bonds and the terminal methyl group of the fatty acid. Thus, as humans, we cannot convert oleic acid into linoleic acid nor linoleic acid into ALA. Therefore, these fatty acids are essential for our well-being and must be provided in the diet. This is normally not a problem as these fatty acids are abundant in plants and also in herbivorous animals.

The first indication that fatty acids were essential components of the diet was provided by Evans and Burr (1927, 1928) following on from the initial work by Osborne and Mendel (1920) that had addressed the nutritive value of certain lipids in rats. Evans and Burr (1928) observed that a dietary deficiency of fat in the diet of rats led to severe and detrimental effects on growth and reproduction. They concluded that '… the favorable substance in fats is possibly a new vitamin (F) which, unlike vitamins A, D and E, is not concentrated in the non-saponifiable fraction of the dietary extract'. This work was followed by further experimental work by Burr and Burr (1929, 1930) who found that rats, having a rigid exclusion of all fats in their diet, produced kidney lesions and eventually died due to kidney failure. Also, male rats appeared to be sterile. Rats only recovered from these symptoms if fed linoleic acid or fats containing it; saturated fatty acids and also oleic acid did not seem to be effective. They concluded that linoleic acid, and possibly other fatty acids, was therefore 'an essential fatty acid'. This was the first use of this term. An early review of the work of these two pioneers and other contributions covering the consequences of essential fatty acid deficiency was by Holman (1971). Some personal details about the Burrs and their early research appeared later (Holman, 1988).

The term *vitamin F* was first given to linoleic acid (18 : 2 *n*-6) but, because linoleic acid could not be distinguished from ALA and indeed it was thought to be a single compound, the term is now accepted as applying to both the 18 : 2(*n*-6) and the 18 : 3(*n*-3) fatty acids. Because the term *vitamin F* is not precise and does not relate to a single compound, it is now no longer used. Sometime after the work of George Burr, 'vitamin FF' was used to describe gamma-linoleic acid (18 : 3 *n*-6) that had been found to be beneficial for the treatment of atopic eczema (Lovell, Burton and Horrobin, 1981; Wright and Burton, 1982) and was also considered to be a possible essential fatty acid. The terms, though, appear to have had only transitory use and the more general and more exact term of *essential fatty acid* is now widely used. Vitamin F and vitamin FF are now no longer used as they lack the precision of descriptors used for other vitamins.

The structures of the principal essential fatty acids are shown in Figure 11.1.

The reason why linoleic acid is 'essential' is that the human body lacks the enzyme required to insert a double bond into oleic acid (18 : 1 *n*-9) between the existing double bond and the terminal methyl group to form linoleic acid (18 : 2 *n*-6). This enzyme, δ-12 desaturase (see Figure 11.2), is, though, found in almost all plants; thus, linoleic acid finds its way into the human food chain by the direct intake of plant materials or by eating herbivorous animals that themselves have fed on plants. Vertebrate animals also lack the δ-15 desaturase for converting linoleic acid into ALA (18 : 3 *n*-3); thus, this PUFA is also considered to be essential. Again, similar to linoleic acid, ALA is a component fatty acid of most plants and also enters the food chain by direct ingestion of plant materials or by eating animals that have fed on plants. In practice, it is therefore almost

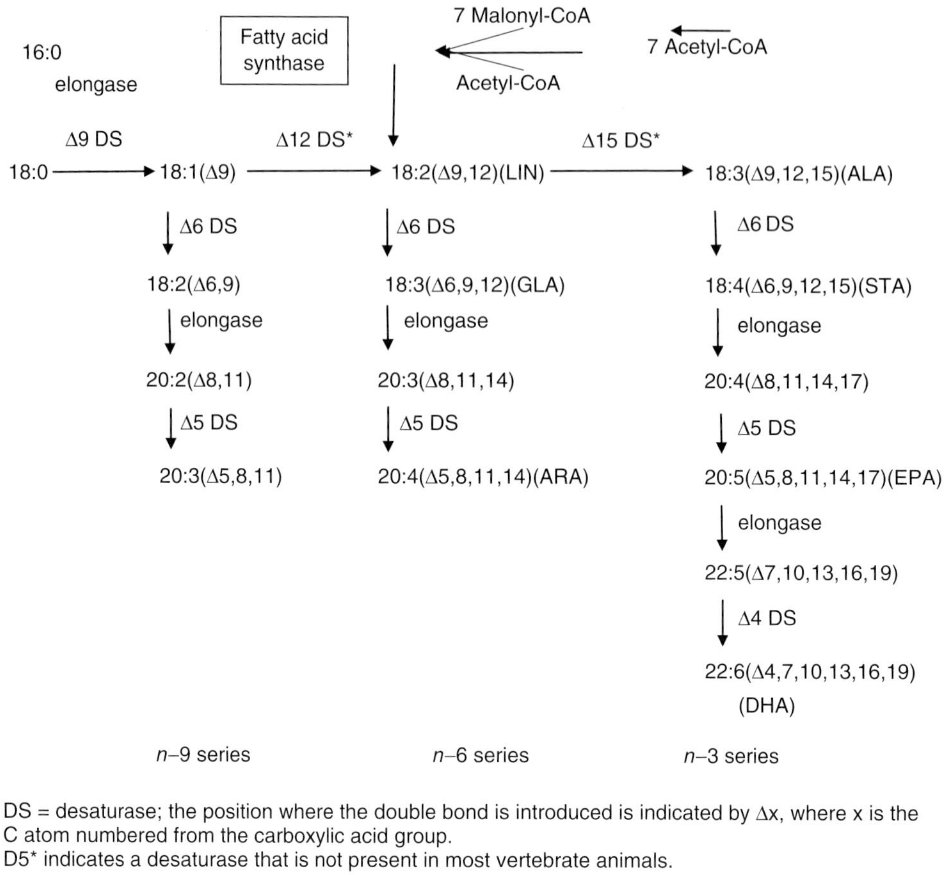

Figure 11.2 Biochemical conversions of fatty acids by fatty acid desaturases (DS) and elongases leading to the production of long-chain polyunsaturated fatty acids of the *n*-3, *n*-6 and *n*-9 series. The reactions are found in some microorganisms and animals, except for the Δ12 and Δ15 desaturases (indicated by *); hence, linoleic acid (LIN) and alpha-linolenic acid (ALA) are regarded as 'essential' fatty acids. In plants, polyunsaturated fatty acids up to ALA and GLA are produced and, in some species, stearidonic acid (STA) but no longer.

impossible to be sufficiently deprived of both of these essential fatty acids to cause physiological problems in the body if a normal and diverse diet is followed.

The pathway of fatty acid biosynthesis is shown in Figure 11.2. All fatty acids are derived from the precursors of acetyl-CoA and malonyl-CoA. In the vast majority of organisms, microbes, plants and animals, the synthesis is via the multi-complex enzyme system known as *fatty acid synthase* (FAS). This produces palmitic acid (16 : 0) which is then converted into all the other fatty acids by a series of elongases and desaturases. In some marine microorganisms [including a few bacteria and some microalgae, principally dinoflagellates and thraustochytrids that produce eicosapentaenoic acid (EPA) and/or docosahexaenoic acid (DHA)], PUFA biosynthesis is by a separate system known as the *polyketide synthase* (PKS) pathway. This pathway is shown in Figure 11.3. This also shows how the pathway in thraustochytrid microorganisms can also produce an n-6 PUFA, docosapentaenoic acid (DPA; 22 : 5), along with DHA – see also Section 11.3.3. This fatty acid (DPA *n*-6) occurs in the human brain tissue along with DHA but has a role that is not yet completely understood.

The role of the various fatty acids is extremely diverse. Fatty acids are the major components of membrane lipids, principally phospholipids, that are found in all living cells. They also form triacylglycerols that are the principal form of oils and fats stored in the adipose tissue of the body as well as in plant oilseeds. Fatty acids also act as precursors for many other materials that have major physiological roles in the body. Principal of these are the prostaglandins, thromboxanes, resolvins, leukotrienes, neuroprotectins, and epoxy and hydroxy fatty acids that are derived from EPA (20 : 5 *n*-3) and DHA (22 : 6 *n*-3) (Serhan, 2005; Cottin, Sanders and Hall, 2011). These are involved in many roles including anti-inflammatory, anti-arrhythmic and anti-aggregatory effects. There have been many publications that have then highlighted the role of the long-chain PUFAs, especially EPA and DHA, that are found in oily fish as being beneficial for the improvement of cardiovascular health. The review by Jump, Depner and Tripathy (2012) recorded over 2300 clinical trials that have been carried out to examine the role of *n*-3 fatty acids on the incidence of cardiovascular disease and stroke. Other recent reviews that may be useful on this topic include those of Minihane (2013) and Nicholson, Khademi and Moghadasian (2013). There is, therefore, considerable interest in providing adequate supplies of these PUFAs in the diet for the improvement of general health (Lands, 2014). It is the recommendation of the American Heart Association that adults should consume oily fish (as sources of long-chain PUFAs) at least twice a week (Lichtenstein *et al.*, 2006).

It is not, however, sufficient to rely on the inclusion of ALA in the diet to provide all the other *n*-3 fatty acids (see Figure 11.2). Even though this is a genuine essential fatty acid, its conversion into the longer chains of EPA and DHA is too slow for it to be of much benefit (Sinclair, Attar-Bashi and Li, 2002). DHA in the diet is a much more effective source of DHA in tissues than ALA; in some cases, there is a 20-fold difference in effectiveness. Thus, the term *essential fatty acid* could be reasonably extended to include both EPA and DHA as neither of these fatty acids can be synthesised in the body at rates that appear adequate for maintenance of our well-being.

PUFAs, especially DHA, can also improve the development of the eye function and memory in newly-born infants (Sinclair *et al.*, 2010) such that DHA, along with arachidonic acid (ARA), is now routinely added to infant formulas throughout the world. This is covered in Section 11.3.

On a somewhat cautionary note, ARA (20 : 4 *n*-6) can produce a number of prostaglandins but these are of the 2- and 4-series and are different from those of the 1-series that are derived from EPA and DHA. Prostaglandins of the 2-series are pro-inflammatory and can potentiate thrombosis function. Although this would be somewhat detrimental if ARA were to be consumed in large amounts by adults,

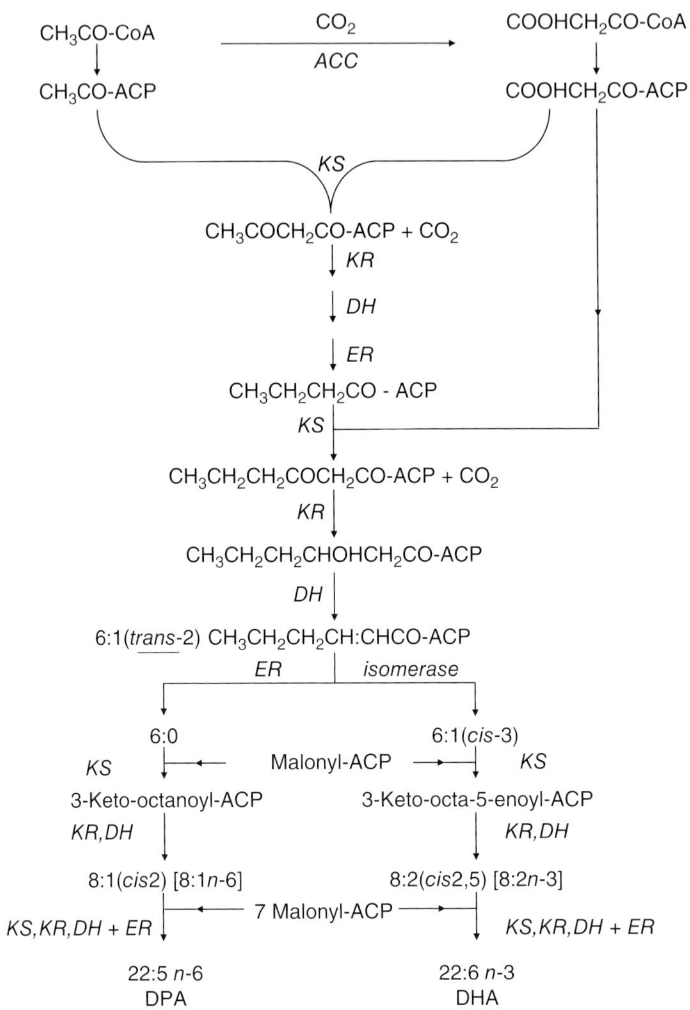
CH3CO-CoA
CO2
ACC
COOHCH2CO-CoA
CH3CO-ACP
COOHCH2CO-ACP
KS
CH3COCH2CO-ACP + CO2
KR
DH
ER
CH3CH2CH2CO - ACP
KS
CH3CH2CH2COCH2CO-ACP + CO2
KR
CH3CH2CH2CHOHCH2CO-ACP
DH
6:1(trans-2) CH3CH2CH2CH:CHCO-ACP
ER
isomerase
6:0
6:1(cis-3)
KS
Malonyl-ACP
KS
3-Keto-octanoyl-ACP
3-Keto-octa-5-enoyl-ACP
KR,DH
KR,DH
8:1(cis2) [8:1n-6]
8:2(cis2,5) [8:2n-3]
7 Malonyl-ACP
KS,KR,DH + ER
KS,KR,DH + ER
22:5 n-6
DPA
22:6 n-3
DHA

Figure 11.3 A suggested scheme to account for the biosynthesis of long-chain polyunsaturated fatty acids in marine microorganisms, and particularly in *Schizochytrium* sp. Other thraustochytrids, using a polyketide synthase (PKS) system (see Metz *et al.*, 2001). The pathway begins with the condensation of acetate, attached to an acyl carrier protein (ACP) and malonyl-ACP [derived from acetyl-CoA using acetyl-CoA carboxylase (ACC)] to give acetoacetyl-ACP. This reaction is carried out by ketosynthetase (KS), sometimes known as *condensing enzyme*. Acetoacetyl-ACP is then reduced by a ketoreductase (KR) to 3-hydroxybutyryl-ACP, which loses water by a dehydratase (DH) reaction to give crotonyl-ACP, which is reduced by an enol reductase (ER) to give butyryl-ACP. Then follows a further addition of malonyl-ACP and the same sequence of reactions, involving KS, KR and DH, up to the formation of 6 : 1(*trans*-2)-ACP. In conventional fatty-acid biosynthesis, this intermediate would be reduced (by ER) to 6 : 0 and the sequence would be repeated a further five times to give 16 : 0 (see Figure 11.2). For the PKS sequence, the 6 : 1(*trans*-2) intermediate might be isomerised to 6 : 1(*cis*-3), which undergoes condensation with malonyl-ACP leading to the formation of 8 : 2(Δ2, Δ5); then follows seven further cycles of condensation with retention of four of the seven double bonds to give DHA as the final product. As *Schizochytrium* spp. and related organisms (but not *Crypthecodinium cohnii*) also synthesise docosapentaenoic acid (DPA) of the ω-6 series, a branched pathway is postulated to occur as the ration of DHA to DPA remains virtually constant under all conditions (Hauvermale *et al.*, 2006) (from Wynn and Ratledge, 2007.).

there appears to be no adverse effects when it is added along with DHA into infant formulas. The reason for adding the ARA is that it prevents the retro-conversion of DHA into EPA (the reverse of the reaction is shown in Figure 11.2) and EPA is not a desirable fatty acid to be given to infants because of its production of undesirable eicosanoids, prostaglandins and hydroxy fatty acids. The background to the addition of DHA + ARA to infant formulas has been reviewed in depth by Sinclair and Jayasooriya (2010).

PUFAs, and *n*-3 fatty acids in particular, have also been implicated to have roles in the nervous system, cognitive development, memory-related learning and neural transmission (see Mazza *et al.*, 2007 for a review). Therefore, inclusion of high levels of these fatty acids in the diet may alleviate numerous neurological and psychiatric disorders (Peet and Stokes, 2005; Mazza *et al.*, 2007; Ross, Seguin and Sieswerda, 2007) including Alzheimer's disease (Hooijmans and Kiliaan, 2008; Ma *et al.*, 2007; Jicha and Markesbery, 2010; Huang, 2010; Daiello *et al.*, 2015) and even chronic bowel disorder (Calder, 2008). A lack of DHA and EPA in the diet has then been suggested as possibly leading to a decline in cognitive functions in patients suffering from Alzheimer's disease (Huang, 2010; see also Raji *et al.*, 2014). There have also been strong claims made to support both *n*-3 and *n*-6 PUFAs having anti-cancer activities (Skender, Vaculova and Hofmanova, 2012; Xu and Qian, 2014; Zheng, Tang and Liu, 2014; Jeong *et al.*, 2014) though clearly these should be considered as adjuncts to the existing chemotherapeutic treatments and not as complete replacements. There is also good evidence to indicate that regular consumption of the long-chain PUFAs (EPA and DHA) can decrease the onset of age-related macular degeneration (SanGiovanni *et al.*, 2008, 2009).

Thus, over many years there has been mounting evidence for the essential role of PUFAs in the diet of both infants and adults. The nutritive value of the long-chain PUFAs, and EPA and DHA in particular, now appears established. The major source of these two fatty acids is fish oils but major questions over the safety of these oils continue to be raised with respect to the possible undesirable levels of organo-mercury compounds as well as other materials such as dioxins and other toxins ingested by the fish from the world's oceans. The following question then arises: what are the likely best and most reliable sources of these fatty acids? Fish oils are undoubtedly the cheapest and most abundant sources but they are always mixtures of EPA and DHA. For maximum nutritional benefits, it is often desirable to consume just one of these fatty acids – normally, it is DHA. This holds true for the addition of DHA to infant formulas. But there are no fish oils that contain DHA alone. No plant source has yet been found that can produce DHA alone or even EPA. For the production of DHA and EPA we must go to microorganisms. Also, there are numerous minority groups who do not wish to consume fish or fish products; for these people alternative, non-animal sources are desired. This chapter now describes the various biotechnological processes that have been developed for the production of the PUFAs as major nutraceuticals.

11.2 General Principles for the Accumulation of Oils and Fats in Microorganisms

Eukaryotic microorganisms, that is yeasts and fungi, have been known for over seven decades to be capable of accumulating lipids up to 70% of their cell dry weights. In all the cases, the process of lipid accumulation depends on growing the cells in a nutritionally unbalanced medium in which one component, not the carbon source but usually the N source, becomes exhausted after the first 24–48 h. The carbon source, which is then in excess, continues to be assimilated by the cells but, as there is now no N available for the synthesis of new proteins and nucleic acids, the cells are unable to continue multiplying. Consequently, the cells are obliged to do something with the excess carbon. This can be in the form of producing polysaccharide materials or, as with the oleaginous microorganisms (i.e. the oil-generating species), it is the accumulation of lipid in the form of triacylglycerols that occurs.

Lipid accumulates in the cells not so much because lipid biosynthetic enzymes become activated after the exhaustion of nitrogen from the culture medium but because other enzyme activities associated with cell growth and replication decline or even cease altogether. The biochemistry of the process has been worked out over a number of years in the author's laboratory (for reviews see Ratledge and Wynn, 2002; Ratledge, 2004, 2014). In general, although there are variations on the theme, when N is exhausted from the medium, this sets up a cascade of metabolic events beginning with the activation of AMP deaminase that deaminates adenosine monophosphate (AMP) to yield inosine monophosphate + NH_4^+ in a seeming attempt to generate additional N for the cells. This, however, has the effect of rapidly decreasing the intracellular concentration of AMP which is an essential cofactor for the activity of isocitrate dehydrogenase (ICDH). ICDH is a component enzyme of the tricarboxylic acid cycle in the mitochondrion of the cell which serves to generate energy (in the form of ATP) for biosynthetic purposes. Without AMP being available, the activity of ICDH decreases or even completely stops. Isocitrate, the substrate of the enzyme, then accumulates, and, because it is in equilibrium with citric acid, this causes a build-up of citric acid also in the mitochondrion. Citric acid is subsequently transported out of the mitochondrion into the cytoplasm of the cell. Here, it is then cleaved by ATP: citrate lyase (ACL) into acetyl-CoA and oxaloacetate. ACL is uniquely found in the oleaginous microbial species and does not occur in those cells that do not accumulate lipid. Thus, there is now a direct production of acetyl-CoA, the major precursor of all fatty acids (see Figure 11.2).

The other product from the action of ACL is oxaloacetate. This is reduced to malic acid which is then oxidatively decarboxylated to pyruvate by malic enzyme with the loss of CO_2 and the generation of NADPH, the essential cofactor used for the reductive reactions involved in fatty acid biosynthesis as carried out by FAS.

A scheme showing these reactions is shown in Figure 11.4.

From an understanding of the biochemical pathways involved in lipid accumulation (rather than just synthesis), we can determine the key activities that make an oleaginous microorganism what it is, as opposed to being a non-oleaginous microorganism such as *Saccharomyces cerevisiae*. A major factor is clearly the possession of the citrate cleavage enzyme, ACL, that generates acetyl-CoA in the cytosol of the organism. However, in addition to providing a copious supply of the carbon precursor for fatty acid biosynthesis, oleaginous microorganisms must also possess the means of reducing the acetyl group, CH_3–CO–, into the growing saturated chain of a fatty acid: –CH_2–CH_2–.

There are two reductive reactions in the cycle of reactions that are carried out by the FAS complex of enzymes, and both require the participation of the reducing cofactor, NADPH. Thus, a supply of this cofactor has to be part of the oleaginicity mechanism. In the majority of oleaginous microorganisms, the generation of NADPH is by the action of malic enzyme (see Figure 11.4). However, in some yeasts, notably *Yarrowia lipolytica*, this enzyme is absent and alternative means of producing NADPH must exist. Other enzymes that can generate NADPH are glucose 6-phosphate dehydrogenase and 6-phosphogluconate dehydrogenase that are part of the pentose phosphate cycle of reactions. A full discussion of the problems that arise in oleaginous cells to generate sufficient reducing power to produce the large amounts of lipid that are found in such cells has been presented by the author Ratledge (2014).

We thus have a clear idea of how oleaginous microorganisms are able to achieve the synthesis of high amounts of lipid, principally in the form of triacylglycerols (see Figure 11.1), which are of the main lipid type of all storage fats and oils and are found not only in microorganisms but also in plants and animals. We can also calculate from the information shown in Figure 11.4 that the theoretical conversion of glucose to a triacylglycerol is 31.6% (Ratledge, 2014). In practice, because some glucose must be used for the production of cell biomass, the yields are somewhat lower: the best yields that have been attained are about 22 g lipid from 100 g

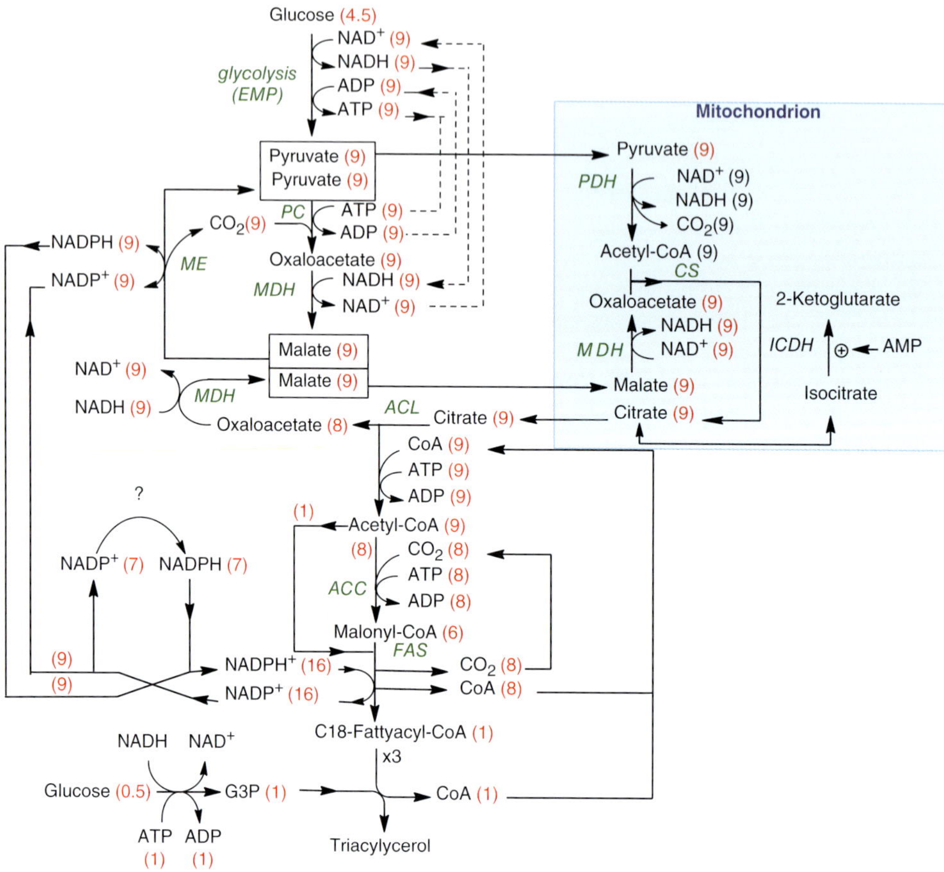

Figure 11.4 Pathway of biosynthesis of fatty acids from glucose in oleaginous microorganisms possessing malic enzyme for NADPH generation. (Adapted from Ratledge, 2014.) Numbers in parentheses indicate the required stoichiometry for the conversion of glucose to 1 mol fatty acyl-CoA. *ACC*, acetyl-CoA carboxylase; *ACL*, ATP:citrate lyase; *CS*, citrate synthase; *EMP* reactions of the Embden–Meyerhof–Parnas pathway; *FAS*, fatty acid synthase; *ICDH*, isocitrate dehydrogenase; *MDH*, malate dehydrogenase; *ME*, malic enzyme (NADP-dependent); *PC*, pyruvate carboxylase; *PDH*, pyruvate dehydrogenase and *G3P*, glycerol 3-phosphate. ? = unspecified reactions, probably using glucose-6-dehydrogenase and 6-phosphogluconate dehydrogenase. It should be noted that when the concentration of AMP falls after N limitation, the activity of ICDH is greatly diminished so that there is little activity of the tricarboxylic acid cycle in the mitochondrion. The requirement of *ICDH* for AMP is indicated by +. Overall conversion: 4.5 glucose + CoA + 9 NAD^+ + 7 NADPH + 17 ATP → C_{18}-fatty acyl-CoA + 9 CO^2 + 9 NADH + 7 $NADP^+$ + 17 ADP + 17P_i.

glucose. If, however, we just consider what happens during the lipid accumulation phase, that is, after nitrogen limitation has been reached, then a conversion of 27% has been recorded for *Y. lipolytica* (Tai and Stephanopoulos, 2013). Thus, in general terms, some 5 t of glucose or equivalent carbohydrate source will be needed to produce 1 t of microbial oil.

11.3 Production of Microbial Oils

Historically, the first production of oleaginous microorganisms occurred in Germany during World War II (1939–1945). The oils were not, however, extracted from the organisms but, instead, the whole biomass was dried and used as direct foodstuff for army horses. This topic has been briefly reviewed (Ratledge, 1992). Although various considerations were made for subsequent developments to produce oils from microorganisms, it was evident from the outset that the costs of producing them would be considerably greater than the costs of producing oils using plants. Agriculture was clearly far cheaper than biotechnology in spite of major developments that had taken place in the 1960s and 1970s for the very large scale production of single-cell proteins (SCPs) using biotechnology.

11.3.1 Production of Gamma-Linolenic Acid (GLA; 18 : 3 n-6)

Although plant oils are generally cheap commodities, there were some that have specialised applications and therefore command premium prices. One such oil is that derived from the seeds of the evening primrose (*Oenothera biennis*) which contains about 9% of its fatty acids as gamma-linolenic acid, GLA (see Figure 11.1). This speciality oil was sold initially as a treatment for multiple sclerosis, a claim that has now been largely discounted, but was subsequently used for the relief of premenstrual tension in women. It is still sold today for this purpose as an over-the-counter (OTC) preparation in the United Kingdom and also in a number of European countries. It is also useful in the treatment of atopic eczema. Other conditions may also respond to it as, although humans can convert linoleic acid (18 : 2) into GLA, the activity of the δ-6 desaturase to carry out this reaction (see Figure 11.2) may diminish in older people and also in certain medical conditions. Hence, evening primrose oil has long been advocated as a fulfilling useful nutritional roles.

In the mid-1980s, as evening primrose oil commanded a price that was about 200 times greater than that commanded by the major commodity plant oils (soybean oil, sunflower oil, palm oil, etc.), there was evidently a niche market that might be exploitable by microorganisms. Fortunately for its commercial development, it was already known that some microorganisms, especially fungi of the order Mucorales, produced GLA in their oils and the task was then to find the most appropriate one that might be used to produce GLA.

Work was carried out in the author's laboratory during the late 1970s that identified *Mucor circinelloides* as the most appropriate organism for the large-scale production of GLA oil out of over 300 strains and species that had been screened for growth, lipid production and GLA contents. It was during this time that the term 'Single Cell Oils' (SCO) was first used to describe microbial oils destined for human or animal consumption in much the same way as SCP had been coined to describe microorganisms with high protein contents that could be used as animal feed materials. The GLA-SCO was produced by J & E Sturge at their fermentation facilities at Selby, North Yorkshire, UK, from 1985 to 1990. The company used its expertise with fungal technology (they were major producers of citric acid using the filamentous fungus, *Aspergillus niger*) to develop SCO production at the 220 m^3 level. The process has been described in some detail (Ratledge, 2006). The fungus could attain cell densities of over 50 g dry weight per litre in about 3 days with an oil content of 25% and a GLA content in the total fatty acids of 18%. A photomicrograph of *M. circinelloides* (although not the actual production strain) is shown in Figure 11.5a. The oil was extracted from the cells

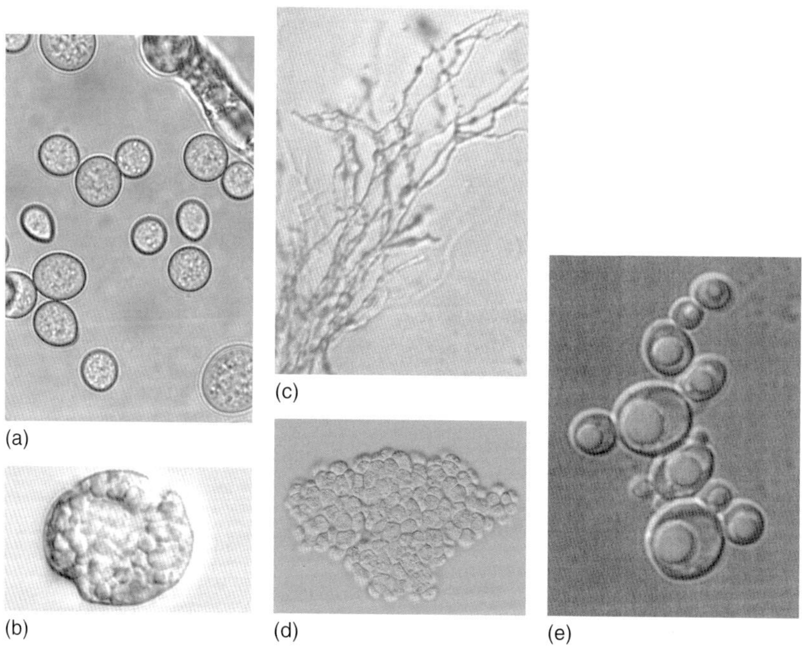

Figure 11.5 Photomicrographs of microorganisms used for the production of SCOs. (a) *Mucor circinelloides* similar to that used for production of GLA-SCO (from Professor Yuanda Song and Mr X. Tang, Jiangnan University, China). (b) *Crypthecodinium cohnii* used for the production of DHA (from Dr Casey Lippmeier, DSM USA). (c) *Mortierella alpina* used for the production of ARA (from Dr C. Lippmeier, DSM USA). (d) *Schizochytrium* sp. used for the production of DHA (from the collection of photographs of the late Dr Kirk Apt and provided by Dr C. Lippmeier, DSM USA). (e) *Yarrowia lipolytica* used for the production of EPA-rich oil – the large droplet in the centre of the yeast is triacylglycerol material (from Dr Quinn Zhu, DuPont USA).

Table 11.1 Fatty acid profiles of fungi and plants used for GLA production.

	Relative % (w/w) of major fatty acids						
	Oil content (% w/w)	16:0	18:0	18:1	18:2 (*n*-6)	18:3 (*n*-6) GLA	18:3 (*n*-3)
Mucor circinelloides[a)]	25	22	6	40	11	18	—
Mortierella isabellina[b)]	~50?	27	6	44	12	8	—
Mortierella ramanniana[b)]	~40?	24	5	51	10	10	—
Evening primrose	16	6	2	8	75	8–10	0.2
Borage	30	10	4	16	40	22	0.5

a) Oil of Javanicus (see text).
b) Production organisms used by Idemitsu Company Ltd, Japan. Oil contents of cells uncertain.

using conventional hexane extraction and, after refinement and purification, was the marketed under the trade name Oil of Javanicus. (The choice of name arose from *M. circinelloides*, originally being named *Mucor javanicus*, as it had been isolated from fermented food produced in Java.)

Before the oil could be sold, it had to undergo stringent toxicological trials as it would be the very first microbial oil that had been sold for human consumption (see Section 11.4). The safety of the oil was considerably helped by the fungus having been originally associated with an oriental food, tempeh, and thus had been consumed by humans for centuries if not millennia. The oil was quickly found to have no toxicological problems and was then sold to the general public under a second trade name of GLA-Forte.

The fatty acid profile of the oil is summarised in Table 11.1 together with that of evening primrose oil. Table 11.1 also lists the fatty acids of two other fungi that were developed by Idemitsu Co. Ltd in Japan in the late 1980s and early 1990s (Nakahara *et al.*, 1992). Although these two fungi, *Mortierella isabellina* and *Mortierella ramanniana*, had higher oil contents than *Mc circinelloides*, their content of GLA in the oil was about half its level in *Mc javanicus* and it is uncertain whether any large-scale production and sales of these oils ever took place.

Production of the GLA-SCO by J & E Sturge, however, ceased in 1990 with the arrival of another GLA-rich plant oil, borage oil, into the market. This oil (see Table 11.1) was produced by *Borago officinalis* with a GLA content of 22% which was slightly higher than the 18% in the GLA-SCO but, importantly, it was cheaper as, amongst other things, it received an EU subsidy for being produced from a non-conventional crop. This meant that the profitability of the biotechnological process was seriously eroded and the company then decided that it was uneconomic to continue production. During the 6 years that the oil was in production, some 50 t was produced.

Although this process was relatively short-lived, it did provide the basic premise that microbial oils were suitable for human consumption and were, in all major respects, as safe as any plant oil. From this small beginning were then laid the foundations on which all further microbial oils could be based. The important principal, though, was established that, if microbial oils were to be commercially successful, then niche markets for them had to be identified, which would then command a sufficiently high price to justify the high costs of the fermentation technology.

11.3.2
Productions of Docosahexaenoic Acid (DHA) and Arachidonic Acid (ARA)

The realisation that mother's milk contained high levels of DHA as the major *n*-3 PUFA and ARA as the major *n*-6 PUFA led to a serious inquiry into the possible benefits of these fatty acids in the diet of newly-born children. It was also established that both these PUFAs were major components of brain lipids (O'Brien and Sampson, 1965; Baker, 1979) and were also found in the membranes of the eye (Anderson, 1970), thereby suggesting their involvement in both neural function and vision. The nutritional roles of both these fatty acids have been reviewed by Sinclair and Jayasooriya (2010). The work of Lucas *et al.* (1992) also established that the IQ values of children who had been breastfed as infants were higher than those of children who had been fed formula milk preparations. This led to the concept that it would be highly beneficial if DHA could be included in the diet of newly-born children. There was also the suggestion that DHA could be regarded as an essential fatty acid as its synthesis in the body seemed to be minimal (Kyle *et al.*, 1992).

With this very strong implication that DHA was an essential fatty acid whose synthesis from ALA (see Figure 11.2) was too slight to be of significance (see Sinclair, Attar-Bashi and Li, 2002), David Kyle in 1985 launched a company, Martek Corporation, that had the production of DHA using a marine alga as its objective. The organism of choice was *Crypthecodinium cohnii*, a non-photosynthetic dinoflagellate, whose lipid composition had been reported by Harrington and Holz (1968) as containing DHA as its sole PUFA. Martek established a process to produce DHA but not without difficulties as most of the available strains of this organism proved to be unsuitable for large-scale cultivation in stirred fermenters. However, by virtue of careful strain selection and process improvement, together with modifications to the culture conditions (see Wynn *et al.*, 2010), a successful oil production process was developed.

The oil content of the cells exceeded 40% and the DHA in the total fatty acids also was above 40%. Extraction and purification of the oil followed standard procedures used in the oil industry for the production of high-purity oils. Once the oil was produced in reasonable quantities, Martek had the foresight to offer free samples of the DHA-SCO to clinicians and nutritionists working with neonatal infants so that the effects of the oil on their development could then be evaluated at first hand. The results from these studies quickly led to a market demand for the DHA oil and, within a few years, the nutritional benefits of including the oil in infant formulas were established. Direct evidence was provided that indicated that DHA could significantly improve memory and visual acuity in both premature babies and neonates (Sinclair *et al.*, 2005; Sinclair and Jayasooriya, 2010). The oil was given GRAS (generally recognised as safe) status by the FDA in 2002. This was then the major factor that established that the oil was entirely safe and could therefore be added as a key nutritional supplement to infant formulas. The recommended level of incorporation is between 0.32% and 0.64% of the total fatty acids (Birch *et al.*, 2010).

The oil is now manufactured in a series of fermenters of over 1000 m^3 total capacity, with individual fermenters of approximately 200 m^3. It is sold and incorporated into infant formulas in over 70 countries of the world, thus creating a huge demand for the product. The profitability of the company was such that when Martek was sold in 2011 to DSM in the Netherlands, it commanded a price of US$ 1.1 billion. Current sales of the oil exceed 2500 t/annum. For the last year of their trading as an independent company, Martek achieved a revenue of US$ 317 million on the sales of oil for infant nutrition. (If we assume that the selling price of the oil is, not unrealistically, about $100 kg^{-1}, then the total amount of oil sold must be over 3000 t.)

A profile of the fatty acids of *C. cohnii* is summarised in Table 11.3, and a photomicrograph of the cells is shown in Figure 11.5b. The oil is now included in infant formulas throughout the world and, in the United States alone, it is included in 99% of all such preparations.

One problem of significance, identified during the early trials of DHA as a supplement for infant formulas, was that it was retro-converted to EPA (see Figure 11.2). This was regarded as undesirable but it was realised that if ARA (20 : 4 *n*-6) were to be added to the DHA oil then it would prevent this from occurring. Thus, a second process had to be developed to produce ARA. This was done independently of Martek's work on DHA production by Gist-brocades in the Netherlands, who later became part of the DSM organisation.

ARA (20 : 4 *n*-6) was initially discovered as a major fatty acid in the fungus *Mortierella alpina* by Totani and Oba (1987), where it represented up to 79% of the total fatty acids. This work, however, had only been carried out using cells grown on solid agar plates. Nevertheless, it proved relatively simple to achieve submerged cultures of the fungus that also produced high contents of ARA (Totani, Watanabe and Oba, 1987). As there are no realistic alternative sources of this PUFA, when it came to considering how ARA might be produced in sufficient quantities to be incorporated into infant formula along with DHA, there was little hesitation in opting for the fungal route of production. Commercial production of ARA was developed simultaneously in Japan and in Europe; in the former case, Suntory Ltd and Lion Corporation had active development programmes whilst in Europe it was Gist-brocades Co. (now DSM) that was primarily active. The main driving force for developing this process, however, came from Martek who needed ARA to be blended with their DHA oil to improve its nutritive value. An agreement was reached between Gist-brocades and Martek that led to the former producing the ARA-SCO using *Mta alpina* with exclusive sales to Martek. The ARA-SCO and the DHA-SCO were, and still are, mixed together at 2 : 1 (v/v) and thus provide the ideal supplement for adding to infant formula.

Table 11.2 Oils derived from *Mortierella alpina* containing arachidonic acid (ARA).

	16:0	18:0	18:1	18:2	18:3 (*n*−6)	20:3 (*n*−6)	ARA 20:4 (*n*−6)	22:0	24:0
ARA-SCO[a)]	8	11	14	7	4	4	49	—	1
CABIO oil[b)]	7.5	6	9	6	2.5	4	43	3	9.5

a) Oil produced by DSM (Netherlands).
b) Oil produced by Cargill Alking Bioengineering (Wuhan) Co. Ltd, from Casterton *et al.* (2009) and Kusumoto *et al.* (2007).
Values are relative % (w/w) total fatty acids.

The process of ARA production has been described in general terms by Streekstra (2010). The organism is somewhat slow growing and requires 8–10 days to reach maximum cell density and lipid levels. It is grown in large (100–150 m^3) fermenters and uses a similar technology to that used for all cultivations of oleaginous microorganisms for SCO production. Contents of ARA in the total lipids can be up to 50%. The fatty acid profile of the oil is summarised in Table 11.2 and a photomicrograph is shown in Figure 11.5c.

A similar process, also using *Mta alpina* for the production of ARA, was developed by Cargill Inc. in conjunction with Wuhan Alking Bioengineering Co. Ltd in Wuhan City in China. In 2010, a dedicated fermentation plant was opened, but sales of the oil were restricted to China in view of the existing patents of Martek/DSM. The profile of the Cargill oil, known as *CABIO oil*, is summarised in Table 11.2 where it can be seen that there are some minor differences in fatty acid composition compared to the DSM oil, notably the higher content of lignoceric acid (24 : 0) in the former. As Cargill's oil has now received authorisation from the European Commission that would allow the ARA-SCO to be sold in Europe (Cargill.com/news/releases/2012/NA3053758), there is likely to be some dispute between the companies as to who has the legal rights to continue ARA production and sales in Europe. It is expected that this matter will be resolved in 2015.

11.3.3 Alternative Sources of DHA

As long ago as 1991, an alternative microbial source of DHA to that produced by *C. cohnii* had been identified by William Barclay who headed a small company, Omega-Tech Inc. in Boulder, CO, USA (Barclay, 1991). This was the use of organisms known as thraustochytrids. Independent work reported by Bajpai, Bajpai and Ward (1991) and Kendrick and Ratledge (1992) had indicated that these organisms appeared to be unlikely sources of DHA and other PUFAs because of their poor growth rates and also their relatively low contents of lipid (<18%) although they did produce DHA. Barclay, however, instead of relying on organisms that had been taken from culture collections (and therefore may have had an uncertain provenance) isolated a number of thraustochytrids directly from marine environments. These proved to be capable of rapid growth and high lipid accumulation.

Thraustochytrids are marine, non-photosynthetic microalgae that originally were considered to be fungi but work by Cavalier-Smith, Allsopp and Chao (1994) placed them as belonging to the heterokont group of algae. They are classified as members of the order of Labyrinthulomycetes and are widely distributed in both temperate and tropical coastal waters. All species appear to produce large amounts of n-3 long-chain PUFAs with DHA being the most abundant.

The work of Barclay established the commercial possibilities for producing an oil rich in DHA using a species of *Schizochytrium* that was a member of the thraustochytrid group. This organism was deposited in the American Type Culture Collection as ATCC 20888. An account of this early work leading up to large-scale production has been provided by Barclay *et al.* (2010). The main problem in developing the process was to create an appropriate culture medium that did not have to use high concentrations of sea salts as these were detrimental to the mild steel used in the construction of the large fermenters. Chloride ions were particularly undesirable but these, it was found, could be replaced by sulfate ions. There was also considerable strain improvement as well as optimisation of the culture medium and conditions of growth. Bailey *et al.* (2003) recorded, in a patent, cell dry weight values of up to 210 g/l being attained in less than 72 h with the cells having lipid contents of 50–73% and a DHA content in the extracted oil of between 35% and 45%. These are, for any oleaginous microorganism, impressive values and probably are the highest yields and productivities ever achieved. Omega-Tech Inc., once it was established as having attained a potentially highly valuable process, was taken over by Martek BioSciences in 2002, and the process, with the acquisition of Martek by DSM, as mentioned earlier, is therefore now the property of DSM.

A profile of the fatty acids in the *Schizochytrium* oil is summarised in Table 11.3 and a photomicrograph of the organism is shown in Figure 11.5d. Of considerable interest, but of some initial concern, was the presence of another long-chain PUFA besides DHA. This was DPA but was of the *n*-6 series (DPA, *n*-6). The concerns about the possible safety of this unusual PUFA were, however, nullified when it was appreciated that DPA was a major component of the human brain tissue along with DHA and could not, therefore, in any way be considered as an undesirable or even toxic fatty acid. Opposition to the inclusion of the oil in materials destined for human consumption and also for animal feedstuffs was then withdrawn. The oil was given GRAS status by the FDA in 2004 (see Barclay, Weaver and Metz, 2005).

The commercially-produced *Schizochytrium* sp. contains up to 60% oil with at least 40% of the total fatty acids being DHA. The presence of DPA is not regarded as a problem. The oil has been used extensively for animal feeding under the trade name DHA-Gold. This has included feeding to poultry for the production of eggs rich in DHA. Interestingly, although the eggs contained high amounts of DHA, DPA was not present even though it was a constituent of the *Schizochytrium* oil.

Other processes for the production of DHA using alternative species of thraustochytrids have been developed by companies such as Nutrinova GmbH,

Table 11.3 Commercially available microbial oils containing docosahexaenoic acid (DHA).

	14:0	16:0	16:1	18:0	18:1	DPA 22:5 (n−6)	DHA 22:6 (n−3)
DHA-SCO[a)]	20	18	2	<0.5	15	—	40
Schizo-SCO[b)]	7	16	<0.5	1	16	16	39
Ulkenia-SCO[c)]	3	30	<0.5	1	—	11	43–46
Schizo-ONC[d)]	13	27	2	1	<1	8	40
Schizo-TK[e)]	6	18	–	0.5	0.5	19	49

Principal fatty acids given as relative % (w/w) of total fatty acids.

a) Oil from *Crypthecodinium cohnii* (as produced by DSM); trade name: DHASCO and sold as *life's* DHA.
b) Oil from *Schizochytrium* sp. ATCC 20888 (as produced by DSM); trade names DHASCO-S and DHA-Gold.
c) Oil from *Ulkenia* sp. (originally manufactured by Lonza, Switzerland).
d) Oil from *Schizochytrium* sp. as produced by Ocean Nutrition Canada Ltd.
e) Oil from *Schizochytrium*-TK as produced by Jiangsu TianKai Biotechnology Co. Ltd (Nanjing, China).

Frankfurt, Germany that was subsequently acquired by Lonza Group AG in Switzerland in 2005. This process used *Ulkenia* sp. and was described somewhat briefly by Kiy, Rusing and Fabritius (2005). A profile of the fatty acids in the oil is summarised in Table 11.3. Similar to the *Schizochytrium* oil, this too contained DPA. Indeed, DPA is now regarded as a signature fatty acid that indicates that the origin of the oil is, in all probability, from a thraustochytrid microorganism. Lonza now sells the DHA oil as DHA-CL (clear liquid) and DHAid but, because of a dispute with DSM over patent rights, the *Ulkenia* oil is no longer produced, and, instead, Lonza buys its DHA oil directly from DSM.

Ocean Nutrition Canada Ltd also developed a similar process using *Thraustochytrium* sp. ONC-T8 (see Burja *et al.*, 2006) with similar characteristics of the oil (see Table 11.3). Because this company had other products derived from fish oils and had significant oil processing and refining facilities, including microencapsulation, in Nova Scotia, Wisconsin, and in Peru, they were an attractive target for acquisition. DSM was again the buyer paying $CAN 540 million for the company in 2012.

Other companies may also be involved in developing similar processes using other strains and species of the thraustochytrids. However, it is likely that the majority of companies offering DHA for sale as derived from microalgae may be obliged to buy their oil from DSM. This does not, however, apply in China, and at least one Chinese company, but there may well be others, is producing such oils: Jiangsu TainKai Biotechnology Co. Ltd, based in Nanjing, has developed a *Schizochytrium*-based process for the production of an oil similar to those listed in Table 11.3 (see Ren *et al.*, 2010). This is used as a supplement for infant formula after being mixed with ARA in the usual ratio (1 : 2 v/v) and is thus presumably judged to be safe for this use even though no company in Europe or North

America has yet taken this step with these thraustochytrid oils to include them in infant formulas. There has also been an announcement (November 2014) of the intended construction of a kiloton production process for a DHA oil in Qingdao City, China, using an unspecified marine microalga. In all probability, this will be using another thraustochytrid as the Chinese press release mentions oil yields of 110 g/l with DHA at 45–50% of the total fatty acids (http://www.most.gov.cn/dfkj/sd/zxdt/201411/t20141102).

11.3.4 Production of Eicosapentaenoic Acid (EPA n-3)

EPA gives rise to a number of prostaglandins, prostacyclins, thromboxanes, leukotrienes and lipoxins, all of which have physiological roles in human metabolism (see Cottin, Sanders and Hall, 2011; Nicholson, Khademi and Moghadasian, 2013). Administration of pure EPA as its ethyl ester has shown some beneficial effects on various neuropsychiatric disorders that include bipolar disorder, depression and schizophrenia (Peet and Stokes, 2005; Riediger, Othaman and Suh, 2009; Lin, Huang and Su, 2010; Sublette *et al.*, 2011). Natural sources of oils rich in EPA as the sole long-chain PUFA, however, are non-existent, and its production relies on the extremely expensive process of fractionating fish oils using very large-scale high-performance liquid chromatography (HPLC). EPA is then usually produced as its ethyl ester. Such preparations are currently sold by Amarin Corp. plc (a company registered both in the United States and Ireland) under the trade name of Vascepa which is an ultra-pure EPA ethyl ester for the treatment of adults with high levels of triacylglycerols in their blood. A similar product to treat the same condition had previously been launched by GSK (GlaxoSmithKline) under the trade name of Lovaza. This, however, is simply a purified fish oil that contains both EPA and DHA. This pharma-grade fish oil is produced by BASF through its earlier acquisition of Equateq, now BASF Pharma Callanish, based in the Outer Hebrides, Scotland. The costs of producing the pure ethyl EPA ester are, however, considerable. The FDA of the United States has also approved (July 2014: http://newhope360.com/breaking-news/fda approves-astrazeneca-s-fish-oil-drug) a mixture of DHA and EPA but as free fatty acids and not their ethyl esters. Again, this mixture is derived directly from fish oils and is sold under the name of Epanova. It is produced by AstraZeneca. Cheaper alternatives to this approach of purifying fish oils and for producing EPA as the sole PUFA therefore would be clearly desirable. But, here, EPA would be considered a medical food rather than a simple dietary supplement or nutraceutical as its use would be directed to the treatment of a specific clinical condition.

EPA-rich oils are not found in naturally-occurring microorganisms, including algae, in the same way that oils with high contents of ARA or DHA have been identified. An original approach was, though, developed by DuPont, USA, to genetically engineer a yeast to produce an oil with a high content of EPA. This work began in the early 2000s and has now led to the production of an oil with 57% EPA

Table 11.4 Fatty acid profiles of oils from *Yarrowia lipolytica* and *Schizochytrium* sp. that contain high levels of EPA.

Fatty acid	Y. l. WT[a)]	Y4305[b)]	Ovega[c)]
14 : 1	—	—	2
16 : 0	18	3	21.5
16 : 1	16	0.7	<0.1
18 : 0	6	1	1
18 : 1	45	4	2.5
18 : 2	15	17	—
18 : 3(*n*-3)	—	2	—
20 : 2	—	3.5	—
20 : 3(*n*-6)	—	2	—
20 : 4(*n*-6)	—	0.6	1.6
20 : 3(*n*-3)	—	0.7	0.2
20 : 4(*n*-3)	—	2	—
20 : 5(*n*-3)	—	56.6	21.7
22 : 5(*n*-6)	—	—	1.6
22 : 5(*n*-3)	—	—	3.5
22 : 6(*n*-3)	—	—	40

a) Y.l. WT = *Yarrowia lipolytica* wild-type (ATCC 20362); from Xue *et al.* (2013) and (Dr Q. Zhu, DuPont Co. Ltd (personal communication).
b) Y4305 = *Yarrowia lipolytica* recombinant strain derived from the wild-type; from Xue *et al.* (2013).
c) Ovega oil produced by *Schizochytrium* sp; from submission to FDA (USA) by DSM: 'DHA SCO-B produced from a new strain of *Schizochytrium* for use as an ingredient in infant formula (pre-term and term)', 2014; from (Dr C. Lippmeier, DSM, USA (personal communication).

(see Table 11.4). The yeast of choice was *Y. lipolytica*. It was selected, because at the time of the commencement of the work, it was the only oleaginous yeast whose genome had been sequenced. Also, being a haploid yeast, it was also amenable to genetic manipulation. The final recombinant yeast (Y4305) contains 30 copies of 9 different genes (Xue *et al.*, 2013). When grown under appropriate conditions to engender lipid accumulation, the yeast contains about 30% of its biomass as lipid. It has been in commercial production for approximately 3 years. The biotechnology behind the work carried out has been described in some detail by Xue *et al.* (2013); some 90 patents and applications have been filed for the process and the product. A photomicrograph of the oleaginous yeast is shown in Figure 11.5e.

The first commercial product from the yeast was marketed through New Harvest (a wholly owned subsidiary company of DuPont) as an OTC, vegetarian/vegan ω-3 fatty acid supplement. The subsequent use of the product has been as a fish food supplement for rearing sustainably farmed salmon. Here, however, the whole yeast is used rather than the extracted oil. This clearly decreases the downstream processing costs as no extraction and refinement of the oil need to be carried out. The company concerned is Verlasso which is a joint venture between AquaChile, which raises salmon, and DuPont. Although the final salmon sells for \$4–10 kg^{-1}

more than other salmon varieties, it appears to have an improved taste and other characteristics that justify this additional cost. It would not be unreasonable, however, to consider that this is an under-use of the oil as it would appear to have some potential for being considered a medical food because of its high EPA content. Its possible clinical use as an equivalent to the highly purified ethyl EPA, as sold by Amarin, for the treatment of hyperglycerolaemia would appear to be worthy of some investigation. At the very least, the oil would seem to be a much more appropriate starting material than fish oils for the manufacture of 98% purity ethyl EPA.

Of some possible interest as a source of EPA, although with DHA still being present, comes from another oil derived from another *Schizochytrium* sp. that has been launched by DSM to stand alongside its other oils (see Table 11.3). This oil is called *Ovega-3* and is now available via a number of outlets including online sales and health food stores as an OTC neutraceutical. The fatty acid profile of the oil is summarised in Table 11.4 where it can be seen that EPA is about 22% of the total fatty acids with DHA at 40%. The content of DPA (20 : 5 *n*-6), which is the usual 'fingerprint' fatty acid for all thraustochytrid oils, is now only 1.6% in the fatty acid profile. Although the content of EPA is much higher than is normally found in these microorganisms, the final oil should be regarded as a vegetarian/vegan substitute for fish oil and not as a source of high levels of EPA. This, indeed, is how it is being marketed. The fact, though, that *Schizochytrium* spp. can produce EPA (*n*-3) is of obvious interest as it indicates that the pathway of PUFA biosynthesis by the PKS pathway (see Figure 11.3) is capable of considerable variation to account for the variety of fatty acids found in this particular species.

11.3.5
Prospects of Photosynthetic Microalgae for Production of PUFAs

Notwithstanding the classification of both *C. cohnii* and the thraustochytrids as algae, none of these organisms is capable of growing photosynthetically. Their growth is, therefore, in large, stirred fermenters using a fixed carbon source (glucose or sucrose).

The potential of using photosynthetically-grown alga for the production of PUFAs, on first inspection, seems to be an extremely attractive proposition. There is the prospect of free carbon as CO_2 and free energy in the form of sunlight. Unfortunately, however, putting this into practice is not that easy. Firstly, algae tend not to produce large amounts of triacylglycerols (although there are some exceptions, see Leu and Boussiba, 2014) which form the basis of most neutraceutical oils to date. Algae produce complex lipids in the form of various phospho- and glycolipids associated with the photosynthetic apparatus. Extraction and production of a triacylglycerol oil is thus not feasible and, instead, all the lipids, both neutral and polar, have to be extracted from the cells. Some hydrolysis of the lipids may be necessary followed by either enzymatic or chemical reformulation of the released fatty acids into acceptable products. Alternatively, the polar lipid fraction might be an acceptable alternative to a triacylglycerol oil as this would be readily digested in the body and the released fatty acids would then be taken up in the same way that they are when derived from triacylglycerol oils.

Secondly, algae are similar to other oleaginous microorganisms and require a surfeit of carbon in the medium to ensure a high accumulation of lipids in their cells. This means that it is not sufficient for the algae to rely on atmospheric CO_2 as the provider of the essential carbon for fatty acid biosynthesis. Instead, for engendering high contents of lipid, one must enrich the culture system with additional CO_2 coming from an alternative source (and therefore necessarily more expensive) than the atmosphere. The source must perforce be clean and plentiful and the usual source of such CO_2 would be from commercial bottled gas.

Thirdly, there is the choice of cultivation system. Most protagonists for algal cultures for lipid production have based their conclusions on using data derived from photobioreactors. Even the simplest of these, polythene tubing arrays held on outdoor frames, are fraught with difficulties. There is the problem of cell adhesion to the tube walls that then creates shading within the rest of the tube thereby restricting growth. These systems are, moreover, prohibitively expensive to run for all but the highest value products. Nevertheless, such systems are used but are for the production of very expensive products such as astaxanthin which is produced in Israel using *Haematococcus pluvialis* as production organism (see Ratledge and Cohen, 2008; Leu and Boussiba, 2014). Ratledge and Cohen (2008) calculated that the cost of producing 1 kg of biomass is about US\$ 40, although this figure has been recently revised to \$100 kg^{-1} (Leu and Boussiba, 2014). If the alga being grown in a tubular array contained 40% lipid, then this would imply that the oil would cost at least \$250 kg^{-1}. But this does not include the costs of downstream processing, oil extraction, refinement and purification. These costs could therefore increase the price of the final oil to well over \$350 kg^{-1} which is far higher than any of the currently produced SCOs. Clearly, such costs would not allow for a reasonable return on investment for the production of even the most expensive PUFA oils using photobioreactors.

Thus, consideration of cheaper alternatives for algal cultivation is necessary. Such systems include the use of lagoons or ponds (raceways) of suitable size but these, being outdoors, do not allow the essential additional CO_2 to be fed in an economic manner to the growing alga. Such systems, though, are regarded as good if they can produce more than 400 kg biomass/hectare per year. For production of, say, 100 t of oil per year, some 2500 ha (about 6250 acres) would be needed, an area too large to consider placing under cover even of the simplest type. It would therefore seem to be an impossible situation: you either use closed tubular reactors so that CO_2 can be introduced to the culture to boost lipid accumulation, but this is too expensive, or use outdoor lagoons or ponds but then the alga would not accumulate lipid at much over 15–20% of the biomass. In addition, as such systems are open to the atmosphere, there comes the problem of verifying that this is a good manufacturing system as a monoculture of the alga in question cannot be guaranteed. Pollution of the culture water from the air or surrounding land areas would seem to be inevitable.

Table 11.5 EPA contents of photosynthetically grown algae.

Organism	Percentage of EPA in total fatty acids
Pavlova spp.	16–29
Asterionella sp.	26
Chaetoceros constrictus	19
Nannochloropsis oceania	23
Nannochloropsis oculata	30–36
Nannochloropsis salina	26
Nannochloropsis spp.	31–33
Phaeodactylum tricornutum	14–30
Porphyridium cruentum	20–38

It should be noted that the total lipid contents of these algae were not given. The lipids will necessarily be complex phospho- and glycolipids with only small amounts of triacylglycerols.
From Bellou *et al.* (2014).

Nevertheless, a number of companies have declared interests in pursuing algae as sources of PUFAs for, undoubtedly, algae are potentially very useful sources of these materials. Table 11.5 summarises a selection of algae that have been considered for the production of EPA. (Production of DHA is now regarded as the province of the thraustochytrids and *C. cohnii* which are not likely to be bettered either in terms of amounts of DHA in the oil or in the overall costs of production.) An authoritative review of the potential of microalgal lipids has recently appeared (Bellou *et al.*, 2014), from which the data in Table 11.5 is taken.

Companies that are actively pursuing the production of EPA/DHA oils using photosynthetic microalgae include Aurora Algae, Qponics, AlgaeBio, BioProcess Algae, Algae Biotechnologies, Renewable Algae Energy and Qualitas Health. All companies have developed technologies for the cultivation of various algae for a variety of products. Of the ones mentioned, only Qualitas Health (with its head office in Jerusalem, Israel), however, has currently achieved commercial production of an EPA-rich, polar lipid-structured ω-3 oil. The company uses *Nannochloropsis oculata* and is the first company to manufacture an ω-3 oil using phototropic algal technology; (press release: 2 December 2014: (http://money.cnn.com/news/newsfeeds/articles/prnewswire/LA76287.htm). The oil is sold under the trade name of Almega PL (where PL stands for polar lipids).

The fatty acid profile of the oil is summarised in Table 11.6 where it is compared to the fatty acids in krill oil and it can be seen that the EPA content is a very reasonable 25% of the total fatty acid with a small amount of ARA (20 : 4 *n*-6) as the only other major PUFA. Although the oil is composed of various phospholipids and glycolipids and, thus, is typical of oils found in the majority of photosynthetically-grown algae, it is considered that these may be more easily digested than the more usual triacylglycerols found in heterotrophically-grown yeasts and fungi (see www.almegaPL.com). An additional benefit of the algal oil is that it may avoid the burping that is often found with the ingestion of fish oils, including

Table 11.6 A comparison of fatty acid profiles from *Nannochloropsis oculata* (Almega PL oil) and krill oil (from Kagan *et al.*, 2013).

Fatty acid[a]	14:0	16:0	16:1	18:0	18:1	18:2	20:4 (n−6)	20:5 (n−3)	22:6 (n−3)	Others
Almega oil	2.5	9.6	12	0.1	1	1	5	25	—	13.6
Krill oil	5	12	3	0.6	6	1	0.4	13.6	7	16.4

a) Values are given as % of oil. In addition to the fatty acids, the algal oil contained 17.3% polar lipids and the krill oil contained 33.4% polar lipids plus 2% cholesterol.

krill oil. So far, two clinical trials of the algal oil have been carried out (Kagan *et al.*, 2013; Kagan, Levy and Leikin-Frenkel, 2015) comparing its performance with krill oil (the only other fish oil that has a predominance of EPA in its fatty acids). Both studies showed similar absorptions in humans and tissue uptake in animals with some evidence that the algal oil may be superior to krill oil. The alga is grown phototrophically in large open raceways, in West Texas (http://omega3.supplysideinsights.com/documentaries/2014/08/omega3.aspx).

Although the lipid content of the cells has not been declared, it is likely for the reasons advanced that it will not be over 20% (w/w) of the cell dry weight. How this oil will then compare with the EPA oil produced using the genetically-modified *Y. lipolytica* in terms of nutritional benefits, customer acceptability and, importantly, overall costs of manufacture will be of considerable future interest.

11.4 Safety Issues

It is axiomatic that all microbial oils have to undergo stringent trials before they can be offered for sale to the general public. These trials will include feeding trials with laboratory animals for periods of up to 90 days or even longer. Trials in humans are also necessary if claims for improvements in specific conditions or for the alleviation of an illness are being claimed for the oil. The data assembled concerning the safety of the oil and its effectiveness must then be submitted to appropriate regulatory authorities to gain approval for the oil to be sold to the general population. Single-cell oils, however, unlike food materials such as SCPs, have the advantage that they are purified entities that can be defined to less than 0.1% of their total weight so that it is technically possible to identify all the components in the oil besides the principal triacylglycerols. The adverse effects of any of the minor components on human health will probably be known as, almost invariably, these components are already known to occur in other food substances.

With each new microbial oil arriving at the market, establishing its safety for use becomes a little easier. Each oil can be compared with all the previous oils to show that it too has a similar lipid profile and that all the fatty acids and lipid components in the oil are known and have already been ingested by humans, often for many centuries if not millennia. The first microbial oil that was offered for sale was Oil of Javanicus, being the GLA-rich oil from *Mucor circinelloides* – see Section 11.3.1. Initially, the oil was found to be toxic to the first experimental animal on which it was tested – brine shrimps – and this appeared to be a major setback for the oil until it was realised that the oil contained a high proportion of free fatty acids. These had arisen during the harvesting of the cells and their subsequent extraction using hexane. The naturally-occurring lipases within the cell had continued to be active and, as a result, caused partial hydrolysis of the lipid leading to the appearance of free fatty acids in the final oil. Once this was appreciated, it was a relatively simple matter to heat the entire culture broth, albeit 220 m^3, before the cells were recovered and further processed. (This heating step is now a standard procedure used in the production of all SCOs.) The oil then obtained from the heat-treated cells was completely free from any toxicological effects on brine shrimps or any other animal on which it was tested. After feeding trials had been completed against various rodents for 90 days, the oil was found to be completely free of any undesirable effect and no objections were raised by the then regulatory body in the United Kingdom: the UK Advisory Committee on Novel Foods and Processes for the oil to be sold to the general public. The stringency of testing of this oil was absolute as it was the first microbial oil to be offered for sale. It was obviously compared to plant oils for its safety evaluation and, in some respects, was found to be superior. For example, the content of herbicides, insecticides and fungicidal agents was much less in the microbial oil than in plant oils, such as sunflower oil or soybean oil. The reason was simple: microbial oils are not sprayed with a variety of chemical agents as are used with the vast majority of commercially-grown plants. Minute traces of these residues were, though, found in the GLA-SCO and these had arisen from the plant-derived glucose used as the principal carbon source. Other plant products, such as soybean meal, may have also been used in the culture medium as supplementary sources of amino acids and vitamins and, again, these contained residues of the chemical sprays that had been used on them when growing in the field. Suffice it to say that the amounts of these residues were well below the nationally and internationally agreed limits.

Although production of Oil of Javanicus ceased in 1990, it was by then firmly established that microbial oils were safe and free from any undesirable effects. The oil could be tolerated by humans and compared favourably with any conventional plant oil. There were no allergic reactions or other adverse immunological consequences.

As further microbial oils came on to the market, the need for lengthy feeding trial for humans or infants did not seem to be necessary although for the oil from *C. cohnii* (see Section 11.3.2), being included in infant formula, there was clearly a need to establish its complete safety. Establishing the safety of this oil (see Zeller, 2005) was considerably helped as the production organism was known to be non-pathogenic for humans or animals and did not produce any toxins, nor was it related to any toxin-producing alga. Studies on the safety of this DHASCO have been reviewed in depth by Kyle and Arterburn (1998), Zeller (2005) and Ryan, Zeller and Nelson (2010). These have included testing for *in vitro* mutagenicity and geno-toxicological trials plus numerous studies with a variety of rodents and non-rodent animals. Effects on development and reproduction have also been examined. No evidence has been found that this oil, nor indeed any of the other SCOs in current production, has caused the slightest problem to any consumer or indeed in any animal, including fish, that have received them. Trials have also been carried out using very high doses of some of the oils where laboratory animals have been given DHASCO-S (the oil from *Schizochytrium* sp.) was at 5% (w/w) of the diet (Fedorova-Dahms *et al.*, 2011) which was the equivalent of a human dose of about 30 g oil per day. No adverse effects were noted in any of the numerous parameters that were evaluated. Similar results have been reported for trials of both the DHA oils from *Schizochytrium* and *C. cohnii* in pre-weaned farm piglets and showed that both oils were well tolerated during the 3 weeks of the feeding trial that commenced immediately after birth (Fedorova-Dahms *et al.*, 2014). The only problem identified when adult humans were fed 7.3 g DHASCO per day, which is about 25 times the recommended daily dose, was a complaint of 'fishy burps' (Wynn and Ratledge, 2006).

Single-cell oils are probably some of the most extensively tested microbial products. Their safety has been examined for over three decades without any substantial adverse report of their effects being noted. None of the regulatory authorities have ever prohibited the use of any SCO even though the most stringent requirements for the oils are demanded. For further details of the regulatory aspects governing the acceptance of microbial oils, the erudite reader is referred to the authoritative reviews of Zeller (2005) and Ryan, Zeller and Nelson (2010).

11.5 Future Prospects

Microbial oils have found their niche markets as high-value nutraceuticals for the supply of PUFAs into the diet of humans and also animals. All the currently available oils, with the exception of the phospholipid preparation derived from *Nannochloropsis* (see Section 11.3.5), are derived from heterotrophically-grown organisms, that is organisms using a fixed carbon source (glucose or sometimes sucrose) in large stirred and aerated fermentation vessels. Research continues to improve the productivity of all the major species currently in commercial production. Also, the search for new microorganisms offering higher contents of the major PUFAs still continues. It is more than likely that more members of the thraustochytrids, which have already proved exceptionally good at producing DHA and, in some cases, EPA + DHA, will be found to add to the list of important oleaginous organisms. Genetic engineering of appropriate yeasts and possibly other oleaginous organisms is also likely to continue to be a major method of improving yields and desirable fatty acid profiles. As the marketplace is now secure for microbial oils, at least for the foreseeable future, we can expect the major commercial players in this field to achieve considerable improvements in overall yields and productivities. It is also not unreasonable to expect that the commercial companies will develop oils that could be directly useful as medical foods for the treatment of a variety of clinical disorders, including various neurological and psychiatric disorders – see Section 11.1. Clearly, work on the nutritional value of PUFAs will continue apace and will be aided by the availability of a number of microbial oils having just a single PUFA in their fatty acids. Such oils should help considerably in determining which fatty acids are the most efficacious for the treatment of which particular disorder. Treatment of various cancers may also involve PUFAs as a means of supporting and optimising the responses to conventional chemotherapy thereby opening up further opportunities for the development of single cell oils.

Of major future significance is likely to be the arrival in the marketplace of a variety of oils derived from photosynthetic algae. The advent of appropriate technology for the large-scale cultivation of photosynthetic algae in outdoor ponds and raceways has seemingly opened up opportunities for algal oils now to be produced for the same market. The first product being launched, AlmegaPL from Qualitas Health, now presents new possibilities of exploiting other photosynthetically-grown algae for PUFA production. However, AlmegaPL differs significantly from the other SCOs as it is not a triacylglycerol oil but is a mixture of polar lipids derived by extraction of the production organism, *N. oculata*. There would seem no valid reason why such an oil would not be able to provide the body with the desired PUFAs as polar lipids will be digested as readily as triacylglycerols. With there being numerous other algae of potential interest for the production of PUFAs (see, e.g. Bellou *et al.*, 2014; Ryckebosch *et al.*, 2014; Leu and Boussiba, 2014), it would seem only to be a question of time before oils from other algae appear on the market. However, it is more than likely that these algal lipids will be mixtures of the main long-chain PUFAs: EPA + DHA with perhaps ARA also being present in some cases. Algae may therefore be a major alternative source of EPA and DHA mixtures to rival fish oils (Ryckebosch *et al.*, 2014). They will, though, be much more expensive than fish oils, and their take-up in the marketplace will probably be by minority groups of people wishing, for whatever reason, to avoid ingestion of any animal product.

The prospects of deriving EPA and DHA from genetically-modified (GM) plants continue to be enticing and, although considerable effort has been expanded on this pursuit over the past 20 years and more, the realistic view is that this is still at least 10 years away. It is still regarded as something of an achievement if an oleaginous plant can be transformed to produce small amounts of a long-chain PUFA. Thus, work to produce ARA in transgenic oil seeds resulted in amounts of ARA at almost 25% of the total fatty acids in *Arabidopsis thaliana* and at about 10% in *Brassica napus* (Petrie *et al.*, 2011). Unfortunately, neither of these plants is known for its high lipid content so, at best, these results could be taken as proof-of-principle that a genetically-modified plant *could* be created that may be appropriate for large-scale cultivation and produce sufficient quantities of PUFAs to be economically viable. Further work by the same group (Petrie *et al.*, 2012) reported the production of DHA at up to 15% of the total fatty acids in another series of genetically-engineered *Arabidopsis* plants and this Australian group at CSIRO has then used this technology to engineer an established oilseed crop plant *Camelina sativa*, to produce DHA at approximately 10% (w/v) in the triacylglycerol oil fraction of the seeds (Petrie *et al.*, 2014). Similar genetic engineering work with *C. sativa* has been carried out at Rothamsted in the United Kingdom. Ruiz-Lopez *et al.* (2014, 2015) described recombinant plants with EPA and DHA at 11% and 8%, respectively, with the claim that these are equivalent levels found in fish oils (Ruiz-Lopez *et al.*, 2014). In a review of the publications on this topic, Ruiz-Lopez *et al.* (2015) concluded that their claims for high levels of production of EPA/DHA by *C. sativa* were, in fact, amongst the highest achieved. These initial findings have been extended to growing trial crops of the GM-*C. sativa* in the United Kingdom. The crop was sown in mid-May 2014 and, after 4 months, it had matured and set seed. The results of the analysis of the lipid content of the seed and the fatty acid profile of the oil have been published (Usher *et al.*, 2015). The average content of EPA/DHA was 14.5% of the total fatty acids; the oil content of the transgenic seeds was slightly diminished – 27% down from 30% in the wild-type seeds. No indication was given, however, of the yield of oil per unit area but this is probably not surprising for a relatively small field trial. The final crop, however, could eventually be grown in Europe or in the United Kingdom but, according to the research group, this is might not be until the end of this decade (Napier *et al.*, 2015). The overall aim of the work both in Australia and in the United Kingdom, however, is to feed the oil to farmed salmon as a substitute for fish oil that is currently used (Betancor *et al.*, 2015a, 2015b). As fish oil itself is not particularly expensive, the economics of this route of production still need to be established but clearly the route is sustainable. It is, though, clear that the content of EPA and DHA in the transgenic plants is insufficient to rival the amounts being produced by the currently available microorganisms.

The use of genetically-modified plants, of course, raises questions of their acceptability by the general public in spite of clear evidence that such crops pose no threat whatsoever to health or the environment. This is particularly acute in Europe and a useful summary of the current status has been provided by Halford *et al.* (2014). Objections to the use of GM crops in Europe has been sufficiently serious for at least one major company, BASF, to have declared in 2012 that it was withdrawing from all plant biotechnology in Europe. Thus, it is against this background that the future of PUFAs from GM crops has to be viewed and, it is therefore for these reasons, the view that EPA + DHA oils from *C. sativa* were at least 10 years away from commercial realisation was expressed. Persuading the EU regulatory authority that the crop is entirely safe to grow, thereby posing no threats to the environment, and is safe to be fed to salmon could take even longer than a decade if past performance of the authority is anything to go by.

Where genetic engineering of a plant for improved oil production has been a success is with the creation of a transgenic safflower (*Carthamus tinctorius*) for the production of a high level of GLA (Nykiforuk *et al.*, 2012; Knauf *et al.*, 2011). As safflower oil already has a high content (~80%) of linoleic acid (18 : 2), by introducing a gene from *Saprolegnia diclina* coding for a δ-6 desaturase, it was then possible for the plant to convert the linoleic acid into GLA with final levels of it exceeding 70% of the total fatty acids. This oil has now been commercialised by Arcadia Biosciences Inc and is sold under the trade name Sonova. Needless to say, as this is a company registered in the United States and the product is aimed at the US market having already received FDA approval, there are apparently no serious objections to this GM oil being sold to the general public. Consumption of the oil would presumably be aimed at decreasing blood cholesterol levels and reducing the risk of atherosclerosis as well as being useful in the treatment of a wide range of other conditions (Knauf *et al.*, 2011). It certainly should be able to supplant all other sources of GLA oil currently available: evening primrose oil, borage oil and blackcurrant oil.

As with many other plant-derived materials, the acceptability of GM products continues to be a major obstacle in Europe although other countries, notably the United States, China and India, seem much more relaxed about this issue. This concern, though, also impinges upon other microbial products but no European consumer apparently objects to the consumption of the huge number of antibiotics that have been derived from engineered microorganisms. However, the key issue for the future, as far as the topic of this chapter is concerned, is likely to be the acceptability of oils derived from genetically-modified organisms – whether plants or microorganisms. If GM microorganisms with improved contents of PUFAs in them are then fed to fish, such as salmon, the interesting question is whether these fish would then be accepted by those people who are vehemently opposed to consuming GM products *per se*.

The future for microbial oils looks secure and it would appear very unlikely that, within the next 10 years, alternative sources will be produced in significant amounts from genetically-modified plants that are the only other realistic sources of the key PUFAs. Thus, from very modest beginnings in the 1980s, microbials oils are now major materials for inclusion in infant formulas and in numerous neutraceutical preparations aimed at improving our health and well-being. The market for them is, therefore, set to increase steadily for the foreseeable future.

Acknowledgements

I am indebted to the advice and help from many people. In particular, I would like to thank Dr Ethel Jackson and Dr Quinn Zhu of DuPont, Wilmington, USA, for their generous help in providing information regarding the oil from *Y. lipolytica* and its current uses and also for providing Figure 11.5e. I also thank Dr Casey Lippmeier for his advice and help with the various SCOs produced by DSM and for providing Figures 11.5b – d. He was also extremely helpful in clarifying the current status of the newly-launched oil from *Schizochytrium* sp. known as Ovega oil (see Table 11.4). Dr David Hart of Qualitas Health was particularly helpful for providing information regarding AlmegaPL, the polar lipid oil derived from *N. oculata* (see Table 11.6), and I am indebted to him for his generous help and advice. I would also like to thank Professor Yuanda Song and his colleagues of Jiangnan University, Wuxi City, China, for providing the photograph of *Mucor circinelloides* used in Figure 11.5a. I am also extremely grateful to Dr Xiao-Jun Ji of Nanking University for providing information of the oil that is produced by Jiangsu TianKai Biotechnology Co. Ltd using a *Schizochytrium* sp. (see Table 11.3). Finally, but by no means least, I would like to thank my secretary, Mrs. Elizabeth Sparke, for the assembly of the figures and tables.

References

Anderson, R.E. (1970) Lipids of the ocular tissues. IV. A comparison of the phospholipids from the retina of six mammalian species. *Exp. Eye Res.*, **10**, 339–344.

Bailey, R.B., DiMasi, D., Hansen, J.M., Mirrasoul, P.J., Ruecker, C.M., Veeder, G.M., Kaneko, T. and Barclay, W.R. (2003) Enhanced production of lipids containing polyenoic fatty acid by very high density cultures of eukaryotic microorganisms in fermentors. US Patent 6,607,900.

Bajpai, P.K., Bajpai, P., and Ward, O.P. (1991) Production of docosahexaenoic acid by Thraustochytrium aureum. *Appl. Microbiol. Biotechnol.*, **35**, 706–710.

Baker, R.R. (1979) The fatty acid composition of phosphoglycerides of nerve cell bodies isolated in bulk from rabbit cerebral cortex: changes during development and positional distribution. *Can. J. Biochem.*, **57**, 378–384.

Barclay, W.R. (1991) Process for the heterotrophic production of long chain omega-3 highly unsaturated fatty acids. World Patent W091/07498.

Barclay, W.R., Weaver, C., and Metz, J. (2005) in *Single Cell Oils*, 1st edn (eds Z. Cohen and C. Ratledge), AOCS Press, Champaign, IL, pp. 36–52.

Barclay, W., Weaver, C., Metz, J., and Hansen, J. (2010) in *Single Cell Oils: Microbial and Algal Oils*, 2nd edn (eds Z. Cohen and C. Ratledge), AOCS Press, Urbana, IL, pp. 75–96.

Bellou, S., Baeshen, M.N., Elazzazy, A.M., Aggeli, D., Sayegh, F., and Aggelis, G. (2014) Microalgal lipids biochemistry and biotechnological perspectives. *Biotechnol. Adv.*, **32**, 1476–1493.

Betancor, M.B., Sprague, M., Sayanova, O., Usher, S., Campbell, P.J., Napier, J., Caballero, M.J., and Tocher, D.R. (2015a) Evaluation of a high-EPA oil from transgenic *Camelina sativa* in feeds for Atlantic salmon (*Salmo salar* L.): effects on tissue fatty acid composition, histology and gene expression. *Aquaculture*, **444**, 1–12.

Betancor M.B., Sprague M., Usher S., Sayanova O., Campbell P.J., Napier J.A. and Tocher D.R. (2015b) A nutritionally-enhanced oil from transgenic *Camelina sativa* effectively replaces fish oil as a source of eicosapentaenoic acid for fish. *Sci. Rep.* **5**:8014 (doi: 10.1038/srep08104).

Birch, E.E., Carlson, S.E., Hoffman, R.R., Fitzgerald-Gustafson, K.M. *et al.* (2010) The DIAMOND (DHA Intake And Measurement Of Neural Development) study: a double-masked, randomized controlled clinical trial of the maturation of infant visual acuity as a function of the dietary level of docosahexaenoic acid. *Am. J. Clin. Nutr.*, **91**, 848–859.

Burja, A.M., Radianingtyas, H., Windust, A., and Barrow, C.J. (2006) Isolation and characterisation of polyunsaturated fatty acid producing *Thraustochytrium* species; screening of strains and optimization of omega-3 production. *Appl. Microbiol. Biotechnol.*, **72**, 1161–1169.

Burr, G.O. and Burr, M.M. (1929) A new deficiency disease produced by rigid exclusion of fat from the diet. *J. Biol. Chem.*, **82**, 345–367.

Burr, G.O. and Burr, M.M. (1930) On the nature and role of the fatty acids essential in nutrition. *J. Biol. Chem.*, **86**, 587–621.

Calder, P.C. (2008) Polyunsaturated fatty acids, inflammatory processes and inflammatory bowel diseases. *Mol. Nutr. Food Res.*, **52**, 885–897.

Casterton, P.L., Curry, L.L., Lina, B.A.R., Wolterbeek, A.P.M., and Kruger, C.L. (2009) 90-Day feeding and genotoxicity studies on a refined arachidonic acid-rich oil. *Food Chem. Toxicol.*, **47**, 2407–2418.

Cavalier-Smith, T., Allsopp, M.T.E.P., and Chao, E.E. (1994) Thraustochytrids are chromists not fungi: 18sRNA signatures of heterokonta. *Philos. Trans. R. Soc. London, Ser. B*, **346**, 387–397.

Cottin, S.C., Sanders, T.A., and Hall, W.L. (2011) The differential effects of EPA and DHA on cardiovascular risk factors. *Proc. Nutr. Soc.*, **70**, 215–231.

Daiello, L.A., Gongvatana, A., Dunsiger, S., Cohen, R.A., and Ott, B.R. (2015) Association of fish oil supplement use with preservation of brain volume and cognitive function. *Alzheimers Demet.*, **11**, 226–235.

Evans, H.M. and Burr, G.O. (1927) A new dietary deficiency with highly purified diets. *Proc. Soc. Exp. Biol. Med.*, **24**, 740–743.

Evans, H.M. and Burr, G.O. (1928) A new dietary deficiency with highly purified diets III. The beneficial effect of fat in the diet. *Proc. Soc. Exp. Biol. Med.*, **25**, 390–397.

Fedorova-Dahms, I., Marone, P.A., Bailey-Hall, E., and Ryan, A.S. (2011) Safety evaluation of algal oil from *Schizochytrium* sp. *Food Chem. Toxicol.*, **49**, 70–77.

Fedorova-Dahms, I., Thorsrud, B.A., Bailey, E., and Salem, N. (2014) A 3-week dietary bioequivalence study in preweaning farm piglets of two sources of docosahexaenoic acid produced from two different organisms. *Food Chem. Toxicol.*, **65**, 43–51.

Halford, N.G., Hudson, E., Gimson, A., Weightman, R., Shewry, P.R., and Tompkins, S. (2014) Safety assessment of genetically modified plants with deliberately altered composition. *Plant Biotechnol. J.*, **12**, 651–654.

Harrington, G.W. and Holz, G.G. (1968) The monoenoic acid docosahexaenoic fatty acids of a heterotrophic dinoflagellate. *Biochim. Biophys. Acta*, **164**, 137–139.

Hauvermale, A., Kuner, J., Rosenzweig, B., Guerra, D., Diltz, S., and Metz, J.G. (2006) Fatty acid production in *Schizochytrium* sp.: involvement of a polyunsaturated fatty acid synthase and a type I fatty acid synthase. *Lipids*, **41**, 739–747.

Holman, R.T. (1971) Essential fatty acid deficiency. *Prog. Chem. Fats Other Lipds*, **9**, 279–348.

Holman, R.T. (1988) George O. Burr and the discovery of essential fatty acids. *J. Nutr.*, **118**, 535–540.

Hooijmans, C.R. and Kiliaan, A.J. (2008) Fatty acids, lipid metabolism and Alzheimer pathology. *Eur. J. Pharmacol.*, **585**, 176–196.

Huang, T.L. (2010) Omega-3 fatty acids, cognitive decline, and Alzheimer's disease: a critical review and evaluation of the literature. *J. Alzheimers Dis.*, **21**, 673–690.

Jeong, S., Jing, K., Kim, N., Shin, S., Kim, S., Song, K.-S., Heo, J.-Y., Park, J.-H., Seo, K.-S., Han, J., Kweon, G.-R., Park, S.-K., Park, J.-I., and Lim, K. (2014) Docosahexaenoic acid-induced apoptosis is mediated by activation of mitogen-activated protein kinases in human cancer cells. *BMC Cancer*, **14**, 481–491.

Jicha, G.A. and Markesbery, W.R. (2010) Omega-3 fatty acids: potential role in the management of early Alzheimer's disease. *Clin. Interv. Aging*, **5**, 45–61.

Jump, D.B., Depner, C.M., and Tripathy, S. (2012) Omega-3 fatty acid supplementation and cardiovascular disease. *J. Lipid Res.*, **53**, 2525–2545.

Kagan, M.L., Levy, A., and Leikin-Frenkel, A. (2015) Comparative study of tissue deposition of omega-3 fatty acids from polar-lipid rich oil of the microalgae *Nannochloropsis oculata* with krill oil in rats. *Food Funct.*, **6**, 185–191.

Kagan, M.L., West, A.L., Zante, C., and Calder, P.C. (2013) Acute appearance of fatty acids in human plasma – a comparative study between polar-lipid rich oil from the microalgae *Nannochloropsis oculata* and krill oil in healthy young males. *Lipids Health Dis.*, **12**, 102.

Kendrick, A. and Ratledge, C. (1992) Lipids of selected molds grown for production of n-3 and n-6 polyunsaturated fatty acids. *Lipids*, **27**, 15–20.

Kiy, T., Rusing, M., and Fabritius, D. (2005) in *Single Cell Oils*, 1st edn (eds Z. Cohen and C. Ratledge), AOCS Press, Champaign, IL, pp. 99–106.

Knauf, V.C., Shewmaker, C., Flider, F., Emlay, D. and Rey, E. (2011) Safflower with elevated gamma-linolenic acid. US Patent 7893321 B2.

Kusumoto, A., Ishikura, Y., Kawashima, H., Kiso, Y., Takai, S., and Miyazaki, M. (2007) Effects of arachidonate-enriched triacylglycerol supplementation on serum fatty acids and platelet aggregation in healthy male subjects with a fish diet. *Br. J. Nutr.*, **98**, 626–635.

Kyle, D.J. and Arterburn, L.M. (1998) in *The Return of n-3 Fatty Acids into the Food Supply. I. Land Based Animal Food Products and Their Health Effects*, World Review of Nutrition and Dietetics, vol. **83** (ed A.P. Simopoulos), Kruger, Basel, pp. 116–131.

Kyle, D.J., Sicotte, V.J., Singer, J.J., and Reeb, S.E. (1992) in *Industrial Applications of Single Cell Oils* (eds D.J. Kyle and C. Ratledge), AOCS Press, Champaign, IL, pp. 287–300.

Lands, B. (2014) Historical perspective on the impact of n-3 and n-6 nutrients on health. *Prog. Lipid Res.*, **55**, 17–29.

Leu, S. and Boussiba, S. (2014) Advances in the production of high-value products by microalgae. *Ind. Biotechnol.*, **10**, 169–183.

Lichtenstein, A.H., Appel, L.J., Brands, M., Danield, S., Franch, H.A., Franklin, B., Fris-Etherton, P., Harris, W.H., Howard, B., Karanja, N., Lefevre, M., Rudel, L., Sack, F., Van Horn, L., Winston, M., and Wylie-Rosett, J. (2006) Diet and lifestyle recommendations revision 2006; a scientific statement from the American Heart Association Nutrition Committee. *Circulation*, **114**, 82–96.

Lin, P.-Y., Huang, S.-Y., and Su, K.-P. (2010) A meta-analytic review of polyunsaturated fatty acid compositions in patients with depression. *Biol. Psychiatry*, **68**, 140–147.

Lovell, C.R., Burton, J.L., and Horrobin, D.F. (1981) Treatment of atopic eczema with evening primrose oil. *Lancet*, **1**, 278.

Lucas, A., Morley, R., Cole, T.J., Lister, G., and Leeson-Payne, C. (1992) Breast milk and subsequent intelligence quotient in children born preterm. *Lancet*, **339**, 261–264.

Ma, Q.-L., Teter, B., Ubeda, O.J., Morihara, T., Dhoot, D., Nyby, M.D., Tuck, M.L., Frautschy, S.A., and Cole, G.M. (2007) Omega-3 fatty acid docosahexaenoic acid increases SorLA/LR11, a sorting protein with reduced expression in sporadic Alzheimer's disease(AD): relevance to AD prevention. *J. Neurosci.*, **27**, 14299–14307.

Mazza, M., Pomponi, M., Janiri, L., Bria, P., and Mazza, S. (2007) Omega-3 fatty acids and antioxidants in neurological and psychiatric diseases: an overview. *Prog. Neuropsychopharmacol. Biol. Psychiatry*, **31**, 12–26.

Metz, J.G., Roessler, P., Facciotti, D., Levering, C., Dittrich, F., Lassner, M., Valentine, R., Lardizabal, K., Domergue, F., Yamada, A., Yazawa, K., Knauf, V., and Browse, J. (2001) Production of polyunsaturated fatty acids by polyketide synthases in both prokaryotes and eukaryotes. *Science*, **293**, 290–293.

Minihane, A.M. (2013) Fish oil omega-3 fatty acids and cardio-metabolic health, alone or with statins. *Eur. J. Clin. Nutr.*, **676**, 536–540.

Nakahara, T., Yokocki, T., Kamisaka, Y., and Suzuki, O. (1992) in *Industrial Applications of Single Cell Oils* (eds D.J. Kyle and C. Ratledge), American Oil Chemists' Association, pp. 61–97.

Napier, J.A., Usher, S., Haslam, R.P., Ruiz-Lopez, N., and Sayanova, O. (2015) Transgenic plants as a sustainable, terrestrial source of fish oils. *Eur. J. Lipid Sci. Technol.*, **117**, 000–000.

Nicholson, T., Khademi, H., and Moghadasian, M.H. (2013) The role of marine n-3 fatty acids in improving cardiovascular health: a review. *Food Funct.*, **4**, 357–365.

Nykiforuk, C.L., Shewmaker, C., Harry, I., Yurchenko, O.P., Zhang, M., Reed, C., Oinam, G.S., Zaplachinski, S., Fidantsef, A., Boothe, J.G., and Moloney, M.M. (2012) High level accumulation of gamma linolenic acid (C18:3 delta6,9,12) in transgenic safflower (*Carthamus tinctorius*) seeds. *Transgenic Res.*, **21**, 367–381.

O'Brien, J.S. and Sampson, E.L. (1965) Fatty acid and fatty aldehyde composition of the major brain lipids in normal human gray matter, white matter and myelin. *J. Lipid Res.*, **6**, 545–551.

Osborne, T.B. and Mendel, L.B. (1920) Growth on diets poor in true fats. *J. Biol. Chem.*, **45**, 145–152.

Peet, M. and Stokes, C. (2005) Omega-3 fatty acids in the treatment of psychiatric disorders. *Drugs*, **65**, 1051–1059.

Petrie J.R., Shrestha P., Beide S., Kennedy Y., Lester G., Liu Q., Divi U.K., Mulder R.J., Mansour M.P., Nichols P.D. and Singh S.P. (2014) Metabolic engineering *Camelina sativa* with fish oil-like levels of DHA. *PLoS One* **9**:e85061 (doi:10.1371/journal.pne.0085061).

Petrie, J.R., Shrestha, P., Belide, S., Mansour, M.P., Liu, Q., Horne, J., Nichols, P.D., and Singh, S.P. (2011) Transgenic production of arachidonic acid in oilseeds. *Transgenic Res.*, **21**, 139–147.

Petrie, J.R., Shrestha, P., Zhou, X.-R., Mansour, M.P., Liu, Q., Belide, S., Nichols, P.D., and Singh, S.P. (2012) Metabolic engineering of plant seeds with fish oil-like levels of DHA. *PLoS One*, **7**, e49165.

Raji, C.A., Erickson, K.I., Lopez, O.L., Kuller, L.H., Gach, M., Thompson, P.M., Riverol, M., and Becker, J.T. (2014) Regular fish consumption and age-related brain gray matter loss. *Am. J. Prev. Med.*, **47**, 444–451.

Ratledge, C. (1992) in *Industrial Applications of Single Cell Oils* (eds D.J. Kyle and C. Ratledge), American Oil Chemists' Association, pp. 1–15.

Ratledge, C. (2004) Fatty acid biosynthesis in microorganisms being used for single cell oil production. *Biochimie*, **86**, 807–815.

Ratledge, C. (2006) In *Handbook of Functional Lipids* (ed A.A. Akoh), Taylor & Francis Group, Boca Raton, FL, pp. 19–45.

Ratledge, C. (2014) The role of malic enzyme as the provider of NADPH in oleaginous microorganisms: a reappraisal and unsolved problems. *Biotechnol. Lett.*, **36**, 1557–1568.

Ratledge, C. and Cohen, Z. (2008) Microbial and algal oils: do they have a future for biodiesel or as commodity oils? *Lipid Technol.*, **20**, 155–160.

Ratledge, C. and Wynn, J.P. (2002) The biochemistry and molecular biology of lipid accumulation in oleaginous microorganisms. *Adv. Appl. Microbiol.*, **51**, 1–51.

Ren, L.-J., Ji, X.J., Huang, H., Qu, L., Feng, Y., Tong, Q.Q., and Ouyang, P.K. (2010) Development of a stepwise aeration control strategy for efficient docosahexaenoic acid production by *Schizochytrium* sp. *Appl. Microbiol. Biotechnol.*, **87**, 1649–1656.

Riediger, N.D., Othaman, R.A., and Suh, M. (2009) A systemic review of the roles of n-3 fatty acids in health and disease. *J. Am. Diet. Assoc.*, **109**, 668–679.

Ross, B.M., Seguin, J., and Sieswerda, L.E. (2007) Omega-3 fatty acids as treatments for mental illness: which disorder and which fatty acid? *Lipids Health Dis.*, **6**, 21.

Ruiz-Lopez, N., Haslam, R.P., Napier, J.A., and Sayanova, O. (2014) Successful high-level accumulation of fish oil omega-3 long chain polyunsaturated fatty acids in a transgenic oilseed crop. *Plant J.*, **77**, 198–208.

Ruiz-Lopez, N., Usher, S., Sayanova, O.V., Napier, J.A., and Haslam, R.P. (2015) Modifying the lipid content and composition of plant seeds: engineering the production of LC-PUFA. *Appl. Microbiol. Biotechnol.*, **99**, 143–154.

Ryan, A., Zeller, S., and Nelson, E.B. (2010) In *Single Cell Oils*, 2nd edn (eds Z. Cohen and C. Ratledge), AOCS Press, Urbana, IL, pp. 317–350.

Ryckebosch, E., Bruneel, C., Termote-Verhalle, R., Goiris, K., Kuylaert, K., and Foubert, I. (2014) Nutritional evaluation of microalgae oils rich in omega-3 long chain polyunsaturated fatty acids as an alternative to fish oil. *Food Chem.*, **160**, 393–400.

SanGiovanni, J.P., Agron, E., Meleth, A.D., Reed, G.R., Sperduto, R.D., Clemons, T.E., and Chew, E.Y. (2009) Omega-3 long-chain polyunsaturated fatty acid intake and 12-y incidence of neovascular age-related macular degeneration and central geographic atrophy: AREDS report 30, a prospective cohort study from the age-related eye disease study. *Am. J. Clin. Nutr.*, **90**, 1601–1607.

SanGiovanni, J.P., Chew, E.M., Agron, E., Clemons, T.E., Ferris, F.L., Gensler, G., Lindblad, A.S., Milton, R.C., Seddon, J.M., Klein, R., and Sperduto, R.D. (2008) The relationship of dietary omega-3 long-chain polyunsaturated fatty acid intake with incident age-related macular degeneration. *Arch. Ophthalmol.*, **126**, 1274–1279.

Serhan, C.N. (2005) Novel eicosanoid and docosaniod mediators: resolvins, docosatrienes, and neuroprotectins. *Curr. Opin Clin. Nutr. Metab. Care*, **8**, 115–121.

Sinclair, A., Attar-Bashi, N., Jayasooriya, A., Gibson, R., and Makrides, M. (2005) In *Single Cell Oils*, 1st edn (eds Z. Cohen and C. Ratledge), AOCS Press, Champaign, IL, pp. 182–201.

Sinclair, A.J., Attar-Bashi, N.M., and Li, D. (2002) What is the role of alpha-linolenic acid for mammals? *Lipids*, **37**, 1113–1123.

Sinclair, A.J. and Jayasooriya, A. (2010) In *Single Cell Oils: Microbial and Algal Oils*, 2nd edn (eds Z. Cohen and C. Ratledge), AOCS Press, Urbana, IL, pp. 351–368.

Skender, B., Vaculova, A.H., and Hofmanova, J. (2012) Docosahexaenoic acid (DHA) in the regulation of colon cell growth and cell death: a review. *Biomed Pap.*, **156**, 186–199.

Streekstra, H. (2010) in *Single Cell Oils: Microbial and Algal Oils*, 2nd edn (eds Z. Cohen and C. Ratledge), AOCS Press, Urbana, IL, pp. 97–114.

Sublette, M.E., Ellis, S.P., Geant, A.L., and Mann, J.J. (2011) Meta-analysis: effect of eicosapentaenoic acid in clinical trials in depression. *J. Clin. Psychiatry*, **72**, 1577–1584.

Tai, M. and Stephanopoulos, G. (2013) Engineering the push and pull of lipid biosynthesis in oleaginous yeast *Yarrowia lipolytica* for biofuel production. *Metab. Eng.*, **15**, 1–9.

Totani, N. and Oba, K. (1987) The filamentous fungus *Mortierella alpina*, high in arachidonic acid. *Lipids*, **22**, 1060–1062.

Totani, N., Watanabe, A., and Oba, K. (1987) An improved method of arachidonic acid production by *Mortierella* sp. S-17. *J. Jpn. Oil Chem. Soc.*, **36**, 328–331.

Usher, S., Haslam, R.P., Ruiz-Lopez, N., Sayanova, O., and Napier, J.A. (2015) Field trial evaluation of the accumulation of omega-3 long chain polyunsaturated fatty acids in transgenic *Camelina sativa*: making fish oil substitutes in plants. *Metab. Eng. Commun.*, **2**, 93–98.

Wright, S. and Burton, J.L. (1982) Oral evening primrose-seed oil improves atopic eczema. *Lancet*, **21**, 1120–1122.

Wynn, J., Behrens, P., Sundararajan, A., Hansen, J., and Apt, K. (2010) In *Single Cell Oils: Microbial and Algal Oils*, 2nd edn (eds Z. Cohen and C. Ratledge), AOCS Press, Urbana, IL, pp. 115–129.

Wynn, J.P. and Ratledge, C. (2006) in *Food Biotechnology*, 2nd edn (eds K. Shetty, G. Paliyath, A. Pometto, and R.E. Levin), Taylor & Francis Group, Boca Raton, FL, pp. 443–472.

Wynn, J.P. and Ratledge, C. (2007) in *Long-Chain Omega-3 Specialty Oils* (ed H. Brevivik), The Oily Press, Bridgwater, pp. 43–76.

Xu, Y. and Qian, S.Y. (2014) Anti-cancer activities of omega-6 polyunsaturated fatty acids. *Biomed J.*, **37**, 112–119.

Xue, Z., Sharpe, P.L., Hong, S.-P., Yadav, N.S. *et al.* (2013) Production of omega-3 eicosapentaenoic acid by metabolic engineering of *Yarrowia lipolytica*. *Nat. Biotechnol.*, **31**, 734–741.

Zeller, S. (2005) in *Single Cell Oils*, 1st edn (eds Z. Cohen and C. Ratledge), AOCS Press, Champaign, IL, pp. 161–181.

Zheng, H., Tang, H., and Liu, M. (2014) Inhibition of endometrial cancer by n-3 polyunsaturated fatty acids (PUFAs) in preclinical models. *Cancer Prev. Res.*, **7**, 824–834.

12
Vitamin Q_{10}: Property, Production and Application

Joong K. Kim, Eun J. Kim, and Hyun Y. Jung

12.1
Background of Vitamin Q_{10}

12.1.1
Historical Aspects

In 1957, Dr Fred Crane isolated coenzyme Q_{10} (CoQ_{10}) from beef heart mitochondria while working at the Enzyme Research Center in the University of Wisconsin (Crane *et al.*, 1957). After about a year, the chemical structure of CoQ_{10} was determined to be a benzoquinone by Dr Karl Folkers' group working at Merck Sharp and Dohme. In 1968, Dr Peter Mitchell first depicted the possible use of CoQ_{10} as an electron–proton carrier in the mitochondria inner membrane during the synthesis of ATP (Adenosine triphosphate). For this effort, he was awarded the Nobel Prize in Biology in 1978. In the meantime, Dr Karl Folkers studied CoQ_{10} and vitamin B complexes continuously, and he was awarded both the Priestly Award by the International Chemical Society in the mid-1980s and the National Medal of Science in the late 1980s. Since then, scientific efforts on CoQ_{10} have continued, building an amazing knowledge base of more than 8000 scientific publications, and currently, more than 600 scientists are working daily on CoQ_{10} basic science and clinical research (PWS, 2014).

12.1.2
Definition

Coenzyme Q (CoQ) molecules, 2,3-dimethoxy-5-methyl-6-multiprenyl-1,4-benzoquinone, are located in the hydrophobic domain of cellular membranes (Battino *et al.*, 1990; Lenaz *et al.*, 1999). The diverse CoQ molecules have the following structure: a quinone head capable of transferring electrons and a long isoprenoid side chain (Choi *et al.*, 2005). The CoQ molecules are distinguished by their number of isoprenoid side chains (Jeya *et al.*, 2010). As shown in Figure 12.1, CoQ_{10} contains 10 isoprenoid units on the side chain, also called ubiquinone-10. This molecule is a naturally occurring oil-soluble material found abundantly in

Industrial Biotechnology of Vitamins, Biopigments, and Antioxidants, First Edition.
Edited by Erick J. Vandamme and José L. Revuelta.

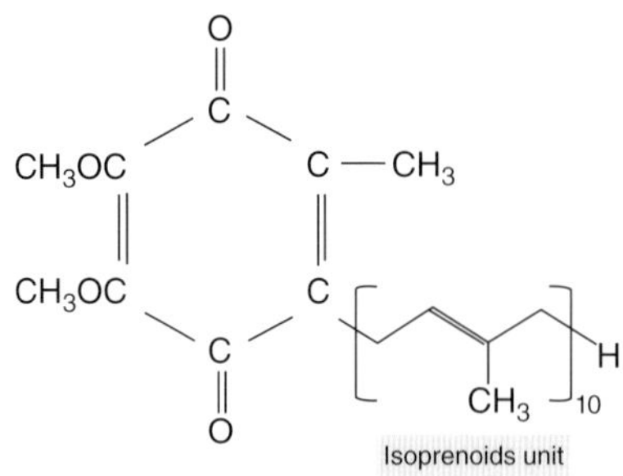

Figure 12.1 Chemical structure of CoQ_{10}.

animals, plants and microorganisms, and it functions as a coenzyme involved in many electron-transferring reactions (Yen and Shih, 2009).

12.1.3 Occurrence

12.1.3.1 In Nature

CoQ compounds are widely found in nature, from microorganisms to human beings. In some species, including humans, the side chain of CoQ compounds is comprised of 10 isoprene units, that is, CoQ_{10} (Crane, 2001). Particularly, animal products such as beef, pork and chicken are relatively good sources of CoQ_{10}, and the best sources are organ meats such as the heart and muscle. In general, tissues with high energy demands contain relatively high amounts of CoQ_{10}. In plants, broccoli and spinach have been reported to contain significant amounts of CoQ_{10} (Natural Products Insider, 2014). Unrefined vegetable oils such as soybean oil and palm oil are also reported to be good sources of CoQ_{10} (Pravst *et al.*, 2010).

12.1.3.2 In Food Sources

Nutritional sources containing an abundant amount of CoQ_{10} include meat, fish, nuts and some oils. In most dairy products, vegetables, fruits and cereals, much lower levels of CoQ_{10} are available. Foods and food products of different geographical origins have large variations of CoQ_{10} content, as summarised in Table 12.1 for food sources and Table 12.2 for processed food sources. As seen in Table 12.1, the highest content of CoQ_{10} is found in beef, pork and chicken hearts. In these animals, the CoQ_{10} content ranges from 11 to 282 mg/kg. Their levels are much higher than those in fish, shellfish, fruits, nuts/seeds and vegetables. Variations in CoQ_{10} content also occur depending on the part of each species. While the CoQ_{10} content of raw materials obtained from beef is high, dairy products such as processed foods reveal a much lower CoQ_{10} content (Table 12.2). However, the content of CoQ_{10} in soybean oil (221–279 mg/kg) and Jeotgal (288.2 mg/kg), a Korean fermented food, is comparable to that found in pork heart. On average, the dietary intake of CoQ_{10} is only 3–6 mg, and about half of it is absorbed in its reduced form (Hoppe *et al.*, 1999).

Table 12.1 CoQ_{10} contents of diverse food sources.

Foods	CoQ_{10} (mg/kg)	References	Foods	CoQ_{10} (mg/kg)	References
Meats			Fish		
Beef			Horse mackerel	3.6–130	Kubo *et al.* (2008), Prošek *et al.* (2007) and Passi *et al.* (2002)
Heart	113.3	Mattila and Kumpulainen (2001)	Sardine	5.1–64.3	
Liver	39.2–50.5	Mattila and Kumpulainen (2001), Kubo *et al.* (2008) and Kraszner-Berndorfer and Telegdy Kováts (1972)	Pollack	14.1	Mattila and Kumpulainen (2001)
Shoulder	40.1	Kubo *et al.* (2008)	Eel	7.4–11.1	Mattila and Kumpulainen (2001) and Passi *et al.* (2002)
Sirloin	30.6	Mattila and Kumpulainen (2001)	Salmon	4.3–7.6	Mattila and Kumpulainen (2001), Kubo *et al.* (2008) Weber *et al.* (1997)
Thigh	30.3	Kubo *et al.* (2008)	Flat fish	1.8–5.5	Kubo *et al.* (2008) and Kamei *et al.* (1986)
Tenderloin	26.5	Mattila and Kumpulainen (2001)	Tuna	4.9	Kubo *et al.* (2008)
Pork			Yellowtail	12.8–20.7	Kubo *et al.* (2008) and Kamei *et al.* (1986)
Heart	118.1–282	Mattila and Kumpulainen (2001), Weber *et al.* (1997) and Mattila *et al.* (2000)	Pike	5.4	Passi *et al.* (2002)
Liver	22.7–54	Mattila and Kumpulainen (2001) and Kraszner-Berndorfer and Telegdy Kováts (1972)	Rainbow trout	8.5–11	Mattila and Kumpulainen (2001) and Weber, Bysted and Hølmer (1997)
Shoulder	45	Mattila and Kumpulainen (2001)	Cuttle fish	4.7–8.2	Kubo *et al.* (2008) and Passi *et al.* (2002)
Sirloin	14	Kubo *et al.* (2008)	**Shellfish**		
Thigh	13.8	Mattila and Kumpulainen (2001)	Oyster	3.4–4.3	Kubo *et al.* (2008) and Passi *et al.* (2002)

(*continued overleaf*)

Table 12.1 *(Continued)*

Foods	CoQ_{10}(mg/kg)	References	Foods	CoQ_{10} (mg/kg)	References
Chicken			Scallop	5	Kubo *et al.* (2008)
Heart	92.3–192	Mattila and Kumpulainen (2001) and Kubo *et al.* (2008) and Prošek *et al.* (2007)	Shrimp	1.7–2.8	Kubo *et al.* (2008) and Passi *et al.* (2002)
Liver	116.2–132.2	Mattila and Kumpulainen (2001), Kraszner-Berndorfer and Telegdy Kováts (1972) and Prošek *et al.* (2007)	**Nuts/Seeds**		
Thigh	24.2–25	Kubo *et al.* (2008) and Prošek *et al.* (2007)	Peanut	26.7	Kamei *et al.* (1986)
			Pistachio nuts	20.1	
Breast/chest	7.8–17.1	Mattila and Kumpulainen (2001), Kubo *et al.* (2008) and Prošek *et al.* (2007)	Walnuts	19	
			Chestnuts	6.3	
Wing	11	Prošek *et al.* (2007)	Almond	5–13.8	Kubo *et al.* (2008) and Kamei *et al.* (1986)
			Sesame seed	17.6–23	
Fruits			**Vegetables**		
Apple	1.1–1.3	Hoppe *et al.* (1999), Kubo *et al.* (2008) and Weber *et al.* (1997)	Sweet potato	3–3.6	Kubo *et al.* (2008) and Kamei *et al.* (1986)
Orange	1–2.2		Carrot	2.2	Mattila and Kumpulainen (2001) , Weber *et al.* (1997) and Kamei *et al.* (1986)
Strawberry	0.5	Mattila and Kumpulainen, 2001)	Potato	0.5–1.1	
Banana	0.8	Kubo *et al.* (2008)	Onion	0.7–1	Kubo *et al.* (2008) and Kamei *et al.* (1986)
Kiwi	0.5	Weber *et al.* (1997)	Garlic	2.7–3.5	

Table 12.2 CoQ_{10} contents of processed food sources.

Foods	Fat (%)	CoQ_{10}(mg/kg)	References	Foods	CoQ_{10}(mg/kg)	References
Dairy products		**Oils**				
Butter	—	7.1	Kamei *et al.* (1986)	Peanut	77	Cabrini *et al.* (2001)
Emmental cheese	—	1.3	Mattila and Kumpulainen (2001)	Coconut	n.d	Kamei *et al.* (1986)
Edam cheese	—	1.2		Rapeseed	63.5–73.4	Mattila and Kumpulainen (2001) and Kamei *et al.* (1986)
Cow milk	3.6	1.9	Strazisar *et al.* (2005)	Sesame	32	Kamei *et al.* (1986)
	3.5	1.3		Cotton seed	17.3	
	1.5–1.6	0.7–1.2	Mattila and Kumpulainen (2001) and Cabrini *et al.* (2001)	Safflower	4	
	3.5	1.7	Strazisar *et al.* (2005)	Sunflower	10–15	Cabrini *et al.* (2001) and Pregnolato *et al.* (1994)
	1.5	1.2		Soybean	221–279	Pregnolato *et al.* (1994)
	0.5	0.5		Corn	113–139	
Yogurt	3.2	0.7–1.1		Olive	109	
	0	0.1		**Korean fermented foods**		
Cream	20–22	0.5–0.9		Doenjang	189.7	Pyo and Oh (2011)
	35	0.9		Gochujang	26.5	
Curd	35	0.7		Cheonggukjang	123.8	
	13	0.7		Jeotgal	288.2	
Cereals		Kimchi	121.5			
Corn	—	7	Kraszner-Berndorfer and Telegdy Kováts (1972)	Soy sauce	3.1	
Rice	—	4.9	Strazisar *et al.* (2005)			
Wheat	—	3.5–6.8	Kraszner-Berndorfer and Telegdy Kováts (1972) and Strazisar *et al.* (2005)			

12.1.3.3 In Microorganisms

Various microorganisms have been reported as CoQ_{10} producers. The representative yeast species are *Candida, Rhodotorula* and *Saitoella* (Matsuda *et al.*, 2000), and the strains *Agrobacterium tumefaciens* (ATCC 4452), *Rhodobacter sphaeroides* (FERM-P4675) and *Paracoccus denitrificans* (ATCC 19367) have also been reported to be excellent CoQ_{10} producers (Choi, Ryu and Seo, 2005).

12.1.4 Functions

Ubiquinones including CoQ_{10} are obligatory cofactors in aerobic respiratory electron transfer for the production of ATP. They are membrane-bound redox-active molecules and participate in several cellular functions such as the formation of disulfide bonds in proteins, detoxification of harmful oxygen radicals, control of cellular redox status, generation of cell signals and gene expression (Battino *et al.*, 1990; Bader *et al.*, 1999; Kawamukai, 2002; Turunen, Olsson and Dallner, 2004; Groneberg *et al.*, 2005; Okada *et al.*, 1998).

CoQ_{10} plays an important role as an important rate-limiting cofactor in the electron transport chain during cellular respiration. Because cellular activities are dependent on energy, the role of CoQ_{10} is pivotal for the efficient functioning of almost all cells. The function of CoQ_{10} located in the inner mitochondrial membrane of eukaryotes or in the plasma membrane of prokaryotes can be found in the transport of electrons from Complex I or II to the cytochrome bc1 complex (Ernster and Dallner, 1995). In humans, CoQ_{10} boosts energy and improves the immune system (Okada *et al.*, 1998). In recent years, CoQ_{10} has become the centre of public interest regarding its antioxidant properties. CoQ_{10} forms in all cellular membranes as well as in blood serum and in serum lipoproteins. CoQ_{10}, as a lipid-soluble antioxidant, efficiently protects membrane phospholipids and serum low-density lipoproteins from lipid peroxidation. It has been reported that CoQ_{10} also protects both mitochondrial membrane proteins and DNA from free-radical-induced oxidative damage (Martin *et al.*, 2007). Furthermore, CoQ_{10} is also involved in cellular metabolism. Nowadays, the use of CoQ_{10} has increased in therapeutic applications for several diseases such as heart disease, high blood pressure, high cholesterol, immune deficiencies and Alzheimer's disease (Boreková *et al.*, 2008).

12.2 Chemical and Physical Properties of CoQ_{10}

12.2.1 Chemical Properties

Biologically active ubiquinone is chemically known as *2,3-dimethoxy-5-methyl-6-polyisoprene* parabenzoquinone. It is composed of a benzoquinone ring with an isoprenoid side chain, and its structure is similar to those of vitamin K and vitamin E. The various CoQs can be distinguished by the number of isoprenoid subunits in their side chains. CoQ_{10} is the most prevalent form in humans and most mammals, while CoQ_9, having nine isoprenoid units, is the primary form found in rats and mice (Barker and Frost, 2001). Other coenzymes CoQ_6, CoQ_7 and CoQ_8 are found in yeasts and bacteria (Overvad *et al.*, 1999). The main chemical characteristic of CoQ_{10} is its existence in three alternate redox states (Battino *et al.*, 1990; Boreková *et al.*, 2008; James *et al.*, 2004): (i) fully oxidised ubiquinone form, (ii) semiquinone radical form and (iii) fully reduced ubiquinol form. When two hydrogen atoms are added in a row, ubiquinone converts first into a semiquinone radical, a partially reduced form, and then the semiquinone radical further converts into ubiquinol. CoQ_{10} can exist in either the *cis* or the *trans* form. Although only the *trans* form is found in nature, both forms can be produced via fermentation or a chemical process (West, 2001). The CoQ_{10} can be reduced by $Na_2S_2O_4$, KBH_4, Zn in acid and more slowly by ascorbic acid–HCl. The reduced solution of CoQ_{10} in ethanol is slowly reoxidised in air, while the reduced CoQ_{10} can be rapidly reoxidised in the presence of $FeCl_3$ or Ag_2O. When the reduced CoQ_{10} is dissolved in cyclohexane, less rapid reoxidation can be achieved.

12.2.2 Physical Properties

CoQ_{10} occurs naturally and forms an orange lipophilic powder. The orange colour is bleached when it is reduced. The oil-soluble, vitamin-like CoQ_{10} is odourless and tasteless. The molecular formula and weight of CoQ_{10} are $C_{59}H_{90}O_4$ and 863.34 g per molecule, respectively. CoQ_{10} is not very stable and deteriorates at about 46 °C (Udell *et al.*, 2003). Compared CoQ_6 to CoQ_9, the physical values of CoQ_{10} vary, reflecting the change in the length of the isoprenoid side chain (Youssef, 1963). The melting point of CoQ_{10} is 49.9 °C (Bhandari *et al.*, 2007), and its solubility increased with increasing concentrations of poloxamer 188 in water (Bhandari *et al.*, 2007). During this solubilisation, Gibbs free energy change was negative, indicating the spontaneous nature of CoQ_{10} solubilisation.

12.3 Biosynthesis and Metabolic Regulation of CoQ_{10}

12.3.1 Biosynthesis of CoQ_{10}

12.3.1.1 Microorganisms

Living organisms synthesise different types of CoQ depending on the length of the isoprenoid side chain. Interestingly, in the case of yeasts, *Schizosaccharomyces pombe*, *Candida albicans*, *Candida utilis* and *Saccharomyces cerevisiae* produce CoQ_{10}, CoQ_9, CoQ_7 and CoQ_6, respectively. Diverse microorganisms produce CoQ_{10}; bacteria such as *Pseudomonas*, *Agrobacterium* and *Paracoccus* (Yen and Shih, 2009; Kuratsu *et al.*, 1984; Ha *et al.*, 2007a; Bule and Singhal, 2011; Natori *et al.*, 1978; Zhong *et al.*, 2009, 2013; Urakami and Yoshida, 1993; Yen and Chiu, 2007; Kien *et al.*, 2010; Yoshida *et al.*, 1998; Sakato *et al.*, 1992; Tian *et al.*, 2010a), moulds such as *Aureobasidium* and *Trichosporon* (Kondo *et al.*, 1971) and yeasts such as *Candida* and *Torulopsis* (Kondo *et al.*, 1971). As presented in Table 12.3, the reported specific CoQ_{10} contents of prokaryotes showed higher values than those of eukaryotes. These values were 0.48–12.5 mg/g dry cell weight (DCW) in prokaryotes and 0.075–0.45 mg/g DCW in eukaryotes. The highest CoQ_{10} content (12.5 mg/g DCW) was obtained from the strain *R. sphaeroides*. Thus, the variance in specific CoQ_{10} content was significant and dependent on cell species. For commercial production of CoQ_{10}, this result has to be referred in the development of superior cells.

Table 12.3 Specific CoQ_{10} contents produced from different cell types.

Cell type	Microorganism	Specific CoQ_{10} content (mg/g DCW)	References
Prokaryotes	*Agrobacterium* sp.	5.10	Kuratsu *et al.* (1984)
	Agrobacterium tumefaciens KCCM 10413	8.54	Ha *et al.* (2007a)
	Paracoccus denitrificans NRRL B-3785	0.81	Bule and Singhal (2011)
	Protaminobacter ruber	1.52	Natori *et al.* (1978)
	Proteus penneri	11.5	Zhong *et al.* (2013)
	Pseudomonas N842	1.20	Natori *et al.* (1978)
	Rhodobacter capsulatus	4.61	Urakami and Yoshida (1993)
	Rhodobacter sphaeroides	4.60	Yen and Shih (2009)
		4.61	Yen and Chiu (2007)
		6.34	Kien *et al.* (2010)
		8.70	Yoshida *et al.* (1998)
		12.5	Sakato *et al.* (1992)
	Rhodospirillum rubrum	5.66	Tian *et al.* (2010a)
	Sphingomonas sp.ZUTEO3	0.48	Zhong *et al.* (2009)
Eukaryotes	*Aureobasidium pullulans* FERM P-852	0.075	Kondo *et al.* (1971)
	Candida bogoriensis FERM P-666	0.3	
	Candida japonica FERM P-662	0.4	
	Torulopsis ingeniosa FERM P-665	0.45	
	Trichosporon cutaneum FERM P-850	0.31	

12.3.1.2
Biosynthetic Pathways

The synthesis of CoQ_{10} is performed endogenously in all cells. However, the biosynthetic pathway of CoQ_{10} has not been completely defined. Most of the available information on CoQ_{10} pathways has been derived from studies on bacteria and yeast. From the results of the genetic analysis using *Escherichia coli* and *S. cerevisiae* mutants, the metabolic pathway in the synthesis of CoQ_{10} was proposed (Kawamukai, 2002; Choi *et al.*, 2005; Cluis *et al.*, 2007; Meganathan, 2001a). Although there are some differences in supplying precursors for prokaryotes and eukaryotes, the biosynthetic pathway of CoQ_{10} is representatively composed of three major steps: the synthesis of decaprenyl diphosphate, the synthesis of a quinonoid ring and quinonoid ring modification (Choi *et al.*, 2005). Biosynthetic pathways of CoQ_{10} proposed in most prokaryotes and eukaryotes are presented in Figure 12.2, and the major enzymes involved in these biosyntheses are tabulated in Table 12.4.

Synthesis of Decaprenyl Diphosphate First of all, isoprenoids are synthesised from common building units, which are the universal metabolic precursors, isopentenyl diphosphate (IPP) and its isomer dimethylallyl diphosphate (DMAPP). The IPP precursor is produced via two distinct pathways: the well-known mevalonate (MVA) pathway and the relatively new 2-*C*-methyl-D-erythritol 4-phosphate (MEP) pathway or non-MVA pathway (Rohmer *et al.*, 1993). In general, bacteria possessing CoQ appear to use the MEP pathway, while the eukaryotes use the MVA pathway (de Dieu Ndikubwimana and Lee, 2014). However, there are some exceptions. Some eukaryotic microbes such as green algae and the malarial parasite *Plasmodium falciparum* appear to undergo the MEP pathway, while a few bacteria undergo the MVA pathway (Meganathan, 2001a). In addition, plants and *Streptomycetes* use both pathways (Hamano *et al.*, 2002; Lichtenthaler *et al.*, 1997).

In the MEP pathway of *E. coli*, pyruvate and D-glyceraldehyde 3-phosphate (GA3P) are the starting precursors for the biosynthesis of IPP and DMAPP. The pyruvate is decarboxylated in a thiamine pyrophosphate (TPP)-requiring reaction. The resulting hydroxyethyl TPP anion condenses with the aldehyde group of GA3P and forms 1-deoxy-D-xylulose 5-phosphate (DXP). This reaction is catalysed by DXP synthase (DXS) The DXP undergoes a benzylic-type rearrangement and is then reduced to MEP by nicotinamide adenine dinucleotide phosphate (NADPH)-dependent DXP reductoisomerase (DXR). To convert MEP into the building blocks of IPP and DMAPP, a series of additional MEP pathway enzymes are involved. MEP is converted to 4-diphosphocytidyl-2-*C*-methylerythritol (DCME) in a novel CTP-dependent reaction. This reaction is catalysed by 4-diphosphocytidyl-2-*C*-methylerythritol (CDP-me) synthase. Subsequently, DCME is phosphorylated by an ATP-dependent kinase. The resulting product is 4-diphosphocytidyl-2-*C*-methyl-D-erythritol 2-phosphate (DCME-2-P). In the next step of the pathway, 2-*C*-methyl-D-erythritol 2, 4-cyclodiphosphate (MECDP) synthase eliminates CMP from DCME-2-P, and forms MECDP

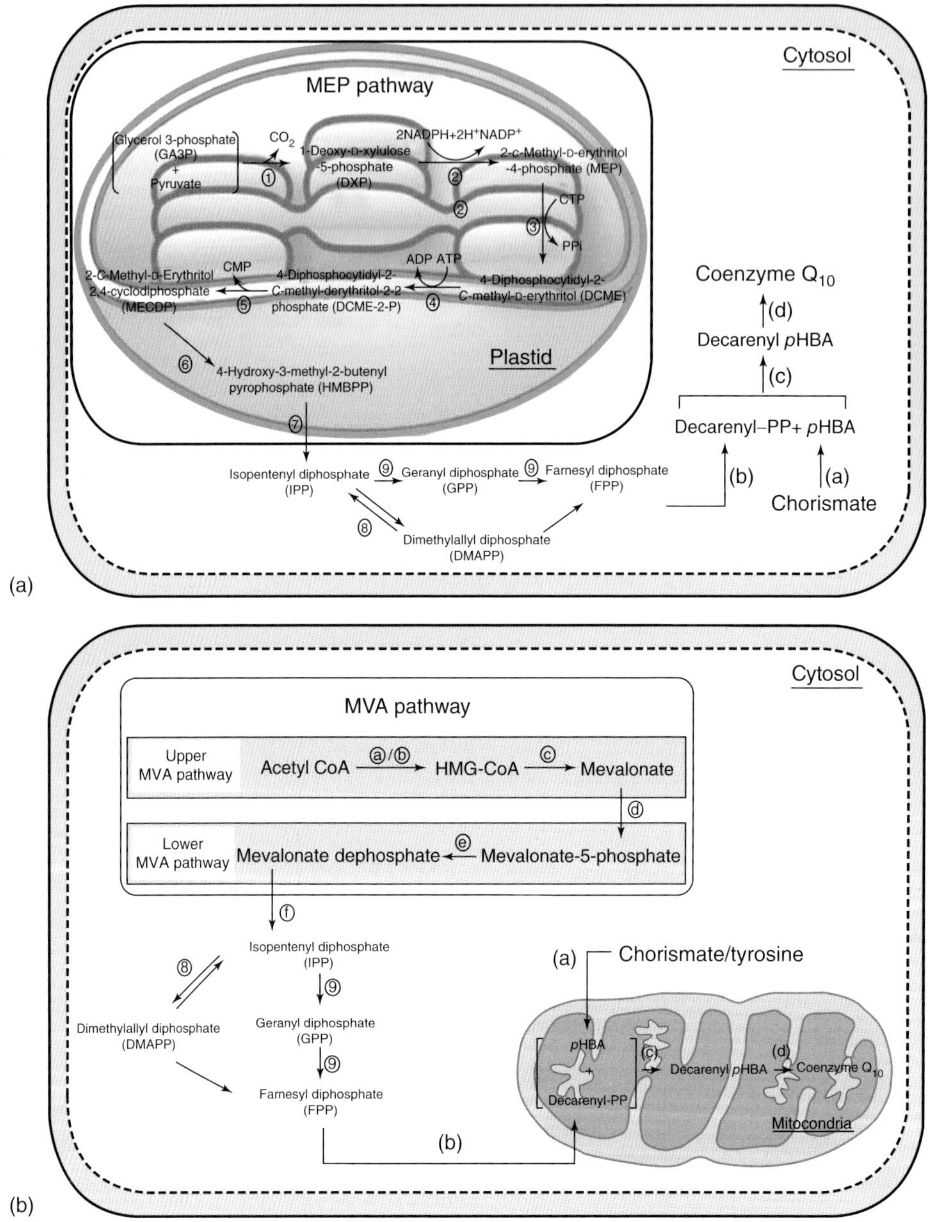

Figure 12.2 Biosynthetic pathways of CoQ_{10} proposed in most prokaryotes (a) and eukaryotes (b).

Table 12.4 Major enzymes involved in CoQ_{10} biosynthesis.

Pathway	No.	Enzymes	References
MEP pathway	①	1-Deoxy-D-xylulose-5-phosphate synthase (DXS)	Kim and Keasling (2001)
	②	1-Deoxy-D-xylulose-5-phosphatereductoisomerase (DXR)	Kuzuyama (2002)
	③	4-Diphosphocytidyl-2-*C*-Methyl-D-erythritol synthase (IspD)	
	④	4-Diphosphocytidyl-2-*C*-methyl-D-erythritol kinase (IspE)	
	⑤	2-*C*-Methyl-D-erythritol-2,4-cyclodiphosphate synthase (IspF)	
	⑥	4-Hydroxy-3-methyl-but-2-enyl pyrophosphate synthase (IspG)	
	⑦	4-Hydroxy-3-methyl-but-2-enyl pyrophosphate reductase (IspH)	
	⑧	Isopentenyl pyrophosphate isomerase (IdI)	Lee, Mijts and Schmidt-Dannert (2004)
	⑨	Farnesyl diphosphate synthase (IspA)	
	⑩	Polyprenyl diphosphate synthase (IspB)	
MVA pathway	ⓐ	Acetoacetyl-CoA thiolase (PhbA)	Nichols and Green (1992)
	ⓑ	HMG-CoA reductase (MvaS)	
	ⓒ	HMG-CoA synthase (MvaA)	
	ⓓ	Mevalonate kinase (MvaK1)	
	ⓔ	Phosphomevalonate kinase (MvaK2)	
	ⓕ	Mevalonate-5-pyrophosphate decarboxylase (MvaD)	
Chorismate pathway	(a)	Chorismate lyase (UbiC)	Nichols and Green (1992)
	(b)	Decaprenyl diphosphate synthase (DdsA or DPS)	Suzuki *et al.* (1997)
Ubiquinone pathway	(c)	pHB-polyprenyltransferase (UbiA)	Siebert *et al.* (1992)
	(d)	*O*-Methyltransferase (UbiG)	Wu *et al.* (1992)
		C-Methyltransferase (UbiE)	Lee *et al.* (1997)
		Monooxygenase (UbiH)	Nakahigashi *et al.* (1992)
		Monooxygenase (UbiF)	Kwon *et al.*, (2000)
		Monooxygenase (UbiB)	Poon *et al.* (2000)
		Chorismate lyase (UbiC)	Nichols and Green (1992)
		Decarboxylase (UbiD/UbiX)	Cox *et al.* (1969)

(Meganathan, 2001a; Eisenreich *et al.*, 2001). It was reported that MECDP was accumulated in bacteria under oxidative stress (Meganathan, 2001a). The next step of the pathway, the conversion of MECDP to IPP and DMAPP, is not yet clearly elucidated. Thus, genes, enzymes, intermediates and reactions involved in this reaction are yet to be determined (Urakami and Yoshida, 1993; Eisenreich *et al.*, 2001). It has been postulated that the formation of DMAPP and IPP from MECDP may involve a ring-opening reaction, two dehydrations and two reduction steps (Eisenreich *et al.*, 2001). An IPP isomerase interconverts IPP and DMAPP (Nakahigashi *et al.*, 1992; Eisenreich *et al.*, 2001). Fungi and yeasts lack the MEP pathway and generally rely on the MVA pathway. In the MVA pathway, acetyl-CoA is a starting precursor. The transfer of an acetyl group from one acetyl-CoA to the methyl carbon of a second acetyl-CoA initiates the formation of acetoacetyl-CoA. After condensing with another molecule of acetyl-CoA, 3-hydroxy-3-methylglutaryl-CoA (HMG-CoA) is formed. The MVA is finally formed when HMG-CoA is reduced with two moles of NADPH. Thus, three molecules of acetyl-CoA convert to MVA through acetoacetyl-CoA and HMG-CoA (Maury *et al.*, 2005). With two phosphorylation reactions mediated by MVA kinase and phospho-MVA kinase, MVA is converted to MVA diphosphate. Sequentially, MVA diphosphate further undergoes dehydration–decarboxylation in the presence of ATP, resulting in the formation of IPP. Then, IPP converts to DMAPP by isomerisation mediated by the IPP isomerase (Jeya *et al.*, 2010). Farnesyl diphosphate (FPP) synthase further catalyses the sequential 1′-4 coupling of IPP with DMAPP and geranyl diphosphate (GPP), resulting in the formation of GPP and FPP. After this reaction, FPP is elongated by decaprenyl diphosphate synthase (DPS). DPS has been reported to be cloned from various organisms, such as *A. tumefaciens, Gluconobacter suboxydans, P. denitrificans* and *S. pombe* (Barker and Frost, 2001; Suzuki *et al.*, 1997; Takahashi *et al.*, 2003; Lee *et al.*, 2004).

Synthesis of Quinonoid Ring The formation of 4-hydroxybenzoate (*p*HBA) from chorismate is the first step in the synthesis of the quinonoid ring. In *E. coli*, this reaction is catalysed by chorismate lyase encoded by the UbiC gene (Nichols and Green, 1992). In bacteria, *p*HBA as a precursor of the quinonoid ring is derived from the shikimate pathway. The shikimate pathway is a key pathway in the synthesis of aromatic amino acids through chorismate. *p*HBA is formed from tyrosine in mammals, because of their lack of the shikimate pathway. In yeast, *p*HBA is formed in two different ways, either directly from chorismate via the chorismate pyruvate lyase reaction, similar to *E. coli*, or alternately from tyrosine, similar to mammals (Jeya *et al.*, 2010). In animal cells, *p*HBA is formed from the essential amino acid tyrosine, which was proposed by Booth *et al.* (1960) from urinary excretion studies on animals administered with radio-labelled phenolic acids. From an *in vitro* study of rat liver and yeast extracts, the following pathway was proposed: Tyrosine → 4-hydroxyphenyl pyruvate (HPP) → 4-hydroxyphenyl lactate (HPLA) → 4-hydroxyphenyl cinnamate → *p*HBA. Recently, evidence in support of this pathway has also been obtained from higher plants (Loscher and Heide, 1994).

In UbiC-mutant *E. coli*, the formation of CoQ_{10} was poor, and its ability to grow aerobically on oxidisable substrates such as succinate was lost (Pennock and Threlfall, 1983). In the mean time, the presence of two alternate routes was possibly obtained using shikimate-pathway yeast mutants. Mutants blocked in the formation of shikimate or chorismate were deficient in the formation of CoQ_{10}, because of their inability to form *p*HBA. The ability of these mutants to form *p*HBA and CoQ_{10} was restored by the addition of tyrosine to the growth medium. It was found that tyrosine fully provided *p*HBA in these yeast mutants, although wild-type yeasts normally use the conversion of chorismate to *p*HBA as the source of precursor for CoQ_{10} (Booth *et al.*, 1960; Pennock and Threlfall, 1983).

Quinonoid Ring Modification The biosynthesis of CoQ_{10} begins with prenylation of *p*HBA, followed by several steps of ring modification reactions (Meganathan, 2001a). The prenylation is mediated by a membrane-bound enzyme, *p*HBA decaprenyl transferase (Ashby *et al.*, 1992; Melzer and Heide, 1994). This enzyme is characterised to be non-specific and, thus, can react with aromatic substrates, letting it tolerate substitutions by various groups at different positions on the benzene ring of CoQ_{10}. In addition, the enzyme can use a variety of prenyl diphosphates as side-chain precursors. The other ring modification reactions include decarboxylation, three hydroxylation reactions and three methylation reactions, and they take place in different orders in prokaryotes and eukaryotes (Meganathan, 2001a).

In decarboxylation, 3-octaprenyl-4-hydroxybenzoate is decarboxylated to 2-octaprenylphenol by the enzyme, 3-octaprenyl-4-hydroxybenzoate decarboxylase. After decarboxylation, 2-octaprenylphenol undergoes three hydroxylation reactions alternating with three methylation reactions, resulting in the formation of ubiquinol and then CoQ_{10}. When wild-type *E. coli* was aerobically grown on glycerol (an oxidisable substrate) with fumarate as an electron acceptor, considerable quantity of CoQ (50–70% of cells) was formed (Meganathan, 1996, 2001b). Mutants blocked in the non-hydroxylating reactions of the pathway were CoQ_{10} deficient under both aerobic and anaerobic conditions. The final three methylation reactions in the pathway are as follows: *O*-methylation (from 2-octaprenyl-6-hydroxyphenol to 2-octaprenyl-6-methoxyphenol); *C*-methylation (from 2-octaprenyl 6-methoxy-1,4-benzoquinol to 2-octaprenyl-3-methyl-6-methoxy-1,4-benzoquinol) and *O*-methylation (from 2-octaprenyl-3-methyl-5-hydroxy-6-methoxy-1,4-benzoquinol to ubiquinol). These reactions introduce methyl groups at the 6-OH, at the ring C-3 and at the 5-OH, respectively. A non-specific UbiE methylates the ring C-3, resulting in the conversion of 2-octaprenyl-6-methoxy-1,4-benzoquinol to 2-octaprenyl-3-methyl-6-methoxy-1,4-benzoquinol. In addition, this methyl transferase also methylates demethylmenaquinone (DMK) to menaquinone (MK). The two *O*-methylations are carried out by the same UbiG. In addition to carrying out the methylation of 2-octaprenyl-6-hydroxyphenol to 2-octaprenyl-6-methoxyphenol and 2-octaprenyl-3-methyl-5-hydroxy-6-methoxy-1,4-benzoquinol to ubiquinol, this non-specific UbiG also undergoes the *O*-methylation of the *S. cerevisiae* CoQ_{10} biosynthetic intermediate (from 3,4-dihydroxy-5-polyprenylbenzoate to 3-methoxy-4-hydroxy-5-polyprenylbenzoate) (Meganathan, 2001b; Jonassen and Clarke, 2001). In yeast, the side-chain precursor used for the prenyltransferase is hexaprenyl diphosphate. The product of the reaction, 3-hexaprenyl-4-hydroxybenzoate, undergoes further ring modification reactions in a different manner than that of *E. coli*. Accordingly, the 3-hexaprenyl-4-hydroxybenzoate first undergoes hydroxylation to 3,4-dihydroxy-5-polyprenylbenzoate, followed by methylation to 3-methoxy-4-hydroxy-5-polyprenylbenzoate and finally decarboxylation to 2-polyprenyl-6-methoxyphenol.

12.3.2
Metabolic Regulation

Metabolic regulation is necessary to improve CoQ_{10} biosynthesis. Key enzymes are involved in the main way to regulate the metabolic pathway, and the concentrations of nutrients, waste products and hormones can control the metabolic rates. Six enzymes are considered a key for CoQ_{10} biosynthesis via the MEP pathway. Overexpression of DXP synthase exhibited many positive results, and DXR is responsible for this pathway. IPP synthase and FPP synthase promote the biosynthesis of all-E-FPP that is the allylic substrate of decaprenyl DPS. The decaprenyl DPS regulates the chain length and thereby provides the long hydrophobic tail. Finally, p-hydroxybenzoate polyprenyl transferase combines the head and tail groups and transfers the reaction product to membrane. Therefore, functionally active coexpression of the six enzymes can facilitate systematic biological approaches to CoQ_{10} biosynthesis, with quantitative analyses of the metabolites and metabolic fluxes (Jeya *et al.*, 2010). In the MEP pathway, DXS, DXR, IDI and IspD are generally identified as rate-limiting enzymes, and metabolic engineering of this pathway targeting the four enzymes has successfully improved the metabolic flux and CoQ_{10} production (Ajikumar *et al.*, 2010; Kim *et al.*, 2006; Lv *et al.*, 2013; Yuan *et al.*, 2006). In the ubiquinone pathway, the major enzymes involved in the ubiquinone pathway are UbiC, UbiA, UbiB, UbiG and UbiH (Overvad *et al.*, 1999; Zhang *et al.*, 2007). It was reported that UbiA activity was a rate-limiting factor in CoQ_{10} synthesis, and UbiC was responsible for the derivation of CoQ_{10} from the precursor chorismate (Zhang *et al.*, 2007). Consequently, the strategy to regulate key enzymes involved in the rate-limiting steps of the CoQ_{10} biosynthesis is necessary and thereby has been efficiently used.

In the wild-type microorganism strains, growth conditions have been optimised to modulate their CoQ_{10} biosynthesis, and their cellular-regulatory mechanisms have also been altered. One important factor to regulate the CoQ_{10} biosynthesis during the fermentation is the oxidation–reduction potential (ORP) of the fermentation medium (Sakato *et al.*, 1992). In the case of mutant *Rhodobacter sphaeroides* strains, a higher CoQ_{10} content (up to 8.7 mg/g DCW) was achieved when the aeration rate was reduced during the fermentation (Yoshida *et al.*, 1998).

Choi *et al.* (2005) reported that the increase in the intracellular CoQ_{10} content is closely related to limitation of the oxygen supply and addition of azide as an electron flux inhibitor.

12.3.3
Strain Development

For the commercial production of CoQ_{10}, wild-type strains of microorganisms have been isolated and their chemical mutants have been successfully developed. Recently, opportunities have arisen for the metabolic engineering of CoQ_{10} production in microorganisms as the knowledge on both the biosynthetic enzymes and the regulatory mechanisms of CoQ_{10} production has accumulated from many related studies. Currently, the strain *E. coli* is the most frequently used to engineer a CoQ_{10} pathway, because it is well suited for genetic modifications and large-scale fermentation. However, several biochemical steps are required to be modified and/or optimised in this metabolic engineering of *E. coli* for a high yield of CoQ_{10} (Cluis *et al.*, 2007).

12.3.3.1
Mutagenesis

To improve the CoQ_{10} content in cells, superior mutants were developed based on several indirect phenotypes, such as growth on structural-analogue inhibitors of the pathway or alteration in pigment production. The trial was to develop mutants with a better regulated CoQ_{10} pathway by selection for growth on pathway or respiration inhibitors. As seen in Table 12.5, the CoQ_{10} contents of *A. tumefaciens* mutants were higher than that of the wild-type strain, because of their ability to grow in the presence of the structural analogues of ubiquinone, daunomycin and menadione. Mutants were also selected based on growth in the presence of L-ethionine. The L-ethionine is an analogue of methionine, which is a precursor supplying three methyl groups capable of modifying the ubiquinone of the benzoic ring (Yoshida *et al.*, 1998). It was also reported that analogues of aromatic amino acids and inhibitors of respiration could be used to improve CoQ_{10} content in *A. tumefaciens* mutants (Yoshida *et al.*, 1998). In the case of *R. phaeroides* mutants, they were selected based on the appearance of green colonies, indicating reduced carotenoid content (Yoshida *et al.*, 1998). Thus, the change of pigment-production phenotype could increase CoQ_{10} content in mutants. The same result was reported in the mutant strain of *Protomonas extorquens* associated with low carotenoid content (Urakami and Hori-Okubo, 1988). Although the correlation between the reduced carotenoid content and improved CoQ_{10} content is not clearly understood, excess isoprenoid precursors resulting from mutations in the carotenoid pathway might be redirected to CoQ_{10} synthesis.

12.3.3.2
Genetic Modification

Genetic engineering or genetic modification is the direct manipulation of a microorganism's genome. New DNA can be inserted in the host genome

Table 12.5 Specific CoQ_{10} contents produced from chemically modified mutants.

Mutant strain	Mutagen	Specific CoQ_{10} content (mg/g DCW)	References
Agrobacterium tumefaciens M-37	Ethionine	4.50	Yoshida *et al.* (1998)
Agrobacterium tumefaciens AU-55	Daunomycin, ethionine and vitamin K_3	5.10	
Agrobacterium tumefaciens KCCM 10413	*N*-Methyl-*N′*-nitro-*N*-nitrosoguanidine	8.50	
Rhodobacter sphaeroides Co-22–11	Carotenoid-deficiency	2.60	Yoshida *et al.* (1998)
Rhodobacter sphaeroides KY-8598	Carotenoid-deficiency	14.50	Sakato *et al.* (1992)
Rhodobacter sphaeroides	Menadione	2.94	Jeong *et al.* (2008)
	Ethionine	1.76	
	Daunomycin	1.90	
	Menadione and ethionine	2.10	

by isolating and copying the target genetic material using molecular cloning methods to generate a DNA sequence or by synthesising the DNA and then inserting it into the host cell. To improve CoQ_{10} content in recombinant *E. coli*, overexpression of the major genes involved in MEP, chorismate and ubiquinone pathways has been proposed.

Overexpression of the MEP Pathway Enzymes In *E. coli*, the overexpression of the DXS enzyme isolated from *Pseudomonas aeruginosa*, *Bacillus subtilis* or *Synechocystis* sp. 6803 was reported to result in small increases in CoQ_{10} content (Kim *et al.*, 2006; Harker and Bramley, 1999). The DXS enzyme is the first step of the MEP pathway, which catalyses the condensation reaction between pyruvate and GA3P (Lois *et al.*, 1998). The DXS enzyme plays an important regulatory role in the MEP pathway for the synthesis of IPP (de Dieu Ndikubwimana and Lee, 2014). The rise in DXS level was reported to increase lycopene production significantly (de Dieu Ndikubwimana and Lee, 2014; Harker and Bramley, 1999; Matthews and Wurtzel, 2000). Thus, the enlargement of the IPP pool by the overexpression of DXS seems to increase the production of CoQ_{10}, which uses IPP as a precursor, similarly to lycopene. For this reason, several studies have focussed on the metabolic engineering of IPP biosynthesis via the overexpression of rate-limiting genes of the MEP pathway (de Dieu Ndikubwimana and Lee, 2014; Harker and Bramley, 1999; Albrecht *et al.*, 1999).

Overexpression of Chorismate Pathway Enzymes To obtain a high yield of *p*HBA in recombinant *E. coli*, overexpression of the chorismate pathway enzymes has been attempted. In the study performed by Barker and Frost (2001), a high yield of *p*HBA was achieved by enhancing the carbon flow to *p*HBA. They overexpressed chorismate pathway enzymes, while eliminating the production of the aromatic amino acids competing with the same precursors. This strategy resulted in higher levels of *p*HBA, compared with those produced normally by *E. coli*. Zhu *et al.* (1995) also reported CoQ_{10} production in *E. coli* by overexpressing UbiA and IspB genes. In their study, *E. coli* harbouring plasmids containing UbiA and IspB produced a higher amount of CoQ_{10} than the wild type.

Overexpression of Ubiquinone Pathway Enzymes Of the biosynthetic pathways of CoQ_{10}, the ubiquinone pathway after the isoprenoid and the chorismate pathways are considered a potential metabolic bottleneck that limits CoQ_{10} biosynthesis. In *E. coli*, the major enzymes involved in the ubiquinone pathway are UbiC, UbiA, UbiB, UbiG and UbiH, and the improvement of CoQ_{10} content by overexpressing the genes encoding them has been reported (Overvad *et al.*, 1999; Zhang *et al.*, 2007). Among these enzymes, the overexpression of UbiA resulted in the most impact on CoQ_{10} content, suggesting that UbiA activity is a rate-limiting factor in CoQ_{10} synthesis (Albrecht *et al.*, 1999). This result was somewhat predicted because *p*HBA prenylation catalysed by UbiA represents the branch point at which the isoprenoid precursor becomes committed to CoQ_{10} biosynthesis. The effect of UbiC overexpression on CoQ_{10} yields was also important because this enzyme is responsible for the derivation of CoQ_{10} from the precursor chorismate. However, the overexpression of UbiB, UbiG and UbiH had little effect on CoQ_{10} production, indicating that they don't involve in limiting steps in the CoQ_{10} synthetic pathway (Zhang *et al.*, 2007; Zhu *et al.*, 1995). Other candidates for the overexpression of ubiquinone enzymes could include UbiD and UbiX, as a certain correlation was reported between their expression and CoQ_{10} production in *E. coli* (Zhang and Javor, 2003).

12.3.3.3 Metabolic Engineering

To increase CoQ_{10} content in production, optimising genetic and regulatory processes within cells have been performed. These processes are chemical networks that use a series of biochemical reactions and enzymes. Specifically, metabolic engineering seeks to pinpoint parts of the network that constrain CoQ_{10} production. To relieve these constraints, gentic engineering techniques can be used for the modification of the networks. Thus, the final goal of metabolic engineering is to be able to use the modified microorganisms for the commercial production of CoQ_{10} in a cost-effective manner. In this section, two techniques for the modification of the networks are described. Other pathways competing for chorismate and isoprenoid precursors could be considered as a metabolic engineering strategy to improve the CoQ_{10} production in *E. coli*.

Introduction of Foreign MVA Pathway MVA is a key intermediate in the MVA pathway. However, it is not produced or consumed by wild-type *E. coli* (Tabata and

Hashimoto, 2004). Thus, a foreign MVA pathway can be alternatively introduced into *E. coli* using the MEP pathway. The introduction of the foreign MVA pathway results in an increase of the IPP supply in the CoQ_{10}-producing *E. coli* strains. It was reported that higher productions of lycopene (Yoon *et al.*, 2006) and β-carotene (Yoon *et al.*, 2007) in *E. coli* were found when the MVA pathway of *Streptococcus pneumoniae* was introduced with exogenous supplementation of MVA. Similarly, Zahiri *et al.* (2006) reported that the introduction of a foreign MVA pathway into *E. coli* increased the IPP supply in the CoQ_{10}-producing *E. coli* strains.

Blocking of CoQ_{10} Biosynthesis Blocking the biosynthesis of competing molecules, particularly MK and DMK, was also attempted to improve CoQ_{10} yields in *E. coli*, because MK and DMK share the quinone pool with CoQ_{10} and are formed of octaprenyl diphosphate and a naphtoquinone derived from chorismate (Poon *et al.*, 2000). The biosynthesis of CoQ_{10} is favoured over that of DMK and MK under aerobic conditions, leading to an abundant pool of quinone (Shestopalov *et al.*, 1997). However, the ratio between the different types of quinones is significantly dependent on oxygen availability (Bekker *et al.*, 2007). Therefore, blocking or down-regulating the biosynthesis of MK and DMK could be used as a metabolic engineering strategy to improve CoQ_{10} yields in *E. coli*.

Analysis of Interactions among Protein Complexes To develop a successful metabolic-engineering strategy for CoQ_{10} production in microbes, it is necessary to investigate the cellular location of the enzymes and to determine if they are located in protein complexes that are necessary for their activity. All intermediates generated in the biosynthetic CoQ_{10} pathway are localised to the cytoplasmic membrane, because they possess a hydrophobic isoprenoid side chain. Based on this fact, several coenzyme Q-biosynthetic enzymes have demonstrated this feature (Young *et al.*, 1972; Leppik *et al.*, 1976a, 1976b). Furthermore, strains with mutations in UbiB, UbiG or UbiH accumulate 2-octaprenylphenol as an intermediate (Siebert *et al.*, 1992; Urakami and Hori-Okubo, 1988; Young *et al.*, 1973), indicating that an activity loss of any of these three enzymes can block the hydroxylation of this intermediate. Thus, this phenotype suggests interdependency or a complex among these proteins. The existence of such a protein complex was supported by a study reporting that a soluble complex isolated from the cytoplasmic membrane of *E. coli* could convert 2-octaprenylphenol to CoQ_{10} in the presence of cytosolic proteins (Knoell, 1979). Moreover, the evidence for the existence of such a complex is increasing in *S. cerevisiae*, implying that a related complex might be functional in *E. coli* (Baba *et al.*, 2004; Marbois *et al.*, 2005; Gin and Clarke, 2005; Hsu *et al.*, 2000). To engineer this biosynthetic pathway of CoQ_{10} successfully in *E. coli*, a careful analysis of interactions among CoQ_{10} biosynthetic enzymes and additional proteins is needed.

12.3.4
Fermentation Process

CoQ_{10} can be produced by chemical (Negishi *et al.*, 2002), semi-chemical (Lipshutz *et al.*, 2002) or biological synthetic methods, and the biological synthesis is more diversely used than the other methods. This is because the starting materials used for the chemical synthesis of CoQ_{10} are different from those synthesised in microorganisms and human beings (Ha *et al.*, 2007a). For the commercial production of CoQ_{10}, the biological synthesis using microorganisms has attracted increasing attention (Choi *et al.*, 2005). For this purpose, the construction of genetically engineered microorganisms and their metabolic modification have been attempted to improve CoQ_{10} production (Cluis *et al.*, 2007; Yoon *et al.*, 2007; Young *et al.*, 1973; Park *et al.*, 2005).

To date, various fermentation systems have been used to produce CoQ_{10}, because fermentation has some advantages over the production of CoQ_{10} through chemical synthesis. One of the advantages is easy scale-up. To date, CoQ_{10} production on the 50-100-kl scale ranges from 350 to 770 mg/l, and the commercial price of CoQ_{10} is in the range of US$ 600–800 kg^{-1} (Bule and Singhal, 2011). In the fermentation of CoQ_{10}, various species of microorganisms have been used: *A. tumefaciens* KCCM 10413 (Ha *et al.*, 2007a); *Paracoccus denitrificans* NRRL B-3785 (Bule and Singhal, 2011); *Protaminobacter ruber* (Natori *et al.*, 1978); *Pseudomonas* N842 (Natori *et al.*, 1978); *Proteus penneri* CA8 (Zhong *et al.*, 2013); *Rhodobacter capsulatus* (Urakami and Yoshida, 1993); *R. sphaeroides* FERM-P4675 (Yen and Chiu, 2007); *R. sphaeroides* (Kien *et al.*, 2010; Yoshida *et al.*, 1998); *Sphingomonas* sp. ZUTEO3 (Zhong *et al.*, 2009); and *A. tumefaciens* ATCC 4452 (Tokdar *et al.*, 2013).

A traditional process of CoQ_{10} production was reported by Zhong *et al.* (2009). The batch-type fermentation was executed using *Sphingomonas* sp. ZUTEO3, with advantages of process simplification and integration. However, an improvement of CoQ_{10} yield was required, suggesting strain development by mutation or metabolic modification. Alternatively, a fed-batch operation using an appropriate feeding strategy was proposed to obtain better CoQ_{10} production, because it is typically used in bio-industrial processes to reach a high cell density in the fermenter (Gu *et al.*, 2006). The controlled addition of the nutrient directly affects the cell growth rate and helps to avoid overflow metabolism such as formation of side metabolites, resulting in a higher value of the specific CoQ_{10} content. In this operation, the concentration of each component in the concentrated culture medium was determined by considering the change of working volume, and thereby, the concentration in the fed-batch operation was the same as that of the culture medium used in batch fermentation.

For the commercial CoQ_{10} production, lab-scale data should be adequately transferred to pilot-plant-scale or commercial-scale processes. Recently, interest in CoQ_{10} production was renewed because of the growing demands of the pharmaceutical industry (Mattila *et al.*, 2000). However, not many scale-up studies of CoQ_{10} production have been performed using microorganisms other than photosynthetic bacteria. The main bottleneck in the scale-up of phototrophic fermentation is the low efficiency of light energy conversion to the desired product, because of excessive dissipation of light energy to heat (Strazisar *et al.*, 2005). In commercial-scale CoQ_{10} production, thermodynamics and microbial kinetics are scale independent, while transport phenomena are scale dependent. Nutrients have to be adequately supplied to the cells by transport processes, and major parameters have to be monitored and controlled.

In Table 12.6, specific contents of CoQ_{10} produced from various fermentation systems with working volumes of 1–300 l are shown. In the batch mode, the values of specific CoQ_{10} contents ranged from 1.20 to 11.5 mg/g DCW depending on both species and characteristics of the strains used in fermentation. The highest specific content of CoQ_{10} (11.5 mg/g DCW) was obtained from the fermentation of *P. penneri* CA8 with a 0.5 l working volume. Surprisingly, this wild-type strain had higher specific CoQ_{10} content than other mutants reported. In addition, a fairly high specific CoQ_{10} content (6.34 mg/g DCW) was reported from the fermentation of *R. sphaeroides* operated in a 150 l scale fermenter at 30 °C and 50 lux. In this photosynthetic fermentation, it was reported that the aeration shift from an adequate supply at the early cell growth phase to a limited supply at the active cell growth phase resulted in a high CoQ_{10} content (Kien *et al.*, 2010). To improve specific CoQ_{10} contents further, fermentations were conducted in the fed-batch mode using mutants. The values of specific CoQ_{10} contents were 8.12–9.25 mg/g DCW, which were higher than those produced from the wild-type cells. The highest specific content of CoQ_{10} (9.25 mg/g DCW) was obtained from fermentation of *A. tumefaciens* KCCM 10413 with a 250 l working volume (Ha *et al.*, 2007b). The result was obtained using the advantage of a fed-batch mode in which a high cell density in the fermenter was possible.

12.3.5 Upstream and Downstream Processing

12.3.5.1 Upstream Processing

CoQ_{10} can be produced from microorganisms such as *A. tumefaciens, P. denitrificans, Rhizobium radiobacter* and *Rhodobacter sphaeroides* (Yoshida *et al.*, 1998; Seo, Kook and Kim, 2012). However, CoQ_{10} production from the wild-type strains is not satisfactory because of the limits of CoQ_{10} accumulation in cells, implying the necessity of strain improvements using chemical mutagenesis, genetic engineering, UV radiation, high hydrostatic pressure (HHP) treatment and so on (Yuan, Tian and Yue, 2012; Ranadive *et al.*, 2014). Therefore, strain improvement plays a central role in commercial-scale microbial fermentation processes. To secure superior CoQ_{10}-producing strains, various methods have been used. Yoshida *et al.* (1998) developed a higher CoQ_{10}-producing mutant

Table 12.6 Specific contents of CoQ_{10} produced from various fermentation systems.

Mode of fermentation	Strain	Working volume (l)	Specific CoQ_{10} content (mg/g DCW)	References
Batch	*Agrobacterium tumefaciens* KCCM 10413 (mutant)	5	6.61	Ha *et al.* (2007a)
	Paracoccus denitrifican NRRL B-3785	0.1	0.81	Bule and Singhal (2011)
	Protaminobacter ruber	15	1.52	Natori *et al.* (1978)
	Proteus penneri CA8	0.5	11.5	Zhong *et al.* (2013)
	Pseudomonas N842	15	1.20	Natori *et al.* (1978)
	Rhodobacter capsulatus	1	4.61	Urakami and Yoshida (1993)
	Rhodobacter sphaeroides	0.3	2.70	Yoshida *et al.* (1998)
	R. sphaeroides FERM-F4675	5	4.61	Yen and Chiu (2007)
	R. sphaeroides (mutant)	150	6.34	Kien *et al.* (2010)
	Sphingomonas sp. ZUTEO3	0.5	0.48	Zhong *et al.* (2009)
Fed-batch	*A. tumefaciens* ATCC 4452	7	1.87	Tokdar *et al.* (2013)
	A. tumefaciens KCCM 10413 (mutant)	2	9.05	Ha *et al.* (2007b)
		5	8.54	Ha *et al.* (2007a)
		250	9.25	Ha *et al.* (2007b)
	R. sphaeroides (mutant)	150	8.12	Kien *et al.* (2010)

from the wild-type *A. tumefaciens* A-9 strain by 100 mg/l *N*-Methyl-*N′*-nitro-*N*-nitrosoguanidine treatment at 30 °C for 50 min with shaking. This improvement could result from the mutant's ability to overcome growth inhibition by inhibitors during CoQ_{10} biosynthesis or from its related metabolism overproducing CoQ_{10} (Kim *et al.*, 2014). Mutants showing a better CoQ_{10} production were selected by the standard of blue colours using Craven's method (Kim *et al.*, 2014). The other chemicals used in mutation were L-ethionine, daunomycin and menadione. For mutation, cells ($10^8 - 10^9$ cells/ml) were suspended in 0.5 M Tris-maleate buffer (pH 6.2) containing 2 mg/ml of each chemical for 20 min and washed twice with 0.85% saline (Yoshida *et al.*, 1998). After cells were resuspended in Tris-maleate buffer, 10 ml solution was incubated for 24 h. Then, the cells were harvested by centrifugation and spread on M-medium containing 20 μg/ml 5-bromo-1-chloro-3-indolyl-*β*-D-galactoside (X-gal). The composition of the M-medium was (per litre): 100 g glucose, 5 g $(NH_4)_2SO_4$, 0.5 g KH_2PO_4, 0.5 g K_2HPO_4, 0.25 g $MgSO_4 \cdot 7H_2O$, 0.1 g $MnCl_2 \cdot 4H_2O$, 1 ml trace element solution, 1 ml of vitamin mixture and 20 g agar. Finally, blue colonies formed on the agar plates were isolated (Jeong *et al.*, 2008).

The use of recombinant DNA technology is another way to improve the CoQ_{10}-producing abilities of wild-type strains. *E. coli* is frequently used in metabolic engineering areas, but the recombinant *E. coli* strains sometimes accumulate CoQ_8 and CoQ_9 in addition to CoQ_{10} (Park *et al.*, 2005). From an industrial point of view, this is undesirable because several downstream processing steps are needed. Accordingly, the development of strains with stringent product specificity of CoQ_{10} might be a good approach to improve the yields and to reduce the costly steps for purification of CoQ_{10} from unwanted CoQ types. *E. coli* is engineered by recombinant DNA technology for the synthesis of CoQ_{10} via the introduction of decaprenyl DPS from various microorganisms including *A. tumefaciens* and *G. suboxydans* (Okada *et al.*, 1998; Zahiri *et al.*, 2006). These recombinant strains are further modified by the overexpression of several upstream enzymes to improve the metabolic flux towards CoQ_{10} synthesis. Some instances are as follows: Overexpression of DXS in the non-MVAe pathway exhibited positive effects on CoQ_{10} production (Kim *et al.*, 2006; Seo *et al.*, 2007); coexpression of the UbiA and UbiC enzymes involved in the synthesis of a phenolic ring constituting CoQ_{10} also improved the specific content of CoQ_{10} (Cluis *et al.*, 2007; Zhang *et al.*, 2007).

Another method to secure superior CoQ_{10}-producing mutants from wild-type strains was reported to be the combined method of UV radiation and DES (diethyl sulfate) (Yuan *et al.*, 2012). In this method, UV radiation treatment was initiated, followed by chemical treatment. After the cell suspension was spread onto a presterilised plate (9 cm diameter), the cells were irradiated by a UV lamp (254 nm, 30 W) at a distance of 30 cm from the plate for 60 s. Then, the cell suspension was treated with 1% (v/v) DES for 20 min.

The HHP is another method to secure superior CoQ_{10}-producing mutants (Wuytack *et al.*, 2002). The HHP equipment contained a 2 l working pressure chamber (UHPF-750MP, BaoTou KeFa New Type Hi-Tech Food Machine Limited

Company, China). For HHP treatment, cells suspended in potassium phosphate buffer were first transferred aseptically into sterile polyethylene pouches, and the cell suspension was heat-sealed. After the prepared pouches were placed into the HHP equipment, the HHP treatments were carried out at 25 °C and constant pressures (100–400 MPa) for 10–30 min. Control samples were maintained at 1 atm, and the effect of HHP treatment was confirmed by counting the colonies formed on the agar plates.

12.3.5.2
Downstream Processing

After the superior strains are developed and secured, downstream processing is required for CoQ_{10} production. Downstream processing for CoQ_{10} production refers to fermentation operation with optimisation, recovery and purification and the disposal of waste after fermentation. This is an essential step in the manufacture of CoQ_{10} for commercialisation. One of the downstream processes is the optimisation of fermentation conditions to improve CoQ_{10} production. During the fermentation, major environmental parameters are pH, temperature, carbon or/and nitrogen sources, aeration, viscosity and minerals (Choi *et al.*, 2005; Jeya *et al.*, 2010).

In the fermentation of *A. tumefaciens* operated in 5 l jar fermenter, the maximum specific CoQ_{10} content and CoQ_{10} concentration were obtained at pH 7.0 and 32 °C (Zhong *et al.*, 2013). The specific CoQ_{10} content varied little with the cultivation temperature, but culture pH had a significant effect. In a study to examine the effect of the carbon source on *Rhodospirillum rubrum* growth and CoQ_{10} production, the best carbon source was reported to be malate (Tian *et al.*, 2010b). The specific CoQ_{10} contents differed depending on the carbon sources. This was because *R. rubrum* cells adapted from chemotrophic to phototrophic conditions, resulting in an increase of the cellular CoQ_{10} content (Carr and Exell, 1965). Thus, malate as a photosynthetic reductant may play a critical role in the noncyclic photophosphorylation of *R. rubrum* (Stenn, 1968). Using malic acid as carbon source, the effects of nitrogen sources on CoQ_{10} production by *R. rubrum* were investigated, and complex nitrogen sources (a mixture of yeast extract and ammonium sulfate) were found to be more desirable than a single nitrogen source, probably because of the rich nitrogen, vitamins and other growth-stimulating compounds in the yeast extract (Kalil *et al.*, 2008). Ammonium sulfate also showed a positive effect on CoQ_{10} production by *Azotobacter vinelandii* (Knowles and Redfearn, 1968) and by *Agrobacterium* sp. (Kuratsu *et al.*, 1984a).

The dissolved oxygen (DO) level is another key factor in the optimisation of CoQ_{10} production. A low level of DO rate was found to be very effective in enhancing CoQ_{10} production by an *Agrobacterium* species (Kuratsu *et al.*, 1984; Ha *et al.*, 2007a; Choi *et al.*, 2005), by *R. sphaeroides* (Yoshida *et al.*, 1998) and by *P. denitrificans* (Matsumura *et al.*, 1983). Thus, these results show that DO levels should be kept low to increase CoQ_{10} production. During the fermentation, the culture broth becomes highly viscous because of the formation of extracellular polysaccharides (EPSs) when cultivated on a sucrose-based medium. This significantly affects not only the mixing of the medium but also oxygen transfer in the fermenter (Kuratsu and Inuzuka, 1985). In addition, the highly viscous broth makes the purification process difficult during the downstream processing (Ha *et al.*, 2007b). Thus, the decrease in EPS production is important to increase CoQ_{10} production on an industrial scale. Other than the aforementioned factors affecting CoQ_{10} production, mineral concentration is also an important factor. Among the various ion sources, $CaCl_2$ resulted in the highest CoQ_{10} production by increasing the specific CoQ_{10} content of the cell without the inhibition of cell growth (Ha *et al.*, 2009). The increase of cellular CoQ_{10} content was reported to be ascribed to the oxidative stress induced by Ca^{2+} supplementation, because of the defensive role of CoQ_{10} against oxidative stress (Schroeder and Johnson, 1995; Søballe and Poole, 2000).

After the fermentation of CoQ_{10} is completed under optimal culture conditions, the recovery of CoQ_{10} contained in the cells is initiated. The cells are first harvested and separated by centrifugation. Then, cells are disrupted, and CoQ_{10} is generally extracted by solvents from the disrupted cell solution. After solvent extraction, a CoQ_{10}-rich phase solution is evaporated before injecting into a high-performance liquid chromatography (HPLC) instrument for analysis and purification. The details of these processes are described in the following section.

During the manufacturing process for CoQ_{10} production, large amounts of the fermentation broth are ultimately left over. The fermentation broth normally contains a variety of amino acids, organic matter and nutrients (Xinqiang *et al.*, 2011). Thus, inefficient treatment of the fermentation broth, such as conventional wastewater treatment, not only causes environmental problem, but also causes further treatment expenses. If the fermentation broth can be reutilised, resource utilisation will be improved, and this also avoids environmental pollution. In addition, the amino-acid liquid fertiliser as a foliar fertiliser can promote plant growth, increase crop yield, improve crop quality, enhance crop resistance and therefore can be widely used. Hence, the preparation of amino-acid liquid fertiliser from CoQ_{10} fermentation broth can provide not only a variety of amino acids needed for crop growth, but also nutrients for crop nitrogen, potassium and ammonia. At present, residues from CoQ_{10} production is manufactured as organic fertiliser after dehydration (Xinqiang *et al.*, 2011). From the waste treatment viewpoint, a physical cell disruption method has been suitably proposed for the efficient manufacture of CoQ_{10} (Matsuda *et al.*, 2000). Waste cells could be effectively utilised through this method.

To improve the solubility, photostability and thermal stability of CoQ_{10} as a product, many methods have been reported: the encapsulation of CoQ_{10} in carbohydrate matrices such as gum Arabic, β-cyclodextrin and γ-cyclodextrin (Yang and Song, 2006; Fir *et al.*, 2009; Bule, Singhal and Kennedy, 2010); solid self-emulsifying drug delivery system of CoQ_{10} (Onoue *et al.*, 2012); and the incorporation of CoQ_{10} with poly(methyl methacrylate) nanoparticles (Kwon *et al.*, 2002). For the microencapsulation of CoQ_{10} in carbohydrate matrices, CoQ_{10} enclosed in gum Arabic was undertaken with coconut oil and sodium stearoyl lactylate as an emulsifier while the encapsulation of CoQ_{10} with cyclodextrins was prepared by an inclusion complex with water. In this method, the cyclodextrins were used as stabilisers to preserve CoQ_{10}, and their use was simple and with ease of operation. The solid self-emulsifying drug delivery system of CoQ_{10} could be prepared by spray-drying an emulsion pre-concentrate containing CoQ_{10}, medium-chain triglyceride, sucrose ester of fatty acid and hydroxypropyl cellulose. To improve the photostability of CoQ_{10}, the incorporation of CoQ_{10} with poly(methyl methacrylate) nanoparticles was performed under a micro-fluidisation and solvent evaporation method. In this method, both the types of surfactants used and the recycling number of the micro-fluidisation process significantly affected the mean diameter of the nanoparticles (Chung and Han, 2010).

12.4 Chemical Synthesis and Separation of CoQ_{10}

12.4.1 Chemical Synthesis

Because of the high value of ubiquinones, several approaches of their synthesis have been proposed over the past three to four decades. Such representative approaches were Lewis acid–induced prenoidal stannane additions to quinones, (Naruta, 1980) reiterative Pd(0)-catalysed couplings of doubly activated prenoidal chains with allylic carbonates bearing the required aromatic nucleus in a protected form (Eren and Keinan, 1988), and a retro Diels-Alder route to arrive directly at the quinone oxidation state (van Liemt *et al.*, 1994; Rüttimann and Lorenz, 1990). Nonetheless, all these approaches had disadvantages, such as lengthy, linear rather than convergent and inefficient aspects. In addition, there were some problems in controlling double-bond stereochemistry, which led to complicated mixtures of geometrical isomers which were difficult to separate given the hydrocarbon nature of the side chains (Yanagisawa *et al.*, 1991). To overcome these problems, a convergent method for the synthesis of the ubiquinones which originates with a simple benzenoid precursor and proceeds with retention of the double-bond stereochemistry was developed, causing significant advancement in the synthesis of ubiquinones (Lipshutz, 2000). Several major improvements have been made in the synthesis of CoQ_{10}. A new sequence was developed leading to the substituted para-quinone head group, resulting in both the reduction in the extent of manipulation of the side chain and the elimination of two synthetic steps positioning in the late synthesis (Lipshutz *et al.*, 2005). With a shortened route to CoQ_{10} synthesis, it was expected that opportunities for its potential industrial scale-up would be enhanced.

Because CoQ_{10} has biological significance and thus commercial importance, its chemical synthesis has attracted considerable interest. To synthesise CoQ_{10}, solanesol isolated from tobacco leaves was used as a starting material for isoprenoid side-chain synthesis, and the quinonoid head was prepared from

2,3,4-methoxy-6-methyl benzaldehyde (Yamamura *et al.*, 1981) The overall yield was 64% without impurities. Another way to synthesise CoQ_{10} was reported by development of a highly stereo-selective process, which is important for the all-E-conformation of the CoQ_{10} tail (Negishi *et al.*, 2002). Without stereo-isomeric separation, the overall yield of this process was 26%, with 98% selectivity. This highly regio- and stereo-selective methodology for the synthesis of terpenoids critically involves Pd-catalysed homoallyl– and homopropargyl–alkenyl coupling and Zr-catalysed carboalumination of alkynes. However, these two methods are not only environmentally unfavourable, but also suffering from low yields (Lipshutz *et al.*, 2002). To date, various synthetic routes to CoQ_{10} have been developed. CoQ_{10} can be synthesised using trimethoxy toluene and solanesol (Lipshutz *et al.*, 2002). During the synthetic steps, the benzyl chloride intermediate and vinyl alane are generated from trimethoxy toluene and solanesol, respectively. By coupling these benzyl chloride and vinyl alane intermediates, CoQ_{10} is finally synthesised, with only a 50% yield from the two starting materials (Hatakeyama *et al.*, 2006). In the synthesis of CoQ_{10}, the development of an efficient and highly regio- and stereo-selective preparation of CoQ_{10} is still in demand for scale-up processes. For this reason, an improved route to synthesise CoQ_{10} was reported starting from commercially available coenzyme Q_1 via SeO_2-mediated oxidation of protected isoprenyl hydroquinone into the (*E*)-allyl alcohol (Yu *et al.*, 2006). This process resulted in a stereo-selective synthesis of CoQ_{10} with a 38% overall yield. Although this is a well-known organic synthesis method for CoQ_{10}, its viable commercial-scale production has to be studied further. Alternatively, fermentation using superior microbes is another means for the commercial-scale production of CoQ_{10}.

12.4.2
Solvent Extraction

After cell disruption, CoQ_{10} contained in the cell can be extracted by solvents. Among the solvents, 1-propanol, dichloromethanol and the mixture of 1-propanol and hexane revealed a higher yield of CoQ_{10} than other solvents (Wu and Tsai, 2013). From this result, it was presumed that water content affects CoQ_{10} solubility in the organic phase. A method for the extraction and assay of CoQ_{10} was reported by Ha *et al.* (2007a). The overall flowchart for CoQ_{10} separation by solvent extraction and HPLC is shown in Figure 12.3. The extraction of CoQ_{10} was initiated by the addition of cell lysis solution (0.5 ml) to the cell pellet prepared by centrifugation of the 0.5 ml culture broth (Figure 12.3). After 30 min incubation at 25 °C, a solvent mixture of propanol and hexane (3 : 5) was added to the cell lysate and mixed vigorously. The solvent phase was dried, and the resulting residue was applied to an HPLC system to separate CoQ_{10}. The HPLC system was equipped with a Capcell Pak C_{18} column (Shodex, Showa, Japan), coupled to a UV detector. The mobile phase consisted of methanol and ethanol (13 : 7) at a flow rate of 1.0 ml/min, and CoQ_{10} was detected at 275 nm.

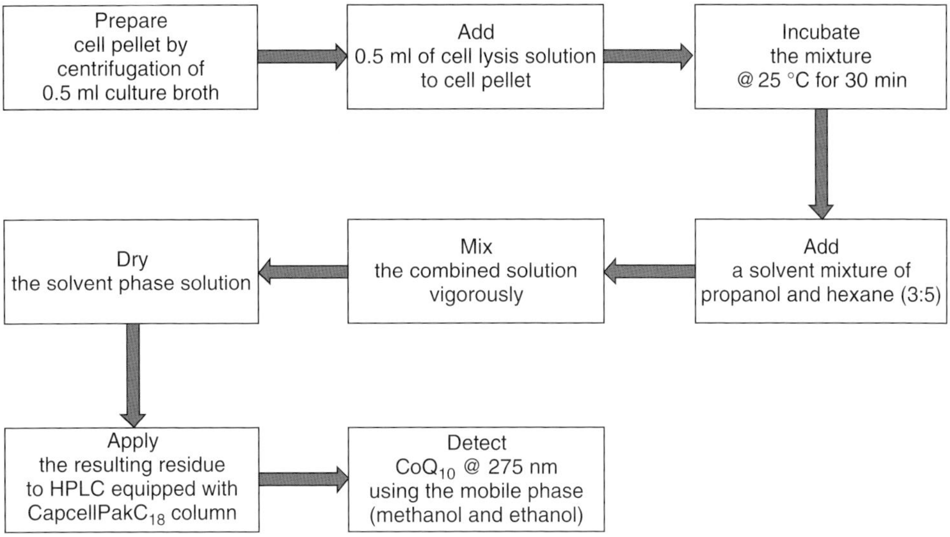

Figure 12.3 Flowchart for CoQ_{10} separation by solvent extraction and HPLC.

In Figure 12.4, a flowchart to separate CoQ_{10} from the wet cells is shown. It was reported that *Pseudomonas* N842 was first cultivated in a jar fermenter using 15 l of culture medium at 30 °C and 200 rpm (Natori *et al.*, 1978). The wet cells (equivalent to 80 g of dry cells) collected by centrifugation were refluxed for 2 h in 80% methanol containing 60 g of pyrogallol and 320 g of sodium hydroxide. From the cooled saponification mixture, unsaponifiable material was extracted twice with 2 l of hexane, and the combined hexane extract was washed with water. After two phases were separated, the hexane phase was evaporated *in vacuo*. Then, the residue was dissolved in 50 ml of acetone and allowed to stand overnight at 5 °C to remove the waxy substance. The remaining red paste was dissolved in 2 ml of benzene after acetone was evaporated *in vacuo*. The sample solution was finally applied to a silica gel column (2 cm× 25 cm) and eluted with benzene to separate CoQ_{10}.

Another method for the separation of CoQ_{10} and its analysis by solvent extraction was reported by Cluis *et al.* (2011). As shown in Figure 12.5, the cells of 1 ml aliquots of *E. coli* cultures were first harvested from centrifugation at 16 100*g*. Then, the cells were washed with 1 ml of 50 mM Tris-HCl (pH 7.5) and resuspended in 450 μl Cell Lytic B (Sigma-Aldrich Co., Saint Louis, USA). After 30 min of tempering at 25 °C, CoQ_{10} contained in the sample solution was extracted with 900 μl of hexane and 2-propanol (5 : 3). After the two phases separated, the upper phase was transferred into a new tube, followed by a second extraction using 500 μl of hexane. Finally, the extracts were dried under an air stream. To analyse the concentration of CoQ_{10}, the dried extract was resuspended in 50 μl acetone and a 10 μl aliquot was injected into a 1200 Series HPLC system equipped with a ZORBAX Eclipse XDB-C_{18} (4.6 mm× 150 mm, 5 μm, Agilent technologies, USA).

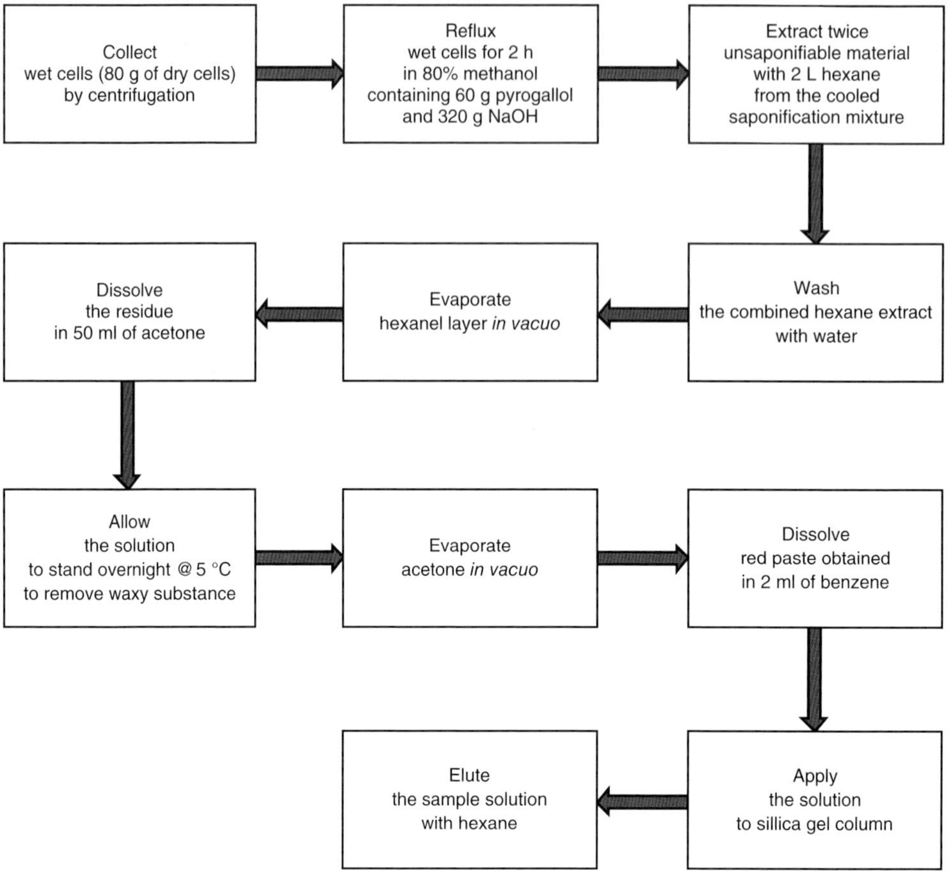

Figure 12.4 Flowchart for CoQ_{10} separation by solvent extraction and silica gel column.

The mobile phase consisted of acetonitrile and ethanol, and CoQ_{10} was detected at 275 nm. CoQ_{10} was quantified by a standard curve, and the specific CoQ_{10} content was estimated by dividing the obtained CoQ_{10} concentration by the sample DCW.

Another solvent extraction using the mixture of 1-propanol and hexane was reported to separate CoQ_{10} from the sample (Wu and Tsai, 2013). As shown in Figure 12.6, the prepared wet cells were mixed with D.W. and disrupted by freezing and subsequent heating. Then, 4 N HCl (3 ml) was added to the mixture, and the mixture was shaken in a water bath. A solvent mixture of 1-propanol and hexane (28 ml) was added to the sample solution and vortexed. After the mixture was centrifuged, the supernatant was withdrawn and introduced into a flask. The solution containing CoQ_{10} was finally concentrated by evaporating the organic solvent using a vacuum evaporator (Wu and Tsai, 2013).

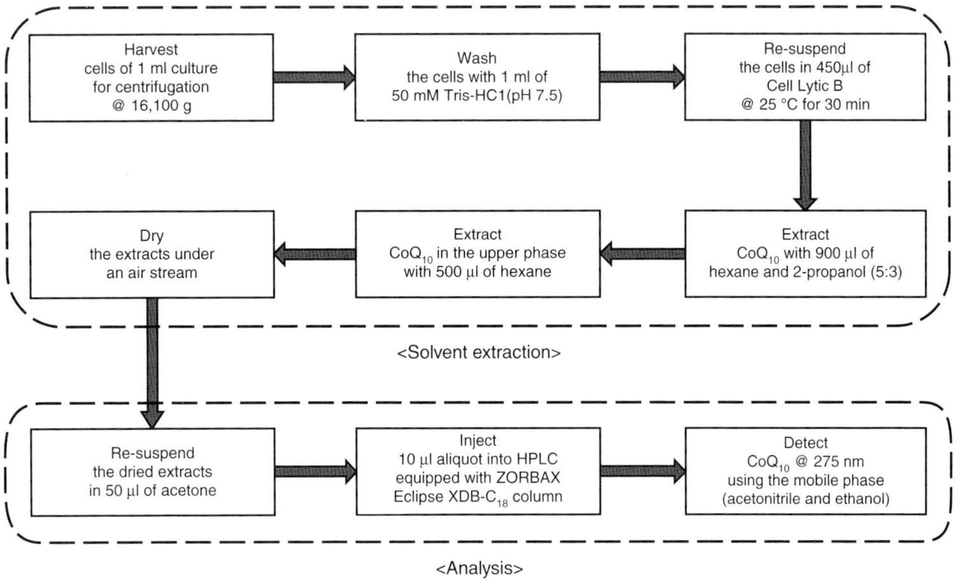

Figure 12.5 Flowchart for CoQ_{10} separation by solvent extraction and its analysis.

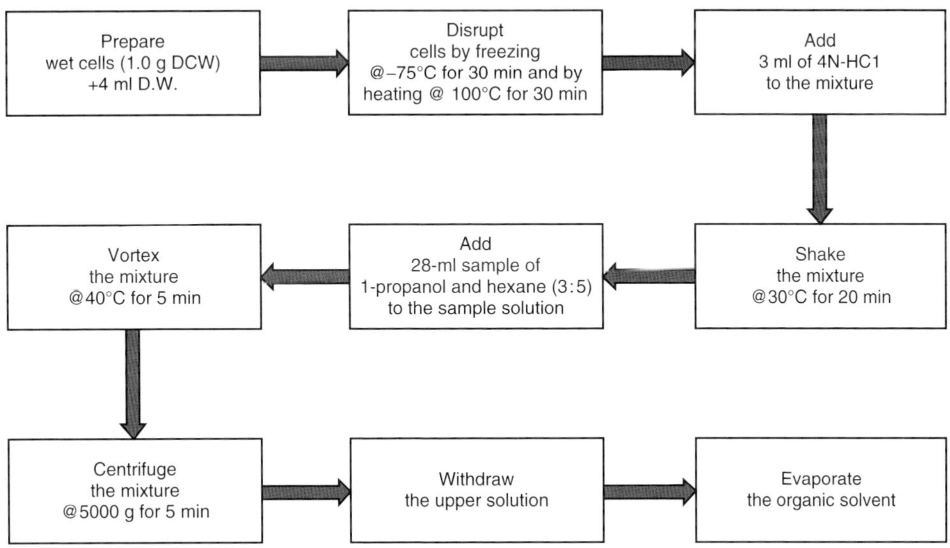

Figure 12.6 Flowchart for CoQ_{10} separation by solvent extraction.

12.4.3
Purification

After CoQ_{10} is extracted from the cell suspension, purification is the next step. Purification is a process of rendering pure CoQ_{10} and needs to produce a product suitable for use in humans with reliability and predictability. Impurities existing in the sample solution and those introduced during the purification process must be removed (Liu *et al.*, 2010). Crystallisation is a chemical purification method, and it separates a product from a liquid stream by cooling the liquid stream or adding precipitants, which lowers the solubility of the desired product. Then, the desired product forms crystals, often in extremely pure form with uniform size (Belter *et al.*, 1988). The pure solid crystals are separated from the remaining liquor by filtration or centrifugation. More than 80% of the substances used in pharmaceuticals, fine chemicals, agrochemicals, food and cosmetics are generally isolated or formulated in their solid form (Novasep, 2014). Thus, monitoring and controlling the isolation of solids via crystallisation are important for diverse applications (Novasep, 2014).

It has been reported that pure CoQ_{10} can be obtained from the solution eluted through a silica gel column (Natori *et al.*, 1978). The eluted solution was crystallised by solvents, and pure CoQ_{10} crystal was obtained once from acetone and twice from ethanol. To confirm that the crystals contain only CoQ_{10} and not any other CoQ homologues, mass, IR, UV and NMR spectra and reversed-phase chromatograms are useful.

To purify CoQ_{10} from the cell extract, 1 g/ml sample solution was prepared by dissolving 6.5 g of cell extract in 6.5 ml of light petroleum, followed by loading it onto the top of a silica gel column (Cao *et al.*, 2006). The column was eluted with light petroleum–diethyl ether (1 : 1) at a flow rate of 2 ml/min, and the orange-yellow fractions containing CoQ_{10} were collected. The CoQ_{10}-rich fractions (more than 94%) were selected by analysis of HPLC using a Waters C_{18} column (150 mm × 4.6 mm i.d., 5 μm) at 25 °C. The peak was detected at 275 nm, and the mobile phase consisting of methanol and ethanol (5 : 95) was isocratically eluted at a flow rate of 1 ml/min. After the CoQ_{10}-rich fractions were concentrated, they were dissolved in a small amount of ethanol at 45 °C and crystallised by gradually cooling down to 4 °C.

To concentrate CoQ_{10} present in the extracted solution, the organic solvent used in the extraction was evaporated with a vacuum evaporator, thus obtaining the crude CoQ_{10} product. After evaporation, the crude product was further purified by silica gel chromatography, liquid–liquid extraction or crystallisation. When CoQ_{10} was purified by silica gel chromatography, 0.05 g of the crude product was first dissolved in 1 ml of elution solvent to make a sample solution of 0.05 g/ml. This sample solution was then loaded on top of the silica gel (3 cm length and 60–200 mesh) packed in a glass column (130 mm × Ø14 mm). The column was eluted with a mixture of hexane and diethyl ether at a flow rate of 1 ml/min, and the effluent was collected in 1 ml fractions. Finally, the orange-yellow fractions containing CoQ_{10} were monitored using HPLC. When CoQ_{10} was purified by liquid–liquid extraction, CoQ_{10} contained within wet cells was first extracted with ethanol, followed by the extraction of CoQ_{10} from the ethanol solution by hexane. Then, the sample solution was separated into two layers. The upper hexane phase was collected, and the solution was evaporated to obtain the solid sample. This partially solid sample was dissolved in ethanol and crystallised after cooling from 25 to 1 °C (Wu and Tsai, 2013). Finally, the yellow crystals were collected using either a filter or centrifuge and then washed with cold ethanol. During these purification and crystallisation processes, the percentages of CoQ_{10} purity and recovery were reported to be 65% and 98% after purification and 96% and 88% after crystallisation, respectively.

12.5 Applications and Economics of CoQ_{10}

In the past, it was merely thought that CoQ_{10} is a key component for oxidative phosphorylation in the respiratory chain. However, its existence was then found in other subcellular fractions and in plasma lipoproteins as an antioxidant. Besides that, the effect of CoQ_{10} on gene expression was also demonstrated (Ernster and Dallner, 1995). Nowadays, these functions of CoQ_{10} extend its use in many clinical practices.

12.5.1 Applications

12.5.1.1 In Diseases

It has already been reported that CoQ_{10} slows the progression of diseases when it is given at high dosages (Shults *et al.*, 2002). The reduced form of CoQ_{10} as an antioxidant protects against oxidative stress and prevents the initiation and propagation of lipid peroxidation. CoQ_{10} was known to be much more effective than vitamin E at inhibiting the lipid peroxidation of low-density lipoproteins (LDLs), (Ingold *et al.*, 1993), and plays an important role in the regeneration of vitamin E (Thomas *et al.*, 1995).The role of CoQ_{10} in bioenergetics and its antioxidant properties extend its clinical applications ranging from cardiovascular disease to neuromuscular degenerative diseases and infertility (Littarru and Tiano, 2010). With no known toxicity or side effects, CoQ_{10} has been widely and successfully used as prophylactic and therapy for a variety of diseases: congestive heart failure, ischemic heart disease and diastolic dysfunction of the left ventricle (Choi *et al.*, 2005). In therapeutic applications, CoQ_{10} has also been used for treating breast cancer (Portakal *et al.*, 2000) and Alzheimer's and Parkinson's diseases (Beal, 2004), because it plays an important role in cellular energy production and in free-radical scavenging in the human body (Singh *et al.*, 1999).

The cardiovascular effects of CoQ_{10} were reported to possibly be ascribed to its bioenergetic role, its capability of antagonising the oxidation of plasma LDL and its effect in a meliorating endothelial function (Belardinelli *et al.*, 2006). When patients affected by coronary artery disease were treated with CoQ_{10}, a significant improvement was found in the activity of endothelium-bound extracellular superoxide dismutase that is thought to protect the blood vessels against oxidant-induced damage (Tiano *et al.*, 2007). In addition, this effect was accompanied by an increase of maximal oxygen uptake and of flow mediated dilation, demonstrating an endothelial function.

In neurodegenerative diseases, CoQ_{10} has been used because it reveals some effects on biochemical features such as oxidative stress and damage and mitochondrial respiratory chain dysfunction. Friedreich's ataxia is one of these conditions, and the treatment with CoQ_{10} and vitamin E exhibited a prolonged improvement in the cardiac and skeletal muscle bioenergetics and clinical scores (Cooper *et al.*, 2008). Another reported effect of CoQ_{10} was that it can improve the oxidative phosphorylation of the occipital cortex (Stamelou *et al.*, 2008). When patients suffering from progressive supranuclear palsy were treated with CoQ_{10}, the ratio of high energy to low energy phosphates increased significantly.

The positive effect of CoQ_{10} has also been reported on migraines, a condition where some inflammatory components may produce reactive oxygen species, leading to the overconsumption of CoQ_{10} (Sándor *et al.*, 2005; Hershey *et al.*, 2007). The promising result of CoQ_{10} has also been found in children suffering from Down's syndrome, in an attempt to counteract the oxidative imbalance present in this condition (Tiano *et al.*, 2008; Miles *et al.*, 2007). Furthermore, patients with maternally inherited diabetes and deafness favourably responded to CoQ_{10} (Bergamin *et al.*, 2008).

12.5.1.2
In Cosmetics

CoQ_{10} is the only lipophilic cellular antioxidant synthesised in humans. This molecule is able to reduce photoaging *in vivo* with a reduction in wrinkle depth (Hoppe *et al.*, 1999). These effects could be caused by the ability of CoQ_{10} to increase the production of basal membrane components and fibroblast proliferation and to protect cells against oxidative damage (Muta-Takada *et al.*, 2009). It has found that CoQ_{10} synthesis decreases during the aging process, leading to lower plasma and tissue levels in aged people (Kalén *et al.*, 1989; Söderberg *et al.*, 1990). This fact has a close relation to dermal tissue, because CoQ_{10} can play a key role in the remodelling process of the dermal tissue in skin aging (Ma *et al.*, 2001). Wrinkle improvement was also reported when a 1% CoQ_{10} cream was used for 5 months (Inui *et al.*, 2008). In addition, *in vitro* CoQ_{10} could reduce the ultraviolet-A-induced production of matrix metalloproteinase in human dermal fibroblasts (Hoppe *et al.*, 1999). Recently, the inhibitory mechanism of CoQ_{10} on ultraviolet-B-induced wrinkle formation was proposed (Inui *et al.*, 2008). The result showed that CoQ_{10} could inhibit the production of interleukin-6 and metalloproteinase. The collagen-fibre-degrading collagenase is a metalloproteinase inhibited by CoQ_{10}, which could lead to the rejuvenation of wrinkled skin.

Another effect of CoQ_{10} on skin is increased suppleness and moisture. As we grow older, skin becomes deflated and fine lines are further enhanced. This is because older skin has less ability to retain moisture and stay hydrated. In addition, less production of CoQ_{10} is also related to less production of collagen and elastin which are essential to keep skin soft and elastic. The supplementation of CoQ_{10} encourages the skin to produce more collagen and elastin, leading to moisturising the skin and making the skin softer. Accordingly, CoQ_{10} as an anti-aging product has recently attracted big interest from the cosmetic industry (Pardeike *et al.*, 2010).

Unlike most additional ingredients present in soap, CoQ_{10} showed a significant beneficial effect to skin cells through its ability to quench free radicals (Lipshutz and Dolnick, 2006). The highly lipophilic characteristics of CoQ_{10} could facilitate the cleaning action of soaps and facial cleansers. As a result, a CoQ_{10}-formulated soap bar has been invented by combination with a solubilising agent, providing increased bioavailability of this important bio-nutrient. CoQ_{10} can also be formulated into facial masks, enhancing the quality of the cleansing process through absorption into the skin. Because CoQ_{10} is far more lipophilic than other vitamins frequently found in facial cleansers such as vitamin C or vitamin E, CoQ_{10} is more absorbed into the skin, improving the health and appearance of the skin (Lipshutz and Dolnick, 2006).

12.5.1.3
In Foods and Others

CoQ_{10} has been also used as a food supplement because of its various physiological functions (Yuan *et al.*, 2012). In general, individuals obtain adequate amounts of CoQ_{10} through a balanced diet, but supplementation of CoQ_{10} is particularly useful for individuals with specific health conditions (Saini, 2011). Primary dietary sources of CoQ_{10} are oily fish, such as salmon and tuna, organ meats such as liver and whole grains. CoQ_{10} as a supplement is available in commercial forms, such as soft gel capsules, oral spray, hard-shell capsules and tablets.

It has also been reported that CoQ_{10} plays a significant role in boosting the immune system including phagocytic rate, circulating antibody level, neoplasia, viral and parasitic infections (Folkers and Wolaniuk, 1985). This is because tissues and cells involved with the immune function are highly energy dependent and thus require an adequate supply of CoQ_{10} for optimal function. CoQ_{10} at the mitochondrial level was found to be essential for optimal function of the immune system, indicating its role as an immunomodulating agent.

Because CoQ_{10} has been recognised as a potent gene regulator (Groneberg *et al.*, 2005), it is considered not only as a bioactive compound targeted at clinical practice uses, but also as a food supplement (Overvad *et al.*, 1999; Sándor *et al.*, 2005; Kikkawa *et al.*, 2007). It was reported that oral CoQ_{10} supplementation could result in a reduction in human LDL cholesterol oxidation (Kaikkonen *et al.*, 1997). Besides its use as a nutraceutical supplement (Tarnopolsky and Beal, 2001; Hermann, 2002), the use of CoQ_{10} has been extended further to energy drinks (Nutraingredients-USA, 2014). CoQ_{10} is increasingly being used in new food and beverage products, and the market for CoQ_{10} as a functional ingredient is gradually opening up. Furthermore, CoQ_{10} as an athletic performance enhancer has been marketed for the oral supplementation of endurance athletes, and is encouraged for its use in the enhancement of cardiac and/or skeletal muscle aerobic energy production at a mitochondrial level (Braun *et al.*, 1991; Snider *et al.*, 1992). Recently, CoQ_{10} nano-sized particles were revealed because it is poorly soluble in water and decreases in stability according to changes in light, heat or pH (Chung and Han, 2010).

12.5.2
Economics

So far, three methods have been representatively developed for the production of CoQ_{10}: (i) extraction from animal or plant tissues, (ii) chemical synthesis and (iii) fermentation using microorganisms. Because CoQ_{10} is an important ingredient in formulations used widely, large-scale processes have been developed for its commercial production. One such process is the chemical synthesis from solanesol. However, it is not only a multi-step reaction, but also not cost-effective (Keinan and Eren, 1988). For the safe and economical production of CoQ_{10}, the biological process is becoming widespread. CoQ_{10} production by fermentation is considered the most viable approach because this approach is able to produce biofunctional CoQ_{10} without optical isomers and at reduced costs as well (Choi *et al.*, 2005; Cluis *et al.*, 2007). Besides, there is growing demand from the pharmaceutical industries for the biologically potent CoQ_{10} (Choi *et al.*, 2005). However, the key point of this biological process is the development of superior microbes producing high levels of CoQ_{10}.

CoQ_{10} is sold as a pharmaceutical or dietary supplement at reasonable prices. Nowadays, quantities of CoQ_{10} are available via well-established fermentation and extraction processes, although its cost for research purposes is still rather high (\$479.74 g^{-1}) (Sigma-Aldrich, USA). The marketable CoQ_{10} extracted from bacterial cells has been produced from many companies, such as Kyowa Co., Nisshin Flourmilling Co., Kaneta, Ajinomoto and Merck (Cheong *et al.*, 2005). However, the CoQ_{10} products manufactured from these companies exhibited low productivity with high cost, because the concentrations of CoQ_{10} content in cells were too low for commercial sale. Therefore, it is necessary to isolate the bacterial strains overexpressing the key enzymes for high CoQ_{10} production. Furthermore, it is necessary to optimise the fermentation conditions for maximum biomass containing high CoQ_{10} content by controlling the fermentation conditions such as temperature, pH, aeration, stirring and DO level on an industrial scale.

In recent years, the use of microbial fermentation for fat-soluble CoQ_{10} production has a greater competitive edge, in terms of product quality or costs. A company occupying 20% of the global CoQ_{10} production is the Zhejiang Medicine Company. Another company (Kyowa Hakko, Japan) successfully developed a fatty alcohol production process to achieve its industrial production (Armmedicine, 2014). In the microbial production of CoQ_{10}, the cost was about \$375 g^{-1} and was much lower than that of the synthetic production. The Kyowa Company has further developed a patented non-yeast fermentation technology to provide a unique and cost-effective CoQ_{10} product.

To date, the most successful strategy proven to improve CoQ_{10} yields is the isolation of strains by mutagenesis and selection on inhibitors. However, the prospect for further improvement using this approach seems limited because the mutations enabling growth on the selection media do not necessarily result in higher CoQ_{10} yields (Cluis *et al.*, 2007). To improve CoQ_{10} production, higher throughput screening strategies have to be developed to pursue a random mutagenesis approach towards strain improvement. Furthermore, metabolic engineering enables the targeting of genetic modifications to specific biochemical pathways. Despite promising results obtained from metabolic engineering, current yields are not sufficient for commercial production, indicating the need for a careful assessment of the physiological and metabolic bottlenecks limiting CoQ_{10} biosynthesis. In host cells such as in *E. coli*, there could be some additional, yet unidentified, physiological factors limiting CoQ_{10} accumulation or the CoQ_{10} yields could be strictly limited by the flux through the pathways. To develop a better cost-effective large-scale process, constant efforts have been made.

References

Ajikumar, P.K., Xiao, W.H., Tyo, K.E., Wang, Y., Simeon, F., Leonard, E., Mucha, O., Phon, T.H., Pfeifer, B., and Stephanopoulos, G. (2010) Isoprenoid pathway optimization for taxol precursor overproduction in *Escherichia coli*. *Science*, **330** (6000), 70–74.

Albrecht, M., Misawa, N., and Sandmann, G. (1999) Metabolic engineering of the terpenoid biosynthetic pathway of *Escherichia coli* for production of the carotenoids b-carotene and zeaxanthin. *Biotechnol. Lett*, **21**, 791–795.

Armmedicine (2014) http://www.armmedicine.info/losses-of-domestic-producers-of-coenzyme-q10-why-changing-faces-coenzyme-medicine-medicine-pha/ (accessed 25May 2014).

Ashby, M.N., Kutsunai, S.Y., Ackerman, S., Tzagoloff, A., and Edwards, P.A. (1992) CoQ_2 is a candidate for the structural gene encoding parahydroxybenzoate: polyprenyltransferase. *J. Biol. Chem.*, **267** (6), 4128–4136.

Baba, S.W., Belogrudov, G.I., Lee, J.C., Lee, P.T., Strahan, J., Shepherd, J.N., and Clarke, C.F. (2004) Yeast Coq5 C-methyltransferase is required for stability of other polypeptides involved in coenzyme Q biosynthesis. *J. Biol. Chem.*, **279** (11), 10052–10059.

Bader, M., Muse, W., Ballou, D.P., Gassner, C., and Bardwell, J.C. (1999) Oxidative protein folding is driven by the electron transport system. *Cell*, **98** (2), 217–227.

Barker, J.L. and Frost, J.W. (2001) Microbial synthesis of p-hydroxybenzoic acid from glucose. *Biotechnol. Bioeng.*, **76** (4), 376–390.

Battino, M., Ferri, E., Gorini, A., Federico, V.R., Rodriguez, H.J.F., Fiorella, P., Genova, M.L., Lenaz, G., and Marchetti, M. (1990) Natural distribution and occurrence of coenzyme Q homologues. *Membr. Biochem.*, **9** (3), 179–190.

Beal, M.F. (2004) Mitochondrial dysfunction and oxidative damage in Alzheimer's and Parkinson's diseases and coenzyme Q_{10} as a potential treatment. *J. Bioenerg. Biomembr.*, **36** (4), 381–386.

Bekker, M., Kramer, G., Hartog, A.F., Wagner, M.J., de Koster, C.G., Hellingwerf, K.J., and de Mattos, M.J. (2007) Changes in the redox state and composition of the quinone pool of Escherichia coli during aerobic batch-culture growth. *Microbiology*, **153**(Pt. 6), 1974–1980.

Belardinelli, R., Muçaj, A., Lacalaprice, F., Solenghi, M., Seddaiu, G., Principi, F., Tiano, L., and Littarru, G.P. (2006) Coenzyme Q_{10} and exercise training in chronic heart failure. *Eur. Heart J.*, **27** (22), 2675–2681.

Belter, P.A., Cussler, E.L., and Hu, W.S. (eds) (1988) *Bioseparations: Downstream Process for Biotechnology*, John Wiley & Sons, Inc., New York.

Bergamin, C.S., Rolim, L.C., Dib, S.A., and Moisés, R.S. (2008) Unusual occurrence of intestinal pseudo obstruction in a patient with maternally inherited diabetes and deafness (MIDD) and favorable outcome with coenzyme Q_{10}. *Arq. Bras. Endocrinol. Metabol.*, **52** (8), 1345–1349.

Bhandari, K.H., Newa, M., Kim, J.A., Yoo, B.K., Woo, J.S., Lyoo, W.S., Lim, H.T., Choi, H.G., and Yong, C.S. (2007) Preparation, characterization and evaluation of coenzyme Q_{10} binary solid dispersions for enhanced solubility and dissolution. *Biol. Pharm. Bull.*, **30** (6), 1171–1176.

Booth, A.N., Masri, M.S., Robbins, D.J., Emerson, O.H., Jones, F.T., and DeEds, F. (1960) Urinary phenolic metabolites of tyrosine. *J. Biol. Chem.*, **235** (9), 2649–2652.

Boreková, M., Hojerová, J., Koprda, V., and Bauerová, K. (2008) Nourishing and health benefits of coenzymeQ_{10} – a review. *Czech J. Food Sci.*, **26** (4), 229–241.

Braun, B., Clarkson, P.M., Freedson, P.S., and Kohl, R.L. (1991) Effects of coenzyme Q_{10} supplementation on exercise performance, VO_2max, and lipid peroxidation in trained cyclists. *J. Int. Soc. Sports Nutr.*, **1** (4), 353–365.

Bule, M.V. and Singhal, R.S. (2011) Fermentation kinetics of production of ubiquinone-10 by *Paracoccus denitrificans* NRRL B-3785: effect of type and concentration of carbon and nitrogen sources. *Food Sci. Biotechnol.*, **20** (3), 607–613.

Bule, M.V., Singhal, R.S., and Kennedy, J.F. (2010) Microencapsulation of ubiquinone-10 in carbohydrate matrices for improved stability. *Carbohydr. Polym.*, **82** (4), 1290–1296.

Cabrini, L., Barzanti, V., Cipollone, M., Fiorentini, D., Grossi, G., Tolomelli, B., Zambonin, L., and Landi, L. (2001) Antioxidants and total peroxyl radical-trapping ability of olive and seed oils. *J. Agric. Food. Chem.*, **49** (12), 6026–6032.

Cao, X.L., Xu, Y.T., Zhang, G.M., Xie, S.M., Dong, Y.M., and Ito, Y. (2006) Purification of coenzyme Q_{10} from fermentation extract: high-speed counter-current chromatography versus silica gel column chromatography. *J. Chromatogr. A*, **1127** (1-2), 92–96.

Carr, N.G. and Exell, G. (1965) Ubiquinone concentrations in *Athiorhodaceae* grown under various environmental conditions. *Biochem. J.*, **96** (3), 688–692.

Cheong, S.R., Kim S.Y., Lee, J.K., Lee, H.C., Ha, S.J., Koo, B.S. and Yoo, J.H. (2005) Fermentation process for preparing coenzyme Q_{10} by the recombinant *Agrobacterium tumefaciens*. US Patent 20050181490 A1, filed Jan. 26, 2005 and issued Aug. 18, 2005.

Choi, G.S., Kim, Y.S., Seo, J.H., and Ryu, Y.W. (2005a) Restricted electron flux increases coenzyme Q_{10} production in *Agrobacterium tumefaciens* ATCC4452. *Process Biochem.*, **40** (10), 3225–3229.

Choi, J.H., Ryu, Y.W., and Seo, J.H. (2005b) Biotechnological production and applications of coenzyme Q_{10}. *Appl. Microbiol. Biotechnol.*, **68** (1), 9–15.

Chung, B.H. and Han, J.H. (2010) Coenzyme Q10 nanoparticles, preparation method thereof and composition containing said nanoparticles. US Patent 20120041178 A1, filed Apr. 6, 2010 and issued Feb. 16, 2012.

Cluis, C.P., Burja, A.M., and Martin, V.J. (2007) Current prospects for the production of coenzyme Q_{10} in microbes. *Trends Biotechnol.*, **25** (11), 514–521.

Cluis, C.P., Ekins, A., Narcross, L., Jiang, H., Gold, N.D., Burja, A.M., and Martin, V.J. (2011) Identification of bottlenecks in *Escherichia coli* engineered for the production of CoQ_{10}. *Metab. Eng.*, **13** (6), 733–744.

Cooper, J.M., Korlipara, L.V., Hart, P.E., Bradley, J.L., and Schapira, A.H. (2008) Coenzyme Q_{10} and vitamin E deficiency in Friedreich's ataxia: predictor of efficacy of vitamin E and coenzyme Q_{10} therapy. *Eur. J. Neurol.*, **15** (12), 1371–1379.

Cox, G.B., Young, I.G., McCann, L.M., and Gibson, F. (1969) Biosynthesis of ubiquinone in *Escherichia coli* K-12: location of genes affecting the metabolism of 3-octaprenyl-4-hydroxybenzoic acid and 2-octaprenylphenol. *J. Bacteriol.*, **99** (2), 450–458.

Crane, F.L. (2001) Biochemical functions of coenzyme Q_{10}. *J. Am. Coll. Nutr.*, **20** (6), 591–598.

Crane, F.L., Hatefi, Y., Lester, R.L., and Widmer, C. (1957) Isolation of a quinine from beef heart mitochondria. *Biochim. Biophys. Acta*, **25** (1), 220–221.

de Dieu Ndikubwimana, J. and Lee, B.H. (2014) Enhanced production techniques, properties and uses of coenzyme Q_{10}. *Biotechnol. Lett.*, **36** (10), 1917–1926.

Eisenreich, W., Rohdich, F., and Bacher, A. (2001) Deoxyxylulose phosphate pathway to terpenoids. *Trends Plant Sci.*, **6** (2), 78–84.

Eren, D. and Keinan, E. (1988) Total synthesis of linear polyprenoids. 3.[1] Synthesis of ubiquinones via palladium—catalyzed oligomerization of monoterpene monomers. *J. Am. Chem. Soc.*, **110**, 4356–4362.

Ernster, L. and Dallner, G. (1995) Biochemical, physiological and medical aspects of ubiquinone function. *Biochim. Biophys. Acta*, **1271** (1), 195–204.

Fir, M.M., Smidovnik, A., Milivojevic, L., Zmitek, J., and Prosek, M. (2009) Studies of CoQ_{10} and cyclodextrin complexes: solubility, thermo- and photo-stability. *J. Inclusion Phenom. Macrocyclic Chem.*, **64**, 225–232.

Folkers, K. and Wolaniuk, A. (1985) Research on coenzyme Q_{10} in clinical medicine and in immunomodulation. *Drugs Exp. Clin. Res.*, **11** (8), 539–545.

Gin, P. and Clarke, C.F. (2005) Genetic evidence for a multi-subunit complex in coenzyme Q biosynthesis in yeast and the role of the Coq1hexaprenyl diphosphate synthase. *J. Biol. Chem.*, **280** (4), 2676–2681.

Groneberg, D.A., Kindermann, B., Althammer, M., Klapper, M., Vormann, J., Littarru, G.P., and Döring, F. (2005) Coenzyme Q_{10} affects expression of genes involved in cell signalling, metabolism and transport in human CaCo-2 cells. *Int. J. Biochem. Cell Biol.*, **37** (6), 1208–1218.

Gu, S.B., Yao, J.M., Yuan, Q.P., Xue, P.J., Zheng, Z.M., and Yu, Z.L. (2006) Kinetics of *Agrobacterium tumefaciens* ubiquinone-10 batch production. *Process Biochem.*, **41** (8), 1908–1912.

Ha, S.J., Kim, S.Y., Seo, J.H., Jeya, M., Zhang, Y.W., Ramu, T., Kim, I.W., and Lee, J.K. (2009) Ca^{2+} increases the specific coenzyme Q_{10} content in *Agrobacterium tumefaciens*. *Bioprocess. Biosyst. Eng.*, **32** (5), 697–700.

Ha, S.J., Kim, S.Y., Seo, J.H., Oh, D.K., and Lee, J.K. (2007a) Optimization of culture conditions and scale-up to pilot and plant scales for coenzyme Q_{10} production by *Agrobacterium tumefaciens*. *Appl. Microbiol. Biotechnol.*, **74** (5), 974–980.

Ha, S.J., Kim, S.Y., Seo, J.H., Moon, H.J., Lee, K.M., and Lee, J.K. (2007b) Controlling the sucrose concentration increases Coenzyme Q_{10} production in fed-batch culture of *Agrobacterium tumefaciens*. *Appl. Microbiol. Biotechnol.*, **76** (1), 109–116.

Hamano, Y., Dairi, T., Yamamoto, M., Kuzuyama, T., Itoh, N., and Seto, H. (2002) Growth-phase dependent expression of the mevalonate pathway in a terpenoid antibiotic-producing *Streptomyces* strain. *Biosci. Biotechnol., Biochem.*, **66** (4), 808–819.

Harker, M. and Bramley, P.M. (1999) Expression of prokaryotic 1- deoxy-D-xylulose-5-phosphatases in *Escherichia coli* increases carotenoid and ubiquinone biosynthesis. *FEBS Lett.*, **448** (1), 115–119.

Hatakeyama, S., Kawase, S., and Yoshimura, I. (2006) Comparative oral toxicity of coenzyme Q_{10} and its (2Z)-isomer in rats: single and four-week repeated dose toxicity studies. *J. Nutr. Sci. Vitaminol.*, **52**, 9–20.

Hermann, D.D. (2002) Naturoceutical agents in the management of cardiovascular disease. *Am. J. Cardiovasc. Drugs*, **2** (3), 173–196.

Hershey, A.D., Powers, S.W., Vockell, A.L., Lecates, S.L., Ellinor, P.L., Segers, A., Burdine, D., Manning, P., and Kabbouche, M.A. (2007) Coenzyme Q_{10} deficiency and response to supplementation in pediatric and adolescent migraine. *Headache*, **47** (1), 73–80.

Hoppe, U., Bergemann, J., Diembeck, W., Ennen, J., Gohla, S., Harris, I.,

Jacob, J., Kielholz, J., Mei, W., Pollet, D., Schachtschabel, D., Suermann, G., Schreiner, V., Stäb, F., and Steckel, F. (1999) Coenzyme Q, a cutaneous antioxidant and energizer. *Biofactors*, **9** (2-4), 371–378.

Hsu, A.Y., Do, T.Q., Lee, P.T., and Clarke, C.F. (2000) Genetic evidence for a multi-subunit complex in the O-methyltransferase steps of coenzyme Q biosynthesis. *Biochim. Biophys. Acta*, **1484** (2-3), 287–297.

Ingold, K.U., Bowry, V.W., Stocker, R., and Walling, C. (1993) Autoxidation of lipids and antioxidation by alpha-tocopherol and ubiquinol in homogeneous solution and in aqueous dispersions of lipids: unrecognized consequences of lipid particle size as exemplified by oxidation of human low density lipoprotein. *Proc. Natl. Acad. Sci. U.S.A.*, **90** (1), 45–49.

Inui, M., Ooe, M., Fujii, K., Matsunaka, H., Yoshida, M., and Ichihashi, M. (2008) Mechanisms of inhibitory effects of CoQ_{10} on UVB-induced wrinkle formation *in vitro* and *in vivo*. *Biofactors*, **32** (1-4), 237–243.

James, A.M., Smith, R.A.J., and Murphy, M.P. (2004) Antioxidant and prooxidant properties of mitochondrial Coenzyme Q. *Arch. Biochem. Biophys.*, **423** (1), 47–56.

Jeong, S.K., Ahn, S.C., Kong, I.S., and Kim, J.K. (2008) Isolation and identification of a photosynthetic bacterium containing high content of coenzyme Q_{10}. *J. Fish. Sci. Technol.*, **11**, 172–176.

Jeya, M., Moon, H.J., Lee, J.L., Kim, I.W., and Lee, J.K. (2010) Current state of coenzyme Q_{10} production and its applications. *Appl. Microbiol. Biotechnol.*, **85** (6), 1653–1663.

Jonassen, T. and Clarke, C. (2001) in *Coenzyme Q: Molecular Mechanisms in Health and Disease* (eds V.E. Kagan and P.J. Quinn), CRC Press, Boca Raton, FL, pp. 185–208.

Kaikkonen, J., Nyyssönen, K., Porkkala-Sarataho, E., Poulsen, H.E., Metsä-Ketelä, T., Hayn, M., Salonen, R., and Salonen, J.T. (1997) Effect of oral coenzyme Q_{10} supplementation on the oxidation resistance of human VLDL + LDL fraction: absorption and antioxidative properties of oil and granule-based preparations. *Free Radical Biol. Med.*, **22** (7), 1195–1202.

Kalén, A., Appelkvist, E.L., and Dallner, G. (1989) Age-related changes in the lipid compositions of rat and human tissues. *Lipids*, **24** (7), 579–584.

Kalil, M.S., Alshiyab, H.S., and Yusoff, W.M.W. (2008) Effect of nitrogen source and carbon to nitrogen ratio on hydrogen production using *C. acetobutylicum*. *Am. J. Biochem. Biotechnol.*, **4** (4), 393–401.

Kamei, M., Fujita, T., Kanbe, T., Sasaki, K., Oshiba, K., Otani, S., Matsui-Yuasa, I., and Morisawa, S. (1986) The distribution and content of ubiquinone in foods. *Int. J. Vitam. Nutr. Res.*, **56** (1), 57–63.

Kawamukai, M. (2002) Biosynthesis, bioproduction and novel roles of ubiquinone. *J. Biosci. Bioeng.*, **94** (6), 511–517.

Keinan, E. and Eren, D. (1988) Total synthesis of polyprenoid natural products via Pd (0)-catalyzed oligomerization. *Pure Appl. Chem.*, **60** (1), 89–98.

Kien, N.B., Kong, I.S., Lee, M.G., and Kim, J.K. (2010) Coenzyme Q_{10} production in a 150-l reactor by a mutant strain of *Rhodobacter sphaeroides*. *J. Ind. Microbiol. Biotechnol.*, **37** (5), 521–529.

Kikkawa, K., Takehara, I., Miyakoshi, T., and Miyawaki, H. (2007) Safety of high dose supplementation of coenzyme Q_{10} in healthy human adults. *Jpn. J. Food Chem.*, **14** (2), 76–81.

Kim, S.W. and Keasling, J.D. (2001) Metabolic engineering of the nonmevalonate isopentenyl diphosphate synthesis pathway in *Escherichia coli* enhances lycopene production. *Biotechnol. Bioeng.*, **72** (4), 408–415.

Kim, S.J., Kim, M.D., Choi, J.H., Kim, S.Y., Ryu, Y.W., and Seo, J.H. (2006) Amplification of 1-deoxy-D-xyluose 5-phosphate (DXP) synthase level increases coenzyme Q_{10} production in recombinant *Escherichia coli*. *Appl. Microbiol. Biotechnol.*, **72** (5), 982–985.

Kim, T.S., Yoo, J.H., Kim, S.Y., Pan, C.H., Kalia, V.C., Kang, Y.C., Lee, J.K. (2014) Screening and characterization of an *Agrobacterium tumefaciens* mutant strain producing high level of coenzyme Q_{10}. *Process Biochem.*, **50**, 33–39 DOI: 10.1016/j.procbio.2014.10.024.

Knoell, H.E. (1979) Isolation of a soluble enzyme complex comprising the ubiquinone-8 synthesis apparatus from the cytoplasmic membrane of *Escherichia coli*. *Biochem. Biophys. Res. Commun.*, **91** (3), 919–925.

Knowles, C.J. and Redfearn, E.R. (1968) The effect of combined-nitrogen sources on the synthesis and function of the electron transport system of *Azotobacter vinelandii*. *Biochim. Biophys. Acta – Bioenerg.*, **162** (3), 348–355.

Kondo, K., Yamada, Y., Mitzugi, K. and Otsuka, S.I. (1971).Method of producing coenzyme Q_{10} by microorganisms. US Patent 3,769,170, filed Mar. 8, 1971 and issued Oct. 30, 1973.

Kraszner-Berndorfer, E. and Telegdy Kováts, L. (1972) Importance and presence of several bioquinones in foods. *Eur. Food Res. Technol.*, **148** (1), 7–15.

Kubo, H., Fujii, K., Kawabe, T., Matsumoto, S., Kishida, H., and Hosoe, K. (2008) Food content of ubiquinol-10 and ubiquinone-10 in the Japanese diet. *J. Food Compos. Anal.*, **21** (3), 199–210.

Kuratsu, Y., Hagino, H., and Inuzukah, K. (1984a) Effect of ammonium ion on coenzyme Q_{10} fermentation by *Agrobacterium* species. *Agric. Biol. Chem.*, **48** (5), 1347–1348.

Kuratsu, Y. and Inuzuka, K. (1985) Factors affecting broth viscosity and coenzyme Q_{10} production by *Agrobacterium* species. *Appl. Microbiol. Biotechnol.*, **21**, 55–59.

Kuratsu, Y., Sakurai, M., Inuzuka, K., and Hagino, H. (1984b) Aeration–agitation effect on coenzyme Q_{10} production by *Agrobacterium* species. *J. Ferment. Technol.*, **62** (3), 305–308.

Kuzuyama, T. (2002) Mevalonate and non-mevalonate pathways for the biosynthesis of isoprene units. *Biosci. Biotechnol., Biochem.*, **66** (8), 1619–1627.

Kwon, O., Kotsakis, A., and Meganathan, R. (2000) Ubiquinone (coenzyme Q) biosynthesis in Escherichia coli: identification of the ubiF gene. *FEMS Microbiol. Lett.*, **186** (2), 157–161.

Kwon, S.S., Nam, Y.S., Lee, J.S., Ku, B.S., Han, S.H., Lee, J.Y., and Chang, I.S. (2002) Preparation and characterization of coenzyme Q_{10}-loaded PMMA nanoparticles by a new emulsification process based on microfluidization. *Colloids Surf., A*, **210** (1), 95–104.

Lee, J.K., Her, G., Kim, S.Y., and Seo, J.H. (2004a) Cloning and functional expression of the dps gene encoding decaprenyl diphosphatesynthase from *Agrobacterium tumefaciens*. *Biotechnol. Progr.*, **20** (1), 51–56.

Lee, P.C., Mijts, B.N., and Schmidt-Dannert, C. (2004b) Investigation of factors influencing production of the onocylic carotenoid torulene in metabolically engineered *Escherichia coli*. *Appl. Microbiol. Biotechnol.*, **65** (5), 538–546.

Lee, P.T., Hsu, A.Y., Ha, H.T., and Clarke, C.F. (1997) A C-methyltransferase involved in both ubiquinone and menaquinone biosynthesis: isolation and identification of the *Escherichia coli* ubiE gene. *J. Bacteriol.*, **179** (5), 1748–1754.

Lenaz, G., Fato, R., Di, B.S., Jarreta, D., Costa, A., Genova, M.L., and Parenti Castelli, G. (1999) Localization and mobility of coenzyme Q in lipid bilayers and membranes. *Biofactors*, **9** (2-4), 87–93.

Leppik, R.A., Stroobant, P., Shineberg, B., Young, I.G., and Gibson, F. (1976a) Membrane-associated reactions in ubiquinone biosynthesis. 2-Octaprenyl-3-methyl-5-hydroxy-6-methoxy-1,4-benzoquinone methyltransferase. *Biochim. Biophys. Acta*, **428** (1), 146–156.

Leppik, R.A., Young, I.G., and Gibson, F. (1976b) Membrane-associated reactions in ubiquinone biosynthesis in *Escherichia coli*. 3-Octaprenyl-4- hydroxybenzoate carboxy-lyase. *Biochim. Biophys. Acta*, **436** (4), 800–810.

Lichtenthaler, H.K., Rohmer, M., and Schwender, J. (1997) Two independent biochemical pathways for isopentenyl diphosphate and isoprenoid biosynthesis in higher plants. *Physiol. Plant.*, **101** (3), 643–652.

van Liemt, W.B.S., Steggerda, W.F., Esmeijer, R., and Lugtenburg, J. (1994) Synthesis and spectroscopic characterisation of ^{13}C-labelled ubiquinone-0 and ubiquinone-10. *Recl. Trav. Chim. Pays-Bas*, **113** (3), 153–161.

Lipshutz, B.H. (2000) Practical, cost-effective synthesis of CoQ_{10}. US Patent 6,545,184

B1, filed Aug. 15, 2000 and issued Apr. 8, 2003.

Lipshutz, B.H. and Dolnick, D. (2006) Skin enrichment using CoQ10 as the delivery system. US Patent 0,251,690 A1, filed Apr. 3, 2006 and issued Nov. 9, 2006.

Lipshutz, B.H., Lower, A., Berl, V., Schein, K., and Wetterich, F. (2005) An improved synthesis of the "miracle nutrient" coenzyme Q_{10}. *Org. Lett.*, **7** (19), 4095–4097.

Lipshutz, B.H., Mollard, P., Pfeiffer, S.S., and Chrisman, W. (2002) A short, highly efficient synthesis of coenzyme Q_{10}. *J. Am. Chem. Soc.*, **124** (48), 14282–14283.

Littarru, G.P. and Tiano, L. (2010) Clinical aspects of coenzyme Q_{10}: an update. *Nutrition*, **26** (3), 250–254.

Liu, H.F., Ma, J., Winter, C., and Bayer, R. (2010) Recovery and purification process development for monoclonal antibody production. *mAbs*, **2** (5), 480–499.

Lois, L.M., Campos, N., Putra, S.R., Danielsen, K., Rohmer, M., and Boronat, A. (1998) Cloning and characterization of a gene from Escherichia coli encoding a transketolase-like enzyme that catalyzes the synthesis of D-1-deoxyxylulose 5-phosphate, a common precursor for isoprenoid, thiamin, and pyridoxol biosynthesis. *Proc. Natl. Acad. Sci. U.S.A.*, **95** (5), 2105–2110.

Loscher, R. and Heide, L. (1994) Biosynthesis of *p*-hydroxybenzoate from *p*-coumarate and *p*-coumaroyl-coenzyme A in cell-free extracts of *Lithospermum erythrorhizon* cell cultures. *Plant Physiol.*, **106** (1), 271–279.

Lv, X., Xu, H., and Yu, H. (2013) Significantly enhanced production of isoprene by ordered coexpression of genes dxs, dxr, and idi in *Escherichia coli*. *Appl. Microbiol. Biotechnol.*, **97** (6), 2357–2365.

Ma, W., Wlaschek, M., Tantcheva-Poór, I., Schneider, L.A., Naderi, L., Razi-Wolf, Z., Schüller, J., and Scharffetter-Kochanek, K. (2001) Chronological ageing and photoageing of the fibroblasts and the dermal connective tissue. *Clin. Exp. Dermatol.*, **26** (7), 592–599.

Marbois, B., Gin, P., Faull, K.F., Poon, W.W., Lee, P.T., Strahan, J., Shepherd, J.N., and Clarke, C.F. (2005) Coq3 and Coq4 define a polypeptide complex in yeast mitochondria for the biosynthesis of coenzyme Q. *J. Biol. Chem.*, **280** (21), 20231–20238.

Martin, S.F., Burón, I., Espinosa, J.C., Castilla, J., Villalba, J.M., and Torres, J.M. (2007) Coenzyme Q and protein/lipid oxidation in a BSE infected transgenic mouse model. *Free Radical Biol. Med.*, **42** (11), 1723–1729.

Matsuda, H., Kawamukai, M., Yajima, K., Ikenaka, Y., Hasegawa, J. and Takahashi, S. (2000) Process for producing coenzyme Q_{10}. US Patent 6,762,037 B1, filed Aug. 24, 2000 and issued Jul. 13, 2004.

Matsumura, M., Kobayashi, T., and Aiba, S. (1983) Anaerobic production of ubiquinone-10 by *Paracoccus denitrificans*. *Eur. J. Appl. Microbiol. Biotechnol.*, **17**, 85–89.

Matthews, P.D. and Wurtzel, E.T. (2000) Metabolic engineering of carotenoid accumulation in *Escherichia coli* by modulation of the isoprenoid precursor pool with expression of deoxyxylulose phosphate synthase. *Appl. Microbiol. Biotechnol.*, **53** (4), 396–400.

Mattila, P. and Kumpulainen, J. (2001) Coenzymes Q_9 and Q_{10}: contents in foods and dietary intake. *J. Food Compos. Anal.*, **14**, 409–417.

Mattila, P., Lehtonen, M., and Kumpulainen, J. (2000) Comparison of in-line connected diode array and electrochemical detectors in the high-performance liquid chromatographic analysis of coenzymes Q_9 and Q_{10} in food materials. *J. Agric. Food. Chem.*, **48** (4), 1229–1233.

Maury, J., Asadollahi, M.A., Møller, K., Clark, A., and Nielsen, J. (2005) Microbial isoprenoid production: an example of green chemistry through metabolic engineering. *Adv. Biochem. Eng. Biotechnol.*, **100**, 19–51.

Meganathan, R. (1996) in *Escherichia coli and Salmonella: Cellular and Molecular Biology* (eds F.C. Neidhardt, R. Curtiss III,, J.L. Ingraham, E.C.C. Lin, K.B. Low, B. Magasanik, W.S. Reznikoff, M. Riley, M. Schachter, and H.E. Umbarger), American Society for Microbiology, Washington, DC, pp. 642–656.

Meganathan, R. (2001a) Ubiquinone biosynthesis in microorganisms. *FEMS Microbiol. Lett.*, **203** (2), 131–139.

Meganathan, R. (2001b) Biosynthesis of menaquinone (vitamin K_2) and ubiquinone (coenzyme Q): a perspective on enzymatic mechanisms. *Vitam. Horm.*, **61**, 173–218.

Melzer, M. and Heide, L. (1994) Characterization of polyprenyldiphosphate: 4-hydroxybenzoate polyprenyl transferase from *Escherichia coli. Biochim. Biophys. Acta*, **1212** (1), 93–102.

Miles, M.V., Patterson, B.J., Chalfonte-Evans, M.L., Horn, P.S., Hickey, F.J., Schapiro, M.B., Steele, P.E., Tang, P.H., and Hotze, S.L. (2007) Coenzyme Q_{10} (ubiquinol-10) supplementation improves oxidative imbalance in children with trisomy 21. *Pediatr. Neurol.*, **37** (6), 398–403.

Muta-Takada, K., Terada, T., Yamanishi, H., Ashida, Y., Inomata, S., Nishiyama, T., and Amano, S. (2009) Coenzyme Q_{10} protects against oxidative stress-induced cell death and enhances the synthesis of basement membrane components in dermal and epidermal cells. *Biofactors*, **35** (5), 435–441.

Nakahigashi, K., Miyamoto, K., Nishimura, K., and Inokuchi, H. (1992) Isolation and characterization of a light-sensitive mutant of *Escherichia coli* K-12 with a mutation in a gene that is required for the biosynthesis of ubiquinone. *J. Bacteriol.*, **174** (22), 7352–7359.

Naruta, Y. (1980) Regio- and stereoselective synthesis of coenzymes Q_n (n = 2-10), vitamin K, and related polyprenylquinones. *J. Org. Chem.*, **45** (21), 4097–4104.

Natori, Y., Nagasaki, T., Kobayashi, A., and Fukawa, H. (1978) Production of coenzyme Q_{10} by *Pseudomonas* N842. *Agric. Biol. Chem.*, **42** (9), 1799–1800.

Natural Products Insider (2014) http://www.naturalproductsinsider.com/ (accessed 5 October 2014).

Negishi, E.I., Liou, S.Y., Xu, C., and Huo, S. (2002) A novel, highly selective, and general methodology for the synthesis of 1,5-diene-containing oligoisoprenoids of all possible geometrical combinations exemplified by an iterative and convergent synthesis of coenzyme Q_{10}. *Org. Lett.*, **4** (2), 261–264.

Nichols, B.P. and Green, J.M. (1992) Cloning and sequencing of *Escherichia coli* ubiC and purification of chorismate lyase. *J. Bacteriol.*, **174** (16), 5309–5316.

Novasep (2014) http://www.novasep.com/technologies/Crystallization.asp (accessed 26 November 2014).

Nutraingredients-USA (2014) http://www.nutraingredients-usa.com/Suppliers2/CoQ10-creeps-into-US-energy-drinks (accessed 8 August 2014).

Okada, K., Kainou, T., Tanaka, K., Nakagawa, T., Matsuda, H., and Kawamukai, M. (1998) Molecular cloning and mutational analysis of the ddsA gene encoding decaprenyl diphosphate synthase from *Gluconobacter suboxydans. Eur. J. Biochem.*, **255** (1), 52–59.

Onoue, S., Uchida, A., Kuriyama, K., Nakamura, T., Seto, Y., Kato, M., Hatanaka, J., Tanaka, T., Miyoshi, H., and Yamada, S. (2012) Novel solid self-emulsifying drug delivery system of coenzyme Q_{10} with improved photochemical and pharmacokinetic behaviors. *Eur. J. Pharm. Sci.*, **46** (5), 492–499.

Overvad, K., Diamant, B., Holm, L., Holmer, G., Mortensen, S.A., and Stender, S. (1999) Coenzyme Q_{10} in health and disease. *Eur. J. Clin. Nutr.*, **53** (10), 764–770.

Pardeike, J., Schwabe, K., and Müller, R.H. (2010) Influence of nanostructured lipid carriers (NLC) on the physical properties of the Cutanova Nanorepair Q_{10} cream and the *in vivo* skin hydration effect. *Int. J. Pharm.*, **396** (1-2), 166–173.

Park, Y.C., Kim, S.J., Choi, J.H., Lee, W.H., Park, K.M., Kawamukai, M., Ryu, Y.W., and Seo, J.H. (2005) Batch and fed-batch production of coenzyme Q_{10} in recombinant *Escherichia coli* containing the decaprenyl diphosphate synthase gene from *Gluconobacter suboxydans. Appl. Microbiol. Biotechnol.*, **67** (2), 192–196.

Passi, S., Cataudella, S.F., Di Marco, P., De Simone, F., and Rastrelli, L. (2002) Fatty acid composition and antioxidant levels in muscle tissue of different Mediterranean marine species of fish and shellfish. *J. Agric. Food. Chem.*, **50** (25), 7314–7322.

Pennock, J.F. and Threlfall, D.R. (1983) in *Biosynthesis of Isoprenoid Compounds* (eds J.W. Porter and S.L. Spurgeon), John Wiley & Sons, Inc., New York, pp. 191–303.

Poon, W.W., Davis, D.E., Ha, H.T., Jonassen, T., Rather, P.N., and Clarke, C.F. (2000) Identification of *Escherichia coli* ubiB, a gene required for the first monooxygenase step in ubiquinone biosynthesis. *J. Bacteriol.*, **182** (18), 5139–5146.

Portakal, O., Özkaya, Ö., Inal, M.E., Bozan, B., Koşan, M.B., and Sayek, I. (2000) Coenzyme Q_{10} concentrations and antioxidant status in tissues of breast cancer patients. *Clin. Biochem.*, **33** (4), 279–284.

Pravst, I., Zmitek, K., and Zmitek, J. (2010) Coenzyme Q_{10} contents in foods and fortification strategies. *Crit. Rev. Food Sci. Nutr.*, **50** (4), 269–280.

Pregnolato, P., Maranesi, M., Mordenti, T., Turchetto, E., Barzanti, V., and Grossi, G. (1994) Coenzyme Q_{10} and Q_9 content in some edible oils. *Riv. Ital. Delle Sostanze Grasse*, **71**, 503–505.

Prošek, M., Šmidovnik, A., Fir, M., Golc-Wondra, A., Žmitek, J., Kostanjevec, B., Donša, B. and Vindi š-Zelenko, B. (2007) Use of coenzyme Q_{10} for improved effectiveness of animal husbandry and production of animal tissues with an increased content of the said coenzyme. Patent WO 2008, 082369, filed Nov. 22, 2007 and issued Jul. 10, 2008.

PWS (2014) http://www.pws.org.au/coq10 .html (accessed 25 November 2014).

Pyo, Y.H. and Oh, H.J. (2011) Ubiquinone contents in Korean fermented foods and average daily intakes. *J. Food Compos. Anal.*, **24** (8), 1123–1129.

Ranadive, P., Mehta, A., Chavan, Y., Marx, A., and George, S. (2014) Morphological and molecular differentiation of *Sporidiobolus johnsonii* ATCC 20490 and its coenzyme Q_{10} overproducing mutant strain UF16. *Indian J. Microbiol.*, **54** (3), 343–357.

Rohmer, M., Knani, M., Simonin, P., Sutter, B., and Sahm, H. (1993) Isoprenoid biosynthesis in bacteria: a novel pathway for the early steps leading to isopentenyl diphosphate. *Biochem. J.*, **295**(Pt. 2), 517–524.

Rüttimann, A. and Lorenz, P. (1990) 80. Ein neuer syntheischer Zugang Zu Ubichinonen. *Helv. Chim. Acta*, **73**, 790–796.

Saini, R. (2011) Coenzyme Q_{10}: the essential nutrient. *J. Pharm. Bioallied Sci.*, **3** (3), 466–467.

Sakato, K., Tanaka, H., Shibata, S., and Kuratsu, Y. (1992) Agitation-aeration studies on coenzyme Q_{10} production using *Rhodopseudomonas spheroides*. *Biotechnol. Appl. Biochem.*, **16**, 19–28.

Sándor, P.S., Di Clemente, L., Coppola, G., Saenger, U., Fumal, A., Magis, D., Seidel, L., Agosti, R.M., and Schoenen, J. (2005) Efficacy of coenzyme Q_{10} in migraine prophylaxis: a randomized controlled trial. *Neurology*, **64** (4), 713–715.

Schroeder, W.A. and Johnson, E.A. (1995) Singlet oxygen and peroxyl radicals regulate carotenoid biosynthesis in *Phaffia rhodozyma*. *J. Biol. Chem.*, **270** (31), 18374–18379.

Seo, M.J., Im, E.M., Nam, J.Y., and Kim, S.O. (2007) Increase of CoQ_{10} production level by coexpression of decaprenyl diphosphate synthase and 1-deoxy-d-xylulose 5-phosphate synthase isolated from *Rhizobium radiobacter* ATCC 4718 in recombinant *Escherichia coli*. *J. Microbiol. Biotechnol.*, **17** (6), 1045–1048.

Seo, M.J., Kook, M.C., and Kim, S.O. (2012) Association of colony morphology with coenzyme Q_{10} production and its enhancement from *Rhizobium radiobacter* T6102W by addition of isopentenyl alcohol as a precursor. *J. Microbiol. Biotechnol.*, **22** (2), 230–233.

Shestopalov, A.I., Bogachev, A.V., Murtazina, R.A., Viryasov, M.B., and Skulachev, V.P. (1997) Aeration-dependent changes in composition of the quinone pool in *Escherichia coli*. Evidence of post-transcriptional regulation of the quinone biosynthesis. *FEBS Lett.*, **404** (2-3), 272–274.

Shults, C.W., Oakes, D., Kieburtz, K., Beal, M.F., Haas, R., Plumb, S., Juncos, J.L., Nutt, J., Shoulson, I., Carter, J., Kompoliti, K., Perlmutter, J.S., Reich, S., Stern, M., Watts, R.L., Kurlan, R., Molho, E., Harrison, M., Lew, M., and Parkinson Study Group (2002) Effects of coenzyme Q_{10} in early Parkinson disease: evidence of slowing of the functional decline. *Arch. Neurol.*, **59** (10), 1541–1550.

Siebert, M., Bechthold, A., Melzer, M., May, U., Berger, U., Schröder, G., Schröder, J., Severin, K., and Heide, L. (1992) Ubiquinone biosynthesis. Cloning of the genes coding for chorismate pyruvate-lyase and 4-hydroxybenzoate octaprenyl transferase from *Escherichia coli*. *FEBS Lett.*, **307** (3), 347–350.

Singh, R.B., Niaz, M.A., Rastogi, S.S., and Verma, S.P. (1999) Coenzyme Q_{10} and its role in heart disease. *J. Clin. Biochem. Nutr.*, **26** (2), 109–118.

Snider, I.P., Bazzarre, T.L., Murdoch, S.D., and Goldfarb, A. (1992) Effects of coenzyme athletic performance system as an ergogenic aid on endurance performance to exhaustion. *J. Int. Soc. Sports Nutr.*, **2** (3), 272–286.

Søballe, B. and Poole, R.K. (2000) Ubiquinone limits oxidative stress in *Escherichia coli*. *Microbiology*, **146** (Pt. 4), 787–796.

Söderberg, M., Edlund, C., Kristensson, K., and Dallner, G. (1990) Lipid compositions of different regions of the human brain during aging. *J. Neurochem.*, **54** (2), 415–423.

Stamelou, M., Reuss, A., Pilatus, U., Magerkurth, J., Niklowitz, P., Eggert, K.M., Krisp, A., Menke, T., Schade-Brittinger, C., Oertel, W.H., and Höglinger, G.U. (2008) Short-term effects of coenzyme Q_{10} in progressive supranuclear palsy: a randomized, placebo-controlled trial. *Mov. Disord. J.*, **23** (7), 942–949.

Stenn, K.S. (1968) Cation transport in a photosynthetic bacterium. *J. Bacteriol.*, **96** (3), 862–864.

Strazisar, M., Fir, M., Golc-Wondra, A., Milivojevic, L., Prosek, M., and Abram, V. (2005) Quantitative determination of coenyzme Q_{10} by liquid chromatography and liquid chromatography/mass spectrometry in dairy products. *J. AOAC Int.*, **88** (4), 1020–1027.

Suzuki, K., Okada, K., Kamiya, Y., Zhu, X.F., Nakagawa, T., Kawamukai, M., and Matsuda, H. (1997) Analysis of the decaprenyl diphosphate synthase (dps) gene in fission yeast suggests a role of ubiquinone as an antioxidant. *J. Biochem.*, **121** (3), 496–505.

Tabata, K. and Hashimoto, S. (2004) Production of mevalonate by a metabolically-engineered *E. coli*. *Biotechnol. Lett.*, **26** (19), 1487–1491.

Takahashi, S., Nishino, T., and Koyama, T. (2003) Isolation and expression of *Paracoccus denitrificans* decaprenyl diphosphate synthase gene for production of ubiquinone-10 in *Escherichia coli*. *Biochem. Eng. J.*, **16** (2), 183–190.

Tarnopolsky, M.A. and Beal, M.F. (2001) Potential for creatine and other therapies targeting cellular energy dysfunction in neurological disorders. *Ann. Neurol.*, **49** (5), 561–574.

Thomas, S.R., Neuzil, J., Mohr, D., and Stocker, R. (1995) Coantioxidants make alpha-tocopherol an efficient antioxidant for low-density-lipoprotein. *Am. J. Clin. Nutr.*, **62** (6), 1357–1364.

Tian, Y., Yue, T., Soma, P.K., Williams, P.D., Machado, P.A., Fu, H., Kratochvil, R.J., Wei, C.I., and Lo, Y.M. (2010a) Tobacco biomass hydrolysate enhances coenzyme Q_{10} production using photosynthetic *Rhodospirillum rubrum*. *Bioresour. Technol.*, **101** (20), 7877–7881.

Tian, Y., Yue, T., Yuan, Y., Soma, P.K., and Martin Lo, Y. (2010b) Improvement of cultivation medium for enhanced production of coenzyme Q_{10} by photosynthetic *Rhodospirillum rubrum*. *Biochem. Eng. J.*, **51** (3), 160–166.

Tiano, L., Belardinelli, R., Carnevali, P., Principi, F., Seddaiu, G., and Littarru, G.P. (2007) Effect of coenzyme Q_{10} administration on endothelial function and extracellular superoxide dismutase in patients with ischaemic heart disease: a double-blind, randomized controlled study. *Eur. Heart J.*, **28** (18), 2249–2255.

Tiano, L., Padella, L., Carnevali, P., Gabrielli, O., Bruge, F., Principi, F., and Littarru, G.P. (2008) Coenzyme Q_{10} and oxidative imbalance in down syndrome: biochemical and clinical aspects. *Biofactors*, **32** (1-4), 161–167.

Tokdar, P., Wani, A., Kumar, P., Ranadive, P., George, S. (2013) Process and strain development for reduction of broth viscosity with improved yield in coenzyme Q_{10} fermentation by *Agrobacterium tumefaciens* ATCC 4452. *Ferment. Technol.*, **2**(1), 110, doi: 10.4172/2167-7972.1000110.

Turunen, M., Olsson, J., and Dallner, G. (2004) Metabolism and function of coenzyme Q. *Biochim. Biophys. Acta*, **1660** (1-2), 171–199.

Udell, R.G., Naguib, Y.M.A. and Hari, S.P. (2003) Formulation and manufacturing process for coenzyme Q_{10} soft gel capsules. US Patent 6,855,733, filed Feb. 18, 2003 and issued Feb. 15, 2005.

Urakami, T. and Hori-Okubo, M. (1988) Production of isoprenoid compounds in the facultative methylotroph *Protomonas extorquens*. *J. Ferment. Technol.*, **66** (3), 323–332.

Urakami, T. and Yoshida, T. (1993) Production of ubiquinone and bacteriochlorophyll *a* by *Rhodobacter sphaeroides* and *Rhodobacter sulfidophilus*. *J. Ferment. Bioeng.*, **76** (3), 191–194.

Weber, C., Bysted, A., and Hølmer, G. (1997) Coenzyme Q_{10} in the diet-daily intake and relative bioavailability. *Mol. Aspects Med.*, **18**, 251–254.

West, D.D. (2001) Synthesis of coenzyme Q_{10} ubiquinone. US Patent 6,506,915 B1, filed Jun. 14, 2001 and issued Jan. 14, 2003.

Wu, H.S. and Tsai, J.J. (2013) Separation and purification of coenzyme Q_{10} from *Rhodobacter sphaeroides*. *J. Taiwan Inst. Chem. Eng.*, **44** (6), 872–878.

Wu, G., Williams, H.D., Zamanian, M., Gibson, F., and Poole, R.K. (1992) Isolation and characterization of *Escherichia coli* mutants affected in aerobic respiration: the cloning and nucleotide sequence of ubiG. Identification of an S-adenosylmethionine-binding motif in protein, RNA, and small molecule methyltransferases. *J. Gen. Microbiol.*, **138** (10), 2101–2112.

Wuytack, E.Y., Diels, A.M.J., and Michiels, C.W. (2002) Bacterial inactivation by high-pressure homogenisation and high hydrostatic pressure. *Int. J. Food Microbiol.*, **77** (3), 205–212.

Xinqiang, S., Yan, H., Yueyong, W., Weiyong, S., Yajun, X. and Liang, N. (2011) Method for preparing amino acid trace fertilizer with coenzyme Q_{10} fermentation liquid. CN Patent 102432364 A, filed Sep. 16, 2011 and issued May 2, 2012.

Yamamura, Y., Folkers, K., and Ito, Y. (eds) (1981) *Biomedical and Clinical Aspects of Coenzyme Q_{10}*, Elsevier, Amsterdam.

Yanagisawa, A., Nomura, N., Noritake, Y., and Yamamoto, H. (1991) Highly S_N2'-,(E)-, and antiselective alkylation of allylic phosphates. Facile synthesis of coenzyme Q_{10}. *Synthesis*, **12**, 1130–1136.

Yang, H.Y. and Song, J.F. (2006) High-sensitive determination of coenzyme Q_{10} in iodinate-β-cyclodextrin medium by inclusion reaction and catalytic polarography. *Anal. Biochem.*, **348** (1), 69–74.

Yen, H.W. and Chiu, C.H. (2007) The influences of aerobic-dark and anaerobic-light cultivation on CoQ_{10} production by *Rhodobacter sphaeroides* in the submerged fermenter. *Enzyme Microb. Technol.*, **41** (5), 600–604.

Yen, H.W. and Shih, T.Y. (2009) Coenzyme Q_{10} production by *Rhodobacter sphaeroides* in stirred tank and in airlift bioreactor. *Bioprocess. Biosyst. Eng.*, **32** (6), 711–716.

Yoon, S.H., Kim, J.E., Lee, S.H., Park, H.M., Choi, M.S., Kim, J.Y., Lee, S.H., Shin, Y.C., Keasling, J.D., and Kim, S.W. (2007) Engineering the lycopene synthetic pathway in *E. coli* by comparison of the carotenoid genes of *Pantoea agglomerans* and *Pantoea ananatis*. *Appl. Microbiol. Biotechnol.*, **74** (1), 131–139.

Yoon, S.H., Lee, Y.M., Kim, J.E., Lee, S.H., Lee, J.H., Kim, J.Y., Jung, K.H., Shin, Y.C., Keasling, J.D., and Kim, S.W. (2006) Enhanced lycopene production in *Escherichia coli* engineered to synthesize isopentenyl diphosphate and dimethylallyl diphosphate from mevalonate. *Biotechnol. Bioeng.*, **94** (6), 1025–1032.

Yoshida, H., Kotani, Y., Ochiai, K., and Araki, K. (1998) Production of ubiquinone-10 using bacteria. *J. Gen. Appl. Microbiol.*, **44** (1), 19–26.

Young, I.G., Leppik, R.A., Hamilton, J.A., and Gibson, F. (1972) Biochemical and genetic studies on ubiquinone biosynthesis in *Escherichia coli* K-12:4-hydroxybenzoate octaprenyltransferase. *J. Bacteriol.*, **110** (1), 18–25.

Young, I.G., Stroobant, P., Macdonald, C.G., and Gibson, F. (1973) Pathway for ubiquinone biosynthesis in *Escherichia coli* K-12: gene–enzyme relationships and intermediates. *J. Bacteriol.*, **114** (1), 42–52.

Youssef, H. (1963) in *Advances in Enzymology and Related Areas of Molecular Biology* (ed F.F. Nord), John Wiley & Sons, Inc., New York and London, pp. 275–328.

Yu, X.J., Dai, H.F., and Chen, F.E. (2006) Synthetic studies on coenzyme Q_{10}, part 2: an efficient and improved synthesis of coenzyme Q_{10} *via* the C5+C45 approach. *Helv. Chim. Acta*, **89**, 1317–1321.

Yuan, L.Z., Rouvière, P.E., Larossa, R.A., and Suh, W. (2006) Chromosomal promoter replacement of the isoprenoid pathway for enhancing carotenoid production in *E. coli*. *Metab. Eng.*, **8** (1), 79–90.

Yuan, Y., Tian, Y., and Yue, T. (2012) Improvement of coenzyme Q_{10} production: mutagenesis induced by high hydrostatic pressure treatment and optimization of fermentation conditions. *J. Biomed. Biotechnol.*, **2012**, 1–8.

Zahiri, H.S., Yoon, S.H., Keasling, J.D., Lee, S.H., Kim, S.W., Yoon, S.C., and Shin, Y.C. (2006) Coenzyme Q_{10} production in recombinant *Escherichia coli* strains engineered with a heterologous decaprenyl diphosphate synthase gene and foreign mevalonate pathway. *Metab. Eng.*, **8** (5), 406–416.

Zhang, H. and Javor, G.T. (2003) Regulation of the isofunctional genes ubiD and ubiX of the ubiquinone biosynthetic pathway of *Escherichia coli*. *FEMS Microbiol. Lett.*, **223** (1), 67–72.

Zhang, D., Shrestha, B., Li, Z., and Tan, T. (2007) Ubiquinone-10 production using *Agrobacterium tumefaciens* dps gene in *Escherichia coli* by coexpression system. *Mol. Biotechnol.*, **35** (1), 1–14.

Zhong, W., Fang, J., Liu, H., and Wang, X. (2009) Enhanced production of CoQ_{10} by newly isolated *Sphingomonas* sp. ZUTEO3 with a coupled fermentation-extraction process. *J. Ind. Microbiol. Biotechnol.*, **36** (5), 687–693.

Zhong, L., Kong, Z., Bian, T., Cheng, C., Shu, M., Shi, Y., Pan, J., Therese, B.A.M., and Zhong, W. (2013) Identification and characterization of a novel coenzyme Q_{10}-producing strain, *Proteus penneri* CA8, newly isolated from Ebolowa, Cameroon. *J. Chem. Pharm. Res.*, **5** (12), 201–209.

Zhu, X., Yuasa, M., Okada, K., Suzuki, K., Nakagawa, T., Kawamukai, M., and Matsuda, H. (1995) Production of ubiquinone in *Escherichia coli* by expression of various genes responsible for ubiquinone biosynthesis. *J. Ferment. Bioeng.*, **79** (5), 493–495.

13
Pyrroloquinoline Quinone (PQQ)

Hirohide Toyama

13.1
Introduction and Historical Outline

PQQ (Pyrroloquinoline quinone: 4,5-dihydro-4,5-dioxo-1*H*-pyrrolo-[2,3-f] quinoline-2,7,9-tricarboxylic acid, Figure 13.1) is a prosthetic group of certain dehydrogenases, a redox active substance that can accept one or two electron(s) in redox cycle and exhibits several different bioactivities on bacteria, plants and mammals.

The research history of PQQ is well summarised by Duine (1999). Hauge (1964) found this compound as the third redox cofactor after nicotinamide and flavin cofactors in quinoprotein glucose dehydrogenase of bacteria (although he hypothesised that it was naphthoquinone). Afterwards, Anthony and Zatman (1967) also found the unknown redox cofactor in quinoprotein methanol dehydrogenase (MDH). Salisbury *et al.* (1979) as well as Duine and colleagues (Westerling, Frank and Duine, 1979) extracted this prosthetic group from MDH of methylotrophs and identified its molecular structure by X-ray crystallographic analysis and named it methoxatin and PQQ, respectively. Ameyama *et al.* (1981) identified that PQQ was also found in other oxidative bacteria. These enzymes containing PQQ are called *quinoproteins* (Adachi *et al.*, 2007). Glucose dehydrogenase (GDH), one of the quinoproteins, is used for a glucose biosensor. Subsequently, PQQ was found to stimulate growth in bacteria (Ameyama *et al.*, 1988). Effect of PQQ on mammals was first reported by Killgore *et al.* (1989), although no enzymes having PQQ as a cofactor in mammals have been identified yet. In addition, several physiological activities including antioxidant and neuro-protective effects were also reported (Rucker, Chowanadisai and Nakano, 2009; Akagawa, Nakano, Ikemoto, 2015).

13.2
Occurrence in Natural/Food Sources

As mentioned earlier, methylotrophic bacteria which are grown on methanol as sole carbon and energy sources produce and secrete PQQ in large amounts.

Industrial Biotechnology of Vitamins, Biopigments, and Antioxidants, First Edition.
Edited by Erick J. Vandamme and José L. Revuelta.

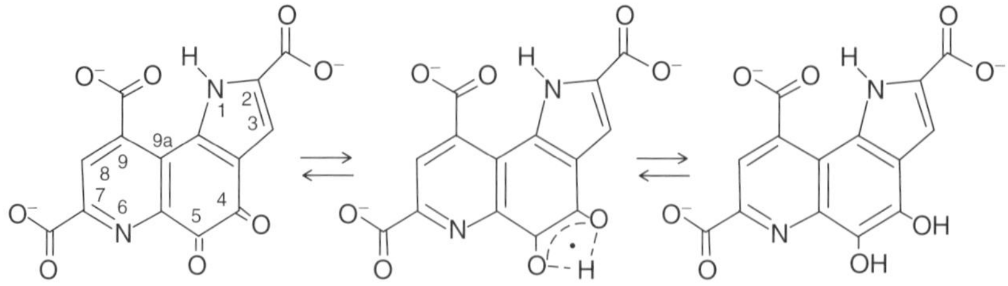

Figure 13.1 Structure of PQQ and its redox activity. PQQ is normally in oxidised state under aerobic conditions and accepts one electron in a semiquinone form or two electrons in a reduced form.

PQQ is distributed ubiquitously in nature and found in numerous dietary sources, including fermented soy beans (natto), tea, green peppers, parsley, kiwi fruit and human milk (Kumazawa *et al.*, 1995; Mitchell *et al.*, 1999). Various methods for instrumental analyses and bioassays for PQQ have been developed, but the PQQ content in foods varies according to different reports because PQQ is chemically reactive and prone to form derivatives or condensation products with other nutrients (Ameyama *et al.*, 1985; Bergethon, 1990; Noji *et al.*, 2007, see Figure 13.3). Kumazawa *et al.* (1992) have developed a method based on gas chromatography/mass spectrometry (GC/MS) with isotopic dilution for free PQQ after derivatization with phenyltrimethylammonium hydroxide. Using this analytical method, the levels of free PQQ in various foods, including vegetables, fruits and teas, were determined to be in the range of 3.7–61 ng/g wet weight or ng/ml in liquid foods (Kumazawa *et al.*, 1995). Recent analyses of PQQ using a reliable liquid chromatography/electrospray-ionisation tandem mass spectrometry (LC/ESI-MS/MS) method elucidated that free PQQ was present in various food samples in the range of 0.19–7.02 ng/g fresh weight or ng/ml in liquid foods including beer (Noji *et al.*, 2007). They also found PQQ in yeast and *Arabidopsis thaliana* cells (Noji, Kasahara and Asami, 2008). Based on the available food composition data, it is estimated that humans consume 0.1 – 1.0 mg PQQ and its derivatives per day (Harris *et al.*, 2013).

13.3
Physiological Role as Vitamin or as Bioactive Substance

In 1989, Killgore *et al.* reported that nutritional importance in rodents which show various systemic responses is observed, including growth impairment, immune dysfunction, decreased reproductive performance and reduced respiratory quotient, when PQQ is omitted from a chemically defined diet fed to rodents (Killgore *et al.*, 1989). After that, many efforts were made to find target enzymes having PQQ as a prosthetic group in animals. Many enzymes isolated from mammals are once claimed to have PQQ as the prosthetic group (Duine, 1999); however, all of them are shown to have the other quinone cofactors, the so-called built-in cofactors (Klinman and Bonnot, 2014). Later, Kasahara and Kato (2003) found one gene showing homology in part to PQQ-dependent quinoproteins from among mammal genes and proposed that the gene product is a dehydrogenase oxidising 2-aminoadipic acid semialdehyde (AAS) to 2-aminoadipic acid (AAA) in a metabolic pathway of lysine degradation. In PQQ-deprived mice, AAA concentrations were significantly decreased, whereas lysine levels remained the same, indicating that PQQ is important in the degradation of lysine. According to these results, they claimed that PQQ is a novel B-type vitamin (Kasahara and Kato, 2003). The cDNA of the gene is reported to be cloned (Wang *et al.*, 2005), but there is no report describing about AAS-dehydrogenase activity or PQQ-binding of the gene product. Very recently, PQQ-dependent quinoprotein is found in eukaryotic cells, and the homologues are found in genome sequences of mammals (Matsumura *et al.*, 2014), implying that it is a possible target enzyme to show vitamin-like effects in rodents.

Although at this moment no enzymes have been identified to contain PQQ as a cofactor in animals, PQQ has been shown to be essential for normal growth and development in animals. Oral supplementation of PQQ improves reproduction and enhances neonatal rates of growth compared with the response from diets devoid of PQQ (Steinberg, Gershwin and Rucker, 1994). More recently, dietary supplementation of PQQ in broiler chickens is shown to improve their growth, carcass yield, immunity and plasma status (Samuel *et al.*, 2015). Thus, PQQ is characterised as an important growth factor or putative essential nutrient in animals, whereas the nutritional benefits of PQQ for human are not yet completely understood. Although the detailed mechanism of PQQ action in animals still remains unclear, the ability to carry out continuous redox cycling suggests a role for PQQ as a redox signalling molecule, or antioxidant, as well as a cofactor of a certain enzyme.

In the cultured cells of humans and mice, PQQ also functions as a potential growth factor to promote cell proliferation when added to the culture media at concentrations as low as 3 nM (Naito *et al.*, 1993; Kumazawa *et al.*, 1993; Kimura *et al.*, 2012). Kumazawa *et al.* (2007) observed that PQQ treatment stimulated activation of extracellular signal-regulated kinase 1/2 (ERK 1/2) in c-Ha-ras transformed NIH/3T3 mouse fibroblasts, resulting in increased cell proliferation. ERK, one of the mitogen-activated protein kinases (MAPKs), activates transcription in the ras-signalling pathway and plays a pivotal role in cell proliferation and survival. This signal transduction by sequential phosphorylation is often initiated by the binding of peptide growth factors to receptor tyrosine kinases (RTKs). Recently, Kimura et al. (2012) showed that PQQ also significantly enhanced proliferation of human epithelial A431 cells. Moreover, they showed that PQQ activated epidermal growth factor receptor (EGFR) in a ligand-independent manner, that is, autophosphorylation of tyrosine residue occurred in the presence of PQQ. The activation of the ERK pathway accompanying EGFR phosphorylation via binding of epidermal growth factor (EGF) plays a prominent role in the proliferation of epithelial cells (Figure 13.2a). On the other hand, EGFR signalling is negatively regulated by protein tyrosine phosphatase 1B (PTP1B), which catalyses tyrosine dephosphorylation of the activated (phosphorylated) EGFR, and the inhibition of PTP1B has been reported to evoke a ligand-independent activation of EGFR (Figure 13.2b). Recent findings also indicate that PTP1B activity is modulated by post-translational modification, such as oxidation and alkylation of an extremely reactive cysteine residue at the catalytic centre (Chiarugi and Cirri, 2003). On the basis of these results, it is elucidated that PQQ inhibits PTP1B through the oxidation of catalytic cysteinyl thiol by H_2O_2 produced during its redox cycling of PQQ, and then active (phosphorylated) forms of EGFR were accumulated without the ligand binding and inducing the ligand-independent activation of sequential phosphorylation (Figure 13.2c). PTP1B has a substrate-specific ability to dephosphorylate RTKs, including the insulin receptor (Kenner *et al.*, 1996), insulin-like growth factor-I receptor (Kenner *et al.*, 1996), platelet-derived growth factor receptor (Chang *et al.*, 2006), vascular endothelial growth factor receptor (Nakamura *et al.*, 2008) and nerve growth factor (NGF) receptor (Shibata *et al.*, 2008), indicating that the multiple signalling pathways involved in these growth factors should be activated. Moreover, it is also suggested that similar inhibition of other tyrosine phosphatases or other enzymes with reactive cysteinyl thiol to that of PTP1B via redox cycling by PQQ might occur and more diverse range of physiological effects through potentiated RTK-mediated signalling and gene expression may occur, which would explain the broad effects of PQQ on mammal systems.

Recently, dietary PQQ supplementation has been revealed to enhance mitochondrial function and biogenesis and improve metabolic homeostasis in mice and rats (Stites *et al.*, 2006; Tao *et al.*, 2007; Bauerly *et al.*, 2011). PQQ deficiency in young mice increases the plasma glucose level, reduces hepatic mitochondrial content by 20–30% and suppresses mitochondrial respiration (Stites *et al.*, 2006). Similarly, rats fed a diet deficient in PQQ exhibit elevated plasma lipid and ketone bodies owing to lower mitochondrial content and decreased energy expenditure (Bauerly *et al.*, 2011). Moreover, PQQ supplementation recovers the mitochondrial alterations and metabolic impairment and significantly improves the lipid profile in UC Davis-type 2 diabetes mellitus (UCD-T2DM) rats. Mitochondrial biogenesis and function are stimulated by the transcriptional coactivator, peroxisome proliferator-activated receptor gamma coactivator-1 alpha (PGC-1α), through activation of the nuclear respiratory factors, NRF-1 and NRF-2 (Handschin and Spiegelman, 2006). The transcription factor cAMP-responsive element-binding protein (CREB) increases transcription of PGC-1α via a conserved CREB-binding site in the proximal promoter and is activated by exercise or fasting (Feige and Auwerx, 2007). Indeed, the exposure of mouse Hepa 1–6 hepatocytes to PQQ elevates PGC-1α promoter activity by enhancing CREB transcriptional activity and stimulating mitochondrial biogenesis (Chowanadisai *et al.*, 2010; Bauerly *et al.*, 2011). Exposure of PQQ also increases the levels of NRF-1 and NRF-2, resulting in the up-regulation of the mitochondrial transcription factor A (Tfam) and mitochondrial gene expression. However, the molecular mechanism underlying the activation of CREB-PGC-1α signalling pathway by PQQ remains unclear.

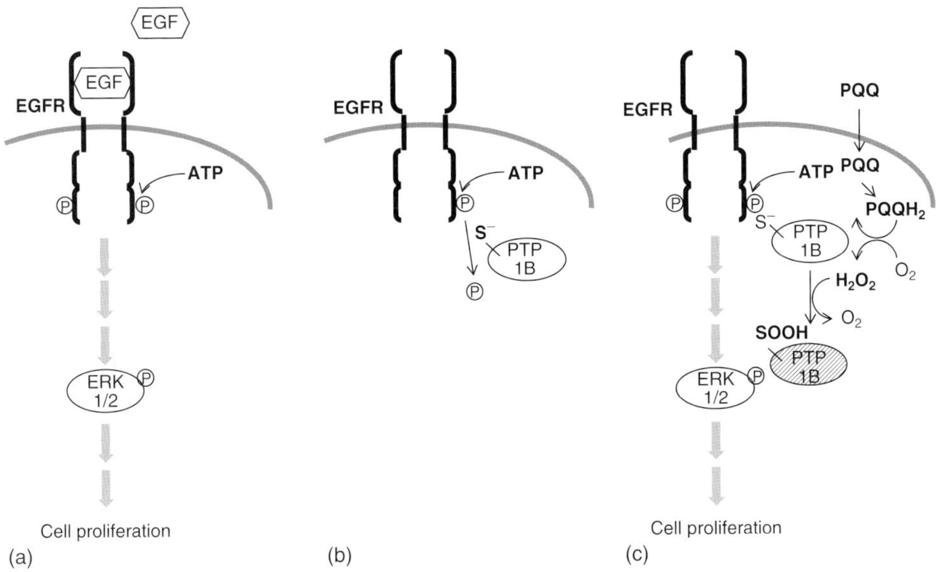

Figure 13.2 Proposed mechanism for the ligand-independent activation of epidermal growth factor receptor (EGFR) signalling through redox cycling of PQQ. (a) When epidermal growth factor (EGF) binds to EGFR, phosphorylation of the cytosolic domain occurs, then signalling pathway starts through sequential phosphorylation via activation of extracellular signal-regulated kinase 1/2 (ERK 1/2). (b) The activated (phosphorylated) EGFR is dephosphorylated by protein tyrosine phosphatase 1B (PTP1B), which has catalytic cysteinyl thiol. (c) PQQ is incorporated into the cytoplasm and reduced by reductants existing in cytosol to become $PQQH_2$, which is sequentially oxidised by O_2 to produce H_2O_2 (redox cycling). Substantial amount of H_2O_2 is accumulated, and the catalytic cysteinyl thiol in PTRP1B is oxidised and inactivated. Inhibition of PTRP1B leads to accumulation of phosphorylated (activated) EGFR which is derived by autophosphorylation without EGF binding at low level. Thus, signalling pathway is activated in the ligand-independent manner.

In vitro, PQQ showed inhibitory activity on fibril formation of alpha-synuclein (Kobayashi *et al.*, 2006) and amyloid formation and cytotoxicity of truncated alpha-synuclein (Kim *et al.*, 2010). PQQ also showed protection for neuroblastoma cells from oxidative stress by 6-hydroxydopamine (6-OHDA) or H_2O_2, in a dose-dependent manner, and the inhibitory activity for oxidative stress was much higher than that of vitamin C or vitamin E (Hara, Hiramatsu and Adachi, 2007; Nunome *et al.*, 2008). These results suggest that the protection effect of PQQ against 6-OHDA or H_2O_2-induced neurotoxicity is involved in its function as a radical scavenger, especially against O_2^-. Furthermore, PQQ enhanced production and secretion of NGF in astroglial cells and in fibroblast cells without cytotoxicity *in vitro* and *in vivo* (Yamaguchi *et al.*, 1993; Murase *et al.*, 1993), although precise mechanism of enhancement of NGF production by PQQ is not yet clear. Cyclooxygenase activation is supposed to be an essential process, because the induction of NGF is inhibited by cyclooxygenase inhibitor or dexamethasone (Yamaguchi *et al.*, 1996).

The effects of PQQ on the learning and memory function of young rats were reported (Ohwada *et al.*, 2008). The rats fed with a diet supplemented with PQQ showed significantly better learning ability than the control rats. In addition, after receiving hyperoxia to induce oxidative stress, rats fed with PQQ-supplemented diet showed better memory function than the control rats did. The combination of PQQ with Coenzyme Q_{10} (CoQ_{10}) showed synergistic effects on memory function. Similar effects of PQQ were also observed in aged rats (Takatsu *et al.*, 2009). These results suggest that PQQ is potentially effective for preventing neurodegradation caused by oxidative stress.

In human studies, a placebo-controlled, double-blinded study using the repeatable battery for the assessment of neuropsychological status (RBANS) was conducted with the participation of old Japanese people who presented with self-identified forgetfulness or forgetfulness identified by a family member, a colleague or an acquaintance (Koikeda, Nakano and Masuda, 2011). RBANS is a neuropsychological battery, and the neuropsychological battery questions allow repeated and quick evaluation of higher brain function disorders with a variety of brain disease complications. Although no significant difference was seen between groups in the high-scoring subgroup, the PQQ + CoQ_{10} group in the low-scoring subgroup showed a significantly better score than the placebo group did. These findings showed that individuals with lower RBANS scores may achieve a better improvement in response to PQQ supplementation than individuals with higher scores do.

In relation to cognitive functions, PQQ had an effect on stress, fatigue and sleep (Nakano *et al.*, 2012). The effectiveness of PQQ was examined using an open-label trial, and the results in the Profile of Mood States-Short Form showed that all six measures of vigour, fatigue, tension–anxiety, depression, anger–hostility and confusion significantly improved following PQQ supplementation compared with scores for those measures before supplementation of PQQ. The results of the Oguri–Shirakawa–Azumi Sleep Inventory (Middle Aged and Aged version) also showed significant improvement in drowsiness at awakening, sleep onset and maintenance and sleep duration. For validation, the Japanese version of the Pittsburgh Sleep Quality Index also showed significant improvement in sleep-related behaviour. Furthermore, the changes in these global scores were correlated with the changes in the cortisol awakening response; that is, the effects of PQQ on the improvement of sleep quality are supported by a biomarker.

PQQ is also reported to affect bone metabolism (Gallop *et al.*, 1993). Recently, it has been reported that PQQ inhibited osteolysis (Kong *et al.*, 2013), and effects on mineralisation in osteoblastic cells were examined (Higa and Tachibana, 2014).

Recently, it has been reported that PQQ is helpful for the improvement of skin conditions and lipid metabolism (Nakano *et al.*, 2015a,2015b), which might be the results of activation of EGFR (Kimura *et al.*, 2012) described earlier. PQQ may be useful not only for the improvement of brain functions but also for various health benefits. The underlying mechanisms of the effects of PQQ should be elucidated further.

PQQ also showed positive effects on plant growth (Choi *et al.*, 2008). Previously, it was shown that PQQ contributes to plant growth promotion to help solubilisation of poorly soluble mineral phosphates with gluconic acid produced by PQQ-dependent membrane-bund glucose dehydrogenase (mGDH); however, PQQ was shown to have effects on plant growth promotion (summarised in Misra, Rajpurohit and Khairnar, 2012). PQQ was also reported to increase tolerance to high salinity in *A. thaliana* (Noji, Kasahara and Asami, 2008).

13.4 Physiological Role as a Cofactor

Enzymes containing PQQ as a prosthetic group are called quinoproteins (Matsushita, Toyama and Adachi, 2002). In quinoproteins, PQQ is tightly but non-covalently bound to the apo-enzyme with calcium or magnesium ion essential for enzyme activity. These enzymes contain 1 mol Ca^{2+} or Mg^{2+}/mol of PQQ that is coordinated both by PQQ and by several amino-acid side-chain atoms (Toyama *et al.*, 2004). In the case of MDH of *Methylobacterium extorquens* AM1, several genes (*mxaAKLD*) in a large gene cluster *mxaFJGIRSACKLDEHB* are found to be required for insertion of Ca^{2+} (Goodwin and Anthony, 1998). Previously, MDH consisting of large and small subunits, which are encoded by *mxaF* and *mxaI*, respectively, could accept Ba^{2+} and Sr^{2+} instead of Ca^{2+} (Goodwin and Anthony, 1998), whereas, recently, another gene *xoxF* has been reported to be responsible for MDH with lanthanide-group rare earth metals (Keltjens *et al.*, 2014). It is suggested that *mxaAKLD*-like machinery for metal ion insertion is not required for XoxF-type enzyme. The three dimensional structure of MxaF-type MDH is well studied (Goodwin and Anthony, 1998), and many of PQQ-dependent quinoproteins have similar super-barrel structure with eight β-propeller blades consisting of four anti-parallel β-sheets (Toyama *et al.*, 2004; Yakushi and Matsushita, 2010).

Some of these quinoproteins have, together with PQQ, an additional prosthetic group, haem *c*, within a single polypeptide; these are called *quinohaemoproteins* to distinguish them from the quinoproteins which contain only the quinone cofactor as the prosthetic group (Matsushita, Toyama and Adachi, 1994). Alcohol dehydrogenase (ADH) is widely distributed in many different types of organisms ranging from bacteria to mammals. It is most often an NAD(P)-dependent enzyme localised in the cytoplasm. In contrast, PQQ-dependent ADHs are rather unique and are found in only a narrow range of species of bacteria, the α-, β- and γ-proteobacteria, and are localised only in the periplasmic fraction. PQQ-dependent ADH includes both quinoprotein- and quinohaemoprotein-type enzymes as described earlier. Some are soluble form in the periplasm, and the others are bound to the outer surface of the cytoplasmic membrane. Thus, similar to other quinoprotein dehydrogenases, quino(haemo)protein ADH forms a so-called periplasmic alcohol oxidase system together with the membrane-bound respiratory chain. The quino(haemo)protein ADH functions as a primary dehydrogenase, which transfers reducing equivalents directly to the bacterial aerobic respiratory chain, in the periplasm. As such, they have a truncated, and thus less energy-efficient respiratory chain that leads to a direct oxidation of substrate without any energy-consuming uptake of the substrate or excretion of the oxidised products (Matsushita, Toyama and Adachi, 1994). PQQ-dependent ADH is the largest enzyme group within the quinoprotein family having a super-barrel structure with eight propeller blades similarly to PQQ-dependent MDHs, and the group is classified into three subgroups, type I, II and III ADHs (Toyama *et al.*, 2004). Type I ADH found in a limited number of Proteobacteria are very similar to quinoprotein MDH in methylotrophs or XoxF-type enzymes and are simple quinoproteins having PQQ as the only prosthetic group; they can be differentiated from MDH with respect to substrate specificity (Goodwin and Anthony, 1998). Unlike type I ADH, type II and III ADHs are quinohaemoproteins or contain a quinohaemoprotein as one of its subunits, respectively. Type II ADH is an enzyme soluble in the periplasm and has a relatively wide distribution among several Proteobacteria. Type III ADH is membrane-bound, working on the periplasmic surface and is unique to acetic acid bacteria. It consists of three subunits, two of which comprise a quinohaemoprotein catalytic subunit and a trihaem cytochrome *c* subunit (Yakushi and Matsushita, 2010).

Another group of quinoproteins is membrane-bound PQQ quinoprotein such as mGDH and glycerol dehydrogenase (GLDH) having five membrane-spanning α-helices in N-terminal region and in the membrane-anchoring subunit, respectively (Yamada *et al.*, 1993, Miyazaki, 2002). In the genome of *Gluconobacter oxydans*, more quinoproteins, PQQ1, PQQ3 and PQQ4 having five membrane-spanning α-helices in the N-terminal region are found (Prust *et al.*, 2005). Therefore, these sequences are classified as mGDH-type quinoprotein. PQQ1 is shown to be inositol dehydrogenase (Hölscher, Weinert-Sepalage and Görisch, 2007), and it is mainly distributed in α-proteobacteria. PQQ3 and PQQ4 are specific in acetic acid bacteria. PQQ3 is only conserved in *Gluconobacter* and *Gluconacetobacter*, while PQQ4 is conserved in *Gluconobacter*, *Acetobacter* and *Gluconacetobacter*. In addition to the difference of repertoire of PQQ quinoproteins among genera, it was reported that there is a strain-specific quinoprotein (quinate dehydrogenase) in *Gluconobacter* (Vangnai *et al.*, 2004). Thus, these differences of repertoire of PQQ quinoproteins may define the species/strain-specific phenotype. PQQ3 and PQQ4 are very closely related to each other and belong to almost the same clade: the two groups should be re-defined as the same new group. The mGDH-type quinoproteins are mainly distributed in α- and γ-proteobacteria. Before genome sequences are available, *Escherichia coli*, *Pseudomonas aeruginosa*, *Acinetobacter calcoaceticus* and *Gluconobacter* specie are already shown to have mGDH (Matsushita, Toyama and Adachi, 1994). GLDH found in acetic acid bacteria has strict preference of (*R*)-configuration of secondary alcohol in polyols (Matsushita *et al.*, 2003) but a broad-range substrate specificity and is able to oxidise various polyols to the corresponding ketones (Adachi *et al.*, 2001; Matsushita *et al.*, 2003). The GLDH homologues are distributed in α- and γ-proteobacteria. The mGDH-type enzymes also have the super-barrel structure consisting of eight blades in its molecular structure (Goodwin and Anthony, 1998).

One more group of quinoproteins is a soluble form of glucose dehydrogenase (sGDH) and is completely different from mGDH. This type of GDH was used in the test strips for the determination of glucose in blood produced by Boehringer/Mannheim (now Roche Diagnostics); however, high reactivity to maltose is a problem. Thus, several attempts to reduce the reactivity of sGDH to maltose were examined (Lau *et al.*, 2007; Stredanský *et al.*, 2013). sGDH was first discovered in *A. calcoaceticus* and specifically found in this species (Cleton-Jansen *et al.*, 1988; Matsushita *et al.*, 1989); however, recent accumulation of genome sequence data revealed that sGDH-type enzyme with low homology is also found in other bacteria including *E. coli* (Oubrie, 2003). It is shown to have the ability to bind PQQ in the presence of Ca^{2+} and broad substrate specificity (Southall *et al.*, 2006). The enzyme is named as aldose sugar dehydrogenase (Asd). The X-ray structure has been determined in the apo-form and as the PQQ-bound active holoenzyme, and the super-barrel structure having six β-propeller folds is conserved between *E. coli* Asd and *A. calcoaceticus* sGDH (Oubrie, 2003), with major structural differences lying in loop and surface-exposed regions (Southall *et al.*, 2006). Asd is widely identified in phylogenetically diverse prokaryotic genera spanning Bacteria and Archaea (Oubrie, 2003). One of the Archea enzymes is cloned from *Pyrobaculum aerophilum* and overexpressed in *E. coli* (Sakuraba *et al.*, 2010). It is shown that the enzyme could bind PQQ, and its overall structure is similar to that of *E. coli* Asd. It was extremely thermostable and active even after incubation at 100 °C for 10 min.

Sorbosone dehydrogenase (SNDH) of *Gluconacetobacter liquefaciens* (formerly *Acetobacter liquefaciens*) is an L-sorbosone:ubiquinone oxidoreductase which oxidises position C1 of L-sorbosone to yield 2-keto-L-gulonate, which is the important intermediate in vitamin C production industry (Pappenberger and Hohmann, 2014). The enzyme has no discrete signature for transmembrane region or signal sequence, but likely functions on the periplasmic surface of the cytoplasmic membrane. SNDH does not have any binding motif for PQQ (Shinjoh *et al.*, 1995); however, by using a recent bioinformatics tool, the enzyme likely belongs to a family of sGDH and Asd. Another enzyme L-Sorbose/L-SNDH of *Ketogulonicigenium vulgare* (formally *G. oxydans*) DSM 4025, which is a soluble enzyme and directly converts L-sorbosone to L-ascorbic acid, is shown to contain PQQ (Asakura and Hoshino, 1999), showing homology to sGDH (Miyazaki, Sugisawa and Hoshino, 2006).

Only recently, a novel PQQ-dependent sugar oxidoreductase was found in a mushroom, the basidiomycete *Coprinopsis cinerea* (Matsumura *et al.*, 2014). This is the first discovery of PQQ-dependent quinoprotein in eukaryotic cells. The enzyme, named as CcSDH, has cytochrome *b* and cellulose-binding domains, besides a domain for PQQ-dependent dehydrogenase having six β-propeller folds. The PQQ-dependent dehydrogenase domain showed very low sequence identity to Asd, and the enzyme showed activity with D-glucosone but not with D-glucose. Homology search revealed that many genes which encode enzymes similar to CcSDH are found in bacteria, archaea, amoebozoa and fungi (Matsumura *et al.*, 2014), indicating that they are distinct from other classes of known quinoproteins. The homologue is found even in *Homo sapiens*; thus, it is a possible target enzyme to show vitamin-like effects in rodents.

13.5
Chemical and Physical Properties; Technical Functions

PQQ is an ortho-quinone tricyclic compound in which all three carboxyl groups are deprotonated under neutral conditions (Figure 13.1), and it is water-soluble and heat-stable. It is a redox active substance which can accept one or two electron(s) in redox cycle, and this character makes it a prosthetic group in several bacterial dehydrogenases as described earlier. The redox potentials of the PQQ/PQQ_{red} couple were reported as 0.090 V at pH 7.0 and −0.230 V versus SHE at pH 13.0 (Duine, Frank and Verwiel, 1981). It is reactive with nucleophiles, especially amino acids (Adachi *et al.*, 1988), to form imidazopyrroloquinoline (IPQ) compounds (Urakami, Sugamura and Niki, 1995/1996, Urakami *et al.* 1996, Figure 13.3). Other compounds such as H_2O, alcohols or amines make complex at C5 position of PQQ (Dekker *et al.*, 1982, Figure 13.3). Under alkaline condition, two molecules of H_2O are added to form PQQ-$2H_2O$ (Figure 13.3).

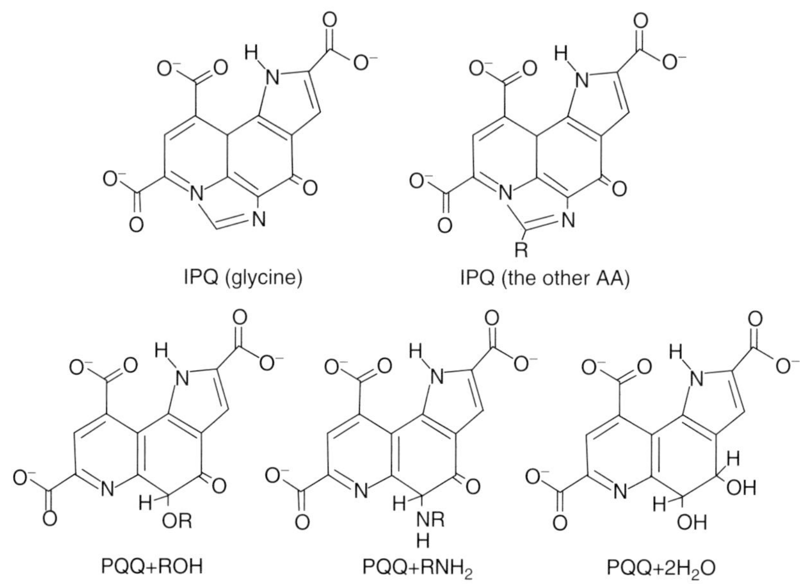

Figure 13.3 Several adducts of PQQ with nucleophilic compounds. IPQ (glycine): glycine adduct of PQQ, and it is also obtained after reaction with other amino acids. IPQ (other AA): adduct of PQQ reacted with amino acids other than glycine and the side chain is represented as R. PQQ + ROH: adduct of PQQ with alcohols (ROH). PQQ + RNH_2: adduct of PQQ with amines (RNH_2). PQQ + $2H_2O$: adduct of PQQ with two molecules of water in alkaline conditions.

Cyclic voltammetric analysis of PQQ was reported (Kano *et al.*, 1990), showing that reduced form of PQQ is stabilised under acid conditions, and under basic conditions, stable formation of semiquinone form was observed.

PQQ is reduced easily to $PQQH_2$ by reaction with reducing agents such as NADPH, sodium borohydride, glutathione or cysteine. A couple of *in vitro* studies demonstrated that $PQQH_2$ exhibits antioxidative capacity (Miyauchi *et al.*, 1994, He *et al.*, 2003; Ouchi *et al.*, 2009; Mukai, Ouchi and Nakano, 2011). The radical-scavenging activity of $PQQH_2$ was 7.4-fold higher than that of vitamin C, which is known as the most active water-soluble physiological antioxidant (Ouchi *et al.*, 2009). The singlet-oxygen-quenching activity of $PQQH_2$ was found to be 6.3-fold higher than that of vitamin C (Mukai, Ouchi and Nakano, 2011). Interestingly, $PQQH_2$ works as a catalyst in the singlet-oxygen-quenching reactions. Moreover, it has been clarified that $PQQH_2$ may rapidly convert two molecules of α-tocopheroxyl radicals to α-tocopherol (Ouchi *et al.*, 2013). These results demonstrate that the pro-oxidant effect of α-tocopheroxyl radicals is suppressed by the coexistence of $PQQH_2$.

One of the notable properties of PQQ is a redox cycling activity in the presence of excess glycine, oxygen and nitro blue tetrazolium at pH 10 (Paz *et al.*, 1996). This activity is not observed in other redox active compounds such as ascorbate (Paz, Flückiger and Gallop, 1990).

13.6
Assay Methods

PQQ is conveniently detected and quantified enzymatically with apo-form of mGDH from *E. coli* (Ameyama *et al.*, 1985) or with apo-form of recombinant sGDH expressed in *E. coli* (Toyama, Chistoserdova and Lidstrom, 1997).

Kumazawa *et al.* (1992) have developed a method based on GC/MS with isotopic dilution for free PQQ after derivatization with phenyltrimethylammonium hydroxide. PQQ and its derivatives are easily quantified by using LC/ESI-MS/MS (Noji *et al.*, 2007) and LC-MS/MS analysis using a reverse-phase column with the ion-pair reagent, dibutylammonium acetate (Ando *et al.*, 2014).

Redox cycling activity described earlier can also possibly detect PQQ; however, separation from other reductants is required before applying this method for quantification.

13.7
Biotechnological Synthesis

13.7.1
Producing Microorganisms

Methylotrophic bacteria which can grow on methanol as energy and carbon sources produce PQQ in culture medium in a large amount (Ameyama *et al.*,

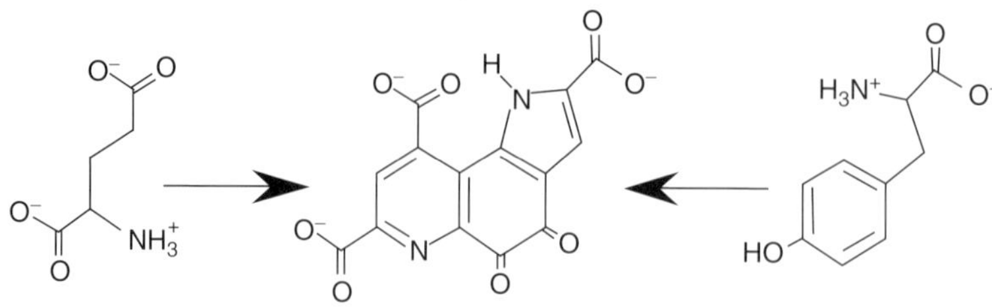

Figure 13.4 PQQ backbone structure is derived from two amino acids, glutamate and tyrosine.

1988; Van Kleef and Duine, 1989). Conditions for higher production were examined (Urakami *et al.*, 1992).

13.7.2
Biosynthesis and Metabolic Regulation

The biosynthetic pathway leading to PQQ has not yet been solved completely, although several genes required for biosynthesis of PQQ are known. The first step in the elucidation of PQQ biosynthesis was the identification of the amino acid precursors, glutamate and tyrosine (Figure 13.4) by ^{13}C-labelling experiments and nuclear magnetic resonance (NMR). *M. extorquens* AM1 was grown on 1-^{13}C- or 2-^{13}C-ethanol or ^{13}C-methanol, and the resulting ^{13}C enrichments in PQQ were compared to the labelling patterns in amino acids (Houck, Hanners and Unkefer, 1988; van Kleef and Duine, 1988; Houck, 1989). From such studies, PQQ is concluded to be formed from the cross-linking of a glutamate and a tyrosine side chain (Figure 13.4), although direct incorporation of labelled glutamate into PQQ was not observed in *M. extorquens* AM1 (Unkefer *et al.*, 1995).

13.8
Strain Development: Genetic Modification, Molecular Genetics and Metabolic Engineering

Six or seven genes are shown to be involved in PQQ biosynthesis (Figure 13.5, Velterop *et al.*, 1995; Goodwin and Anthony, 1998; Shen *et al.*, 2012), although some reports claimed that *E. coli* introduced with a single gene could produce PQQ (Liu *et al.*, 1992; Babu-Khan *et al.*, 1995). Only one compound was isolated as a biosynthetic intermediate of PQQ: 3a-(2-amino-2-carboxyethyl)-4,5-dioxo-4,5,6,7,8,9-hexahydroquinoline-7,9-dicarboxylic acid (AHQQ, Figure 13.5, Magnusson *et al.*, 2002b). This compound was successfully converted to PQQ *in vitro* with the purified PqqC from *Klebsiella pneumoniae* (Magnusson *et al.*, 2002a) or PqqC/D from *M. extorquens* AM1 (Toyama *et al.*, 2002), both of which were purified from recombinant *E. coli* cells. Site directed mutagenetic analysis of PqqC is done (Magunusson et. al, 2007). PqqA is a small peptide having around 20–30 amino acid residues, with conserved glutamate and tyrosine residues,

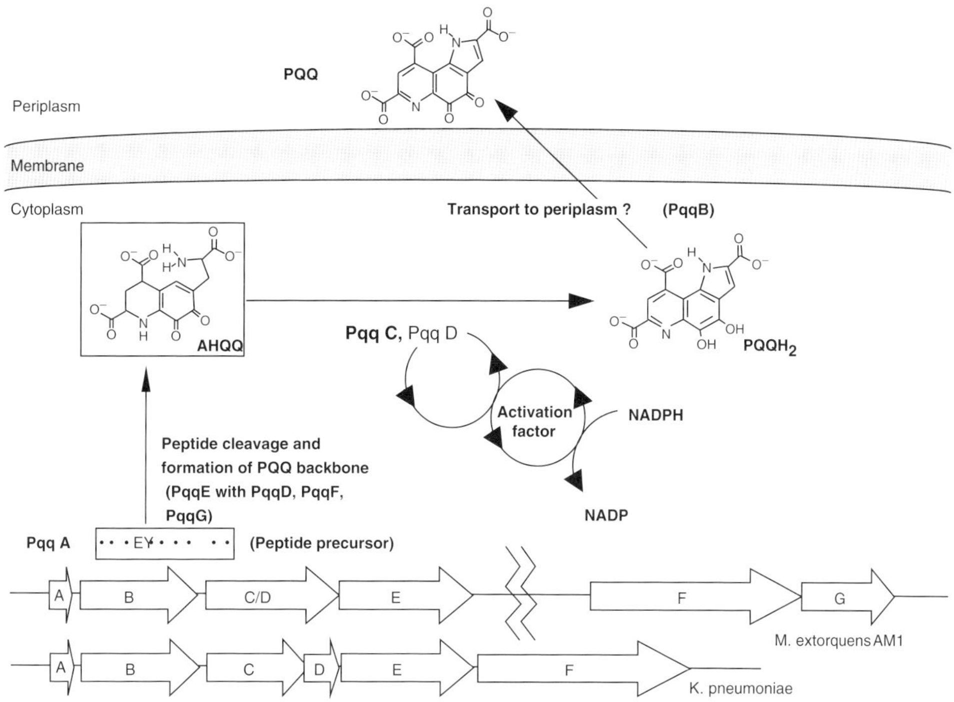

Figure 13.5 Illustrated scheme of PQQ biosynthesis. Six (*K. pneumoniae*) or seven (*M. extorquens* AM1) genes are shown to be involved in synthesis. PqqA has conserved glutamate and tyrosine residues and is believed to be the starting material of PQQ. PqqB is thought to be involved in transport of PQQ to periplasm or in introduction of quinone moiety. PqqC is the enzyme in the last step of the PQQ synthesis, producing $PQQH_2$ from AHQQ which is the biosynthetic intermediate and accumulated in the mutant disrupted in *pqqC*. PqqD forms a fusion protein with PqqC or PqqE in some genomes and was recently shown to be able to form a complex with PqqA. PqqF and PqqG are homologous to zinc proteases and are believed to cleave four peptide bonds in PqqA to release AHQQ or a precursor of AHQQ.

indicating that it is the precursor of PQQ. Multiple existence of the gene was reported in methylotrophs (Ge *et al.*, 2015). The importance of PqqB is ambiguous because significant amount of PQQ production was observed inside the cells but much less outside the cells, even after disruption of *pqqB* gene (Velterop *et al.*, 1995). Thus, PqqB was thought to be involved in transport of PQQ from cytosol to periplasm, but recently, PqqB was suggested to be a member of the metallo-β-lactamase superfamily (Shen *et al.*, 2012) and speculated to be an oxygenase to introduce oxygen onto the aromatic ring of the tyrosine residue in PqqA (Klinman and Bonnot, 2014). PqqE belongs to radical SAM (S-adenosyl methionine) superfamily, and recently, it was shown that the recombinant PqqE actually had a SAM cleavage activity; however, any modification of PqqA peptide was not observed (Wecksler *et al.*, 2009). More recently, PqqE and PqqD formed complex with PqqA, making a ternary complex, and it is proposed that PqqD has a chaperone function (Latham *et al.*, 2015). PqqE is supposed to catalyse C–C bond formation between C9 and C9a in PQQ (Figure 13.1, Klinman and Bonnot, 2014). PqqF is most likely a metalloendopeptidase with a zinc centre in its active site, involved in the processing of the tyrosine and glutamate residues of PqqA (Figure 13.5). It is shown to be dispensable in several bacteria (Velterop *et al.*, 1995) and is absent from 53 of the identified PQQ operons (Shen *et al.*, 2012), indicating that its role is able to be substituted by non-specific, cell-associated proteases. PqqG is a homologue of PqqF and shown to be essential for PQQ production in *M. extorquens* AM1 (Springer, Ramamoorthi and Lidstrom, 1996, PqqF in the reference).

The most recent review on biosynthesis of PQQ is published by Klinman and Bonnot (2014).

13.9
Up- and Down-stream Processing; Purification and Formulation

The starting material for industrial production of PQQ is methanol, and production of PQQ by methylotrophs was reported (Ameyama *et al.*, 1984, 1988; Urakami *et al.*, 1992). The methods for purification of PQQ from culture medium were reported in the aforementioned references and several Japanese patents.

13.10
Chemical Synthesis or Extraction Technology

The first total synthesis of PQQ was published by Corey and Tramontano (1981) with an overall yield of 10% in 10 steps. Chemical synthesis of PQQ and its analogues was also reported by Sleath *et al.* (1985), and its analogue preparation is summarised by Ohshiro (1995). Such PQQ analogues synthesised were used for reconstitution experiments with apo-mGDH, revealing that the 9-carboxyl group of PQQ was essential for the reconstitution of the enzyme activity (Shinagawa *et al.*, 1986). Trimethyl ester of pyrroloquinoline quinone (PQQTME) and its derivatives are synthesised and used for reactivity with phenylhydrazine in organic solvents, revealed the significant role of the pyrrole nucleus in conducting the intramolecular general base catalysis in the amine oxidation (Itoh *et al.*, 1992).

Chemical synthesis of imidazole analogues of PQQ is reported by Fouchard, Tillekeratne and Hudson (2004). Kimachi *et al.* (1995) reported the synthesis of 5-deazaflavoquinone (hybrid model compound of 5-deazaflavin and coenzyme PQQ).

13.11
Application and Economics

Chemically synthesised PQQ is available from Nascent Health Sciences, and PQQ prepared by bacterial fermentation is available from Mitsubishi Gas Chemical.

Supplementary capsules containing PQQ are already in market in the United States, and cosmetics containing PQQ are commercialised in Japan.

PQQ-dependent sGDH was used for test strips or biosensor for glucose determination in blood (Duine, 1999); however, it has been substituted by another FAD-dependent GDH (Tsujimura *et al.*, 2006).

References

Adachi, O., Ano, Y., Toyama, H., and Matsushita, K. (2007) in *Modern Biooxidation. Enzymes, Reactions and Applications* (eds R.F. Schmid and V.B. Urlacher), Wiley-VCH Verlag GmbH & Co. KGaA, Weinheim, pp. 1–41.

Adachi, O., Fujii, Y., Ghaly, M.F., Toyama, H., Shinagawa, E., and Matsushita, K. (2001) Membrane-bound quinoprotein D-arabitol dehydrogenase of *Gluconobacter suboxydans* IFO 3257: a versatile enzyme for the oxidative fermentation of various ketoses. *Biosci. Biotechnol., Biochem.*, **65**, 2755–2762.

Adachi, O., Okamoto, K., Shinagawa, E., Matsushita, K., and Ameyama, M. (1988) Adduct formation of pyrroloquinoline quinone and amino acid. *Biofactors*, **1**, 251–254.

Akagawa, M., Nakano, M., Ikemoto, K. (2015) Recent progress in studies on the health benefits of pyrroloquinoline quinone. *Biosci. Biotechnol. Biochem.*, **80**, 13–22.

Ameyama, M., Hayashi, M., Matsushita, K., Shinagawa, E., and Adachi, O. (1984) Microbial production of pyrroloquinoline quinone. *Agric. Biol. Chem.*, **48**, 561–565.

Ameyama, M., Matsushita, K., Ohno, Y., Shinagawa, E., and Adachi, O. (1981) Existence of a novel prosthetic group, PQQ, in membrane-bound, electron transport chain-linked, primary dehydrogenases of oxidative bacteria. *FEBS Lett.*, **130**, 179–183.

Ameyama, M., Matsushita, K., Shinagawa, E., Hayashi, M., and Adachi, O. (1988) Pyrroloquinoline quinone: excretion by methylotrophs growth stimulation for microorganisms. *Biofactors*, **1**, 51–53.

Ameyama, M., Nonobe, M., Shinagawa, E., Matsushita, K., and Adachi, O. (1985) Method of enzymatic determination of pyrroloquinoline quinone. *Anal. Biochem.*, **151**, 263–267.

Ando, A., Takeda, M., Kato, S., Kimura, F., Nakagawa, K., Nakano, M., and Miyazawa, N. (2014) LC-MS/MS analysis of pyrroloquinoline quinone (PQQ). *Vitamin (Japan)*, **88**, 601–609 (Japanese).

Anthony, C. and Zatman, L.J. (1967) The microbial oxidation of methanol. The prosthetic group of the alcohol dehydrogenase of *Pseudomonas* sp. M27: a new oxidoreductase prosthetic group. *Biochem. J.*, **104**, 960–969.

Asakura, A. and Hoshino, T. (1999) Isolation and characterization of a new quinoprotein dehydrogenase, L-sorbose/L-sorbosone dehydrogenase. *Biosci. Biotechnol. Biochem.*, **63**, 46–53.

Babu-Khan, S., Yeo, T.C., Martin, W.L., Duron, M.R., Rogers, R.D., and Goldstein, A.H. (1995) Cloning of a mineral phosphate-solubilizing gene from *Pseudomonas cepacia*. *Appl. Environ. Microbiol.*, **61**, 972–978.

Bauerly, K., Harris, C., Chowanadisai, W., Graham, J., Havel, P.J., Tchaparian, E., Satre, M., Karliner, J.S., and Rucker, R.B. (2011) Pyrroloquinoline quinone modulates mitochondrial quantity and function in mice. *PLoS One*, **6**, e21779.

Bergethon, P.R. (1990) Amperometric electrochemical detection of pyrroloquinoline quinone in high-performance liquid chromatography. *Anal. Biochem.*, **186**, 324–327.

Chang, Y., Ceacareanu, B., Zhuang, D., Zhang, C., Pu, Q., Ceacareanu, A.C., and Arterioscler, H.A. (2006) Counter-regulatory function of protein tyrosine phosphatase 1B in platelet-derived growth factor- or fibroblast growth factor-induced motility and proliferation of cultured smooth muscle cells and in neointima formation. *Arterioscler. Thromb. Vasc. Biol.*, **26**, 501–507.

Chiarugi, P. and Cirri, P. (2003) Redox regulation of protein tyrosine phosphatases during receptor tyrosine kinase signal transduction. *Trends Biochem. Sci.*, **28**, 509–514.

Choi, O., Kim, J., Kim, J.-G., Jeong, Y., Moon, J.S., Park, C.S., and Hwang, I. (2008) Pyrroloquinoline quinone is a plant growth promotion factor produced by *Pseudomonas fluorescens* B16. *Plant Physiol.*, **146**, 657–668.

Chowanadisai, W., Bauerly, K.A., Tchaparian, E., Wong, A., Cortopassi, G.A., and Rucker, R.B. (2010) Pyrroloquinoline quinone stimulates mitochondrial biogenesis through cAMP response element-binding protein phosphorylation and increased PGC-1alpha expression. *J. Biol. Chem.*, **285**, 142–152.

Cleton-Jansen, A.M., Goosen, N., Wenzel, T.J., and van de Putte, P. (1988) Cloning of the gene encoding quinoprotein glucose dehydrogenase from *Acinetobacter calcoaceticus*: evidence for the presence of a second enzyme. *J. Bacteriol.*, **170**, 2121–2125.

Corey, E.J. and Tramontano, A. (1981) Total synthesis of the quinonoid alcohol dehydrogenase coenzyme (1) of methylotrophic bacteria. *J. Am. Chem. Soc.*, **103**, 5599–5600.

Dekker, R.H., Duine, J.A., Frank, J., Verwiel, P.E., and Westerling, J. (1982) Covalent addition of H_2O, enzyme substrates and activators to pyrrolo-quinoline quinone, the coenzyme of quinoproteins. *Eur. J. Biochem.*, **125**, 69–73.

Duine, J.A. (1999) The PQQ story. *J. Biosci. Bioeng.*, **88**, 231–236.

Duine, J.A., Frank, J., and Verwiel, P.E.J. (1981) Characterization of the second prosthetic group in methanol dehydrogenase from *Hyphomicrobium* X. *Eur. J. Biochem.*, **118**, 395–399.

Feige, J.N. and Auwerx, J. (2007) Transcriptional coregulators in the control of energy homeostasis. *Trends Cell Biol.*, **17**, 292–301.

Fouchard, D.M., Tillekeratne, L.M., and Hudson, R.A. (2004) Synthesis of imidazolo analogues of the oxidation – reduction cofactor pyrroloquinoline quinone (PQQ). *J. Org. Chem.*, **69**, 2626–2629.

Gallop, P.M., Paz, M.A., Flückiger, R., and Henson, E. (1993) Is the antioxidant, anti-inflammatory putative new vitamin, PQQ, involved with nitric oxide in bone metabolism? *Connect. Tissue Res.*, **29**, 153–161.

Ge, X., Wang, W., Du, B., Wang, J., Xiong, X., and Zhang, W. (2015) Multiple *pqqA* genes respond differently to environment and one contributes dominantly to pyrroloquinoline quinone synthesis. *J. Basic Microbiol.*, **55**, 312–323.

Goodwin, P.M. and Anthony, C. (1998) The biochemistry, physiology and genetics of PQQ and PQQ-containing enzymes. *Adv. Microb. Physiol.*, **40**, 1–80.

Handschin, C. and Spiegelman, B.M. (2006) Pyrroloquinoline quinone preserves mitochondrial function and prevents oxidative injury in adult rat cardiac myocytes. *Endocr. Rev.*, **27**, 728–735.

Hara, H., Hiramatsu, H., and Adachi, T. (2007) Pyrroloquinoline quinone is a potent neuroprotective nutrient against 6-hydroxydopamine-induced neurotoxicity. *Neurochem. Res.*, **32**, 489–495.

Harris, C.B., Chowanadisai, W., Mishchuk, D.O., Satre, M.A., Slupsky, C.M., and Rucker, R.B. (2013) Dietary pyrroloquinoline quinone (PQQ) alters indicators of inflammation and mitochondrial-related metabolism in human subjects. *J. Nutr. Biochem.*, **24**, 2076–2084.

Hauge, J.G. (1964) Glucose dehydrogenase of *Bacterium anitratum*: an enzyme with a novel prosthetic group. *J. Biol. Chem.*, **239**, 3630–3639.

He, K., Nukada, H., Urakami, T., and Murphy, M.P. (2003) Antioxidant and pro-oxidant properties of pyrroloquinoline quinone (PQQ): implications for its function in biological systems. *Biochem. Pharmacol.*, **65**, 67–74.

Higa, Y. and Tachibana, S. (2014) Effects of pyrroloquinoline quinone on minelization in murine osteoblastic MC3T3-E1 cells. *Vitamins (Japan)*, **88**, 396–404 (Japanese).

Hölscher, T., Weinert-Sepalage, D., and Görisch, H. (2007) Identification of membrane-bound quinoprotein inositol

dehydrogenase in *Gluconobacter oxydans* ATCC 621H. *Microbiology*, **153**, 499–506.
Houck, D.R., Hanners, J.L., and Unkefer, C.J. (1988) Biosynthesis of pyrroloquinoline quinone. 1. Identification of biosynthetic precursors using carbon-13 labeling and NMR spectroscopy. *J. Am. Chem. Soc.*, **110**, 6920–6921.
Houck, D.R., Hanners, J.L., Unkefer, C.J., van Kleef, M.A., and Duine, J.A. (1989) PQQ: biosynthetic studies in *Methylobacterium* AM1 and *Hyphomicrobium* X using specific ^{13}C labeling and NMR. *Antonie Van Leeuwenhoek*, **56**, 93–101.
Itoh, S., Fukui, Y., Haranou, S., Ogino, M., Komatsu, M., and Ohshiro, Y. (1992) Synthesis and characterization of dimethyl 9,10-dihydro-9,10-dioxobenzo[f]quinoline-2,4-dicarboxylate. Effect of the pyrrole nucleus on the reactivity of coenzyme PQQ. *J. Org. Chem.*, **57**, 4452–4457.
Kano, K., Mori, K., Uno, B., Kubota, T., Ikeda, T., and Senda, M. (1990) Voltammetric and spectroscopic studies of pyrroloquinoline quinone coenzyme under neutral and basic conditions. *Bioelectrochem. Bioenerg.*, **23**, 227–238.
Kasahara, T. and Kato, T. (2003) Nutritional biochemistry: a new redox-cofactor vitamin for mammals. *Nature (London)*, **422**, 832.
Keltjens, J.T., Pol, A., Reimann, J., and Op den Camp, H.J. (2014) PQQ-dependent methanol dehydrogenases: rare-earth elements make a difference. *Appl. Microbiol. Biotechnol.*, **98**, 6163–6183.
Kenner, K.A., Anyanwu, E., Olefsky, J.M., and Kusari, J. (1996) Protein-tyrosine phosphatase 1B is a negative regulator of insulin- and insulin-like growth factor-I-stimulated signaling. *J. Biol. Chem.*, **271**, 19810–19816.
Killgore, J., Smidt, C., Duich, L., Romero-Chapman, N., Tinker, D., Reiser, K., Melko, M., Hyde, D., and Rucker, R.B. (1989) Nutritional importance of pyrroloquinoline quinone. *Science*, **245**, 850–852.
Kim, J., Harada, R., Kobayashi, M., Kobayashi, N., and Sode, K. (2010) The inhibitory effect of pyrroloquinoline quinone on the amyloid formation and cytotoxicity of truncated alpha-synuclein. *Mol. Neurodegener.*, **5**, 20.
Kimachi, T., Sugita, K., Bessho, K., and Yoneda, F. (1995) Synthesis of a new type of 5-deazaflavoquinone. (hybrid model compound of 5-deazaflavin and coenzyme PQQ). *Bioorg. Med. Chem. Lett.*, **5**, 31–34.
Kimura, K., Takada, M., Ishii, T., Tsuji-Naito, K., and Akagawa, M. (2012) Pyrroloquinoline quinone stimulates epithelial cell proliferation by activating epidermal growth factor receptor through redox cycling. *Free Radic. Biol. Med.*, **53**, 1239–1251.
van Kleef, M.A.G. and Duine, J.A. (1988) L-tyrosine is the precursor of PQQ biosynthesis in *Hyphomicrobium* X. *FEBS Lett.*, **237**, 91–97.
van Kleef, M.A. and Duine, J.A. (1989) Factors relevant in bacterial pyrroloquinoline quinone production. *Appl. Environ. Microbiol.*, **55**, 1209–1213.
Klinman, J.P. and Bonnot, F. (2014) Intrigues and intricacies of the biosynthetic pathways for the enzymatic quinocofactors: PQQ, TTQ, CTQ, TPQ, and LTQ. *Chem. Rev.*, **114**, 4343–4365.
Kobayashi, M., Kim, J., Kobayashi, N., Han, S., Nakamura, C., Ikebukuro, K., and Sode, K. (2006) Pyrroloquinoline quinone (PQQ) prevents fibril formation of alpha-synuclein. *Biochem. Biophys. Res. Commun.*, **349**, 1139–1144.
Koikeda, T., Nakano, M., and Masuda, K. (2011) Pyrroloquinoline quinone disodium salt improves higher brain function. *Med. Consult. New Remedies*, **48**, 519–527.
Kong, L., Yang, C., Yu, L., Smith, W., Zhu, S., Zhu, J., and Zhu, Q. (2013) Pyrroloquinoline quinine inhibits RANKL-mediated expression of NFATc1 in part via suppression of c-Fos in mouse bone marrow cells and inhibits wear particle-induced osteolysis in mice. *PLoS One*, **8**, e61013.
Kumazawa, T., Hiwasa, T., Takiguchi, M., Suzuki, O., and Sato, K. (2007) Activation of Ras signaling pathways by pyrroloquinoline quinone in NIH3T3 mouse fibroblasts. *Int. J. Mol. Med.*, **19**, 765–770.
Kumazawa, T., Sato, K., Seno, H., Ishii, A., and Suzuki, O. (1995) Levels of pyrroloquinoline quinone in various foods. *Biochem. J.*, **307**, 331–333.
Kumazawa, T., Seno, H., and Suzuki, O. (1993) Failure to verify high levels of pyrroloquinoline quinone in eggs and skim

milk. *Biochem. Biophys. Res. Commun.*, **193**, 1–5.

Kumazawa, T., Seno, H., Urakami, T., Matsumoto, T., and Suzuki, O. (1992) Trace levels of pyrroloquinoline quinone in human and rat samples detected by gas chromatography/mass spectrometry. *Biochim. Biophys. Acta*, **1156**, 62–66.

Latham, J.A., Iavarone, A.T., Barr, I., Juthani, P.V., and Klinman, J.P. (2015) PqqD is a novel peptide chaperone that forms a ternary complex with the radical S-adenosylmethionine protein PqqE in the pyrroloquinoline quinone biosynthetic pathway. *J. Biol. Chem.*, **290**, 12908–12918.

Lau, C., Borgmann, S., Maciejewska, M., Ngounou, B., Gründler, P., and Schuhmann, W. (2007) Improved specificity of reagentless amperometric PQQ-sGDH glucose biosensors by using indirectly heated electrodes. *Biosens. Bioelectron.*, **22**, 3014–3020.

Liu, S.T., Lee, L.Y., Tai, C.Y., Hung, C.H., Chang, Y.S., Wolfram, J.H., Rogers, R., and Goldstein, A.H. (1992) Cloning of an *Erwinia herbicola* gene necessary for gluconic acid production and enhanced mineral phosphate solubilization in *Escherichia coli* HB101: nucleotide sequence and probable involvement in biosynthesis of the coenzyme pyrroloquinoline quinone. *J. Bacteriol.*, **174**, 5814–5819.

Magnusson, O.T., RoseFigura, J.M., Toyama, H., Schwarzenbacher, R., and Klinman, J.P. (2007) Pyrroloquinoline quinone biogenesis: characterization of PqqC and its H84N and H84A active site variants. *Biochemistry*, **46**, 7174–7186.

Magnusson, O.T., Toyama, H., Saeki, M., Rojas, A., Reed, J.C., Liddington, R.C., Klinman, J.P., and Schwarzenbacher, R. (2002a) Quinone biogenesis: Structure and mechanism of PqqC, the final catalyst in the production of pyrroloquinoline quinone. *Proc. Natl. Acad. Sci. U.S.A.*, **101**, 7913–7918.

Magnusson, O.T., Toyama, H., Saeki, M., Schwarzenbacher, R., and Klinman, J.P. (2002b) The structure of a biosynthetic intermediate of pyrroloquinoline quinone (PQQ) and elucidation of the final step of PQQ biosynthesis. *J. Am. Chem. Soc.*, **126**, 5342–5343.

Matsumura, H., Umezawa, K., Takeda, K., Sugimoto, N., Ishida, T., Samejima, M., Ohno, H., Yoshida, M., Igarashi, K., and Nakamura, N. (2014) Discovery of a eukaryotic pyrroloquinoline quinone-dependent oxidoreductase belonging to a new auxiliary activity family in the database of carbohydrate-active enzymes. *PLoS One*, **9**, e104851.

Matsushita, K., Fujii, Y., Ano, Y., Toyama, H., Shinjoh, M., Tomiyama, N., Miyazaki, T., Sugisawa, T., Hoshino, T., and Adachi, O. (2003) 5-keto-D-gluconate production is catalyzed by a quinoprotein glycerol dehydrogenase, major polyol dehydrogenase, in *Gluconobacter* species. *Appl. Environ. Microbiol.*, **69**, 1959–1966.

Matsushita, K., Shinagawa, E., Adachi, O., and Ameyama, M. (1989) Quinoprotein D-glucose dehydrogenase of the *Acinetobacter calcoaceticus* respiratory chain: membrane-bound and soluble forms are different molecular species. *Biochemistry*, **28**, 6276–6280.

Matsushita, K., Toyama, H., and Adachi, O. (1994) in *Advances in Microbial Physiology*, vol. **36** (eds A.H. Rose and D.W. Tempest), Academic Press, London, pp. 247–301.

Matsushita, K., Toyama, H., and Adachi, O. (2002) Quinoproteins: structure, function, and biotechnological applications. *Appl. Microbiol. Biotechnol.*, **58**, 13–22.

Misra, H.S., Rajpurohit, Y.S., and Khairnar, N.P. (2012) Pyrroloquinoline-quinone and its versatile roles in biological processes. *J. Biosci.*, **37** (2), 313–325.

Mitchell, A.E., Jones, A.D., Mercer, R.S., and Rucker, R.B. (1999) Characterization of pyrroloquinoline quinone amino acid derivatives by electrospray ionization mass spectrometry and detection in human milk. *Anal. Biochem.*, **269**, 317–325.

Miyauchi, K., Urakami, T., Abeta, H., Shi, H., Noguchi, N., and Niki, E. (1994) Action of pyrroloquinolinequinol as an antioxidant against lipid peroxidation in solution. *Antioxid. Redox Signal.*, **1**, 547–554.

Miyazaki, T., Sugisawa, T., and Hoshino, T. (2006) Pyrroloquinoline quinone-dependent dehydrogenases from *Ketogulonicigenium vulgare* catalyze the direct conversion of L-sorbosone to L-ascorbic acid. *Appl. Environ. Microbiol.*, **72**, 1487–1495.

Miyazaki, T., Tomiyama, N., Shinjoh, M., and Hoshino, T. (2002) Molecular cloning and functional expression of D-sorbitol dehydrogenase from *Gluconobacter suboxydans* IFO3255, which requires pyrroloquinoline quinone and hydrophobic protein SldB for activity development in *E. coli*. *Biosci. Biotechnol., Biochem.*, **66**, 262–270.

Mukai, K., Ouchi, A., and Nakano, M. (2011) Kinetic study of the quenching reaction of singlet oxygen by pyrroloquinolinequinol ($PQQH_2$, a reduced form of pyrroloquinolinequinone) in micellar solution. *J. Agric. Food Chem.*, **59**, 1705–1712.

Murase, K., Hattori, A., Kohno, M., and Hayashi, K. (1993) Stimulation of nerve growth factor synthesis/secretion in mouse astroglial cells by coenzymes. *Biochem. Mol. Biol. Int.*, **30**, 615–621.

Naito, Y., Kumazawa, T., Kino, I., and Suzuki, O. (1993) Effects of pyrroloquinoline quinone (PQQ) and PQQ-oxazole on DNA synthesis of cultured human fibroblasts. *Life Sci.*, **52**, 1909–1915.

Nakamura, Y., Patrushev, N., Inomata, H., Mehta, D., Urao, N., Kim, H.W., Razvi, M., Kini, V., Mahadev, K., Goldstein, B.J., McKinney, R., Fukai, T., and Ushio-Fukai, M. (2008) Role of protein tyrosine phosphatase 1B in vascular endothelial growth factor signaling and cell-cell adhesions in endothelial cells. *Circ. Res.*, **102**, 1182–1191.

Nakano, M., Kamimura, A., Watanabe, F., Kamiya, T., Watanabe, D., Yamamoto, E., Fukagawa, M., Hasumi, K., and Suzuki, E. (2015a) Effects of orally administered pyrroloquinoline quinone disodium salt on dry skin conditions in mice and healthy female subjects. *J. Nutr. Sci. Vitaminol.*, **61**, 241–246.

Nakano, M., Kawasaki, Y., Suzuki, N., and Takara, T. (2015b) Effects of pyrroloquinoline quinone disodium salt intake on the serum cholesterol levels of healthy Japanese adults. *J. Nutr. Sci. Vitaminol.*, **61**, 233–240.

Nakano, M., Yamamoto, T., Okumura, H., Tsuda, A., and Kowatari, Y. (2012) Effects of oral supplementation with pyrroloquinoline quinone on stress, fatigue, and sleep. *J. Funct. Foods Health Dis.*, **2**, 307–324.

Noji, N., Kasahara, K., and Asami, T. (2008) PQQ vitamin setsu he no apurohchi. *Kagaku to Seibutsu*, **46**, 339–345 (in Japanese).

Noji, N., Nakamura, T., Kitahata, N., Taguchi, K., Kudo, T., Yoshida, S., Tsujimoto, M., Sugiyama, T., and Asami, T. (2007) Simple and sensitive method for pyrroloquinoline quinone (PQQ) analysis in various foods using liquid chromatography/electrospray-ionization tandem mass spectrometry. *J. Agric. Food Chem.*, **55**, 7258–7263.

Nunome, K., Miyazaki, S., Nakano, M., Iguchi-Ariga, S., and Ariga, H. (2008) Pyrroloquinoline quinone prevents oxidative stress-induced neuronal death probably through changes in oxidative status of DJ-1. *Biol. Pharm. Bull.*, **3**, 1321–1326.

Ohshiro, Y. (1995) Development of novel synthetic method for polyfunctionalized heterocycles. *J. Synth. Org. Chem. Jpn.*, **53**, 846–857 (in Japanese).

Ohwada, K., Takeda, H., Yamazaki, M., Ishogai, H., Nakano, M., Shimomura, M., Fukui, K., and Urano, S. (2008) Pyrroloquinoline quinone (PQQ) prevents cognitive deficit caused by oxidative stress in rats. *J. Clin. Biochem. Nutri.*, **42**, 29–34.

Oubrie, A. (2003) Structure and mechanism of soluble glucose dehydrogenase and other PQQ-dependent enzymes. *Biochim. Biophys. Acta*, **1647**, 143–151.

Ouchi, A., Ikemoto, K., Nakano, M., Nagaoka, S., and Mukai, K. (2013) Kinetic study of aroxyl radical scavenging and α-tocopheroxyl regeneration rates of pyrroloquinolinequinol ($PQQH_2$, a reduced form of pyrroloquinolinequinone) in dimethyl sulfoxide solution: finding of synergistic effect on the reaction rate due to the coexistence of α-tocopherol and $PQQH_2$. *J. Agric. Food Chem.*, **61**, 11048–11060.

Ouchi, A., Nakano, M., Nagaoka, S., and Mukai, K. (2009) Kinetic study of the antioxidant activity of pyrroloquinolinequinol ($PQQH_2$, a reduced form of pyrroloquinolinequinone) in micellar solution. *J. Agric. Food Chem.*, **57**, 450–456.

Pappenberger, G. and Hohmann, H.-P. (2014) in *Biotechnology of Food and Feed Additives*, vol. **143** (eds H. Zorn and P. Czermak), Springer, Berlin and Heidelberg, pp. 143–188.

Paz, M.A., Flückiger, R., and Gallop, P.M. (1990) Redox-cycling is a property of PQQ but not of ascorbate. *FEBS Lett.*, **264**, 283–284.

Paz, M.A., Martin, P., Flückiger, R., Mah, J., and Gallop, P.M. (1996) The catalysis of redox cycling by pyrroloquinoline quinone (PQQ), PQQ derivatives, and isomers and the specificity of inhibitors. *Anal. Biochem.*, **238**, 145–149.

Prust, C., Hoffmeister, M., Liesegang, H., Wiezer, A., Fricke, W.F., Ehrenreich, A., Gottschalk, G., and Deppenmeier, U. (2005) Complete genome sequence of the acetic acid bacterium *Gluconobacter oxydans*. *Nat. Biotechnol.*, **23**, 195–200.

Rucker, R., Chowanadisai, W., and Nakano, M. (2009) Potential physiological importance of pyrroloquinoline quinone. *Altern. Med. Rev.*, **14**, 179–183.

Sakuraba, H., Yokono, K., Yoneda, K., Watanabe, A., Asada, Y., Satomura, T., Yabutani, T., Motonaka, J., and Ohshima, T. (2010) Catalytic properties and crystal structure of quinoprotein aldose sugar dehydrogenase from hyperthermophilic archaeon *Pyrobaculum aerophilum*. *Arch. Biochem. Biophys.*, **502**, 81–88.

Salisbury, S.A., Forrest, H.S., Cruse, W.B., and Kennard, O. (1979) A novel coenzyme from bacterial primary alcohol dehydrogenases. *Nature*, **280**, 843–844.

Samuel, K.G., Zhang, H.J., Wang, J., Wu, S.G., Yue, H.Y., Sun, L.L., and Qi, G.H. (2015) Effects of dietary pyrroloquinoline quinone disodium on growth performance, carcass yield and antioxidant status of broiler chicks. *Animal*, **9**, 409–416.

Shen, Y.Q., Bonnot, F., Imsand, E.M., RoseFigura, J.M., Sjölander, K., and Klinman, J.P. (2012) Distribution and properties of the genes encoding the biosynthesis of the bacterial cofactor, pyrroloquinoline quinone. *Biochemistry*, **51**, 2265–2275.

Shibata, T., Nakahara, H., Kita, N., Matsubara, Y., Han, C., Morimitsu, Y., Iwamoto, N., Kumagai, Y., Nishida, M., Kurose, H., Aoki, N., Ojika, M., and Uchida, K. (2008) Role of protein tyrosine phosphatase 1B in vascular endothelial growth factor signaling and cell-cell adhesions in endothelial cells. *J. Neurochem.*, **107**, 1248–1260.

Shinagawa, E., Matsushita, K., Nonobe, M., Adachi, O., Ameyama, M., Ohshiro, Y., Itoh, S., and Kitamura, Y. (1986) The 9-carboxyl group of pyrroloquinoline quinone, a novel prosthetic group, is essential in the formation of holoenzyme of D-glucose dehydrogenase. *Biochem. Biophys. Res. Commun.*, **139**, 1279–1284.

Shinjoh, M., Tomiyama, N., Asakura, A., and Hoshino, T. (1995) Cloning and nucleotide sequencing of the membrane-bound L-sorbosone dehydrogenase gene of *Acetobacter liquefaciens* IFO 12258 and its expression in *Gluconobacter oxydans*. *Appl. Environ. Microbiol.*, **61**, 413–420.

Sleath, P.R., Noar, J.B., Eberlein, G.A., and Bruice, T.C. (1985) Synthesis of 7,9- didecarboxymethoxatin (4,5-dihydro-4,5-dioxo-1*H*-pyrrolo[2,3-*f*]quinoline-2-carboxylic acid) and comparison of its chemical properties with those of methoxatin and analogous o-quinones. Model studies directed toward the action of PQQ requiring bacterial oxidoreductases and mammalian plasma amine oxidase. *J. Am. Chem. Soc.*, **107**, 3328–3338.

Southall, S.M., Doel, J.J., Richardson, D.J., and Oubrie, A. (2006) Soluble aldose sugar dehydrogenase from *Escherichia coli*. A highly exposed active site conferring broad substrate specificity. *J. Biol. Chem.*, **281**, 30650–30659.

Springer, A.L., Ramamoorthi, R., and Lidstrom, M.E. (1996) Characterization and nucleotide sequence of *pqqE* and *pqqF* in *Methylobacterium extorquens* AM1. *J. Bacteriol.*, **178**, 2154–2157.

Steinberg, F.M., Gershwin, M.E., and Rucker, R.B. (1994) Dietary pyrroloquinoline quinone: growth and immune response in BALB/c mice. *J. Nutr.*, **124**, 744–753.

Stites, T., Storms, D., Bauerly, K., Mah, J., Harris, C., Fascetti, A., Rogers, Q., Tchaparian, E., Satre, M., and Rucker, R.B. (2006) Pyrroloquinoline quinone modulates mitochondrial quantity and function in mice. *J. Nutr.*, **136**, 390–396.

Streďanský, M., Monošík, R., Mastihuba, V., and Sturdík, E. (2013) Monitoring of PQQ-dependent glucose dehydrogenase substrate specificity for its potential use in biocatalysis and bioanalysis. *Appl. Biochem. Biotechnol.*, **171**, 1032–1041.

Takatsu, H., Owada, K., Abe, K., Nakano, M., and Urano, S. (2009) Effect of vitamin E on learning and memory deficit in aged rats. *J. Nutr. Sci. Vitaminol.*, **55**, 389–393.

Tao, R., Karliner, J.S., Simonis, U., Zheng, J., Zhang, J., Honbo, N., and Alano, C.C. (2007) Pyrroloquinoline quinone preserves mitochondrial function and prevents oxidative injury in adult rat cardiac myocytes. *Biochem. Biophys. Res. Commun.*, **363**, 257–262.

Toyama, H., Chistoserdova, L., and Lidstrom, M.E. (1997) Sequence analysis of *pqq* genes required for biosynthesis of pyrroloquinoline quinone in *Methylobacterium extorquens* AM1 and the purification of a biosynthetic intermediate. *Microbiology*, **143**, 595–602.

Toyama, H., Fukumoto, H., Saeki, M., Matsushita, K., Adachi, O., and Lidstrom, M.E. (2002) PqqC/D, which converts a biosynthetic intermediate to pyrroloquinoline quinone. *Biochem. Biophys. Res. Commun.*, **299**, 268–272.

Toyama, H., Mathews, F.S., Adachi, O., and Matsushita, K. (2004) Quinohemoprotein alcohol dehydrogenases: Structure, function and physiology. *Arch. Biochem. Biophys.*, **428**, 10–21.

Tsujimura, S., Kojima, S., Kano, K., Ikeda, T., Sato, M., Sanada, H., and Omura, H. (2006) Novel FAD-dependent glucose dehydrogenase for a dioxygen-insensitive glucose biosensor. *Biosci. Biotechnol. Biochem.*, **70**, 654–659.

Unkefer, C.J., Houck, D.R., Britt, B.M., Sosnick, T.R., and Hanners, J.L. (1995) Biogenesis of pyrroloquinoline quinone from 3C-labeled tyrosine. *Methods Enzymol.*, **258**, 227–235.

Urakami, T., Sugamura, K., and Niki, E. (1995/1996) Characterization of imidazopyrroloquinoline compounds synthesized from coenzyme PQQ and various amino acids. *Biofactors*, **5**, 75–81.

Urakami, T., Tanaka, A., Tanimoto, T., and Niki, E. (1996) Synthesis and aldose reductase-inhibitory activity of imidazopyrroloquinoline esters. *Chem. Pharm. Bull. (Tokyo)*, **44**, 1493–1497.

Urakami, T., Yashima, K., Kobayashi, H., Yoshida, A., and Ito-Yoshida, C. (1992) Production of pyrroloquinoline quinone by using methanol-utilizing bacteria. *Appl. Environ. Microbiol.*, **58**, 3970–3976.

Vangnai, A.S., Toyama, H., De-Eknamkul, W., Yoshihara, N., Adachi, O., and Matsushita, K. (2004) Quinate oxidation in *Gluconobacter oxydans* IFO3244: purification and characterization of quinoprotein quinate dehydrogenase. *FEMS Microbiol. Lett.*, **241**, 157–162.

Velterop, J.S., Sellink, E., Meulenberg, J.J., David, S., Bulder, I., and Postma, P.W. (1995) Synthesis of pyrroloquinoline quinone in vivo and in vitro and detection of an intermediate in the biosynthetic pathway. *J. Bacteriol.*, **177**, 5088–5098.

Wang, L., Jil, C., Xu, Y., Xu, J., Dai, J., Wu, Q., Wu, M., Zou, X., Sun, L., Gu, S., Xie, Y., and Mao, Y. (2005) Cloning and characterization of a novel human homolog of mouse U26, a putative PQQ-dependent AAS dehydrogenase. *Mol. Biol. Rep.*, **32**, 47–53.

Wecksler, S.R., Stoll, S., Tran, H., Magnusson, O.T., Wu, S.P., King, D., Britt, R.D., and Klinman, J.P. (2009) Pyrroloquinoline quinone biogenesis: demonstration that PqqE from *Klebsiella pneumoniae* is a radical S adenosyl-L-methionine enzyme. *Biochemistry*, **48**, 10151–10161.

Westerling, J., Frank, J., and Duine, J.A. (1979) The prosthetic group of methanol dehydrogenase from *Hyphomicrobium* X: electron spin resonance evidence for a quinone structure. *Biochem. Biophys. Res. Commun.*, **87**, 719–724.

Yakushi, T. and Matsushita, K. (2010) Alcohol dehydrogenase of acetic acid bacteria: structure, mode of action, and applications in biotechnology. *Appl. Microbiol. Biotechnol.*, **86**, 1257–1265.

Yamada, M., Sumi, K., Matsushita, K., Adachi, O., and Yamada, Y. (1993) Topological analysis of quinoprotein glucose dehydrogenase in *Escherichia coli* and its ubiquinone-binding site. *J. Biol. Chem.*, **268**, 12812–12817.

Yamaguchi, K., Sasano, A., Urakami, T., Tsuji, T., and Kondo, K. (1993) Stimulation of nerve growth factor production by pyrroloquinoline quinone and its derivatives in vitro and in vivo. *Biosci. Biotechnol. Biochem.*, **57**, 1231–1233.

Yamaguchi, K., Tsuji, T., Uemura, D., and Kondo, K. (1996) Cyclooxygenase induction is essential for NGF synthesis enhancement by NGF inducers in L-M cells. *Biosci. Biotechnol. Biochem.*, **60**, 92–94.

Part III
Other Growth Factors, Biopigments and Antioxidants

Industrial Biotechnology of Vitamins, Biopigments, and Antioxidants, First Edition.
Edited by Erick J. Vandamme and José L. Revuelta.

14
L-Carnitine, the Vitamin B_T: Uses and Production by the Secondary Metabolism of Bacteria

Vicente Bernal, Paula Arense, and Manuel Cánovas

14.1
Introduction and Historical Outline

L-Carnitine ((*R*)-3-hydroxy-4-trimethylaminobutyrate) is a natural component of tissues with high physiological relevance. It takes its name from the Latin term *carnis* (meat). Carnitine was originally found as a growth factor for mealworm *Tenebrio molitor*, playing an essential role in the metamorphosis of beetles (Fraenkel, 1948). Carnitine was thus named vitamin B_T, although further investigations have demonstrated that it is not a true vitamin, since most organisms are able to produce it themselves (Vaz and Wanders, 2002).

Today, L-carnitine is considered an important factor in the metabolism of long-chain fatty acids, although it also exerts several effects on various organ systems. Many clinical applications have been found for L-carnitine and its derivatives in many fields of medicine, especially in the treatment of cardiovascular diseases (Löster, 2003; Mescka *et al.*, 2011; Seim, Eichler and Kleber, 2001) Moreover, other uses are for health and social welfare, for example, as dietary supplement or nutraceutical to improve weight management and exercise performance (Jeppesen and Kiens, 2012; Johri *et al.*, 2014; Wall *et al.*, 2011).

In 1905, carnitine was discovered almost simultaneously by Gulewitsch and Krimberg (1905) in Moscow, Russia, and by Kutscher (1905), in Marburg, Germany (Löster, 2003). It took more than 20 years to elucidate its chemical structure and almost 60 years to determine the absolute configuration of its chiral centre. In 1927, the β-position of the hydroxyl group was established (Tomita and Sendju, 1927), and in 1962, the naturally occurring enantiomer was identified (Kaneko and Yoshida, 1962).

Although the presence of carnitine in different tissues was known, its actual role was a mystery for years. Initially, carnitine was seen as choline substituted with acetic acid and expected to produce similar effects to choline. It was Strack who detected that an unknown compound was contaminating choline preparations from smooth muscle (Strack and Fosterling, 1937). During the investigations of the vitamin requirements of the mealworm *T. molitor*, Fraenkel found that a 'growth factor' which they considered to be a vitamin (vitamin B_T) played an essential role

Industrial Biotechnology of Vitamins, Biopigments, and Antioxidants, First Edition.
Edited by Erick J. Vandamme and José L. Revuelta.

in the metamorphosis of that beetle (Fraenkel, 1948). In 1952, it was established that this factor was, in fact, L-carnitine (Carter *et al.*, 1952). It is known today that L-carnitine is not a vitamin for most organisms.

In 1955, Fritz reported that muscle extracts added to rat liver homogenates increased the palmitate oxidation rate (Fritz, 1955). L-Carnitine was this stimulating molecule, and Bremer observed that acetyl-carnitine was an effective supplier of acetyl groups for β-oxidation in the mitochondria (Bremer, 1962, 1963).

Today, the role of L-carnitine in the metabolism of fatty acids is well known. It plays an important role in metabolism, regulating the levels of fats in serum. In fact, carnitine deficiencies result in abnormally elevated levels of fats. Several clinical applications have been identified, which impelled research in the pharmaceutical and food industry in order to develop applicable production processes and new and safe supplements (Bernal *et al.*, 2007c; Löster, 2003).

14.2 Occurrence in Natural/Food Sources

In human beings and other animals, L-carnitine is obtained by biosynthesis and diet. Biosynthesis requires lysine and methionine (both essential amino acids), vitamin C, vitamin B_3, vitamin B_6 and iron. Consequently, biosynthesis also depends on a well-balanced diet. Tunable tubular reabsorption by the kidney contributes to efficiently maintain its plasma levels. Although animals are capable of synthesising carnitine endogenously, L-carnitine is primarily obtained from the diet (Vaz and Wanders, 2002).

Carnitine content in food is highly variable, depending on the source, its content varying in up to 2 orders of magnitude. Animal products contain more carnitine than vegetables, which may be an issue for vegetarians, whose plasma carnitine levels are often lowered (Krajcovicová-Kudlácková *et al.*, 2000). In omnivorous humans, approximately 75% of L-carnitine comes from the diet and 25% comes from *de novo* biosynthesis. In strict vegetarians, biosynthesis contributes to more than 90% of their L-carnitine (Vaz and Wanders, 2002). Dietary supply of carnitine may also be limiting in subjects on total parenteral nutrition (Duran *et al.*, 1990).

Carnitine content in foods also depends on the tissue and the species considered. Due to its role in mitochondrial fatty-acid metabolism, its concentration is especially high in the skeletal and cardiac muscle (content in the muscle: from 203 mg/100 g in sheep to 9.7 mg/100 g in chicken). It is also present in milk (from 14.1 mg/100 g in sheep milk to 2.6 mg/100 g in cow milk) and eggs (0.012 mg/100 g). Carnitine is thermostable, and, therefore, its content is hardly altered by food cooking or freezing (Jayasena *et al.*, 2014; Rigault *et al.*, 2008), although it may be partially extracted from products due to its high solubility in water (Löster, 2003).

Normal dietary intake of L-carnitine supplies enough carnitine to maintain serum levels of carnitine and excludes food-related deficiencies. Limited intestinal resorption or transportation may appear in ill or elderly people (Löster, 2003).

Patients with Crohn's disease and ulcerative colitis with adapted/special diets can obtain sufficient daily amounts. Only in the cases of severe malabsorption, intravenous administration of carnitine would be necessary (Bene *et al.*, 2006; Iwamoto *et al.*, 2014).

14.3 Physiological Role as Vitamin or as Coenzyme

L-Carnitine plays a significant role at various stages of intermediary metabolism of lipids:

- In the β-oxidation of medium- and long-chain fatty acids in the mitochondria.
- In the α- and β-oxidation of fatty acids in the peroxisomes.
- In the exchange of acyl and acetyl groups with CoA in the mitochondria and, thus, in the alteration of the acyl-CoA/CoA and acyl-carnitine/carnitine ratios.
- In the production of ketone bodies.

L-Carnitine is essential for fatty acid catabolism and closely related to acetyl-CoA. In fact, acetyl-carnitine can be seen as a second form of activated acetic acid, an acetyl buffer or a depot of acetyl group (Löster, 2003).

14.3.1 Physiological Role of Carnitine in the Mitochondria

Fatty acids are transported through the inner mitochondrial membrane into the mitochondrial matrix by the carnitine system (Eaton, Bartlett and Pourfarzam, 1996). Usually, short- and medium-chain fatty acids are able to pass the inner mitochondria membrane without concurrence of this system, although ATP shortages in the mitochondria may make the passage carnitine-dependent (Löster, 2003).

This transport mechanism comprises the following steps:

1) Long-chain fatty acids are activated by an acyl-CoA synthetase located on the inside of the outer mitochondrial membrane, but also in peroxisomes and the microsome fraction.
2) Carnitine palmitoyl transferases (CPTs) catalyse the reversible transfer of activated fatty acids between CoA and L-carnitine. These are membrane-bound enzymes which need lipid interactions for activity. Two enzymes exist: CPT I and II, which are located on the inner side of the outer mitochondrial membrane and on the inner side of the inner mitochondrial membrane, respectively.
3) The transport through the inner mitochondrial membrane is facilitated by the carnitine-acylcarnitine translocase.

β-Oxidation of fatty acids provides a major source of energy for the heart and skeletal muscle. In the liver, β-oxidation provides ketone bodies to the peripheral circulation, to be used as fuel by other organs, especially the brain, when blood glucose levels are low (Drynan, Quant and Zammit, 1996; Eaton, Bartlett and Pourfarzam, 1996). CPT I is the primary regulator of fatty acid oxidation and ketogenesis. The existing isozymes in the liver and muscle differ by post-translational modifications. In liver, CPT I is inhibited by malonyl-CoA, levels of which are, in turn, regulated by insulin. This regulation limits β-oxidation during lipogenic conditions. Muscle cells have high carnitine levels, and the carnitine system exerts a limiting effect only under lipolytic conditions, when energy consumption in the cell is high. Under normal conditions, the substrate flow is limited by the availability of free fatty acids and by NAD^+/NADP or ADP/ATP ratios (Eaton, Bartlett and Pourfarzam, 1996; Löster, 2003).

14.3.2 Physiological Role of Carnitine in the Peroxisomes

The role of L-carnitine in peroxisomes is not as well understood. In peroxisomes, very long-chain fatty acids, branched-chain fatty acids, bile acids and fatty dicarboxylic acids are metabolised (Violante *et al.*, 2013). In peroxisomes, L-carnitine has a role in the export of chain-shortened products produced during β-oxidation of very long-chain fatty acids. Acyl-CoA oxidase produces acetyl-CoA and medium-chain acyl-CoAs of about eight carbon atoms. Carnitine acyl-transferases exchange the acyl groups onto carnitine and regenerate the CoA for β-oxidation. These carnitine esters are then transported to mitochondria for further degradation (Ramsay, 1999; Wanders *et al.*, 2001).

14.3.3 Other Functions of Carnitine

The involvement of carnitine in the metabolism of the heart, liver, muscle, brain and lipid, as well as a certain role in sperm maturation, the immune system and connecting tissue, has been described (Löster, 2003).

14.4 Chemical and Physical Properties

This amphoteric compound belongs to the family of betaines (neutral compounds with a positively charged functional group). L-Carnitine free base is a crystalline, white powder which is extremely hygroscopic and suitable for all liquid formulations (Löster, 2003; Meyer and Robins, 2005). In solution, carnitine is a zwitterion and is highly soluble. With few exceptions (picrate, tetra-chloroaurate, tetraphenylboronate, triphenylcyanoboranate, Reinecke salt, etc.), its salts are soluble in water. Salts such as L-carnitine tartrate are free flowing, which makes them suitable for use in solid products such as tablets and capsules (Meyer and Robins, 2005).

14.5
Assay Methods and Units

The development of analytical methods to detect and quantify carnitine and/or its derivatives in biological samples has been key to identify natural sources of this compound, to provide information about metabolic disorders, to show the effect of carnitine therapy or to establish levels of biosynthesis and absorption of carnitine. Both free and total carnitines are clinically relevant for the follow-up of patients with abnormalities in carnitine metabolism, since deficiency is diagnosed from either a low free carnitine or a high proportion of esterified carnitine (Dąbrowska and Starek, 2014). Moreover, enantioselective analysis of D- and L-carnitine is essential in quality control of chemical and biological production processes, since L-carnitine is the only biologically active enantiomer (Gross and Henderson, 1984; Jung, Jung and Kleber, 1993; Stieger, O'Neill and Krähenbühl, 1995).

Selection of an analytical method is related to its features of precision, accuracy, cost or time consumption. On the other hand, sample matrix (biological, pharmaceutical or food) and application of results are important parameters involved in the selection of an efficient analytical method. Analytical methods for L-carnitine and related compounds have been recently reviewed (Dbrowska and Starek, 2014). Here, we outline the major considerations in this field.

14.5.1
Chromatographic Methods

Chromatographic methods have been widely used. High-performance liquid chromatography (HPLC) and gas chromatography (GC) allow automated and easy separation of carnitine and acyl-carnitines from several types of samples (plasma, human seminal plasma, urine, tissues, blood, dietary supplements, pharmaceutical and infant formulation samples, etc.) with high resolution. Enantioseparation and GC technique require derivatization of the analytes to enable separation and to generate volatile compounds, respectively. Some examples of chromatographic methods are listed in Table 14.1.

14.5.2
MS-Based Methods

MS-based methods are more sensitive, specific and reproducible and require lower sample amounts to assay carnitine and its short-, medium- and long-chain esters (Table 14.2). When coupled to separation techniques as monitoring detector, MS improves the capacity of determination of the technique. In capillary electrophoresis, MS detection allows the analysis of non-derivatised carnitine and reduces the analysis time per sample (Heinig and Henion, 1999). For the determination of D-and L-carnitine from commercial vegetable oils (Sánchez-Hernández *et al.*, 2011), pharmaceutical formulations (Sánchez-Hernández

Table 14.1 Chromatographic methods for the analysis of L-carnitine and its derivatives.

Sample	Separation mode	Stationary phase	Mobile phase	Derivatizing reagent	Detection	References
Plasma	Normal phase	—	Acetonitrile/water (80 : 20, v/v)	—	Tandem mass spectrometry	Talián *et al.* (2007)
Human seminal plasma	Normal phase	LiChrospher SiO_2	6.4 mM citrate buffer (pH 5) 12 mM triethanolamine acetonitrile (86:14, v/v)	*p*-Bromophenacyl bromide	Spectrophotometry; $\lambda = 260$ nm	Li and Huang (2007)
Human urine	Normal phase	—	—	3-Bromomethyl-6,7-dimethoxy-1-methyl-2(1*H*)-quinoxalinone	Fluorometry	Kamimori, Hamashima and Konishi (1994)
Dietary supplements, pharmaceuticals	Enantioseparation	C_8 Ascentis Express (150 mm × 4.6 mm; 2.7 μm)	Methanol/2.5 mM sodium 1-octane sulfonate/50 mmol/l sodium phosphate buffer (pH 3) (2.5:97.5, v/v)	—	Spectrophotometry; $\lambda = 232$ nm	De Andrés, Castañeda and Ríos (2010)

Milk-based infant formula, health-care products	Reverse phase	Inertsil ODS-3 (250 mm × 3 mm; 5 µm)	Dissolve 1115 mg ammonium acetate in about 500 ml add 1420 µl heptafluorobutyric acid (pH 4.20–4.25), transfer into a 1 l volumetric flask, add 200 ml methanol and dilute to volume with water	—	Mass spectrometry; *m/z* 162.2	Andrieux *et al.* (2008)
Food supplements	Ion-pair chromatography	Waters Nova Pak C_{18} (150 mm × 3.9 mm; 4 µm)	0.64 mmol/l sodium octanesulfonate acidified with 5.2 mmol/l trifluoroacetic acid	—	Indirect conductometry	Kakou, Megoulas and Koupparis (2005)
Biological materials	GC	Carbowax 20 M (20%): 3% H3PO4 on Chromosorb *W*	Hydrogen	—	Flame ionisation detector	Lewin, Peshin and Sklarz (1975)

Table 14.2 Mass spectrometry (MS) methods of analysis of L-carnitine and its derivatives.

Sample	Method	References
Dried plasma	Isotope dilution electrospray tandem	Osorio and Pourfarzam (2002)
Blood spot	Electrospray tandem MS	Johnson (1999)
Plasma	Isotope-dilution tandem MS	Stevens *et al.* (2000)
Plasma	LC-MS/MS	Johnson (2010)

et al., 2010) and infant formulae (Castro-Puyana *et al.*, 2009) by capillary electrophoresis, previous derivatization of carnitine and acylcarnitines was needed.

14.5.3 Enzymatic Methods

These methods take advantage of the enantioselectivity of enzymes metabolising D- or L-carnitine. Thus, they are especially useful to determine the D-carnitine content of pharmaceutical preparations or to monitor D-carnitine levels during microbial biotransformations (Table 14.3). Various methods based on D- or L-carnitine dehydrogenases have been implemented (Obón *et al.*, 1999a; Takahashi *et al.*, 1994). L-Carnitine acetyl-transferase has been exploited to quantify L-carnitine in serum and urine (Cederblad and Lindstedt, 1972; Seccombe *et al.*, 1976; Wieland, Deufel and Paetzke-Brunner, 1985).

Table 14.3 Enzymatic methods.

Sample	Enzyme	Detection	References
Bacterial culture and pharmaceutical preparation	D-Carnitine dehydrogenase	NAD^+ cycling fluorometer	Obón *et al.* (1999a)
Serum	Carnitine dehydrogenase	Thio-NADH at 415 nm UV/VIS spectrometer	Takahashi *et al.* (1994)
Serum and urine	Carnitine acetyl transferase	TNB at 550 nm UV/VIS spectrometer	Seccombe *et al.* (1976)
Tissue	Carnitine acetyl transferase	CoA-DNTB at 412 nm UV/VIS spectrometer	Alhomida *et al.* (1995)
Serum	Carnitine dehydrogenase	CoA-NAM at 362 nm fluorometer	Serdar *et al.* (2001)

14.5.4
Automated Methods

Flow Injection Analysis (FIA) of L-carnitine in pharmaceutical preparations using L-carnitine dehydrogenase covalently immobilised to Eupergit C and quantifying the NADH produced (which is proportional to the L-carnitine concentration) by fluorescence detection (Manjón, Obón and Iborra, 2000) has been reported. A Sequential Injection Analysis (SIA) system using amperometric detection was applied to determine the two enantiomers (D- and L-carnitine) (Stefan *et al.*, 2003).

14.6
Biotechnological Synthesis of L-Carnitine Microbial Metabolism of L-Carnitine and Its Regulation

14.6.1
Biotechnological Methods for L-Carnitine Production

Production of enantiomerically pure compounds challenges the chemical industry. The requirements of enantiospecific processes have an impact on overall economics (Breuer *et al.*, 2004). The chiral nature of L-carnitine makes biotechnological production an attractive alternative to chemical methods, encouraging extensive research into the microbial metabolism of L-carnitine and its derivatives (Hoeks, Kulla and Meyer, 1992; Jung, Jung and Kleber, 1993; Kleber, 1997; Naidu *et al.*, 2000).

In principle, three biological process schemes have been considered for the biotechnological production of L-carnitine: *de novo* synthesis, biological resolution of racemic mixtures and enantioselective biotransformation of achiral substrates into L-carnitine.

14.6.1.1
De novo Biosynthesis of L-Carnitine

The biosynthetic pathway in eukaryotes, which uses lysine as metabolic precursor (Vaz and Wanders, 2002), has not been observed in prokaryotes. Moreover, production with eukaryotes has several disadvantages, since only very low concentrations have been achieved in cultures of few fungal species such as *Neurospora crassa* and always using complex media (Caspi *et al.*, 2014; Kaufman and Broquist, 1977). These processes cannot compete with the current technologies since extensive strain improvement programmes would be necessary (Meyer and Robins, 2005).

14.6.1.2
Biological Resolution of Racemic Mixtures

Selective hydrolysis of racemic carnitine amide by nitrilases from *Pseudomonas* (Nakayama *et al.*, 1988), racemic acetyl-D, L-carnitine (Dropsy and Klibanov, 1984) or acetyl or betaine esters using bacterial betaine ester hydrolases (Bornscheuer *et al.*, 2002) has been reported. D-Carnitine can be racemised into L-carnitine by *Escherichia coli* O44K74 (Castellar *et al.*, 1998a). In general, these processes are not industrially attractive (Meyer and Robins, 2005).

14.6.1.3
Biotransformation from Non-Chiral Substrates

Chemical synthesis of non-chiral precursors is less expensive, and they can be transformed into L-carnitine using enzymes or whole cells. This was the strategy selected by the current worldwide L-carnitine producers to develop their own processes.

Biotechnological methods preferentially use industrial waste products as substrates. The most commonly used starting materials are achiral precursors (mostly crotonobetaine, γ-butyrobetaine and 3-dehydrocarnitine) or racemic mixtures (such as D,L-acyl-carnitine, D,L-carnitinamide and D,L-carnitine) (Cavazza, 1981; Hoeks, Kulla and Meyer, 1992; Jung, Jung and Kleber, 1993; Kleber, 1997; Kulla, 1991; Meyer and Robins, 2005; Naidu *et al.*, 2000). A great variety of microorganisms can be used for these biotransformations (Kleber, 1997).

Since the early 1980s, many companies worldwide have patented bioprocesses for L-carnitine production. We focus on the processes developed by two companies: Lonza (Switzerland) and Sigma-Tau (Italy). The bioprocesses developed for commercial production of L-carnitine by Sigma-Tau are based on the biotransformation of crotonobetaine by *E. coli* and *Proteus mirabilis* strains. Lonza (Switzerland) uses γ-butyrobetaine as starting material and a derivative of the HK4 strain, which is phylogenetically related to Rhizobium.

Production of L-Carnitine by Lonza (Switzerland) Scientists at Lonza developed a process for L-carnitine production from γ-butyrobetaine with a bacterial strain isolated from a soil sample. The pathway is analogous but not identical to fatty acid degradation. Although this HK4 strain has not been characterised, it was phylogenetically described as related to *Agrobacterium* and *Rhizobium* (Kulla, 1991). This strain was able to grow on L-carnitine as the sole source of carbon and nitrogen under aerobic conditions. The degradation of L-carnitine was blocked by frameshift mutagenesis, giving rise to a derivative strain, HK13, lacking L-carnitine dehydrogenase (Kulla and Lehky, 1985). The strain development programme followed by scientists at Lonza led to a strain with high productivity, a high precursor uptake rate and a high L-carnitine tolerance (Meyer and Robins, 2005). At the process development stage, various schemes were tested at a large scale. Volumetric productivity of continuous processes (130 g/l/day) was higher than in fed-batch culture (30 g/l/day), but biotransformation yields were lower (91% vs. 99.5%). This lower conversion resulted in additional unit operations for downstream operation to separate γ-butyrobetaine. This greatly affected overall process costs, which were 40% lower for the fed-batch process (Meyer and Robins, 2005). Moreover, compared with classical chemical processes, biotechnological production has several advantages. For instance, production by fed-batch cultivation of the *Rhizobium*-like strain developed by Lonza involves 50% less total organic waste, 25% less waste water and 90% less waste for incineration (Kulla and Lehky, 1985; Meyer and Robins, 2005; Naidu *et al.*, 2000).

Production of L-Carnitine by Enterobacteria While there is limited information on the characteristics of the HK4 strain, which was industrially protected, there is a wealth of information available on the metabolism of trimethylammonium compounds in Enterobacteria. Productivities of *E. coli* strains are not the highest reported for L-carnitine production, but the genetic and metabolic constraints controlling its biotransformations are well described (Bernal *et al.*, 2007c). Additionally, there is a strong track record in bioprocess development in *E. coli* due to its well-demonstrated capacity for high-density cultivation (Lee, 1996; Obón *et al.*, 1999b), resistance to immobilisation (Bernal *et al.*, 2007a; Obón *et al.*, 1997) and environmental stresses (Arense *et al.*, 2010; Cánovas *et al.*, 2007b). Finally, the availability of well-established molecular biology techniques allows the development of Systems Biology Bottom-Up strategies for bioprocess optimisation (Arense *et al.*, 2013; Bernal *et al.*, 2007c; Sevilla *et al.*, 2005a,b).

14.6.2 Roles of L-Carnitine in Microorganisms

The role of L-carnitine in bacteria is not clear (Kleber, 1997). Three major roles have been described: as a protectant agent, as a carbon and/or nitrogen source and as an electron acceptor.

14.6.2.1 Protectant Agent

Similar to other osmoprotectants, L-carnitine exerts several protective effects in bacteria. Microorganisms counteract the outflow of water that occurs when exposed to high-osmolarity environments by increasing their intracellular solute pool. Many bacteria accumulate in their cytosol large amounts of organic osmolytes, the so-called compatible solutes, which protect cells from water outflow and desiccation. Many bacterial species possess uptake systems with different degrees of specificity which allow them to scavenge osmoprotectants from the environment (Jung, Jung and Kleber, 1990; Kempf and Bremer, 1998; Verheul *et al.*, 1998).

In addition, in some species, betaine uptake systems have also been related to the ability to grow and survive in foods and to provoke infections *in vivo* (Angelidis and Smith, 2003; Kempf and Bremer, 1998; Sleator *et al.*, 2003).

14.6.2.2 Carbon and Nitrogen Source

Some bacterial species metabolise trimethylammonium compounds under certain conditions (Kleber, 1997). Different pathways may be involved in the metabolism of L-carnitine depending on the species and the cultivation conditions (e.g. presence of carbon and nitrogen sources, aerobic or anaerobic conditions, etc.). These catabolic pathways are induced by L-carnitine and, partly, also by other trimethylammonium compounds.

Different genera are able to degrade L-carnitine under aerobic conditions, assimilating both carbon and nitrogen in the molecule backbone. This is the case of certain *Pseudomonas* species, such as *Pseudomonas aeruginosa* A7244 and *Pseudomonas* sp. AK1, which grow aerobically on L-carnitine as the sole source of carbon and nitrogen. In these species, L-carnitine degradation starts by oxidation of the hydroxyl group with the concomitant formation of 3-dehydrocarnitine by an L-carnitine dehydrogenase (EC 1.1.1.108) (Aurich, Kleber and Schöpp, 1967). *Pseudomonas* sp. AK1 also grows on γ-butyrobetaine, which is an intermediate in the degradation pathway (Lindstedt, Lindstedt and Nordin, 1977). This pathway is similar to the eukaryotic pathway of L-carnitine biosynthesis.

Some species such as *Acinetobacter calcoaceticus* 69/V do not assimilate nitrogen from the L-carnitine skeleton and degradation occurs with stoichiometric formation of trimethylamine (Miura-Fraboni, Kleber and Englard, 1982). *A. calcoaceticus* is able to metabolise D- and L-carnitine, L-*O*-acylcarnitines and γ-butyrobetaine as the sole carbon source, but only L-carnitine induces this metabolism. Moreover, two separate transport systems for the D- and L-isomers exist in *A. calcoaceticus* ATCC 39647 (Ditullio *et al.*, 1994).

Microorganisms of the gastrointestinal tract may play a role in lowering the concentration of dietary L-carnitine (Kleber, 1997; Seim *et al.*, 1982). More recently, it has been proposed that production of trimethylamine-*N*-oxide from gut microbiota-derived trimethylamine may enhance cardiovascular risk via promoting atherosclerotic lesion development. Therefore, metabolism of L-carnitine and other trimethylamine-forming compounds such as choline and phosphatidylcholine, which are often found in large quantities in red meat, may be critical. However, a number of studies demonstrate beneficial properties for L-carnitine consumption against metabolic diseases including skeletal muscle insulin resistance and ischemic heart disease (Ussher, Lopaschuk and Arduini, 2013).

14.6.2.3
Electron Acceptor: Carnitine Respiration

Enterobacteria, such as *E. coli, Salmonella typhimurium, Proteus vulgaris* and *P. mirabilis,* do not assimilate the carbon and nitrogen skeleton of trimethylammonium compounds, but use D- and L-carnitine as electron acceptors, reducing it to γ-butyrobetaine via crotonobetaine (Kleber, 1997) (Figure 14.1). This process is frequently referred to as *carnitine respiration* (Seim *et al.*, 1982). In *E. coli,* this anaerobic process is considered a respiration variant, which sustains bacterial growth in the presence of adequate carbon and nitrogen sources. Similar anaerobic respiration processes occur on nitrate, fumarate and trimethylamine-*N*-oxide (Haddock and Jones, 1977). Although the respiration pathway is only expressed under anaerobic conditions, the biotransformation also occurs in the presence

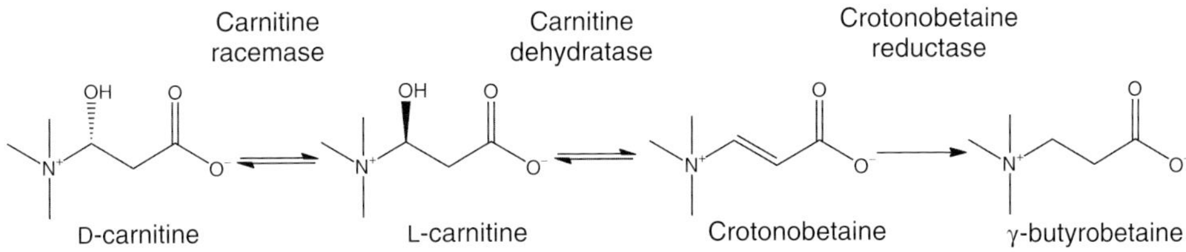

Figure 14.1 Biotransformation of trimethylammonium compounds in *E. coli*. Crotonobetaine is transformed into L-carnitine by carnitine dehydratase activity. D- and L-carnitine are transformed through carnitine racemase activity. Crotonobetaine reductase activity leads to the production of the by-product γ-butyrobetaine.

of oxygen as electron acceptor, and even in the absence of nutrients, which has been exploited for the production of L-carnitine with resting cells (Castellar *et al.*, 1998b).

14.6.3 L-Carnitine Metabolism in Enterobacteria and Its Regulation

As mentioned previously, Enterobacteria have the ability to use L-carnitine as electron acceptor. Prof. Hans-Peter Kleber's group in the University of Leipzig, Germany, analysed this pathway in detail, elucidating its genetic and biochemical constraints in *E. coli, Proteus* sp. and other bacteria.

14.6.3.1 Metabolism of L-Carnitine in E. coli

Initially, a two-step pathway was proposed to explain the conversion of L-carnitine into γ-butyrobetaine, which included two enzyme activities (Figure 14.1): L-carnitine dehydratase (CDH) (EC 4.2.1.89) and crotonobetaine reductase (CR) (EC 1.3.99) (Eichler *et al.*, 1994b; Roth *et al.*, 1994). Interconversion of the D-and L-isomers was explained by carnitine racemase activity (CRac) (Jung and Kleber, 1991).

The molecular cloning of the *cai* operon in *E. coli* (Eichler *et al.*, 1994a) demonstrated a high degree of complexity (Figure 14.2). Six ORFs were found. Gene functions were first assigned on the basis of sequence homology and experimentally demonstrated later on.

The first ORF of the operon, *caiT*, encodes a highly specific transporter working as antiporter, allowing substrate and product exchange (Jung *et al.*, 2002). The biotransformation occurs at the level of CoA derivatives (Elssner *et al.*, 2000, 2001). A betainyl-CoA ligase (EC 6.2.1) encoded by *caiC* catalyses the synthesis of the CoA derivatives of trimethylammonium compounds (Bernal *et al.*, 2008) (Figure 14.3). The initially described CDH and CR activities depend on two proteins: enoyl-CoA hydratase (EC 4.2.1.89) encoded by *caiD* and crotonobetainyl-CoA reductase (EC 1.3.99) encoded by *caiA*. Both enzymes need the crotonobetainyl-CoA:carnitine CoA transferase (EC 2.8.3) encoded by

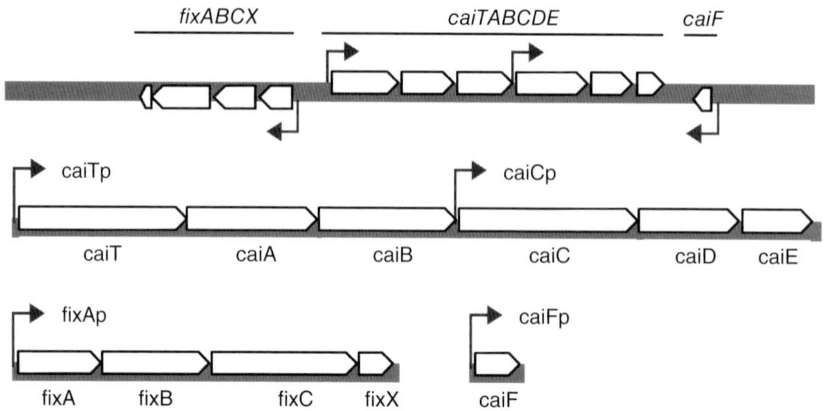

Figure 14.2 Genetic organisation of carnitine-metabolising genes in *E. coli*. Two structural operons are expressed from the same promoter/operator region. Carnitine-metabolising enzymes are encoded by *caiTABCDE* operon. The *fixABCX* operon-encoded proteins are necessary for crotonobetaine reduction. A seventh ORF located at the 3′-end of the *cai* operon, *caiF*, encodes for a transcriptional regulator which specifically regulates the expression of *cai* and *fix* operons.

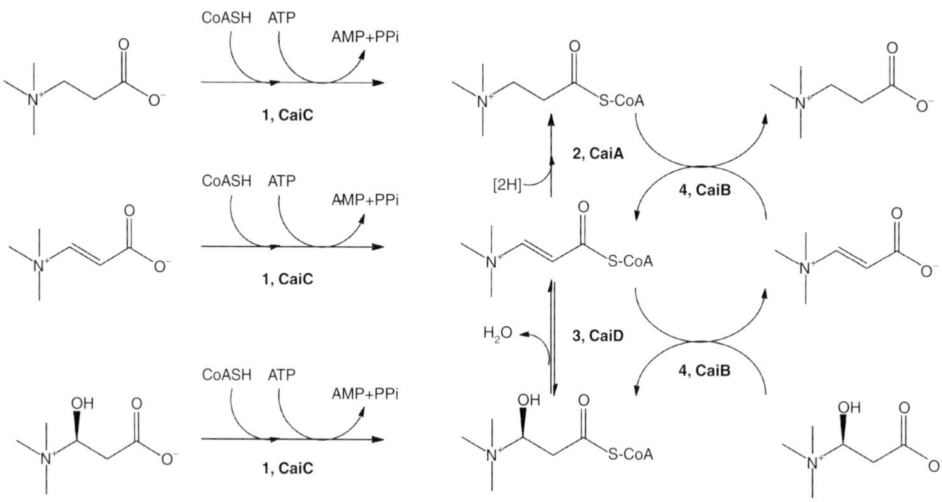

Figure 14.3 Metabolism of trimethylammonium compounds in *Escherichia coli*. Trimethylammonium compounds γ-butyrobetaine, crotonobetaine and L-carnitine are activated to form the corresponding thioesters, γ-butyrobetainyl-CoA, crotonobetainyl-CoA and L-carnitinyl-CoA. The reaction proceeds at the expense of ATP hydrolysis (1, betaine:CoA ligase, CaiC). Crotonobetainyl-CoA is reduced to γ-butyrobetainyl-CoA (2, crotonobetainyl-CoA reductase, CaiA). Crotonobetainyl-CoA is enantiospecifically hydrated to L-carnitinyl-CoA (3, carnitinyl-CoA dehydratase, CaiD). The coenzyme A moiety is exchanged from products to substrates (4, crotonobetainyl-CoA:L-carnitine coenzyme A transferase, CaiB). See the main text for details.

caiB, to recycle the CoA moiety between products and substrates of the biotransformation (Elssner *et al.*, 2001) (Figure 14.3). The only gene whose function remains unconfirmed is *caiE*, although early studies observed an activation of the CaiD/CaiB function (CDH activity), suggesting a possible role in the metabolism of CoA derivatives (Eichler *et al.*, 1994a).

A seventh ORF, the *caiF* gene, was found downstream of the *cai* operon (Eichler *et al.*, 1996) (Figure 14.2). The product is a transcriptional factor which regulates the expression of the *cai* operon genes.

The *cai* operon is highly regulated by both general and specific transcriptional regulators which respond to environmental signals. Transcription of the *cai* operon is induced during anaerobic growth in the presence of L-carnitine (Buchet, Eichler and Mandrand-Berthelot, 1998). The activator of carbon catabolic operons, the cAMP Receptor Protein (CRP) and the global regulator of anaerobiosis (FNR) are required for induction, while the histone-like H-NS protein and the σ^S factor (RpoS), which are involved in gene regulation in the stationary phase and in stress responses, exert a repressive effect on carnitine metabolism (Buchet *et al.*, 1999; Eichler *et al.*, 1994a). CRP and FNR bind to the promoter region of both the *caiTABCDE* and *caiF* transcriptional units. In *E. coli*, *caiF* is expressed under anaerobic conditions and in the absence of glucose. In the presence of L-carnitine, a *caiF* dimer binds to the promoter region of the *cai* operon, inducing its expression (Buchet *et al.*, 1999; Buchet, Eichler and Mandrand-Berthelot, 1998).

The four ORFs' operon located at the 5′ end of the *cai* locus in *E. coli* is co-transcribed from the same promoter/operator region (Eichler *et al.*, 1995). Sequence homology studies demonstrated significant sequence homology with polypeptides encoded by the *fix*ABCX operon from *Azorhizobium caulinodans* and *Rhizobium meliloti* and were therefore named *fix*. This operon is involved in electron transfer to crotonobetainyl-CoA reductase, and deletion studies demonstrated that it is essential for the reduction of L-carnitine to γ-butyrobetaine (Walt and Kahn, 2002).

14.6.3.2
Metabolism of L-Carnitine in Proteus sp.

This biotransformation pathway also exists in other Enterobacteria. This is the case of *Proteus* sp. in which molecular characterisation of the *cai* operon revealed a high conservation (Engemann *et al.*, 2005). Two are the most remarkable characteristics of the metabolism of L-carnitine in *Proteus* sp. Contrary to what occurs in *E. coli*, expression of *cai* operon is aerobic (Engemann and Kleber, 2001). Aerobic metabolism of L-carnitine has also been described in other Enterobacteria (Elssner *et al.*, 1999). It is worth mentioning that *caiF* is the gene having the lowest homology between both species and that alterations in its promoter region have been proposed as being responsible for the very different regulation of this pathway in *Proteus* sp. (Engemann *et al.*, 2005). Second, this bacterium transforms crotonobetaine into L-carnitine with high yields, the amount of γ-butyrobetaine produced being almost negligible. This occurs even though the bacteria possess the crotonobetainyl-CoA reductase activity encoded by *caiA*. In fact, low amounts of this compound are found in cell extracts (Engemann, Elssner and Kleber, 2001) and in high-density cultures (probably as a result of cell decay) (Cánovas *et al.*, 2003b). It has been proposed that this is due to the inability of *caiT* to antiport γ-butyrobetaine outside of the cell.

14.6.4
Expression of Metabolising Activities: Effect of Inducers, Oxygen and Substrates

The metabolism of trimethylammonium compounds in Enterobacteria is induced by crotonobetaine and D- or L-carnitine (Kleber, 1997). As mentioned previously, expression of carnitine metabolism is anaerobic in *E. coli* and *Salmonella,* and aerobic in other Enterobacteria, such as *Proteus* sp. (Elssner *et al.*, 1999).

The two major enzyme activities of L-carnitine metabolism (CDH and CR) are co-expressed under proper growth conditions (Figure 14.1).

When induced cells are incubated with D- or L-carnitine, this is transformed to γ-butyrobetaine, while when incubated with crotonobetaine, a mixture of L-carnitine and γ-butyrobetaine is produced (Bernal *et al.*, 2007c; Kleber, 1997). This drawback of natural carnitine metabolism can be avoided by inhibiting CR with alternative electron acceptors such as fumarate or oxygen. This is precisely the reason why the use of aerobic resting cells leads to important productivity improvements (Castellar *et al.*, 1998b, 2001; Miura-Fraboni, Kleber and Englard, 1982). The resting cell strategy consists in the separation of the metabolic induction and biotransformation phases. In a first step, cells are grown under conditions optimal for the full expression of the metabolism to be used. Then, cells are transferred to a 'biotransformation medium', which typically consists of a buffered solution of the biotransformation substrate. In the absence of nutrients, cells cannot grow and all cellular resources are allocated for the biotransformation. In the case of carnitine metabolism in *E. coli,* best results were obtained when carnitine metabolism was expressed anaerobically and induced cells were subsequently used in an aerobic resting biotransformation medium. In the presence of oxygen, the formation of the side product γ-butyrobetaine is avoided, since crotonobetainyl-CoA reductase activity is inhibited (Castellar *et al.*, 1998b, 2001).

14.6.5
Biotransformation with D-Carnitine or Crotonobetaine as Substrates

Both D-carnitine and crotonobetaine can be used as biotransformation substrates. Nevertheless, the different levels of CDH and CRac activities explain why the highest productivities are achieved with crotonobetaine (Cánovas *et al.*, 2005). In addition, the production of L-carnitine from D-carnitine is only possible under aerobic or microaerobic conditions with growing or resting cells, since the reduction of D,L-carnitine to γ-butyrobetaine is preferred under anaerobic conditions (Cánovas *et al.*, 2005; Castellar *et al.*, 1998b).

Increased productivity was obtained using a genetically modified strain overproducing the multifunctional protein CaiD, which is part of both CDH and CRac activities (Castellar *et al.*, 2001).

14.6.6 Transport Phenomena for L-Carnitine Production

The kinetics of membrane transport processes often limit the actual biotransformation rates in processes with whole cells (Burkovski and Krämer, 2002). Substrate uptake and product efflux consist of diffusion through the outer membrane (OM) and the cellular envelope and active or facilitated transport through cellular membrane. Various strategies have allowed engineering transport phenomena in microbial processes for L-carnitine production.

14.6.6.1 Membrane Permeabilisation

Membrane integrity is necessary for maintaining the chemical independence of the cell from the extracellular medium and is essential for cell survival. Nevertheless, in bioprocesses in which growth and biotransformation are uncoupled, cell proliferation is not necessary to maintain productivity. In fact, permeabilisation may increase the efficiency of cells, considered as biotransformation units with integrated routes for cofactor regeneration. Cell permeabilisation greatly affects membrane structure, even provoking leakage of cellular components (Cánovas, Torroglosa and Iborra, 2005; Flores, Voget and Ertola, 1994).

Gram-negative bacteria possess a mechanically strong cell envelope made up largely of peptidoglycan, which gives the cells shape and protects them from sudden changes in osmolarity. Outside this layer, there is an OM, with a mainly protective role and which is less selective and more permeable than the cytoplasmic membrane. The OM is an efficient barrier against hydrophilic macromolecules and hydrophobic substances due to a lipopolysaccharide layer. Permeabilisers, such as detergents (Triton X-100 and Tween), EDTA, organic solvents (toluene, lactic acid and alcohols) and certain polycationic substances (such as polyethylenimine, polymyxin and its derivatives, polylysines and protamine) disturb the integrity of cellular envelop (Ayres, Furr and Russell, 1999; Cánovas, Torroglosa and Iborra, 2005; Felix, 1982).

Permeabilisation of *E. coli* and *Proteus* sp. cells disrupts cell wall and increases L-carnitine yields by more than 50% in growing or resting biotransformation media (Cánovas and Iborra, 2005). For *E. coli* O44K74, polyethylenimine (PEI) was the best permeabiliser, with an almost 100% increase in yield (Cánovas, Torroglosa and Iborra, 2005). For *Proteus* sp., Triton X-100 resulted in higher conversion and productivity values than those of the control (Cánovas and Iborra, 2005). Permeability of *E. coli* can be engineered by deletion of *lpp* gene, which encodes Braun's lipoprotein, a major component of the OM. This mutation increases the rate of whole cell-catalysed reactions in which substrate diffusion is a limiting factor (Ni, Reye and Chen, 2007).

14.6.6.2
Osmotic Stress Induction of Transporters

L-Carnitine, as well as other betaines, are osmoprotectants, and bacteria accumulate them in their cytosol in response to high osmolarity of the growth medium (Arense *et al.*, 2010; Cánovas *et al.*, 2003c, 2007b; Csonka and Hanson, 1991; Verheul *et al.*, 1998). Under osmotic stress, the expression of several transport systems devoted to the uptake of compatible solutes, such as ProU and ProP, is induced (Cánovas *et al.*, 2003c; Kempf and Bremer, 1998; Verheul *et al.*, 1998). These active transport systems are in contrast with *caiT*, the transporter associated to carnitine pathway, which is a substrate facilitator functioning as antiporter (Jung *et al.*, 2002), and is not involved in stress adaptation (Verheul *et al.*, 1998).

The adaptations to osmotic stress improve L-carnitine production in resting cells. The biotransformation of crotonobetaine is enhanced when carnitine metabolism is induced under standard conditions and the biotransformation is carried out in 0.5 M NaCl. Biotransformation yields reached 85–90% with the recombinant *E. coli* pT7-5KE32 strain (Cánovas *et al.*, 2003c).

14.6.6.3
Overexpression of the Transporter caiT

When the *caiT* gene, encoding a carnitine/γ-butyrobetaine antiporter was overexpressed in *E. coli* LMG194 (a standard laboratory strain), an almost threefold increase in L-carnitine molar yield was obtained with both growing and resting cells (Cánovas *et al.*, 2007a). However, the level of production of L-carnitine in the engineered strain was still far from that of the wild-type *E. coli* O44K74 strain. Thus, although transport limits L-carnitine production, it is not the main factor controlling the biotransformation in laboratory strains.

14.6.7
Metabolic Engineering for High-Yielding L-Carnitine Producing Strains

Metabolic engineering is the purposeful modification of intermediary metabolism using recombinant DNA technology (Nielsen, 2001). Although molecular biology is the workhorse of metabolic engineers, the choice of proper cultivation techniques, ensuring the availability of substrates or avoiding the inhibitory effects of products, may confine the cells to well-defined physiological states and may also be considered part of this discipline. Quantitative description of growth and metabolic kinetics, detailed metabolic and physiological description at the enzyme, proteomic and transcriptomic levels and metabolic modelling are key tools for the rational design of strain optimisation strategies.

14.6.7.1
Link between Central and Secondary Metabolism during Biotransformation

Although metabolic engineering usually deals with the redirection of fluxes for complex pathways related with carbon and nitrogen central metabolism, its methods and concepts are also useful to improve a biotransformation performed by secondary metabolism. Understanding the link of the biotransformation of trimethylammonium compounds with central metabolism provides clues of which are the main factors to be considered for strain engineering. Carnitine and central carbon metabolisms are connected at the level of cofactors (Cánovas *et al.*, 2003a; Elssner *et al.*, 2000). Detailed analysis of metabolic alterations under biotransformation conditions with resting and growing cells demonstrated that ATP turnover, the redox cellular state and the regeneration of the cofactors required in the biotransformation correlate with biotransformation yields (Cánovas *et al.*, 2003a). This is similar to what has been observed in other processes (Lopez de Felipe *et al.*, 1998; San *et al.*, 2002).

Metabolic pathways associated with the maintenance of the acetyl-CoA/CoA ratio were analysed as indicators of the metabolic state of the cell. The TCA cycle, the glyoxylate shunt and acetate metabolism are closely interrelated and exercise a control on the biotransformation efficiency (Bernal *et al.*, 2007b; Cánovas *et al.*, 2003a). Similar conclusions were drawn from the analysis of the dynamic response to metabolic perturbations affecting the level of carbon sources, biotransformation substrate, salinity or electron acceptors. This underlines that ATP levels and the metabolism of acetyl-CoA are highly dependent on cellular metabolism under production conditions (Cánovas *et al.*, 2006, 2007b).

Metabolic Flux Analysis further supported the importance of the cellular energetic state for the biotransformation (Sevilla *et al.*, 2005a). ATP is devoted to the activation of trimethylammonium compounds by the betaine:CoA ligase CaiC (Bernal *et al.*, 2008), but it is also expedited because of a feasible futile cycle which was detected through the analysis of the network's topology. This cycle is the consequence of the simultaneous operation of the two trimethylammonium compound carriers, *CaiT* and *ProU* (Sevilla *et al.*, 2005a; Verheul *et al.*, 1998), which leads to energy dissipation. Cofactor engineering is defined as the purposeful modification of the availability of cofactors involved in a certain metabolic pathway (San *et al.*, 2002). In the case of L-carnitine metabolism, overexpression of the enzymes involved in substrate activation (betaine:CoA ligase, CaiC) and cofactor recycling (crotonobetainyl-CoA:L-carnitine CoA-transferase, CaiB) demonstrated that activation of substrates limited the biotransformation rate (Bernal *et al.*, 2007b, 2008).

14.6.7.2 Metabolic Engineering for Strain Engineering: Feedback between Modelling and Experimental Analysis of Cell Metabolism

Strain engineering for enhanced L-carnitine production was guided by mathematical modelling of cellular metabolism. Several approaches exist to model cellular metabolism. The best modelling strategy is chosen on the basis of the type of data and knowledge available and the biological questions to be answered.

An unstructured model of L-carnitine production with *E. coli* strains described the effect of oxygen on substrate consumption and on the expression of L-carnitine metabolism during bioprocess (Cánovas *et al.*, 2002). The model was applicable to continuous and batch bioprocesses, both with growing and resting cells, representing the link between cellular metabolic productivity and the macrokinetics of the material mass balance for the reactor. This simple model was the basis to establish an S-system description of the cell–bioreactor combined system, which could, in turn, be used to optimise the biotechnological set-up (Álvarez-Vásquez *et al.*, 2002). Using optimisation techniques, process operation and cellular variables such as the dilution rate and the initial crotonobetaine concentration in continuous cell recycle cultures and the CDH activity of cells were identified as critical factors to maximise L-carnitine production. In fact, a 74% increase in L-carnitine production rate was experimentally assessed, which was in close agreement with the predictions of the model (Álvarez-Vásquez *et al.*, 2002).

The application of optimisation strategies to unstructured models can only predict improvement strategies regarding the cultivation method. A connection with central and secondary metabolism was still needed to fully describe the activity of the microorganism. A combined model of reactor and metabolism linking the macrokinetics of the reactor with microkinetics of cellular processes was developed and genetic engineering strategies were identified based on this model (Sevilla *et al.*, 2005b). The optimised solution suggested the overexpression of the carnitine transferase activity, *caiB* and the protein carrier, *caiT* as the main targets to improve the L-carnitine production rate, predicting an enhancement of up to three times the initial productivity (Sevilla *et al.*, 2005b). In fact, overexpression of *caiB* and *caiT* proteins in the low-producing *E. coli* LMG194 laboratory strain increased L-carnitine production threefold to fourfold (Cánovas *et al.*, 2007a). Nevertheless, overexpression of *caiC* enhanced L-carnitine production 50-fold in *E. coli* LMG194 (a 2–3-fold increase compared with the overproducing *E. coli* O44K74 strain) (Bernal *et al.*, 2007b, 2008). This effect was not predicted by this model. The main reason was that, at the time the model was built, the kinetic characteristics of this protein were not known. Meaningful models strongly depend on the completeness and goodness of the data available. A continuous feedback between *in silico* and *in vivo* experimentation is necessary for the application of Metabolic Engineering and Systems Biology approaches to living systems (Cánovas *et al.*, 2007a). In this context, analysing the effect of deletion of central metabolism genes on L-carnitine production (Bernal *et al.*, 2007b) allowed improving our understanding of the link between central and carnitine metabolisms, a crucial aspect in the cyclic process of model refining.

Finally, a production strain was engineered for high-yielding L-carnitine production in both growing and resting cell systems. Two gene deletions were performed, which led to increasing flux through the C cycle and eliminating the major side reaction for the transformation of crotonobetaine into L-carnitine. This was accomplished by deletion of *aceK* (which encodes a bifunctional kinase/phosphatase which inhibits isocitrate dehydrogenase activity) and deletion of *caiA* (which encodes a crotonobetainyl-CoA reductase, preventing formation of γ-butyrobetaine). Both mutations improved L-carnitine production by 20% and 42%, respectively. In addition, the expression of the *cai* operon gene products was enhanced by substituting the highly regulated promoter by a constitutive synthetic promoter (Arense *et al.*, 2013). This strategy allowed the aerobic expression of carnitine metabolism in *E. coli*, although best results were obtained under anaerobic conditions.

Best results were obtained in a triple mutant strain carrying all three modifications. Resting cells of the BW △*aceK* △*caiA* p37*cai* strain produced 59.6 mmol/l/h of L-carnitine, doubling the productivity of the wild-type strain. In addition, almost total conversion was attained in less than 2 h without concomitant production of the side product γ-butyrobetaine, thus improving downstream processing. These metabolic engineering strategies yielded a robust and high-yielding *E. coli* strain. So far, this is the best performing L-carnitine production *E. coli* strain ever reported.

14.7 Other Methods for L-Carnitine Production: Extraction from Natural Sources and Chemical Synthesis

14.7.1 Isolation of L-Carnitine from Natural Sources

Although animal meat (muscles) is rich in L-carnitine, extraction from natural sources is not a commercially viable option (Meyer and Robins, 2005).

14.7.2 Chemical Synthesis

Much research effort has been focused on the development of methods for the industrial-scale production of L-carnitine. Racemic D,L-carnitine is obtained by chemical synthesis from cheap raw materials such as epichlorohydrin and trimethylamine. The racemate is separated by fractionated crystallisation (Voeffray *et al.*, 1987).

Chemical synthesis of enantiomerically pure L-carnitine and biological racemic resolution are poorly attractive alternatives for industrial-scale L-carnitine production. Numerous chemical procedures can be found in the literature involving asymmetric synthesis (Giannessi *et al.*, 1994; Kitamura *et al.*, 1988; Kolb, Bennani and Sharpless, 1993); chemical multi-step racemisation (Giannessi *et al.*, 1994); resolution through diastereoisomeric derivatives (Cavazza, 1981; Jakob, Hutchmacher and Klenk, 1993; Voeffray *et al.*, 1987); entrainment crystallisation of chemically synthesised precursors (Pallavicini *et al.*, 2008); microbiological or enzymatic techniques (Hashiguchi, Kawada and Natsugari, 1992; Jung, Jung and Kleber, 1993; Kasai and Sakaguchi, 1992; Nakahama *et al.*, 1995; Tian, Wang and Zhang, 2009) and the use of chiral starting materials (Bellamy, Bondoux and Dodey, 1990; Bols, Lundt and Pedersen, 1992; Takano *et al.*, 1987). For instance, the method developed by Bellamy, Bondoux and Dodey (1990) consisted of six steps which, using as starting material (*R*)- and (*S*)-malic acid, respectively, specifically obtained both enantiomers. More recently, Marzi *et al.* (2000) described an enantioselective synthesis using achiral glycerol as starting material and a chiral auxiliary. However, few of these chemical procedures are of practical use on an industrial scale because of the number of steps and the need to use chiral starting materials or chiral auxiliaries. The classical industrial method for the synthesis of carnitine generates a racemic mixture, with D-carnitine as waste product (Cavazza, 1981; Voeffray *et al.*, 1987).

Acknowledgement

This work was supported by MICCIN BIO2011-29233-C02-01, CTQ2012-33717, MINECO BIO2014-54411-C2-1-R (including FEDER funds) and partially by Seneca Foundation CARM 029287PI/05.

References

Alhomida, A.S., Duhaiman, A.S., Al-Jafari, A.A., and Junaid, M.A. (1995) Determination of L-carnitine, acylcarnitine and total carnitine levels in plasma and tissues of camel (Camelus dromedarius). *Comp. Biochem. Physiol. Part B: Biochem. Mol. Biol.*, **111**, 441–445.

Álvarez-Vásquez, F., Cánovas, M., Iborra, J.L., and Torres, N.V. (2002) Modeling, optimization and experimental assessment of continuous L(-)-carnitine production by *Escherichia coli* cultures. *Biotechnol. Bioeng.*, **80**, 794–805.

Andrieux, P., Kilinc, T., Perrin, C., and Campos-Giménez, E. (2008) Simultaneous determination of free carnitine and total choline by liquid chromatography/mass spectrometry in infant formula and health-care products: single-laboratory validation. *J. AOAC Int.*, **91**, 777–785.

Angelidis, A.S. and Smith, G.M. (2003) Role of the glycine betaine and carnitine transporters in adaptation of Listeria monocytogenes to chill stress in defined medium. *Appl. Environ. Microbiol.*, **69**, 7492–7498.

Arense, P., Bernal, V., Charlier, D., Iborra, J.L., Foulquié-Moreno, M.R., and Cánovas, M. (2013) Metabolic engineering for high yielding L(-)-carnitine production in *Escherichia coli. Microb. Cell Fact.*, **12**, 56.

Arense, P., Bernal, V., Iborra, J.L., and Cánovas, M. (2010) Metabolic adaptation of *Escherichia coli* to long-term exposure to salt stress. *Process Biochem.*, **45**, 1459–1467.

Aurich, H., Kleber, H.P., and Schöpp, W.D. (1967) An inducible carnitine dehydrogenase from *Pseudomonas aeruginosa. Biochim. Biophys. Acta*, **139**, 505–507.

Ayres, H.M., Furr, J.R., and Russell, A.D. (1999) Effect of permeabilizers on antibiotic sensitivity of *Pseudomonas aeruginosa. Lett. Appl. Microbiol.*, **28**, 13–16.

Bellamy, F.D., Bondoux, M., and Dodey, P. (1990) A new, short and efficient synthesis of both enantiomers of carnitine. *Tetrahedron Lett.*, **31**, 7323–7326.

Bene, J., Komlósi, K., Havasi, V., Talián, G., Gasztonyi, B., Horváth, K., Mózsik, G., Hunyady, B., Melegh, B., and Figler, M. (2006) Changes of plasma fasting carnitine ester profile in patients with ulcerative colitis. *World J. Gastroenterol.*, **12**, 110–113.

Bernal, V., Arense, P., Blatz, V., Mandrand-Berthelot, M.A., Cánovas, M., and Iborra, J.L. (2008) Role of betaine:CoA ligase (CaiC) in the activation of betaines and the transfer of coenzyme A in *Escherichia coli. J. Appl. Microbiol.*, **105**, 42–50.

Bernal, V., González-Veracruz, M., Cánovas, M., and Iborra, J.L. (2007a) Plasmid maintenance and physiology of a genetically engineered *Escherichia coli* strain during continuous L-carnitine production. *Biotechnol. Lett.*, **29**, 1549–1556.

Bernal, V., Masdemont, B., Arense, P., Cánovas, M., and Iborra, J.L. (2007b) Redirecting metabolic fluxes through cofactor engineering: role of CoA-esters pool during L(-)-carnitine production by *Escherichia coli. J. Biotechnol.*, **132**, 110–117.

Bernal, V., Sevilla, A., Cánovas, M., and Iborra, J.L. (2007c) Production of L-carnitine by secondary metabolism of bacteria. *Microb. Cell Fact.*, **6**, 31.

Bols, M., Lundt, I., and Pedersen, C. (1992) Simple synthesis of (R)-carnitine from D-galactono-1,4-lactone. *Tetrahedron*, **48**, 319–324.

Bornscheuer, U., Musidlowska, A., Werlen, J., Zimmermann, T. and Tscherry, B. (2002) Microbiological method for producing L-carnitine. PCT Application WO2002061094.

Bremer, J. (1962) Carnitine in intermediary metabolism. The metabolism of fatty acid esters of carnitine by mitochondria. *J. Biol. Chem.*, **237**, 3628–3632.

Bremer, J. (1963) Carnitine in intermediary metabolism. The biosynthesis of palmitylcarnitine by cell subfractions. *J. Biol. Chem.*, **238**, 2774–2779.

Breuer, M., Ditrich, K., Habicher, T., Hauer, B., Kesseler, M., Stürmer, R., and Zelinski, T. (2004) Industrial methods for the production of optically active intermediates. *Angew. Chem. Int. Ed.*, **43**, 788–824.

Buchet, A., Eichler, K., and Mandrand-Berthelot, M.A. (1998) Regulation of the carnitine pathway in *Escherichia coli*: investigation of the cai-fix divergent promoter region. *J. Bacteriol.*, **180**, 2599–2608.

Buchet, A., Nasser, W., Eichler, K., and Mandrand-Berthelot, M.A. (1999) Positive co-regulation of the *Escherichia coli* carnitine pathway cai and fix operons by CRP and the CaiF activator. *Mol. Microbiol.*, **34**, 562–575.

Burkovski, A. and Krämer, R. (2002) Bacterial amino acid transport proteins: occurrence, functions, and significance for biotechnological applications. *Appl. Microbiol. Biotechnol.*, **58**, 265–274.

Cánovas, M., Bernal, V., González, M., Kleber, H.P., and Iborra, J.L. (2005) Factors affecting the biotransformation of trimethylammonium compounds into L-carnitine by *Escherichia coli. Biochem. Eng. J.*, **26**, 145–154.

Cánovas, M., Torroglosa, T., and Iborra, J.L. (2005) Permeabilization of *Escherichia coli* cells in the biotransformation of trimethylammonium compounds into L-carnitine. *Enzyme Microb. Technol.*, **37**, 300–308.

Cánovas, M., Bernal, V., Sevilla, A., and Iborra, J.L. (2007a) Role of wet experiment design in data generation: from in vivo to in silico and back. *Silico Biol.*, **7**, S3–S16.

Cánovas, M., Bernal, V., Sevilla, A., Torroglosa, T., and Iborra, J.L. (2007b) Salt stress effects on the central and carnitine metabolisms of *Escherichia coli. Biotechnol. Bioeng.*, **96**, 722–737.

Cánovas, M., Bernal, V., Torroglosa, T., Ramirez, J.L., and Iborra, J.L. (2003a) Link between primary and secondary metabolism in the biotransformation of trimethylammonium compounds by *Escherichia coli. Biotechnol. Bioeng.*, **84**, 686–699.

Cánovas, M., Maiquez, J., Diego, T., Buendia, B., Espinosa, G., and Iborra, J.L. (2003b) Membrane cell retention systems for continuous production of L-carnitine using Proteus sp. *J. Membr. Sci.*, **214**, 101–111.

Cánovas, M., Torroglosa, T., Kleber, H.P., and Iborra, J.L. (2003c) Effect of osmotic stress in the uptake and biotransformation of crotonobetaine and D-carnitine into L-carnitine by resting cells of *Escherichia coli. J. Basic Microbiol.*, **43**, 259–268.

Cánovas, M. and Iborra, J.L. (2005) Whole cell biocatalysts stabilization for L-carnitine production. *Biocatal. Biotransform*, **23**, 149–158.

Cánovas, M., Máiquez, J.R., Obón, J.M., and Iborra, J.L. (2002) Modeling of the biotransformation of crotonobetaine into L(-)-carnitine by *Escherichia coli* strains. *Biotechnol. Bioeng.*, **77**, 764–775.

Cánovas, M., Sevilla, A., Bernal, V., Leal, R., and Iborra, J.L. (2006) Role of energetic coenzyme pools in the production of L-carnitine by *Escherichia coli. Metab. Eng.*, **8**, 603–618.

Carter, H.E., Bhattacharyya, P.K., Weidman, K.R., and Fraenkel, G. (1952) Chemical studies on vitamin BT isolation and characterization as carnitine. *Arch. Biochem. Biophys.*, **38**, 405–416.

Caspi, R., Altman, T., Billington, R., Dreher, K., Foerster, H., Fulcher, C.A., Holland, T.A., Keseler, I.M., Kothari, A., Kubo, A., Krummenacker, M., Latendresse, M., Mueller, L.A., Ong, Q., Paley, S., Subhraveti, P., Weaver, D.S., Weerasinghe, D., Zhang, P., and Karp, P.D. (2014) The MetaCyc database of metabolic pathways and enzymes and the BioCyc collection of pathway/genome databases. *Nucleic Acids Res.*, **42**, D459–D471.

Castellar, M.R., Canovas, M., Kleber, H.P., and Iborra, J.L. (1998a) Biotransformation of D(+)-carnitine by resting cells of *Escherichia coli* O44K74. *J. Appl. Microbiol.*, **85**, 883–890.

Castellar, M.R., Cánovas, M., Kleber, H.P., and Iborra, J.L. (1998b) Biotransformation of D(+)-carnitine into L(-)-carnitine by resting cells of *Escherichia coli* O44 K74. *J. Appl. Microbiol.*, **85**, 883–890.

Castellar, M.R., Obón, J.M., Marán, A., Cánovas, M., and Iborra, J.L. (2001) L(-)-carnitine production using a recombinant *Escherichia coli* strain. *Enzyme Microb. Technol.*, **28**, 785–791.

Castro-Puyana, M., García-Ruiz, C., Crego, A.L., and Marina, M.L. (2009) Development of a CE-MS(2) method for the enantiomeric separation of L/D-carnitine: application to the analysis of infant formulas. *Electrophoresis*, **30**, 337–348.

Cavazza, M. (1981) D-camphorate of L-carnitinamide and D-camphorate of D-carnitinamide. BE Patent 877609A1.

Cederblad, G. and Lindstedt, S. (1972) A method for the determination of carnitine in the picomole range. *Clin. Chim. Acta*, **37**, 235–243.

Csonka, L.M. and Hanson, A. (1991) Prokaryotic osmoregulation: genetics and physiology. *Annu. Rev. Microbiol.*, **45**, 569–606.

Dbrowska, M. and Starek, M. (2014) Analytical approaches to determination of carnitine in biological materials, foods and dietary supplements. *Food Chem.*, **142**, 220–232.

De Andrés, F., Castañeda, G., and Ríos, A. (2010) Achiral liquid chromatography with circular dichroism detection for the determination of carnitine enantiomers in dietary supplements and pharmaceutical formulations. *J. Pharm. Biomed. Anal.*, **51**, 478–483.

Ditullio, D., Anderson, D., Chen, C.S., and Sih, C.J. (1994) L-carnitine via enzyme-catalyzed oxidative kinetic resolution. *Bioorg. Med. Chem.*, **2**, 415–420.

Dropsy, E.P. and Klibanov, A.M. (1984) Cholinesterase-catalyzed resolution of D,L-carnitine. *Biotechnol. Bioeng.*, **26**, 911–915.

Drynan, L., Quant, P.A., and Zammit, V.A. (1996) Flux control exerted by mitochondrial OM carnitine palmitoyltransferase over beta-oxidation, ketogenesis and tricarboxylic acid cycle activity in hepatocytes isolated from rats in different metabolic states. *Biochem. J.*, **317** (Pt. 3), 791–795.

Duran, M., Loof, N.E., Ketting, D., and Dorland, L. (1990) Secondary carnitine deficiency. *J. Clin. Chem. Clin. Biochem.*, **28**, 359–363.

Eaton, S., Bartlett, K., and Pourfarzam, M. (1996) Mammalian mitochondrial beta-oxidation. *Biochem. J.*, **320**, 345–357.

Eichler, K., Bourgis, F., Buchet, A., Kleber, H.P., and Mandrand-Berthelot, M.A. (1994a) Molecular characterization of the cai operon necessary for carnitine metabolism in *Escherichia coli. Mol. Microbiol.*, **13**, 775–786.

Eichler, K., Schunck, W.H., Kleber, H.P., and Mandrand-Berthelot, M.A. (1994b) Cloning, nucleotide sequence, and expression of the *Escherichia coli* gene encoding carnitine dehydratase. *J. Bacteriol.*, **176**, 2970–2975.

Eichler, K., Buchet, A., Bourgis, F., Kleber, H.P., and Mandrand-Berthelot, M.A. (1995) The fix *Escherichia coli* region contains four genes related to carnitine metabolism. *J. Basic Microbiol.*, **35**, 217–227.

Eichler, K., Buchet, A., Lemke, R., Kleber, H.P., and Mandrand-Berthelot, M.A. (1996) Identification and characterization of the caiF gene encoding a potential transcription activator of carnitine metabolism in *Escherichia coli. J. Bacteriol.*, **178**, 1248–1257.

Elssner, T., Engemann, C., Baumgart, K., and Kleber, H.P. (2001) Involvement of coenzyme A esters and two new enzymes, an enoyl-CoA hydratase and a CoA-transferase, in the hydration of crotonobetaine to L-carnitine by *Escherichia coli*. *Biochemistry*, **40**, 11140–11148.

Elssner, T., Hennig, L., Frauendorf, H., Haferburg, D., Kleber, H.P., and Henning, L. (2000) Isolation, identification and synthesis of butyrobetanyl-CoA and crotonobetainyl-CoA, compounds involved in carnitine metabolism of *E. coli*. *Biochemistry*, **39**, 10761–10769.

Elssner, T., Preusser, A., Wagner, U., and Kleber, H.P. (1999) Metabolism of L(-)-carnitine by Enterobacteriaceae under aerobic conditions. *FEMS Microbiol. Lett.*, **174**, 295–301.

Engemann, C., Elssner, T., and Kleber, H.P. (2001) Biotransformation of crotonobetaine to L(-)-carnitine in Proteus sp. *Arch. Microbiol.*, **175**, 353–359.

Engemann, C., Elssner, T., Pfeifer, S., Krumbholz, C., Maier, T., and Kleber, H. (2005) Identification and functional characterisation of genes and corresponding enzymes involved in carnitine metabolism of *Proteus sp. Arch. Microbiol.*, **183**, 176–189.

Engemann, C. and Kleber, H.P. (2001) Epigenetic regulation of carnitine metabolising enzymes in Proteus sp. under aerobic conditions. *FEMS Microbiol. Lett.*, **196**, 1–6.

Felix, H. (1982) Permeabilized cells. *Anal. Biochem.*, **120**, 211–234.

Flores, M.V., Voget, C.E., and Ertola, J.J. (1994) Permeabilization of yeast cells (*Kluyveromyces lactis*) with organic solvents. *Enzyme Microb. Technol.*, **16**, 340–346.

Fraenkel, G. (1948) Bt, a new vitamin of the B-group and its relation to the folic acid group, and other anti-anaemia factors. *Nature*, **161**, 981–983.

Fritz, I. (1955) The effect of muscle extracts on the oxidation of palmitic acid by liver slices and homogenates. *Acta Physiol. Scand.*, **34**, 367–385.

Giannessi, F., Castagnani, R., Tinti, M.O., De Angelis, F., De Witt, P. and Misiti, D. (1994) Improved process for manufacturing L-(-)-carnitine from waste products having opposite configuration. European Patent Application EP EP0623588.

Gross, C.J. and Henderson, L.M. (1984) Absorption of D- and L-carnitine by the intestine and kidney tubule in the rat. *Biochim. Biophys. Acta*, **772**, 209–219.

Gulewitsch, W., Krimber, R. (1905) Zur kenntnis der extraktionsstoffe der muskeln. 2. Mitteilung über das carnitin. *Hoppe-Seyler's Z Physiol Chem*, **45**, 326–330.

Haddock, B.A. and Jones, C.W. (1977) Bacterial respiration. *Bacteriol. Rev.*, **41**, 47–99.

Hashiguchi, S., Kawada, A., and Natsugari, H. (1992) Bakers-yeast reduction of N-protected methyl 4-amino-3-oxobutanoates and 4-amino-3-oxopentanoates. *Synthesis (Stuttg)*, **4**, 403–408.

Heinig, K. and Henion, J. (1999) Determination of carnitine and acylcarnitines in biological samples by capillary electrophoresis–mass spectrometry. *J. Chromatogr. B: Biomed. Sci. Appl.*, **735**, 171–188.

Hoeks, F., Kulla, H., and Meyer, H.P. (1992) Continuous cell-recycle process for L-carnitine production: performance, engineering and downstream processing aspects compared with discontinuous processes. *J. Biotechnol.*, **22**, 117–127.

Iwamoto, J., Honda, A., Miyamoto, Y., Miyazaki, T., Murakami, M., Saito, Y., Ikegami, T., Miyamoto, J., and Matsuzaki, Y. (2014) Serum carnitine as an independent biomarker of malnutrition in patients with impaired oral intake. *J. Clin. Biochem. Nutr.*, **55**, 221–227.

Jakob, H., Hutchmacher, K. and Klenk, H. (1993) Method of preparation of L-carnitine from D,L-carnitine nitrile salts. European Patent Application EP0508133.

Jayasena, D.D., Jung, S., Bae, Y.S., Kim, S.H., Lee, S.K., Lee, J.H., and Jo, C. (2014) Changes in endogenous bioactive compounds of Korean native chicken meat at different ages and during cooking. *Poult. Sci.*, **93**, 1842–1849.

Jeppesen, J. and Kiens, B. (2012) Regulation and limitations to fatty acid oxidation during exercise. *J. Physiol.*, **590**, 1059–1068.

Johnson, D.W. (1999) Inaccurate measurement of free carnitine by the electrospray tandem mass spectrometry screening method for blood spots. *J. Inherit. Metab. Dis.*, **22**, 201–202.

Johnson, D.W. (2010) An acid hydrolysis method for quantification of plasma free and total carnitine by flow injection tandem mass spectrometry. *Clin. Biochem.*, **43**, 1362–1367.

Johri, A.M., Heyland, D.K., Hétu, M.-F., Crawford, B., and Spence, J.D. (2014) Carnitine therapy for the treatment of metabolic syndrome and cardiovascular disease: evidence and controversies. *Nutr. Metab. Cardiovasc. Dis.*, **24**, 808–814.

Jung, H., Buchholz, M., Clausen, J., Nietschke, M., Revermann, A., Schmid, R., and Jung, K. (2002) CaiT of *Escherichia coli*, a new transporter catalyzing L-carnitine/gamma -butyrobetaine exchange. *J. Biol. Chem.*, **277**, 39251–39258.

Jung, H., Jung, K., and Kleber, H.P. (1990) L-carnitine metabolization and osmotic stress response in *Escherichia coli*. *J. Basic Microbiol.*, **30**, 409–413.

Jung, H., Jung, K., and Kleber, H.P. (1993) Synthesis of L-carnitine by microorganisms and isolated enzymes. *Adv. Biochem. Eng. Biotechnol.*, **50**, 21–44.

Jung, H. and Kleber, H.P. (1991) Metabolism of D(+)-carnitine by *Escherichia coli*. *Appl. Microbiol. Biotechnol.*, **35**, 393–395.

Kakou, A., Megoulas, N.C., and Koupparis, M.A. (2005) Determination of L-carnitine in food supplement formulations using ion-pair chromatography with indirect conductometric detection. *J. Chromatogr. A*, **1069**, 209–215.

Kamimori, H., Hamashima, Y., and Konishi, M. (1994) Determination of carnitine and saturated-acyl group carnitines in human urine by high-performance liquid chromatography with fluorescence detection. *Anal. Biochem.*, **218**, 417–424.

Kaneko, T. and Yoshida, R. (1962) On the absolute configuration of L-carnitine (vitamin BT). *Bull. Chem. Soc. Jpn.*, **35**, 1153–1155.

Kasai, N. and Sakaguchi, K. (1992) An efficient synthesis of (R)-carnitine. *Tetrahedron Lett.*, **33**, 1211–1212.

Kaufman, R.A. and Broquist, H.P. (1977) Biosynthesis of carnitine in *Neurospora crassa*. *J. Biol. Chem.*, **252**, 7437–7439.

Kempf, B. and Bremer, E. (1998) Uptake and synthesis of compatible solutes as microbial stress responses to high-osmolality environments. *Arch. Microbiol.*, **170**, 319–330.

Kitamura, M., Ohkuma, T., Takaya, H., and Noyori, R. (1988) A practical asymmetric-synthesis of carnitine. *Tetrahedron Lett.*, **29**, 1555–1556.

Kleber, H.P. (1997) Bacterial carnitine metabolism. *FEMS Microbiol. Lett.*, **147**, 1–9.

Kolb, H.C., Bennani, Y.L., and Sharpless, K.B. (1993) Short and practical synthesis of (R)-(-)-carnitine and (R)-(-)-gamma-amino-beta-hydroxybutyric acid (GABOB). *Tetrahedron Asymmetry*, **4**, 133–141.

Krajcovicová-Kudlácková, M., Simoncic, R., Béderová, A., Babinská, K., and Béder, I. (2000) Correlation of carnitine levels to methionine and lysine intake. *Physiol. Res.*, **49**, 399–402.

Kulla, H.G. (1991) Enzymatic hydroxylations in industrial applications. *Chim. Int. J. Chem.*, **45**, 81–85.

Kulla, H.G. and Lehky, P. (1985) Verfahren zur herstellung von L-carnitin auf mikrobiologischem. Weg European Patent Application EP 0158194.

Kutscher, F. (1905) Über liebigs fleischextrakt. *Mitteilung I. Z Unters nahr GenuBm*, **10**, 528–537.

Lee, S.Y. (1996) High cell-density culture of *Escherichia coli*. *Trends Biotechnol.*, **14**, 98–105.

Lewin, L.M., Peshin, A., and Sklarz, B. (1975) A gas chromatographic assay for carnitine. *Anal. Biochem.*, **68**, 531–536.

Li, K., Li, W., and Huang, Y. (2007) Determination of free L-carnitine in human seminal plasma by high performance liquid chromatography with pre-column ultraviolet derivatization and its clinical application in male infertility. *Clin. Chim. Acta*, **378**, 159–163.

Lindstedt, G., Lindstedt, S., and Nordin, I. (1977) Purification and properties of gamma-butyrobetaine hydroxylase from Pseudomonas sp AK 1. *Biochemistry*, **16**, 2181–2188.

Lopez de Felipe, F., Kleerebezem, M., de Vos, W.M., and Hugenholtz, J. (1998) Cofactor engineering: a novel approach to metabolic engineering in Lactococcus lactis by controlled expression of NADH oxidase. *J. Bacteriol.*, **180**, 3804–3808.

Löster, H. (2003) *Carnitine and Cardiovascular Diseases*, Ponte Press Verlags GmbH, Bochum.

Manjón, A., Obón, J.M., and Iborra, J.L. (2000) Determination of L-carnitine by flow injection analysis with NADH fluorescence detection. *Anal. Biochem.*, **281**, 176–181.

Marzi, M., Minetti, P., Moretti, G., Tinti, M.O., and De Angelis, F. (2000) Efficient enantioselective synthesis of (R)-(-)-carnitine from glycerol. *J. Org. Chem.*, **65**, 6766–6769.

Mescka, C., Moraes, T., Rosa, A., Mazzola, P., Piccoli, B., Jacques, C., Dalazen, G., Coelho, J., Cortes, M., Terra, M., Regla Vargas, C., and Dutra-Filho, C.S. (2011) In vivo neuroprotective effect of L-carnitine against oxidative stress in maple syrup urine disease. *Metab. Brain Dis.*, **26**, 21–28.

Meyer, H.-P. and Robins, K.T. (2005) Large scale bioprocess for the production of optically pure L-carnitine. *Monatsh. Chem. –Chem. Mon.*, **136**, 1269–1277.

Miura-Fraboni, J., Kleber, H.P., and Englard, S. (1982) Assimilation of gamma-butyrobetaine and D- and L-carnitine by resting cell suspensions of *Acinetobacter calcoaceticus* and Pseudomonas putida. *Arch. Microbiol.*, **133**, 217–221.

Naidu, G.S.N., Lee, I.Y., Lee, E.G., Kang, G.H., and Park, Y.H. (2000) Microbial and enzymatic production of L-carnitine. *Bioprocess Eng.*, **23**, 627–635.

Nakahama, K., Izawa, M., Kanamaru, T. and Shinagawa S. (1995) Production of L-carnitine. Japanese Patent Office JP7170990.

Nakayama, K., Honda, H., Ogawa, Y., Ohta, T. and Ozawa, T. (1988) Method of pro ducing carnitine. US Patent Application US5041375 A

Ni, Y., Reye, J., and Chen, R.R. (2007) lpp deletion as a permeabilization method. *Biotechnol. Bioeng.*, **97**, 1347–1356.

Nielsen, J. (2001) Metabolic engineering. *Appl. Microbiol. Biotechnol.*, **55**, 263–283.

Obón, J.M., Buendía, B., Cánovas, M., and Iborra, J.L. (1999a) Enzymatic cycling assay for D-carnitine determination. *Anal. Biochem.*, **274**, 34–39.

Obón, J.M., Máiquez, J.R., Cánovas, M., Kleber, H.P., and Iborra, J.L. (1999b) High-density *Escherichia coli* cultures for continuous L(-)-carnitine production. *Appl. Microbiol. Biotechnol.*, **51**, 760–764.

Obón, J.M., Máiquez, J.R., Cánovas, M., Kleber, H.P., and Iborra, J.L. (1997) L(-)-carnitine production with immobilized *Escherichia coli* cells in continuous reactors. *Enzyme Microb. Technol.*, **21**, 531–536.

Osorio, J.H. and Pourfarzam, M. (2002) Plasma free and total carnitine measured in children by tandem mass spectrometry. *Braz. J. Med. Biol. Res.*, **35**, 1265–1271.

Pallavicini, M., Bolchi, C., Binda, M., Ferrara, R., Fumagalli, L., Piccolo, O., and Valoti, E. (2008) Entrainment resolution of carnitinamide chloride. *Tetrahedron: Asymmetry*, **19**, 1637–1640.

Ramsay, R.R. (1999) The role of the carnitine system in peroxisomal fatty acid oxidation. *Am. J. Med. Sci.*, **318**, 28–35.

Rigault, C., Mazué, F., Bernard, A., Demarquoy, J., and Le Borgne, F. (2008) Changes in L-carnitine content of fish and meat during domestic cooking. *Meat Sci.*, **78**, 331–335.

Roth, S., Jung, K., Jung, H., Hommel, R.K., and Kleber, H.P. (1994) CR from *Escherichia coli*. A new inducible enzyme of anaerobic metabolization of L(-)-carnitine. *Antonie Van Leeuwenhoek*, **65**, 63–69.

San, K.Y., Bennett, G.N., Berrios-Rivera, S.J., Vadali, R.V., Yang, Y.T., Horton, E., Rudolph, F.B., Sariyar, B., and Blackwood, K. (2002) Metabolic engineering through cofactor manipulation and its effects on metabolic flux redistribution in *Escherichia coli. Metab. Eng.*, **4**, 182–192.

Sánchez-Hernández, L., Castro-Puyana, M., Luisa Marina, M., and Crego, A.L. (2011) Determination of betaines in vegetable oils by capillary electrophoresis tandem mass spectrometry--application to the detection of olive oil adulteration with seed oils. *Electrophoresis*, **32**, 1394–1401.

Sánchez-Hernández, L., García-Ruiz, C., Crego, A.L., and Marina, M.L. (2010) Sensitive determination of D-carnitine as enantiomeric impurity of levo-carnitine in pharmaceutical formulations by capillary electrophoresis-tandem mass spectrometry. *J. Pharm. Biomed. Anal.*, **53**, 1217–1223.

Seccombe, D.W., Dodek, P., Frohlich, J., Hahn, P., Skala, J.P., and Campbell, D.J. (1976) Automated method for

L-carnitine determination. *Clin. Chem.*, **22**, 1589–1592.

Seim, H., Eichler, K., Kleber, H. (2001) L(-)-Carnitine and its precursor, gamma-butyrobetaine. *Nutraceuticals in Health and Disease Prevention*, (eds Kramer, K., Hoppe, P., Packer, L.), Marcel Dekker, Inc., New York, 217–256.

Seim, H., Löster, H., Claus, R., Kleber, H.P., and Strack, E. (1982) Stimulation on the anaerobic growth of *Salmonella typhimurium* by reduction of L-carnitine, carnitine derivatives and structure-related trimethylammonium compounds. *Arch. Microbiol.*, **132**, 91–95.

Serdar, M.A., Kurt, İ., Bayraktar, İ., Kenar, L., Özçelik, F., Onur, İ., and Kutluay, T. (2001) Optimization of fluorometric measurement of free and total carnitine in serum. *Turk. J Med. Sci.*, **31**, 309–316.

Sevilla, A., Schmid, J.W., Mauch, K., Iborra, J.L., Reuss, M., and Cánovas, M. (2005a) Model of central and trimethylammonium metabolism for optimizing L-carnitine production by *E. coli*. *Metab. Eng.*, **7**, 401–425.

Sevilla, A., Vera, J., Díaz, Z., Cánovas, M., Torres, N.V., and Iborra, J.L. (2005b) Design of metabolic engineering strategies for maximizing L-(-)-carnitine production by *Escherichia coli*. Integration of the metabolic and bioreactor levels. *Biotechnol. Prog.*, **21**, 329–337.

Sleator, R.D., Francis, G.A., O'Beirne, D., Gahan, C.G., and Hill, C. (2003) Betaine and carnitine uptake systems in *Listeria monocytogenes* affect growth and survival in foods and during infection. *J. Appl. Microbiol.*, **95**, 839–846.

Stefan, R.-I., Bokretsion, R.G., van Staden, J.F., and Aboul-Enein, H.Y. (2003) Simultaneous determination of L- and D-carnitine using a sequential injection analysis/amperometric biosensors system. *J. Pharm. Biomed. Anal.*, **33**, 323–328.

Stevens, R.D., Hillman, S.L., Worthy, S., Sanders, D., and Millington, D.S. (2000) Assay for free and total carnitine in human plasma using tandem mass spectrometry. *Clin. Chem.*, **46**, 727–729.

Stieger, B., O'Neill, B., and Krähenbühl, S. (1995) Characterization of L-carnitine transport by rat kidney brush-border-membrane vesicles. *Biochem. J.*, **309** (2), 643–647.

Strack, E. and Fosterling, K. (1937) Über die biologische wirkung des carnitins und acetylcarnitins. *Naunyn-Schmiedeberg's Arch. Exp. Pathol. Pharmakol.*, **185**, 612–621.

Takahashi, M., Ueda, S., Misaki, H., Sugiyama, N., Matsumoto, K., Matsuo, N., and Murao, S. (1994) Carnitine determination by an enzymatic cycling method with carnitine dehydrogenase. *Clin. Chem.*, **40**, 817–821.

Takano, S., Yanase, M., Sekiguchi, Y., and Ogasawara, K. (1987) Practical synthesis of (R)-gamma-amino-beta-hydroxybutanoic acid (GABOB) from (R)-epichlorohydrin. *Tetrahedron Lett.*, **28**, 1783–1784.

Talián, G.C., Komlósi, K., Decsi, T., Koletzko, B., and Melegh, B. (2007) Determination of carnitine ester patterns during the second half of pregnancy, at delivery, and in neonatal cord blood by tandem mass spectrometry: complex and dynamic involvement of carnitine in the intermediary metabolism. *Pediatr. Res.*, **62**, 88–92.

Tian, J., Wang, Q., and Zhang, Z. (2009) A novel strategy to improve the bioconversion of l-carnitine from crotonobetaine. *Eur. Food Res. Technol.*, **229**, 721–724.

Tomita, M. and Sendju, Y. (1927) Über die Oxyaminoverbindungen, welche die Biuretreaktion zeigen. III. Spaltung der γ -Amino- β -oxy-buttersäure in die optisch-aktiven Komponenten. *Hoppe-Seyler's Z. Physiol. Chem.*, **169**, 263–277.

Ussher, J.R., Lopaschuk, G.D., and Arduini, A. (2013) Gut microbiota metabolism of L-carnitine and cardiovascular risk. *Atherosclerosis*, **231**, 456–461.

Vaz, F.M. and Wanders, R.J.A. (2002) Carnitine biosynthesis in mammals. *Biochem. J.*, **361**, 417–429.

Verheul, A., Wouters, J.A., Rombouts, F.M., and Abee, T. (1998) A possible role of ProP, ProU and CaiT in osmoprotection of *Escherichia coli* by carnitine. *J. Appl. Microbiol.*, **85**, 1036–1046.

Violante, S., Ijlst, L., Te Brinke, H., Koster, J., Tavares de Almeida, I., Wanders, R.J.A., Ventura, F.V., and Houten, S.M. (2013) Peroxisomes contribute to the acylcarnitine production when the carnitine shuttle is deficient. *Biochim. Biophys. Acta*, **1831**, 1467–1474.

Voeffray, R., Perlberger, J.C., Tenud, L., and Gosteli, J. (1987) L-carnitine-novel synthesis and determination of the optical purity. *Helv. Chim. Acta*, **70**, 2058–2064.

Wall, B.T., Stephens, F.B., Constantin-Teodosiu, D., Marimuthu, K., Macdonald, I.A., and Greenhaff, P.L. (2011) Chronic oral ingestion of L-carnitine and carbohydrate increases muscle carnitine content and alters muscle fuel metabolism during exercise in humans. *J. Physiol.*, **589**, 963–973.

Walt, A. and Kahn, M.L. (2002) The fixA and fixB genes are necessary for anaerobic carnitine reduction in *Escherichia coli*. *J. Bacteriol.*, **184**, 4044–4047.

Wanders, R.J., Vreken, P., Ferdinandusse, S., Jansen, G.A., Waterham, H.R., van Roermund, C.W., and Van Grunsven, E.G. (2001) Peroxisomal fatty acid alpha- and beta-oxidation in humans: enzymology, peroxisomal metabolite transporters and peroxisomal diseases. *Biochem. Soc. Trans.*, **29**, 250–267.

Wieland, O., Deufel, T., and Paetzke-Brunner, J. (1985) in *Methods Enzymatic Analysis*, 3rd edn (ed H. Bergmeyer), VCH Verlagsgesellschaft, Weinheim, pp. 481–488.

15
Application of Carnosine and Its Functionalised Derivatives

Isabelle Chevalot, Elmira Arab-Tehrany, Edouard Husson, and Christine Gerardin

15.1
Introduction and Historical Outline

Carnosine was isolated for the first time from Liebig's meat extract in 1900 (Gulewitsch and Amiradzibi, 1900). The structure of this compound was then identified as β-alanyl-L-histidine (Figure 15.1) in 1918 and constituted the first peptide ever isolated from biomaterial (Barger and Tutin, 1918; Baumann and Ingvaldsen, 1918). Carnosine is accumulated in the animal tissue particularly at high concentrations in the muscle tissue (Schönherr, 2002). This polar dipeptide exhibits various biological activities such as agents that act as hepatoprotection, neuroprotection or regulation of enzymatic activity (Babizhayev, 2006; Fouad, El-Rehany and Maghraby, 2007; Shen *et al.*, 2007). More precisely, carnosine was shown to contribute to the protection of biological membranes from lipid-peroxidation-induced damages, the inactivation of reactive oxygen species (ROS) or the reversal of normal features of senescent fibroblasts (Babizhayev, 2006). Some studies have reported on its neuroprotective effect against ROS (Boldyrev *et al.*, 2004; Shen *et al.*, 2007; Liu, Liu and Yin, 2008; Mehmetçik *et al.*, 2008; Tomonaga *et al.*, 2008). These antioxidant properties have been deeply investigated and described in the literature, suggesting the preponderant implication of its constitutive histidine. The antioxidant mechanism of carnosine has been attributed to its chelation of metal ions, its superoxide dismutase (SOD)-like activity and its ability to scavenge ROS and other free radicals. Because of these properties, carnosine has been used to prevent damage caused by radiation therapy, as well as necrotic and apoptotic cell death in hepatic tissue caused by ischemia/reperfusion (Guney *et al.*, 2006; Badr, 2007). All of these properties represent only a non-exhaustive panel of multiple biological potentials of this singular dipeptide. Furthermore, studies showed that this peptide can be extracted from meat tissues of low economic value such as poultry tissues and can be exploited to increase the economy of the meat industry (Manhiani *et al.*, 2013).

In this context, carnosine constitutes an undeniable challenging substrate for chemical or enzymatic modifications leading to pharmaceutical, cosmetic and nutraceutical applications.

Industrial Biotechnology of Vitamins, Biopigments, and Antioxidants, First Edition.
Edited by Erick J. Vandamme and José L. Revuelta.

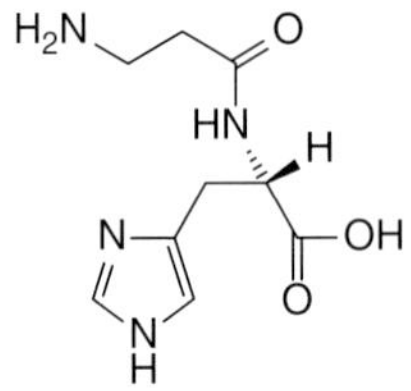

Figure 15.1 Molecular structure of carnosine.

15.2 Sources and Synthesis

15.2.1 Occurrence in Natural/Food Sources

Carnosine is a natural dipeptide present in large amounts in the skeletal muscle. A lot of studies concern the production of extract from beef or chicken with high percentage of carnosine. Several methods were tested to produce a carnosine-containing antioxidant extract from beef or chicken muscle (Chan, Decker and Means, 1993).

Heat treatment, ultrafiltration and demineralisation increased antioxidant activity of beef extract. The white muscle is known to contain a higher amount of carnosine than does the red muscle tissue. Some efforts have been made to identify the optimum extraction conditions of this dipeptide. Studies attempted heat and ultrafiltration processes with acidic extraction (Intarapichet and Maikhunthod, 2005) and others used membrane separation (Nabetani *et al.*, 2012). These different methods involve, in general, acid extraction followed by deproteinisation by ethanol treatment and have been applied by other researchers. For instance, a method of extraction from chicken breast was recently developed for preserving the maximum of physiological activities, through response surface methodology using multiple objective optimisations (Kim *et al.*, 2014).

15.2.2 Chemical Synthesis of Carnosine

The first chemical synthesis of carnosine dates back to the early twentieth century with that proposed in 1918 (Baumann and Ingvaldsen, 1918) to verify the hypothesis of the supposed chemical structure of the carnosine extracted from beef muscle extract. Carnosine was prepared from histidine and β-iodo propionic acid followed by aminolysis by ammonia (Figure 15.2).

A new synthesis has been proposed in 1934 by a classical method used for peptide synthesis. Carbobenzyloxy-β-alanine was converted to acylazide by the treatment of ester by sodium nitrite (Sifferd and Du Vignaud, 1935). The condensation of the methylester of histidine on the acylazide led to the carbobenzyloxy carnosine which was the converted to the free crystalline base by saponification of the ester and by scission of the carbobenzyloxy group by catalytic hydrogenation.

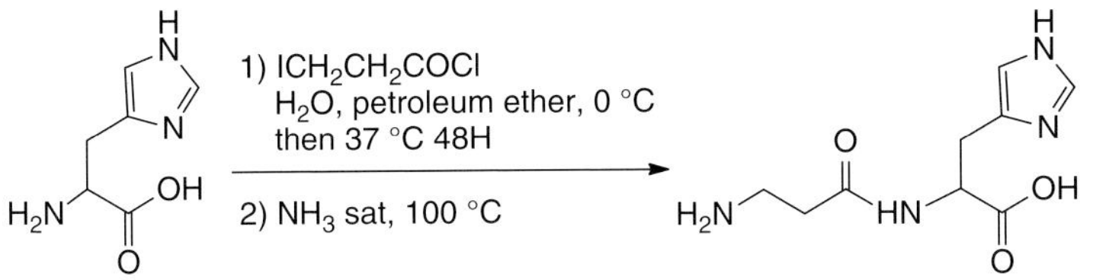

Figure 15.2 First chemical synthesis of carnosine (Baumann and Ingvaldsen, 1918).

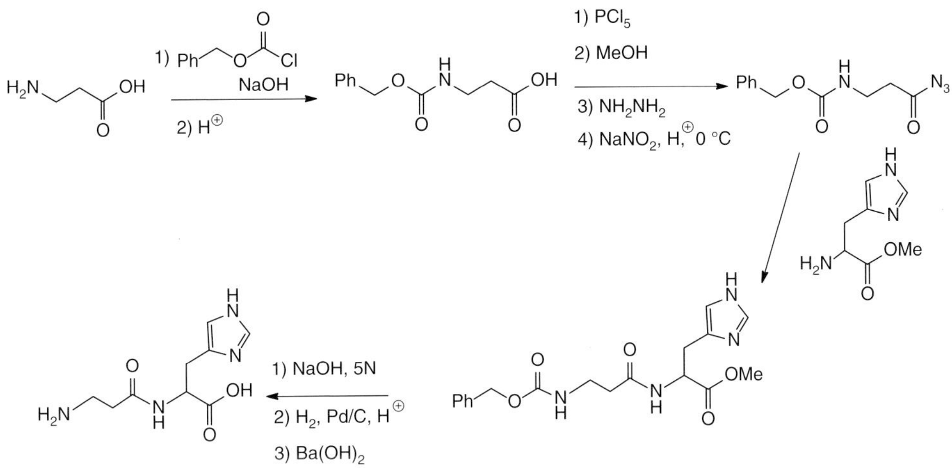

Figure 15.3 Chemical synthesis of carnosine by Curtius method (Sifferd and Du Vignaud, 1935).

This method, named 'method of Curtius', allows, in this case, a better yield than with acyl chloride activation (Figure 15.3).

More recently, studies proposed the synthesis of carnosine from the *N*-trifluoroacetyl-β-alanine (Cherevin *et al.*, 2007). The ease of introduction and removal of trifluoroacetyl protection makes such protection quite interesting in those cases where no racemisation is possible. Different activations were tested from this protected β-alanine (Figure 15.4).

Otherwise, carnosine could be synthetised by other chemical methods used in peptide coupling (Kimmerlin and Seebach, 2005).

15.2.3 Enzymatic Synthesis of Carnosine

Alternative enzymatic synthesis strategies can be used to overcome limitations in chemical synthesis, such as lack of specificity, use of organic solvents and necessity to block or activate functional groups.

Carnosine is known to be synthesised from β-Ala and L-His by carnosine synthetase (EC 6.3.2.11), which has been characterised from various animals (Bauer

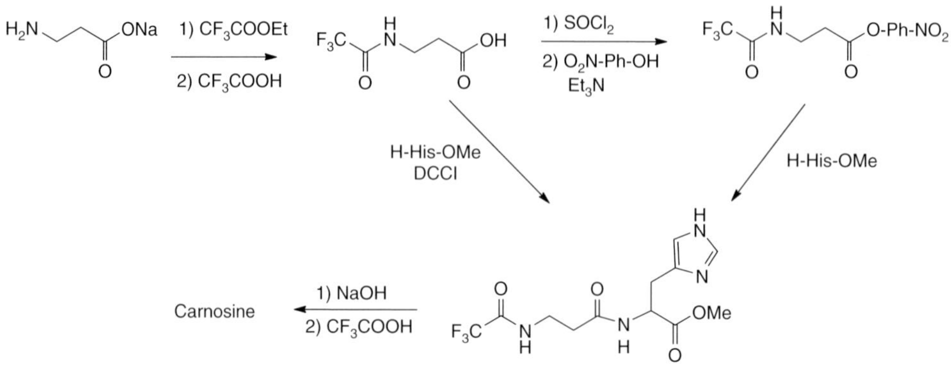

Figure 15.4 Chemical synthesis of carnosine (Cherevin *et al.*, 2007).

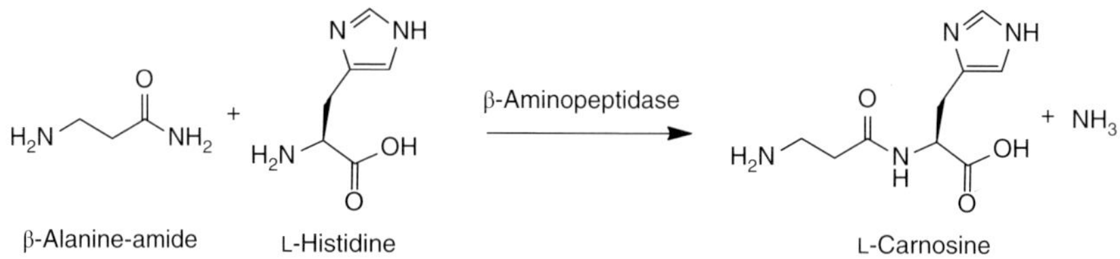

Figure 15.5 Use of bacterial β-aminopeptidase for the synthesis of carnosine.

and Schulz, 1994), but this enzyme exhibited limited substrate specificity and required expensive additives such as ATP. Other enzymes have been described such as the family of S9 peptidases that showed oligopeptidase activity synthesis of various β-alanyl-dipeptides. For instance, the conversion rate of β-Ala-OBzl to carnosine-OMe was estimated as greater than 30% using S9 aminopeptidase from *Streptomyces thermocyaneoviolaceus* NBRC14271 (S9AP-St) (Arima *et al.*, 2010).

Other bacterial β-amino-peptidases DmpA from *Ochrobactrum anthropi* and BapA from *Sphingosinicella xenopeptidilytica* were described for the preparation of L-carnosine in two aqueous enzymatic reaction systems (Heck *et al.*, 2010) (Figure 15.5). The results showed that the reaction is under kinetic control and the yield reached more than 50%, depending on the acyl donor used. The application of these two enzymes in whole cells for the coupling of H-β-Ala-NH_2 and L-histidine to L-carnosine was also studied, and the optimisation of reaction conditions has led to an improved whole-cell process using recombinant *Escherischia coli*, obtaining high yields of 70% L-carnosine. The studies of activity stability showed enzymatically active cells for more than 8 h (Heyland *et al.*, 2010).

Other studies reported the synthesis of L-carnosine using various whole-cell biocatalyst systems. In particular, recombinant yeast and bacteria strains can overexpress a β-peptidase and were shown to be used directly as whole-cell biocatalysts for the synthesis of L-carnosine.

The synthesis of carnosine from nonprotected amino acids as substrates was also performed by cloning the carnosinase (CN1) gene. A whole-cell biocatalyst displaying CN1 on the yeast cell surface was constructed with α-agglutinin as the anchor protein. This CN1-displaying yeast cells catalysed carnosine synthesis in hydrophobic organic solvents and hydrophobic ionic liquids, from non-protected amino acids in only one step (Inaba *et al.*, 2010). However, the synthesis efficiency of carnosine was ~5% which appeared to be not sufficient for industrial applications. The limiting factor may probably be the amount of water in the system. Using ionic liquids that can dissolve amino acids, carnosine could be efficiently synthesised without the addition of water.

15.3
Physico-Chemical and Biological Properties of Carnosine

15.3.1
Physico-Chemical Properties

Carnosine is a molecule exhibiting three ionisable groups: the carboxylic group with a pKa of 2.75, the primary amino group of β-alanine with a pKa of 9.3 and nitrogen present on the imidazole ring with a pKa of 6.75. This dipeptide is highly soluble in water. Some physical properties of carnosine such as equilibrium geometries, harmonic vibrational frequencies, electronic transition energies and NMR chemical shifts were determined to characterise carnosine from a theoretical point of view (Pis-Diez and Baran, 2003).

Numerous functions of this dipeptide have been described, such as buffering activity, metal-ion-chelating activity, especially for Cu II and Zn II, antioxidant and anti-glycating activity (Kang *et al.*, 2002; Di Paola *et al.*, 2011), all being important under physiological and pathological conditions (Renner *et al.*, 2010). The presence of the imidazole ring in carnosine is quite important as it plays a major role in the regulation of the buffering activity of carnosine.

Furthermore, carnosine inhibited the oxidative hydroxylation of deoxyguanosine induced by ascorbic acid and copper ions (Kohen *et al.*, 1988). Other roles of carnosine, such as quenching of singlet oxygen, binding of hydroperoxides and regulation of enzyme activity, have already been described (Babizhayev, 2006).

Carnosine decreases the levels of free radicals through its direct interaction with ROS. It has been found that the metabolites of carnosine, such as anserine, ophidine and *N*-acetyl-carnosine, also function as antioxidants (Boldyrev and Abe, 1999).

Considering the ability of carnosine to chelate heavy metal ions (copper, iron, cobalt, cadmium), the protection by carnosine against lipid peroxidation was probably due to a decrease in active concentrations of variable valence metals. The binding constant of iron ions by carnosine calculated from values of the chemiluminescent response of phospholipid liposomes on their oxidation was $2.3 \times 10^{-3}\,M^{-1}$ (Vladimirov, 1994). However, in some cases, the effect of carnosine was observed also at lower concentrations (0.1–0.5 mM) (Klebanov *et al.*, 1997). Thus, it was concluded that the effect of carnosine was not limited only to its ability to chelate iron ions.

15.3.2
Physiological Properties

Carnosine is the first peptide ever isolated from natural material (Quinn, Boldyrev and Formazuyk, 1992) and reaches very high concentrations (up to 20 mM) in the muscle and nervous tissues of several animal species (Decker, Livisay and Zhou, 2000). This dipeptide (β-Ala-His) is known to present a large diversity of functions.

The research on the physiological role of carnosine was first directed towards the skeletal muscle, as the molecule was initially discovered in meat extract. It was shown that carnosine presents a role in the contractile function of skeletal muscle. The role of carnosine as a pH buffer in the skeletal muscle was also demonstrated (Abe, 2000), and it was supposed that this property explains its predominant association with white, glycolytic, muscles possessing relatively few mitochondria and generating lactic acid. Thus, the physiological role of carnosine in contracting muscle is mainly related to homeostasis, including protection against acidosis (Begum, Cunliffe and Leveritt, 2005). In humans, the degree of acidosis in the blood during high-intensity exercise can be limited when carnosine content in the muscle is increased by nutritional intervention (Baguet *et al.*, 2010). L-Carnosine has an important role as a pH buffer to protect against damage caused by oxidation under anaerobic respiration conditions. During exercise, carnosine suppresses acidification of intracellular environments and maintains muscle activity (Kim *et al.*, 2014).

Another possible role of carnosine in skeletal muscle tissue concerns its physiological role of Cu^{2+} chelation. By proton NMR spectroscopy, studies showed the presence of Cu^{2+}-carnosine chelates in the human calf muscle *in vivo* (Schroder, Schmitz and Bachert, 2008).

As the muscle contains the main carnosine present in the organism due to the high expression of the enzyme responsible for carnosine synthesis, it was supposed that carnosine is stored in the skeletal muscle before it is released to other places where it can be used for the delivery of L-histidine or β-alanine. For instance, studies in rats showed that when the skeletal muscles contract, they release carnosine into the circulation. Consequently, the carnosine concentration in plasma was twice for rats accessing a running wheel in comparison with sedentary rats (Nagai *et al.*, 2003).

With regard to brain function, carnosine is located in the olfactory receptor neurons and was supposed to be involved in sensory neurotransmission, either as a neurotransmitter or as a neuromodulator. Its putative function and mechanisms are still unclear. Carnosine could serve as a neurotransmitter in neuron-to-glia communication (Baslow, 2010). Carnosine could be also indirectly involved in neurotransmission through its capacity to chelate transition metals such as copper and zinc (Trombley, Horning and Blakemore, 1998).

In cardiovascular function, carnosine was shown to increase contractility when perfused into isolated rat heart, increasing myoplasmic Ca^{2+} concentration. Consequently, carnosine was considered as a modulator of calcium-regulated proteins in cardiac muscle cells displaying an important role in contractility and cardiac function (Roberts and Zaloga, 2000).

15.4 Biotechnological Synthesis of Carnosine Derivatives: Modification, Vectorisation and Functionalisation

The main limitation on therapeutic use of carnosine is due to increased oxidative stress associated with the hydrolysis by the specific dipeptidase carnosinase. Several attempts have been made to overcome this drawback. For instance, studies reported that the carnosine glycoconjugation protects the dipeptide moiety from carnosinase hydrolysis that can lead to an increase in carnosine availability (Lanza *et al.*, 2011). Several other carnosine derivatives with saccharides (Bonomo *et al.*, 2003), such as β-cyclodextrin (Mineo *et al.*, 2004) and trehalose, have been synthesised. All these compounds are able to scavenge hydroxyl radicals, and their copper (II) complexes exhibit SOD activity (Amorini *et al.*, 2007; La Mendola *et al.*, 2002). Furthermore, they are resistant to the hydrolysis of carnosinase (Bellia *et al.*, 2008) and exhibit antioxidant efficacy at concentrations 10–20 times lower than that reported for other synthetic derivatives (Boldyrev *et al.*, 2004).

Studies investigated the impact of the covalent attachment of an acyl group on this dipeptide. It appeared that the acetyl derivative of carnosine exhibited improved antioxidant activities in the liposome peroxidation system and its transfer throughout the biological membranes was increased due to a higher hydrophobicity compared with carnosine (Babizhayev *et al.*, 2001).

Other derivatives of carnosine were synthesised; for instance, a series of carnosine amides bearing the amido group alkyl substituents was produced and characterised: they exhibited different lipophilicities and displayed carnosine-like properties. Furthermore, these derivatives were demonstrated to be stable over 3 h of incubation in the human serum at 37 °C (Bertinaria *et al.*, 2011).

15.4.1 Chemical Functionalisation

The chemical derivatisation of carnosine is a very promising approach to obtain therapeutic carnosinase-resistant molecules based on carnosine. In general, the dipeptide has been modified either at the amino group of the β-alanine or at the carboxyl group of the histidine, depending on the importance of these two groups for the biological activity.

The principal *N*-grafted derivatives are shown in Figure 15.6. Different kinds of derivatisation have been proposed. The amino group has been grafted by various groups such as acetyl groups (*N*-acetylcarnosine, NAC), by acylation with a fatty

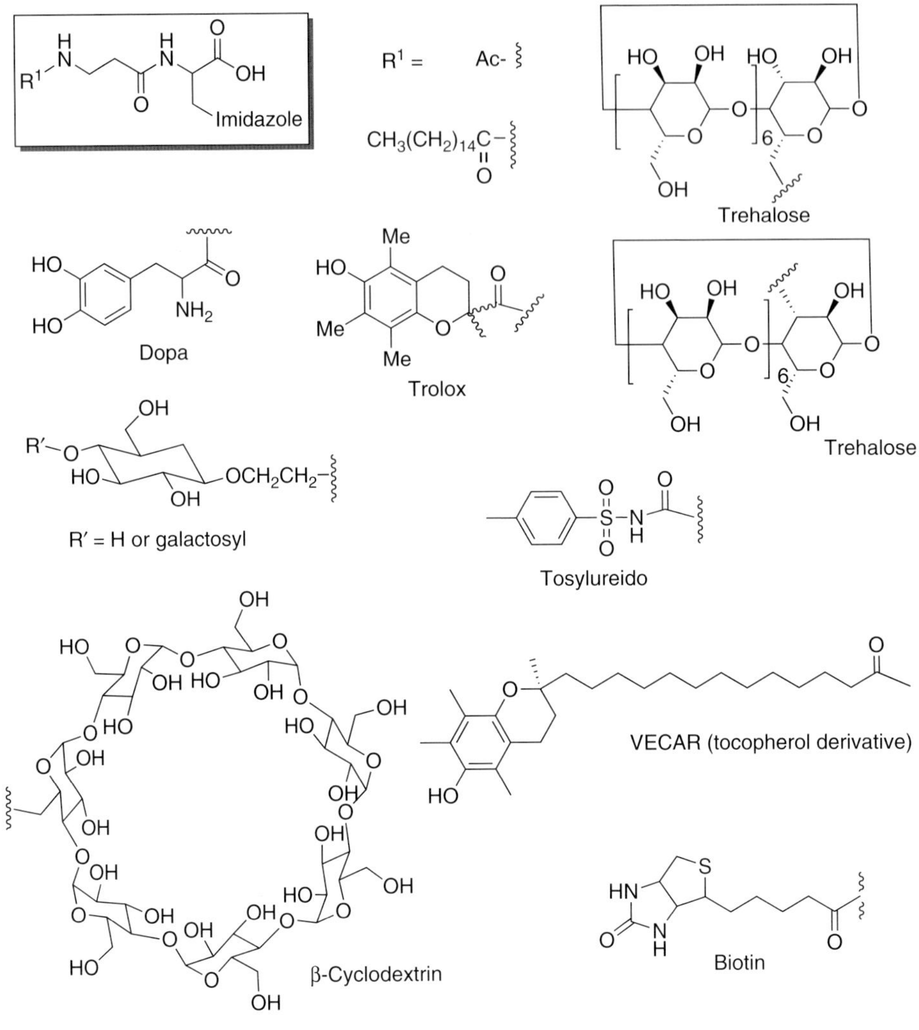

Figure 15.6 N-derivatives of carnosine.

acid chain or modified by sugar, oligosugar, cyclodextrin, L-Dopa, tocopherol and so on. A widely used synthetic procedure consists in the alkylation of carnosine using tosylate, iodide or bromide in a nucleophilic substitution reaction. For acylation, the condensation reaction was used, in the presence of an activating agent.

The NAC is a well-known derivative because of its capacity to be an ophthalmic drug in the treatment of human cataract (Babizhayev *et al.*, 2001; Babizhayev, 2006; Babizhayev, Guiotto and Kasus-Jacobi, 2009). Acylation with longer chain acetyl, such as palmitoyl chain, is performed to obtain amphiphilic derivatives. These types of compounds have been founded to self-assemble, and the addition of them, to DPPC-multilamellar vesicles, leads to a transition to well-defined unilamellar vesicles with a potential for applications in the delivery to cells (Castelletto *et al.*, 2012). Acylation by benzoic acid with a fatty acid chain in para position led to hydrogelators, which could be interesting for drug delivery (Pal, Shrivastava and Dey, 2009).

The association of carnosine with the other different moieties has been made to exploit the beneficial activity of each constituent. In general, the conjugation shows a better activity due to a synergistic effect. For example, a higher quenching activity against the DPPH radical has been reported for the *R*-Trolox or the VECAR with respect to that of the constituents separately (Astete *et al.*, 2013; Stvolinsky *et al.*, 2010). Glycoconjugate derivatives obtained by functionalisation with carnosine in different position of the sugar or the cyclodextrin are also promising because of their protected effect from the degradation by carnosinase (Bellia *et al.*, 2008; Lanza *et al.*, 2011). Moreover, cyclodextrins are particularly used in pharmaceutical science for their ability to include or stabilise drugs, and trehalose is appreciated for its protective and moisturising functions in cosmetics (La Mendola *et al.*, 2002). 4-Tosylureido carnosine has also been shown to be stable with respect to carnosinase and has been tested as a target moiety to the delivery to tumour cells. One of the widely used molecules for selective delivery is vitamin H, better known as *biotin*. Functionalisation of gold nanoparticles (NPs) with biotin derivative has been also exploited (Bellia *et al.*, 2013).

The principal derivatives modified at the carboxyl group are presented in Figure 15.7. A number of these derivatives are amides of carnosine. The most used strategy for the synthesis of carnosine amide involves the coupling between the Boc-protected β-alanine and a histidine amide with classical agent peptidic coupling.

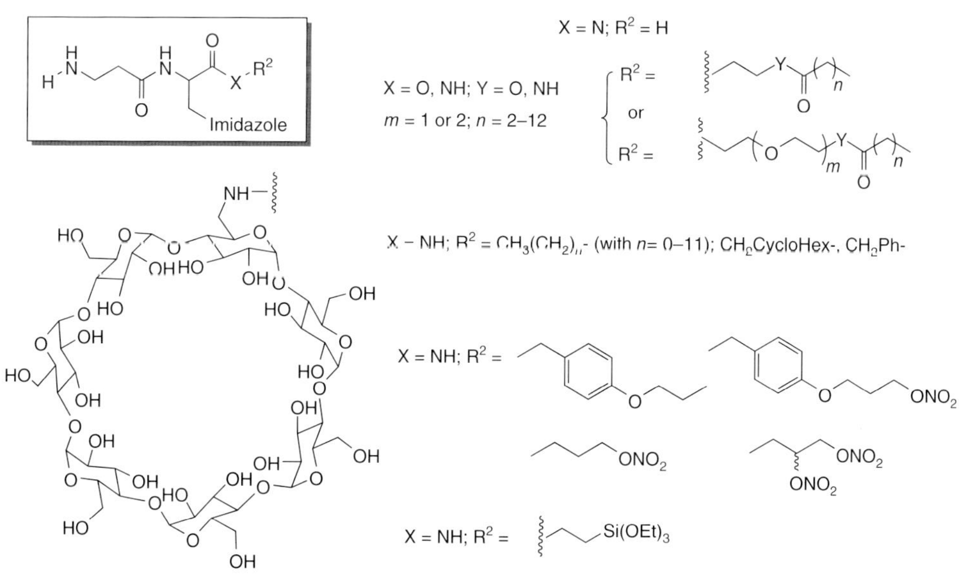

Figure 15.7 Derivatives of carnosine at the carboxylic moiety.

The first example corresponds to the primary amide ($R^2 = NH_2$). The functionalisation by an amide on the carboxylic-moiety seems to be beneficial for the protection against carnosinase, while maintaining some important biological functions of the dipeptides (Bertinaria *et al.*, 2011). Indeed, it has been reported that the carboxylic moiety is important in the recognition by the carnosinase enzymes. Other amides or esters have been synthetised. Amino-β-cyclodextrin has also been used (La Mendola *et al.*, 2002). Some carnosine derivatives also include NO-donor nitro-oxy substructure. This class of carnosine derivatives could be a potential tool for treating a wide range of chronic vascular and neurodegenerative diseases in which NO-bioavailability is reduced (Bertinaria *et al.*, 2012).

Other derivatives have been prepared to exploit the physicochemical properties of the carnosine. Indeed, some derivatives have been obtained by introduction of a polyoxyethylenic moiety and a fatty acid chain by various links between the different parts, in order to modify the lipophilic and hydrophilic balance of the carnosine and to obtain amphiphilic compounds with complexing properties and surfactive or gelator properties (Gizzi *et al.*, 2009).

Finally, silica-hybrid material has been obtained by co-condensation of tetraethoxysilane with an organotriethoxysilane containing a carnosine dipeptide. The hybrid material has propensity for binding CU(II) species from aqueous solutions (Hamdoune *et al.*, 2000; Sayen *et al.*, 2003; Hamdoune *et al.*, 2001; Walcarius *et al.*, 2004).

Some analogues of carnosine have also been proposed with the objective either to strengthen the carnosine against carnosinases or for exploiting the physicochemical properties of carnosine in other compounds. Two examples are presented in Figure 15.8.

A study reported the synthesis of $\Psi(SO_2NH)$ isostere of the carnosine-dipeptide (Calcagni *et al.*, 1999). The sulfamido pseudopeptides have shown stability to carnosinase activity.

Moreover, perfluorinated and hybrid amphiphilic surfactants have been proposed and their surfactant and complexing ability have been investigated (Cosgun *et al.*, 2001, 2004).

15.4.2 Enzymatic Functionalisation: Enzymatic N-Acylation of Carnosine

The enzymatic functionalisation of carnosine mainly concerned acylation reactions. The classical way to acylate peptides usually followed the chemical

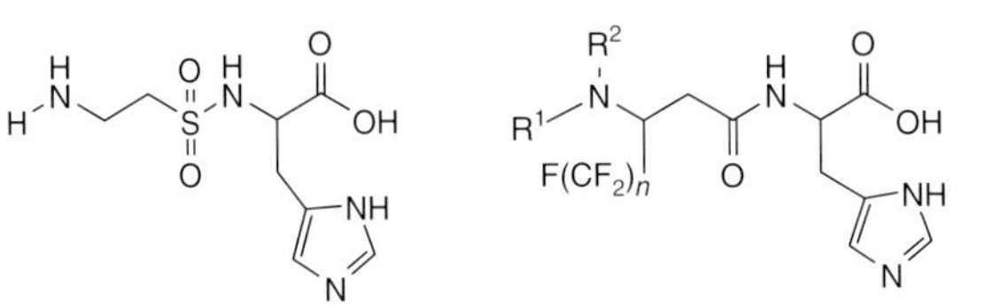

Figure 15.8 Analogues of carnosine.

procedure of acylation requiring organic solvent and protection/activation of functional groups. Furthermore, some drastic reaction conditions are often required for these chemical modifications, inconsistent with the current environmental requirements. Another way can be the use of enzyme-catalysed reactions in organic solvents (Zaks and Klibanov, 1985; Fernandez-Perez and Otero, 2001). Some studies demonstrated the feasibility of the enzymatic *N*-acylation of amino acids or dipeptides catalysed by lipases in organic solvents (Montet *et al.*, 1990; Gardossi, Bianchi and Klibanov, 1991).

The impact of the covalent attachment of an acyl group to carnosine on its *in vivo* and *in vitro* properties has been described in some studies. For instance, the *N*-acetyl derivative of carnosine has been demonstrated to induce the improvement of antioxidant activities in the liposome peroxidation system. Its transfer throughout the biological membranes was also increased due to its higher hydrophobicity compared with pristine carnosine (Babizhayev, 2006). In addition, the *N*-acetyl derivative of carnosine exhibited a higher stability than carnosine to carnosinase-catalysed hydrolysis, suggesting a better bioavailability (Pegova, Abe and Boldyrev, 2000). Another study showed that *N*-acylation of carnosine with long acyl chains considerably improved its capacity to suppress phosphatidylcholine liposome oxidation induced by ferrous ions and ascorbic acid (Murase, Nagao and Terao, 1993). Biocatalysts used for such acylation reactions are usually hydrolytic enzymes such as lipases (EC 3.1.1.3) which are commonly used in non-conventional (non-aqueous) solvents to catalyse *N*-acylation of amino acids and dipeptides (Montet *et al.*, 1990; Gardossi, Bianchi and Klibanov, 1991; Soo *et al.*, 2004). Similarly, the use of other enzymes, such as lipase/acyltransferase from *Candida parapsilosis*, has also been considered. These enzymes can indeed catalyse *N*-acylation reactions in aqueous media where polar substrates are completely solubilised (Vaysse *et al.*, 2002). Aminoacylases from *Streptomyces mobaraensis* have also been described for their ability to catalyse amino acid acylation in aqueous media (Koreishi *et al.*, 2005, 2006). Advantages of these enzymatic pathways are the use of mild conditions avoiding the consumption of hazardous chemical reagents such as thionyl chloride ($SOCl_2$), the specificity of reaction and the decrease in by-product production. Two major constraints due to the intrinsic properties of carnosine may govern its enzymatic *N*-acylation: a low solubility in apolar medium which could considerably limit its availability for enzyme and the reactivity/accessibility of its amino group. Furthermore, a suitable reaction medium has to be selected to ensure activity/stability of the enzyme. For overcoming these difficulties, medium and processes engineering approaches coupled with a rational choice of biocatalysts have been developed for finding appropriate solutions allowing efficient enzymatic *N*-acylation of carnosine.

15.4.2.1 Lipase-Catalysed N-Acylation of Carnosine in Non-Aqueous Medium

The feasibility of carnosine acylation by enzymatic way was firstly investigated using the lipase B of *Candida antarctica* in organic medium (Husson *et al.*, 2009). The aim of this work was to study both the reactivity of carnosine primary amino group and the influence of its solubility in different media on the performance of the reaction.

Based on the previous studies about *N*-acylation of model amino-alcohols (Husson *et al.*, 2008, 2010), the enzymatic acylation of carnosine (0.12 M) was performed in 2-methyl-2-butanol at 55 °C with free oleic acid (0.24 M) as acyl donor and the immobilised lipase B from *C. antarctica* as biocatalyst. Mass spectrometric analysis of the reaction medium at the thermodynamic equilibrium (72 h) of the reaction indicated the synthesis of a monoacylated derivative of carnosine identified as the *N*-oleoyl-β-Ala-His. However, quantitative monitoring of the reaction showed very weak carnosine conversion (9.5%) and synthesis initial rate of 5 $mM \cdot h^{-1}$. These very low performances were attributed to the very limited solubility value of carnosine (0.29 $g \cdot l^{-1}$) under such conditions. To improve the solubility of carnosine, the reaction was performed at higher temperature (80 °C) in the same organic solvent, but the performances obtained were not satisfactory despite the carnosine solubility value being slightly increased (0.33 $g \cdot l^{-1}$). Another non-conventional medium was then tested: 1-butyl-3-methyl imidazolium hexafluorophosphate. This ionic liquid constituted an interesting solvent with the capacity to solubilise a wide range of biomolecules in maintaining and/or improving synthesis activity of lipase (Sheldon *et al.*, 2002; van Rantwijk and Sheldon, 2007). However, the solubility of carnosine in this solvent remained very weak, at both 55 and 80 °C and the substrate conversion remained very low. This first investigation evidenced the limitation of the lipase-catalysed strategy to *N*-acylate polar dipeptide as carnosine in heterogeneous system (Husson *et al.*, 2009). The major presence of carnosine particles as solid aggregates in these reaction systems, whatever may be the temperature, would induce indeed a very low availability of carnosine for lipase.

Thus, high-pressure homogenisation pretreatments (0–2.5 kbar) were applied on the carnosine suspension in 2-methyl-2-butanol (before enzyme and acyl donor addition) to improve the availability of carnosine during further reaction. With this physical pretreatment, the maximum value of carnosine solubility in 2-methyl-2-butanol at 80 °C was only 0.45 g/l, but the reaction medium was more homogeneous with smaller sized carnosine particles. Decreased particle size resulted in a significant increase in both initial rate of reaction and carnosine conversion (32% versus 16% without pretreatment). This result could be explained by an increase in the substrate/enzyme exchange surface allowing higher kinetic and conversion performances as already admitted for another enzymatic conversions (Sakakibara *et al.*, 1996; Xiao *et al.*, 2005). The overall enzymatic process is presented in Figure 15.9. By enhancing the temperature and the dispersibility of substrate particles, the possibility to reach a significant yield of lipase-catalysed *N*-acylation of carnosine in organic solvent (Husson *et al.*, 2011) was demonstrated for the first time.

15.4.2.2
Acyltransferase-Catalysed N-Acylation of Carnosine in Aqueous Medium

A second enzymatic alternative to produce *N*-oleyl carnosine was described (Husson *et al.*, 2011). The retained strategy was the use of a reaction system in

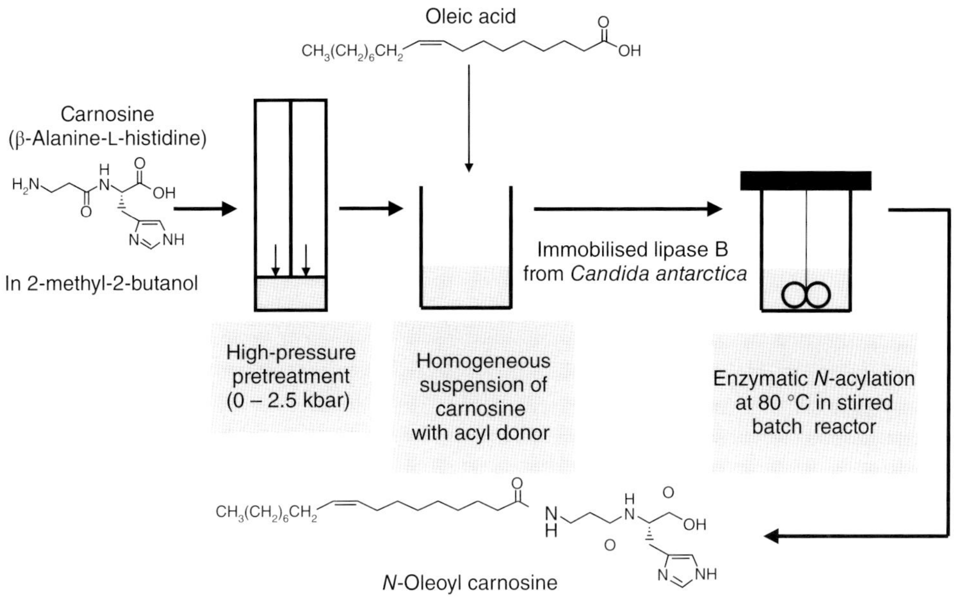

Figure 15.9 Enzymatic process of N-oleoyl carnosine synthesis (Husson *et al.*, 2011).

which the solubility of carnosine is not a limiting factor. Enzymatic acylation was thus performed in an aqueous medium (phosphate buffer/acyl donor emulsion) using the acyltransferase from *C. parapsilosis*. The maximum solubility of carnosine at 45 °C in this medium was 228.5 g/l (1.01 M). Under the conditions of the enzymatic synthesis (0.12 M carnosine), the peptide was completely solubilised in the reaction medium and thus available for the enzyme. In this aqueous medium, the acyl donor was emulsified by sonication to favour its homogeneous distribution in the system. Acylation reactions catalysed by the acyltransferase from *C. parapsilosis* were first performed with oleic acid as the acyl donor. The synthesis of *N*-oleyl carnosine was confirmed by mass spectrometry analysis. This work demonstrated for the first time the feasibility of the enzymatic *N*-acylation of carnosine in aqueous medium (Husson *et al.*, 2011). Influence of acyl donor was then studied in comparing the kinetic and conversion performances obtained with free oleic acid and its corresponding ethyl ester. With oleic acid, the carnosine conversion yield was 37% at the thermodynamic equilibrium (120 h), whereas the initial rate of the reaction was 39 $mM{\cdot}h^{-1}$. This result might be limited by a competition between the synthetic and hydrolytic pathways during the reaction. The kinetic and conversion performances obtained with ethyl oleate as acyl donor were better (48% and 116 $M{\cdot}h^{-1}$, respectively) in comparison to those with oleic acid. However, the presence of free oleic acid in the reaction medium of transacylation indicated that hydrolytic events of ethyl oleate occurred, which may explain the limitation of the *N*-oleyl carnosine synthesis. The carnosine acylation may also be limited by the pKa value (9.36) of the primary amino group of carnosine (Nielsen *et al.*, 2002). In fact, the optimal pH for the acyltransferase activity is lower than this pKa (6.5). Consequently, a large portion of carnosine molecules are under protonated form, unable to perform nucleophilic attack, thereby limiting substrate conversion. This innovative emulsified aqueous system constituted a promising eco-friendly alternative for *N*-acylation of carnosine.

15.4.2.3 Impact of Enzymatic Oleylation of Carnosine on Some Biological Properties

The xanthine oxidase inhibition property of both carnosine and *N*-oleyl carnosine synthesised by enzymatic way was evaluated in a previous study (Husson *et al.*, 2011). The *N*-oleyl carnosine IC_{50} value in the xanthine/xanthine oxidase system was similar to those of pristine carnosine, suggesting that carnosine acylation does not affect xanthine oxidase inhibition potential of carnosine. On the contrary, it has been demonstrated that *N*-acylation of carnosine increased the superoxide scavenging activity of the free dipeptide (IC_{50} increased by a factor 1.3). As suggested by other authors, this result might be explained by the higher hydrophobicity and the resulting higher availability of the dipeptide derivative in the reaction system (Murase, Nagao and Terao, 1993). These results showed that enzymatic *N*-acylation should provide derivatives with improved bioavailability and higher resistance towards endogenous proteases without affecting some of the interesting biological activities of carnosine.

15.4.3 Vectorisation

One approach that can be used for the selective delivery of carnosine consisted in the use of NPs. They are attracting considerable and growing interest because of their unique physical and chemical properties. The integration of nanotechnology with biology and medicine has led to further development of a new emerging research area, nanobiotechnology, which offers opportunities for discovering new materials, processes and phenomena (Bellia *et al.*, 2013).

NPs are a new class of carriers in nanomedicine, with special applications in the case of cancer. Recently, the interest on NPs has been increasing and also NPs based on carnosine have been obtained. For instance, carnosine has been functionalised with L-lipoic acid to synthesise gold NPs (Winum and Supuran, 2011). These NPs have been synthesised as activators of carbonic anhydrase. The role of carnosine could be related to the presence of histidine, and the activity of carnosine NPs is very similar to that of histidine NPs. Recently, carnosine-coated iron-oxide and gold NPs have been obtained. L-carnosine-coated iron-oxide nanoparticles (CCIO NPs) have been prepared via co-precipitation of iron oxide in the presence of L-carnosine.

Magnetic NPs have been surface-modified for the assembly of molecules whose function, when attached, often defines their application. For example, magnetic NPs have been reported in magnetofection and drug delivery as well as in biosensing by attaching molecules including chemicals, oligonucleotides and proteins (Ma *et al.*, 2004; Nidumolu *et al.*, 2006). These NPs have been investigated as activators of carbonic anhydrase and nickel sensors.

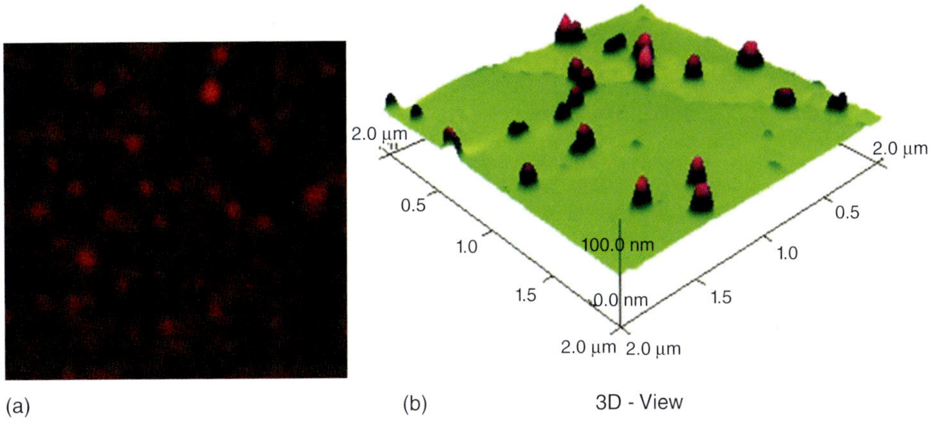

Figure 15.10 Rhodamine – labelled POPC in STED multiphoton (a) and 3D reconstruction of DPPC – liposomal membrane by atomic force microscopy (b).

Liposomes are widely used as biocompatible vehicles for the delivery of drugs and gene-based therapy. The effectiveness of liposomes as delivery vehicles depends on the concentration of the entrapped species, and much development effort has been devoted to increase vesicular loading. The encapsulation efficiency, which is a measure of the percentage of the total compound entrapped within the liposome, is an important parameter in liposomal characterisation (Perkins *et al.*, 1993; Vance *et al.*, 1985).

The entrapment of carnosine into nanoliposomes might represent an alternative to overcome the problems related to the direct application of these antioxidant peptides in food (Maherani *et al.*, 2011). Encapsulation of antioxidants in nanoliposomes offers a potential solution to not only protect antioxidants but also enhance their efficacy and stability in food applications. Another advantage of liposomal delivery systems is the ability to release components on demand (Were *et al.*, 2003).

Maherani *et al.* (2012) showed that encapsulation efficiency tended to increase by augmentation of the saturation degree of the lipids use in liposome membrane. Membrane fluidity was also one of the major factors affecting the encapsulation efficiency of the carnosine. The average height of individual surface-bound DPPC-liposome was 48 nm (Figure 15.10).

15.5 Applications of Carnosine and Its Derivatives

15.5.1 Nutraceutics and Food Supplementation

Recently, carnosine has gained increasing attention as a functional ingredient for human food because of its high antioxidant activity (Mozdzan *et al.*, 2005), high buffering capacity to maintain intracellular pH change (Abe, 2000) and anti-glycating and anti-aldehyde effects (Guiotto *et al.*, 2005). Previous experiments indicated that carnosine concentrations in animal products and human body could be increased by dietary manipulation.

Ma *et al.* (2010) showed the effect of dietary carnosine supplementation on antioxidant capacity and quality of pig meat. They showed that addition of 100 mg carnosine per kilogram diet increased glutathione peroxidase (GSH-Px), SOD and catalase (CAT) activities in plasma, liver or muscle, as well as SOD and GSH-Px gene expression in muscle. As an endogenous peptide, carnosine is safe for oral delivery. It is absorbed as the intact dipeptide with hydrolysis in the intestinal mucosa readily saturable. However, human plasma carnosine levels are low due to rapid hydrolysis by carnosinase. Synthetic derivatives are less susceptible to hydrolysis and may have therapeutic potential. In contrast, chronic supplementation of rodents results in persistent and significant elevations of both plasma and tissue (e.g. aorta, heart, kidney, liver and spleen) levels; this, therefore, provides a suitable model to examine the potential of carnosine as an anti-glycation/anti-dyslipidaemic agent (Brown *et al.*, 2014).

15.5.2 Cosmetics

Due to its various biological activities such as antioxidant, carnosine can also find applications in cosmetics and can be found in a range of skin-care preparations. As carnosine can inhibit the lipid peroxidation within the cell membrane due to the ability of imidazole in histidine to trap the lipid peroxy-radical, the benefits of antioxidants such as carnosine and carnosine derivatives in skin care were already reported. Thus, it was shown that they help to protect and repair cell membrane damage and lead to younger looking skin. The role of imidazole-containing dipeptide (NAC, carcinine, carnosine) on the mechanisms of skin diseases, ageing, cellular and signalling pathways mediated by targeting molecular chaperone protein(s) in skin-care formulations was also demonstrated (Babizhayev *et al.*, 2001).

Studies were carried out on the topical effects on the skin: experiments with carnosine applied on mice skin showed that this dipeptide has immunomodulating properties and interesting immunoprotective action from UVB radiation (Reeve, Bosnic and Rozinova, 1993). Other results showed that carnosine and NAC can present an effect of UVB photoprotection when applied on human skin (Nino *et al.*, 2011).

15.5.3 Pharmaceuticals

Since its discovery in the beginning of twentieth century, carnosine has crossed the decades carrying with it an abundance of promising therapeutic applications but encountering to date very few real applications (Guiotto *et al.*, 2005; Sale *et al.*, 2013). Studies carried out on laboratory animals and humans have shown that carnosine can have a beneficial influence on the organism. Due to its antioxidant, protective, chelating, anti-glycation activity, this dipeptide can be used to prevent and treat diseases such as diabetes, neurodegenerative diseases, diseases of the sense organs and cancers.

Many studies have indeed underlined the pharmacological properties and the potential therapeutic applications of the carnosine (Boldyrev, 1992; Budzeń and Rymaszewska, 2013). The first therapeutic use postulated for the carnosine dates from 1935 for the treatment of polyarthritis and the last corresponds to protection against diabetes in 2012. Indeed, the administration of carnosine seems to protect humans against diabetic nephropathy.

Some other therapeutic actions have also been proposed. Indeed, in rats, carnosine probably inhibits splenic sympathetic nerve activity and proliferation of cancer cells. It has been shown that in humans, carnosine may inhibit the growth of glioblastoma multiform, a malignant tumour of the brain. Other studies showed that L-carnosine inhibits the proliferation of human colorectal carcinoma cells by affecting the ATP and ROS production and by inducing the cell cycle arrest in G1 phase (Iovine *et al.*, 2012). The Hypoxia-Inducible Factor 1a (HIF-1a) was suggested as a possible target of L-carnosine in HCT-116 cell line. Experimental data showed that L-carnosine reduces the HIF-1a protein level affecting its stability and decreases the HIF-1 transcriptional activity (Iovine *et al.*, 2014).

The hepatoprotective effect of carnosine is related to the antioxidant mechanism of carnosine due to its chelating effect against metal ions, its SOD-like activity, its ROS and free radicals scavenging ability. Thanks to these properties, carnosine has been used to prevent damage due to radiation therapy, necrotic and apoptotic cell death in hepatic tissue due to ischemia/reperfusion (Badr, 2007). It has also been proven that polaprezinc, the zinc (II) – carnosine complex, is effective for the recovery of ulcers and other lesions in the alimentary tract (Katayama *et al.*, 2000). Moreover, human consumption of natural products with a high carnosine content can inhibit the growth of influenza virus infections.

The metal binding ability of L-carnosine especially for copper (II) and zinc (II) ions has been extensively studied (Baran, 2000). The copper- and zinc-mediated neurotoxicity involved in several pathologies, such as amyotrophic lateral sclerosis, Alzheimer's and Parkinson's diseases (Barnham and Bush 2008), might be reduced or prevented by endogenous metal-chelating agents, such as L-carnosine. For instance, an effective treatment of the Parkinson's disease has been presented with the combination of the standard therapy with carnosine (Boldyrev *et al.*, 2008). The transition metal-chelating effect of carnosine may assist in the transport of these elements into the body. Disturbance in copper metabolism may be associated with particular pathological states, for example, Wilson's disease. The antioxidant properties of carnosine give it a role in the protection of cells and tissues against peroxidative processes which induce a number of diseases (Guiotto *et al.*, 2005; Orioli *et al.*, 2011; Xie *et al.*, 2013).

The principal limitations of carnosine use are certainly due to the issues surrounding bioavailability. Finally, the most realistic application until now is the formulation of ophthalmic solutions. Indeed, carnosine exhibited optimal pharmaceutical characteristics for use as a buffer in chronically administered topical ocular formulations (Singh *et al.*, 2009; Sunkireddy *et al.*, 2013).

In conclusion, carnosine exhibits many biological activities and is a very promising therapeutic molecule. However, it has not been widely used until now because of insufficient knowledge about its mechanism of action and the basis for its use in medical treatment (Boldyrev, Aldini and Derave, 2013).

References

Abe, H. (2000) Role of histidine-related compounds as intracellular proton buffering constituents in vertebrate muscle. *Biochemistry (Moscow)*, **65**, 757–765.

Amorini, A.M., Bellia, F., Di Pietro, V., Giardina, B., La Mendola, D., Lazzarino, G., Sortino, S., Tavazzi, B., Rizzarelli, E., and Vecchio, G. (2007) Synthesis and antioxidant activity of new homocarnosine beta-cyclodextrin conjugates. *Eur. J. Med. Chem.*, **42**, 910–920.

Arima, J., Morimoto, M., Usuki, H., Mori, N., and Hatanaka, T. (2010) β-Alanyl peptide synthesis by Streptomyces S9 aminopeptidase. *J. Biotechnol.*, **147**, 52–58.

Astete, C.E., Songe Meador, D., Spivak, D., and Sabliov, C. (2013) Synthesis of Vitamin E-Carnosine (VECAR): new antioxidant molecule with potential application in atherosclerosis. *Synth. Commun.*, **43**, 1299–1313.

Babizhayev, M.A., Deyev, A.I., Yermakova, V.N., Semiletov, Y.A., Davydova, N.G., Kurysheva, N.I., Zhukotskii, A.V., and Goldman, I.M. (2001) N-acetylcarnosine, a natural histidine-containing dipeptide, as a potent ophthalmic drug in treatment of human cataract. *Peptides*, **22**, 979–994.

Babizhayev, M.A. (2006) Biological activities of the natural imidazole-containing peptidomimetics N-acetylcarnosine, carcinine and L-carnosine in ophthalmic and skin care products. *Life Sci.*, **78**, 2343–2357.

Babizhayev, M.A., Guiotto, A., and Kasus-Jacobi, A. (2009) N-acetylcarnosine and histidyl-hydrazine are potent agents for multitargeted ophthalmic therapy of senile cataract and diabetic ocular complications. *J. Drug Targeting*, **17** (1), 36–63.

Badr, H.M. (2007) Antioxidative activity of carnosine in gamma irradiated ground beef and beef patties. *Food Chem.*, **104**, 665–679.

Baguet, A., Koppo, K., Pottier, A., and Derave, W. (2010) Beta-alanine supplementation reduces acidosis but not oxygen uptake response during high-intensity cycling exercise. *Eur. J. Appl. Physiol.*, **108**, 495–503.

Baran, E.J. (2000) Metal complexes of carnosine. *Biochemistry (Mosc)*, **65**, 789–797.

Barger, G. and Tutin, F. (1918) Carnosine, constitution and synthesis. *Biochemistry*, **12**, 402–405.

Barnham, K.J. and Bush, A.I. (2008) Metals in Alzheimer's and Parkinson's diseases. *Current Opinion in Chemical Biology*, **12**, 222–228.

Baslow, M.H. (2010) A novel key-lock mechanism for inactivating amino acid neurotransmitters during transit across extracellular space. *Amino Acids*, **38**, 51–55.

Baumann, L. and Ingvaldsen, T. (1918) Concerning histidine and carnosine. The synthesis of carnosine. *J. Biol. Chem.*, **35**, 263–276.

Bauer, K. and Schulz, M. (1994) Biosynthesis of carnosine and related peptides by skeletal muscle cells in primary culture. *Eur. J. Biochem.*, **219** (1-2), 43–47.

Bellia, F., Amorini, A.M., La Mendola, D., Vecchio, G., Tavazzi, B., Giardina, B., Di Pietro, V., Lazzarino, G., and Rizzarelli, E. (2008) New glycosidic derivatives of histidine-containing dipeptides with antioxidant properties and resistant to carnosinase activity. *Eur. J. Med. Chem.*, **43**, 373–380.

Begum, G., Cunliffe, A., and Leveritt, M. (2005) Physiological role of carnosine in contracting muscle. *Int. J. Sport Nutr. Exercise Metab.*, **15**, 493–514.

Bellia, F., Oliveri, V., Rizzarelli, E., and Vecchio, G. (2013) New derivatives of carnosine for nanoparticules assemblies. *Eur. J. Med. Chem.*, **70**, 225–232.

Bertinaria, M., Rolando, B., Giorgis, M., Montanaro, G., Guglielmo, S., Buonsanti, M.F., Carabelli, V., Gavello, D., Daniele, P.G., Fruttero, R., and Gasco, A. (2011) Synthesis, physicochemical characterization, and biological activities of new carnosine derivatives stable in human serum as potential neuroprotective agents. *J. Med. Chem.*, **54**, 611–621.

Bertinaria, M., Rolando, B., Giorgis, M., Montanaro, G., Marini, E., Collino, M., Benetti, E., Daniele, P.G., Fruttero, R., and Gasco, A. (2012) Carnosine analogues containing NO-donor substructures: synthesis, physico-chemical characterization and preliminary pharmacological profile. *Eur. J. Med. Chem.*, **54**, 103–112.

Boldyrev, A.A. (1992) Carnosine: biological role and potential applications in medicine. *Biochemistry (Moscow)*, **57**, 898–904.

Boldyrev, A.A. and Abe, H. (1999) Metabolic transformation of neuropeptide carnosine modifies its biological activity. *Cell. Mol. Neurobiol.*, **19**, 163–175.

Boldyrev, A., Bulygina, E., Leinsoo, T., Petrushanko, I., Tsubone, S., and Abe, H. (2004) Protection of neuronal cells against reactive oxygen species by carnosine and related compounds. *Comp. Biochem. Physiol. B: Biochem. Mol. Biol.*, **137**, 81–88.

Boldyrev, A., Fedorova, T., Boldanova, N., Bagyeva, G., Ivanova Smolenskaya, I., and Illarioshkin, S. (2008) Carnosine increases efficiency of DOPA therapy of Parkinson's disease. *Rejuvenation Res.*, **11**, 988–994.

Boldyrev, A.A., Aldini, G., and Derave, W. (2013) Physiology and pathophysiology of carnosine. *Physiol. Rev.*, **93** (4), 1803–1845.

Bonomo, R.P., Bruno, V., Conte, E., De Guidi, G., La Mendola, D., Maccarrone, G., Nicoletti, F., Rizzarelli, E., Sortino, S., and Vecchio, G.J. (2003) Potentiometric, spectroscopic and antioxidant activity studies of SOD mimics containing carnosine. *Dalton Trans.*, **23**, 4406–4415.

Brown, B.E., Kim, C.H., Torpy, F.R., and Bursill, C.A. (2014) Supplementation with carnosine decreases plasma triglycerides and modulates atherosclerotic plaque composition in diabetic apo E(-/-) mice. *Atherosclerosis*, **232** (2), 403–409.

Budzeń, S. and Rymaszewska, J. (2013) The biological role of carnosine and its possible applications in medicine. *Adv. Clin. Exp. Med.*, **22** (5), 739–744.

Calcagni, A., Ciattini, P.G., Di Stefano, A., Duprè, S., Luisi, G., Pinnen, F., Rossi, D., and Spirito, A. (1999) $\Psi(SO_2NH)$ transition state isosteres of peptides. Synthesis and bioactivity of sulfonamido pseudopeptides related to carnosine. *Il Farmaco*, **54**, 673–677.

Castelletto, V., Cheng, G., Stain, C., Connon, C.J., and Hamley, I.W. (2012) Self-assembly of a peptide amphiphile containing L-carnosine and its mixtures with a multilamellar vesicle forming lipid. *Langmuir*, **28**, 11599–11608.

Chan, K.M., Decker, E.A., and Means, W.J. (1993) Extraction and activity carnosine, a naturally occurring antioxidant in beef muscle. *J. Food Sci.*, **58** (1), 1–4.

Cherevin, M.S., Zubreichuk, Z.P., Popova, L.A., Gulevich, T.G., and Knizhnikov, V.A. (2007) Trifluoroacetyl-β-alanine in the synthesis of carnosine. *Russ. J. Gen. Chem.*, **7** (9), 1576–1579.

Cosgun, S., Ozer, M., Hamdoune, F., Gerardin, C., Thiebaut, S., Henry, B., Amos, J., Rodehüser, L., and Selve, C. (2001) Amphiphilic analogues of peptidoamines with perfluorinated side chains. Synthesis and preliminary investigations of their surfactant and complexing abilities. *J. Fluorine Chem.*, **107**, 375–386.

Cosgun, S., Gérardin-Charbonnier, C., Amos, J., and Selve, C. (2004) Efficient synthesis of new perfluorinated or hybrid amphiphilic surfactants. *J. Fluorine Chem.*, **125**, 55–61.

Decker, E.A., Livisay, S.A., and Zhou, S. (2000) A re-evaluation of the antioxidant activity of purified carnosine. *Biochemistry*, **65**, 766–770.

Di Paola, R., Impellizzeri, D., Salinaro, A.T., Mazzon, E., Bellia, F., Cavallaro, M., Cornelius, C., Vecchio, G., Calabrese, V., Rizzarelli, E., and Cuzzocrea, S. (2011) Administration of carnosine in the treatment of acute spinal cord injury. *Biochem. Pharmacol.*, **82**, 1478–1489.

Fernandez-Perez, M. and Otero, C. (2001) Enzymatic synthesis of amide surfactants from ethanolamine. *Enzyme Microb. Technol.*, **28** (6), 527–536.

Fouad, A.A., El-Rehany, M.A.-A., and Maghraby, H.K. (2007) The hepatoprotective effect of carnosine against ischemia/reperfusion liver injury in rats. *Eur. J. Pharmacol.*, **572**, 61–68.

Gardossi, L., Bianchi, D., and Klibanov, A.M. (1991) Selective acylation of peptides catalyzed by lipases in organic solvents. *J. Am. Chem. Soc.*, **113**, 6328–6329.

Gizzi, P., Pasc, A., Dupuy, N., Parant, S., Henry, B., and Gérardin, C. (2009) Molecular tailored histidine-based complexing surfactants: from micelles to hydrogels. *Eur. J. Org. Chem.*, **2009** (23), 3953–3963.

Guiotto, A., Calderan, A., Ruzza, P., and Borin, G. (2005) Carnosine and carnosine-related antioxidants: a review. *Curr. Med. Chem.*, **12**, 2293–2315.

Gulewitsch, W. and Amiradzibi, S. (1900) Ueber das Carnosin, eine neue organische Base des Fleischextraktes. *Ber. Dtsch. Chem. Ges.*, **33**, 1902–1903.

Guney, Y., Ozel Turkcu, U., Hicsonmez, A., Nalca Andrieu, M., Guney, H.Z., Bilgihan, A., and Kurtman, C. (2006) Carnosine may reduce lung injury caused by radiation therapy. *Med. Hypotheses*, **66**, 957–959.

Hamdoune, F., El Moujahid, C., Rodehüser, L., Gérardin, C., Henry, B., Stébé, M.-J., Amos, J., Marraha, M., Asskali, A., and Selve, C. (2000) Amphiphilic and cation-complexing compounds based on peptidoamines. *New J. Chem.*, **24**, 1037–1042.

Hamdoune, F., El Moujahid, C., Stébé, M.-J., Gérardin, C., Tekely, P., Selve, C., and Rodehüser, L. (2001) Interactions of colloidal grafted silica with transition-metal cations. *Prog. Colloid Polym. Sci.*, **118**, 107–109.

Heck, T., Makam, V.S., Lutz, J., Blank, L.M., Schmid, A., Seebach, D., Kohler, H.-P.E., and Geueke, B. (2010) Kinetic analysis of L-carnosine formation by β–aminopeptidases. *Adv. Synthesis Catal.*, **352**, 407–415.

Heyland, J., Antweiler, N., Lutz, J., Heck, T., Geueke, B., Kohler, H.P.E., Blank, L.M., and Schmid, A. (2010) Simple enzymatic procedure for L-carnosine synthesis: whole-cell biocatalysis and efficient biocatalyst recycling. *Microb. Biotechnol.*, **3** (1), 74–78.

Husson, E., Humeau, C., Blanchard, F., Framboisier, X., Marc, I., and Chevalot, I. (2008) Chemo selectivity of the N,O-enzymatic acylation in organic media and in ionic liquids. *J. Mol. Catal. B: Enzym.*, **55**, 110–117.

Husson, E., Humeau, C., Paris, C., Vanderesse, R., Framboisier, X., Marc, I., and Chevalot, I. (2009) Enzymatic acylation of polar dipeptides: influence of reaction media and molecular environment of functional groups. *Process Biochem.*, **44**, 428–434.

Husson, E., Garcia-Matilla, V., Humeau, C., Chevalot, I., Fournier, F., and Marc, I. (2010) Enzymatic acylation of a bifunctional molecule in 2-methyl-2-butanol: kinetic modelling. *Enzyme Microb. Technol.*, **46**, 338–346.

Husson, E., Humeau, C., Harscoat, C., Framboisier, X., Paris, C., Dubreucq, E., Marc, I., and Chevalot, I. (2011) Enzymatic acylation of the polar dipeptide, carnosine: reaction performances in organic and aqueous media. *Process Biochem.*, **46**, 945–952.

Inaba, C., Higuchi, S., Morisaka, H., Kuroda, K., and Ueda, M. (2010) Synthesis of functional dipeptide carnosine from nonprotected amino acids using carnosinase-displaying yeast cells. *Appl. Microbiol. Biotechnol.*, **86**, 1895–1902.

Intarapichet, K.O. and Maikhunthod, B. (2005) Genotype and gender difference in carnosine extract and antioxidant activities of chicken breast and thigh meats. *Meat Sci.*, **71**, 634–642.

Iovine, B., Iannella, M.L., and Nocella, F. (2012) Carnosine inhibits KRAS-mediated HCT116 proliferation by affecting ATP and ROS production. *Cancer Lett.*, **315**, 122–128.

Iovine, B., Oliviero, G., Garofalo, M., and Orefice, M. (2014) The Anti-Proliferative Effect of L-Carnosine Correlates with a Decreased Expression of Hypoxia Inducible Factor 1 alpha in Human Colon Cancer Cells. *PLoS ONE*, **9** (5), e96755.

Kang, J.H., Kim, K.S., Choi, S.Y., Kwon, H.Y., Won, M.H., and Kang, T.C. (2002) Carnosine and related dipeptides protect human ceruloplasmin against peroxyl radical-mediated modification. *Mol. Cells*, **13**, 498–502.

Katayama, S., Nishizawa, K., Hirano, M., Yamamura, S., and Momose, Y. (2000) Effect of polaprezinc on healing of acetic acid-induced stomatitis in hamsters. *Journal of Pharmacy and Pharmaceutical Sciences*, **3**, 114–117.

Kim, S.-K., Kwon, D., Kwon, D.-A., Paik, I.K., and Auh, J.-H. (2014) Optimizing carnosine extract preparation from chicken breast for anti-glycating agents. *Korean J. Food Sci. Technol.*, **34** (1), 127–132.

Kimmerlin, T. and Seebach, D. (2005) 100 years of peptide synthesis: ligation methods for peptide and protein synthesis with application to β-peptide assemblies. *J. Pept. Res.*, **65** (2), 229–260.

Klebanov, G.I., Veselkin, Y.O., Babenkova, I.V., Popov, I.N., Levin, G., Tyvlina, O.V., Boldyrev, A.A., and Vladimirov, Y.A. (1997) Evidence for a direct interaction of superoxide anion radical with carnosine. *Biochem. Mol. Biol. Int.*, **43**, 99–106.

Kohen, R., Yamamoto, Y., Cundy, K.C., and Ames, B.N. (1988) Antioxidant activity of carnosine, homocarnosine, and anserine present in muscle and brain. *Proc. Natl. Acad. Sci. U.S.A.*, **85** (9), 3175–3179.

Koreishi, M., Kawasaki, R., Imanaka, H., Imamura, K., and Nakanishi, K. (2005) A novel lysine acylase from *Streptomyces mobaraensis* for synthesis of N-acyl-L-lysines. *J. Am. Oil Chem. Soc.*, **82** (9), 631–637.

Koreishi, M., Zhang, D., Imanaka, H., Imamura, K., Adachi, S., Matsuno, R., and Nakanishi, K. (2006) A novel acylase from *Streptomyces mobaraensis* that efficiently catalyzes hydrolysis/synthesis of capsaicins as well as N-acyl-L-amino acids and N-acyl-peptides. *J. Agric. Food. Chem.*, **54** (1), 72–78.

La Mendola, D., Sortino, S., Vecchio, G., and Rizzarelli, E. (2002) Synthesis of new carnosine derivatives of beta-cyclodextrine and their hydroxyl radical savenger ability. *Helv. Chim. Acta*, **85**, 1633–1643.

Lanza, V., Bellia, F., D'Agata, R., Grasso, G., Rizzarelli, E., and Vecchio, G. (2011) New glycoside derivatives of carnosine and analogs resistant to carnosinase hydrolysis: synthesis and characterization of their copper(II) complexes. *J. Inorg. Chem.*, **105**, 181–188.

Liu, W.-H., Liu, T.-C., and Yin, M.-C. (2008) Beneficial effects of histidine and carnosine on ethanol-induced chronic liver injury. *Food Chem. Toxicol.*, **46**, 1503–1509.

Ma, Y.H., Manolache, S., Denes, F.S., Thamm, D.H., Kurzman, I.D., and Vali, D.M. (2004) Plasma synthesis of carbon magnetic nanoparticles and immobilization of doxorubicin for targeted drug delivery. *J. Biomater. Sci. Polym. Ed.*, **15** (8), 1033–1049.

Ma, X.Y., Jiang, Z.Y., Lin, Y.C., Zheng, C.T., and Zhou, G.L. (2010) Dietary supplementation with carnosine improves antioxidants capacity and meat quality of finishing pigs. *J. Anim. Physiol. Anim. Nutr.*, **94**, 286–295.

Maherani, B., Arab-Tehrany, E., Mozafari, M.R., Gaiani, C., and Linder, M. (2011) A liposomes: review of manufacturing techniques and targeting strategies. *Curr. Nanosci.*, 7, 436–452.

Maherani, B., Arab-Tehrany, E., Kheirolomoom, A., Cleymand, F., and Linder, M. (2012) Influence of lipid composition on physicochemical properties of natural dipeptide antioxidant (L-carnosine) loaded Liposome. *Food Chem.*, **134** (2), 632–640.

Manhiani, P.S., Morthcutt, J.K., Han, I., Bridges, W.C., and Dawson, P.L. (2013) Antioxidant activity of carnosine extracted from various poultry tissues. *Poult. Sci.*, **92** (2), 444–453.

Mehmetçik, G., Özdemirler, G., Koçak-Toker, N., Çevikbaş, U., and Uysal, M. (2008) Role of carnosine in preventing thioacetamide-induced liver injury in the rat. *Peptides*, **29**, 425–429.

Mineo, P., Vitalini, D., La Mendola, D., Rizzarelli, E., Scamporrino, E., and Vecchio, G. (2004) Coordination features of difunctionalized beta-cyclodextrins with carnosine: ESI-MS and spectroscopic investigations on 6A, 6D-di-(beta-alanyl-L-histidine)-6A, 6D-dideoxy-beta-cyclodextrin and 6A,

6C-di-(beta-alanyl-L-histidine)-6A, 6C-dideoxy-beta-cyclodextrin and their copper (II) complexes. *J. Inorg. Biochem.*, **98**, 254–265.

Montet, D., Servat, F., Pina, M., Graille, J., Galzy, P., Arnaud, A., Ledon, H., and Marcou, L. (1990) Enzymatic synthesis of N-ε-acyllysines. *J. Am. Oil Chem. Soc.*, **67**, 771–774.

Mozdzan, M., Szemraj, J., Rysz, J., and Nowak, D. (2005) Antioxidant properties of carnosine re-evaluated with oxidizing systems involving iron and copper ions. *Basic and Clinical Pharmacology*, **96**, 352–360.

Murase, H., Nagao, A., and Terao, J. (1993) Antioxidant and emulsifying activity of N-(long-chain-acyl)histidine and N-(long-chain-acyl)carnosine. *J. Agric. Food. Chem.*, **41**, 1601–1604.

Nabetani, H., Hagiwara, S., Yanai, N., Shiotani, S., Baljinnyam, J., and Nakazima, M. (2012) Purification and concentration of antioxidative dipeptides obtained from chicken extract and their application as functional food. *J. Food Drug Anal.*, **20**, 179–183.

Nagai, K., Niijima, A., Yamano, T., Otani, H., Okumra, N., Tsuruoka, N., Nakai, M., and Kiso, Y. (2003) Possible role of L-carnosine in the regulation of blood glucose through controlling autonomic nerves. *Exp. Biol. Med.*, **228**, 1138–1145.

Nidumolu, B.G., Urbina, M.C., Hormes, J., Kumar, C.S., and Monrore, W.T. (2006) Functionalization of gold and glass surfaces with magnetic nanoparticles using biomolecular interactions. *Biotechnol. Progr.*, **22** (1), 91.

Nielsen, C., Supuran, C., Scozzafava, A., Frokjaer, S., Steffansen, B., and Brodin, B. (2002) Transport characteristics of L-carnosine and the anticancer derivative 4-toluenesulfonylureido-carnosine in a human epithelial cell line. *Pharm. Res.*, **19**, 1337–1344.

Nino, M., Iovine, B., and Santoianni, P. (2011) Carnosine and N-Acetylcarnosine Induce Inhibition of UVB Erythema in Human Skin. *Journal of Cosmetics, Dermatological Sciences and Applications*, **1**, 177–179.

Orioli, M., Vistoli, G., Regazzoni, L., Pedretti, A., Lapolla, A., Rossoni, G., Canevotti, R., Gamberoni, L., Previtali, M., Carini, M., and Aldini, G. (2011) Design, synthesis, ADME properties, and pharmacological activities of beta-alanyl-D-histidine (D-carnosine) prodrugs with improved bioavailability. *ChemMedChem*, **6**, 1269–1282.

Pal, A., Shrivastava, S., and Dey, J. (2009) Salt, pH and thermoresponsive supramolecular hydrogel of N-(4-n-tetradecyloxybenzoyl)-L-carnosine. *Chem. Commun.*, **45**, 6997–6999.

Pegova, A., Abe, H., and Boldyrev, A. (2000) Hydrolysis of carnosine and related compounds by mammalian carnosinases. *Comp. Biochem. Physiol. B: Biochem. Mol. Biol.*, **127**, 443–446.

Perkins, W.R., Minchey, S.R., Ahl, P.L., and Janoff, A.S. (1993) The determination of liposome captured volume. *Chem. Phys. Lipids*, **64**, 197–217.

Pis-Diez, E. and Baran, E.J. (2003) A density functional study of some physical properties of carnosine (N-β-alanyl-L-histidine). *J. Mol. Struct. (Theochem)*, **621**, 245–251.

Quinn, P.J., Boldyrev, A.A., and Formazuyk, V.E. (1992) Carnosine: its properties, functions and potential therapeutic applications. *Mol. Aspects Med.*, **13**, 379–444.

Reeve, V.E., Bosnic, M., and Rozinova, E. (1993) Carnosine (beta-alanylhistidine) protects from the suppression of contact hypersensitivity by ultraviolet B (280–320 nm) radiation or by cis-urocanic acid. *Immunology*, **78** (1), 99–104.

Renner, C., Zemitzsch, N., Fuchs, B., Geiger, K.D., Hermes, M., Hengstler, J., Gebhardt, R., Meixensberger, J., and Gaunitz, F. (2010) Carnosine retards tumor growth in vivo in an NIH3T3-HER2/neu mouse model. *Mol. Cancer*, **9** (2), 1–7.

Roberts, P.R. and Zaloga, G.P. (2000) Cardiovascular effects of carnosine. *Biochemistry*, **65**, 856–861.

Sakakibara, M., Wang, D., Takahashi, R., Takahashi, K., and Mori, S. (1996) Influence of ultrasound irradiation on hydrolysis of sucrose catalyzed by invertase. *Enzyme Microb. Technol.*, **18**, 444–448.

Sale, C., Artioli, G.G., Gualamo, B., Saunders, B., Hobson, R.M., and Harris, R.C. (2013) Carnosine: from exercice performance to health. *Amino Acids*, **44**, 1477–1491.

Sayen, S., Gérardin, C., Rodehüser, L., and Walcarius, A. (2003) Electrochemical detection of copper(II) at an electrode modified by a carnosine-silica hybrid material. *Electroanalysis*, **15** (5-6), 422–430.

Schönherr, J. (2002) Analysis of products of animal origin in feeds by determination of carnosine and related dipeptides by high-performance liquid chromatography. *J. Agric. Food. Chem.*, **50**, 1945–1950.

Schroder, L., Schmitz, C.H., and Bachert, P. (2008) Carnosine as molecular probe for sensitive detection of Cu(II) ions using localized (1)H NMR spectroscopy. *J. Inorg. Biochem.*, **102**, 174–183.

Sheldon, R.A., Lau, R.M., Sorgedrager, M.J., van Rantwijk, F., and Seddon, K.R. (2002) Biocatalysis in ionic liquids. *Green Chem.*, **4**, 147–151.

Shen, Y., Fan, Y., Dai, H., Fu, Q., Hu, W., and Chen, Z. (2007) Neuroprotective effect of carnosine on necrotic cell death in PC12 cells. *Neurosci. Lett.*, **414**, 145–149.

Sifferd, R.H. and Du Vignaud, V. (1935) A new synthesis of carnosine, with some observations on the splitting of the benzyl group from carbobenzoxy derivatives and from benzylthio ethers. *J. Biol. Chem.*, **108**, 753–761.

Singh, S.R., Carreiro, S.T., Chu, J., Prasanna, G., Niesman, M.R., Collette, W.W., Younis, H.S., Sartnurak, M.R., and Gukasyan, H.J. (2009) L-carnosine: multifunctional dipeptide buffer for sustained-duration topical ophthalmic formulations. *J. Pharm. Pharmacol.*, **61**, 733–742.

Soo, E.L., Salleh, A.B., Basri, M., Rahman, R.N.Z.A., and Kamaruddin, K. (2004) Response surface methodological study on lipase-catalyzed synthesis of amino acid surfactants. *Process Biochem.*, **39**, 1511–1518.

Stvolinsky, S.L., Bulygina, E.R., Fedorova, T.N., Meguro, K., Sato, T., Tyulina, O.V., Abe, H., and Boldyrev, A.A. (2010) Biological activity of novel synthetic derivatives of carnosine. *Cell. Mol. Neurobiol.*, **30**, 395–404.

Sunkireddy, P., Jha, S.N., Kanwar, J.R., and Yadav, S. (2013) Natural antioxidant biomolecules promises future nanomedicine based therapy for cataract. *Colloids Surf., B*, **112**, 554–562.

Tomonaga, S., Yamane, H., Onitsuka, E., Yamada, S., Sato, M., Takahata, Y., Morimatsu, F., and Furuse, M. (2008) Carnosine-induced antidepressant-like activity in rats. *Pharmacol. Biochem. Behav.*, **89**, 627–632.

Trombley, P.Q., Horning, M.S., and Blakemore, L.J. (1998) Carnosine modulates zinc and copper effects on amino acid receptors and synaptic transmission. *Neuroreport*, **9**, 3503–3507.

Vance, D.E. and Vance, J.E. (1985) (Eds) Biochemistry of Lipids and Membranes, The Benjamin/Cummings Publishing Company, Menlo Park, pp. 116–142.

van Rantwijk, F. and Sheldon, R.A. (2007) Biocatalysis in ionic liquids. *Chem. Rev.*, **107**, 2757–2785.

Vaysse, L., Ly, A., Moulin, G., and Dubreucq, E. (2002) Chain-length selectivity of various lipases during hydrolysis, esterification and alcoholysis in biphasic aqueous medium. *Enzyme Microb. Technol.*, **31**, 648–655.

Vladimirov, Y.A. (1994) in *Natural Antioxidant* (eds L. Packer, M.G. Trabe, and W. Xin), AOCS Press, Champaign, pp. 125–144.

Walcarius, A., Sayen, S., Gérardin, C., Hamdoune, F., and Rodehüser, L. (2004) Dipeptide-functionalized mesoporous silica spheres. *Colloids Surf., A*, **234**, 145–151.

Were, L.M., Bruce, B.D., Davidson, P.M., and Weiss, J (2003) Size, stability, and entrapment efficiency of phospholipid nanocapsules containing polypeptide antimicrobials. *J. Agric. Food. Chem.*, **51** (27), 8073–8079.

Winum, J.Y. and Supuran, C.T. (2011) Carbonic anhydrase activators: gold nanoparticles coated with derivatized histamine, histidine, and carnosine show enhanced activatory effects on several mammalian isoforms. *J. Med. Chem.*, **54**, 1170–1177.

Xiao, Y.-M., Wu, Q., Cai, Y., and Lin, X.-F. (2005) Ultrasound-accelerated enzymatic synthesis of sugar esters in nonaqueous solvents. *Carbohydr. Res.*, **340**, 2097–2103.

Xie, Z., Baba, S.P., Sweeney, B.R., and Barski, O.A. (2013) Detoxification of aldehyde by histidine-containing dipeptides: from chemistry to clinical implications. *Chem. Biol. Interact.*, **202**, 288–297.

Zaks, A. and Klibanov, A.M. (1985) Enzyme-catalyzed processes in organic solvents. *Proc. Natl. Acad. Sci. U.S.A.*, **82** (10), 3192–3196.

16
Metabolism and Biotechnological Production of Gamma-Aminobutyric Acid (GABA)

Feng Shi, Yalan Ni, and Nannan Wang

16.1
Introduction

Gamma-aminobutyric acid (GABA) is a non-protein amino acid widely distributed in various organisms, from microorganisms to plants and animals (Ueno, 2000). It is a major inhibitory neurotransmitter in mammals and has many physiological properties, such as anti-anxiety, hypotensive, tranquilising, diuretic and analgesic properties (Wong, Bottiglieri and Snead, 2003; Hayakawa *et al.*, 2004). Thus, it is recognised as the twenty-first century's health supplement and has been applied in food, feed and pharmaceutical fields. Moreover, GABA can be used as the precursor of a biodegradable plastic, polyamide 4, also known as Nylon-4 (Saskiawan, 2008). However, direct addition of chemically prepared GABA in the food and pharmaceutical fields is considered unsafe. So, it is necessary to prepare GABA by safe methods. Microbial fermentation and whole-cell or enzymatic conversion have a particular advantage.

The dominant GABA-producing strains include *Escherichia coli* (Fonda, 1985) and lactic acid bacteria (LAB) (Li and Cao, 2010). These strains possess high activity of glutamate decarboxylase (GAD), which catalyses the irreversible α-decarboxylation of L-glutamate (Glu) to GABA with the concomitant consumption of a proton. As LABs are generally regarded as safe, they have been well studied for GABA production. Various GABA-producing LABs have been screened from as many as possible traditional fermented foods, such as kimchi, Chinese traditional pao cai, Japanese traditional fermented fish, cheese, yoghurt and fresh milk, alcohol distillery lees as well as black raspberry juice (Dhakal, Bajpai and Baek, 2012). Moreover, many factors including pH, temperature, culture time and media additives have been optimised to achieve the maximum GAD activity and GABA production. Among them, the highest yield of GABA reached 1005.81 ± 47.88 mM (Li *et al.*, 2010). Besides the production of GABA by microbial fermentation and whole-cell conversion, production of GABA by enzymatic conversion has also been well studied. GADs derived from different microorganisms such as *E. coli* (Wang *et al.*, 2011), *Lactobacillus brevis* (Hiraga, Ueno and Oda, 2008) and *Streptococcus salivarius* ssp. *thermophilus* (Yang

Industrial Biotechnology of Vitamins, Biopigments, and Antioxidants, First Edition.
Edited by Erick J. Vandamme and José L. Revuelta.

et al., 2008) have been investigated. However, Glu or monosodium glutamate (MSG) must be added as the precursor during GABA production by these microorganisms or GAD enzymes. Recently, in order to synthesise GABA more cost-effectively, a single-step GABA fermentation system by recombinant *Corynebacterium glutamicum* expressing GAD gene(s) derived from LAB or *E. coli* had been created (Shi and Li, 2011; Takahashi *et al.*, 2012; Zhang *et al.*, 2014). This system can directly convert the endogenous Glu into GABA and thus is a promising method for GABA production.

This chapter focusses on the properties and occurrence of GABA in natural source, metabolism and regulatory mechanisms of GABA in different organisms, production of GABA by whole-cell and enzymatic conversion of Glu or MSG and *de novo* fermentation of *C. glutamicum*, physiological functions of GABA in mammals and current applications of GABA.

16.2 Properties and Occurrence of GABA in Natural Sources

The molecular formula of GABA is $C_4H_9NO_2$ with a relative molecular weight of 103.2. The structural formula is shown in Figure 16.1.

GABA has an amino group on the γ-carbon rather than on the α-carbon and is highly soluble (Shelp, Bown and McLean, 1999). It is zwitterionic at physiological pH values, carrying both a positive charge (NH_3^+ with a pK value of 10.56) and a negative charge (COO^- with a pK value of 4.03).

GABA was artificially synthesised as a chemical substance in the early 1880s and then was discovered as a natural substance in some fungi, bacteria and plants. It was until 1950 that Roberts and Frankel (1950) and Awapara *et al.* (1950) first individually identified large quantities of this amino acid in mammalian brain and hence attracted huge attention. It was demonstrated that GABA mimics the action of an endogenous neurotransmitter and plays an important role in regulating overall nervous system function in the late 1960s (Krnjevic and Schwartz, 1967). Later, three GABA receptors, that is, ionotropic $GABA_A$ receptor (Curtis *et al.*, 1970), metabotropic $GABA_B$ receptor (Bowery *et al.*, 1980) and ionotropic $GABA_C$ receptor (Johnston, 1996) were discovered and studied. Through research on the pharmacology, structure, genetics and cellular localisation of these GABA receptors, a large number of physiological functions of GABA have been discovered in succession. Due to these important functions, GABA shows great potentials in foods and medicines; therefore, safe and effective methods for GABA production were researched and developed. Besides mammals, more and more microorganisms have been found to exhibit higher GAD activity and accumulate GABA. Therefore, GABA productions

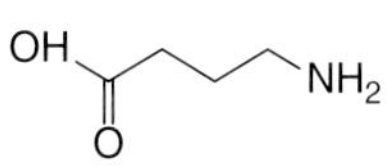

Figure 16.1 The structure of GABA.

by microorganisms have become a focus of research after 1990s, as mainly represented by *E. coli* and LAB.

16.3
Metabolism of GABA

GABA is widely distributed in nature from microorganisms to plants and mammals, in which GABA is synthesised from Glu by the enzyme GAD (Figure 16.2). In some of these living organisms, the synthesised GABA can be exported out of cells by a Glu/GABA antiporter (usually named as GadC) with concomitant importing of Glu, therefore, contributing to the acid resistance of these organisms. Besides the biosynthesis of GABA, some species can uptake the extracellular GABA from environment by a GABA-specific transporter/permease (usually named as GabP). The synthesised or obtained GABA can be degraded into succinate, which then flow into tricarboxylic acid (TCA) cycle; therefore, it can be utilised as carbon and/or nitrogen source in some living organisms. Degradation of GABA into succinate contains two reactions: conversion of GABA to succinic semialdehyde (SSA) by gamma-aminobutyric acid transaminase (GABA-T)

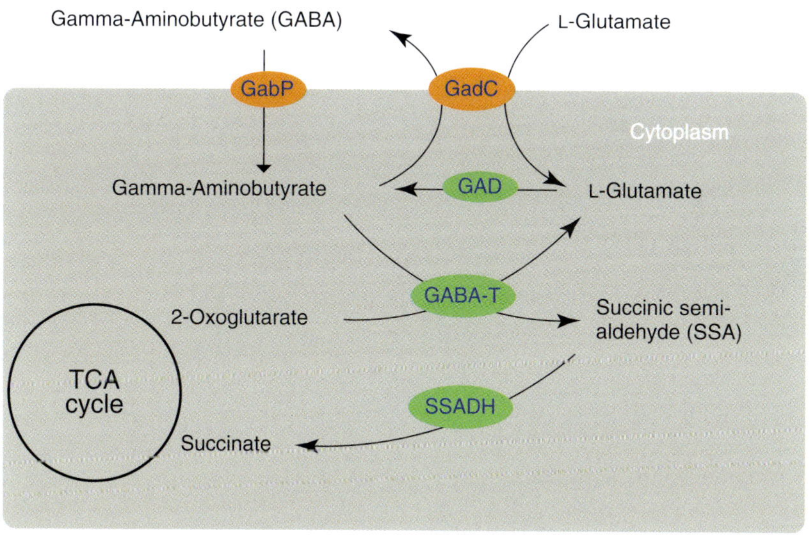

Figure 16.2 The metabolism pathway of GABA in bacteria. The pathway consists of biosynthesis and catabolic pathway, as well as export and uptake systems of GABA. The biosynthesis pathway is catalysed by L-glutamate decarboxylase (GAD). The catabolic pathway consists of two enzymatic steps. They are catalysed by GABA transaminase (GABA-T) and succinic semialdehyde dehydrogenase (SSADH). The export and uptake of GABA are carried out by Glu/GABA antiporter (GadC) and GABA permease (GabP). GadC mediates the exchange of intracellular GABA with the extracellular Glu. GabP transports GABA into cytoplasm. Other abbreviation: TCA, tricarboxylic acid cycle.

followed by conversion of SSA to succinate by succinic semialdehyde dehydrogenase (SSADH). This metabolic pathway relating to conversion of Glu to succinate via GABA is termed as GABA shunt. Recently, a model of the intracellular compartmentation of the GABA shunt has been introduced in *Arabidopsis* (Figure 16.3) where GAD is located in the cytosol, but GABA-T and SSADH are located in mitochondria (Shelp, Mullen and Waller, 2012). However, in some bacteria, only part of GABA shunt is found; some such as *L. brevis* possess only GABA biosynthesis enzyme (GAD), while some such as *C. glutamicum* possess only GABA degradation enzymes (GABA-T and SSADH) (Table 16.1).

Besides the biosynthesis and degradation of GABA, the export and uptake of GABA are also involved in GABA metabolism. Export of GABA is carried out by Glu/GABA antiporter (GadC) that mediates the exchange of intracellular GABA with the extracellular Glu. GadC mainly exists in microorganisms. The gene encoding GAD and GadC is often flanking and forms as an operon. The GAD/GadC system is involved in acid resistance (AR) system in some bacteria

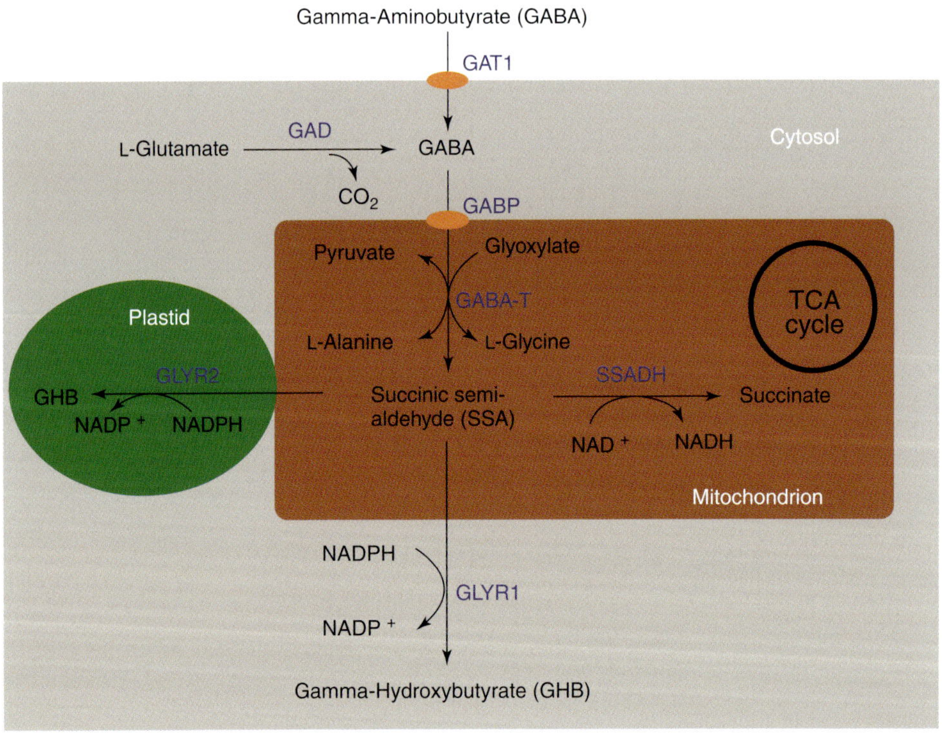

Figure 16.3 Compartmentation of GABA metabolism in *Arabidopsis*. The key enzymes involved are coloured blue. GAD is located in the cytosol, but GABA-T and SSADH are located in mitochondria. Two transporters, GAT1 (GABA transporter 1) and GABP (GABA permease), are located at the plasma and mitochondrial membrane, respectively. GLYR1 and GLYR2 (glyoxylate reductase isoforms 1 and 2) are located at cytosol and plastid, respectively. The published information by Shelp, Mullen and Waller (2012).

Table 16.1 GABA metabolic genes and their encoded proteins in several species.

Organism	Gene	Encoded protein
Escherichia coli K-12	*gadA, gadB*	GAD
	gadC	Glu/GABA antiporter
	gadX	GAD regulator
	gabT	GABA-T
	gabD	SSADH
	gabP	GABA transporter
Lactobacillus brevis ATCC 367	*LVIS_2213, LVIS_1847, LVIS_0079*	GAD
	LVIS_0078	Glu/GABA antiporter
	LVIS_0077	Transcriptional regulator
Corynebacterium glutamicum ATCC 13032	*NCgl0462*	GABA-T
	NCgl0463	NAD-dependent aldehyde dehydrogenase
	NCgl0464	Amino acid permease
Saccharomyces cerevisiae	*GAD1*	GAD
	UGA1	GABA-T
	UGA2	SSADH
	UGA4	GABA permease
	GAP1	Amino acid permease
	PUT4	Proline permease
	UGA3, UGA35	Positive regulator
	UGA43	Negative regulator
Aspergillus nidulans FGSC A4	*AN5447.2, AN7278.2*	GAD
	gatA	GABA-T
	ssuA	SSADH
	intA	Positive regulator
Neurospora crassa OR74A	*NCU06803, NCU11190, NCU06112*	GAD
	NCU08998	GABA-T
	NCU00936	SSADH
	gabA	GABA permease
Arabidopsis thaliana	*GAD1, GAD2, GAD3, GAD4, GAD5*	GAD
	POP2	GABA-T
	ALDH5F1	SSADH
	BAT1	GABA permease
	AT1G08230	GABA transporter 1
Homo sapiens (human)	*GAD1, GAD2*	GAD
	ABAT	GABA-T
	ALDH5A1	SSADH
	GAT1, GAT2, GAT3, GAT4	GABA transporter

such as *E. coli* (Gut *et al.*, 2006) and *Listeria monocytogenes* (Karatzas *et al.*, 2010). Unlike GABA export system, GABA uptake system is widely spread in microorganisms, plants and mammals. The gene encoding GABA transporter/permease, GABA-T and SSADH usually forms as an operon in microorganisms. Uptake of GABA by GABA permease followed by conversion of GABA to succinate enables GABA to be assimilated as nitrogen and/or carbon source in microorganisms such as *Saccharomyces cerevisiae* (Andre *et al.*, 1993), *E. coli* (Niegemann, Schulz and Bartsch, 1993), *Bacillus subtilis* (Ferson, Wray and Fisher, 1996), *Rhizobium leguminosarum* (Prell *et al.*, 2009) and *C. glutamicum* (Zhao *et al.*, 2012). *Arabidopsis* is the main object of research on GABA transporters in plants. A member of amino acid transporter (GABA transporter 1, GAT1) and a member of amino acid/polyamine/organocation (APC) transporter family (GABA permease, GABP) have been confirmed to transport GABA in *Arabidopsis* (Figure 16.3) (Shelp, Mullen and Waller, 2012). In mammals, GABA transporters are present in neurons and astrocytes, and their activity is crucial to regulate the extracellular concentration of GABA under basal conditions and during ongoing synaptic events (Scimemi, 2014).

Metabolism of GABA consists of GABA shunt along with its export and uptake systems, represented by bacteria in Figure 16.2. It is proposed to be involved in a legion of cellular processes ranging from acid resistance in microbes to environmental stress resistance in plants and neuronal inhibition in mammals.

16.3.1 Biosynthesis and Export of GABA

Biosynthesis of GABA is fulfilled by the direct and irreversible α-decarboxylation of Glu by the enzyme GAD. In some bacteria, the synthesised GABA can be secreted out of cells by the Glu/GABA antiporter which is usually named as GadC; therefore, it contributes to the acid resistance. While in most of other living organisms, the Glu/GABA antiporter is not found.

16.3.1.1 Biosynthesis of GABA

Biosynthesis of GABA widely exists in microorganisms, plants and mammals. Due to the accumulation of synthesised GABA, some microorganisms such as *E. coli*, LAB, yeast and many moulds are used as GABA-producing strains. GABA was first isolated from acid-treated yeast extract (Reed, 1950) and later detected while investigating the amino acid composition of red yeast, *Rhodotorula glutinis* (Krishnaswamy and Giri, 1953). An important GABA pool was observed in the early phase of spore germination of *Neurospora crassa* (Schmit and Brody, 1975) in which GAD is well studied (Kumar and Punekar, 1997). Other filamentous fungi such as *Aspergillus niger* accumulate GABA during acidogenesis (Kubicek, Hampel and Rohr, 1979). But now, *E. coli* and LAB, especially LAB, have become the most potential bacteria for GABA production. Several strains of *Lactobacillus* (*Lb.*) and *Lactococcus (Lc.)* (Li and Cao, 2010) have been screened, such as *Lb.*

brevis, *Lc. lactis*, *Lb. paracasei*, *Lb. delbrueckii* subsp. *bulgaricus*, *Lb. buchneri*, *Lb. plantarum* and *Lb. helveticus*. However, vast difference of GABA-producing ability is displayed among them. Up to date, *Lb. brevis* NCL912 produced the highest amount of GABA (Li *et al.*, 2010).

In plants, GABA biosynthesis was first reported in potato tuber (Dent, Stepka and Steward, 1947). Its rapid accumulation in plants is largely due to biotic and abiotic stresses (Kinnersley and Turano, 2000). Recently, due to the recognition of the important physiological functions of GABA in mammals, development of plant-based functional foods containing GABA has been actively pursued, such as GABA-enriched brown rice (Miwako *et al.*, 1999), GABA-enriched rice germ (Okada *et al.*, 2000; Zhang, Yao and Chen, 2006), GABA-enriched green tea (Wang *et al.*, 2006) and GABA-enriched tempeh-like fermented soybean (Aoki *et al.*, 2003). Improvement of GABA biosynthesis in these plants has been researched.

In higher animals, the two isozymes of GAD, that is, GAD67 and GAD65, function to produce the inhibitory neurotransmitter GABA from Glu constitutively and transiently, respectively (Fenalti *et al.*, 2007) and control fundamental processes such as neurogenesis (Ge *et al.*, 2006). The structure of GAD67 shows a tethered loop covering the active site, providing a catalytic environment that sustains GABA production. In contrast, GAD65 is transiently activated in response to the demand for extra GABA in neurotransmission and cycles between an active holo form and an inactive apo form. GABA generated by GAD65 at the synaptic vesicles membrane is taken up preferentially over the GABA generated by GAD67 in the cytoplasm (Jin *et al.*, 2003). GABA generated by GAD67 is purposed for synaptogenesis during early development, protection after neuronal injury, source of energy via the GABA shunt and regulator of redox potential during oxidative stress (Lamigeon *et al.*, 2001; Pinal and Tobin, 1998; Waagepetersen *et al.*, 2001).

16.3.1.2
Essential Enzyme for GABA Biosynthesis – GAD

GAD is the only essential enzyme for GABA biosynthesis. It is pyridoxal 5′-phosphate (PLP)-dependent; hence, it is a vitamin-B_6-dependent enzyme (Capitani *et al.*, 2003). The active site of GADs, where PLP covalently binds, is highly conserved. But, structures and properties of GADs vary in different organisms.

Bacterial GADs exhibit a sharp acidic pH optimum (3.8–4.6) and are expressed in response to environmental stresses. Most of them consist of 1–6 subunits with molecular mass ranging from 54 to 62 kD (Blankenhorn, Phillips and Slonczewski, 1999; De Biase *et al.*, 1996). GAD isoforms with hexamer formation have been confirmed in the GadA and GadB of a Gram-negative bacterium, *E. coli* (Smith *et al.*, 1992). Both of the two distinct GAD structural genes, *gadA* and *gadB*, encode the 466-residue polypeptide with the difference of only five amino acid residues. The two isoenzymes show identical kinetic and physico-chemical properties. In LAB, the tetrameric GadB2 of *Lb. brevis* IFO 12005 (Hiraga, Ueno and Oda, 2008), the dimeric GAD of *Lb. paracasei* (Komatsuzaki *et al.*, 2005), the monomeric GadB1 of *Lb. brevis* Lb85 (Shi *et al.*, 2014) and homologous GAD of *Lb. brevis* CGMCC 1306 (Yu *et al.*, 2012) have been identified. The optimal temperatures of *E. coli* and LAB GADs range from 30 to 50 °C, and the optimum pH values are in the range of 4.0–5.0 (Li and Cao, 2010). At pHs lower than 4.0 and higher than 5.0, GAD activity decreases sharply. Broadening the active pH range of GAD towards near-neutral pH may be important for effective production of GABA by GAD (Ho *et al.*, 2013; Kang, Ho and Pack, 2013). Therefore, several efforts to extend GAD activity towards near-neutral pH have been made. Yu *et al.* (2012) deleted the C-terminal residues of GAD of *Lb. brevis* CGMCC 1306, and the activity is significantly extended towards pH 6.0. The activity range of a GAD, that is, GadB1 from *Lb. brevis* Lb85, is broadened towards a near-neutral pH by directed evolution and site-specific mutagenesis (Shi *et al.*, 2014).

Unlike bacterial GADs, GADs from various plants possess a calmodulin (CaM)-binding domain in the C-terminal region and exhibit an acidic pH optimum of about 5.8 (Ueno, 2000). These GADs are regulated by pH and binding of Ca^{2+}/CaM to the C-terminal domain and are involved in normal development and stress response (Baum *et al.*, 1996). In tobacco (*Nicotiana tabacum*), removal of the CaM-binding domain can result in altered Glu and GABA metabolism and developmental problems (Baum *et al.*, 1996). Moreover, no activity of recombinant petunia GAD can be recovered when both Ca^{2+} and CaM were added (Baum *et al.*, 1993). However, GAD derived from rice (*Oryza sativa*) does not depend on Ca^{2+}/CaM for activity (Akama *et al.*, 2001).

Sequence homology of mammalian GAD with non-mammalian counterpart is significantly lower; for example, mammalian GAD versus *E. coli* GAD shows around 24% identity. In mammals, the two isozymes of GAD, GAD65 and GAD67 are active as dimmers at neutral pH (Soghomonian and Martin, 1998; Fenalti *et al.*, 2007). The first 100 amino acid residues in the two isoforms differ substantially, but residues within the middle and the C-terminal domain of two isoforms show 74% sequence identity. Within the N-terminal region, GAD65 contains more cysteine (Cys) residues than GAD67. Three of such Cys residues of GAD65 were found to be palmitoylated, which further suggests the affinity of this isozyme to the membrane and their involvement in anchoring mechanism for vesicles (Christgau *et al.*, 1992; Shi, Veit and Baekkeskov, 1994). However, the N-terminal segment of GAD67 may not be required for GAD activity (Chu and Metzler, 1994).

16.3.1.3
Export of GABA

After synthesis by GADs in some bacteria, GABA can be excreted into the extracellular medium by a Glu/GABA antiporter. The Glu/GABA antiporter is usually encoded by *gadC* gene; therefore, it is named as GadC, as in the case of *E. coli* (Table 16.1). But in *L. monocytogenes*, the Glu/GABA antiporter is encoded by *gadT1* and *gadT2* (Cotter *et al.*, 2005). GadC is mainly present in several bacteria, such as the Gram-negative bacteria *E. coli*, *Salmonella enterica*, *Shigella flexneri* and the Gram-positive bacteria *Lb. brevis*, *Lc. lactis* and *Bifidobacterium* (De Biase and Pennacchietti, 2012). The GAD-Glu/GABA antiporter system functions as an important AR system in these bacteria.

To better explain the acid-resistant function of GAD-GadC system, the crystal structure of *E. coli* GadC at pH 8.0 has been resolved recently, which provides insights into the mechanism of substrate transport (Ma *et al.*, 2012). GadC is a representative member of APC transporter superfamily and consists of 12 transmembrane (TM) segments (Figure 16.4a). It contains an extended C-terminal fragment and seems to adopt an inward-open conformation (Figure 16.4a). The open path leads to a negatively charged environment

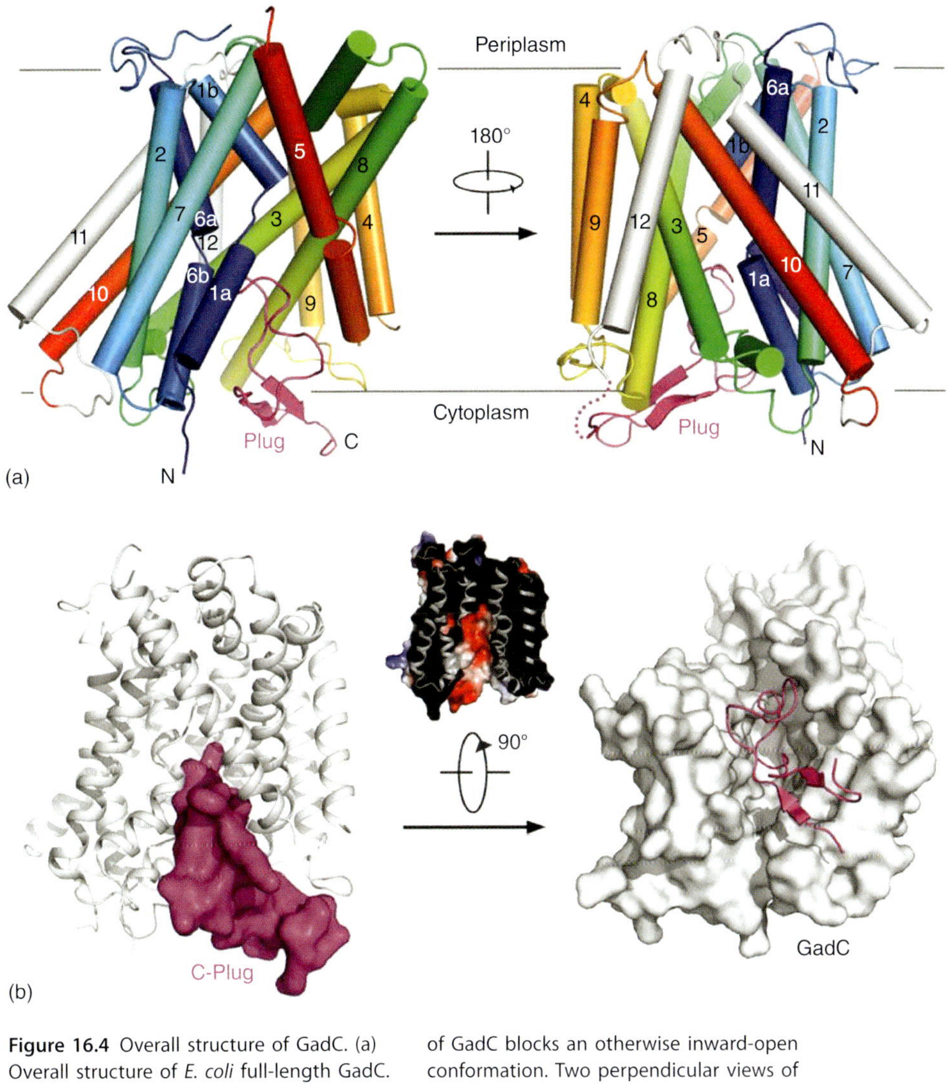

Figure 16.4 Overall structure of GadC. (a) Overall structure of *E. coli* full-length GadC. TM1–TM10 are rainbow-coloured, with TM1 in blue and TM10 in red. TM11 and TM12 are shown in grey. The C-terminal fragment (C-plug) is coloured magenta. (b) The C-plug of GadC blocks an otherwise inward-open conformation. Two perpendicular views of GadC are shown. The C-plug blocks access to the negatively charged substrate-binding cleft (inset). The published information by Ma *et al.* (2012).

(Figure 16.4b), where substrate-binding residues are likely to be located. However, the C-terminal fragment forms a folded domain and completely blocks the path to the putative substrate-binding site at pH 8.0. Therefore, the C-terminal fragment is referred to as the C-plug. Deletion of the C-plug in GadC shifted its acidic pH-dependent substrate transport towards a higher pH. With the use of an *in vitro* proteoliposome-based assay, researchers show that GadC transports GABA/Glu only under acidic conditions and no activity can be detected at pH values higher than 6.5. It is also found that GadC can efficiently transport not only Glu and GABA but also L-glutamine and L-methionine.

16.3.2 Uptake and Catabolism of GABA

Besides the biosynthesis of GABA, many species can uptake the extracellular GABA from the environment by a GABA-specific transporter/permease. Subsequently, the synthesised or obtained GABA can be degraded into succinate by GABA-T and SSADH, which then flows into TCA cycle for further metabolism. Therefore, GABA can be utilised as a carbon and/or nitrogen source in these species. However, the enzymes for GABA catabolism are not found in some species such as *Lb. brevis*.

16.3.2.1 The Uptake System of GABA

The extracellular GABA can be transported into cells by a GABA-specific transporter (sometimes named as GABA permease). In bacteria, the GABA permease is usually encoded by *gabP* gene; therefore, it is named as GabP, as in the case of *E. coli* (Table 16.1). GABA transporters are distributed widely in microbes, plants and mammals. For most of the species which possess GABA catabolic enzymes (GABA-T and SSADH), GABA permease/transporter also exists. However, this is not all the case.

All previously identified microbial GABA transporters, including GabP in *E. coli* (Niegemann, Schulz and Bartsch, 1993), GabP in *B. subtilis* (Brechtel and King, 1998) and UGA4, GAP1 and PUT4 *in S. cerevisiae* (Andre *et al.*, 1993), are members of the APC transporter superfamily and function as solute/cation symporters and solute/solute antiporters. However, *R. leguminosarum* has two GABA transporters that belong to ATP-binding cassette type (Hosie *et al.*, 2002; White *et al.*, 2009). *C. glutamicum* has one GABA-specific transporter that belongs to the hydroxy/aromatic amino acid permease (HAAAP) family (Zhao *et al.*, 2012). Among them, GABA transporter of *E. coli* has been sequenced and expressed. It consists of 466 amino acids and is believed to have 12 TM α-helices, with both N- and C-termini facing the cytosol (Li *et al.*, 2001). The substrate translocation pathway is thought to be formed by a conserved consensus amphipathic region in the sequence, which has been identified for members of the APC transporter family, ranging from bacteria to mammals (Reizer *et al.*, 1993). GABA transport in *B. subtilis* is mediated by one GABA permease, which exhibits 47%

sequence identity to the GABA transporter of *E. coli* but is not expressed in the form of *gab* cluster (Ferson, Wray and Fisher, 1996). *R. leguminosarum* contains two GABA transporters, the Bra (branched-chain amino acid permease) and the Gts (GABA transport system) (White *et al.*, 2009). Transport of GABA in *S. cerevisiae* is mediated by three transport systems, that is, the general amino acid permease encoded by *GAP1*, the proline permease encoded by *PUT4* and a specific GABA permease encoded by *UGA4* which is induced in the presence of GABA and depends on cell growth conditions (Moretti, Garcia and Batlle, 1998). Additionally, efficient growth of *Arabidopsis* on GABA has provided evidence of GABA transporters in plants which include plasma-membrane-located AtGAT1 belonging to amino acid transporter family and mitochondria-located GABA permease (Michaeli *et al.*, 2011).

In mammalian brains, four different GABA transporters, GAT1, GAT2, GAT3 and GAT4, can transport GABA. They are part of a family of neurotransmitter transporter sharing similar structure and amino acid sequences but are different in pharmacological specificities and substrate specificities (Tamura *et al.*, 1995). GAT1 exclusively transports GABA, but the others are also able to transport β-alanine and taurine.

16.3.2.2
The Catabolism of GABA

In some species, GABA can be decomposed to succinate by two-step reactions catalysed by GABA-T and SSADH under some certain conditions. The catabolism of GABA is present in microorganism, plants and mammals. But, plants are the most thoroughly studied targets. GABA-T of most plants such as soybean, *Petunia* and *N. tabacum*, exhibits a broad pH optimum from 8 to 10. It can use either pyruvate or 2-oxoglutarate as amino acceptor to reversibly catalyse the conversion of GABA to SSA (Shelp *et al.*, 1995; Van Cauwenberghe and Shelp, 1999). Use of the former leads to the formation of L-alanine, whereas use of the latter leads to the formation of Glu and thus would potentially set up a futile cycle, since at least part of the Glu recycled by the transamination of GABA would eventually feed back into the GABA shunt (Fernie *et al.*, 2001; Geigenberger and Stitt, 1993). But, the GABA-T of *Arabidopsis* uses glyoxylate instead of 2-oxoglutarate as amino acceptor and produces L-glycine (Figure 16.3). SSADH of plants irreversibly oxidises SSA to succinate, the intermediate of TCA cycle. It has an alkaline pH optimum of about 9; activity is up to 20 times greater with NAD^+ than with $NADP^+$ as cofactor (Satyanarayan and Nair, 1990). A branch point for SSA catabolism to γ-hydroxybutyrate (GHB) catalysed by glyoxylate reductase (GLYR) (Figure 16.3) has been demonstrated in *Arabidopsis* (Hoover *et al.*, 2007).

The genes involved in GABA catabolic pathway usually exist in the form of *gab* gene cluster (Metzner, Germer and Hengge, 2004). In *E. coli*, the *csiD-ygaF-gabDTP* gene cluster forms a complex operon controlled by δ^S factor (Metzner, Germer and Hengge, 2004). But, in *B. subtilis*, the operon which comprises two genes, *gabT* (encoding GABA-T) and *gabD* (encoding SSADH), is regulated by a divergent gene, *gabR* (Belitsky and Sonenshein, 2002). The structure and regulation of its *gab* gene cluster are controlled by a δ^{54} factor (Zhu *et al.*, 2010). Besides, genes encoding those enzymes have also been identified in *R. leguminosarum* and *C. glutamicum* which carry a putative *gabTDP* operon (Zhao *et al.*, 2012).

16.4 Regulation of GABA Biosynthesis

Regulation of GABA biosynthesis occurs in all of the GABA metabolic process including GABA synthesis, GABA decomposition, GABA export and GABA uptake. In mammals, biosynthesis and secretion of GABA are a critical step in ensuring normal neural function, and loss of each protein in this pathway has been linked to human genetic diseases (Jakobs, Jaeken and Gibson, 1993). But the specific mechanisms that regulate the activity of these GABA metabolic proteins are not been well described except for those in plants and microorganisms.

In plants, GABA metabolism plays a role in signalling, herbivore deterrence, pH regulation, redox regulation, energy production and maintenance of carbon/nitrogen balance (Bouche and Fromm, 2004). GABA levels can be regulated by biotic and abiotic stresses and GABA export and uptake (Shelp, Bown and McLean, 1999). Stresses can increase cytosolic Ca^{2+}/CaM, H^+ or Glu levels, which in turn stimulate the production of GABA by GAD. Stresses also can increase the import of GABA and increase the ratio of NADH to NAD^+, thereby limiting or competitively inhibiting SSADH activity and causing the accumulation of SSA, the feedback of which in turn inhibits GABA-T activity. In addition, intracellular compartmentation of GABA metabolism (Figure 16.3) is another mechanism for regulating GABA levels (Shelp, Mullen and Waller, 2012). This model will implicate the transport of GABA across the mitochondrial membranes which further affects cytosolic GABA levels. The transport of GABA across plasma membrane also affects cytosolic GABA levels.

The molecular mechanisms about the regulation of GABA biosynthesis and export in *E. coli* have been fully elucidated than in other strains. In *E. coli*, the amount of GadA and GadB proteins increases in response to stationary phase and exponential phase under acidic conditions. Both exponential and stationary phase inductions of *gadA*, *gadB* and *gadC* expression are the outcome of a complex interplay between specific regulators, such as GadE, RcsB, GadX and GadW, and global regulators such as HNS (the histone-like nucleoid-structuring protein) and RpoS (the stationary-phase sigma factor of the RNA polymerase) (De Biase and Pennacchietti, 2012). Besides, many stress conditions, such as starvation, acidic pH, hyper- or hypo-osmolar stress and anaerobiosis, can also stimulate the expression of these Glu-dependent AR structural genes (*gadA*, *gadB* and *gadC*). As outlined in Figure 16.5, *gadA* and 13 additional genes which contribute at various levels to AR compose the acid fitness island (AFI). Most of the AFI genes are organised into operons (Figure 16.5), that is, *slp-yhiF*, *hdeAB-yhiD*, *gadE-mtdEF*, *gadXW* and *gadAX*. The genes *gadE*, *gadX*, *gadW*, *yhiF* (*dctR*), *arrS* and *gadY* code for

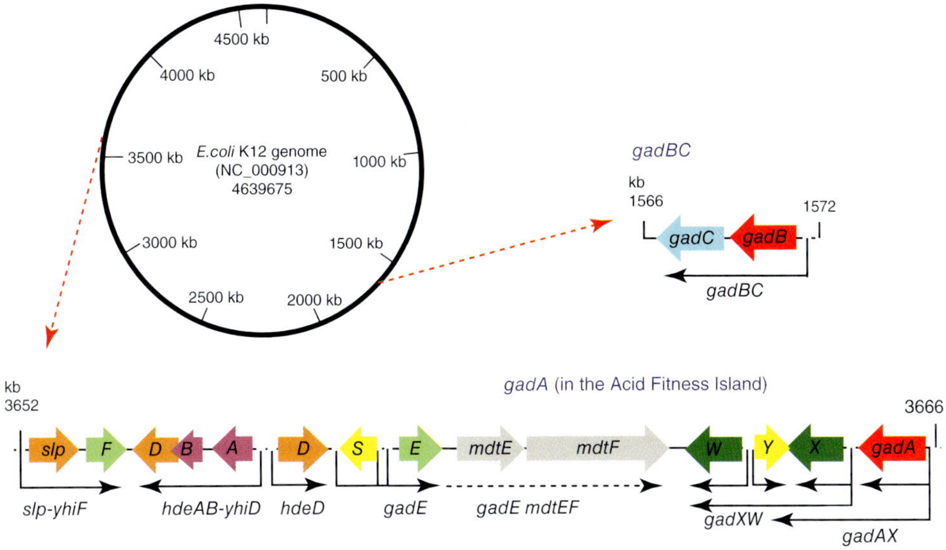

Figure 16.5 Representation of the *E. coli* genome map with the location of the gene loci relating to GABA biosynthesis and regulation. Genes encoding GAD (red), Glu/GABA antiporter (blue), additional membrane proteins (orange), multi-drug exporters (grey), acid-stress periplasmic chaperons (magenta), LuxR-like transcriptional regulators (light green), AraC-like transcriptional regulators (dark green) and small regulatory RNAs (yellow). The origin of each arrow is where transcripts start. The published information by De Biase *et al.* (2012).

four specific transcriptional regulators and two small RNAs, respectively; *hdeA* and *hdeB* code for pH-regulated acid-stress periplasmic denatured proteins in this cellular compartment; *slp*, *yhiD* and *hdeD* code for membrane proteins required for protection from acidic metabolites such as lactate, succinate and formate and for AR at high cell densities; *mdtE* and *mdtF* code for multidrug exporters which are unique among the 20 known drug exporters in that their expression is induced in stationary phase and is GadX-dependent. The one GAD gene *gadA* is either independently transcribed or transcribed with *gadX* (Figure 16.5) to generate a *gadAX* transcript, whereas the other GAD gene *gadB* is typically cotranscribed with the downstream Glu/GABA antiporter gene *gadC*.

16.5 Biotechnological Production of GABA

Currently, biotechnological production of GABA is mainly performed by whole-cell and enzymatic conversion of Glu or MSG. During the whole-cell conversion, LAB cells possessing high GAD activity are usually used as the whole-cell catalysts and before or during the GABA conversion process, LAB cells must be cultured and fermented. Therefore, this method is also classified as microbial fermentation of GABA. During the enzymatic conversion of GABA, the isolated GAD enzyme with high activity is used as the catalyst, and GAD is usually isolated from *E. coli* or LAB cells. For whatever whole-cell or enzymatic conversion, the GABA precursors Glu or MSG must be added. Recently, to make the GABA production process more effectively, a novel GABA fermentation method based on recombinant *C. glutamicum* has been reported. The recombinant *C. glutamicum* does not need the addition of GABA precursor and can produce GABA using its own synthesised Glu.

16.5.1
Fermentative Production of GABA by LAB

During the fermentation of GABA by LAB, LAB cells are generally pre-cultured at first, then MSG or Glu is added and GABA is produced by the living LAB cells under certain fermentation conditions. The ability of LAB cells for producing GABA is varied among species and strains, mostly due to their different GAD activity and Glu/GABA antiporter activity. In addition, fermentation conditions such as pH, medium components or additives and temperature also affect the rate of GABA production, attributing to the enzymatic properties of GAD in these LAB cells. The fermentation conditions can be optimised based on these properties.

Regulation of pH is crucial for GABA production by LABs. GADs derived from LABs are only active under acidic conditions and sharply lose activity at pH values higher than 5.0. However, as the GAD reaction proceeds, a proton is consumed; meanwhile, decarboxylation of the acidic substrate Glu results in the stoichiometric release of the neutral product GABA; these cause the pH to rise and inactivate GAD. Such property enables GAD to function as an important member of AR system in microbes. But this has negative effect on the GABA production. Therefore, acid must be added continuously to maintain the optimum acidic pH during GABA fermentation by LABs (Li *et al.*, 2010). After fermentation by *Lb. brevis* NCL912, the GABA production can be improved to 1005.81 ± 47.88 mM by regulating pH of culture medium to 5.0 instead of 4.0 or 6.0 (Li *et al.*, 2010). Similarly, GABA production by the growing cells of *Lb. brevis* TCCC13007 reached a higher level of 38 g/l under a controlled pH of 4.6, which improved by about 40% compared to the production without pH control (Zhang, Song and Gao, 2012).

Besides pH, medium composition also affects the GABA production. A research indicated that after optimisation of medium composition, the maximum GABA yield of *Lc. lactis* improved from 3.68 to 6.41 g/l (Lu *et al.*, 2008). Among these components, Glu and PLP are the major factors affecting the production of GABA (Li *et al.*, 2010), with the former being the substrate and the latter being the cofactor of GAD enzyme.

Temperature also affects GABA production. Fermentation of *Lb. brevis* NCL912 indicated that the GABA production had a positive correlation with the cell density, which was dependent on the culturing temperature (Li *et al.*, 2010).

16.5.2
Production of GABA by Enzymatic Conversion

During the enzymatic conversion of GABA, the isolated GAD enzyme with high activity is used as the catalyst and GAD is usually isolated from *E. coli* or LAB cells. Because GAD accounts for the major cost for GABA production, in order for the enzyme to be used repeatedly and steadily, immobilisation of enzymes has been adopted in GABA production. Furthermore, GADs derived from *E. coli* or LABs exhibit maximum enzymatic activity at acidic pH and exhibit little activity at near-neutral to alkaline pH values. Therefore, pH increase resulting from the progress of the GAD reaction will eventually inactivate the enzyme and limit the conversion of GABA. To make GADs more suitable for producing GABA, research studies have been conducted to broaden the active pH range of GADs.

16.5.2.1
Production of GABA by Immobilised GAD

Immobilisation of enzymes has been applied to many biochemical reactions owing to the advantages of repeated usage of enzyme, process stability, less inhibition, relative easiness of product separation and so on (Yao *et al.*, 2013). Effective enzyme immobilisation can be achieved using physical adsorption, entrapment and covalent binding to carriers or cross-linking. During immobilisation, the characteristics of the carrier govern the performance of an immobilised enzyme. The carriers, such as ion-exchange resin (Ling *et al.*, 2000), crystalline cellulose (Park *et al.*, 2012) and Ni-Sepharose (Lee *et al.*, 2013; Lee and Jeon, 2014), were used in the production of GABA by immobilised GAD. Recently, as a new and environment-friendly support carrier, bacteria cellulose membrane (BCM) has been used to immobilise GAD (Yao *et al.*, 2013). BCM is generated by bacteria due to a protection mechanism. It is a promising polymer material that possesses the ultra-fine network structure, biocompatibility and unique mechanical strength. Catalysed by this BCM-immobilised GAD, the productivity of GABA reached 6.03 g/l/h, higher than that from other reported processes. So, it is a potential method for GABA production.

16.5.2.2
Improving GAD Activity by Rational and Irrational Designs

To make GADs more suitable for GABA production, rational design using site-specific mutation and irrational design using directed evolution have been applied to broaden GAD activity to more alkaline pH values. Recently, the crystal structures of two bacterial GADs, *E. coli* GadB and GadA, were solved (Capitani *et al.*, 2003; Dutyshev *et al.*, 2005). The crystal structures of GadB at acidic and neutral pH values revealed the molecular details of its pH-dependent conformational change and the structural basis for its optimal activity at acidic pH values (Gut *et al.*, 2006). Based on the *E. coli* GadB crystal structure, a series of site-specific mutants have been constructed that broaden the range of its activity towards more alkaline pH values; these mutants include the His465 mutants

(Pennacchietti *et al.*, 2009), Δ452-466, Δ465-466, His465Ala, Glu89Gln/Δ452-466, Glu89Gln/Δ465-466 and Glu89Gln/His465Ala (Ho *et al.*, 2013), as well as the Δ466 and extension 467Ala mutants (Kang, Ho and Pack, 2013). Recently, a C-terminal truncated mutant of GAD derived from *Lb. brevis* CGMCC 1306, GADΔC, was constructed according to the monomer homology model of wild-type GAD deduced from the *E. coli* GadB structure and exhibited higher activity at pH 6.0 (Yu *et al.*, 2012). Quite recently, irrational design using directed evolution and rational design using site-specific mutagenesis were performed to broaden the active range of the GAD, that is, GadB1 from *Lb. brevis* Lb85 towards a near-neutral pH (Shi *et al.*, 2014). The combination of the beneficial mutations generated by directed evolution (Thr17Ile/Asp294Gly/Gln346His) and site-specific mutation (Glu312Ser) enhanced the GAD activity at near-neutral pH. These GAD variants can be used as potent candidates for GABA production.

16.5.3
Fermentation of GABA by Recombinant *C. glutamicum*

Either for whole-cell or enzymatic conversion, the GABA precursors Glu or MSG must be added. Glu-producing strain, *C. glutamicum* shows potential to produce GABA after heterologous expressing GAD gene(s). The recombinant *C. glutamicum* can use its own synthesised Glu and therefore combine the two procedures (cell fermentation and GABA conversion) into one procedure (Shi and Li, 2011; Takahashi *et al.*, 2012).

C. glutamicum ATCC 13032 can produce Glu, but no GABA was detected in its fermented culture. Recently, by expressing the GAD gene from *Lb. brevis* Lb85, *C. glutamicum* ATCC13032 was engineered to produce GABA spontaneously, but production was low (2.15 ± 0.16 g/l) (Shi and Li, 2011). A later study indicated that, by expressing the GAD gene from *E. coli* W3110, *C. glutamicum* ATCC13032 was able to produce 12.37 g/l of GABA after 72-h fermentation (Takahashi *et al.*, 2012). Later, a recombinant *C. glutamicum* strain co-expressing two GAD genes derived from *Lb. brevis* Lb85 was constructed. After optimising fermentation condition, GABA production increased to a high level of 18.66 g/l after 84-h flask cultivation and 26.32 g/l after 60-h fed-batch fermentation, while the conversion ratio of Glu to GABA reached 0.60–0.74 mol/mol (Shi *et al.*, 2013). Quite recently, a better producer of Glu, *C. glutamicum* G01, was engineered by expressing GAD and pyridoxal kinase genes derived from *Lb. plantarum* CCTCC M209102 and blocking by-product pools of L-arginine, L-proline and L-lysine, resulting in much higher production of GABA (70.6 g/l) after 70-h fermentation (Zhang *et al.*, 2014). These research studies indicate that the recombinant *C. glutamicum* strains are useful candidates for producing GABA.

To further enhance GABA production in recombinant *C. glutamicum*, the intracellular concentration of Glu was increased recently by deleting *pknG* gene which encodes serine/threonine protein kinase G. PknG catalyses the phosphorylation of OdhI and renders OdhI inactive. The unphosphorylated, active OdhI binds to the E1 subunit (OdhA) of 2-oxoglutarate dehydrogenase complex (ODHC) and inhibits its activity. The activity of ODHC can be reduced by deleting *pknG*; hence, the production of Glu is improved. After deletion of *pknG*, GABA production in recombinant *C. glutamicum* strain expressing *E. coli gadB* gene increased to 31.1 g/l after 120-h fermentation, 2.29-fold higher than that in recombinant *C. glutamicum* strain retaining *pknG* gene (Okai *et al.*, 2014).

16.6 Physiological Functions and Applications of GABA

16.6.1 Physiological Functions of GABA

In mammalian brains, GABA acts as a major inhibitory neurotransmitter. It can also regulate other psychological and physiological processes. The most important one is the hypotensive effect that has been demonstrated in animals and in human intervention trials (Diana, Quilez and Rafecas, 2014). This works mainly by reducing sympathetic nerve activity (Kajekar *et al.*, 2002) and dilating blood vessels. GABA may also be used for treating cardiovascular disease. Kelly and Saravanan (2008) reported that GABA may reduce inflammation in rheumatoid arthritis and attenuate the metabolic response to ischemic incidents (Abel and McCandless, 1992).

Meanwhile, GABA is associated with several typical brain diseases and many psychiatric diseases. Alterations in GABA levels are associated with many brain diseases, including Alzheimer's disease, Huntington's disease, stiff person syndrome and schizophrenia (Wong, Bottiglieri and Snead, 2003). Other physiological dysfunctions such as relaxation (Wong, Bottiglieri and Snead, 2003), alcoholism (Oh, Soh and Cha, 2003), mood disorders (Bjork *et al.*, 2001; Krystal *et al.*, 2002), sleeplessness and depression (Okada *et al.*, 2000; Mohler, 2012) can be treated with GABA.

Additionally, GABA can delay or inhibit the invasion and metastasis of various types of cancers, such as mammary gland, colon, hepatic cancers and small-airway-derived lung adenocarcinoma (Kleinrok *et al.*, 1998; Minuk, 2000; Opolski *et al.*, 2000; Schuller, Al-Wadei and Majidi, 2008). GABA can also act on other vital organs. It can potentially protect against chronic kidney disease, activate liver function (Sun, 2004), improve visual function (Leventhal *et al.*, 2003) and increase the rate of protein synthesis in brain (Tujioka *et al.*, 2009). It can also effectively control asthma (Xu and Xia, 1999) and breathing (Kazemi and Hoop, 1991).

Finally, there is evidence that GABA can act as a hormonal regulator (Parkash and Kaur, 2007) and cell regulator. It can increase the concentration of growth hormone, promote the secretion of insulin, regulate secretion of progesterone and thyroid hormone and prevent obesity. As a cell regulator, it is involved in maintaining cell volume homeostasis under UV radiation (Warskulat *et al.*, 2004), in the synthesis of hyaluronic acid and in enhancing the rate of dermal fibroblasts (Han *et al.*, 2007).

16.6.2 Applications of GABA

Due to its numerous physiological functions, GABA has been widely applied in foods, pharmaceuticals and stockbreeding. The applications of GABA-enriched foods, as represented by Gabaron tea, began in the middle of 1980s. But now, GABA-enriched functional foods have covered cereal-based products, dairy products and Chinese tea (Diana, Quilez and Rafecas, 2014). In the development of these foods, the germinated brown rice with high content of GABA has attracted the most extensive attention (Jannoey *et al.*, 2010). Recently, GABA-enriched dairy products fermented by LAB are generally regarded as safe and, hence, have become the focus of research (Li and Cao, 2010). In addition to the aforementioned *in situ* accumulated GABA in GABA-enriched foods, GABA has also been used as food additive in many foods, such as bread, chocolates, biscuits, and some beverages. As the ideal health factor and medicine, GABA can also be combined with other functional factors, producing all kinds of nutritional supplements such as capsules and soft capsules. Recently, GABA has been reported to be the precursor of one of the most promising heat-resistant biopolymers, polyamide 4, also known as *Nylon-4*, which is biodegradable and composed of a repeating GABA unit. Thus, GABA shows huge economic potential in the synthesis of biodegradable plastics.

16.7 Conclusion

Over the past years, more and more research studies have revealed the mechanisms of GABA metabolism and regulation in various organisms. Meanwhile, some physiological functions of GABA in mammals have been discovered and explained, demonstrating it as a bioactive molecule for mammals. This expands the applications of GABA in foods, pharmaceuticals and stockbreeding. Therefore, biotechnological production of GABA with high efficiency is highly desired and pursued.

Acknowledgement

This work was supported by the Program of State Key Laboratory of Food Science and Technology (SKLF-ZZB-201405) and the Fundamental Research Funds for the Central Universities (JUSRP 21109).

References

Abel, M.S. and McCandless, D.W. (1992) Elevated γ-aminobutyric acid levels attenuate the metabolic response to bilateral ischemia. *J. Neurochem.*, **58**, 740–744.

Akama, K., Akihiro, T., Kitagawa, M. *et al.* (2001) Rice (*Oryza sativa*) contains a novel isoform of glutamate decarboxylase that lacks an authentic calmodulin-binding domain at the C-terminus. *Biochim. Biophys. Acta*, **1522**, 143–150.

Andre, B., Hein, C., Grenson, M. *et al.* (1993) Cloning and expression of the UGA4 gene coding for the inducible GABA-specific transport protein of *Saccharomyces cerevisiae*. *Mol. Gen. Genet.*, **237**, 17–25.

Aoki, H., Uda, I., Tagami, K. *et al.* (2003) The production of a new tempeh-like fermented soybean containing a high level of γ-aminobutyric acid by anaerobic incubation with Rhizopus. *Biosci. Biotechnol., Biochem.*, **67**, 1018–1023.

Awapara, J., Landua, A.J.R., Fuerst, J. *et al.* (1950) Free amino γ-aminobutyric acid in brain. *J. Biol. Chem.*, **187**, 35–39.

Baum, G., Chen, Y., Arazi, T. *et al.* (1993) A plant glutamate decarboxylase containing a calmodulin-binding domain. *J. Biol. Chem.*, **268**, 19610–19617.

Baum, G., Lev-Yadun, S., Fridmann, Y. *et al.* (1996) Calmodulin binding to glutamate decarboxylase is required for regulation of glutamate and GABA metabolism and normal development in plants. *EMBO J.*, **15**, 2988–2996.

Belitsky, B.R. and Sonenshein, A.L. (2002) GabR, a member of a novelprotein family, regulates utilization of γ-aminobutyrate in *Bacillus subtilis*. *Mol. Microbiol.*, **45**, 569–583.

Bjork, J.M., Moeller, F.G., Kramer, G.L. *et al.* (2001) Plasma GABA levels correlate with aggressiveness in relatives of patients with unipolar depressive disorder. *Psychiatry Res.*, **101**, 131–136.

Blankenhorn, D., Phillips, J., and Slonczewski, J.L. (1999) Acid- and base-induced proteins during aerobic and anaerobic growth of *Escherichia coli* revealed by two-dimensional gel electrophoresis. *J. Bacteriol.*, **181**, 2209–2216.

Bouche, N. and Fromm, H. (2004) GABA in plants: just a metabolite? *Trends Plant Sci.*, **9**, 110–115.

Bowery, N.G., Hill, D.R., Hudson, A.L. *et al.* (1980) Baclofen decreases neurotransmitter release in the mammalian CNS by an action at a novel GABA receptor. *Nature*, **283**, 92–94.

Brechtel, C.E. and King, S.C. (1998) 4-aminobutyrate (GABA) transporter from the amine-polyamine-choline superfamily:substrate specificity and ligand recognition profile of the 4-aminobutyrate permease from *Bacillus subtilis*. *Biochem. J.*, **333**, 565–571.

Capitani, G., De Biase, D., Aurizi, C. *et al.* (2003) Crystal structure and functional analysis of *Escherichia coli* glutamate decarboxylase. *EMBO*, **22**, 4027–4037.

Christgau, S., Aanstoot, H.J., Schierbeck, H. *et al.* (1992) Membrane anchoring of autoantigen GAD65 to microvesicles in pancreatic β-cells by palmitoylation in the NH2-terminal domain. *J. Cell Biol.*, **118**, 309–320.

Chu, W.C. and Metzler, D.E. (1994) Enzymatically active truncated cat brain glutamate decarboxylase: expression, purification, and absorption spectrum. *Arch. Biochem. Biophys.*, **313**, 287–295.

Cotter, P.D., Ryan, S., Gahan, C.G. *et al.* (2005) Presence of GadD1 glutamate decarboxylase in selected *Listeria monocytogenes* strains is associated with an ability to grow at low pH. *Appl. Environ. Microbiol.*, **71**, 2832–2839.

Curtis, D.R., Duggan, A.W., Felix, D. *et al.* (1970) GABA, bicuculline and central inhibition. *Nature*, **226**, 122–124.

De Biase, D. and Pennacchietti, E. (2012) Glutamate decarboxylase-dependent acid resistance in orally acquired bacteria: function, distribution and biomedical implications of the gadBC operon. *Mol. Microbiol.*, **86**, 770–786.

De Biase, D., Tramonti, A., John, R.A. *et al.* (1996) Isolation overexpression and biochemical characterization of the two isoforms of glutamic acid decarboxylase from *Escherichia coli*. *Protein Expr. Purif.*, **8**, 430–438.

Dent, C.E., Stepka, W., and Steward, F.C. (1947) Detection of the free amino acids of plant cells by partition chromatography. *Nature*, **160**, 682–683.

Dhakal, R., Bajpai, V.K., and Baek, K.H. (2012) Production of GABA (γ-aminobutyric acid) by microorganisms: a review. *Braz. J. Microbiol.*, **43**, 1230–1241.

Diana, M., Quilez, J., and Rafecas, M. (2014) Gamma-aminobutyric acid as a bioactive compound in foods: a review. *J. Funct. Foods*, **10**, 407–420.

Dutyshev, D.I., Darii, E.L., Fomenkova, N.P. *et al.* (2005) Structure of *Escherichia coli* glutamate decarboxylase (GAD) in complex with glutarate at 2.05 angstroms resolution. *Acta Crystallogr., Sect. D: Biol. Crystallogr.*, **61**, 230–235.

Fenalti, G., Law, R.H., Buckle, A.M. *et al.* (2007) GABA production by glutamic acid decarboxylase is regulated by a dynamic catalytic loop. *Nat. Struct. Mol. Biol.*, **14**, 280–286.

Fernie, A.R., Roessner, U., Trethewey, R.N. *et al.* (2001) The contribution of plastidial phosphoglucomutase to the control of starch synthesis within the potato tuber. *Planta*, **213**, 418–426.

Ferson, A.E., Wray, L.V. Jr.,, and Fisher, S.H. (1996) Expression of the *Bacillus subtilis* gabP gene is regulated independently in response to nitrogen and amino acid availability. *Mol. Microbiol.*, **22**, 693–701.

Fonda, M.L. (1985) L-glutamate decarboxylase from bacteria. *Methods Enzymol.*, **113**, 1–10.

Ge, S.Y., Goh, E.L.K., Sailor, K.A. *et al.* (2006) GABA regulates synaptic integration of newly generated neurons in the adult brain. *Nature*, **439**, 589–593.

Geigenberger, P. and Stitt, M. (1993) Sucrose synthase catalyses a readily reversible reaction in vivo in developing potato tubers and other plant tissues. *Planta*, **189**, 329–339.

Gut, H., Pennacchietti, E., John, R.A. *et al.* (2006) *Escherichia coli* resistance: pH-sensing, activation by chloride and autoinhibition in GadB. *EMBO J.*, **25**, 2643–2651.

Han, D., Kim, H.Y., Lee, H.J. *et al.* (2007) Wound healing activity of gamma-aminobutyric acid (GABA) in rats. *J. Microbiol. Biotechnol.*, **17**, 1661–1669.

Hayakawa, K., Kimura, M., Kasaha, K. *et al.* (2004) Effect of a gamma-aminobutyric acid-enriched dairy product on the blood pressure of spontaneously hypertensive and normotensive Wistar-Kyoto rats. *Br. J. Nutr.*, **92**, 411–417.

Hiraga, K., Ueno, Y., and Oda, K. (2008) Glutamate decarboxylase from *Lactobacillus brevis*: activation by ammonium sulfate. *Biosci. Biotechnol., Biochem.*, **72**, 1299–1306.

Ho, N.A.T., Hou, C.Y., Kim, W.H. *et al.* (2013) Expanding the active pH range of *Escherichia coli* glutamate decarboxylase by breaking the cooperativeness. *J. Biosci. Bioeng.*, **115**, 154–158.

Hoover, G.J., Van Cauwenberghe, O.P., Breitkreuiz, K.E. *et al.* (2007) Characteristics of an Arabidopsis glyoxylate reductase: general biochemical properties and substrate specificity for the recombinant protein, and developmental expression and implications of glyoxylate and succinic semialdehyde metabolism in plants. *Can. J. Bot.*, **85**, 883–895.

Hosie, A.H.F., Allaway, D., Galloway, C.S. *et al.* (2002) *Rhizobium leguminosarum* has a second general amino acid permease with unusually broad substrate specificity and high similarity to branched-chain amino acid transporters (Bra/LIV) of the ABC Family. *J. Bacteriol.*, **184**, 4071–4080.

Jakobs, C., Jaeken, J., and Gibson, K.M. (1993) Inherited disorders of GABA metabolism. *J. Inherit. Metab. Dis.*, **16**, 704–715.

Jannoey, P., Niamsup, H., Lumyong, S. *et al.* (2010) Gamma-aminobutyric acid accumulations in rice during germination. *Chiang Mai J. Sci.*, **37**, 124–133.

Jin, H., Wu, H., Osterhaus, G. *et al.* (2003) Demonstration of functional coupling between gamma-aminobutyric acid (GABA) synthesis and vesicular GABA transport into synaptic vesicles. *Proc. Natl. Acad. Sci. U.S.A.*, **100**, 4293–4298.

Johnston, G.A. (1996) GABAC receptors: relatively simple transmitter-gated ion channels? *Trends Pharmacol. Sci.*, **17**, 319–323.

Kajekar, R., Chen, C.Y., Mutoh, T. *et al.* (2002) GABA (A) receptor activation at medullary sympathetic neurons contributes to post exercise hypotension.

Am. J. Physiol. Heart Circ. Physiol., **282**, H1615–H1624.

Kang, T.J., Ho, N.A.T., and Pack, S.P. (2013) Buffer-free production of gamma-aminobutyric acid using an engineered glutamate decarboxylase from *Escherichia coli*. *Enzyme Microb. Technol.*, **53**, 200–205.

Karatzas, K.A., Brennan, O., Heavin, S. *et al.* (2010) Intracellular accumulation of high levels of γ-aminobutyrate by *Listeria monocytogenes*10403S in response to low pH: uncoupling of γ-aminobutyrate synthesis from efflux in a chemically defined medium. *Appl. Environ. Microbiol.*, **76**, 3529–3537.

Kazemi, H. and Hoop, B. (1991) Glutamic acid and gamma-aminobutyric acid neurotransmitters in central control of breathing. *J. Appl. Physiol.*, **70**, 1–7.

Kelly, C. and Saravanan, V. (2008) Treatment strategies for a rheumatoid arthritis patient with interstitial lung disease. *Expert Opin. Pharmacother.*, **9**, 3221–3230.

Kinnersley, A.M. and Turano, F.J. (2000) Gamma aminobutyric acid (GABA) and plant responses to stress. *Crit. Rev. Plant Sci.*, **19**, 479–509.

Kleinrok, Z., Matuszek, M., Jesipowicz, J. *et al.* (1998) GABA content and GAD activity in colon tumors taken from patients with colon cancer or from human colon cancer cells growing as S.C. tumors in a thymic nu/nu mice. *J. Physiol. Pharmacol.*, **49**, 303–310.

Komatsuzaki, N., Shima, J., Kawamoto, S. *et al.* (2005) Production of γ-aminobutyric acid (GABA) by *Lactobacillus paracasei* isolated from traditional fermented foods. *Food Microbiol.*, **22**, 497–504.

Krishnaswamy, P.R. and Giri, K.V. (1953) The occurrence of 4-aminobutyric acid and glutamic acid decarboxylase in red yeast (*Rhodotorula glutinis*). *Curr. Sci.*, **22**, 143–144.

Krnjevic, K. and Schwartz, S. (1967) The action of γ-aminobutyric acid on cortical neurones. *Exp. Brain Res.*, **3**, 320–326.

Krystal, J.H., Sanacora, G., Blumberg, H. *et al.* (2002) Glutamate and GABA systems as targets for novel antidepressant and mood-stabilizing treatments. *Mol. Psychiatry*, **7**, 71–80.

Kubicek, C.P., Hampel, W., and Rohr, M. (1979) Manganese deficiency leads to elevated amino acid pools in critic acid accumulating *Aspergillus niger*. *Arch. Microbiol.*, **123**, 73–79.

Kumar, S. and Punekar, N.S. (1997) The metabolism of 4-aminobutyrate in fungi. *Mycol. Res.*, **101**, 403–409.

Lamigeon, C., Bellier, J.P., Sacchettoni, S. *et al.* (2001) Enhanced neuronal protection from oxidative stress by coculture with glutamic acid decarboxylase-expressing astrocytes. *J. Neurochem.*, **77**, 598–606.

Lee, S., Ahn, J., Kim, Y.G. *et al.* (2013) Gamma-aminobutyric acid production using immobilized glutamate decarboxylase followed by downstream processing with cation exchange chromatography. *Int. J. Mol. Sci.*, **14**, 1728–1739.

Lee, J.Y. and Jeon, S.J. (2014) Characterization and immobilization on nickel-chelated Sepharose of a glutamate decarboxylase A from *Lactobacillus brevis* BH2 and its application for production of GABA. *Biosci. Biotechnol., Biochem.*, **78**, 1656–1661.

Leventhal, A., Wang, Y., Pu, M. *et al.* (2003) GABA and its agonists improved visual cortical function in senescent monkeys. *Science*, **300**, 812–815.

Li, H. and Cao, Y. (2010) Lactic acid bacterial cell factories for gamma-aminobutyric acid. *Amino Acids*, **39**, 1107–1116.

Li, H., Qiu, T., Huang, G. *et al.* (2010) Production of gamma-aminobutyric acid by *Lactobacillus brevis* NCL912 using fed-batch fermentation. *Microb. Cell Fact.*, **9**, 85.

Li, X.D., Villa, A., Gownley, C. *et al.* (2001) Monomeric state and ligand binding of recombinant GABA transporter from *Escherichia coli*. *FEBS Lett.*, **494**, 165–169.

Ling, D., Wu, G., Wang, C. *et al.* (2000) The preparation and characterization of an immobilized l-glutamic decarboxylase and its application for determination of l-glutamic acid. *Enzyme Microb. Technol.*, **27**, 516–521.

Ma, D., Lu, P., Yan, C. *et al.* (2012) Structure and mechanism of a glutamate-GABA antiporter. *Nature*, **483**, 632–636.

Metzner, M., Germer, J., and Hengge, R. (2004) Multiple stress signal integration in the regulation of the complex δS-dependent csiD-ygaF-gabDTP operon in *Escherichia coli*. *Mol. Microbiol.*, **51**, 799–811.

Michaeli, S., Fait, A., Lagor, K. *et al.* (2011) A mitochondrial GABA permease connects the GABA shunt and the TCA cycle, and is essential for normal carbon metabolism. *Plant J.*, **67**, 485–498.

Minuk, G.Y. (2000) GABA and hepatocellular carcinoma. *Mol. Cell. Biochem.*, **207**, 105–108.

Miwako, K., Miyuki, S., Akira, Y. *et al.* (1999) Accumulation of GABA in brown rice by high pressure treatment. *J. Jpn. Soc. Food Sci. Technol.*, **46**, 329–333.

Mohler, H. (2012) The GABA system in anxiety and depression and its therapeutic potential. *Neuropharmacology*, **62**, 42–53.

Moretti, M.B., Garcia, S.C., and Batlle, A. (1998) UGA4 gene expression in *Saccharomyces cerevisiae* depends on cell growth conditions. *Cell Mol. Biol.*, **44**, 585–590.

Niegemann, E., Schulz, A., and Bartsch, K. (1993) Molecular organization of the *Escherichia coli* gab cluster: nucleotide sequence of the structural genes gabD and gabP and expression of the GABA permease gene. *Arch. Microbiol.*, **160**, 454–460.

Oh, S.H., Soh, J.R., and Cha, Y.S. (2003) Germinated brown rice extract shows a nutraceutical effect in the recovery of chronic alcohol-related symptoms. *J. Med. Food*, **6**, 115–121.

Okada, T., Sugishita, T., Murakami, T. *et al.* (2000) Effect of the defatted rice germ enriched with GABA for sleeplessness depression, autonomic disorder by oral administration. *J. Jpn. Soc. Food Sci.*, **47**, 596–603.

Okai, N., Takahashi, C., Hatada, K. *et al.* (2014) Disruption of pknG enhances production of gamma-aminobutyric acid by *Corynebacterium glutamicum* expressing glutamate decarboxylase. *AMB Express*, **4**, 20.

Opolski, A., Mazurkiewicz, M., Wietrzyk, J. *et al.* (2000) The role of GABA-ergic system in human mammary gland pathology and in growth of transplantable murine mammary cancer. *J. Exp. Clin. Cancer Res.*, **19**, 383–390.

Park, H., Ahn, J., Lee, J. *et al.* (2012) Expression, immobilization and enzymatic properties of glutamate decarboxylase fused to a cellulose-binding domain. *Int. J. Mol. Sci.*, **13**, 358–368.

Parkash, J. and Kaur, G. (2007) Potential of PSA-NCAM in neuronglial plasticity in the adult hypothalamus: role of noradrenergic and GABAergic neurotransmitters. *Brain Res. Bull.*, **74**, 317–328.

Pennacchietti, E., Lammens, T.M., Capitani, G. *et al.* (2009) Mutation of His465 alters the pH-dependent spectroscopic properties of *Escherichia coli* glutamate decarboxylase and broadens the range of its activity toward more alkaline pH. *J. Biol. Chem.*, **115**, 154–158.

Pinal, C.S. and Tobin, A.J. (1998) Uniqueness and redundancy in GABA production. *Perspect. Dev. Neurobiol.*, **5**, 109–118.

Prell, J., Bourdes, A., Karunakaran, R. *et al.* (2009) Pathway of gamma-aminobutyrate metabolism in *Rhizobium leguminosarum* 3841 and its role in symbiosis. *J. Bacteriol.*, **191**, 2177–2186.

Reed, L.J. (1950) The occurrence of γ-aminobutyric acid in yeast extract; its isolation and identification. *J. Biol. Chem.*, **183**, 451–458.

Reizer, J., Finley, K., Kakuda, D. *et al.* (1993) Mammalian integral membrane receptors are homologous to facilitators and antiporters of yeast, fungi, and eubacteria. *Protein Sci.*, **2**, 20–30.

Roberts, E. and Frankel, S. (1950) γ-aminobutyric acid in brain: its formation from glutamic acid. *J. Biol. Chem.*, **187**, 55.

Saskiawan, I. (2008) Biosynthesis of polyamide 4, a biobased and biodegradable polymer. *Microbiology*, **2**, 119–123.

Satyanarayan, V. and Nair, P.M. (1990) Metabolism, enzymology and possible roles of 4-aminobutyrate in higher plants. *Phytochemistry*, **29**, 367–375.

Schmit, J.C. and Brody, S. (1975) *Neurospora crassa* conidial germination: role of endogenous amino acid pools. *J. Bacteriol.*, **124**, 232–242.

Schuller, H.M., Al-Wadei, H.A.N., and Majidi, M. (2008) Gammaaminobutyric acid, a potential tumor suppressor for small airway-derived lung adenocarcinoma. *Carcinogenesis*, **29**, 1979–1985.

Scimemi, A. (2014) Structure, function and plasticity of GABA transporters. *Front. Cell. Neurosci.*, **8**, 1–14.

Shelp, B.J., Bown, A.W., and McLean, M.D. (1999) Metabolism and functions of gamma-aminobutyric acid. *Trends Plant Sci.*, **4**, 446–452.

Shelp, B.J., Mullen, R.T., and Waller, J.C. (2012) Compartmentation of GABA metabolism raises intriguing questions. *Trends Plant Sci.*, **17**, 57–59.

Shelp, B.J., Walton, C.S., Snedden, W.A. *et al.* (1995) GABA shunt in developing soybean seeds is associated with hypoxia. *Physiol. Plant.*, **94**, 219–228.

Shi, F., Jiang, J.J., Li, Y.X. *et al.* (2013) Enhancement of gamma-aminobutyric acid production in recombinant *Corynebacterium glutamicum* by co-expressing two glutamate decarboxylase genes from *Lactobacillus brevis*. *J. Ind. Microbiol. Biotechnol.*, **40**, 1285–1296.

Shi, F. and Li, Y. (2011) Synthesis of γ-aminobutyric acid by expressing *Lactobacillus brevis*-derived glutamate decarboxylase in the *Corynebacterium glutamicum* strain ATCC 13032. *Biotechnol. Lett.*, **33**, 2469–2474.

Shi, Y., Veit, B., and Baekkeskov, S. (1994) Amino acid residues 24-31 but not palmitoylation of cysteines 30 and 45 are required for membrane anchoring of glutamic acid decarboxylase, GAD65. *J. Cell Biol.*, **6**, 927–934.

Shi, F., Xie, Y., Jiang, J. *et al.* (2014) Directed evolution and mutagenesis of glutamate decarboxylase from *Lactobacillus brevis* Lb85 to broaden the range of its activity toward a near-neutral pH. *Enzyme Microb. Technol.*, **61–62**, 35–43.

Smith, D.K., Kassam, T., Singh, B. *et al.* (1992) *Escherichia coli* has two homologous glutamate decarboxylase genes that map to distinct loci. *J. Bacteriol.*, **174**, 5820–5826.

Soghomonian, J.J. and Martin, D.L. (1998) Two isoforms of glutamate decarboxylase: why? *Trends Pharmacol. Sci.*, **19**, 500–505.

Sun, B.S. (2004) Research of some physiological active substance by fermentation of Monascus spp. Dissertation for Master's Degree. Zhejiang Industry University, China, pp. 40–55.

Takahashi, C., Shirakawa, J., Tsuchidate, T. *et al.* (2012) Robust production of gamma-amino butyric acid using recombinant *Corynebacterium glutamicum* expressing glutamate decarboxylase from *Escherichia coli*. *Enzyme Microb. Technol.*, **51**, 171–176.

Tamura, S., Nelson, H., Tamura, A. *et al.* (1995) Short external loops as potential substrate binding site of γ-aminobutyric acid transporters. *J. Biol. Chem.*, **270**, 28712–28715.

Tujioka, K., Ohsumi, M., Horie, K. *et al.* (2009) Dietary gamma-aminobutyric acid affects the brain protein synthesis rate in ovariectomized female rats. *J. Nutr. Sci. Vitaminol.*, **55**, 75–80.

Ueno, H. (2000) Enzymatic and structural aspects on glutamate decarboxylase. *J. Mol. Catal. B: Enzym.*, **10**, 67–79.

Van Cauwenberghe, O.R. and Shelp, B.J. (1999) Biochemical characterization of partially purified GABA: pyruvate transaminase from *Nicotiana tabacum*. *Phytochemistry*, **52**, 575–581.

Waagepetersen, H.S., Sonnewald, U., Gegelashvili, G. *et al.* (2001) Metabolic distinction between vesicular and cytosolic GABA in cultured GABAergic neurons using 13C magnetic resonance spectroscopy. *J. Neurosci. Res.*, **6**, 347–355.

Wang, H.F., Tsai, Y.S., Lin, M.L. *et al.* (2006) Comparison of bioactive components in GABA tea and green tea produced in Taiwan. *Food Chem.*, **96**, 648–653.

Wang, Q., Xin, Y.Q., Zhang, F. *et al.* (2011) Enhanced γ-aminobutyric acid-forming activity of recombinant glutamate decarboxylase (gadA) from *Escherichia coli*. *World J. Microbiol. Biotechnol.*, **27**, 693–700

Warskulat, U., Reinen, A., Grether-Beck, S. *et al.* (2004) The osmolyte strategy of normal human keratinocytes in maintaining cell homeostasis. *J. Investig. Dermatol.*, **123**, 516–521.

White, J.P., Prell, J., Ramachandran, V.K. *et al.* (2009) Characterization of a γ-aminobutyric acid transport system of *Rhizobium leguminosarum* bv.viciae 3841. *J. Bacteriol.*, **191**, 1547–1555.

Wong, C.G., Bottiglieri, T., and Snead, O.C. (2003) GABA, gamma-hydroxybutyric acid, and neurological disease. *Ann. Neurol.*, **54**, S3–S12.

Xu, C.W. and Xia, Y.H. (1999) Clinical observations on the control acute attack of deficiency-syndrome asthma with γ-aminobutyric acid. *Chin. J. Binzhou Med. Coll.*, **22**, 181.

Yang, S.Y., Lin, Q., Lu, Z.X. *et al.* (2008) Characterization of a novel glutamate decarboxylase from *Streptococcus salivarius* ssp. thermophilus Y2. *J. Chem. Technol. Biotechnol.*, **83**, 855–861.

Yao, W., Wu, X., Zhu, J. *et al.* (2013) In vitro enzymatic conversion of γ-aminobutyric acid immobilization of glutamate decarboxylase with bacterial cellulose membrane (BCM) and non-linear model establishment. *Enzyme Microb. Technol.*, **52**, 258–264.

Yu, K., Lin, L., Hu, S. *et al.* (2012) C-terminal truncation of glutamate decarboxylase from *Lactobacillus brevis* CGMCC 1306 extends its activity toward near-neutral pH. *Enzyme Microb. Technol.*, **50**, 263–269.

Zhang, Y., Song, L., and Gao, Q. (2012) The two-step biotransformation of monosodium glutamate to GABA by *Lactobacillus brevis* growing and resting cells. *Appl. Microbiol. Biotechnol.*, **94**, 1619–1627.

Zhang, R.Z., Yang, T.W., Rao, Z.M. *et al.* (2014) Efficient one-step preparation of γ-aminobutyric acid from glucose without an exogenous cofactor by the designed *Corynebacterium glutamicum*. *Green Chem.*, **16**, 4190–4197.

Zhang, H., Yao, H.Y., and Chen, F. (2006) Accumulation of gamma-aminobutyric acid in rice germ using protease. *Biosci. Biotechnol., Biochem.*, **70**, 1160–1165.

Zhao, Z., Ding, J.Y., Ma, W.H. *et al.* (2012) Identification of characterization of γ-aminobutyric acid uptake system GabPCg (NCgl0464) in *Corynebacterium glutamicum*. *Appl. Environ. Microbiol.*, **78**, 2596–2601.

Zhu, L., Peng, Q., Song, F. *et al.* (2010) Structure and regulation of gab gene cluster, involved in the γ-aminobutyric acid shunt, are controlled by a δ54 factor in *Bacillus thuringiensis*. *J. Bacteriol.*, **192**, 346–355.

17
Flavonoids: Functions, Metabolism and Biotechnology

Celestino Santos-Buelga and Ana M. González-Paramás

17.1
Introduction

Flavonoids constitute one of the largest groups of plant secondary metabolites. They are phenolic compounds synthesised through the phenylpropanoid pathway that are widespread in higher plants. Flavonoids play relevant roles in plant ecology and plant physiology, contributing to plant structural integrity, UV photoprotection, reproduction and internal regulation of plant cell signalling. They are involved in the mechanisms of natural resistance against biotic and abiotic stresses and act as key chemical modulators of plant communication with insects and microbes and as phytoalexins against pathogens and herbivores. They also contribute to seed dispersal, attract pollinators via flower colour and induce root nodulation when excreted by symbiotic nitrogen-fixing rhizobia. Some flavonoid groups such as proanthocyanidins are major determinants of seed-coat-imposed dormancy and also prevent feeding by herbivores by providing a bitter taste to leaves (Hichri *et al.*, 2011; Falcone, Ferreyra and Casati, 2012). Flavonoids are also widely distributed in the human diet through cereals, pulses, fruits, vegetables and their derived products such as wine, tea or chocolate, in which they contribute to sensory, technological and health properties.

The first observations about the effects of food flavonoids in the human organism were published by Szent-Gyorgyi and co-workers in the mid-1930s. They observed that preparations from lemon juice and paprika could restore the normal capillary resistance and permeability and prevent spontaneous bleeding associated to the deficiency of ascorbic acid in human beings. The substances responsible for that activity were identified as flavonoids (flavones or flavonols) and tentatively termed 'vitamin P' (for permeability) (Rusznyak and Szent-Gyorgyi, 1936). Further support was found in studies in guinea pigs, where it was observed that the vascular symptoms of the ascorbic acid deficiency could be counteracted by administration of a flavone fraction (called *citrin*) isolated from lemon juice, leading to the conclusion that experimental scurvy was a deficiency caused by the combined lack of vitamins C and P (Benthsath, Rusznyak and Szent-Györgyi, 1936, 1937). Although citrin was initially thought to be a pure

Industrial Biotechnology of Vitamins, Biopigments, and Antioxidants, First Edition.
Edited by Erick J. Vandamme and José L. Revuelta.

substance, further analyses revealed that it consisted of a mixture of flavonoids, namely hesperidin and an eriodictyol glycoside (Bruckner and Szent-Györgyi, 1936). At that time, there was much debate about the actual role of flavonoids on scurvy (Zilva, 1937; Scarborough, 1939, 1940; Rusznyak and Benko, 1941), but in the end, the studies failed to substantiate that they were indispensable. In 1950, following the recommendation of the Joint Committee on Biochemical Nomenclature of the American Society of Biological Chemists and the American Institute of Nutrition, the term *vitamin P* was dropped (Anonymous, 1950). Despite it seemed clear that only vitamin C was actually curative for scurvy, some authors (Cotereau *et al.*, 1948; Crampton and Lloyd, 1950; Douglass and Kamp, 1959) continued claiming for the existence of possible synergistic effects between ascorbic acid and flavonoids, which would increase the antiscorbutic potency of the vitamin, and the name vitamin C_2 was also proposed for flavonoids. However, it was never demonstrated whether this vitamin-C-sparing activity was an expression of an essential role of these compounds or it was rather associated to unspecific antioxidant effects owing to the antioxidant, free-radical scavenging and metal-chelating capacities demonstrated *in vitro* for flavonoids (Clemetson and Andersen, 1966). In 1968, the Food and Drug Administration withdrew approval of the use of 'bioflavonoids' (another denomination proposed for flavonoids, considering their biological activities) as drugs, since it was considered that they have no proven efficacy in humans for any clinical purpose (Singleton, 1981).

In recent years, the interest in the health properties of flavonoids has renewed, especially after the publication of the epidemiological observations by Hertog and coworkers in the early 1990s (Hertog *et al.*, 1993, 1995), which pointed out the existence of an inverse correlation between their dietary consumption and reduced incidence and mortality from cardiovascular disease (CVD). Since then, a large number of epidemiological studies have attempted to correlate the dietary intake of flavonoids, and phenolic compounds in general, with health-promoting effects. Many of these studies have indicated some degree of inverse associations between dietary phenolic/flavonoid intake and the incidence of degenerative diseases (Jaganath and Crozier, 2010). Nowadays, there is accumulating evidence that modest long-term intakes of flavonoids could have favourable effects on the prevalence of not only CVD but also other important diseases that represent major health challenges in developed countries, such as type II diabetes, some types of cancers or neurodegenerative disorders such as Alzheimer's and Parkinson's diseases (Del Rio *et al.*, 2013). Despite the fact that they are not essential nutrients, similar to vitamins, flavonoids are now considered as components responsible, at least in part, for the protective effects of a fruit- and vegetable-rich diet, and the study of their role in human nutrition has become a major issue of interest in food research. Furthermore, owing to the variety of biological activities reported for flavonoids, including antioxidant, anti-inflammatory, estrogenic, antimicrobial, antiproliferative or antitumour abilities, they are also considered as promising compounds for the development of new drugs. Efforts have now to be made on establishing the nature of actual bioactive molecules and provide industrial users the potential to develop methods for their production, either by chemical synthesis or by the use of biotechnological tools (Ververidis *et al.*, 2007a).

17.2
Structure and Occurrence in Food

The flavonoid family comprises several thousand compounds sharing a common C_6–C_3–C_6 phenylchromane skeleton (Figure 17.1). Based on the oxidation level of the ring C, different flavonoids classes are distinguished such as flavones, flavonols, flavanones, flavan-3-ols, anthocyanins, dihydroflavonols and isoflavones (Table 17.1). In a few cases, the heterocyclic ring C occurs in an open form (Figure 17.1), such as in chalcone and dihydrochalcone structures; also, other minor flavonoid groups can be found in plants and foods, such as flavan-3,4-diols or aurones.

Flavonoid molecules are usually hydroxylated in positions 3, 5, 7, 3′, 4′ and/or 5′ and can be further methylated, acetylated, prenylated or sulfated. In their natural sources, they may occur in free forms (aglycones), as glycosylated or acylated derivatives, and as oligomeric and polymerised structures, such as the flavan-3-ol-derived condensed tannins (also called *proanthocyanidins*). But for flavan-3-ols, most flavonoids occur in plants and foods as glycoside derivatives. The sugar residues are usually linked to 3, 7 or 4′ hydroxyl groups in the case of *O*-glycosides and directly to C-6 or C-8 in the case of *C*-glycosides. Sugar substituents can be additionally acylated with aliphatic (e.g. malonyl or acetyl residues) or aromatic acids (e.g. *p*-coumaroyl, caffeoyl or feruloyl residues) (Santos-Buelga and González-Paramás, 2014). This structural diversity gives rise to a large variety of compounds, so that more than 8000 naturally occurring flavonoids have been documented (Andersen and Markham, 2006).

Despite the large variety of existing flavonoids, only a limited number of them are prominent in foods commonly consumed by humans. These include compounds derived from three anthocyanidins (cyanidin, delphinidin, malvidin),

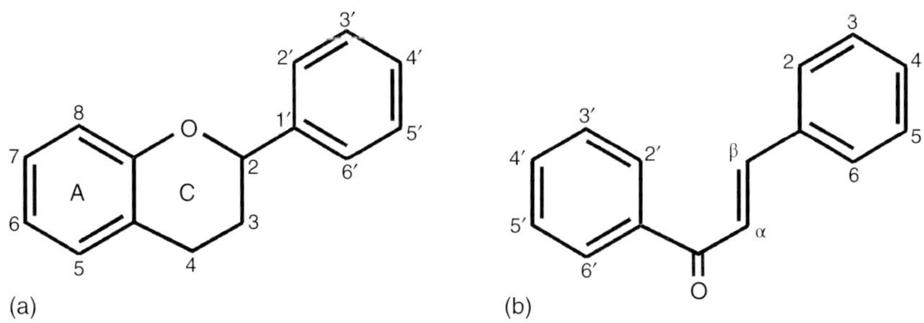

Figure 17.1 Basic structure and numbering of flavonoids: phenylchromane (a) and chalcone (b) forms.

Table 17.1 Basic skeletons of the main flavonoid classes.

Flavonoid class	Core structure	Examples
Flavan-3-ol monomers	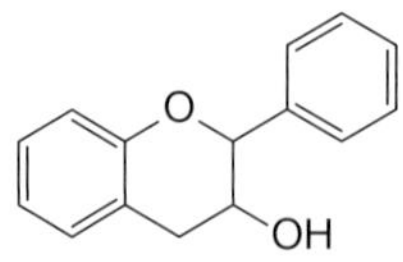	(Epi)catechin, (epi)gallocatechin
Flavanones	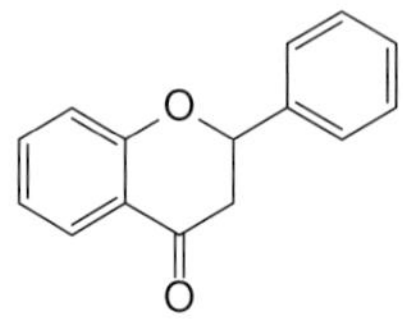	Naringenin, hesperidin, taxifolin, eriodictyol
Isoflavones	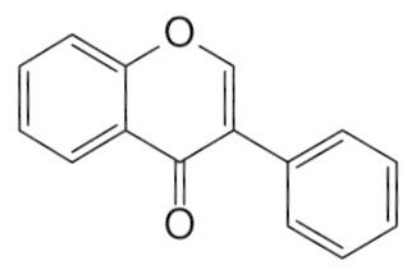	Genistein, daidzein, biochanin A, puerarin
Flavones	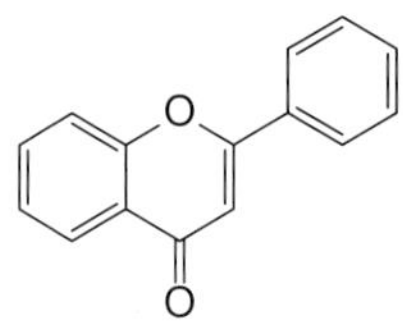	Apigenin, luteolin, chrysin, chrysoeriol
Flavonols	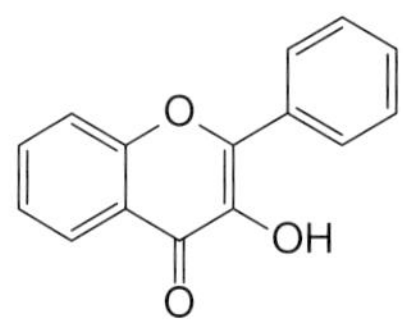	Quercetin, kaempferol, myricetin, isorhamnetin
Anthocyanins	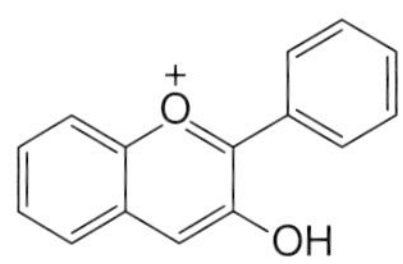	Cyanidin, delphinidin, malvidin, pelargonidin, petunidin, peonidin
Chalcones	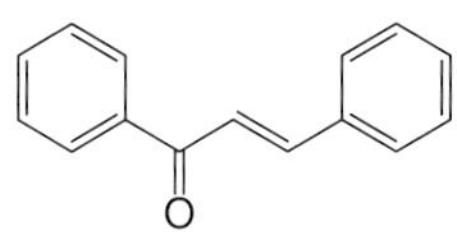	Naringenin chalcone, phloretin, arbutin

(*continued overleaf*)

Table 17.1 (*continued*)

Flavonoid class	Core structure	Examples
Flavan-3-ol oligo- and polymers (condensed tannins or proanthocyanidins)	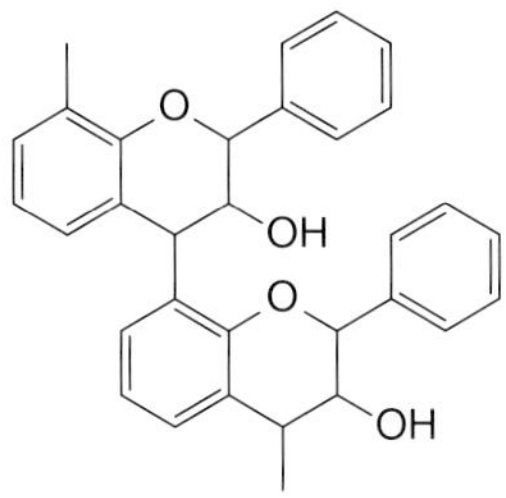	Procyanidins, prodelphinidins

three flavan-3-ols (catechin, epicatechin, epigallocatechin), three flavonols (quercetin, kaempferol, myricetin), two flavanones (hesperetin, naringenin) and two flavones (apigenin, luteolin) (Beecher, 1999). On a global scale, the most important commodities in terms of their contents of flavonoids that are widely consumed are thought to be green and black tea, red wine and cocoa/chocolate. Generally, fruits, and especially vegetables, are on a second level (Crozier, Jaganath and Clifford, 2009).

The interest in the associations between flavonoid consumption and health promotion has made the estimation of their dietary intake a point of interest. It is, however, difficult to calculate flavonoid consumption properly. On the one hand, accurate data on flavonoid composition in foods and beverages are not easy to obtain due to their structural diversity, which makes their analysis problematic. On the other hand, it is difficult to compare content values of flavonoids for a given food, owing to the use of different and non-standardised analytical methods and/or the tremendous variations that may occur in their qualitative and quantitative composition, as influenced by varietal, agronomic and environmental conditions. Furthermore, food processing and storage may involve both processes of degradation and structural transformations leading to the formation of newly derived structures, thus changing flavonoid contents and composition profiles (Santos-Buelga and González-Paramás, 2014). A first estimation of the human intake of flavonoids was made by Kühnau (1976), which calculated their average daily consumption to be around 1 g, although more recent estimates are well below this amount. Flavan-3-ol monomers (catechins) and their polymeric forms (proanthocyanidins) could represent the most common flavonoids consumed in Western diets, followed by anthocyanins and flavonols (Jaganath and Crozier, 2010).

Flavan-3-ols are found in many types of fruits, teas, cocoa, chocolate and red wine, although green tea and chocolate are considered by far the richest sources (Manach *et al.*, 2004). Catechin and epicatechin are the main flavan-3-ol monomers in fruits, whereas gallocatechin, epigallocatechin and epigallocatechin gallate (EGCG) are found in tea and certain seeds of leguminous plants. Tea is probably the most important source of these compounds in many countries; it combines a high level of consumption with a relatively high flavan-3-ol content. Furthermore, it is the only plant-derived product for human consumption that contains relevant amounts of EGCG (Hollman and Arts, 2000). Given the large range of concentrations that may be found in foodstuffs and the insufficient knowledge about the levels of proanthocyanidin polymers, it is difficult to estimate the consumption of flavan-3-ols. Intakes ranging 20–60 mg/day have been calculated for Danish (average intake of catechins of 20–50 mg/day) (Dragsted, Strube and Leth, 1997), Dutch (daily intakes of the six major catechins around 50 mg) (Arts *et al.*, 2001), Spanish (mean intakes of flavan-3-ol monomers to trimers between 17.9 and 30.6 mg/day) (de Pascual-Teresa, Rivas-Gonzalo and Santos-Buelga, 2002) and the US populations (57.7 mg/person/day) (Gu *et al.*, 2004).

Anthocyanins are the most important group of water-soluble pigments in plants and are responsible for most of the red, blue and purple colours of fruits and vegetables. Although more than 25 anthocyanin aglycones (also called *anthocyanidins*) have been identified, only 6 of them are widespread: cyanidin, delphinidin, petunidin, peonidin, pelargonidin and malvidin. Cyanidin glycosides are the most common anthocyanins in foods. These pigments occur in high abundance in berries, reaching values up to 2–6 g/kg fresh weight in blackcurrants, blackberries or red grapes (Manach *et al.*, 2004), although they are also found in other fruits, certain cereals and leafy and root vegetables such as aubergines, cabbage, beans or red onions. In a rough calculation, Andersen and Jordheim (2013) estimated that the average adult intake of anthocyanins in Western countries could be on the scale of 10 mg/day. However, wide variations should exist depending on the country, season and dietary habits, especially related to the consumption of fruits, berries and red wine. Thus, a mean daily intake of 12.5 mg was estimated for the US population (Wu *et al.*, 2006), whereas values of 82.5 mg/day were calculated in Finland, where consumption of berries is high (Heinonen, 2007).

Quercetin glycosides are the most ubiquitous flavonols in the diet, although kaempferol, myricetin and isorhamnetin derivatives are also well represented. They are found in many fruits and vegetables, with especially high concentrations of quercetin, 200–1000 mg/g fresh weight, occurring in onions (Jaganath and Crozier, 2010). Other rich food sources of flavonols are curly kale, leeks, broccoli, beans, apples and different berries (Hollman and Arts, 2000). In plant tissues, flavonols are almost exclusively present as glycosides, located mainly in the leaves, flowers and outer parts of plants, such as skin and peel, and decrease in concentration towards the central core. Hertog *et al.* (1995) estimated a daily intake of quercetin between 3 and 38 mg and indicated onions and apples as its predominant sources in the United States, Finland, Greece and former Yugoslavia, while tea contributed most in the Netherlands and Japan. Although red wine contains relatively low amounts of quercetin, compared with onions, it could be a relevant source of dietary flavonoids in some Mediterranean countries.

Other flavonoid groups have more restricted distribution in food. Flavones are present in herbs, cereals, fruits and vegetables, celery, parsley and artichoke being the only important edible sources of flavones identified (Jaganath and Crozier, 2010). Hollman and Arts (2000) estimated a mean consumption of flavonols and flavones in the Dutch population to be about 23 mg/day, where flavones contributed only a minor fraction (about 7%). Flavanones have restricted occurrence in food, despite being present in significant concentration in citrus fruits. They usually occur as neohesperidosides, which impart a bitter taste (such as naringin in grapefruit), or rutinosides. In citrus fruits, flavanones are present both in juices and in tissues, although concentrations are much higher in the solid tissues compared with juices. Orange juice contains between 200 and 600 mg hesperidin per litre, and the whole fruit may contain up to five times as much as a glass of orange juice. These quantitative data imply a variable but potentially substantial intake of citrus flavanones that would exceed by the intakes calculated for flavonols (Tomás-Barberán and Clifford, 2000). In Finland, average intakes of naringenin and hesperetin of 8.3 and 28.3 mg/day, respectively, were estimated (Kumpulainen, Lehtonen and Mattila, 1999). Nevertheless, because citrus fruits are practically the sole source of flavanones, ingestion of these flavonoids is probably greater in regions where these fruits are produced, such as Southern Europe.

Isoflavones are flavonoids classified as phytoestrogens that have a very limited distribution in the plant kingdom with substantial quantities being found only in leguminous species. Worldwide, soybeans (*Glycine max*) are almost the sole dietary source of isoflavones (Jaganath and Crozier, 2010). The mean daily intake is undoubtedly different among Asian and Western populations. In Asian countries, fermented soy products are part of the traditional diet, which leads to mean daily intake of about 8–50 mg, while in Western countries, the dietary intake of isoflavones is usually lower than 1 mg/day. Although the intake of vegetarians and soy consumers in Western population may range 3–12 mg/day, it is still low compared to intakes in Asian populations (Mortensen *et al.*, 2009).

Given the interest on the health properties of dietary flavonoids, in the past 15 years, different agencies have started to compile databases on their contents in foods. Thus, the Nutrient Data Laboratory (NDL) of the United States Department of Agriculture (USDA) made data on isoflavones available in 1999, which were further updated in 2008 (U.S. Department of Agriculture, Agricultural Research Service, 2008). Further, flavonoid and proanthocyanidin databases were released in 2003 and 2004, respectively, which were combined in 2007 into a unique database containing values for 385 food entries for 50 polyphenols, flavonoids being one of the main four classes of compounds, but also including the data on phenolic acids, lignans and stilbenes (U.S. Department of Agriculture, Agricultural Research Service, 2011). A significant gap in this database was that it only contained data for flavonoid aglycones. Since flavonoids are largely present in foods as glycosides, which are known to vary in their chemical and biological properties, some potentially useful data are lost (Neveu *et al.*, 2010). In Europe, the EuroFIR (European Food Information Resource) project started to develop a harmonised and standardised database that combines data on food composition and biological effects for plant-based bioactive compounds. This database covers various classes of bioactive compounds (flavonoids, isoflavones, carotenoids, phenolic acids, lignans and phytosterols) in 330 major food plants with photographs and detailed description (e.g. colour, size, shape or use) of their edible parts (Gry *et al.*, 2007). Data from the two databases (EuroFIR-BASIS and USDA Database for flavonoids) have their quality evaluated; however, once the systems differ in relation to the evaluation criteria adopted, the same component from an article can be differently classified depending on the system used (Wenzel de Menezes *et al.*, 2011). Another database recently developed is the Phenol-Explorer (Rothwell *et al.*, 2013), a comprehensive Web-based database on polyphenol content in foods that contains more than 37 000 original data points collected from 638 scientific articles published in peer-reviewed journals. The quality of the data was evaluated before they were aggregated to produce final representative mean content values for 502 polyphenols (glycosides, esters and aglycones) in 452 foods. The content values displayed in Phenol-Explorer are expressed in standard units (mg/100 g of fresh weight and mg/100 ml for beverages) after conversion of the original units found in the publications. The Web interface allows raising various queries on the data to identify foods containing a given polyphenol or polyphenols present in a given food (Neveu *et al.*, 2010). Using this database, Pérez-Jimenez *et al.* (2010) identified the 100 richest dietary sources of polyphenols, among which there were various spices and dried herbs, cocoa products, some darkly coloured berries, a few seeds (flaxseed) and nuts (chestnut, hazelnut) and some vegetables, such as olive and globe artichoke heads. More recently, a Brazilian flavonoid database started to be built from the compilation of data from scientific publications, evaluating 773 entries from six flavonoid subclasses (flavonols, flavones, isoflavones, flavanones, flavan-3-ols and anthocyanidins) from 197 Brazilian foods, for future dissemination in the Brazilian Food Composition Database (Wenzel de Menezes *et al.*, 2011).

17.3
Activity and Metabolism

The potential health benefits of flavonoids, and polyphenols in general, have been classically associated to their scavenging/antioxidant activity and interpreted in the framework of the Oxidative Stress theory. This concept, as initially formulated (Sies, 1985), refers to an imbalance in the dynamic equilibrium between pro-oxidants and antioxidants in favour of the formers, potentially leading to damage, which would be the origin of major degenerative diseases. Flavonoids might help counteract oxidative damage, thus contributing to the prevention of those diseases.

The antioxidant properties of flavonoids have been related to their ability to act as effective scavengers of most types of oxidising species, such as superoxide anions and hydroxyl, peroxyl, alkoxyl and NO radicals, through a mechanism that involves the transfer of an H atom to the radical stabilising it. As a consequence, a flavonoid-derived radical (aroxyl radical) is formed that is supposed to be fairly stable (Bors, Michel and Stettmaier, 1997). Structural criteria for optimal scavenging activity are the presence of a catechol group in the B-ring, a 2,3-double bond conjugated with a 4-oxo function in the C-ring and a 3- (and 5-)hydroxy group, as they provide extensive electron delocalisation over the three-ring system and confer higher stability to the produced radical (Bors *et al.*, 1990). It is generally assumed that flavonoids that lack some of these features possess weaker antioxidant activity. The existence of substituents on the hydroxyl groups (e.g. glycosylation or methylation) generally decreases the antioxidant capacity in relation to the parent flavonoids. As for condensed tannins (proanthocyanidins), the degree of polymerisation also appears to have an influence on the radical scavenging properties. In this case, extensive conjugation between the OH at C-3 and catechol groups of the B-ring together with abundant 4-8 linkages confers these oligo/polymers with enhanced radical scavenging ability (Heim, Tagliaferro and Bobilya, 2002).

The stability of the primarily formed flavonoid aroxyl radicals is sometimes questionable, and they may be converted into more reactive secondary radicals that may give rise to undesirable effects. For instance, it has been shown that the flavonol quercetin forms an *o*-semiquinone radical that can be disproportionated to produce *o*-quinones and also react with O_2 to form superoxide (Metodiewa *et al.*, 1999). Similar observations have been made in the oxidation of other polyphenols leading to the formation of mixtures of quinones, semiquinones and reactive oxygen species, ROS (e.g. $O_2^{\bullet-}$, H_2O_2), all of which are pro-oxidants and potentially cytotoxic (Halliwell, 2008). Nevertheless, although high levels of pro-oxidant activity are expected to produce toxic effects, light pro-oxidant effects, as might be provided by the cellular levels of polyphenols, could be beneficial, since, by imposing a mild degree of oxidative stress, the levels of antioxidant defences and xenobiotic-metabolising enzymes might be raised, leading to overall cytoprotection (Halliwell, 2008; Tang and Halliwell, 2010). It has also been suggested that the reduction of oxidatively formed quinones (e.g. by cellular NADPH-cytochrome P450 reductase) could give rise to a regulated metabolically recycling of flavonoids (e.g. quercetin) helping to maintain their cellular pool (Metodiewa *et al.*, 1999). Another point to consider is the interaction of flavonoids with other antioxidants, such as ascorbate, allowing the recycling of the aroxyl radicals. Flavonoids that possess a catechol group in the B-ring and a double bond at position 2-3 in the C-ring, such as quercetin, would have a higher redox potential than ascorbate and would be capable of oxidising it to the ascorbyl radical (Bors, Michel and Schikora, 1995). The ascorbyl radical may be further enzymatically reduced or disproportionate to the non-radical form (Laranjinha, 2010).

Another mechanism that may contribute to the antioxidant activity of flavonoids is their ability to act as chelators of transition metal ions, thus preventing iron- and copper-catalysed formation of initiating radical species. Two possible points of attachment of metal ions to the flavonoid molecule have been proposed: the catechol unit in B-ring and the 4-oxo function in C-ring linked to the presence of hydroxyl groups at position 3 in flavonols or position 5 in the case of flavones (Rice-Evans, Miller and Paganga, 1996).

Although flavonoids may behave as antioxidants *in vitro*, they are, in general, less bioavailable and largely biotransformed in the organism. A variable but usually small fraction of the consumed flavonoids may be absorbed in the small intestine to be further conjugated in the intestinal wall, the liver and the peripheral tissues. It has been estimated that less than 5–10% of the consumed flavonoids follow this way and will be found in the plasma in the form of conjugated metabolites (glucuronides, sulfates and methylated derivatives); the remaining 90–95% go to the large intestine (compounds not absorbed in the small gut or recycled by enterohepatic circulation), where they will be fermented by the colonic microflora (Clifford, 2004). A poorly defined fraction of these latter could also be absorbed and will be found in blood mostly as conjugated forms. Thus, the metabolites present in cells and tissues are chemically and, in many instances, functionally distinct from the dietary form, and such features underlie their bioactivity (Kroon *et al.*, 2004). It is well established that conjugated forms of flavonoids have a significantly lower capacity for donating hydrogen ions and scavenging free radicals compared to the parent compounds (Duenas *et al.*, 2010, 2011). Furthermore, flavonoids probably undergo intracellular metabolism, such as conjugation with thiols (especially glutathione), oxidative metabolism and P450-related metabolism (Spencer, Abd El Mohsen and Rice-Evans, 2004).

The actual concentrations of phenolic metabolites that can be found in the plasma of humans under realistic polyphenol consumption are in the nanomolar to low micromolar range (Manach *et al.*, 2005). Maximum levels in the plasma would be reached 1–2 h after consumption in the case of conjugated metabolites derived from the absorption in the small intestine and between 6 and 24 h for metabolites produced by the gut microflora. Even at the highest levels reported in the plasma (low micromolar values), the concentrations are far below those of other antioxidants such as urate, α-tocopherol or ascorbate, which are present in blood and/or in the intracellular milieu in micromolar and even millimolar ranges. Furthermore, in cells, they should be competing with other relevant antioxidants such as glutathione. In a rough estimation, cells might have 10^7-10^{10} molecules of glutathione and less than 10^3 molecules of flavonoids as a protection from the oxidative injury (Fraga, Celep and Galleano, 2010). In these circumstances, a direct antioxidant effect might be only expected in tissues directly exposed to polyphenols after their consumption, such as the gastrointestinal tract.

Despite the challenges posed by their limited bioavailability, mechanisms based on antioxidant activity have still been considered to explain some *in vivo* effects of flavonoids. Laranjinha and co-workers (Laranjinha, Almeida and Madeira, 1994; Laranjinha *et al.*, 1995) explored the hypothesis that circulating polyphenols might overcome the isotropic dilution in blood plasma by binding to biomembranes and lipoproteins, in view of their capacity to establish H-bonds and hydrophobic interactions. By accumulating at lipid:water interfaces, they could achieve local concentrations high enough to afford a confined antioxidant protection, for example, they could protect low-density lipoproteins (LDLs) from lipid oxidation by recycling α-tocopherol from the α-tocopheroxyl radical at the LDL surface. Actually, flavonoids and other polyphenols are known to interact with lipids and proteins. The interactions with lipids seem to be rather unspecific, based essentially on physical adsorption, mostly dependent on the hydrophobic/hydrophilic characteristics of the molecule. In turn, polyphenol–protein interactions could be either unspecific or specific and have been epitomised as similar to antigen–antibody interactions in that a binding agent and a ligand associate through single or multiple moieties to form a complex. The chemical characteristics that govern such associations are mainly related to (i) the hydrophobicity of the aromatic nuclei of polyphenols and (ii) the availability of multiple phenolic hydroxyl groups that allow hydrogen bonding. Proline-rich proteins, such as saliva proteins, seem preferential targets for these interactions (Fraga, Celep and Galleano, 2010). Also, a significant number of enzyme activities have been reported to be inhibited by flavonoids and other polyphenols, including, among others, cyclooxygenases, lipoxygenases, protein kinases, metalloproteinases, NADPH oxidases, drug metabolism enzymes or telomerase (Frade *et al.*, 2008). Of particular interest, because of its role in inflammatory conditions, is the reductive inactivation of lipoxygenases, cyclooxygenases, myeloperoxidase and xanthine oxidase, which, in view of their pro-oxidant activity, may be regarded as an indirect antioxidative action.

Even though the concept of oxidative stress as a global imbalance between pro-oxidants and antioxidants has been assumed for years, many studies, including intervention trials in humans, failed to show that shifting the balance by providing more dietary antioxidants resulted in an increased protection against oxidative stress in human disease. These and other observations, such as the lack of equilibration between the two key thiol/disulfide systems (i.e. glutathione and thioredoxin) and recognition of the existence of multiple, discrete redox signalling pathways, led Jones (2006) to redefine oxidative stress as a condition that disrupts redox signalling and control. In line with this new concept, in recent years, increasing attention has been paid to the possibility that flavonoids could act as potential modulators of intracellular signalling cascades vital to cellular function independent of their classical antioxidant capacity. Thus, it has been proposed that beneficial properties of flavonoids could be mediated by their ability to modulate the activity of both protein and lipid kinase signalling cascades (e.g. Mitogen-activated protein (MAP) kinases, protein kinase C, Akt/PKB, tyrosine kinases, phosphoinositide 3-kinase) and transcription factors (e.g. Nrf2, AP-1, NF-kB). Inhibitory or stimulatory actions at these pathways would likely profoundly affect cellular function by altering the phosphorylation state of target molecules and/or by modulating gene expression (Williams, Spencer and Rice-Evans, 2004). The concentrations of flavonoids that might be encountered *in vivo* could be sufficient to exert pharmacological activity at receptors, kinases and transcription factors. However, precise sites of action are yet to be established. It is likely that their activity depends on their ability to: (i) bind to ATP sites on enzymes and receptors; (ii) modulate the activity of kinases directly; (iii) affect the function of phosphatases, which act in opposition to kinases; (iv) preserve Ca^{2+} homeostasis, thereby preventing Ca^{2+}-dependent activation of kinases in neurons; and (v) modulate signalling cascades lying downstream of kinases, that is, transcription factor activation and binding to promoter sequences (Spencer, 2007). Whereas these actions may be more important *in vivo* than antioxidant effects, again most of them have only been demonstrated *in vitro* at non-physiologically high levels of these compounds (Jaganath and Crozier, 2010).

The modulation of nitric oxide (•NO) metabolism is another possible mechanism that could contribute to explain the *in vivo* activity of polyphenols. NO is a gaseous free radical and multifunctional messenger that easily permeates cell membranes and that has emerged as a fundamental signalling molecule in cellular functions, being involved in the regulation of major organs and systems, including immune, cardiovascular and nervous systems (Moncada, Palmer and Higgs, 1991). Flavonoids exert complex actions on the synthesis and bioavailability of NO that may result in enhanced or decreased NO levels. In cell-free systems, flavonoids may scavenge NO via its pro-oxidant properties by increasing superoxide; however, under conditions of oxidative stress, they may also protect NO from superoxide-driven inactivation. In intact healthy tissues, some flavonoids increase eNOS activity in endothelial cells. Paradoxically, this effect involves a pro-oxidant effect that results in Ca^{2+}-dependent activation of eNOS. Under conditions of inflammation and oxidative stress, flavonoids may prevent the inflammatory signalling cascades via inhibition of NFkB and thereby down-regulate iNOS. On the other hand, they also prevent the overexpression of ROS generating enzymes, reducing superoxide and peroxynitrite levels and, hence, preventing superoxide-induced NO inactivation and eNOS uncoupling. In the end, the final effect of flavonoids on NO levels will depend on the flavonoid structure and the concentrations used, on the cell type under study and particularly on the presence of inflammatory/oxidative conditions (Duarte, Francisco and Perez-Vizcaino, 2014).

In the human organism, NO is produced not only through enzymatic pathways but also by the reduction of dietary nitrate and nitrite in the stomach. After a meal, high concentrations of both nitrite and flavonoids/polyphenols may occur at this location. By taking into account the reductive activity of flavonoids, nitrite reduction to •NO is likely to occur at the acidic gastric pH, which would constitute a large source of this molecule independent of its enzymatic synthesis from nitric oxide synthase (Laranjinha, 2010). What is more, this effect would be produced by the compounds in the form they are conveyed in food; for instance, the ability of wine polyphenols to promote the production of NO from nitrite was shown both *in vitro* and in human volunteers by measuring NO in the air expelled from the stomach (Gago *et al.*, 2007). In the stomach, dietary polyphenols may not only promote nitrite reduction to NO but also embark in a complex network of chemical reactions to produce higher nitrogen oxides with signalling functions, namely by inducing post-translational modifications in both endogenous and exogenous macromolecules (most notably lipids and proteins). Thus, local and systemic effects of NO could be, in this sense, triggered by dietary flavonoids. Although a direct role of nitrite or of NO derived from nitrite reduction in the tissues has to be equated, it is also reasonable to consider that NO generated in the stomach in a polyphenol-dependent reduction of nitrite can convey such systemic and beneficial effect. By this way, the biochemistry of polyphenols in the stomach and intestine, in connection with the nitrate–nitrite–NO pathway, could constitute a shortcut for the biological effects of these molecules with impact on human health (Rocha *et al.*, 2014).

Overall, the actual mechanisms behind the *in vivo* activity of flavonoids/polyphenols are still under discussion. What seems clear is that the notion of these compounds acting as 'systemic' antioxidants and as conventional hydrogen donators is unlikely to be the sole explanation for their putative health effects. Modulation of redox signalling, entailing the modification of gene expression and of enzymatic activity, as well as interference with nitric oxide metabolism are mechanisms that might be involved in the biological effects of polyphenols, contributing to explain their influence on health, beyond direct antioxidant activity (Laranjinha, 2010).

17.4
Biosynthesis of Flavonoids in Plants

Flavonoid biosynthesis pathway is part of the larger phenylpropanoid pathway, which produces a range of other phenolic metabolites, such as phenolic acids, lignins, lignans and stilbenes. It is probably the best characterised secondary metabolic pathway in plants with approximately 20% of the carbon fixed by photosynthesis believed to be channelled into it (Ververidis *et al.*, 2007a). Over the past three decades, tremendous progresses have been made in the characterisation of the flavonoid biosynthesis derived from studies on different plant models, namely maize (*Zea mays*), snapdragon (*Antirrhinum majus*), petunia (*Petunia hybrida*) and Arabidopsis. Genes or cDNAs for the core steps in various branches of the flavonoid pathway have been identified, especially those leading to anthocyanin, proanthocyanidin, flavone and flavonol formation, allowing extensive knowledge of the encoded enzymes. Several DNA sequences have become available for the secondary enzymes that produce the different structures within each class of compounds, and data are starting to emerge on the subcellular organisation of the enzymes in the cytosol and transport mechanisms of flavonoids within the cell. Targeted manipulation of flavonoid production in transgenic plants and their heterologous production in microbial systems has also been achieved. Nevertheless, there are still major areas where data are lacking. Tertiary structures are available for only a few of the biosynthetic enzymes, little is known about the turnover or degradation of flavonoids and details of post-transcriptional regulatory mechanisms are limited. Furthermore, the range of genes encoding secondary modification enzymes that have been characterised is still limited compared to the great array of known flavonoid structures (Pandey and Sohng, 2013).

The general scheme of the flavonoid biosynthesis is shown in Figure 17.2. The key flavonoid precursors are L-phenylalanine, obtained via the shikimate and arogenate pathways, and malonyl-CoA, derived from citrate produced by the

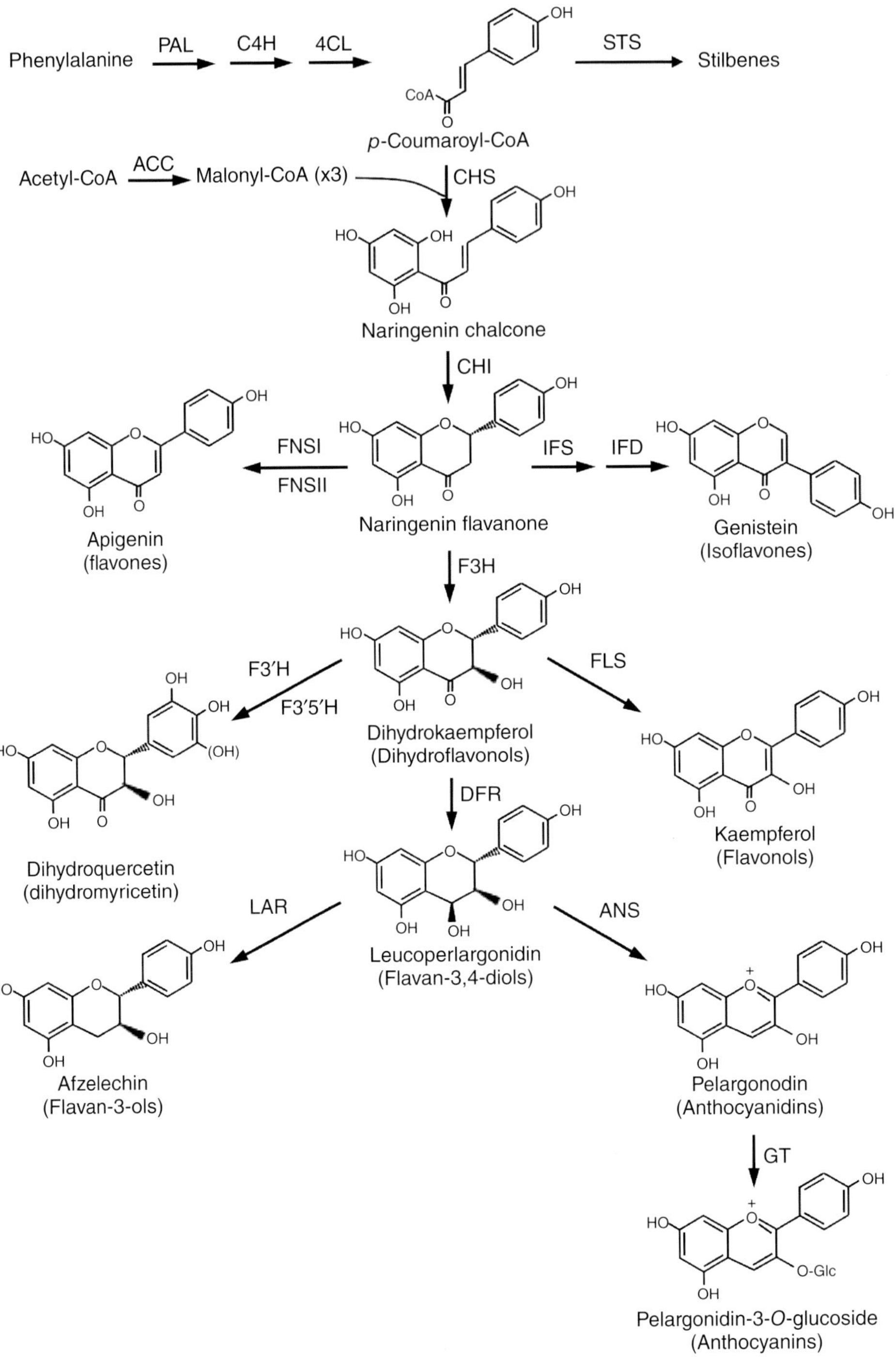

Figure 17.2 Flavonoid biosynthetic pathways. Enzyme abbreviations are described in the text.

Krebs cycle, thus linking the flavonoid biosynthesis with primary metabolism. Phenylalanine undergoes deamination by phenylalanine-ammonia-lyase (PAL) to give cinnamic acid; this activity has been found in many bacteria and fungi in addition to all plants. Further hydroxylation catalysed by a cinnamate-4-hydroxylase (C4H) produces *p*-coumaric acid, which is a common precursor for the biosynthesis of flavonoids, stilbenes, furanocoumarins and lignins. Both PAL and C4H activities can be replaced advantageously by a tyrosine ammonia lyase (TAL) catalysing the deamination of L-tyrosine in cinnamic acid, with potential application in metabolic engineering (Limem *et al.*, 2008). The *p*-coumaric acid is activated into 4-coumaroyl-CoA by a 4-coumaroyl-CoA-ligase activity (4CL). This activity, which is found in all plants and in some bacteria, also allows the conversion of other cinnamate derivatives such as caffeoate and ferulate to their corresponding CoA thiol esters (Limem *et al.*, 2008).

The formation of the C_{15} flavonoid backbone is produced by condensation of 4-coumaroyl-CoA with three molecules of malonyl-CoA catalysed by chalcone synthase (CHS) to form either 4,2′,4′,6′-tetrahydroxychalcone (i.e. naringenin chalcone) or 4,2′,4′-trihydroxychalcone (i.e. isoliquiritigenin). The formation of the latter requires one extra enzymatic step catalysed by a chalcone reductase (CHR) (Pandey and Sohng, 2013). In a few species, caffeoyl-CoA and feruloyl-CoA may also be used as substrates leading to the formation of other differently substituted chalcones. Malonyl-CoA is formed from acetyl-CoA by an acetyl-CoA carboxylase (ACC). Acetyl-CoA may be produced in mitochondria, plastids, peroxisomes and the cytosol by a variety of routes, the cytosolic acetyl-CoA being used for flavonoid biosynthesis (Limem *et al.*, 2008; Davies and Schwinn, 2006). The reaction between 4-coumaroyl-CoA and malonyl-CoA catalysed stilbene synthase (STS) results in the formation of stilbenes (such as resveratrol), a branch of the phenylpropanoid pathway leading to non-flavonoid compounds.

The following step in the flavonoid biosynthesis pathway is the isomerisation of chalcones into (2*S*)-flavanones. This reaction can occur spontaneously but is more efficiently catalysed by a chalcone isomerase (CHI) (Limem *et al.*, 2008). Further on, the pathway diverges into several side branches for the synthesis of the different flavonoid classes, which are produced through the combined actions of functionalising enzymes that hydroxylate, reduce, alkylate, oxidise and glycosylate the phenylpropanoid core structure (Pandey and Sohng, 2013).

Flavanones can be stereospecifically converted to dihydroflavonols by flavanone 3-hydroxylase (F3H), also called flavanone 3-ß-hydroxylase (FHT), or diverted to flavones by the action of flavone synthases I and II (FNS I and II). Also, they can be converted to isoflavones by the action of isoflavone synthase (IFS), which catalyses both C-2 to C-3 aryl migration and hydroxylation of the C-2 of (2*S*)-flavanones to yield (2*R*,3*S*)-2-hydroxyisoflavanones that are further dehydrated, either spontaneously or through the isoflavone dehydratase (IFD). Dihydroflavonols can be transformed into the respective flavan-3,4-diols (leucoanthocyanins) through NADPH-dependent reduction at the 4-carbonyl catalysed by dihydroflavonol reductase (DFR). Moreover, dihydroflavonols can also be converted to flavonols by a flavonol synthase (FLS), which catalyses the introduction of a double bound between carbons 2 and 3 of the C ring. Flavan-3,4-diols are converted to flavan-3-ols by leucoanthocyanidin reductase (LAR) and to anthocyanidins by anthocyanidin synthase (ANS), also referred to as *leucoanthocyanidin dioxygenase* (LDOX). All the previous structures are further modified by various hydroxylases, methyltransferases, reductases and glycosyltransferases to form diverse flavonoids and isoflavonoids. For instance, usual glycosylation at position 3 in C-ring is produced by flavonoid-3-*O*-glucosyltransferase (UFGT), and hydroxylation on the flavonoid B-ring is catalysed by the flavonoid-3′-hydroxylase (F3′H) and flavonoid-3′,5′-hydroxylase (F3′5′H), two members of the P450 monooxygenase family (Limem *et al.*, 2008; Davies and Schwinn, 2006).

The expression of the genes involved in the flavonoid biosynthesis is regulated by the complex of R2R3-MYB, basic helix-loop-helix (bHLH) and WD40 repeats (WDRs) transcription factors. The specific activation of each of the flavonoid pathways relies on the participation of the appropriate R2R3-MYB factors. A simpler mechanism concerns flavonol and phlobaphene biosynthesis and involves R2R3-MYB acting without known bHLH 'cofactors'. However, R2R3-MYB proteins controlling anthocyanin and proanthocyanidin biosynthesis require both bHLH and WDR cofactors acting in a complex interplay (Quattrocchio *et al.*, 2006).

17.5
Biotechnological Production

Increasing evidences about the health-promoting properties of flavonoids have made these compounds a very attractive natural products group, which has awaken the interest of the industry, especially food and pharmaceutical companies, in view of their use as nutraceuticals or as an option for the treatment of different chronic human diseases. Unfortunately, total synthesis of flavonoids poses many challenges for chemists, and many of these secondary metabolites are present in plant material in such small amounts that makes it impossible or very expensive to isolate them in pure form on a large-scale production. To solve this setback, in recent years, biotechnological tools that use techniques of plant and microbial genetic engineering have started to be developed for the production of high-value natural products with biological or technological interest. One of these emerged tools is the so-called *combinatorial biosynthesis* that could be defined as the combination of the product of one species and the enzymes of another species, that in nature would never meet, to yield a desired product. This approach can be applied for the expression of a single gene or for the reconstruction of complete biosynthetic pathways by combining genes of the desired pathway in host organisms. This strategy will deliver compounds that are not influenced by selection pressure, a habitat or the biochemical limitations of an organism (Julsing *et al.*, 2006).

There are some aspects that have contributed to the development of biotechnological approaches for the production of flavonoids, among them, the extensive knowledge about genes that control their biosynthesis, which has facilitated their characterisation and cloning, together with the identification of almost all enzymes and transcription factors involved in the pathways to the different flavonoid classes (Forkmann and Martens, 2001). The first approach in this area was published in 1987 and was focused on the modification of the natural colour of petunia flowers by transformation of a mutant with a maize gene (Meyer *et al.*, 1987). After that, many other strategies have been designed with the objective of increasing the production of flavonoids, such as overexpression of endogenous genes, up- or down-regulation of entire pathways using regulatory factors, inhibition of gene expression or introduction of novel biosynthetic activities from other organism (Ververidis *et al.*, 2007b).

In this chapter, the focus is on the biotechnological applications to increase flavonoid production using combinatorial biosynthesis in plants and microbes.

17.5.1
Reconstruction of Flavonoid Pathways in Plant Systems

In nature, flavonoids are produced exclusively in plants via the phenylpropanoid pathway, a biosynthetic route that has almost been completely elucidated at genetic and enzymatic levels and can be reconstructed in detail. In the literature, a lot of different strategies and achievements are described using genetic engineering and molecular biology techniques to obtain modified plants with enhanced flavonoids content compared to their natural counterparts. In principal, most of the described approaches are based on the same principle, 'heterologous organisms provide precursors from their own primary and secondary metabolism that are metabolised to the desired secondary product by the expression of foreign genes' (Julsing *et al.*, 2006). Furthermore, overexpression of metabolite-specific transcription factors coordinately activates the entire pathway, enhancing the overall activity of the transgene expression. Also, the suppression of competitive metabolic pathways allows redirection of metabolic fluxes, increasing the yield of the desired compounds (Wang, Chen and Yu, 2011).

Initial works on the genetic manipulation of the biosynthetic pathway of flavonoids were focused on the expression of different transcription factors, based on the evidence that one transcription factor typically controls the expression of multiple genes of the same pathway, allowing, in this way, efficient and simple manipulation of multi-enzyme pathways. The first approaches were designed using anthocyanins as target flavonoids and expressing different MYC-type transcription factors from *Z. mays*, *Antirrhinum* or *Perilla frutescens* in tomato (*Lycopersicon esculentum*) so as to produce enhanced pigmentation (Mooney *et al.*, 1995; Goldsbrough, Tong and Yoder, 1996; Gong *et al.*, 1999). Also, the overexpression of orthologs of *ant1*, a gene that encodes an MYB transcription factor, from *L. esculentum* resulted in plants displaying intense purple colour and fruits showing purple spotting on the epidermis and pericarp, indicating that the overexpression of ANT1 caused the up-regulation of genes that encode proteins of anthocyanidin biosynthesis, as well as genes involved in the glycosylation and transport of anthocyanins into the vacuole (Mathews *et al.*, 2003). The co-expression of C_1 and LC transcription factors (MYB and MYC type, respectively) from *Z. mays* in transgenic tomatoes was sufficient to up-regulate the flavonoid pathway in tomato fruit flesh, a tissue that normally does not produce flavonoids, leading to the production of high-flavonol tomatoes (Bovy *et al.*, 2002). There are also examples in the literature of how the expression of some heterologous transcription factors can induce the activation of some branches of the flavonoid pathway, leading to the production of, for example, isoflavones or proanthocyanidins in non-producing plants (Yu *et al.*, 2000; Xie *et al.*, 2006).

Other common strategy to improve the plant ability to synthesise flavonoids in the edible parts consists of the modification of structural genes by introduction of single or multiple heterologous genes in the flavonoids pathway that increase the production of key enzymes, often rate-limiting enzymes. One of the most widespread approaches consist of the overexpression of genes encoding CHI, an enzyme involved in the transformation of chalcones into flavanones, which are the substrate for the different branches of the pathway (Figure 17.2). An early report of this genetic approach was described by Muir *et al.* (2001), who produced a transgenic tomato line by the overexpression of the Petunia *chiA* gene encoding CHI. The resulting modified tomatoes showed increased contents of flavonols in the fruit peel of up to 78-fold, mainly due to the accumulation of rutin (i.e. quercetin-3-*O*-rutinoside) and smaller but still substantial levels of kaempferol glycosides. The same group also demonstrated that using structural flavonoid genes (encoding STS, CHS, CHR, CHI and FNS) from different plant sources to produce transgenic tomatoes resulted in fruits that contained high levels of stilbenes (resveratrol and piceid), deoxychalcones (butein and isoliquiritigenin), flavones (luteolin-7-*O*-glucoside and luteolin aglycone) and flavonols (quercetin and kaempferol glycosides) (Schijlen *et al.*, 2006).

Due to the significant role of isoflavonoids in plant defence and health-related benefits, there is great interest to improve the biosynthesis of this kind of compounds both in leguminous, where isoflavones naturally occur, and in non-leguminous plants. To date, nearly all attempts to produce isoflavonoids are based on the overexpression of the IFS gene from legumes, although some reports indicate that in legumes, increased isoflavonoid synthesis does not necessarily result from higher levels of IFS expression, but enhanced content could also be obtained by increasing the expression of other structural genes such as CHS7 and CHS8 (Dhaubhadel *et al.*, 2007). It has also been suggested that it is possible to generate significant amounts of genistein and daidzein in plants overexpressing HID (2-hydroxyisoflavanone dehydratase) (Shimamura *et al.*, 2007).

Techniques used to redirect endogenous phenylpropanoid pathway have also been applied to different areas of flavonoid engineering in order to focus towards particular branches of the flavonoid biosynthesis. For example, Davies *et al.* (2003) demonstrated that down-regulation of the expression of FLS, a key enzyme in the regulation of the flux into different branches of the flavonoid pathway, increased the production of anthocyanins. RNAi has also been used to knock out metabolic pathways reprogramming the flow from synthesis of the flavone apigenin to the isoflavone genistein (Jiang *et al.*, 2010). More examples of flux control can be found in the reviews by Wang, Chen and Yu (2011) and Mora-Pale *et al.* (2013).

Although the use of plant tissue systems, such as suspension cells and hairy root cultures, may have advantages over cultivated plants for the production of bioactive compounds, there are few examples of their application on flavonoids production. Overexpression of the *Saussurea medusa* CHI gene in *Saussurea involucrata* hairy roots increased the production of flavonoids in general and apigenin in particular (Li *et al.*, 2006). A maize transcription factor transgene was used to generate a potato cell line with high-level production of anthocyanins without the requirement for light induction (Davies and Deroles, 2014).

17.5.2
Reconstruction of Flavonoid Pathways in Microbial Systems

Many high-value metabolites are produced in nature by organisms that are not ideal for large-scale production. The reconstruction of the flavonoid pathway in microorganisms has offered promising results for the industrial production of these compounds by fermentation or bioconversion (Wang, Chen and Yu, 2011). Microbial fermentations have the advantages that their growth is based on inexpensive renewable feedstocks (carbon and nitrogen sources) that can be strictly controlled and allow shorter production times. Different microbial systems have been used as a platform to reconstitute biosynthetic pathways from plants. The first studies testing the microbes potential for the biotransformation of precursor compounds into flavonoids were focused on the expression of specific enzymes such as PAL, CHS or CHI, although nowadays the expression of whole pathways in hosts organisms, such as *E. coli*, *Bacillus subtilis* or *Saccharomyces cerevisiae*, by the application of genetic techniques, has come true (Julsing *et al.*, 2006; Ververidis *et al.*, 2007b).

17.5.2.1
E. coli Platform

Escherichia coli was the first microorganism engineered for the biotechnological production of flavonoids. This bacterium lacks the main genes for flavonoids biosynthesis; hence, recombinant plasmids containing the genes of interest must be used. Although the heterologous expression of diverse phenylpropanoid biosynthetic enzymes in bacteria had been reported before, the production of flavonoids by genetically engineered bacteria was described for the first time in 2003 by Horinouchi's research group (Hwang *et al.*, 2003). An artificial gene cluster including PAL from the yeast *Rhodotorula rubra*, 4CL from the actinomycete *Streptomyces coelicolor* A3(2) and CHS from the plant *Glycyrrhiza echinata*, was constructed in *E. coli*. Starting with the amino-acids tyrosine and phenylalanine as substrates, an *E. coli* strain (BL21(DE3) from Novagen) carrying one of the designed gene clusters (plasmid pET26b-rbs-3GS) was able to produce small amounts of naringenin (from tyrosine) and pinocembrin (from phenylalanine) (Hwang *et al.*, 2003; Kaneko *et al.*, 2003). Further modification of the same system introducing the F3H/FLS and FNS genes into the recombinant *E. coli* cell allowed the production of naringenin, resveratrol, genistein, curcumin, as well as significant amounts of apigenin (2 h; 13 mg/l) from tyrosine, and chrysin (1 h; 9.4 mg/l) from phenylalanine (Horinouchi, 2008). Further models were designed by introducing additional biosynthetic genes to modify the oxygenation pattern of flavonoids, thus establishing a nearly complete biosynthetic pathway from plants in a microorganism (Miyahisa *et al.*, 2005). At the same time, another group described a strategy for synthesising plant anthocyanins in *E. coli,* by the expression of F3H, DFR, ANS and F3GT, producing low concentrations of the 3-*O*-glucosides of pelargonidin and cyanidin (Yan *et al.*, 2005).

Another approach used for high-level production of flavonoids in *E. coli* is to increase the levels of intracellular malonyl-CoA substrate by overexpression of the enzyme ACC, supplementing the culture medium with malonate while expressing a malonyl-CoA synthetase (MatB) or minimising the malonyl-CoA lost to fatty acid synthesis by repressing some genes (Chemler and Koffas, 2008). Computational tools, such as constraint-based flux balance analysis (FBA) algorithms, can be used to predict a set of genetic interventions on the genome scale that redirect the carbon flux to malonyl-CoA or target products (Wu *et al.*, 2014). Similar to malonyl-CoA, it has also been found that NADPH supply can be an important limitation in the engineering of *E. coli* for flavonoid production (Chemler *et al.*, 2010).

Because the addition of expensive phenylpropanoic precursors is one of the impediments to the large-scale microbial production, more economical processes for *de novo* production of naringenin and pinocembrin, without the need to feed precursors, have recently been developed in *E. coli* using glucose as a substrate (Santos, Koffas and Stephanopoulos, 2011). Additional factors must be taken into account, such as the necessity to reduce the accumulation of the final product in the cytoplasmic space to prevent feedback inhibition, or the possibility to use protein engineering to mutate natural enzymes so as to increase their catalytic activity. These and other challenges have been recently reviewed by Wu and coworkers (2014).

In nature, most of the flavonoids are commonly modified through methylation, glycosylation and hydroxylation. Some approaches have been made in this area using transformed *E. coli* to obtain more complex flavonoids such as *O*-methylated or glycosylated structures. For example, 3-*O*-methyl kaempferol was obtained from naringenin by using two *E. coli* transformants, harbouring either FLS (enzyme that converts naringenin to kaempferol) from cottonwood poplar (PFLS) or ROMT-9 (enzyme that transfers the methyl group to the 3′-hydroxyl group) from rice. Independently, grown cells were mixed and incubated with naringenin, which resulted in the production of 3-*O*-methyl kaempferol along with kaempferol (Kim, Joe and Ahn, 2010). A review on glycosyl and methyl transformations can be found in a recent paper (Song *et al.*, 2014).

17.5.2.2
Saccharomyces cerevisiae Platform

The yeast *S. cerevisiae* has also been used to express flavonoid biosynthetic genes. It offers some advantages over *E. coli*, such as it is classified by the FDA as GRAS (generally recognised as safe) for production of consumer goods and that it is also more industrially robust and less susceptible to phage contamination than *E. coli* (Krivoruchko and Nielsen, 2015). In addition, as a eukaryote, it has similar compartmentalisation to plant cells, can post-translationally modify eukaryotic proteins and also has the ability to support functional expression of membrane-bound cytochrome P450 enzymes, including the C4H, which are not expressed in *E. coli* (Wang, Chen and Yu, 2011). Successful expression of PAL and C4H leading to the synthesis of the precursor *p*-coumaric acid from L-phenylalanine was firstly achieved in *S. cerevisiae* by Ro and Douglas (2004). Those authors also demonstrated that no physical interaction of PAL with C4H was required, so co-cultivation of separate strains expressing these genes also resulted in *p*-coumarate production. That work opened the door to generate further strains that channel *p*-coumarate into flavonoid biosynthesis. Similar research but introducing PAL (with TAL activity, too) together with 4CL and CHS led to the partial reconstruction of the flavonoid pathway in the yeast and the production of naringenin from phenylalanine and pinocembrin from tyrosine (Jiang, Wood and Morgan, 2005). A more efficient flavanone producing pathway was reconstructed starting with C4H, followed by 4CL, CHS and CHI (Yan, Kohli and Koffas, 2005).

The flavanone naringenin is a common precursor to many flavonoids and can be used as a substrate by yeasts co-expressing CHI with IFS, F3H or FNSII to obtain isoflavones, dihydroflavonols or flavones, respectively. For instance, apigenin can be obtained from naringenin by the action of FNSII. Similarly, other flavanones can be used as substrates to produce differently substituted flavonoids, such as eriodictyol to obtain luteolin or liquiritigenin for 7,4′-dihydroxyflavone (Martens and Forkmann, 1999). As described for *E. coli*, trying to solve the availability of expensive precursors, which limit the industrial production of flavonoids, *de novo* pathways have been designed using glucose as the exclusive carbon source to obtain naringenin (Koopman *et al.*, 2012). Representative examples of the synthesis of different flavonoids using *S. cerevisiae* as a model organism can be found in recent reviews (Ververidis *et al.*, 2007b; Song *et al.*, 2014). Approaches using *Streptomyces venezuelae* as a heterologous host have also been attempted, although the system has not been fully optimised and shows relatively low productivity compared to *E. coli* and *S. cerevisiae* (Song *et al.*, 2014).

17.6
Concluding Remarks

The engineering of plants aims at increasing the yield of flavonoid production, while the engineering of microbes by reconstructing the flavonoid biosynthetic pathways explores the power of fermentation to continuously produce specific compounds under controlled conditions. Significant advances have been made through metabolic engineering of plant cells to generate a variety of compounds, although improvements are still necessary. A major issue in the application of transgenic approaches to pathway engineering is the time and cost required to generate stably transformed plants. Furthermore, enzymes that catalyse complex chemical substitutions of the flavonoid structure, such as hydroxylation, methylation or conjugation, are yet to be discovered. Microbial production offers an interesting alternative to produce flavonoids on an industrial scale in a more economical and environment-friendly manner. However, at present, microbial production has been achieved only at a laboratory scale, maybe due to obstacles such as the need for addition of expensive phenylpropanoid precursors, the requirement of two different media for cell proliferation and flavonoid production and the low intracellular concentration of malonyl-CoA (Wu *et al.*, 2014). The strategy for the future will be to identify a microbial host in which basic primary pathways can be exploited for the production of biosynthetic precursors for further secondary pathways. In this way, no transfer of genes and promoter sequences for a primary pathway are necessary and genes and expressed enzymes for the desired secondary pathway can just be coupled (Julsing *et al.*, 2006).

References

Andersen, Ø.M. and Jordheim, M. (2013) in *Anthocyanins in Health and Disease* (eds T.C. Wallace and M. Giusti), CRC Press, Boca Raton, FL, pp. 13–90.

Andersen, Ø.M. and Markham, K.R. (eds) (2006) *Flavonoids. Chemistry, Biochemistry and Applications*, Taylor & Francis Group, Boca Raton, FL.

Anonymous1950) Use of the term vitamin P. *Nature*, **166**, 543.

Arts, I.C., Hollman, P.C., Feskens, E.J., Bueno de Mesquita, H.B., and Kromhout, D. (2001) Catechin intake and associated dietary and lifestyle factors in a representative sample of Dutch men and women. *Eur. J. Clin. Nutr.*, **55**, 76–81.

Beecher, G.R. (1999) in *Antioxidant Food Supplements in Human Health* (eds L. Packer, M. Hiramatsu, and T. Yoshikawa), Academic Press, San Diego, CA, pp. 269–281.

Benthsath, A., Rusznyak, S.T., and Szent-Györgyi, A. (1936) Vitamin nature of flavones. *Nature*, **138**, 798.

Benthsath, A., Rusznyak, S.T., and Szent-Györgyi, A. (1937) Vitamin P. *Nature*, **139**, 326–327.

Bors, W., Heller, W., Michel, C., and Saran, M. (1990) Flavonoids as antioxidants: determination of radical scavenging efficiencies. *Methods Enzymol.*, **186**, 343–355.

Bors, W., Michel, C., and Schikora, S. (1995) Interaction of flavonoids with ascorbate and determination of their univalent redox potentials: a pulse radiolysis study. *Free Radic. Biol. Med.*, **19**, 45–52.

Bors, W., Michel, C., and Stettmaier, K. (1997) Antioxidant effects of flavonoids. *BioFactors*, **6**, 399–402.

Bovy, A., de Vos, R., Kemper, M., Schijlen, E., Almenar-Pertejo, M., Muir, S., Collins, G., Robinson, S., Verhoeyen, M., Hughes, S., Santos-Buelga, S., and van Tunen, A. (2002) High-flavonol tomatoes resulting from the heterologous expression of the maize transcription factor genes LC and C1. *Plant Cell*, **14**, 2509–2526.

Bruckner, V. and Szent-Györgyi, A. (1936) Chemical nature of citrin. *Nature*, **138**, 1057 (Letter).

Chemler, J.A., Fowler, Z.L., McHugh, K.P., and Koffas, M.A. (2010) Improving NADPH availability for natural product biosynthesis in *Escherichia coli* by metabolic engineering. *Metab. Eng.*, **12**, 96–104.

Chemler, J.A. and Koffas, M.A.G. (2008) Metabolic engineering for plant natural product biosynthesis in microbes. *Curr. Opin. Biotechnol.*, **19**, 597–605.

Clemetson, C.A. and Andersen, L. (1966) Plant polyphenols as antioxidants for ascorbic acid. *Ann. N.Y. Acad. Sci.*, **136**, 341–376.

Clifford, M.N. (2004) Diet-derived phenols in plasma and tissues and their implications for health. *Planta Med.*, **70**, 1103–1114.

Cotereau, H., Gabe, M., Géro, E., and Parrot, J.L. (1948) Influence of vitamin P (vitamin C2) upon the amount of ascorbic acid in the organs of the guinea pig. *Nature*, **161**, 557.

Crampton, E.W. and Lloyd, L.E. (1950) A quantitative estimation of the effect of rutin on the biological potency of vitamin C. *J. Nutr.*, **41**, 487–498.

Crozier, A., Jaganath, I.B., and Clifford, M.N. (2009) Dietary phenolics: chemistry, bioavailability and effects on health. *Nat. Prod. Rep.*, **26**, 1001–1043.

Davies, K.M. and Deroles, S.C. (2014) Prospects for the use of plant cell cultures in food biotechnology. *Curr. Opin. Biotechnol.*, **26**, 133–140.

Davies, K.M. and Schwinn, K.E. (2006) in *Flavonoids. Chemistry, Biochemistry and Applications* (eds Ø.M. Andersen and K.R. Markham), Taylor & Francis Group, Boca Raton, FL, pp. 143–218.

Davies, K.M., Schwinn, K.E., Deroles, S.C., Manson, D.G., Lewis, D.H., Bloor, S.J., and Bradley, J.M. (2003) Enhancing anthocyanin production by altering competition for substrate between flavonol synthase and dihydroflavonol 4-reductase. *Euphytica*, **131**, 259–268.

Del Río, D., Rodriguez-Mateos, A., Spencer, J.P.E., Tognolini, M., Borges, G., and Crozier, A. (2013) Dietary (poly)phenolics in human health: structures, bioavailability, and evidence of protective effects against chronic diseases. *Antioxid. Redox Signal.*, **8**, 1818–1892.

Dhaubhadel, S., Gijzen, M., Moy, P., and Farhangkhoee, M. (2007) Transcriptome analysis reveals a critical role of CHS7 and CHS8 genes for isoflavonoid synthesis in soybean seeds. *Plant Physiol.*, **143**, 326–338.

Douglass, C.D. and Kamp, G.H. (1959) The effect of orally administered rutin on the adrenal ascorbic acid level in guinea pigs. *J. Nutr.*, **67**, 531–536.

Dragsted, L.O., Strube, M., and Leth, T. (1997) Dietary levels of plant phenols and other non-nutritive components: could they prevent cancer? *Eur. J. Cancer Prev.*, **6**, 522–528.

Duarte, J., Francisco, V., and Perez-Vizcaino, F. (2014) Modulation of nitric oxide by flavonoids. *Food Funct.*, **5**, 1653–1668.

Duenas, M., González-Manzano, S., González-Paramás, A., and Santos-Buelga, C. (2010) Antioxidant evaluation of O-methylated metabolites of catechin, epicatechin and quercetin. *J. Pharm. Biomed. Anal.*, **51**, 443–449.

Duenas, M., Surco-Laos, F., Gonzalez-Manzano, S., Gonzalez-Paramas, A.M., and Santos-Buelga, C. (2011) Antioxidant properties of major metabolites of quercetin. *Eur. Food Res. Technol.*, **232**, 103–111.

Falcone, M.L., Ferreyra, S.P., and Casati, P. (2012) Flavonoids: biosynthesis, biological functions and biotechnological applications. *Front. Plant Sci.*, **3**, 1–15.

Forkmann, G. and Martens, S. (2001) Metabolic engineering and applications of flavonoids. *Curr. Opin. Biotechnol.*, **12**, 155–160.

Frade, J., Ferreira, R., Barbosa, M., and Laranjinha, J. (2008) Mechanisms of neuroprotection by polyphenols. *Curr. Med. Chem.*, **5**, 307–318.

Fraga, C.G., Celep, G.S., and Galleano, M. (2010) in *Plant Phenolics and Human Health: Biochemistry, Nutrition, and Pharmacology* (ed C.G. Fraga), John Wiley & Sons, Inc., Hoboken, NJ, pp. 91–106.

Gago, B., Lundberg, J.O., Barbosa, R.M., and Laranjinha, J. (2007) Red wine-dependent reduction of nitrite to nitric oxide in the stomach. *Free Radic. Biol. Med.*, **43**, 1233–1242.

Goldsbrough, A.P., Tong, Y., and Yoder, J.I. (1996) Lc as a non-destructive visual reporter and transposition excision marker gene for tomato. *Plant J.*, **9**, 927–933.

Gong, Z.Z., Yamagishi, E., Yamazaki, M., and Saito, K. (1999) A constitutively expressed Myc-like gene involved in anthocyanin biosynthesis from *Perilla frutescens*: molecular characterization, heterologous expression in transgenic plants and transactivation in yeast cells. *Plant Mol. Biol.*, **41**, 33–44.

Gry, J., Black, L., Eriksen, F.D., Pilegaard, K., Plumb, J., Rhodes, M., Sheehan, D., Kiely, M., and Kroon, P.A. (2007) EuroFIR-BASIS: a combined composition and biological activity database for bioactive compounds in plant-based foods. *Trends Food Sci. Technol.*, **18**, 434–444.

Gu, L., Kelm, M.A., Hammerstone, J.F., Beecher, G., Holden, J., Haytowitz, D., Gebhardt, S., and Prior, R. (2004) Concentrations of proanthocyanidins in common foods and estimations of normal consumption. *J. Nutr.*, **134**, 613–617.

Halliwell, B. (2008) Are polyphenols antioxidants or pro-oxidants? What do we learn from cell culture and in vivo studies? *Arch. Biochem. Biophys.*, **476**, 107–112.

Heim, K.E., Tagliaferro, A.R., and Bobilya, D.J. (2002) Flavonoid antioxidants: chemistry, metabolism and structure–activity relationships. *J. Nutr. Biochem.*, **13**, 572–584.

Heinonen, M. (2007) Antioxidant activity and antimicrobial effect of berry phenolics – a Finnish perspective. *Mol. Nutr. Food Res.*, **51**, 684–691.

Hertog, M.G., Feskens, E.J., Hollman, P.C., Katan, M.B., and Kromhout, D. (1993) Dietary antioxidant flavonoids and risk of coronary heart disease: the Zutphen elderly study. *Lancet*, **342**, 1007–1011.

Hertog, M.G., Kromhout, D., Aravanis, C., Blackburn, H., Buzina, R., Fidanza, F., Giampaoli, S., Jansen, A., Menotti, A., Nedeljkovic, S., Pekkarinen, M., Simic, B.S., Toshima, H., Feskens, E.J.M., Hollman, P.C.H., and Katan, M.B. (1995) Flavonoid intake and long-term risk of coronary heart disease and cancer in the seven countries study. *Arch. Intern. Med.*, **155**, 381–386.

Hichri, I., Barrieu, F., Bogs, J., Kappel, C., Delrot, S., and Lauvergeat, V. (2011) Recent advances in the transcriptional regulation of the flavonoid biosynthetic pathway. *J. Exp. Bot.*, **62**, 2465–2483.

Hollman, P.C.H. and Arts, I.C.W. (2000) Flavonols, flavones and flavanols – nature, occurrence and dietary burden. *J. Sci. Food Agric.*, **80**, 1081–1093.

Horinouchi, S. (2008) Combinatorial biosynthesis of non-bacterial and unnatural flavonoids, stilbenoids and curcuminoids by microorganisms. *J. Antibiot.*, **61**, 709–728.

Hwang, E.I., Kaneko, M., Ohnishi, Y., and Horinouchi, S. (2003) Production of plant-specific flavanones by *Escherichia coli* containing an artificial gene cluster. *Appl. Environ. Microbiol.*, **69**, 2699–2706.

Jaganath, I.B. and Crozier, A. (2010) in *Plant Phenolics and Human Health: Biochemistry, Nutrition, and Pharmacology* (ed C.G. Fraga), John Wiley & Sons, Inc., Hoboken, NJ, pp. 1–49.

Jiang, Y.N., Wang, B., Li, H., Yao, L.M., and Wu, T.L. (2010) Flavonoid production is effectively regulated by RNAi interference of two flavone synthase genes from *Glycine max*. *J. Plant Biol.*, **53**, 425–432.

Jiang, H., Wood, K.V., and Morgan, J.A. (2005) Metabolic engineering of the phenylpropanoid pathway in *Saccharomyces cerevisiae*. *Appl. Environ. Microbiol.*, **71**, 2962–2969.

Jones, D.P. (2006) Redefining oxidative stress. *Antioxid. Redox Signal.*, **8**, 1865–1879.

Julsing, M.K., Koulman, A., Woerdenbag, H.J., Quax, W.J., and Kayser, O. (2006) Combinatorial biosynthesis of medicinal plant secondary metabolites. *Biomol. Eng.*, **23**, 265–279.

Kaneko, M., Hwang, E.I., Ohnishi, Y., and Horinouchi, S. (2003) Heterologous production of flavanones in *Escherichia coli*: potential for combinatorial biosynthesis of flavonoids in bacteria. *J. Ind. Microbiol. Biotechnol.*, **30**, 456–461.

Kim, B.G., Joe, E.J., and Ahn, J.H. (2010) Molecular characterization of flavonol synthase from poplar and its application to the synthesis of 3-O-methylkaempferol. *Biotechnol. Lett.*, **32**, 579–584.

Koopman, F., Beekwilder, J., Crimi, B., van Houwelingen, A., Hall, R.D., Bosch, D., van Maris, A.J., Pronk, J.T., and Daran, J.M. (2012) De novo production of the flavonoid naringenin in engineered *Saccharomyces cerevisiae*. *Microb. Cell Fact.*, **11**, 155.

Krivoruchko, A. and Nielsen, J. (2015) Production of natural products through metabolic engineering of *Saccharomyces cerevisiae*. *Curr. Opin. Biotechnol.*, **35**, 7–15.

Kroon, P.A., Clifford, M.N., Crozier, A., Day, A.J., Donovan, J.L., Manach, C., and Williamson, G. (2004) How should we assess the effects of exposure to dietary polyphenols in vitro? *Am. J. Clin. Nutr.*, **80**, 15–21.

Kühnau, J. (1976) The flavonoids. A class of semi-essential food components: their role in human nutrition. *World Rev. Nutr. Diet.*, **24**, 117–191.

Kumpulainen, J.T., Lehtonen, M., and Mattila, P. (1999) in *Natural Antioxidants and Anticarcinogens in Nutrition, Health and Disease* (eds J.T. Kumpulainen and J.T. Salonen), The Royal Society of Chemistry, Cambridge, pp. 141–150.

Laranjinha, J. (2010) in *Recent Advances in Polyphenols Research*, vol. **2** (eds C. Santos-Buelga, M.T. Escribano, and V. Lattanzio), Wiley-Blackwell, Chichester, pp. 269–282.

Laranjinha, J.A., Almeida, L.M., and Madeira, V.M. (1994) Reactivity of dietary phenolic acids with peroxyl radicals: antioxidant activity upon low density lipoprotein peroxidation. *Biochem. Pharmacol.*, **48**, 487–494.

Laranjinha, J., Vieira, O., Madeira, V.M., and Almeida, L.M. (1995) Two related phenolic antioxidants with opposite effects on vitamin E content in low density lipoproteins oxidized by ferrylmyoglobin: consumption *vs* regeneration. *Arch. Biochem. Biophys.*, **323**, 373–381.

Li, F.X., Jin, Z.P., Zhao, D.X., Cheng, L.Q., Fu, C.X., and Ma, F. (2006) Overexpression of the *Saussurea medusa* chalcone isomerase gene in S. involucrate hairy root cultures enhances their biosynthesis of apigenin. *Phytochemistry*, **67**, 553–560.

Limem, I., Guedon, E., Hehn, A., Bourgaud, F., Ghedira, L.C., Engasser, J.M., and Ghoul, M. (2008) Production of phenylpropanoid compounds by recombinant microorganisms expressing plant-specific biosynthesis genes. *Process Biochem.*, **43**, 463–479.

Manach, C., Scalbert, A., Morand, C., Rémésy, C., and Jiménez, L. (2004) Polyphenols: food sources and bioavailability. *Am. J. Clin. Nutr.*, **79**, 727–747.

Manach, C., Williamson, G., Morand, C., Scalbert, A., and Remesy, C. (2005) Bioavailability and bioefficacy of polyphenols in humans. I. Review of 97 bioavailability studies. *Am. J. Clin. Nutr.*, **81**, 230S–242S.

Martens, S. and Forkmann, G. (1999) Cloning and expression of flavone synthase II from Gerbera hybrids. *Plant J.*, **20**, 611–618.

Mathews, H., Clendennen, S.K., Caldwell, C.G., Liu, X.L., Connors, K., Matheis, N., Schuster, D.K., Menasco, D.J., Wagoner, W., Lightner, J., and Wagner, D.R. (2003) Activation tagging in tomato identifies a transcriptional regulator of anthocyanin biosynthesis, modification, and transport. *Plant Cell*, **15**, 1689–1703.

Metodiewa, D., Jaiswal, A.K., Cenas, N., Dickancaité, E., and Segura-Aguilar, J. (1999) Quercetin may act as a cytotoxic prooxidant after its metabolic activation to semiquinone and quinoidal product. *Free Radic. Biol. Med.*, **26**, 107–116.

Meyer, P., Heidmann, I., Forkmann, G., and Saedler, H. (1987) A new petunia flower colour generated by transformation of a mutant with a maize gene. *Nature*, **330**, 667–678.

Miyahisa, I., Kaneko, M., Funa, N., Kawasaki, H., Kojima, H., Ohnishi, Y., and Horinouchi, S. (2005) Efficient production of (2S)-flavanones by *Escherichia coli* containing an artificial biosynthetic gene cluster. *Appl. Microbiol. Biotechnol.*, **68**, 498–504.

Moncada, S., Palmer, R.M., and Higgs, E.A. (1991) Nitric oxide: physiology, pathophysiology, and pharmacology. *Pharmacol. Rev.*, **43**, 109–142.

Mooney, M., Desnos, T., Harrison, K., Jones, J., Carpenter, R., and Coen, E. (1995) Altered regulation of tomato and tobacco pigmentation genes caused by the delila gene of *Antirrhinum*. *Plant J.*, 7, 333–339.

Mora-Pale, M., Sanchez-Rodriguez, S.P., Linhardt, R.J., Dordick, J.S., and Koffas, M.A. (2013) Metabolic engineering and in vitro biosynthesis of phytochemicals and non-natural analogues. *Plant Sci.*, **210**, 10–24.

Mortensen, A., Kulling, S.E., Schwartz, H., Rowland, I., Ruefer, C.E., Rimbach, G., Cassidy, A., Magee, P., Millar, J., Hall, W.L., Birkved, F.K., Sorensen, I.K., and

Sontag, G. (2009) Analytical and compositional aspects of isoflavones in food and their biological effects. *Mol. Nutr. Food Res.*, **53**, S266–S309.

Muir, S.R., Collins, G.J., Robinson, S., Hughes, S., Bovy, A., de Vos, R.C.H., van Tunen, A.J., and Verhoeyen, M.E. (2001) Overexpression of petunia chalcone isomerase in tomato results in fruit containing increased levels of flavonols. *Nat. Biotechnol.*, **19**, 470–474.

Neveu, V., Perez-Jimenez, J., Vos, F., Crespy, V., du Chaffaut, L., Mennen, L., Knox, C., Eisner, R., Cruz, J., Wishart, D., and Scalbert, A. (2010) Phenol-explorer: an online comprehensive database on polyphenol contents in foods. *Database*, doi: 10.1093/database/bap024, Version 1.5.2. http://www.phenol-explorer.eu (accessed 22 August 2015).

Pandey, R.P. and Sohng, J.K. (2013) in *Natural Products* (eds K.G. Ramawat and J.M. Merillon), Springer-Verlag, Berlin and Heidelberg, pp. 1617–1645.

de Pascual-Teresa, S., Rivas-Gonzalo, J.C., and Santos-Buelga, C. (2002) in *COST 916 – Bioactive Compounds in Plant Foods. Health Effects and Perspectives for the Food Industry* (eds R. Amadò, B. Abt, L. Bravo, I. Goñi, and F. Saura-Calixto), Office for Official Publications of the European Communities, Tenerife, pp. 317–318.

Perez-Jimenez, J., Neveu, V., Vos, F., and Scalbert, A. (2010) Identification of the 100 richest dietary sources of polyphenols: an application of the Phenol-Explorer database. *Eur. J. Clin. Nutr.*, **64**, S112–S120.

Quattrocchio, F., Baudry, A., Lepiniec, L., and Grotewold, E. (2006) in *The Science of Flavonoids* (ed E. Grotewold), Springer, New York, pp. 97–122.

Rice-Evans, C.A., Miller, N.J., and Paganga, G. (1996) Structure-antioxidant activity relationships of flavonoids and phenolic acids. *Free Radical Biol. Med.*, **20**, 933–956.

Ro, D.K. and Douglas, C.J. (2004) Reconstitution of the entry point of plant phenylpropanoid metabolism in yeast (*Saccharomyces cerevisiae*): implications for control of metabolic flux into the phenylpropanoid pathway. *J. Biol. Chem.*, **279**, 2600–2607.

Rocha, B.S., Nunes, C., Pereira, C., Barbosa, R.M., and Laranjinha, J. (2014) A shortcut to wide-ranging biological actions of dietary polyphenols: modulation of the nitrate-nitrite-nitric oxide pathway in the gut. *Food Funct.*, **5**, 1646–1652.

Rothwell, J.A., Pérez-Jiménez, J., Neveu, V., Medina-Ramon, A., M'Hiri, N., Garcia Lobato, P., Manach, C., Knox, K., Eisner, R., Wishart, D. and Scalbert, A. (2013) Phenol-Explorer 3.0: a major update of the Phenol-Explorer database to incorporate data on the effects of food processing on polyphenol content. *Database*, doi: 10.1093/database/bat070.

Rusznyak, S. and Benko, A. (1941) Experimental vitamin P deficiency. *Science*, **94**, 25.

Rusznyak, S. and Szent-Gyorgyi, A. (1936) Vitamin P: flavonols as vitamins. *Nature*, **138**, 27.

Santos, C.N.S., Koffas, M.A.G., and Stephanopoulos, G. (2011) Optimization of a heterologous pathway for the production of flavonoids from glucose. *Metab. Eng.*, **13**, 392–400.

Santos-Buelga, C. and González-Paramás, A.M. (2014) in *Handbook of Chemical and Biological Plant Analytical Methods*, 1st edn (eds K. Hostettmann, H. Stuppner, A. Marston, and S. Chen), John Wiley & Sons, Ltd., Chichester, 26 pp.

Scarborough, H. (1939) Vitamin P. *Biochem. J.*, **33**, 1400–1407.

Scarborough, H. (1940) Deficiency of vitamin C and vitamin P in man. *Lancet*, **236**, 644–647.

Schijlen, E., de Vos, C.H.R., Jonker, H., van den Broeck, H., Molthoff, J., van Tunen, A., Martens, S., and Bovy, A. (2006) Pathway engineering for healthy phytochemicals leading to the production of novel flavonoids in tomato fruit. *Plant Biotechnol. J.*, **4**, 433–444.

Shimamura, M., Akashi, T., Sakurai, N., Suzuki, H., Saito, K., Shibata, D., Ayabe, S., and Aoki, T. (2007) 2-hydroxyisoflavanone dehydratase is a critical determinant of isoflavone productivity in hairy root cultures of Lotus japonicus. *Plant Cell Physiol.*, **48**, 1652–1657.

Sies, H. (1985) in *Oxidative Stress* (ed H. Sies), Academic Press, London, pp. 1–8.

Singleton, V.L. (1981) in *Advances in Food Research* (eds E.M. Mrak and G.F. Stewart), Academic Press, New York, Chichester, CO, pp. 149–242.

Song, M.C., Kim, E.J., Kim, E., Rathwell, K., Nama, S.-J., and Yoon, Y.J. (2014) Microbial biosynthesis of medicinally important plant secondary metabolites. *Nat. Prod. Rep.*, **31**, 1497–1509.

Spencer, J.P. (2007) The interactions of flavonoids within neuronal signalling pathways. *Genes Nutr.*, **2**, 257–273.

Spencer, J.P.E., Abd El Mohsen, M.M., and Rice-Evans, C. (2004) Cellular uptake and metabolism of flavonoids and their metabolites: implications for their bioactivity. *Arch. Biochem. Biophys.*, **423**, 148–161.

Tang, S.Y. and Halliwell, B. (2010) Medicinal plants and antioxidants: what do we learn from cell culture and *Caenorhabditis elegans* studies? *Biochem. Biophys. Res. Commun.*, **394**, 1–5.

Tomás-Barberán, F.A. and Clifford, M.N. (2000) Flavanones, chalcones and dihydrochalcones – nature, occurrence and dietary burden. *J. Sci. Food Agric.*, **80**, 1073–1080.

U.S. Department of Agriculture, Agricultural Research Service (2008) USDA Database for the Isoflavone Content of Selected Foods. Release 2.0, http://www.ars.usda.gov/Services/docs.htm?docid=6382 (accessed 22 August 2015)

U.S. Department of Agriculture, Agricultural Research Service (2011) USDA Database for the Flavonoid Content of Selected Foods. Release 3.0, http://www.ars.usda.gov/Services/docs.htm?docid=6231 (accessed 22 August 2015).

Ververidis, F., Trantas, E., Douglas, C., Vollmer, G., Kretzschmar, G., and Panopoulos, N. (2007a) Biotechnology of flavonoids and other phenylpropanoid-derived natural products. Part I: chemical diversity, impacts on plant biology and human health. *Biotechnol. J.*, **2**, 1217–1234.

Ververidis, F., Trantas, E., Douglas, C., Vollmer, G., Kretzschmar, G., and Panopoulos, N. (2007b) Biotechnology of flavonoids and other phenylpropanoid-derived natural products. Part II: reconstruction of multienzyme pathways in plants and microbes. *Biotechnol. J.*, **2**, 1235–1249.

Wang, Y., Chen, S., and Yu, O. (2011) Metabolic engineering of flavonoids in plants and microorganisms. *Appl. Microbiol. Biotechnol.*, **91**, 949–956.

Wenzel de Menezes, E., Cardoso Santos, N., Bistriche Giuntini, E., Dan, M.C.T., Genovese, M.I., and Lajolo, F.M. (2011) Brazilian flavonoid database: application of quality evaluation system. *J. Food Compos. Anal.*, **24**, 629–636.

Williams, R.J., Spencer, J.P., and Rice-Evans, C. (2004) Flavonoids: antioxidants or signalling molecules? *Free Radic. Biol. Med.*, **36**, 838–849.

Wu, X., Beecher, G., Holden, J.M., Haytowitz, D.B., Gebhardt, S.E., and Prior, R.L. (2006) Concentrations of anthocyanins in common foods in the United States and estimation of normal consumption. *J. Agric. Food. Chem.*, **54**, 4069–4075.

Wu, J., Du, G., Zhou, J., and Chen, J. (2014) Systems metabolic engineering of microorganisms to achieve large-scale production of flavonoid scaffolds. *J. Biotechnol.*, **188**, 72–80.

Xie, D.Y., Sharma, S.B., Wright, E., Wang, Z.Y., and Dixon, R.A. (2006) Metabolic engineering of proanthocyanidins through co-expression of anthocyanidin reductase and the PAP1 MYB transcription factor. *Plant J.*, **45**, 895–907.

Yan, Y., Chemler, J., Huang, L., Martens, S., and Koffas, M.A. (2005a) Metabolic engineering of anthocyanin biosynthesis in *Escherichia coli*. *Appl. Environ. Microbiol.*, **71**, 3617–3623.

Yan, Y., Kohli, A., and Koffas, M.A.G. (2005b) Biosynthesis of natural flavanones in *Saccharomyces cerevisiae*. *Appl. Environ. Microbiol.*, **71**, 5610–5613.

Yu, O., Jung, W., Shi, J., Croes, R.A., Fader, G.M., McGonigle, B., and Odell, J.T. (2000) Production of the isoflavones genistein and daidzein in non-legume dicot and monocot tissues. *Plant Physiol.*, **124**, 781–794.

Zilva, S.S. (1937) Vitamin P. *Biochem. J.*, **31**, 915–919.

18
Monascus Pigments

Yanli Feng, Yanchun Shao, Youxiang Zhou, Wanping Chen, and Fusheng Chen

18.1
Introduction and History of *Monascus* Pigments

Monascus pigments (MPs) are a group of azaphilonic secondary metabolites produced by *Monascus* spp. They are widely utilised as natural food colourants worldwide, especially in China for more than 1,000 years (Dufossé *et al.*, 2005; Fu, 1997). Besides their use as food colourants, MPs possess a series of physiological functions such as anti-cancer properties (Akihisa *et al.*, 2005a; Su *et al.*, 2005; Izawa *et al.*, 1997), anti-obesity characteristics (Kim *et al.*, 2007a,b) and antimicrobial activities (Kim *et al.*, 2006b; Martínková *et al.*, 1999; Martínková, Jůzlová and Veselý, 1995), and they could even be used for dyeing cotton yarn (Velmurugan *et al.*, 2010b) and leather (Velmurugan *et al.*, 2010a), sensitising solar cells (Sang-aroon, Saekow and Amornkitbamrung, 2012; Ito *et al.*, 2010) and preparing gels (Calvo and Salvador, 2002).

In this chapter, the categories and application prospects, physiological functions, chemical and physical properties, assay methods and units, MP producer—*Monascus* spp., application and economics of MPs are described.

18.2
Categories of MPs

MPs mainly include yellow, orange and red, totally three groups of constituents (Domínguez-Espinosa and Webb, 2003; Yongsmith *et al.*, 1993; Lin and Demain, 1991). With regard to the investigation on structures of MPs, it could be dated back at least to 1932 (Salomon and Karrer, 1932), and a lot of studies had been conducted from the late 1950s to the early 1970s (Whalley *et al.*, 1976; Manchand and Whalley, 1973; Chen, Manchard and Whalley, 1969, 1971; Inouye *et al.*, 1962; Kumasaki *et al.*, 1962; Fielding *et al.*, 1961; Haws *et al.*, 1959). Not until 1973, six MP compounds were identified, of which two were yellow – monascin (Chen, Manchard and Whalley, 1969; Salomon and Karrer, 1932) and ankaflavin (Manchand and Whalley, 1973), two were orange – rubropunctatin (Chen,

Industrial Biotechnology of Vitamins, Biopigments, and Antioxidants, First Edition.
Edited by Erick J. Vandamme and José L. Revuelta.

Manchard and Whalley, 1969) and monascorubrin (Manchand and Whalley, 1973) and two were red – rubropunctamine and monascorubramine (Sweeny *et al.*, 1981; Kumasaki *et al.*, 1962). Later, these six compounds were well known as MPs fundamental compound types (Pattanagul *et al.*, 2007; Jůzlová, Martínková and Křen, 1996; Lin *et al.*, 1992). Nowadays, more and more attention has been paid on MP compounds due to their various biological activities and wide application fields (Velmurugan *et al.*, 2010b; Kim *et al.*, 2007a; Izawa *et al.*, 1997; Martínková, Jůzlová and Veselý, 1995). Until the end of 2014, a total of 65 MP compounds including two yellow compounds without their structure shown (Zheng, Xin and Guo, 2009) had been reported. With the exception of monascin and ankaflavin, many other yellow MP compounds have been isolated, purified and identified from *Monascus*-fermented products. Until November 2014, a total of 34 yellow MP compounds had been isolated (Table 18.1). By contrast, the kinds of orange pigment were fewer than those of yellow and red ones. Besides rubropunctatin and monascorubrin, only four orange MP compounds, monapilol A-D (Table 18.1), were purified and identified (Hsu *et al.*, 2011). With regard to red MPs, more researches were focused on the production of water-soluble pigments (Jeun *et al.*, 2008; Kim *et al.*, 2007a; Jung *et al.*, 2003; Izawa *et al.*, 1997), and most of the red MP compounds listed in Table 18.1 are amino acid derivatives of MPs.

18.3 Physiological Functions of MPs

Besides their usage as food colourants, most of the MPs exhibit multiple physiological functions, such as anti-cancer, antimicrobial activity, and antioxidant, by inhibiting relative enzyme activities (Patakova, 2013).

18.3.1 Anti-Cancer Activities

It has been proven that some MPs and their derivatives as well as analogues possess anti-cancer activities (Jo *et al.*, 2014; Chen *et al.*, 2012a; Akihisa *et al.*, 2005a), and this is mainly attributed to their effects on proliferation and apoptosis of cancer cells or inhibition of the carcinogens (Hong *et al.*, 2008; Izawa *et al.*, 1997). According to the previous reports, the well-known yellow MP, monascin, showed obvious inhibitory activity on mouse skin carcinogenesis induced by ultraviolet light B and peroxynitrite (Akihisa *et al.*, 2005a), without cytotoxicity on Hep G2 (human cancer cell lines) cells (Su *et al.*, 2005). Another yellow MP, ankaflavin, could inhibit Hep G2 and A549 (human cancer cell lines) cells at the concentration of 15 μg/ml but no significant toxicity on normal MRC-5 (diploid fibroblast cell line) and WI-38 (diploid fibroblast cell line) cells at the same concentration (Su *et al.*, 2005). The similar phenomenon was also observed when

Table 18.1 MP constituents isolated from *Monascus*-fermented products.

Colour	No.	Name	Molecular formula	Molecular weight	Chemical structure	Producer	References
Yellow	1	Monascin	$C_{21}H_{26}O_5$	358	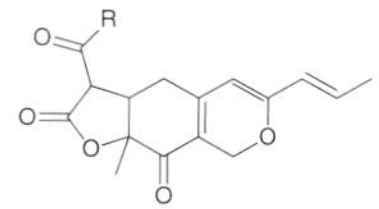 **1** R = C_5H_{11} **2** R = C_7H_{15}	*M. purpureus*	Chen, Manchard and Whalley (1969) and Salomon and Karrer (1932)
	2	Ankaflavin	$C_{23}H_{30}O_5$	386		*M. anka*	Manchand and Whalley (1973)
	3	Xanthomonasin A	$C_{21}H_{24}O_7$	388	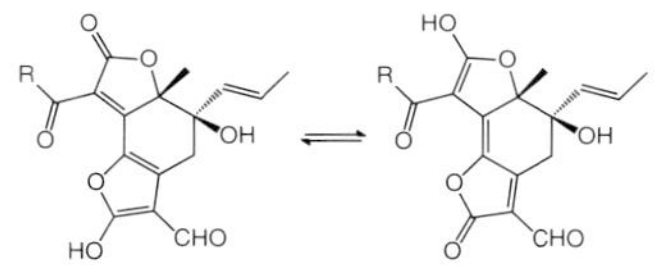 **3** R = C_5H_{11} **4** R = C_7H_{15}	*M. anka* U-1	Sato *et al.* (1992)
	4	Xanthomonasin B	$C_{23}H_{28}O_7$	416			
	5	Yellow II	$C_{22}H_{28}O_5$	372	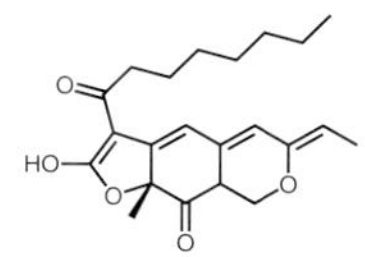 **5**	*Monascus* sp. KB 10	Yongsmith *et al.* (1993)
	6	Monankarin A	$C_{20}H_{22}O_6$	358	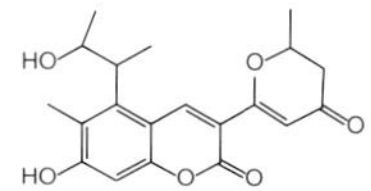 **6** Diastereomers monankarins A **7** Diastereomers monankarins B	*M. anka*	Hossain, Okuyama and Yamazaki (1996)
	7	Monankarin B	$C_{20}H_{22}O_6$	358			

(continued overleaf)

Table 18.1 (Continued)

Colour	No.	Name	Molecular formula	Molecular weight	Chemical structure	Producer	References
	8 9	Monankarin C Monankarin D	$C_{21}H_{24}O_6$ $C_{21}H_{24}O_6$	372 372	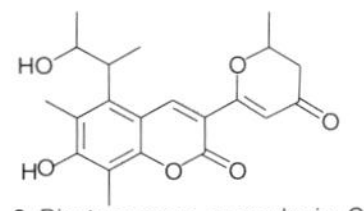 8 Diastereomers monankarin C 9 Diastereomers monankarin D		
	10	Monankarin E	$C_{19}H_{20}O_6$	344	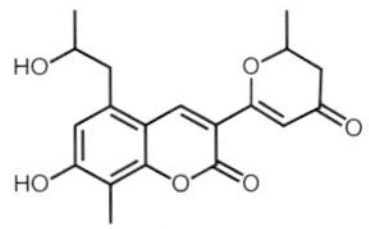 10		
	11	Monankarin F	$C_{20}H_{22}O_6$	358	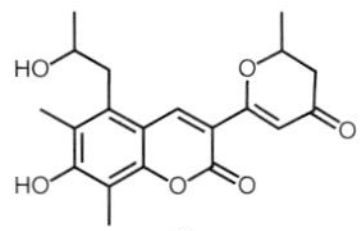 11		
	12	Monascusone A	$C_{13}H_{18}O_5$	254	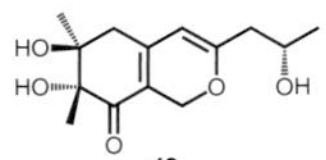 12	*M. kaoliang* KB20M10.2	Jongrungruangchok *et al.* (2004)
	13	Monascusone B	$C_{17}H_{18}O_5$	302	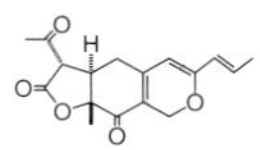 13		
	14	FK17-P2B2	$C_{13}H_{16}O_4$	236	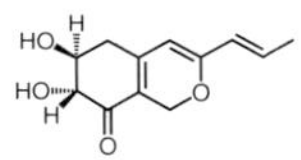 14		

15	Y3	$C_{20}H_{30}O_8S$	430	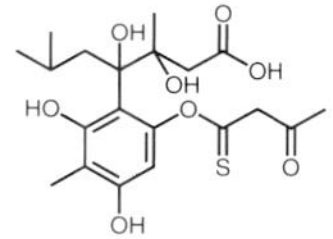 15	*M. purpureus* IB1	Campoy *et al.* (2006)
16	Monasfluore A	$C_{21}H_{24}O_5$	356	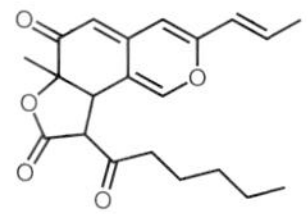 16	*M. pilosus* AS 3.4444	Huang *et al.* (2007)
17	Monasfluore B	$C_{23}H_{28}O_5$	384	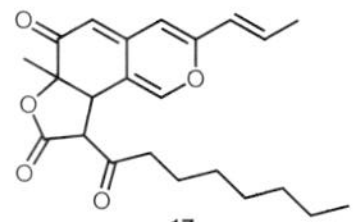 17		
18	Yellow-1	Not reported	356	Not reported	*Monascus* spp.	Zheng, Xin and Guo (2009)
19	Yellow-2		384			
20	Monaphilone A	$C_{22}H_{32}O_4$	360	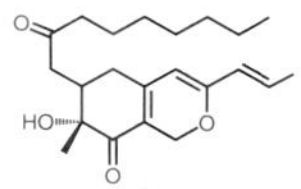 20	*M. purpureus* HM105	Hsu *et al.* (2010)
21	Monaphilone B	$C_{20}H_{28}O_4$	332	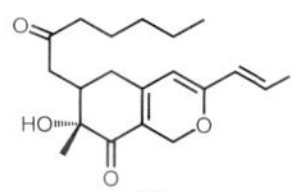 21		
22	Monaphilone C	$C_{20}H_{32}O_4$	336	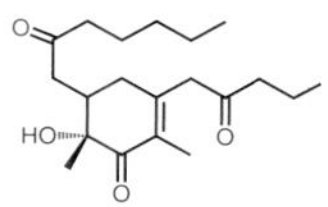 22		

(continued overleaf)

Table 18.1 (Continued)

Colour	No.	Name	Molecular formula	Molecular weight	Chemical structure	Producer	References
	23	Monapurone A	$C_{20}H_{26}O_4$	330	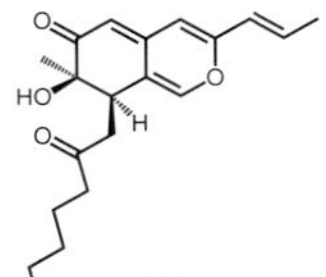 23	*M. purpureus* B0708	Li *et al.* (2010a)
	24	Monapurone B	$C_{21}H_{28}O_4$	344	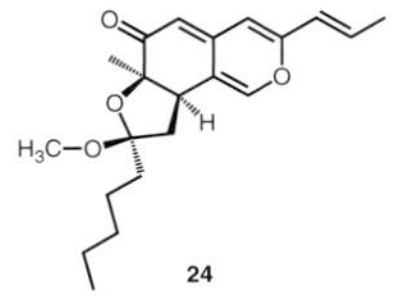 24		
	25	Monapurone C	$C_{21}H_{28}O_4$	344	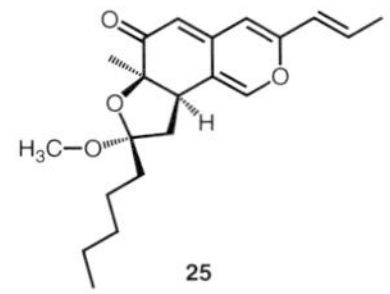 25		
	26	Monarubrin	$C_{20}H_{26}O_4$	330	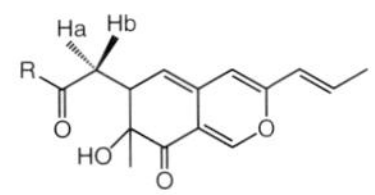 **26** R=C_5H_{11} **27** R=C_7H_{15}	*M. ruber* ATCC 96218	Loret and Morel (2010)
	27	Rubropunctin	$C_{22}H_{30}O_4$	358			
	28	Monaspilosuslin	$C_{15}H_{24}O_6$	300	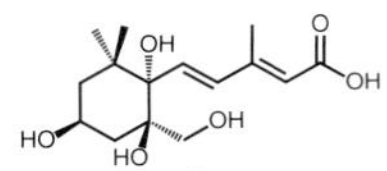 28	*M. pilosus* BCRC 38072	Cheng *et al.* (2010)

29	Purpureusone	$C_{23}H_{34}O_5$	390	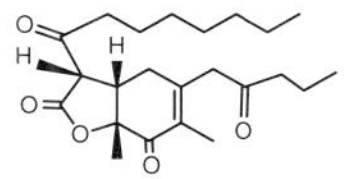 29	*M. purpureus* BCRC 38038	Cheng *et al.* (2011)
30	Monasnicotinate A	$C_{21}H_{27}NO_4$	357	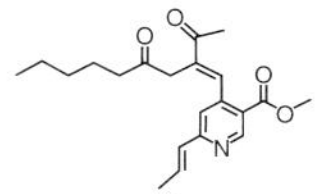 30	*M. pilosus* BCRC 38093	Wu *et al.* (2011b)
31	Monasnicotinate B	$C_{22}H_{29}NO_4$	371	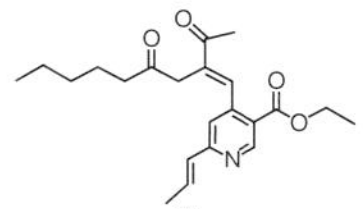 31		
32	Monasnicotinate C	$C_{23}H_{31}NO_4$	386	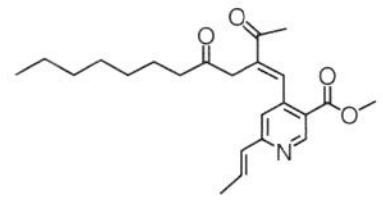 32		
33	Monasnicotinate D	$C_{21}H_{29}NO_4$	359	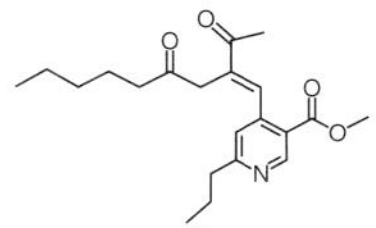 33		
34	Monapilosusazaphilone	$C_{23}H_{32}O_5$	388	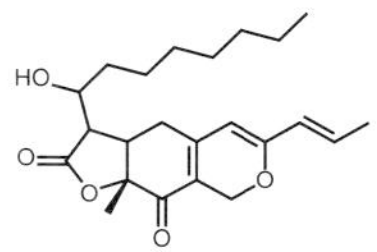 34	*M. pilosus*	Cheng *et al.* (2013)

(*continued overleaf*)

Table 18.1 (*Continued*)

Colour	No.	Name	Molecular formula	Molecular weight	Chemical structure	Producer	References
Orange	35	Rubropunctatin	$C_{21}H_{22}O_5$	354	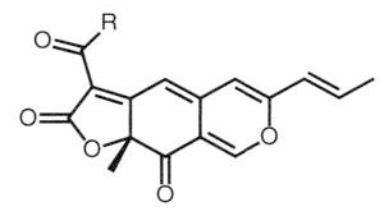 **35** R = C_5H_{11} **36** R = C_7H_{15}	*M. purpureus*	Chen, Manchard and Whalley (1969)
	36	Monascorubrin	$C_{23}H_{26}O_5$	382			Manchand and Whalley (1973)
	37	Monapilol A	$C_{23}H_{28}O_5$	384	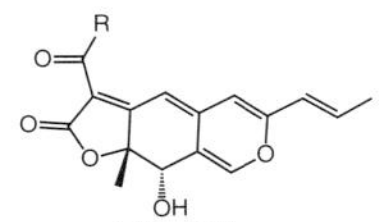 **37** R = C_7H_{15} **38** R = C_5H_{11}	*M. purpureus NTU 568*	Hsu *et al.* (2011)
	38	Monapilol B	$C_{21}H_{24}O_5$	356			
	39	Monapilol C	$C_{26}H_{32}O_6$	440	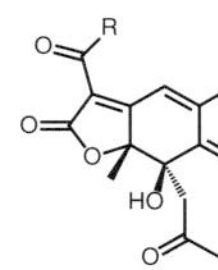 **39** R = C_7H_{15} **40** R = C_5H_{11}		
	40	Monapilol D	$C_{24}H_{28}O_6$	412			
Red	41	Rubropunctamine	$C_{21}H_{23}NO_4$	353	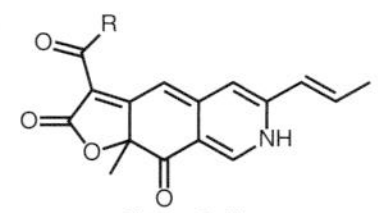 **41** R = C_5H_{11} **42** R = C_7H_{15}	*M. purpureus*	Sweeny *et al.* (1981)
	42	Monascorubramine	$C_{23}H_{27}NO_4$	381			
	43	*N*-Glucosylrubropunctamine	$C_{27}H_{33}NO_9$	515	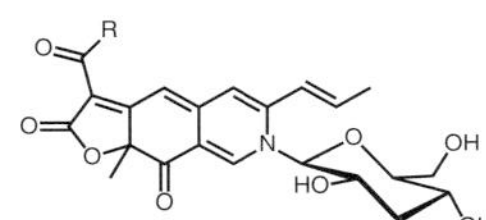 **43** R = C_5H_{11} **44** R = C_7H_{15}		Moll and Farr (1976)
	44	*N*-Glucosylmonascorubramine	$C_{29}H_{37}NO_9$	543			

45	*N*-Glutarylrubropunctamine	$C_{26}H_{29}NO_8$	483	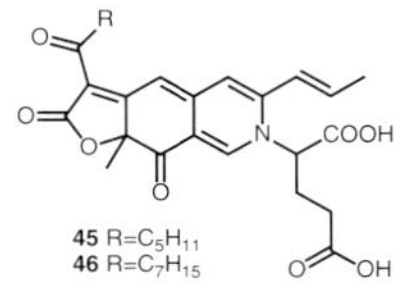 45 R=C_5H_{11} 46 R=C_7H_{15}	*Monascus* sp. TTWMB 6093	Lin *et al.* (1992)
46	*N*-Glutarylmonascorubramine	$C_{28}H_{33}NO_8$	511			
47	Monascorubrin L-Alanine	$C_{26}H_{31}NO_6$	453	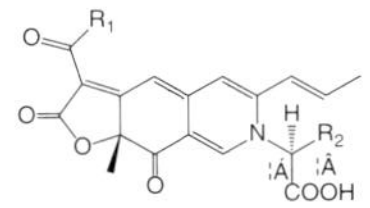 47 $R_1 = C_7H_{15}$, $R_2 = CH_3$ 48 $R_1 = C_5H_{11}$, $R_2 = CH_3$ 49 $R_1 = C_7H_{15}$, $R_2 = CH_2COOH$ 50 $R_2 = C_5H_{11}$, $R_2 = CH_2COOH$	*Monascus* sp.	Sato *et al.* (1997)
48	Rubropunctatin L-Alanine	$C_{24}H_{27}NO_6$	425			
49	Monascorubrin L-Aspartate	$C_{27}H_{31}NO_8$	497			
50	Rubropunctatin L-Aspartate	$C_{25}H_{27}NO_8$	469			
51	Monascorubrin D-Alanine	$C_{26}H_{31}NO_6$	453	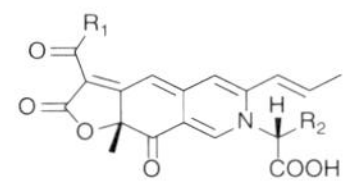 51 $R_1 = C_7H_{15}$, $R_2 = CH_3$ 52 $R_1 = C_5H_{11}$, $R_2 = CH_3$ 53 $R_1 = C_7H_{15}$, $R_2 = CH_2COOH$ 54 $R_2 = C_5H_{11}$, $R_2 = CH_2COOH$		
52	Rubropunctatin D-Alanine	$C_{24}H_{27}NO_6$	425			
53	Monascorubrin D-Aspartate	$C_{27}H_{31}NO_8$	497			
54	Rubropunctatin D-Aspartate	$C_{25}H_{27}NO_8$	469			
55	Glycylrubropunctatin	$C_{23}H_{27}NO_6$	413	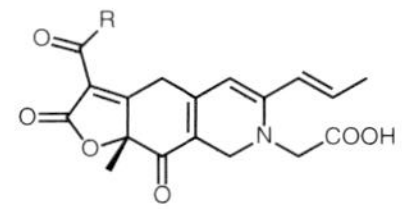 55 R = C_5H_{11} 56 R = C_7H_{15}	*M. anka*	Izawa *et al.* (1997)
56	Glycylmonascorbrin	$C_{25}H_{31}NO_6$	441			
57	Glycine derivative of *Monascus* pigments	$C_{25}H_{29}NO_6$	439	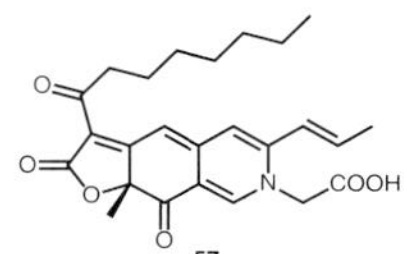 57		Jung *et al.* (2003)

(*continued overleaf*)

Table 18.1 (*Continued*)

Colour	No.	Name	Molecular formula	Molecular weight	Chemical structure	Producer	References
	58	R3	$C_{21}H_{26}O_6$	374	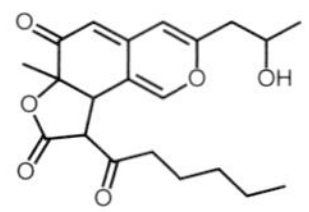 58	*M. purpureus* IB1	(Campoy *et al.*, 2006)
	59	—	$C_{19}H_{28}N_2O_5$	364	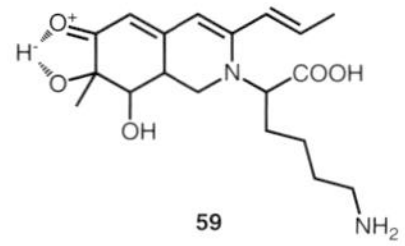 59	*M. ruber* 102w	Lian, Wang and Guo (2007)
	60 61	Rubropunctatin Arginine Monascorubrin Arginine	$C_{27}H_{34}N_4O_6$ $C_{29}H_{38}N_4O_6$	510 538	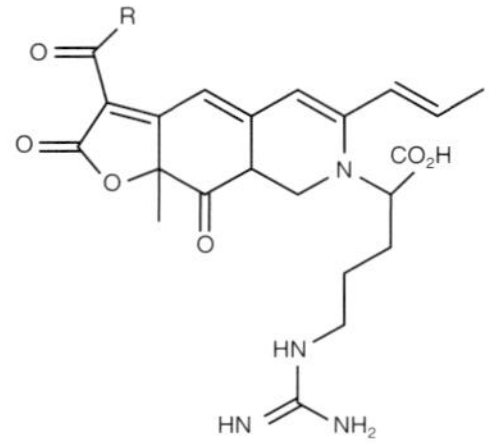 60 R = C_5H_{11} 61 R = C_7H_{15}		
	62	PP-V	$C_{23}H_{25}NO_6$	411	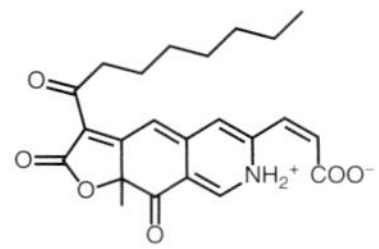 62	*M. ruber* IBT 7904	Mapari *et al.* (2008)

63	—	$C_{21}H_{29}NO_5$	375	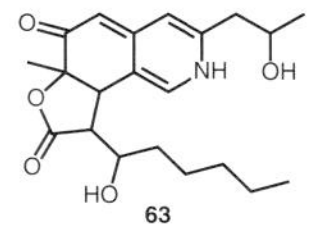 63	*M. purpureus* NFCCI 1756	Mukherjee and Singh (2011)
64	—	$C_{18}H_{25}NO_3$	303	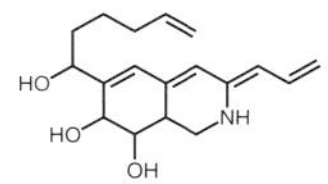 64	*M. ruber*	Lian *et al.* (2015)
65	—	$C_{20}H_{29}NO_3$	331	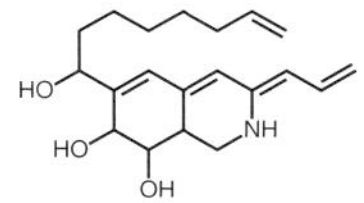 65		

HEp-2 (human laryngeal carcinoma cell line) and WiDr (human colon adenocarcinoma cell line) were treated with two yellow MP compounds, monaphilone A and monaphilone B (Hsu *et al.*, 2010). In addition, a monascin analogue named monascuspiloin could induce autophagy and apoptosis in prostate cancer cells through the PI3K/Akt and AMPK Adenosine 5′-monophosphate-activated protein kinase signalling pathways, which indicated that it is a potential cancer therapeutic agent (Chen *et al.*, 2012a). Orange MPs and their derivatives also exhibited anti-cancer activities. Rubropunctatin could induce the apoptosis mediated by tumour necrosis factor (TNF) and inhibit proliferation of human gastric adenocarcinoma BGC-823 both *in vivo* and *in vitro* at 12.57 μM of IC_{50}, but no significant toxicity to normal gastric epithelial cell GES-1 (gastric epithelial cell) at the same concentration (Zheng *et al.*, 2010a). Its therapeutic effect was the same as that of taxol at equal dose, and its tricyclic structure was recognised as the necessary moiety to the anti-cancer activity (Zheng *et al.*, 2010a,b). Moreover, glutamic acid and (*S*)-(+)-1-amino-2-propanol derivatives of the two well-known orange MPs (rubropunctatin and monascorubrin) displayed high inhibitory activities against melanogenesis through inhibiting the reaction and expression of tyrosinase (an important enzyme for melanin synthesis). From this way, the risk of malignant melanoma could be decreased (Jo *et al.*, 2014). With regard to red MPs, rubropunctamine and monascorubramine showed strong cytotoxicity and antimitotic effects on IHKE (immortalised human kidney epithelial) cells (Knecht and Humpf, 2006). However, two yellow pigments (monascin, ankaflavin) and two orange pigments (rubropunctatin, monascorubrin) showed no significant cytotoxic activity towards rat hepatocytes *in vitro* (Martínková *et al.*, 1999). Monascusone A exhibited no cytotoxicity against breast cancer and human epidermoid carcinoma of cavity cell lines (Jongrungruangchok *et al.*, 2004).

On the other hand, MPs could also display their anti-cancer activities through inhibiting the carcinogens. In the Ames *Salmonella* assay, yellow and red MPs extracted from *Monascus purpureus* and *Monascus anka* inhibited mutagenicity of heterocyclic amines such as 3-hydroxyamino-1-methyl-5*H*-pyrido[4,3-*b*]indole [Trp-P-2(NHOH)], the activated form of Trp-P-2 (3-amino-1-methyl-5*H*-pyrido[4,3-*b*]indole), due to acceleration of the decomposition of the activated mutagen (Izawa *et al.*, 1997).

18.3.2
Antimicrobial Activities

MPs, especially the two well-known orange MPs (rubropunctatin and monascorubrin), and MP derivatives of amino acids have been demonstrated to possess antimicrobial activities, not only against bacteria but also against filamentous fungi and yeasts (Vendruscolo *et al.*, 2014; Kim *et al.*, 2006b; Martínková *et al.*, 1999; Martínková, Jůzlová and Veselý, 1995). However, the effect mechanism of MPs is still unclear. It is inferred that MPs react with enzymes in the germinated spores and vegetative cells to restrict the use of iron and interface with the cell membrane permeability and reduce the transport of nutrients oxygen and metabolites (Vendruscolo *et al.*, 2014). Usually, MP derivatives of amino acids possess higher antimicrobial activities than the original MPs since the derivatives are easier to be adsorbed onto the bacterial cell surface and resulted in limited oxygen transfer (Kim *et al.*, 2006b; Martínková *et al.*, 1999). Another reason for this phenomenon might be the presence of $R\text{-}NH_2$ in the red MP derivatives of amino acids when compared to orange pigments (Jung *et al.*, 2003). For example, the red MP derivatives of L-Phe, D-Phe, L-Tyr and D-Tyr exhibit much higher antimicrobial activities against G^+ and G^- bacteria with minimal inhibitory concentration (MIC) values of 4–8 μg/ml compared to the original red MPs with MIC value of 32 μg/ml. Moreover, the types of amino acids could also affect the antimicrobial activities. The MP derivatives of hydrophobic amino acids have stronger inhibition activities on microorganisms than those of hydrophilic amino acids. For instance, L-Tyr and L-Phe derivatives of MPs exhibited higher antimicrobial activities (MIC, 8 and 16 mg/l, respectively) while L-Glu and L-Asn derivatives exhibited lower activities (MIC, 64 and 128 mg/l, respectively) when *Escherichia coli* was tested (Kim *et al.*, 2006a). With regard to antimicrobial spectrum, red MPs possess a broader antimicrobial spectrum compared to orange MPs (Vendruscolo *et al.*, 2014; Kim *et al.*, 2006b). Furthermore, unlike the original MPs, the red MP derivatives of L-Asp, D-Asp, L-Tyr and D-Tyr could be against *Penicillium citrinum*, *Aspergillus niger* and *Candida albicans* (Kim *et al.*, 2006b).

18.3.3 Anti-Obesity Activities

The extracts of red fermented rice (RFR) by *Monascus* spp. and MP compounds have been proven to display their anti-obesity characteristics through inhibiting lipases (Kim *et al.*, 2007a,b), adipogenesis, adipocyte cell proliferation and so on (Lee *et al.*, 2013; Choe *et al.*, 2012; Jou *et al.*, 2010) or improving glucose tolerance (Yoshizaki *et al.*, 2014).

RFR extracts and MPs could prevent obesity development by inhibiting cell proliferation, adipogenesis, lipolysis and heparin-releasable lipoprotein lipase (HR-LPL) of 3T3-L1 preadipocyte (Choe *et al.*, 2012; Jou *et al.*, 2010). Nowadays, the related studies mainly focus on the anti-obesity activities of amino acid derivatives of MPs (Choe *et al.*, 2012; Kim *et al.*, 2007a, 2010). Results achieved by Choe *et al.* indicated that 16 of 47 amine derivatives of MPs displayed an inhibitory activity against adipogenic differentiation in 3T3-L1 cells. The number and droplet size of fatty cells were reduced after treating with derivatives, and about 40% inhibition was obtained when the derivatives of 4-phenylbutylamine (PBA) (2.5 μM) and 2-(*p*-tolyl)ethylamine (TEA) (12.5 μM) were used (Choe *et al.*, 2012). Besides, the L-Leu-OEt (L-leucinethylester) derivatives of MPs exhibited some specific inhibition to porcine pancreatic lipase but not to other digestive enzymes (Kim *et al.*, 2007a). When mice were fed with L-Trp and L-Leu-OEt derivatives of MPs, the average body weight and the intraperitoneal adipose tissues weights were reduced by 13.6–50.9% and 16.7–30.5%, respectively (Kim *et al.*, 2010). The two well-known yellow pigments, monascin and ankaflavin, also possess anti-obesity characteristics (Lee *et al.*, 2013; Wang *et al.*, 2013; Jou *et al.*, 2010). Monascin and ankaflavin might reduce triglyceride accumulation and suppress expression of adipocyte-specific transcription factors to decrease proliferation and differentiation of preadipocyte related with obesity (Lee *et al.*, 2013; Jou *et al.*, 2010). Both monascin and ankaflavin might also promote mature adipocyte delipidation by releasing glycerol and down-regulating the HR-LPL activities (Jou *et al.*, 2010). Besides, monascin and ankaflavin are also found to suppress Niemann–Pick C_1 like 1 (NPC1L1) protein expression (73.6% and 26.1%), which are associated with small intestine tissue lipid absorption (Lee *et al.*, 2013).

18.3.4
Anti-Inflammation Activities

The MP compounds such as the six well-known MPs (Table 18.1 **1–2, 35–36, 41–42**) exhibited potent inhibitory effects on inflammation induced by 12-*O*-tetradecanoylphorbol-13-acetate (TPA), TNF-α, interleukin-1β (IL-1β) and interleukin 6 (IL-6) (Hsu *et al.*, 2013; Lin *et al.*, 2011; Akihisa *et al.*, 2005b). Lin *et al.* (2011) reported that monascin and ankaflavin reduced endothelial adhesiveness induced by TNF-α. Moreover, monascin could significantly degenerate TNF-α and IL-6 (inflammation-associated cytokines) at both the protein and the mRNA level, and it might also bind peroxisome proliferator-activated receptor-γ (PPAR-γ) and regulate expression of anti-inflammatory genes (Hsu *et al.*, 2012; Lin *et al.*, 2011). In addition, monascin and ankaflavin also exhibited inhibitory activities on nitric oxide (NO) production (Hsu *et al.*, 2013). However, four orange pigments, monaphilols A–D, showed higher activities compared with the aforementioned yellow pigments. The six pigments (monascin and ankaflavin, and orange pigments monaphilols A–D) could not only suppress the TNF-α, 1L-1β and 1L-6 but also down-regulate the protein expression of nitric oxide synthase (iNOS). Animal experiments also verified their anti-inflammatory effects (Hsu *et al.*, 2013). Besides, extracts of *Monascus*-fermented products, such as *Monascus*-fermented dioscorea enriched in MPs, also displayed anti-inflammation activities (Lee, Hsu and Pan, 2012; Shi, Liao and Pan, 2011). So, MPs may be potential anti-inflammation agents and beneficial for reducing the risk of vascular disease associated with inflammation (Lin *et al.*, 2011).

18.3.5
Regulation of Cholesterol Levels

Some MPs or extracts of *Monascus*-fermented products rich in MPs could exhibit activities for regulating cholesterol levels (Shi, Liao and Pan, 2011; Kim *et al.*, 2010; Jeun *et al.*, 2008). The threonine derivative of total MPs and orange MPs decreased the low-density lipoprotein (LDL) level, increased the high-density lipoprotein (HDL) level and the ratio of HDL to LDL in mice sera significantly (Jeun *et al.*, 2008). Moreover, the inhibitory activity against

HMG (hydroxy methylglutaryl)-CoA reductase and lipoprotein lipase (LPL) of L-Leu-OEt derivative of MPs was higher than that of its L-Trp derivative *in vitro* (Kim *et al.*, 2010).

18.3.6
Anti-Diabetes Activities

Yellow pigment, monascin, and extracts of *Monascus*-fermented products rich in MPs have been found to exhibit anti-diabetic activities (Lee *et al.*, 2011). On the one hand, monascin was able to improve insulin sensitivity through the AKt (serine/threonine protein kinases) pathway by stabilising PPAR-γ structure, preventing its phosphorylation and inhibiting JNK (c-Jun *N*-terminal kinase) activation (Lee *et al.*, 2011). On the other hand, monascin could also confer several treatment-oriented properties on diabetic rats through reducing hyperglycaemia, improving antioxidant ability and protecting tissue and a resistance against thermotolerance and oxidative stress on nematode (*Caenorhabditis elegans*) by regulating the FOXO/DAF-16-dependent insulin signalling pathway (Shi, Liao and Pan, 2012). In addition, extracts of *Monascus*-fermented products usually played their anti-diabetes activities through inhibiting α-glucosidase (Srianta *et al.*, 2013) and reducing urine sugar and urine protein levels (Shi and Pan, 2010). These findings suggest that monascin has a therapeutic potential on diabetes and diabetes-associated oxidative stress complications (Shi, Liao and Pan, 2012).

18.4
Chemical and Physical Properties of MPs

18.4.1
Solubility

The six well-known MPs (Table 18.1 **1–2, 35–36, 41–42**) are insoluble in water, but dissolved in ethanol, hexane, acetic acid and so on (Lin *et al.*, 1992; Sweeny *et al.*, 1981). The most commonly used method to improve MPs solubility in water is by adding amino acids such as glutamate, leucine, glycine to the media of *Monascus* spp. (Jeun *et al.*, 2008; Kim *et al.*, 2007a; Jung *et al.*, 2003; Lin *et al.*, 1992). Moreover, chemical modification could also be employed to enhance solubility of MPs in water (Wong and Koehler, 1983).

18.4.2
Stability

18.4.2.1
Effects of Temperature, pH and Solvent on Stability of MPs

Usually, MPs are very stable at 30–60 °C and pH 6.0–8.0 compared to those at high temperature (above 60 °C) and at acidic pH values (4.0–5.0) (Silveira *et al.*,, 2013; Huang *et al.*, 2011). Sometimes, MPs are still stable at pH 11.0 and 150 °C (Li,

Du and Zhang, 2003). Results achieved by Wongjewboot and Kongruang (2011) also indicated that MPs produced by *M. purpureus* TISTR 3002 had greater stability in the basic pH range compared in acidic pH range. Higher pH values could also increase the heat stability when compared to the lower pH values (Vendruscolo *et al.*, 2013). Besides, stabilities of MPs in different solvents are different from each other. So, MP stability is dependent on the producing strains, MP derivatives and solvents (Wongsorn, Wongjewboot and Kongruang, 2011). Results obtained by Lim and Kwak (2004) verified that 80% ethanol was the best solvent for stability of MPs, and this was in agreement with the results that stabilities of amino acid derivatives of MPs decreased in descending order in hexane, ethanol, propanol, methanol, ethyl ether, distilled water, chloroform and acetonitrile (Jung, Kim and Shin, 2005).

18.4.2.2
Effect of Light on Stability of MPs

The MPs are sensitive to lights, especially to sunlight and ultraviolet light, and the yellow MPs constituents are more photostable than the red ones (Mapari, Meyer and Thrane, 2009; Jung, Kim and Shin, 2005; Fabre *et al.*, 1993). Usually, the water-soluble MPs and amino acid derivatives of MPs are more stable than the original MPs (Jung *et al.*, 2011; Sheu, Wang and Shyu, 2000; Lin *et al.*, 1992). It had been demonstrated that L-Phe derivatives of rubropunctamine and monascorubramine were more stable than the original MPs in presence of sunlight irradiation, and the half-lives of MPs amino acid derivatives were 6–25-fold improvement over the original MPs (Jung *et al.*, 2011). Sheu, Wang and Shyu (2000) reported that when nata (a bacterial cellulose produced by *Acetobacter aceti* ssp. *xylinum*) was fermented by *M. purpureus* CCRC3150, MPs in *Monascus*–nata fermentation complex were more stable than those in nata dyed by MPs under ultraviolet irradiation at 366 nm for 36 h. 1,4,6-Trihydroxynaphthalene could also inhibit *N*-glucosylrubropunctamine and *N*-glucosylmonascorubramine fading under sunlight, owing to the formation of a complex (Figure 18.1) (Sweeny *et al.*, 1981). Besides, some results showed that MPs extracted from solid-state fermentation (SSF) products were less sensitive to light than those from liquid-state fermentation (LSF) products (Kaur, Chakraborty and Kaur, 2009).

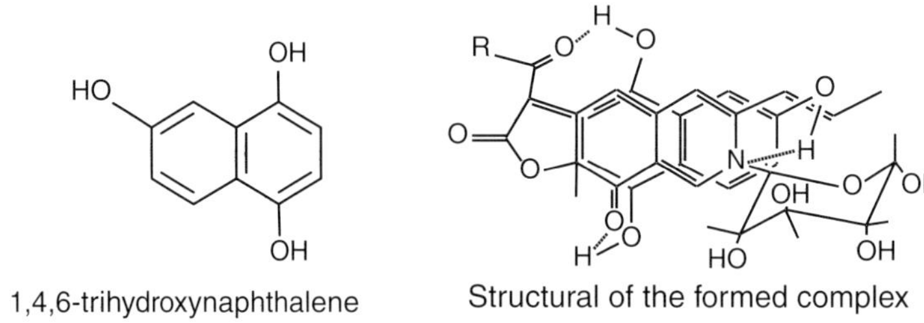

Figure 18.1 The chemical structures of 1,4,6-trihydroxynaphthalene and the complex formed by 1,4,6-trihydroxynaphthalene with *N*-glucosylrubropunctamine or *N*-glucosylmonascorubramine.

18.4.2.3
Effect of Metal Ion on Stability of MPs

Besides the aforementioned factors, metal ions can also affect MP stability. Usually, MPs are stable in the presence of a low concentration of Na^{+}, Mg^{2+}, K^{+}, Al^{3+}, Ca^{2+}, Cu^{2+} and Zn^{2+}, but Fe^{3+} and Fe^{2+} exerted an obvious negative effect on MPs stability at the concentrations of 20, 40, 100 ppm (Zhang *et al.*, 2005; Li, Du and Zhang, 2003; Song, Cui and Si, 1995).

18.4.3
Safety

As natural food colourants, MPs have been used in food industries in China for more than 1,000 years (Jia *et al.*, 2010; Kumari *et al.*, 2009; Lin *et al.*, 2008; Jůzlová, Martínková and Křen, 1996). MPs were used as colourants in more than 20 kinds of foods in China, and no adverse effect of MPs has been reported (Kumari *et al.*, 2009; Gheith *et al.*, 2008; Wang *et al.*, 2007). Kumari *et al.* (2009) reported that there was no significant difference between the experimental rats which were fed with *M. purpureus* MTCC410-fermented rice at different levels for 14 weeks compared to the control rats. However, citrinin, a kind of mycotoxin, was found in *Monascus*-fermented products in 1995 (Blanc, Loret and Goma, 1995). After that, the permitted limited quantity of citrinin in MPs in China and Japan were set at 1 and 0.2 mg/kg, respectively, for the sake of controlling the harmful impact of citrinin (Li *et al.*, 2008; Yang *et al.*, 2007). Besides, investigation of citrinin biosynthesis (Jia *et al.*, 2010), optimization of fermentation parameters to decrease the citrinin yield (Kang *et al.*, 2014) and screening of citrinin-free *Monascus* strains were also carried out (Feng *et al.*, 2015).

18.5
Assay Methods and Units of MPs

18.5.1
Extraction and Detection of MPs

MPs are classified as water-soluble and water-insoluble constituents according to their water-solubility, most of which are water-insoluble (Qian and Wu, 2010; Hajjaj *et al.*, 1997). The total MPs were usually extracted by ethanol at a certain concentration (Lai *et al.*, 2011; Vidyalakshmi *et al.*, 2009b; Babitha, Soccol and Pandey, 2006, 2007b; Johns and Stuart, 1991; Lin and Iizuka, 1982); the water-insoluble MP constituents were obtained by organic solvents such as *n*-hexane, benzene, ethanol, methanol; and the water-soluble MPs were obtained by distilled water (Sato *et al.*, 1997; Lin and Iizuka, 1982; Sweeny *et al.*, 1981).

As we know, MPs could be obtained from SSF or LSF products by *Monascus* (Gong, Chen and Gao, 2002). Both of the MPs from SSF products and mycelia of LSF products could be extracted through solid–liquid extraction (Hu *et al.*, 2012; Kongruang, 2010; Mohamed *et al.*, 2009; Lin and Iizuka, 1982) and micro-extraction (Mapari, Meyer and Thrane, 2009; Mapari *et al.*, 2008; Smedsgaard, 1997); MPs in supernatant of LSF products of *Monascus* could be extracted by liquid–liquid extraction (Hu *et al.*, 2012; Lai *et al.*, 2011; Li *et al.*, 2010b; Velmurugan *et al.*, 2010c; Zhou *et al.*, 2009; Lin and Demain, 1991).

The total MP contents, measured by colour values which are defined as total optical density values at a given wavelength per millilitre or gram of MPs, might be measured at 500 nm (Lin and Demain, 1991; Evans and Wang, 1984), 505 nm (Ding *et al.*, 2008), 480 nm (Santerre, Queinnec and Blanc, 1995), respectively. The red MP constituents were usually detected at the wavelengths between 500 and 510 nm (Hu *et al.*, 2012; Lai *et al.*, 2011; Silveira *et al.*, 2013; Kongruang, 2010; Pisareva and Kujumdzieva, 2010; Mohamed *et al.*, 2009; Nimnoi and Lumyong, 2009; Vidyalakshmi *et al.*, 2009b; Zhou *et al.*, 2009; Yongsmith *et al.*, 2000), sometimes at 495 nm (Domínguez-Espinosa and Webb, 2003) or 480 nm (Hajjaj *et al.*, 2012; Wongjewboot and Kongruang, 2011). The yellow ones were commonly determined at 400–410 nm (Hu *et al.*, 2012; Kongruang, 2010; Pisareva and Kujumdzieva, 2010; Vidyalakshmi *et al.*, 2009b; Zhou *et al.*, 2009), 370 nm (Krairak *et al.*, 2000; Yongsmith *et al.*, 2000) or 340 nm (Wongjewboot and Kongruang, 2011). The orange ones could be analysed at 420 nm (Yongsmith *et al.*, 2000), 440 nm (Wongjewboot and Kongruang, 2011), 460 nm (Kongruang, 2010) or 470 nm (Hu *et al.*, 2012; Domínguez-Espinosa and Webb, 2003; Johns and Stuart, 1991). The extractant or extract-liquor from unfermented substrate was always used as the blank (Nimnoi and Lumyong, 2009; Lee *et al.*, 2001; Johns and Stuart, 1991; Yoshimura *et al.*, 1975).

Moreover, another evaluating indicator of MPs, chroma value, which is expressed as the purity and colour saturation of MPs, was detected by measuring values of L^*, a^* and b^* by a reflectance colorimeter (Jung *et al.*, 2003). Here, L^* value is the luminance or lightness component, which ranges from 0 (black) to 100 (white), whereas a^* value is from green to red and b^* value from blue to yellow, ranging from −120 to +120. As the values of a^* value and b^* value increase, the colour becomes more saturated or chromatic (Kongruang, 2010; Jung *et al.*, 2003).

18.5.2 Isolation and Purification of MPs Components

MP components are mainly isolated and purified by column chromatography (CC), thin-layer chromatography (TLC), high performance liquid chromatography (HPLC), capillary electrophoresis (CE) and high-speed counter-current chromatography (HSCCC). In this section, the key parameters of some important methods such as HPLC for isolating and purifying MP constituents and components are listed.

18.5.2.1 CC and TLC

CC has been widely used for MP isolation and purification, and its isolated MP fractions usually need to be further purified by TLC and HPLC (Vidyalakshmi *et al.*, 2009a; Kim *et al.*, 2006b). In 1973, monascin and ankaflavin from mycelia of *M. anka* were isolated through TLC using 25% ether in benzene as developing agent (Manchand and Whalley, 1973). However, TLC was often used for preliminary isolation of MPs due to its low sensitivity (Sun, Yang and Wang, 2005; Jung *et al.*, 2003), and after TLC analysis of MPs, HPLC is applied to further purify MP compounds.

18.5.2.2
HPLC

HPLC is utilised to isolate, purify and analyse various MP constituents and components owing to its high sensitivity and multiple detection systems including ultraviolet–visible, photodiode array, fluorescence and mass detectors (Turner, Subrahmanyam and Piletsky, 2009; Lin *et al.*, 1992). The main analytical parameters of HPLC used in MP isolation, purification and analysis are presented in Table 18.2 (Feng, Shao and Chen, 2012).

18.5.2.3
CE and the Others

CE is suitable for MP constituents and components analysis because it requires only a little amount of sample and less time and solvent as compared to HPLC. However, few literatures about CE used in MP analysis were found (Watanabe *et al.*, 1997, 1999). For example, micellar electrokinetic chromatography (MEKC) of CE was utilised to separate and analyse the yellow MPs using 50 mM phosphate buffer (pH 7.0) as mobile phase at 10 kV (Watanabe *et al.*, 1997). In addition, xanthomonasin A, glycyl-rubropunctatin and 3-hydroxyamino-1-methyl-5*H*-pyrido[4,3-*b*]indole (Trp-P-2(NHOH)) were analysed by MEKC when 25 mM phosphate buffer (pH 7.0) was used as the running separation solution (Watanabe *et al.*, 1999). Moreover, HSCCC was also used for purification of the six well-known MPs (Table 18.1 **1–2, 35–36, 41–42**), xanthomonasin A and xanthomonasin B (Inoue *et al.*, 2010).

18.5.3
Identification of MPs Components

As far back as the early 1970s, nuclear magnetic resonance (NMR) was used for MP component identification (Yoshimura *et al.*, 1975). Nowadays, although NMR is still a convincing method for identification of MP compounds, it is normally combined with ultraviolet–visible (UV–vis) spectra, infrared (IR) spectra, mass spectra (MS), fluorescence spectra and/or electron paramagnetic resonance (EPR) to identify MPs (Mukherjee and Singh, 2011; Loret and Morel, 2010; Akihisa *et al.*, 2005b). Some MP compounds, which are confirmed by NMR or NMR together with other aforementioned methods, are listed in Table 18.3 (Feng, Shao and Chen, 2012).

Table 18.2 Purification and analysis of MPs by HPLC.

Sample	Column/detector/wavelength	Mobile phase	References
MPs mixture	μ Bondapak C_{18} column/tunable UV–vis absorbance/400, 470 and 500 nm	Acetonitrile:water (70 : 30, v/v) at 1.0 ml/min	Chen and Johns (1993, 1994)
	L-column ODS packed column/programmable solvent module 125 and programmable detector module 166/460 nm	Solution A distilled water containing 0.05% trifluoroacetic acid (TFA), solution B acetonitrile containing 0.05% TFA, the linear-gradient program from 30% to 60% of solution B in 30 min at 1.0 ml/min	Watanabe *et al.* (1997)
	C_{18} column/481 spectrophotometer/233 nm	Acetonitrile:water (80 : 20, v/v), at 0.5 ml/min	Teng and Feldheim (1998, 2000)
	μ Bondapak C_{18} column/tunable UV–vis absorbance/405, 470 and 495 nm	Acetonitrile:water (75 : 25, v/v) at 1.0 ml/min	Domínguez-Espinosa and Webb (2003)
	ODS C_{18} column/UV–vis detector/425 nm	Elution gradient of distilled water:methanol from 100 : 0 to 30 : 70 in 40 min at 0.8 ml/min	Jung *et al.* (2003)
	LiChrospher 100RP-18 pore column/photodiode array detector/390, 470 and 520 nm	Water (A) and acetonitrile (B), the conditions were adapted for optimal resolution of the different MP components	Campoy *et al.* (2006)
	COS-MOSIL 5 C_{18}-MS /—/ 490 and 380 nm	Linear gradient, acetonitrile:water containing 0.1% HCOOH (60 : 40, v/v) to acetonitrile–water containing 0.1% HCOOH (100 : 0, v/v) in 20 min at 0.5 ml/min	Miyake *et al.* (2008)
	Cosmosil C_{18}/photodiode array detector, RF-10AXL fluorescence detector/390 nm, $\lambda_{ex} = 331$ nm and $\lambda_{em} = 500$ nm	Eluent A H_2O:HAC (100 : 10, v/v), eluent B acetonitrile:HAC (100 : 10, v/v), elution gradient: 0 min, 80% A and 20% B; 25 min, 50% A and 50% B; 26 min, 15% A and 85% B, flow rate: 1 ml/min	Zheng, Xin and Guo (2009)

	5 μm Nucleosil C_{18}/fluorimetric detector/λ_{ex} = 340 nm and λ_{em} = 490 nm	Isocratic elution of acetonitrile:water (8 : 2, v/v) at 2 ml/min	Loret and Morel (2010)
	C_{18} column/2487 dual absorbance detector/400 and 500 nm	Acetonitrile:H_2O (8 : 2, v/v) at 0.5 ml/min	Mukherjee and Singh (2011)
	Cosmosil 5C_{18} packing column/—/—	85% MeOH as mobile phase solvent at 7 ml/min	Hsu *et al.* (2011)
	Luna C_{18}/photodiode array detector, intelligent fluorescence detector FP-2020 plus/234 nm	Isocratic elution using 0.05% TFA in acetonitrile-water (62.5 : 37.5, v/v) at 1.0 ml/min	Wu *et al.* (2011a)
Yellow MPs	Pegasil ODS II C_{18} silica column/refractive index detector/—	CH_3OH:H_2O:HAc 75 : 25 : 3 (v/v/v) at 2.5 ml/min	Akihisa *et al.* (2005b)
Orange MPs	Pegasil ODS II C_{18} silica column/refractive index detector/—	CH_3OH:H_2O:HCOOH 75 : 25 : 0.1 (v/v/v) at 2.5 ml/min	Akihisa *et al.* (2005b)
Red MPs	C_{18} column (Techsphere 5ODS 4 μm)/2487 dual absorbance detector/400 and 500 nm	Acetonitrile:H_2O (80:20, v/v) at 0.5 ml/min	Mukherjee and Singh (2011)
	μ Bondapak C_{18} column/Model 450/500 nm	CH_3CN:H_2O (45:55, v/v) at 1.0 ml/min	Sweeny *et al.* (1981)
	μ Bondapak C_{18} column/481 LC spectrophotometer/500 nm	Initial 35% aqueous solution of acetonitrile gradually increasing to 70% within 15 min at 1.0 ml/min	Lin *et al.* (1992)
	Pegasil ODS II C_{18} silica column/refractive index detector/—	CH_3OH:H_2O:AcOH 60 : 40 : 3 (v/v/v) at 2.5 ml/min	Akihisa *et al.* (2005b)

—: Did not present in the literature.

Table 18.3 Identification methods of MP compounds.

Sample	Method	MPs compound:name/molecular weight/molecular formula	References
MPs mixture	NMR	Monascin/358/$C_{21}H_{26}O_5$ Monascorubrin/382/$C_{23}H_{26}O_5$	Yoshimura *et al.* (1975)
	MS and NMR	Y3/430/$C_{20}H_{30}O_8S$ R3/374/$C_{21}H_{26}O_6$	Campoy *et al.* (2006)
	UV–vis, MS and NMR	Monascin/358/$C_{21}H_{26}O_5$ Ankaflavin/386/$C_{23}H_{30}O_5$ Rubropunctatin/354/$C_{21}H_{22}O_5$ Monascorubrin/382/$C_{23}H_{26}O_5$ Rubropunctamine/354/$C_{21}H_{23}NO_4$ Xanthomonasin A/388/$C_{21}H_{24}O_7$ Xanthomonasin B/416/$C_{23}H_{28}O_7$	Akihisa *et al.* (2005b)
	LC-MS	Yellow 1/356/— Yellow 2/384/—	Zheng, Xin and Guo (2009)
	UV, LC-MS, HRMS, IR and NMR	Monarubrin/330/$C_{20}H_{26}O_4$ Rubropunctin/358/$C_{22}H_{30}O_4$	Loret and Morel (2010)
	UV–vis, LC-DAD-MS	Monascin/358/$C_{21}H_{26}O_5$ Ankaflavin/386/$C_{23}H_{30}O_5$ Rubropunctatin/354/$C_{21}H_{22}O_5$ Monascorubrin/382/$C_{23}H_{26}O_5$	Mapari *et al.* (2008)
	UV–vis, IR, GC-MS and NMR	Unnamed/375/$C_{21}H_{29}NO_5$	Mukherjee and Singh (2011)

Yellow MPs	IR, MS and NMR	Monascin/358/$C_{21}H_{26}O_5$	Yongsmith, Krairak and Bavavoda (1994) and Yongsmith *et al.* (1993)
		Ankaflavin/386/$C_{23}H_{30}O_5$	
		Yellow II/372/$C_{22}H_{26}O_5$	
	ESI-MS and MEKC-ESI-MS	Xanthomonasin A/388/$C_{21}H_{24}O_7$	Watanabe *et al.* (1997)
		Xanthomonasin B/416/$C_{23}H_{28}O_7$	
	Polarimeter, IR, UV, ESI-TOF-MS and NMR	Monascusone A/254/$C_{13}H_{18}O_5$	Jongrungruangchok *et al.* (2004)
		Monascusone B/302/$C_{17}H_{18}O_5$	
Orange MPs	UV, IR, Polarimeter, Fluorescence Spectra, ESI-MS, HR-ESI-MS and NMR	Monapilol A/384/$C_{23}H_{28}O_5$	Hsu *et al.* (2011)
		Monapilol B/356/$C_{21}H_{24}O_5$	
		Monapilol C/440/$C_{26}H_{32}O_6$	
		Monapilol D/412/$C_{24}H_{28}O_6$	
Red MPs	UV, MS and NMR	*N*-Glucosylrubropunctamine/515/$C_{27}H_{33}O_9N$	Hajjaj *et al.* (1997)
		N-Glucosylmonascorubramine/543/$C_{29}H_{37}O_9N$	
		N-Glutarylrubropunctamine/483/$C_{26}H_{29}O_8N$	
		N-Glutarylmonascorubramine/511/$C_{28}H_{33}O_8N$	
	IR, UV, ESI-MS and NMR	Unnamed/364/$C_{19}H_{28}O_5N_2$	Lian, Wang and Guo (2007)
	Visible light absorbance, IR, ESI-MS, NMR and X-ray diffraction	303/$C_{18}H_{25}NO_3$	Lian *et al.* (2015)
		331/$C_{20}H_{29}NO_3$	

NMR: nuclear magnetic resonance, UV–vis: ultraviolet–visible spectrometry, IR: infrared spectrometry, TOF-MS: time-of-flight mass spectrometry, LC-MS: liquid chromatogram–mass spectrometry, GC-MS: gas chromatography–mass spectrometry, HRMS: high-resolution mass spectrometry, MEKC-ESI-MS: micellar electrokinetic chromatography electron-impact mass spectrometry, LC-DAD-MS: liquid chromatography–diode array detection–mass spectrometry.

18.6
MPs Producer – *Monascus* spp.

18.6.1
Brief Introduction of *Monascus* Species and Their Applications

The genus of *Monascus* belongs to the family Monascaceae, the order Eurotiales, the subclass Eurotiomycetidae, the class Eurotiomycetes, the subphylum Ascomycota, the phylum Eumycota, the kingdom Fungi (Geiser *et al.*, 2006). Since *Monascus* was first nominated by Van Tieghem (1884), nine species are accepted internationally, including *Monascus pilosus, Monascus ruber, Monascus purpureus, Monascus floridanus, Monascus eremophilus, Monascus pallens, M. sanguineus, Monascus lunisporas* and *M. argentinensis* (Stchigel *et al.*, 2004; Udagawa and Baba, 1998; Cannon, Abdullah and Abbas, 1995; Hocking and Pitt, 1988; Barnard and Cannon, 1987; Hawksworth and Pitt, 1983), and the most common species used for MP production are *Monascus purpureus* (Hsu *et al.*, 2011), *M. ruber* (Lian *et al.*, 2015) and *M. pilosus* (Cheng *et al.*, 2013).

18.6.2
Producing Methods of MPs

Traditionally, MPs are produced by SSF, which means that conidia suspensions are inoculated on the surface of steamed non-sticky rice and cultured for more than 2 weeks at 28–30 °C (Padmavathi and Prabhudessai, 2013). When the fermentation process is completed, the white rice turns into red, which is also called as red fermented rice, red yeast rice, *Hon-chi, Hongqu*, red Chinese rice, red koji, *Anka* and so on. In SSF process, the substrates not only supply the nutrients for the microbial culture, but also serve as an anchorage for the cells producing pigment. The red rice either can be grounded into powder which is directly used as edible colourant, folk medicines and fermentation starters, or can be extracted with ethanol to produce concentrated MPs (Shi and Pan, 2011; Li *et al.*, 2010b). In recent years, agro-industrial residues, such as broken rice, wheat bran, jackfruit seeds, grape pip, palm kernel cake, cassava starch (Babitha, Soccol and Pandey, 2006; Teng and Feldheim, 2001; Pandey, Soccol and Mitchell, 2000; Lee *et al.*, 1995; Lin and Iizuka, 1982), have been applied to replace the non-sticky rice to obtain the pigment product. The usage of RFR in East Asian countries can be dated at least back to the first century AD (Han dynasty in China), but the production by SSF method is a labour-intensive process Chen *et al.*, 2015). So, submerged culture is developed and becomes an attractive method for MP production due to its easier management, shorter cultivation time, lower costs and higher quality (Silveira *et al.*, 2013). During the SSF process for MP production, it is very significant to control moisture content and temperature in the fermentation cylinder. Comparatively, different factors such as carbon/nitrogen sources, dissolved oxygen, pH value, light, physical and chemical stresses, have different effects on pigment yields and the pigment constituents by submerged culture (Kang *et al.*, 2014; Prajapati *et al.*, 2014; Babitha, Soccol and Pandey, 2007a).

18.6.3 Progress of *Monascus* spp. at the Genetic Level

18.6.3.1 DNA Transformation

For a long time, the research studies have mainly concentrated on optimising the fermentation parameters for the purpose of improvement of the bioactive metabolites from *Monascus* spp. With the introduction of genetic transformation into *Monascus* species, a great progress has been made in the secondary metabolism and its modulation in *Monascus* spp. Until now, several kinds of transformation methods are available to transfer exogenous DNA to the genomic DNA of *Monascus* species, including genomic DNA-mediated transformation (Lakrod, Chaisrisook and Skinner, 2003b), biolistic bombardment (Lakrod, Chaisrisook and Daniel, 2003a), polyethylene-glycol-mediated protoplast transformation, *Agrobacterium-tumefaciens*-mediated transformation (Campoy *et al.*, 2003) and restricted-enzyme-mediated integration technique (Chen *et al.*, 2008a; Zhou, Wang and Zhuge, 2006), which facilitate our understanding of the physiological and chemical characteristics of the *Monascus* species at the genetic level. In order to testify the functions of interest genes, *ku*-deleted mutants (*ku70* and *ku80*) and *ligase IV*-deficient mutants were constructed as transformation hosts which could increase gene replacement frequency in *Monascus* spp. due to deletion of the components in the non-homologous end-joining pathway (He, Shao and Chen, 2014; He *et al.*, 2013).

18.6.3.2 Citrinin Synthesis and Its Regulations

In filamentous fungi, secondary metabolic biosynthesis genes are usually physically linked, to form a gene cluster, which minimises the amount of regulatory steps and contributes to physiological economisation (Gacek and Strauss, 2012). In *Monascus* spp., the gene clusters responsible for citrinin, monacolin K (MK) and MPs have been cloned and functionally analysed. The synthesis of these polyketides is catalysed by polyketide synthases consisting of a minimal set of ketosynthase, acyltransferase and acyl carrier protein domains (Lackner *et al.*, 2012; Staunton and Weissman, 2001). The first cloned *pks* gene was *pksCT* (GenBank accession no. AB167465), responsible for the synthesis of citrinin in *M. purpureus*, which has a length of 7838 bp with a single 56 bp intron, and encodeds a 2593-amino acid protein (Shimizu *et al.*, 2005). Following this information, a DNA fragment of 21 kb (GenBank accession no. AB243687) in length was cloned, including four open reading frames (ORFs) in the 5′-end and one ORF in the 3′-end in the vicinity of *pksCT* (Shimizu, Kinoshita and Nihira, 2007). Later, Li, Xu and Huang (2012) isolated a DNA fragment with 43 kb length from *M. aurantiacus*, consisting of *pksCT* homologous gene and other 15 ORFs.

Citrinin is a kind of mycotoxin and is harmful to the kidney, so it is imperative to reduce or restrain its contamination on RFR. Deletion of *pksCT* gene, specific activator gene *ctnA* and other related genes in citrinin synthesis gene cluster could significantly decrease citrinin production, which provides a potential approach to genetically modify *Monascus* strains (Li *et al.*, 2013; Jia *et al.*, 2010).

18.6.3.3
MK Synthesis and Its Regulations

MK, the counterpart of lovastatin and compactin, possesses strong cholesterol-lowering effect. Besides *Aspergillus* spp. and *Penicillium* spp., some *Monascus* species are also the producer of MK and its homologues such as monacolin L, X, J, M, dihydroerinolin and dihydromonalin L (Endo, Komagata and Shimada, 1986; Endo and Hasumi, 1985; Endo, 1979). The gene cluster coding for MK synthesis (GenBank accession no. DQ176595) in *Monascus* species was first isolated from a bacterial artificial chromosome library in *M. pilosus* BCRC38072 (Chen *et al.*, 2008b), of which nine genes shared strong homologues with those in *Aspergillus terreus* and *P. citrinum*. *MokA*-deleted strain (predicted to be responsible for the synthesis of the nonaketide skeleton) caused the complete loss of MK production in *M. pilosus* BCRC38072, suggesting the involvement of *mokA* in MK biosynthesis (Chen *et al.*, 2008b), while *mokB*-deleted mutant (presumed to be responsible for the synthesis of the diketide skeleton) lost the ability to produce MK but accumulated the homologue monacolin J, indicating that *mokB* is responsible for the diketide side-chain synthesis of MK (Sakai, Kinoshita and Nihira, 2009).

Usually, there is a specific activator encoding a transcription factor necessary for expression of the other genes in the secondary metabolite gene cluster. In MK gene cluster, *mokH* gene, predicted as an activator for MK synthesis, was cloned and expressed in *M. pilosus* driven by the promoter of glyceraldehydes-3-phosphate dehydrogenase, and the result suggested that the transcript levels of *mokH* and MK production in the transformants were significantly increased compared to those of the wild-type strain. Simultaneously, gene expression of MK biosynthetic gene cluster in the transformant appeared earlier than that in the wild-type strain. These results indicated that *mokH* could up-regulate the transcription of MK biosynthetic genes and increase its production (Chen *et al.*, 2010).

18.6.3.4
MPs Synthesis and Its Regulation

MPs biosynthesis is generally considered to follow a polyketide pathway, and a putative MP biosynthetic gene cluster is proposed as shown Figure 18.2. Targeted inactivation of *MpPKS5* (*pks* gene) resulted in abolishment of pigment production, suggesting that *MpPKS5* is involved in pigment biosynthesis (Balakrishnan *et al.*, 2013). Similarly, our research group cloned a 53 kb DNA fragment from *M. ruber* M7, predicted as the MP synthesis gene cluster, including the putative PKS gene, fatty acid synthases gene, esterase gene, dehydrogenase gene, reductase gene (*MpigE*), and regulator gene *pigR*. The *pigR*-deleted strain had no ability to

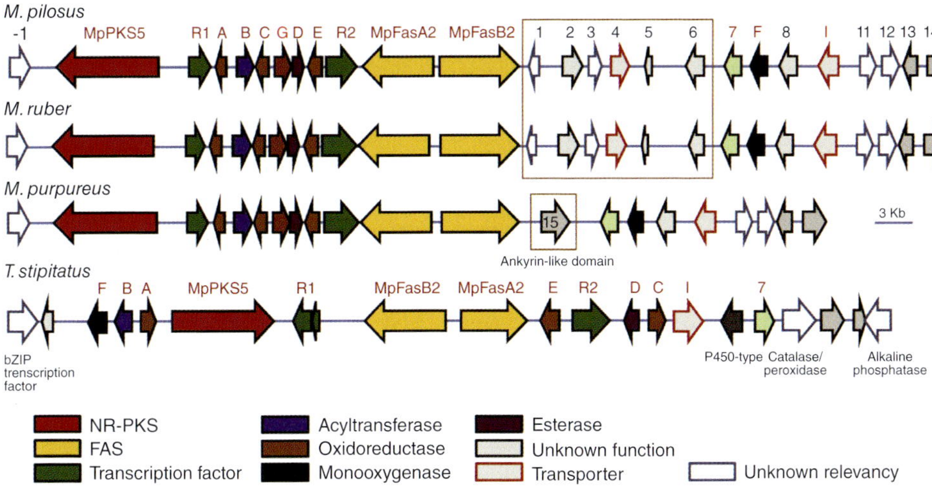

Figure 18.2 Organisation of MP biosynthetic gene clusters from *M. pilosus*, *M. ruber* and *M. purpureus*, with the putative cluster gene from *T. stipitatus* (Balakrishnan *et al.*, 2014a, with kind permission from Elsevier).

produce MPs (Xie, Liu and Chen, 2013). *MpigE*-deleted mutant (Liu *et al.*, 2014) secreted four kinds of yellow MPs but very little red pigments, suggesting that *MpigE* may be involved in the conversion among different pigment compositions. *Mpp7* in *M. pilosus* assists in regioselective Knoevenagel aldol condensation during MP biosynthesis (Balakrishnan *et al.*, 2014a). Later, the product profiles of *mppF*, *mppA* and *mppC* mutants demonstrate that *MppA*-mediated ω-2 ketoreduction is a prerequisite for the synthesis of the pyranoquinone bicyclic core of the MP and that MppC activity determines the regioselectivity of the spontaneous Knoevenagel (Balakrishnan *et al.*, 2014b).

18.6.3.5
The Regulation of Secondary Metabolism in *Monascus* spp.

The regulation of secondary metabolism and its subsequent synthesis pathways are controlled by posttranslational modification. Histone modification, DNA methylation or transcription factors can lead to the induction of secondary metabolite production (Brakhage, 2013). This complex regulatory network responds to various environmental stimuli, including carbon and nitrogen levels, temperature, light, pH and stimuli from other organisms (Brakhage, 2013). Secondary metabolism always couples with developmental process mediated through various signalling pathways (Yu, 2006). In recent years, the regulation of G-protein heterotrimer consisting of α, β and γ subunits (named Mga1, Mgb1 and Mgg1, respectively) and MrflbA (the regulator of Mga1) as well as global regulator LaeA for the development and secondary metabolite production were investigated in *M. ruber* (Li *et al.*, 2010b, 2014; Yang *et al.*, 2012). In *M. ruber*, Mga1 and Mgb1 and Mgg1 have been demonstrated to promote both sexual and

Table 18.4 Information on *Monascus* genomes.

Strains	Genome size (Mb)	GC content (%)	Coding sequence count	Average gene size (bp)	Coding region percentage (%)
M. ruber M7	23.81	48.88	8407	1500	52.97
M. ruber NRRL 1597	24.80	48.91	9650	1769	68.83
M. purpureus NRRL 1596	23.44	49.03	8918	1815	69.05

asexual development and vegetative growth but repress the production of citrinin and pigments (Li *et al.*, 2010b, 2014). *MrflbA*-deleted mutant resulted in less production of pigment and citrinin and autolytic aerial hyphae (Yang *et al.*, 2012). In addition, functions of MpigE gene which involved in MPs biosynthesis in Monascus ruber M7 was also carried out (Liu et al., 2014). Four yellow MPs with little red MPs were produced by MpigE deletion strain while complex mixture MPs yielded by the wild strain.

18.6.4 *Monascus* Genomics

The genome of *M. pilosus* was first sequenced and assembled by the Bioresource Collection and Research Center of Taiwan in 2004, but it has not been released yet (http://www.bcrc.firdi.org.tw/genome_ project_monascus/). In 2010, the genome sequencing of *M. ruber* M7 was completed with *de novo* sequencing technique by our *Monascus* research group (PI: Prof. Fusheng Chen, Huazhong Agricultural University), and its genomic size is 23.81 Mb including 8407 predicted genes (Table 18.4). Thereafter, *M. purpureus* NRRL 1596 and *M. ruber* NRRL 1597 were selected as part of the JGI 1000 Fungal Genomes CSP to represent members of the ascomycete family Monascaceae and their genomic sequencings were finished in 2013, with size of 23.44 and 24.80 Mb, respectively (http://genome.jgi.doe.gov/) (Table 18.4). These data provide the basis for *Monascus* genomics research, helpful in interpreting the biological features of *Monascus* spp. from the view point of whole genomes. In the foreseeable future, *Monascus* genomics will preferentially focus on the research areas in comparative genomics, secondary metabolite biosynthesis and hydrolytic enzymes.

18.7 Application and Economics of MPs

MP production in SSF has been practised in Asian countries for many centuries, and it was first mentioned in a Chinese medical book on herbs which was published in the first century AD (Dufossé, 2006). Nowadays, MPs and

RFR are widely applied in foods and fermentation foods (Mamucod and Dizon, 2014), texture industries (Velmurugan *et al.*, 2010b), medicines and cosmetics (Mostafa and Abbady, 2014), and they could also be used for sensitising solar cells (Sang-aroon, Saekow and Amornkitbamrung, 2012; Ito *et al.*, 2010) and preparing gels (Calvo and Salvador, 2002). Here, only MPs and RFR applications in meat products and fermentation foods are summarised.

MPs can be used not only as colourants in meat products but also as the partial or complete substitute of nitrite, which is a commonly used food additive during meat manufacturing (Mamucod and Dizon, 2014; Liu, Wu and Tan, 2010; Martínez *et al.*, 2006). For example, the sausage products with low-nitrite (25 mg/kg) and 1.5% RFR were acceptable when stored at 4 °C for 28 days (Liu, Wu and Tan, 2010). Experiments in which nitrite was completely replaced by MPs for sausage production were also carried out (Mamucod and Dizon, 2014). Generally, when using MPs to partially or completely replace nitrite, the chemical compositions, moisture contents, titratable acidity and hues (L, a, b values) of the meat products showed no significant differences compared with the control samples.

The MPs and RFR may be used in fermentation foods with two ways. One way is that MPs and RFR are directly used as colour agents of fermentation foods (Chen *et al.*, 2012b; Puttananjaiah, Dhale and Govindaswamy, 2011; Baranova *et al.*, 2004). For instance, when MPs as colour agents were used in yoghurt, the organoleptic qualities and physicochemical characters of yoghurt were stable after storage at 4 °C for 1, 4, 7 and 14 days (Chen *et al.*, 2012b). The other way is that RFR containing active cells of *Monascus* spp. was applied as fermentation starter (Yin *et al.*, 2005). For example, when RFR containing *Monascus* spp. was utilised as the starter in fermented fish products, the volatile base nitrogen of the products was conformed to the limit of food regulation in Taiwan (<25 mg/100 g) after 7 days of fermentation. In addition, the *Monascus*-fermented fish products possessed good flavour and colour which were highly likeable to the sensory panel (Yin *et al.*, 2005). Effect of *M. purpureus*-fermented product extract (MFPE) enriched in MPs on lactic acid bacteria (LAB) was also studied. LABs were not affected after 24 h of incubation in the broth containing MFPE. Besides, the antioxidant activities of isoflavone glycosides by LAB were enhanced after adding MFPE (Puttananjaiah, Dhale and Govindaswamy, 2011).

As for economics of MPs, the production values of the international food colourant market were estimated at about $1.15 billion USD in 2007. The natural food colourants accounted for $465 million USD in 2007 (Mapari, Thrane and Meyer, 2010). In Japan, the annual consumption of MPs increased from 100 t in 1981 to 600 t in 1992 and valued about $12 million (Mostafa and Abbady, 2014). Now, in China, the MP production scale and the actual output are less than 1000 t (colour value 10 000–15 000 U/g), and it is expected to reach more than 1500 t after 2014 (Srianta *et al.*, 2014).

Acknowledgements

In this chapter the authors have summarized some results from their research projects, including the Major Program of National Natural Science Foundation of China (No. 31330059), the National Natural Science Foundation of China (Nos. 31271834, 31171649, 31371824, and 31401631), the Fundamental Research Funds for the Central Universities of China (Nos. 2662014PY034, 2662014BQ051 and 2662015QC003), the International S & T Cooperation Program of Hubei Province in China (No. 2014BHE0016), the Research Project of Education Department of Hubei Province in China (B2015134), and the Open Fund of Hubei Key Laboratory of Edible Wild Plants Conservation & Utilization in China (EWPL201515).

References

Akihisa, T., Tokuda, H., Ukiya, M., Kiyota, A., Yasukawa, K., Sakamoto, N., Kimura, Y., Suzuki, T., Takayasu, J., and Nishino, H. (2005a) Anti-tumor-initiating effects of monascin, an azaphilonoid pigment from the extract of *Monascus pilosus* fermented rice (red-mold rice). *Chem. Biodivers.*, **2**, 1305–1309.

Akihisa, T., Tokuda, H., Yasukawa, K., Ukiya, M., Kiyota, A., Sakamoto, N., Suzuki, T., Tanabe, N., and Nishino, H. (2005b) Azaphilones, furanoisophthalides, and amino acids from the extracts of *Monascus pilosus*-fermented rice (red-mold rice) and their chemopreventive effects. *J. Agric. Food Chem.*, **53**, 562–565.

Babitha, S., Soccol, C.R., and Pandey, A. (2006) Jackfruit seed – a novel substrate for the production of *Monascus* pigments through solid-state fermentation. *Food Technol. Biotechnol.*, **44**, 465–471.

Babitha, S., Soccol, C.R., and Pandey, A. (2007a) Effect of stress on growth, pigment production and morphology of *Monascus* sp. in solid cultures. *J. Basic Microbiol.*, **47**, 118–126.

Babitha, S., Soccol, C.R., and Pandey, A. (2007b) Solid-state fermentation for the production of *Monascus* pigments from jackfruit seed. *Bioresour. Technol.*, **98**, 1554–1560.

Balakrishnan, B., Chen, C.C., Pan, T.M., and Kwon, H.J. (2014a) Mpp7 controls regioselective Knoevenagel condensation during the biosynthesis of *Monascus* azaphilone pigments. *Tetrahedron Lett.*, **55**, 1640–1643.

Balakrishnan, B., Suh, J.-W., Park, S.-H., and Kwon, H.-J. (2014b) Delineating *Monascus* azaphilone pigment biosynthesis: oxidoreductive modifications determine the ring cyclization pattern in azaphilone biosynthesis. *RSC Adv.*, **4**, 59405–59408.

Balakrishnan, B., Karki, S., Chiu, S.H., Kim, H.J., Suh, J.W., Nam, B., Yoon, Y.M., Chen, C.C., and Kwon, H.J. (2013) Genetic localization and in vivo characterization of a *Monascus* azaphilone pigment biosynthetic gene cluster. *Appl. Microbiol. Biotechnol.*, **97**, 6337–6345.

Baranova, M., Mal'a, P., Burdova, O., Hadbavny, M., and Sabolova, G. (2004) Effect of natural pigment of *Monascus purpureus* on the organoleptic characters of processed cheeses. *Bull. Vet. Inst. Pulawy*, **48**, 59–62.

Barnard, E.L. and Cannon, P.F. (1987) A new species of *Monascus* from pine tissues in Florida. *Mycologia*, **79**, 479–484.

Blanc, P.J., Loret, M.O., and Goma, G. (1995) Production of citrinin by various species of *Monascus*. *Biotechnol. Lett.*, **17**, 291–294.

Brakhage, A. (2013) Regulation of fungal secondary metabolism. *Nat. Rev. Microbiol.*, **11**, 21–32.

Calvo, C. and Salvador, A. (2002) Comparative study of the colorants monascus and cochineal used in the preparation of gels made with various gelling agents. *Food Hydrocolloids*, **16**, 523–526.

Campoy, S., Pérez, F., Martín, J., Gutiérrez, S., and Liras, P. (2003) Stable transformants of the azaphilone pigment-producing *Monascus purpureus* obtained by protoplast transformation and Agrobacterium-mediated DNA transfer. *Curr. Genet.*, **43**, 447–452.

Campoy, S., Rumbero, A., Martín, J.F., and Liras, P. (2006) Characterization of an hyperpigmenting mutant of *Monascus purpureus* IB1: identification of two novel pigment chemical structures. *Appl. Microbiol. Biotechnol.*, **70**, 488–496.

Cannon, P.F., Abdullah, S.K., and Abbas, B.A. (1995) Two new species of *Monascus* from Iraq, with a key to known species of the genus. *Mycol. Res.*, **99**, 659–662.

Chen, R.J., Hung, C.M., Chen, Y.L., Wu, M.D., Yuan, G.F., and Wang, Y.J. (2012a) Monascuspiloin induces apoptosis and autophagic cell death in human prostate cancer cells via the Akt and AMPK signaling pathways. *J. Agric. Food Chem.*, **60**, 7185–7193.

Chen, S.H.A., Lv, B.I.N., Du, X., and Chen, F. (2012b) Pigment from red fermented rice as colouring agent for stirred skimmed milk yoghurts. *Int. J. Dairy Technol.*, **65**, 287–292.

Chen, M.H. and Johns, M.R. (1993) Effect of pH and nitrogen source on pigment production by *Monascus purpureus*. *Appl. Microbiol. Biotechnol.*, **40**, 132–138.

Chen, M.H. and Johns, M.R. (1994) Effect of carbon source on ethanol and pigment production by *Monascus purpureus*. *Enzyme Microb. Technol.*, **16**, 584–590.

Chen, F.C., Manchard, P.S., and Whalley, W.B. (1969) The structure of monascin. *J. Chem. Soc. D*, 130–131.

Chen, F.C., Manchand, P.S., and Whalley, W.B. (1971) The chemistry of fungi. LXIV. The structure of monascin: the relative stereochemistry of the azaphilones. *J. Chem. Soc., Perkin Trans. 1*, **21**, 3577–3579.

Chen, W.P., He, Y., Zhou, Y.X., Shao, Y.C., Feng, Y.L., Li, M., and Chen, F.S. (2015) Edible filamentous fungi from the species Monascus: early traditional fermentations, modern molecular biology, and future genomics. *Compr. Rev. Food Sci. F.*, **14**, 555–567.

Chen, Y., Tseng, C., Liaw, L., Wang, C., Chen, I., Wu, W., Wu, M., and Yuan, G. (2008a) Cloning and characterization of monacolin K biosynthetic gene cluster from *Monascus pilosus*. *J. Agric. Food Chem.*, **56**, 5639–5646.

Chen, Y.P., Chen, I.C., Hwang, I.E., Yuan, G.F., Liaw, L.L., and Tseng, C.P. (2008b) Selection of an effective red-pigment producing *Monascus pilosus* by efficient transformation with aurintricarboxylic acid. *Biosci. Biotechnol., Biochem.*, **72**, 3021–3024.

Chen, Y.P., Yuan, G.F., Hsieh, S.Y., Lin, Y.S., Wang, W.Y., Liaw, L.L., and Tseng, C.P. (2010) Identification of the mokH gene encoding transcription factor for the upregulation of monacolin K biosynthesis in *Monascus pilosus*. *J. Agric. Food Chem.*, **58**, 287–293.

Cheng, M.J., Wu, M.D., Chen, Y.L., Chen, I.S., Su, Y.S., and Yuan, G.F. (2013) Chemical constituents of red yeast rice fermented with the fungus *Monascus pilosus*. *Chem. Nat. Compd.*, **49**, 249–252.

Cheng, M.J., Wu, M.D., Chen, I.S., Tseng, M., and Yuan, G.F. (2011) Chemical constituents from the fungus *Monascus purpureus* and their antifungal activity. *Phytochem. Lett.*, **4**, 372–376.

Cheng, M.J., Wu, M.D., Chen, I.S., and Yuan, G.F. (2010) A new sesquiterpene isolated from the extracts of the fungus *Monascus pilosus*-fermented rice. *Nat. Prod. Res.*, **24**, 750–758.

Choe, D., Lee, J., Woo, S., and Shin, C.S. (2012) Evaluation of the amine derivatives of *Monascus* pigment with anti-obesity activities. *Food Chem.*, **134**, 315–323.

Ding, G., Zhao, J.X., Zhang, W., Yao, J.C., Xu, H., Ding, Y.Z., Yang, G.H., Guo, X.G. and Wei, P. (2008) National standard GB4926-2008:Food additive-red kojic rice (powder). General Administration of Quality Supervision, Inspection and Quarantine of the People's Republic of China, pp. 1–5.

Domínguez-Espinosa, R.M. and Webb, C. (2003) Submerged fermentation in wheat substrates for production of *Monascus* pigments. *World J. Microbiol. Biotechnol.*, **19**, 329–336.

Dufossé, L. (2006) Microbial production of food grade pigments. *Food Technol. Biotechnol.*, **44**, 313–321.

Dufossé, L., Galaup, P., Yaron, A., Arad, S.M., Blanc, P., Chidambara Murthy, K.N., and Ravishankar, G.A. (2005) Microorganisms and microalgae as sources of pigments for food use: a scientific oddity or an industrial reality? *Trends Food Sci. Technol.*, **16**, 389–406.

Endo, A. (1979) Monacolin K, a new hypocholesterolemic agent produced by a *Monascus* species. *J. Antibiot.*, **32**, 852–854.

Endo, A. and Hasumi, K. (1985) Dihydromonacolin L and monacolin X, new metabolites those inhibit cholesterol biosynthesis. *J. Antibiot.*, **38**, 321–327.

Endo, A., Komagata, D., and Shimada, H. (1986) Monacolin M, a new inhibitor of cholesterol biosynthesis. *J. Antibiot.*, **39**, 1670–1673.

Evans, P.J. and Wang, H.Y. (1984) Pigment production from immobilized *Monascus* sp. utilizing polymeric resin adsorption. *Appl. Environ. Microbiol.*, **47**, 1323–1326.

Fabre, C.E., Santerre, A.L., Loret, M.O., Baberian, R., Pareilleux, A., Goma, G., and Blanc, P.J. (1993) Production and food applications of the red pigments of *Monascus ruber*. *J. Food Sci.*, **58**, 1099–1102/1110.

Feng, Y.L., Shao, Y.C., and Chen, F.S. (2012) *Monascus* pigments. *Appl. Microbiol. Biotechnol.*, **96**, 1421–1440.

Feng, Y., Shao, Y., Zhou, Y., and Chen, F. (2015) Production and optimization of monacolin K by citrinin-free *Monascus pilosus* MS-1 in solid-state fermentation using non-glutinous rice and soybean flours as substrate. *Eur. Food Res. Technol.*, **240**, 635–643.

Fielding, B.C., Holker, J.S.E., Jones, D.F., Powell, A.D.G., Richmond, K.W., Roberton, A., and Whalley, W.B. (1961) The chemistry of fungi. Part XXXIX. The structure of monascin. *J. Chem. Soc.*, 4579–4589.

Fu, J.Q. (ed) (1997) Chinese Monascus and Its Practical Technology, China Light Industry Press, Beijing.

Gacek, A. and Strauss, J. (2012) The chromatin code of fungal secondary metabolite gene clusters. *Appl. Microbiol. Biotechnol.*, **95**, 1389–1404.

Geiser, D.M., Gueidan, C., Miadlikowska, J., Lutzoni, F., Kauff, F., Hofstetter, V., Fraker, E., Schoch, C.L., Tibell, L., Untereiner, W.A., and Aptroot, A. (2006) Eurotiomycetes: Eurotiomycetidae and Chaetothyriomycetidae. *Mycologia*, **98**, 1053–1064.

Gheith, O., Sheashaa, H., Abdelsalam, M., Shoeir, Z., and Sobh, M. (2008) Efficacy and safety of *Monascus purpureus* Went rice in subjects with secondary hyperlipidemia. *Clin. Exp. Nephrol.*, **12**, 189–194.

Gong, H., Chen, H., and Gao, Q. (2002) The research progress on Hongqu and its pigment. *J. Wuhan Polytech. Univ.*, **1**, 22–24.

Hajjaj, H., François, J.M., Goma, G., and Blanc, P.J. (2012) Effect of amino acids on red pigments and citrinin production in *Monascus ruber*. *J. Food Sci.*, **77**, 156–159.

Hajjaj, H., Klaébé, A., Loret, M.O., Tzedakis, T., Goma, G., and Blanc, P.J. (1997) Production and identification of N-glucosylrubropunctamine and n-glucosylmonascorubramine from *Monascus ruber* and occurrence of electron donor-acceptor complexes in these red pigments. *Appl. Environ. Microbiol.*, **63**, 2671–2678.

Hawksworth, D. and Pitt, J. (1983) A new taxonomy for *Monascus* species based on cultural and microscopical characters. *Aust. J. Bot.*, **31**, 51–61.

Haws, E.J., Holker, J.S.E., Kelly, A., Powell, A.D.G., and Robertson, A. (1959) The chemistry of fungi. Part XXXVII. The structure of rubropunctatin. *J. Chem. Soc.*, **70**, 3598–3610.

He, Y., Liu, Q., Shao, Y., and Chen, F. (2013) ku70 and ku80 null mutants improve the gene targeting frequency in *Monascus ruber* M7. *Appl. Microbiol. Biotechnol.*, **97**, 4965–4976.

He, Y., Shao, Y., and Chen, F. (2014) Efficient gene targeting in ligase IV-deficient *Monascus ruber* M7 by perturbing the non-homologous end joining pathway. *Fungal Biol.*, **118**, 846–854.

Hocking, A.D. and Pitt, J.I. (1988) Two new species of xerophilic fungi and a further record of Eurotium halophilicum. *Mycologia*, **80**, 82–88.

Hong, M.Y., Seeram, N.P., Zhang, Y., and Heber, D. (2008) Anticancer effects of Chinese red yeast rice versus monacolin K alone on colon cancer cells. *J. Nutr. Biochem.*, **19**, 448–458.

Hossain, C.F., Okuyama, E., and Yamazaki, M. (1996) A new series of coumarin derivatives having monoamine oxidase inhibitory activity from *Monascus anka*. *Chem. Pharm. Bull.*, **44**, 1535–1539.

Hsu, Y.W., Hsu, L.C., Liang, Y.H., Kuo, Y.H., and Pan, T.M. (2010) Monaphilones A – C, three new antiproliferative azaphilone derivatives from *Monascus purpureus* NTU 568. *J. Agric. Food Chem.*, **58**, 8211–8216.

Hsu, Y.W., Hsu, L.C., Liang, Y.H., Kuo, Y.H., and Pan, T.M. (2011) New bioactive orange pigments with yellow fluorescence from *Monascus*-fermented dioscorea. *J. Agric. Food Chem.*, **59**, 4512–4518.

Hsu, W.H., Lee, B.H., Liao, T.H., Hsu, Y.W., and Pan, T.M. (2012) *Monascus*-fermented metabolite monascin suppresses inflammation via PPAR-γ regulation and JNK inactivation in THP-1 monocytes. *Food Chem. Toxicol.*, **50**, 1178–1186.

Hsu, L.C., Liang, Y.H., Hsu, Y.W., Kuo, Y.H., and Pan, T.M. (2013) Anti-inflammatory properties of yellow and orange pigments from *Monascus purpureus* NTU 568. *J. Agric. Food Chem.*, **61**, 2796–2802.

Hu, Z.Q., Zhang, X.H., Wu, Z.Q., Qi, H.S., and Wang, Z.L. (2012) Perstraction of intracellular pigments by submerged cultivation of *Monascus* in nonionic surfactant micelle aqueous solution. *Appl. Microbiol. Biotechnol.*, **94**, 81–89.

Huang, L., Cheng, X., Wei, S.J., Tu, X.R., and Li, K.T. (2011) Research on the stability for monascus pigment produced by *Monascus purpureus* JR. *China Condiment*, **36**, 93–96.

Huang, Z., Xu, Y., Li, L., and Li, Y. (2007) Two new *Monascus* metabolites with strong blue fluorescence isolated from red yeast rice. *J. Agric. Food Chem.*, **56**, 112–118.

Inoue, K., Ito, Y., Hattori, Y., Tsutsumiuchi, K., Ito, S., Hino, T., and Oka, H. (2010) Efficient purification of xanthomonasin A and B from *Monascus* yellow colorant by high-speed countercurrent chromatography. *Jpn. J. Food Chem. Saf.*, **17**, 185–191.

Inouye, Y., Nakanishi, K., Nishikawa, H., Ohashi, M., Terahara, A., and Yamamura, S. (1962) Structure of monascoflavin. *Tetrahedron*, **18**, 1195–1203.

Ito, S., Saitou, T., Imahori, H., Uehara, H., and Hasegawa, N. (2010) Fabrication of dye-sensitized solar cells using natural dye for food pigment: *Monascus* yellow. *Energy Environ. Sci.*, **3**, 905–909.

Izawa, S., Harada, N., Watanabe, T., Kotokawa, N., Yamamoto, A., Hayatsu, H., and Arimoto-Kobayashi, S. (1997) Inhibitory effects of food-coloring agents derived from *Monascus* on the mutagenicity of heterocyclic amines. *J. Agric. Food Chem.*, **45**, 3980–3984.

Jeun, J., Jung, H., Kim, J., Kim, Y., Youn, S., and Shin, C. (2008) Effect of the monascus pigment threonine derivative on regulation of the cholesterol level in mice. *Food Chem.*, **107**, 1078–1085.

Jia, X.Q., Xu, Z.N., Zhou, L.P., and Sung, C.K. (2010) Elimination of the mycotoxin citrinin production in the industrial important strain *Monascus purpureus* SM001. *Metab. Eng.*, **12**, 1–7.

Jo, D., Choe, D., Nam, K., and Shin, C. (2014) Biological evaluation of novel derivatives of the orange pigments from *Monascus* sp. as inhibitors of melanogenesis. *Biotechnol. Lett.*, **36**, 1605–1613.

Johns, M.R. and Stuart, D.M. (1991) Production of pigments by *Monascus purpureus* in solid culture. *J. Ind. Microbiol.*, **8**, 23–28.

Jongrungruangchok, S., Kittakoop, P., Yongsmith, B., Bavovada, R., Tanasupawat, S., Lartpornmatulee, N., and Thebtaranonth, Y. (2004) Azaphilone pigments from a yellow mutant of the fungus *Monascus kaoliang*. *Phytochemistry*, **65**, 2569–2575.

Jou, P.C., Ho, B.Y., Hsu, Y.W., and Pan, T.M. (2010) The effect of *Monascus* secondary polyketide metabolites, monascin and ankaflavin, on adipogenesis and lipolysis activity in 3T3 L1. *J. Agric. Food Chem.*, **58**, 12703–12709.

Jung, H., Choe, D., Nam, K.Y., Cho, K.H., and Shin, C.S. (2011) Degradation patterns and stability predictions of the original reds and amino acid derivatives of *Monascus* pigments. *Eur. Food Res. Technol.*, **232**, 621–629.

Jung, H., Kim, C., Kim, K., and Shin, C.S. (2003) Color characteristics of monascus pigments derived by fermentation with various amino acids. *J. Agric. Food Chem.*, **51**, 1302–1306.

Jung, H., Kim, C., and Shin, C.S. (2005) Enhanced photostability of *Monascus* pigments derived with various amino acids via fermentation. *J. Agric. Food Chem.*, **53**, 7108–7114.

Jůzlová, P., Martí, L., and Křen, V. (1996) Secondary metabolites of the fungus *Monascus*: a review. *J. Ind. Microbiol.*, **16**, 163–170.

Kang, B., Zhang, X., Wu, Z., Wang, Z., and Park, S. (2014) Production of citrinin-free *Monascus* pigments by submerged culture at low pH. *Enzyme Microb. Technol.*, **55**, 50–57.

Kaur, B., Chakraborty, D., and Kaur, H. (2009) Production and evaluation of physicochemical properties of red pigment from *Monascus purpureus* MTCC 410. *Internet J. Microbiol.*, **7**, 26. doi: 10.5580/d4a

Kim, C., Jung, H., Kim, J.H., and Shin, C.S. (2006a) Effect of monascus pigment derivatives on the electrophoretic mobility of bacteria, and the cell adsorption and antibacterial activities of pigments. *Colloids Surf., B*, **47**, 153–159.

Kim, C., Jung, H., Kim, Y.O., and Shin, C.S. (2006b) Antimicrobial activities of amino acid derivatives of monascus pigments. *FEMS Microbiol. Lett.*, **264**, 117–124.

Kim, J.H., Kim, H.J., Kim, C., Jung, H., Kim, Y.O., Ju, J.Y., and Shin, C.S. (2007a) Development of lipase inhibitors from various derivatives of monascus pigment produced by *Monascus* fermentation. *Food Chem.*, **101**, 357–364.

Kim, J.H., Kim, H.J., Park, H.W., Youn, S.H., Choi, D.Y., and Shin, C.S. (2007b) Development of inhibitors against lipase and α-glucosidase from derivatives of monascus pigment. *FEMS Microbiol. Lett.*, **276**, 93–98.

Kim, J.H., Kim, Y.O., Jeun, J., Choi, D.Y., and Shin, C.S. (2010) L-Trp and L-Leu-OEt derivatives of the *Monascus* pigment exert high anti-obesity effects on mice. *Biosci. Biotechnol., Biochem.*, **74**, 304–308.

Knecht, A. and Humpf, H.U. (2006) Cytotoxic and antimitotic effects of N-containing *Monascus* metabolites studied using immortalized human kidney epithelial cells. *Mol. Nutr. Food Res.*, **50**, 406–412.

Kongruang, S. (2010) Growth kinetics of biopigment production by Thai isolated *Monascus purpureus* in a stirred tank bioreactor. *J. Ind. Microbiol. Biotechnol.*, **38**, 93–99.

Krairak, S., Yamamura, K., Irie, R., Nakajima, M., Shimizu, H., Chim-Anage, P., Yongsmith, B., and Shioya, S. (2000) Maximizing yellow pigment production in fed-batch culture of *Monascus* sp. *J. Biosci. Bioeng.*, **90**, 363–367.

Kumasaki, S., Nakanishi, K., Nishikawa, E., and Ohashi, M. (1962) Structure of monascorubrin. *Tetrahedron*, **18**, 1171–1184.

Lackner, G., Misiek, M., Braesel, J., and Hoffmeister, D. (2012) Genome mining reveals the evolutionary origin and biosynthetic potential of basidiomycete polyketide synthases. *Fungal Genet. Biol.*, **49**, 996–1003.

Lai, Y., Wang, L., Qing, L., and Chen, F.S. (2011) Effects of cyclic AMP on development and secondary metabolites of *Monascus ruber* M-7. *Lett. Appl. Microbiol.*, **52**, 420–426.

Lakrod, K., Chaisrisook, C., and Daniel, Z. (2003a) Transformation of *Monascus purpureus* to hyromycing B resistance with cosmid pMOcosX reduces fertility. *Electron. J. Biotechnol.*, **6**, 143–147.

Lakrod, K., Chaisrisook, C., and Skinner, D.Z. (2003b) Expression of pigmentation genes following electroporation of albino *Monascus purpureus*. *J. Ind. Microbiol. Biotechnol.*, **30**, 369–374.

Lee, Y.K., Chen, D.C., Chauvatcharin, S., Seki, T., and Yoshida, T. (1995) Production of *Monascus* pigments by a solid-liquid state culture method. *J. Ferment. Bioeng.*, **79**, 516–518.

Lee, B.H., Hsu, W.H., Liao, T.H., and Pan, T.M. (2011) The *Monascus* metabolite monascin against TNF-α-induced insulin resistance via suppressing PPAR-γ phosphorylation in C2C12 myotubes. *Food Chem. Toxicol.*, **49**, 2609–2617.

Lee, B.H., Hsu, W.H., and Pan, T.M. (2012) Red mold rice against hepatic inflammatory damage in Zn-deficient rats. *J. Tradit. Complement. Med.*, **2**, 52–60.

Lee, B.K., Park, N.H., Piao, H.Y., and Chung, W.J. (2001) Production of red pigments by

Monascus purpureus in submerged culture. *Biotechnol. Bioprocess Eng.*, **6**, 341–346.

Lee, C.L., Wen, J.Y., Hsu, Y.W., and Pan, T.M. (2013) *Monascus*-fermented yellow pigments monascin and ankaflavin showed antiobesity effect via the suppression of differentiation and lipogenesis in obese rats fed a high-fat diet. *J. Agric. Food Chem.*, **61**, 1493–1500.

Li, H.R., Du, Z.W., and Zhang, J.R. (2003) Study on the stability of *Monascus* pigment. *Food Sci.*, **24**, 59–62.

Li, L., He, L., Lai, Y., Shao, Y., and Chen, F. (2014) Cloning and functional analysis of the Gbeta gene Mgb1 and the Ggamma gene Mgg1 in *Monascus ruber*. *J. Microbiol.*, **52** (1), 35–43.

Li, Y.P., Pan, Y.F., Zou, L.H., Xu, Y., Huang, Z.B., and He, Q.H. (2013) Lower citrinin production by gene disruption of *ctnB* involved in citrinin biosynthesis in *Monascus aurantiacus* Li AS3.4384. *J. Agric. Food Chem.*, **61**, 7397–7402.

Li, J.J., Shang, X.Y., Li, L.L., Liu, M.T., Zheng, J.Q., and Jin, Z.L. (2010a) New cytotoxic azaphilones from *Monascus purpureus*-fermented rice (red yeast rice). *Molecules*, **15**, 1958–1966.

Li, L., Shao, Y.C., Li, Q., Yang, S., and Chen, F.S. (2010b) Identification of *Mga1*, a G-protein α-subunit gene involved in regulating citrinin and pigment production in *Monascus ruber* M7. *FEMS Microbiol. Lett.*, **308**, 108–114.

Li, Y.P., Xu, Y., and Huang, Z.B. (2012) Isolation and characterization of the citrinin biosynthetic gene cluster from *Monascus aurantiacus*. *Biotechnol. Lett.*, **34**, 131–136.

Li, F., Xu, G., Li, Y. and Chen, X. (2008) GB/T 5009.222-2008: Determination of citrinin in *Monascus* products. AQSIQE, pp. 1–5.

Lian, X., Liu, L., Dong, S., Wu, H., Zhao, J., and Han, Y. (2015) Two new monascus red pigments produced by Shandong Zhonghui Food Company in China. *Eur. Food Res. Technol.*, **240** 719–724.

Lian, X.J., Wang, C.L., and Guo, K.L. (2007) Identification of new red pigments produced by *Monascus ruber*. *Dyes Pigm.*, **73**, 121–125.

Lim, S.I. and Kwak, E.J. (2004) Stability of the pigments from *Monascus purpureus* CBS 281.34. *J. Korean Soc. Food Sci. Nutr.*, **33**, 711–715.

Lin, T.F. and Demain, A.L. (1991) Effect of nutrition of *Monascus* sp. on formation of red pigments. *Appl. Microbiol. Biotechnol.*, **36**, 70–75.

Lin, C.F. and Iizuka, H. (1982) Production of extracellular pigment by a mutant of *Monascus kaoliang* sp. nov. *Appl. Environ. Microbiol.*, **43**, 671–676.

Lin, C.P., Lin, Y.L., Huang, P.H., Tsai, H.S., and Chen, Y.H. (2011) Inhibition of endothelial adhesion molecule expression by *Monascus purpureus*-fermented rice metabolites, monacolin K, ankaflavin, and monascin. *J. Sci. Food Agric.*, **91**, 1751–1758.

Lin, Y.L., Wang, T.H., Lee, M.H., and Su, N.W. (2008) Biologically active components and nutraceuticals in the *Monascus*-fermented rice: a review. *Appl. Microbiol. Biotechnol.*, **77**, 965–973.

Lin, T.F., Yakushijin, K., Büchi, G.H., and Demain, A.L. (1992) Formation of water-soluble *Monascus* red pigments by biological and semi-synthetic processes. *J. Ind. Microbiol.*, **9**, 173–179.

Liu, D.C., Wu, S.W., and Tan, F.J. (2010) Effects of addition of anka rice on the qualities of low-nitrite Chinese sausages. *Food Chem.*, **118**, 245–250.

Liu, Q., Xie, N., He, Y., Wang, L., Shao, Y., Zhao, H., and Chen, F. (2014) MpigE, a gene involved in pigment biosynthesis in *Monascus ruber* M7. *Appl. Microbiol. Biotechnol.*, **98**, 285–296.

Loret, M.O. and Morel, S. (2010) Isolation and structural characterization of two new metabolites from *Monascus*. *J. Agric. Food Chem.*, **58**, 1800–1803.

Mamucod, H.F. and Dizon, E.I. (2014) Potential of biopigments from *Monascus purpureus Went* as natural food colorant for Philippine native sausage (Longganisa). 2014 3rd International Conference on Nutrition and Food Sciences, Vol. 71, pp. 72–76.

Manchand, P.S. and Whalley, W.B. (1973) Isolation and structure of ankaflavin: a new pigment from *Monascus anka*. *Phytochemistry*, **12**, 2531–2532.

Mapari, S.A.S., Hansen, M.E., Meyer, A.S., and Thrane, U. (2008) Computerized screening for novel producers of

Monascus-like food pigments in *Penicillium* species. *J. Agric. Food Chem.*, **56**, 9981–9989.

Mapari, S.A.S., Meyer, A.S., and Thrane, U. (2009) Photostability of natural orange–red and yellow fungal pigments in liquid food model systems. *J. Agric. Food Chem.*, **57**, 6253–6261.

Mapari, S.A.S., Thrane, U., and Meyer, A.S. (2010) Fungal polyketide azaphilone pigments as future natural food colorants? *Trends Biotechnol.*, **28**, 300–307.

Martínez, L., Cilla, I., Beltrán, J.A., and Roncalés, P. (2006) Comparative effect of red yeast rice (*Monascus purpureus*), red beet root (Beta vulgaris) and betanin (E-162) on colour and consumer acceptability of fresh pork sausages packaged in a modified atmosphere. *J. Sci. Food Agric.*, **86**, 500–508.

Martínková, L., Juzlová, P., and Veselý, D. (1995) Biological activity of polyketide pigments produced by the fungus *Monascus*. *J. Appl. Microbiol.*, **79**, 609–616.

Martínková, L., Patáková Jůzlová, P., Krent, V., Kucerová, Z., Havlícek, V., Olsovský, P., Hovorka, O., Ríhová, B., Veselý, D., Veselá, D., Ulrichová, J., and Prikrylová, V. (1999) Biological activities of oligoketide pigments of *Monascus purpureus*. *Food Addit. Contam.*, **16**, 15–24.

Miyake, T., Kono, I., Nozaki, N., and Sammoto, H. (2008) Analysis of pigment compositions in various *Monascus* cultures. *Food Sci. Technol. Res.*, **14**, 194–197.

Mohamed, M.S., Mohamad, R., Manan, M.A., and Ariff, A.B. (2009) Enhancement of red pigment production by *Monascus purpureus* FTC 5391 through retrofitting of helical ribbon impeller in stirred-tank fermenter. *Food Bioprocess Technol.*, **5**, 80–91.

Kumari, H.P., Naidu, K.A., Vishwanatha, S., Narasimhamurthy, K., and Vijayalakshmi, G. (2009) Safety evaluation of *Monascus purpureus* red mould rice in albino rats. *Food Chem. Toxicol.*, **47**, 1739–1746.

Moll, H.R. and Farr, D.R. (1976) Red pigment and process. US Patent 3 993 789 23.

Mostafa, M.E. and Abbady, M.S. (2014) Secondary metabolites and bioactivity of the *Monascus* pigments review article. *Global J. Biotechnol. Biochem.*, **9**, 1–13.

Mukherjee, G. and Singh, S.K. (2011) Purification and characterization of a new red pigment from *Monascus purpureus* in submerged fermentation. *Process Biochem.*, **46**, 188–192.

Nimnoi, P. and Lumyong, S. (2009) Improving solid-state fermentation of *Monascus purpureus* on agricultural products for pigment production. *Food Bioprocess Technol.*, **4**, 1384–1390.

Padmavathi, T. and Prabhudessai, T. (2013) A solid liquid state culture method to stimulate *Monascus* pigments by intervention of different substrates. *Int. Res. J. Biol. Sci.*, **2**, 22–29.

Pandey, A., Soccol, C.R., and Mitchell, D. (2000) New developments in solid state fermentation: I-bioprocesses and products. *Process Biochem.*, **35**, 1153–1169.

Patakova, P. (2013) *Monascus* secondary metabolites: production and biological activity. *J. Ind. Microbiol. Biotechnol.*, **40**, 169–181.

Pattanagul, P., Pinthong, R., Phianmongkhol, A., and Leksawasdi, N. (2007) Review of angkak production (*Monascus purpureus*). *Chiang Mai J. Sci.*, **34**, 319–328.

Pisareva, E.I. and Kujumdzieva, A.V. (2010) Influence of carbon and nitrogen sources on growth and pigment production by *Monascus pilosus* C_1 strain. *Biotechnol. Biotechnol. Equip.*, **24**, 501–506.

Prajapati, V.S., Soni, N., Trivedi, U.B., and Patel, K.C. (2014) An enhancement of red pigment production by submerged culture of *Monascus purpureus* MTCC 410 employing statistical methodology. *Biocatal. Agric. Biotechnol.*, **3**, 140–145.

Puttananjaiah, M.K.H., Dhale, M.A., and Govindaswamy, V. (2011) Non-toxic effect of *Monascus purpureus* extract on lactic acid bacteria suggested their application in fermented foods. *Food Nutr. Sci.*, **2**, 837–843.

Qian, J. and Wu, Q. (2010) Improving water solubility of *Monascus* pigment. *J. Chin. Cereal Oil Assoc.*, **25**, 77–79, 92.

Sakai, K., Kinoshita, H., and Nihira, T. (2009) Identification of mokB involved in monacolin K biosynthesis in *Monascus pilosus*. *Biotechnol. Lett.*, **31**, 1911–1916.

Salomon, H. and Karrer, P. (1932) Pflanzenfarbstoffe XXXVIII. Ein farbstoff aus "rotem" reis, monascin. *Helv. Chim. Acta*, **15**, 18–22.

Sang-aroon, W., Saekow, S., and Amornkitbamrung, V. (2012) Density functional theory study on the electronic structure of *Monascus* dyes as photosensitizer for dye-sensitized solar cells. *J. Photochem. Photobiol., A*, **236**, 35–40.

Santerre, A.L., Queinnec, I., and Blanc, P.J. (1995) A fedbatch strategy for optimal red pigment. *Bioprocess. Eng.*, **13**, 245–250.

Sato, K., Goda, Y., Sakamoto, S.S., and Shibata, H. (1997) Identification of major pigments containing D-amino acid units in commercial *Monascus* pigments. *Chem. Pharm. Bull.*, **45**, 227–229.

Sato, K., Iwakami, S., Goda, Y., and Okuyama, E. (1992) Novel natural colorants from *Monascus anka* U-1. *Heterocycles*, **34**, 2057–2060.

Sheu, F., Wang, C.L., and Shyu, Y.T. (2000) Fermentation of *Monascus purpureus* on bacterial cellulose-nata and the color stability of *Monascus*-nata complex. *J. Food Sci.*, **65**, 342–345.

Shi, Y.C., Liao, J.W., and Pan, T.M. (2011) Antihypertriglyceridemia and anti-inflammatory activities of *Monascus*-fermented dioscorea in streptozotocin-induced diabetic rats. *Exp. Diabetes Res.*, **2011**, 710635.

Shi, Y.C., Liao, V.H.C., and Pan, T.M. (2012) Monascin from red mold dioscorea as a novel antidiabetic and antioxidative stress agent in rats and *Caenorhabditis elegans*. *Free Radical Biol. Med.*, **52**, 109–117.

Shi, Y.C. and Pan, T.M. (2010) Anti-diabetic effects of *Monascus purpureus* NTU 568 fermented products on streptozotocin-induced diabetic rats. *J. Agric. Food Chem.*, **58**, 7634–7640.

Shi, Y.C. and Pan, T.M. (2011) Beneficial effects of *Monascus purpureus* NTU 568-fermented products: a review. *Appl. Microbiol. Biotechnol.*, **90** (4), 1207–1217.

Shimizu, T., Kinoshita, H., Ishihara, S., Sakai, K., Nagai, S., and Nihira, T. (2005) Polyketide synthase gene responsible for citrinin biosynthesis in *Monascus purpureus*. *Appl. Environ. Microbiol.*, **71**, 3453–3457.

Shimizu, T., Kinoshita, H., and Nihira, T. (2007) Identification and in vivo functional analysis by gene disruption of *ctnA*, an activator gene involved in citrinin biosynthesis in *Monascus purpureus*. *Appl. Environ. Microbiol.*, **73**, 5097–5103.

Silveira, S., Daroit, D., Sant'Anna, V., and Brandelli, A. (2013) Stability modeling of red pigments produced by *Monascus purpureus* in submerged cultivations with sugarcane bagasse. *Food Bioprocess Technol.*, **6** (4), 1007–1014.

Smedsgaard, J. (1997) Micro-scale extraction procedure for standardized screening of fungal metabolite production in cultures. *J. Chromatogr. A*, **760**, 264–270.

Song, S.S., Cui, H.X., and Si, S.L. (1995) Study on stability of *Monascus* pigment. *J. Hebei Acad. Sci*, 2, 27–34.

Srianta, I., Kusumawati, N., Nugerahani, I., Artanti, N., and Xu, G.R. (2013) In vitro α-glucosidase inhibitory activity of *Monascus*-fermented durian seed extracts. *Int. Food Res. J.*, **20** (6), 533–536.

Srianta, I., Ristiarini, S., Nugerahani, I., Sen, S.K., Zhang, B.B., Xu, G.R., and Blanc, P.J. (2014) Recent research and development of *Monascus* fermentation products. *Int. Food Res. J.*, **21**, 1–12.

Staunton, J. and Weissman, K.J. (2001) Polyketide biosynthesis: a millennium review. *Nat. Prod. Rep.*, **18**, 380–416.

Stchigel, A.M., Cano, J.F., Abdullah, S.K., and Guarro, J. (2004) New and interesting species of *Monascus* from soil, with a key to the known species. *Stud. Mycol.*, **50**, 299–306.

Su, N.W., Lin, Y.L., Lee, M.H., and Ho, C.Y. (2005) Ankaflavin from *Monascus*-fermented red rice exhibits selective cytotoxic effect and induces cell death on Hep G2 cells. *J. Agric. Food Chem.*, **53**, 1949–1954.

Sun, X.H., Yang, X.R., and Wang, E.K. (2005) Chromatographic and electrophoretic procedures for analyzing plant pigments of pharmacologically interests. *Anal. Chim. Acta*, **547**, 153–157.

Sweeny, J.G., Valdes, M.C.E., Iacobucci, G.A., Sato, H., and Sakamura, S. (1981) Photoprotection of the red pigments of *Monascus anka* in aqueous media by 1,4,6-trihydroxynaphthalene. *J. Agric. Food Chem.*, **29**, 1189–1193.

Teng, S.S. and Feldheim, W. (1998) Analysis of anka pigments by liquid chromatography with diode array detection and tandem mass spectrometry. *Chromatographia*, **47**, 529–536.

Teng, S.S. and Feldheim, W. (2000) The fermentation of rice for anka pigment production. *J. Ind. Microbiol. Biotechnol.*, **25**, 141–146.

Teng, S.S. and Feldheim, W. (2001) Anka and anka pigment production. *J. Ind. Microbiol. Biotechnol.*, **26**, 280–282.

Turner, N.W., Subrahmanyam, S., and Piletsky, S.A. (2009) Analytical methods for determination of mycotoxins: a review. *Anal. Chim. Acta*, **632**, 168–180.

Udagawa, S.I. and Baba, H. (1998) *Monascus lunisporas*: a new species isolated from mouldy feeds. *Cryptogam., Mycol.*, **19**, 269–276.

Van Tieghem, M. (1884) *Monascus*, genre nouveau de l'ordre des Ascomycetes. *Bull. Soc. Bot. France*, **31**, 226–231.

Velmurugan, P., Kamala Kannan, S., Balachandar, V., Lakshmanaperumalsamy, P., Chae, J.C., and Oh, B.T. (2010a) Natural pigment extraction from five filamentous fungi for industrial applications and dyeing of leather. *Carbohydr. Polym.*, **79**, 262–268.

Velmurugan, P., Kim, M.J., Park, J.S., Karthikeyan, K., Lakshmanaperumalsamy, P., Lee, K.J., Park, Y.J., and Oh, B.T. (2010b) Dyeing of cotton yarn with five water soluble fungal pigments obtained from five fungi. *Fibers Polym.*, **11**, 598–605.

Velmurugan, P., Lee, Y.H., Venil, C.K., Lakshmanaperumalsamy, P., Chae, J.C., and Oh, B.T. (2010c) Effect of light on growth, intracellular and extracellular pigment production by five pigment-producing filamentous fungi in synthetic medium. *J. Biosci. Bioeng.*, **109**, 346–350.

Vendruscolo, F., Luise Müller, B., Esteves Moritz, D., de Oliveira, D., Schmidell, W., and Luiz Ninow, J. (2013) Thermal stability of natural pigments produced by *Monascus ruber* in submerged fermentation. *Biocatal. Agric. Biotechnol.*, **2**, 278–284.

Vendruscolo, F., Tosin, I., Giachini, A.J., Schmidell, W., and Ninow, J.L. (2014) Antimicrobial activity of *Monascus* pigments produced in submerged fermentation. *J. Food Process. Preserv.*, **38**, 1860–1865.

Vidyalakshmi, R., Paranthaman, R., Murugesh, S., and Singaravadivel, K. (2009a) Microbial bioconversion of rice broken to food grade pigments. *Global J. Biotechnol. Biochem.*, **4**, 84–87.

Vidyalakshmi, R., Paranthaman, R., Murugesh, S., and Singaravadivel, K. (2009b) Stimulation of *Monascus* pigments by intervention of different nitrogen sources. *Global J. Biotechnol. Biochem.*, **4**, 25–28.

Wang, L.C., Lung, T.Y., Kung, Y.H., Wang, J.J., Tsai, T.Y., Wei, B.L., Pan, T.M., and Lee, C.L. (2013) Enhanced anti-obesity activities of red mold dioscorea when fermented using deep ocean water as the culture water. *Mar. Drugs*, **11**, 3902–3925.

Wang, M.Q., Wang, Z.T., Chen, J.S., Zhang, J.B., Li, X.Y., Chen, Y.J., Luo, X.Y., Fan, Y.X., Wang, J., Zhao, D., Jin, Q.Z., Tian, J., Mao, X.D. and Yang, D.J. (2007) National Standard GB2760-2007: Hygienic Standards for Uses of Food Additives. AQSIQ, pp. 23–24.

Watanabe, T., Mazumder, T.K., Yamamoto, A., Nagai, S., Arimoto-Kobayashi, S., Hayatsu, H., and Terabe, S. (1999) A simple and rapid method for analyzing the *Monascus* pigment-mediated degradation of mutagenic 3-hydroxyamino-1-methyl-5H-Pyrido[4,3-b] indole by in capillary micellar electrokinetic chromatography. *Mutat. Res.*, **444**, 75–83.

Watanabe, T., Yamamoto, A., Nagai, S., and Terabe, S. (1997) Separation and determination of monascus yellow pigments for food by micellar electrokinetic chromatography. *Anal. Sci.*, **13**, 571–575.

Whalley, W.B., Ferguson, G., Marsh, W.C., and Restivo, R.J. (1976) The chemistry of fungi. Part LXVIII. The absolute configuration of (+)-sclerotiorin and of the azaphilones. *J. Chem. Soc., Perkin Trans. 1*, **13**, 1366–1369.

Wong, H.C. and Koehler, P.E. (1983) Production of red water-soluble *Monascus* pigments. *J. Food Sci.*, **48**, 1200–1203.

Wongjewboot, I. and Kongruang, S. (2011) pH stability of ultrasonic Thai Isolated

Monascus purpureus pigments. *Int. J. Biosci. Biochem. Bioinf.*, **1**, 79–83.

Wongsorn, H., Wongjewboot, I., and Kongruang, S. (2011) Solvent stability of ultrasonic mutants of *Monascus purpureus* pigments. *Int. J. Biosci. Biochem. Bioinf.*, **1**, 206–210.

Wu, C.L., Kuo, Y.H., Lee, C.L., Hsu, Y.W., and Pan, T.M. (2011a) Synchronous high-performance liquid chromatography with a photodiode array detector and mass spectrometry for the determination of citrinin, monascin, ankaflavin, and the lactone and acid forms of monacolin K in red mold rice. *J. AOAC Int.*, **94**, 179–190.

Wu, M.D., Cheng, M.J., Yech, Y.J., Chen, Y.L., Chen, K.P., Chen, I.S., Yang, P.H., and Yuan, G.F. (2011b) Monasnicotinates A–D, four new pyridine alkaloids from the fungal strain *Monascus pilosus* BCRC 38093. *Molecules*, **16**, 4719–4727.

Xie, N., Liu, Q., and Chen, F. (2013) Deletion of pigR gene in *Monascus ruber* leads to loss of pigment production. *Biotechnol. Lett.*, **35**, 1425–1432.

Yang, X., Hu, W., Xie, F., and Wang, M. (2007) Cirinin control in processing monascus red. *China Food Addit.*, z1, 209–212, 145.

Yang, Y., Li, L., Li, X., Shao, Y., and Chen, F. (2012) mrflbA, encoding a putative FlbA, is involved in aerial hyphal development and secondary metabolite production in *Monascus ruber* M-7. *Fungal Biol.*, **116**, 225–233.

Yin, L.J., Lu, M.C., Pan, C.L., and Jiang, S.T.S. (2005) Effect of *Monascus* fermentation on the characteristics of mackerel mince. *J. Food Sci.*, **70**, S66–S72.

Yongsmith, B., Kitprechavanich, V., Chitradon, L., Chaisrisook, C., and Budda, N. (2000) Color mutants of *Monascus* sp. KB9 and their comparative glucoamylases on rice solid culture. *J. Mol. Catal. B: Enzym.*, **10**, 263–272.

Yongsmith, B., Krairak, S., and Bavavoda, R. (1994) Production of yellow pigments in submerged culture of a mutant of *Monascus* spp. *J. Ferment. Bioeng.*, **78**, 223–228.

Yongsmith, B., Tabloka, W., Yongmanitchai, W., and Bavavoda, R. (1993) Culture conditions for yellow pigment formation by *Monascus* sp. KB 10 grown on cassava medium. *World J. Microbiol. Biotechnol.*, **9**, 85–90.

Yoshimura, M., Yamanaka, S., Mitsugi, K., and Hirose, Y. (1975) Production of *Monascus*-pigment in a submerged culture. *Agric. Biol. Chem.*, **39**, 1789–1795.

Yoshizaki, Y., Kawasaki, C., Cheng, K.C., Ushikai, M., Amitani, H., Asakawa, A., Okutsu, K., Sameshima, Y., Takamine, K., and Inui, A. (2014) Rice koji reduced body weight gain, fat accumulation, and blood glucose level in high-fat diet-induced obese mice. *PeerJ*, **2**, e540.

Yu, J.H. (2006) Heterotrimeric G protein signaling and RGSs in *Aspergillus nidulans*. *J. Microbiol.*, **44**, 145–154.

Zhang, H.J., Shen, L.J., Xu, G.R., and Chen, Y. (2005) Studies on the extraction and stability of *Monascus* orange pigment. *Food Ferment. Ind.*, **31**, 129–133.

Zheng, Y.Q., Xin, Y.W., and Guo, Y.H. (2009) Study on the fingerprint profile of *Monascus* products with HPLC–FD, PAD and MS. *Food Chem.*, **113**, 705–711.

Zheng, Y.Q., Xin, Y.W., Shi, X.A., and Guo, Y.H. (2010a) Anti-cancer effect of rubropunctatin against human gastric carcinoma cells BGC-823. *Appl. Microbiol. Biotechnol.*, **88**, 1169–1177.

Zheng, Y.Q., Xin, Y.W., Shi, X.A., and Guo, Y.H. (2010b) Cytotoxicity of monascus pigments and their derivatives to human cancer cells. *J. Agric. Food Chem.*, **58**, 9523–9528.

Zhou, B., Wang, J., Pu, Y., Zhu, M., Liu, S., and Liang, S. (2009) Optimization of culture medium for yellow pigments production with *Monascus anka* mutant using response surface methodology. *Eur. Food Res. Technol.*, **228**, 895–901.

Zhou, L.H., Wang, Z.X., and Zhuge, J. (2006) Comparison of different transformation methods for *Monascus* sp. *Hereditas (Chinese)*, **28**, 479–485.

Index

Industrial Biotechnology of Vitamins, Biopigments, and Antioxidants, First Edition.
Edited by Erick J. Vandamme and José L. Revuelta.

p